Materials for Sustainable Energy Storage at the Nanoscale

The book *Materials for Sustainable Energy Storage Devices at Nanoscale* anticipates covering all electrochemical energy storage devices such as supercapacitors, lithium-ion batteries (LIBs), and fuel cells, transformation and enhancement materials for solar cells, photocatalysis, etc. The focal objective of the book is to deliver stunning and current information to the materials application at nanoscale to researchers and scientists in our contemporary time towardthe enhancement of energy conversion and storage devices. However, the contents of the proposed book, *Materials for Sustainable Energy Storage at the Nanoscale*, will cover various fundamental principles and wide knowledge of different energy conversion and storage devices with respect to their advancement due to the emergence of nanoscale materials for sustainable storage devices. This book is targeted to be award-winning as well as a reference book for researchers and scientists working on different types of nanoscale materials-based energy storage and conversion devices.

Features

- Comprehensive overview of, energy storage devices, an important field of interest for researchers worldwide
- Explores the importance and growing impact of batteries and supercapacitors
- Emphasizes the fundamental theories, electrochemical mechanism, and its computational view point and discusses recent developments in electrode designing based on nanomaterials, separators, and fabrication of advanced devices and their performances

Materials for Sustainable Energy Storage at the Nanoscale

Edited by
Fabian I. Ezema
M. Anusuya
Assumpta C. Nwanya

CRC Press
Taylor & Francis Group
Boca Raton London New York

CRC Press is an imprint of the
Taylor & Francis Group, an **informa** business

First edition published 2023
by CRC Press
4 Park Square, Milton Park, Abingdon, Oxon, OX14 4RN

and by CRC Press
6000 Broken Sound Parkway NW, Suite 300, Boca Raton, FL 33487-2742

Library of Congress Cataloging-in-Publication Data
Names: Ezema, Fabian I., editor. | Anusuya, M., editor. | Nwanya, Assumpta C., editor.
Title: Materials for sustainable energy storage at the nanoscale / edited by Fabian I. Ezema, M. Anusaya and Assumpta C. Nwanya.
Description: First edition. | Boca Raton : CRC Press, 2023. | Includes bibliographical references. | Summary: "This book provides a comprehensive overview of the latest, cutting-edge research into electrochemical energy storage devices, including supercapacitors, lithium-ion battery (LIBs), and fuel cells. It also explores transformation and enhancement materials for solar cells and photo catalysis and details their possible material applications and fundamental principles. It explores current knowledge of different energy conversion and storage devices with respect to the possible advancement of the utilization of nanoscale materials for sustainable storage devices. This book will be a valuable reference book for researchers and scientists working with nanoscale material-based energy storage and conversion devices"-- Provided by publisher.
Identifiers: LCCN 2022059034 | ISBN 9781032405438 (hardback) | ISBN 9781032410043 (paperback) | ISBN 9781003355755 (ebook)
Subjects: LCSH: Energy storage--Materials. | Nanostructured materials.
Classification: LCC TK2945.N36 M38 2023 | DDC 621.31/26--dc23/eng/20230307
LC record available at https://lccn.loc.gov/2022059034

ISBN: 978-1-032-40543-8 (hbk)
ISBN: 978-1-032-41004-3 (pbk)
ISBN: 978-1-003-35575-5 (ebk)

DOI: 10.1201/9781003355755

Typeset in Minion
by SPi Technologies India Pvt Ltd (Straive)

Contents

Editors

Fabian I. Ezema is a professor at the University of Nigeria, Nsukka. He earned a Ph.D. in Physics and Astronomy from the University of Nigeria, Nsukka. His research focused on several areas of materials science, from synthesis and characterizations of particles and thin-film materials through chemical routes with an emphasis on energy applications. For the past 15 years, he has been working on energy conversion and storage (cathodes, anodes, supercapacitors, solar cells, among others), including novel methods of synthesis, characterization, and evaluation of electrochemical and optical properties. He has published about 180 papers in various international journals and given over 50 talks at various conferences. His h-index is 21 with over 1500 citations, and he has served as a reviewer for several high impact journals and as an editorial board member.

Dr. M. Anusuya is specialized in materials science, thin-film technology, nanoscience, and crystallography. She is working as a Registrar of Indra Ganesan Group of Institutions, Trichy. Earlier to this, she served as a Vice-Principal at Trichy Engineering College, Trichy. Being an administrator and teacher, with more than 25 years' experience, for her perpetual excellence in academics, she has been recognized with many awards. She has received over 45 awards in academic and social activity. She has published more than 30 research papers in national and international journals, 7 chapters in edited books, and 5 patents, presented 50 papers in conferences, and organized more than 200 webinars, both national and internationally.

Dr. Assumpta C. Nwanya is a lecturer and a FLAIR (Future Leaders – African Independent Research) Scholar at the Department of Physics and Astronomy, University of Nigeria, Nsukka. She obtained her Ph.D. in 2017 (University of Nigeria, Nsukka) with specialization in the synthesis of nanostructured materials for applications in photovoltaics and electrochemical energy storage (batteries and supercapacitors) as well as for sensing. She was a postdoctoral fellow under the UNESCO-University of South Africa (UNISA) Africa Chair in Nanoscience and Nanotechnology (2018–2020). She is a research affiliate with the SensorLab, University of the Western Cape Sensor Laboratories, Cape Town, South Africa. Dr. Nwanya is a very active researcher and has published more than 85 scientific articles in high impact journals and has a Google Scholar's H-index of 24 and 1475 citations.

Contributors

Dr. A. Adaikalam
Anna University BIT Campus
Trichy, India

Ayaz Ahmad
Department of Maths, NIT P
Patna, Bihar, India

K. Anandan
Department of Physics, Assistant Professor,
Academy of Maritime Education and Training
(Deemed to be University)
Kanathur, Chennai, India

P. Anitha
Department of Physics, Roever Engineering
College
Perambalur, Tamilnadu, India

N.K. Anushkannan
Department of ECE, Kathir College
of Engineering
Coimbatore, Tamilnadu, India

Dr. M. Anusuya
Professor, Indra Ganesan College of Engineering
Trichy, Tamil Nadu, India

P. Archana
Institute of Organic and Polymeric Materials,
Research and Development Centre of Smart
Textile Technology
Taiwan

A. Joseph Arockiam
Automobile Engineering, Arasu Engineering
College
Kumbakonam, India

A. Arulmary
Anna University
Chennai, India

V. Aruna
PG and Research Department of
Microbiology, Cauvery College for Women
(Autonomous), Affiliated to Bharathidasan
University
Tiruchirappalli, TamilNadu, India

N. Srinivasan alias Arunsankar
Sri Sai Ram Engineering College
West Tambaram, Chennai, India

Dr. G. Raghu Babu
Department of Mechanical Engineering, VNR
Vignana Jyothi Institute of Engg. & Tech,
Bachupally
Nizampet (So), India

Azhagu Saravana Babu
Department of Biotechnology, Vel Tech
Rangarajan Dr. Sagunthala R&D Institute
of Science and Technology
Tamilnadu, India

Dr. J. Raffiea Baseri
Department of Chemistry, PSG College of Arts
and Science, Coimbatore
Tamilnadu, India

Dr. A. Yasmine Begum
Department of EIE, Sree Vidhya Nikethan Engg.
College
Andhra Pradesh, India

Sumanta Bhattacharya
Department of Textile Technology, MAKAUT
West Bengal, India

Dr. S. Chandra
Department of Chemistry, PSG College of Arts
    and Science, Coimbatore
Tamilnadu, India

Kodanda Rama Rao Chebattina
Mechanical Engineering, GITAM (Deemed to be
    University)
Visakhapatnam, Andhra Pradesh, India

Santhosh Kumar Chinnaiyan
Faculty of pharmacy, Karpagam academy of
    higher education
Coimbatore, Tamilnadu, India

Radhika G. Deshmukh
Department of Physics, Shri Shivaji Science
    College
Amravati, Maharashtra, India

Ramesh Desikan
Department of Renewable Energy Engineering,
    Agricultural Engineering College and
    Research Institute, Tamil Nadu Agricultural
    University
Coimbatore, India

Anitha Rexalin Devaraj
Department of Physics, Academy of Maritime
    Education and Training (AMET)
Kanathur, Chennai, India

M. V. Suganya Devi
Saranathan College of Engineering
Tiruchirappalli, Tamilnadu, India

C. Divya
Centre for Information Technology and
    Engineering, Manonmaniam Sundaranar
    University
Tirunelveli, Tamilnadu, India

J. Eindhumathy
Department of ECE, Saranathan College of
    Engineering
Panjappur, Tiruchirappalli, Tamilnadu, India

Fabian I. Ezema
Africa Centre of Excellence for Sustainable Power
    and Energy Development (ACE-SPED),
    University of Nigeria
Nsukka, Nigeria

Nikhat Farhana
Department of Pharmaceitical Chemistry,
    YPCRC, Yenepoya (Deemed to be University)
Naringana, Mangaluru, Karnataka, India

R. Femi
Electrical Department, G. H. Raisoni College
    of Engineering & Management
Pune, Maharashtra, India

Neelamma Gajji
Department of Pharmaceutics, Vikas college
    of Pharmaceutical Sciences
Rayanigudem, Suryapet, Telangana State, India

K. Gayathri
Department of Physics, Academy of Maritime
    Education and Training
Kanathur, India

R. Gayathri
PG and Research Department of Physics,
    Cauvery College for Women (Autonomous)
Tiruchirappalli, Tamilnadu, India

S. Gayathri
Department of Biotechnology, Karpaga Vinayaga
    College of Engineering and Technology
Chengalpattu, India

S. Gopakumar
Department of Electrical and Electronics
    Engineering, Rohini College of Engineering
    and Technology
Nagarcoil, Tamilnadu, India

Dr. R. Gopal
Department of Mechanical Engineering,
    Trichy Engineering College
Trichy, Tamil Nadu, India

Dr. K. M. Govindaraju
Department of Chemistry, PSG College of Arts
    and Science, Coimbatore
Tamilnadu, India

R. Govindharaju
Department of Chemistry, Thanthai Hans Roever
   College
Perambalur, Tamilnadu, India

Dr. K.V. Gunavathy
Department of Physics, Kongu Engineering
   College
Perundurai, India

Anamika Gupta
Department of Chemistry, Aligarh Muslim
   University
Aligarh, U.P., India

Shubhajit Halder
Department of Chemistry, Hislop College, Civil
   Lines
Nagpur, Maharastra, India

Gururaj Hatti
BLDEA'S V.P. Dr. P. G. Halakatti College of
   Engineering and Technology
Vijayapur, Karnataka, India

Vishwanath V. Hokrani
KLS Vdit, Udyog Vidya Nagar
Dandeli Road, Haliyal, Karnataka, India

Deepa Jaganathan
Department of Food Process Engineering,
   Agricultural Engineering College and
   Research Institute, Tamil Nadu Agricultural
   University
Coimbatore, India

S Jana
Department of ECE, Vel Tech Rangarajan
   Dr. Sagunthala R&D Institute of Science
   and Technology
Avadi, Chennai, India

K. Jayachitra
Physics Department, Oxford Engineering College
Trichy, India

Dr. M. Mahaveer Sree Jayan
Department of Mathematics, Indra Ganesan
   College of Engineering
Trichy, Tamil Nadu, India

Dr. M. Jayapriya
Department of Biotechnology, Pavendhar
   Bharathidasan Institute of Engineering and
   Technology
Trichy, Tamilnadu, India

N. Jeenathunisa
PG and Research Department of Microbiology,
   Cauvery College for Women (Autonomous),
   Affiliated to Bharathidasan University
Tiruchirappalli, Tamil Nadu, India

S. Jeyabharathi
PG and Research Department of Microbiology,
   Cauvery College for Women (Autonomous),
   Affiliated to Bharathidasan University
Tiruchirappalli, Tamil Nadu, India

Rachappa Jopate
Department of Information Technology,
   University of Technology and Applied
   Sciences
Al Musannah, Sultanate of Oman

J. Prakash Arul Jose
Paavai Engineering College (Autonomous)
Namakkal, TamilNadu, India

Gitanjali Jothiprakash
Department of Farm Machinery and Power
   Engineering, Agricultural Engineering
   College and Research Institute, Tamil Nadu
   Agricultural University
Coimbatore, India

Kalaivaani
Department of ECE, Vivekananda College of
   Engineering for Women (Autonomous)
Thiruchengodu, Tamilnadu, India

C. Pearline Kamalini
Department of EEE, Saranathan College of
   Engineering
Panjappur, Trichy, Tamilnadu, India

N. Karthikeyan
Department of Physics, Sri Srinivasa
   Matriculation Higher Secondary School
Orathanadu, Thanjavur, Tamilnadu, India

Dr. T. Kavitha
Veltech Rangarajan Dr. Sagunthala R&D
    Institute of Science and Technology
Chennai, India

S.S. Kerur
KLE Technological University, KLE Society's
    Dr. M. S. Sheshagiri College of Engineering
    and Technology
Belagavi, Karnataka, India

Sarvani Jowhar Khanam
School of Chemistry, University
    of Hyderabad
Hyderabad, India

Ankush Balajirao Khansole
Department of Mechanical Engineering,
    Shreeyash College of Engineering and
    Technology, Satara Parisar
Aurangabad, Maharashtra, India

Dr. S. Kiruthika
Department of Physics, Kongu Engineering
    College
Perundurai, India

Shanthala Kollur
Mechanical Engineeing, RVITM, Chaithanya
    Layout, 8th Phase, J. P. Nagar
Bengaluru, Karnataka, India

Dr. V. Koushick
Department of ECE, Vel Tech Rangarajan
    Dr. Sagunthala R&D Institute of Science and
    Technology
Avadi, Chennai, India

P. Krishnan
Department of Physics, St. Joseph's college of
    Engineering, Rajiv Gandhi salai,
    Kamarajarnagar
Semmancheri, Chennai, India

Dr. A. Arun kumar
Department of Physics (H & Sc), Methodist
    College of Engineering & Technolog
Abids, Hyderabad, India

T. Ch. Anil Kumar
Department of Mechanical Engineering,
    Vignan's Foundation for Science Technology
    and Research
Vadlamudi, Guntur, A.P., India

Dr. V. R. Lenin
Anil Neerukonda Institute of Technology and
    Sciences, Sangivalasa
Visakhapatnam, Andhra Pradesh, India

Dr. M. Malathi
Department of Physics, Kongu Engineering
    College
Perundurai, India

Dr. J. Manikandan
Department of Chemistry, PSG College of Arts
    and Science, Coimbatore
Tamilnadu, India

Kumari Manisha
Department of Electronics and Communication
    Engineering, Gokaraju Rangaraju Institute of
    Engineering and Technology
Bachupally, Hyderabad, India

L. H. Manjunatha
School of Mechanical Engineering, REVA
    University
Bangalore, India

Dr. M. Marimuthu
Department of EEE, Saranathan College of
    Engineering
Panjappur, Trichy, Tamilnadu, India

D. Meenakshi
Department of Physics, Shrimati Indira Gandhi
    College
Trichy, India

Diip Mishra
Department of Mechanical Engineering, Faculty
    of Science & Technology, ICFAI University
Raipur, Chhattisgarh, India

N. Muruganantham
Department of Chemistry, Thanthai Hans Roever
    College
Perambalur, Tamilnadu, India

R. Muthukumaran
Department of Mechanical engineering,
   Kakinada Institute of Technological Sciences
Ramachandrapuram, India

Dr. P. Nagarajan
Rajalakshmi Institute of Technology
Chennai, India

R. Sai Nandhini
Department of Biotechnology, Vel Tech
   Rangarajan Dr. Sagunthala R&D Institute of
   Science and Technology
Tamilnadu, India

N.M. Nandhitha
Sathyabama Institute of Science and Technology
Chennai, Tamil Nadu, India

Sam Nirmala Nisha
Department of Biotechnology, Vel Tech
   Rangarajan Dr. Sagunthala R&D Institute of
   Science and Technology
Tamilnadu, India

R. G. Padmanabhan
Automobile Engineering, Arasu Engineering
   College
Kumbakonam, India

Dr. G. Padmasree
Department of Physics, Stanley College of
   Engineering for Women
Hyderabad, India

Dr. M. Padmavathy
Department of Physics, Shrimati Indira Gandhi
   College
Trichy, India

Jeyanathi Palanivelu
Department of Biotechnology, Vel Tech
   Rangarajan Dr. Sagunthala R&D Institute of
   Science and Technology
Tamilnadu, India

P. Jothi Palavesam
Saranathan College of Engineering
Tiruchirappalli, Tamilnadu, India

A. Pandian
Department of Electrical and Electronics
   Engineering, Koneru Lakshmaiah Education
   Foundation (K L Deemed to be University)
Guntur, Andhra Pradesh, India

C. Pavithra
Department of Physics, Marudhar Kesari Jain
   College for Women
Vaniyambadi, Thirupatur, Tamilnadu, India

Omprakash B. Pawar
Department of Forensic Chemistry, Government
   Institute of Forensic Science
Aurangabad, India

Rajkumar Perumal
Department of Food Process Engineering,
   Agricultural Engineering College and
   Research Institute, Tamil Nadu Agricultural
   University
Coimbatore, India

Dr. N. Prakash
Department of EEE, Kumaraguru College of
   Technology
Coimbatore, India

Dr. P.V. Premalatha
Department of Civil Engineering, M.I.E.T.
   Engineering College
Trichy, India

Dr. N. Sathammai Priya
Cauvery College for women (A), Affiliated to
   Bharathidasan University
Tiruchirappalli, Tamil Nadu, India

Ranjit Kumar Puse
Department of Physical Science Chemistry,
   Rabindranath Tagore University
Bhopal, M.P., India

Arulmari Rajachidambaram
Department of Food Process Engineering,
   Agricultural Engineering College and
   Research Institute, Tamil Nadu Agricultural
   University
Coimbatore, India

K. Rajesh
Department of Physics, Academy of Maritime
Education and Training
Kanathur, India

P. V. Rajesh
Saranathan College of Engineering
Tiruchirappalli, Tamilnadu, India

E. Rajkumar
Senior, Department of School of Mechanical
Engineering (SMEC), VIT University
Vellore, Tamilnadu, India

Dr. K. Rajkumar
Department of EEE, Saranathan College of
Engineering
Panjappur, Trichy, Tamilnadu, India

N. Ramya
Department of Computer Science and
Engineering, Saranathan College of
Engineering
Tiruchirappalli, India

P. Rathidevi
JJ College of Engineering and Technology
Trichy, Tamilnadu, India

Dr. A. Rathika
Department of Physics and Research Centre,
Muslim Arts College, Thiruvithancode.
Affiliated to Manonmaniam Sundaranar
University
Tirunelveli, India

Mr. Kodumuri Veerabhadra Rao
Department of Physics, Methodist College of
Engineering and Technology
Hyderabad, India

D. Ravindran
Department of Physics, Velammal College of
Engineering and Technology
Madurai, India

Dr. K. Anuradha
Department of Physics, Methodist College of
Engineering and Technology
Hyderabad, India

J. Rekha
Department of EEE Part Time Research Scholar,
Sathyabama University
Semancheri, Chennai, India

R. Rekha
Saranathan Engineering College
Trichy, Tamil Nadu, India

L.K.Rex
M.I.E.T. Engineering College
Trichy, India

J. Femila Roseline
Saveetha School of Engineering, Saveetha
Institute of Medical and Technical Sciences,
Saveetha University
Chennai, Tamil Nadu, India

S. Roseline
Department of Mechanical Engineering,
MAM College of Engineering and
Technology
Trichy, India

Dr. P. Janardhan Saikumar
Department of ECE, Audisankara Institute of
Technology
Gudur, India

P. Sakthivel
Department of Physics, Urumu Dhanalakshmi
College
Trichy, Tamilnadu, India

Dr. M. Sangeetha
Department of EEE, M.A.M. School of
Engineering
Siruganur, Trichirappalli, India

Dr. V. Saravanan
Department of Physics, Sri Meenakshi Vidiyal
Arts and Science College
Trichy, Tamilnadu, India

Dr. V. Subha Seethalakshmi
Department of EEE, Sreyas Institute of
Engineering and Technology
Hyderabad, India

S Selvaganesan
Department of AI&DS, JNN Institute of
 Engineering
Tiruvallur, Tamil Nadu, India

R. Selvam
Mechanical Engineering. St. Joseph's College of
 Engineering, OMR
Jeppiaar Nagar, Chennai, India

Dr. R. Selvamani
Department of Mathematics, Karunya Institute
 of Technology and Sciences
Coimbatore, Tamilnadu, India

Sandip Sen
School of Pharmacy, Guru Nanak Institutions
 Technical Campus
Ibrahimpatnam, Hyderabad, India

M.M. Senthamilselvi
Regional Joint Director (Rtd), Directorate of
 Collegiate Education
Trichy, Tamilnadu, India

S Sanjay Sethuganesh
Material Science Engineering, Visvesvaraya
 National Institute of Technology
Nagpur, India

M. Shanmugavalli
Saranathan College of Engineering
Tiruchirappalli, Tamilnadu, India

A.V.K. Shanthi
AIMAN College of Arts and Science
 for Women
K.K. Nagar, Trichy, Taminadu, India

Pradosh Kumar Sharma
Department of Physics, Chinmaya Degree
 College, BHEL
Haridwar, Uttrakhand, India

Dr. R. Shenbagalakshmi
Department of Electrical Engg., G.H. Raisoni
 College of Engineering & Management
Wagholi, Pune, India

Balkeshwar Singh
Department of Mechanical Design &
 Manufacturing Engineering, Adama Science
 and Technology University
Kebele-14, Adama City, Ethiopia

M. Muralidhar Singh
Department of Mechanical Engineering,
 RV Institute of Technology and
 Management
Chaithanya Layout, J. P. Nagar, Bengaluru,
 Karnataka, India

Rahul Singh
Department of Physics, N.B.G.S.M College,
 Gurugram University
Raipur, India

SreeSvarna BhaskaraMohan
Tamil Nadu Agricultural University
Coimbatore, India

R. Sridhar
Department of EEE, Saranathan College of
 Engineering
Panjappur, Trichy, Tamilnadu, India

Sriramajayam Srinivasan
Department of Renewable Energy Engineering,
 Agricultural Engineering College and
 Research Institute, Tamil Nadu Agricultural
 University
Coimbatore, India

Ashish Kumar Srivastava
Department in Mechanical Engineering,
 Galgotias University
Greater Noida, Uttar Pradesh, India

Karthikeyan Subburamu
Department of Renewable Energy Engineering,
 Agricultural Engineering College and
 Research Institute, Tamil Nadu Agricultural
 University
Coimbatore, India

Dr. R. Subramaniyan@Raja
Department of Physics, KPR Institute of
 Engineering & Technology
Coimbatore, India

M. Sudha
Department of Electronics and Communication
Engineering, Paavai Engineering College
(Autonomous)
NH-44, Paavai Nagar, Paachal, Namakkal,
Tamilnadu, India

Dr. K. Sujatha
Department of Physics, Shrimati Indira Gandhi
College
Tiruchirappalli, Tamilnadu, India

A. R. Umayal Sundari
Periyar Maniammai Institute of Science &
Technology (Deemed to be University)
Periyar Nagar, Vallam, Thanjavur, India

K. Suneeta
Electrical and Electronics Engineering,
Pallavi Engineering College
Hyderabad, India

Dr. V. Suresh Kannan
Department of Mechanical Engineering,
Madanapalle Institute of Technology &
Science
Madanapalle, Andhra Pradesh, India

Thirumalvalavan
Department of Mechanical Engineering,
Arunai Engineering College
Velunagar, Tiruvannamalai, India

Dr. V. Vaithiyanathan
Department of Mechanical Engineering,
Indra Ganesan College of Engineering
Trichy, Tamil Nadu, India

Sugumari Vallinayagam
Department of Biotechnology, Vel Tech
Rangarajan Dr. Sagunthala R&D Institute of
Science and Technology
Tamilnadu, India

M.K. Valsakumari
Department of Chemistry, Mookambigai College
of Engineering
Kalamavur, Pudukottai, Tamilnadu, India

E. Veeramanipriya
Periyar Maniammai Institute of Science &
Technology (Deemed to be University)
Periyar Nagar, Vallam, Thanjavur, India

Dr. N. Vengadachalam
Department of EEE, Malla Reddy Engineering
College for Women
Hyderabad, India

P. Vickraman
Department of Physics, Gandhigram Rural
University
Dindigul, Madurai, India

Dr. J. Vidhya
Department of Physics, M. Kumarassamy College
of Engineering (Autonomous)
Thalavapalayam, Karur, Tamilnadu, India
and
PG and Research Department of Physics,
Cauvery College for Women (Autonomous)
Tiruchirappalli, Tamilnadu, India

Dr. S. Vidhya
Department of Physics, Shrimati Indira Gandhi
College
Tiruchirappalli, Tamilnadu, India

Dr. S. Vijayalakshmi
Department of EEE, Saranathan College of
Engineering, Panjappur
Trichy, Tamilnadu, India

D. S. Vijayan
Department of Civil Engineering, Aarupadai
Veedu institute of Technology
Paiyanur, Tamil Nadu, India

S. Vijayaraj
Department of Electrical and Electronics
Engineering, Vels Institute of Science,
Technology & Advanced Studies, Pallavaram
Chennai, Tamil Nadu, India

K. S. Vinod
Department of ECE, Vel Tech Rangarajan Dr.
Sagunthala R&D Institute of Science and
Technology
Avadi, Chennai, India

K. Raju Yadav
Research Scholar, Department of Farm Machinery
and Power Engineering, Agricultural
Engineering College and Research Institute,
Tamil Nadu Agricultural University
Coimbatore, India

Rama Krishna Yellapragada
Department of Computer Science and
Engineering, Koneru Lakshmaiah Education
Foundation, K L University
Greenfields, Vaddeswaram, Andhra Pradesh,
India

1

Prediction and Optimization of Interpulse Tungsten Inert Gas (IPTIG) Arc Welding Process Parameters to Attain Minimum Fusion Zone Area in Ti–6Al–4V Alloy Sheets Used in Energy Storage Devices

Dr. V. Vaithiyanathan and Dr. M. Mahaveer Sree Jayan
Indra Ganesan College of Engineering, Trichy, India

Dr. R. Gopal
Trichy Engineering College, Trichy, India

M. Anusuya
Indra Ganesan College of Engineering, Trichy, India

Ti-6Al-4V titanium alloy has been widely utilized in fabrication of delicate structures like energy storage devices, demanding excellent fatigue strengths, corrosion resistance, and tensile strength. Tungsten inert gas (TIG) welding process is frequently chosen to join these alloys owing to its better economy. Interpulse tungsten inert gas (IPTIG) welding is another variation of the traditional TIG welding process. IPTIG welding method provides more benefits than traditional TIG welding processes such as narrow heat-affected zone and deeper penetration due to arc constrictions. Mechanical properties are not only affected by chemical compositions but also by proper process parameters and fusion zone characteristics. In this analysis, an effort has been completed to develop empirical relationships to

DOI: 10.1201/9781003355755-1

calculate IPTIG-welded fusion zone characteristics (width of bead and fusion zone area) by incorporating IPTIG welding parameters. These parameters were optimized to achieve minimum fusion zone area by response surface methodology (RSM). Additionally, the influence of IPTIG welding parameters on fusion zone physical characteristics was analyzed. From the outcomes, it is found that the delta current has a superior impact on the formation of the fusion zone compared to other parameters.

1.1 Introduction

Titanium and its alloys are key metals in nonferrous metals. Due to their high strength-to-weight ratio, high stiffness, virtuous high temperature, and corrosion resistance, they can be utilized in energy storage devices, automobiles, space laboratory, and medical devices (Yunlian et al. 2000). As a result, the benefits of titanium must be weighed against the additional expense (Caiazzo et al. 2004). Ti-6Al-4V has long been the industry's workhorse, with this grade of titanium accounting for roughly 80–90% of all titanium used in the industry (Balasubramanian et al. 2008). Ti-6Al-4V alloy can be joined through tungsten inert gas (TIG) welding, diffusion bonding, and high-energy density processes such as laser beam welding and electron beam welding (EBW; Balasubramanian et al. 2011). For industrial applications and systems, EBW is regarded as the best possible joining process for titanium alloys (Wang et al. 2003). But drawbacks of the EBW process include X-ray emission (Gao et al. 2013). Titanium alloys can also be joined with traditional TIG welding, but this should be avoided due to the slow process and high heat input. Because of the large heat input, columnar grains form in the fusion zone, resulting in poor mechanical characteristics and deformation (Ahmed & Rack 1998).

Two main issues observed in Ti-6Al-4V laser welding are underfilling and porosity. The fundamental difficulty with porosity is that it shrinks the weld cross-sectional area. The occurrence of porosity always lowers the strength and ductility (Sundaresan & Janaki Ram 1999). Current pulsing techniques primarily affect grain refining in the fusion zone by physically disturbing the arc in the molten weld pool, which is agitated vigorously, thus the optimum frequency minimizes the preceding beta grain boundary, improving tensile ductility (Vaithiyanathan et al. 2019).

The interpulse generated a magnetically restricted columnar profile arc with a frequency of 20 kHz. The magnetic field that surrounds the arc restricts the arc. The interpulse welding machine produces high-frequency pulses with programmable parameters that can be used to change the magnetic field of the arc, allowing for arc constriction control as shown in Figure 1.1. The arc is constricted, resulting in thin but deeper weld beads and a restricted heat-affected zone (Leary et al. 2010).

Based on a review of the literature, it appears that no research into the relationship between interpulse tungsten inert gas (IPTIG) welding process parameters and fusion zone features in Ti-6Al-4V alloy joints has been reported. The majority of investigators have used peak current, frequency, and speed as response welding parameters, but no research on "delta current" has been done.

1.2 Experimental

For welding experiments, Ti-6Al-4V titanium alloy sheets (150 mm × 75 mm × 1.2 mm) of the desired dimensions were fabricated.

1.2.1 Finding the Working Limits of the Parameters

Many trial runs were conducted with various combinations of process parameters in order to obtain defect-free samples for testing. The quality of the joints produced during the testing runs was examined.

The delta current (D), peak current (P), welding speed (S), and delta frequency (F) are the IPTIG welding process factors that determine the fusion zone characteristics. Each process parameter's higher and lower limits were coded as +2 and –2, respectively. Certain process parameters with limits are shown in Table 1.1.

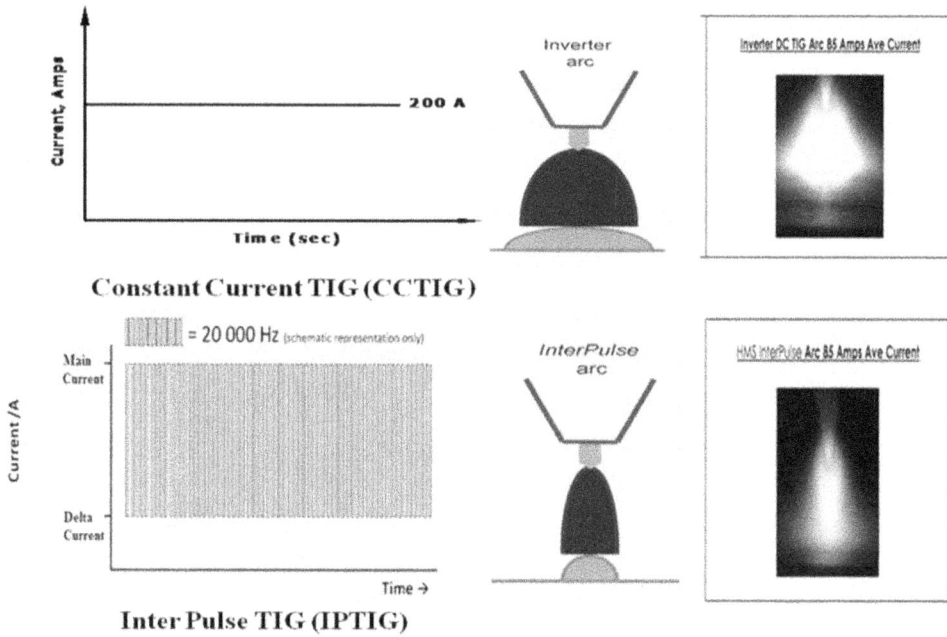

FIGURE 1.1 Schematic diagram of different welding modes.

TABLE 1.1 Feasible IPTIG Welding Parameters and Their Levels

No.	Parameter	Notations	Unit	−2	−1	0	+1	+2
1.	Main current	P	Amp	40	45	50	55	60
2.	Delta current	D	Amp	10	20	30	40	50
3.	Delta frequency	F	kHz	4	8	12	16	20
4.	Welding speed	S	mm/min	40	50	60	70	80

Table 1.2 displays the design matrix. It's a five-level, four-factor rotatable design matrix. As a result of the 30 tentative runs, the quadratic, linear, and two-way interactive influences of the process parameters on bead geometries could be estimated. Figure 1.2 depicts the fusion zone profile. Several essential fusion zone features, including bead width and fusion zone area, were measured.

TABLE 1.2 Design Matrix and Experimental Results

Expt. No.	P	D	F	S	P	D	F	S	Width of Bead, WB (mm)	Fusion Zone Area, FZA (mm²)
1	−1	−1	−1	−1	45	20	8	50	7	8
2	1	−1	−1	−1	55	20	8	50	6.9	7.9
3	−1	1	−1	−1	45	40	8	50	8.59	10.8
4	1	1	−1	−1	55	40	8	50	7.934	9.3
5	−1	−1	1	−1	45	20	16	50	7.26	8.553
6	1	−1	1	−1	55	20	16	50	7.67	8.999

(Continued)

TABLE 1.2   (Continued)

Expt. No.	P	D	F	S	P	D	F	S	Width of Bead, WB (mm)	Fusion Zone Area, FZA (mm²)
7	−1	1	1	−1	45	40	16	50	8.321	9.75
8	1	1	1	−1	55	40	16	50	7.69	8.765
9	−1	−1	−1	1	45	20	8	70	6.5	7.354
10	1	−1	−1	1	55	20	8	70	7	8.23
11	−1	1	−1	1	45	40	8	70	8.332	9.64
12	1	1	−1	1	55	40	8	70	7.836	9.34
13	−1	−1	1	1	45	20	16	70	6.665	7.977
14	1	−1	1	1	55	20	16	70	7.178	8.561
15	−1	1	1	1	45	40	16	70	8	9.15
16	1	1	1	1	55	40	16	70	7.5	8.67
17	−2	0	0	0	40	30	12	60	8.267	9.568
18	2	0	0	0	60	30	12	60	7.98	9.118
19	0	−2	0	0	50	10	12	60	6.994	8.1
20	0	2	0	0	50	50	12	60	8.68	10.345
21	0	0	−2	0	50	30	4	60	6.901	8
22	0	0	2	0	50	30	20	60	6.852	8.089
23	0	0	0	−2	50	30	12	40	7.98	9.15
24	0	0	0	2	50	30	12	80	6.82	7.9
25	0	0	0	0	50	30	12	60	5.882	6.345
26	0	0	0	0	50	30	12	60	5.6	6.543
27	0	0	0	0	50	30	12	60	5.778	6.745
28	0	0	0	0	50	30	12	60	5.8	6.96
29	0	0	0	0	50	30	12	60	5.88	7.056
30	0	0	0	0	50	30	12	60	5.6	6.72

1.3   Development of Empirical Relationships

The bead geometry is a gathering of IPTIG welding parameters such as delta current (D), main current (P), welding speed (S), and delta frequency (F).

$$Y = f(P, D, F, S) \tag{1.1}$$

For the four designated factors and their interface factors, the selected polynomial could be expressed as,

$$Y = b_0 + b_1P + b_2D + b_3F + b_4S + b_{11}P^2 + b_{22}D^2 + b_{33}F^2 \\ + b_{44}S^2 + b_{12}PD + b_{13}PF + b_{14}PS + b_{23}DS + b_{23}DS + b_{24}FS \tag{1.2}$$

The values of these coefficients are calculated and presented in Table 1.3.

Expt.No	Parameters	Macrograph	
		Top Surface	Cross-Section
2	P= 55 A D= 20 A F= 8 kHz S= 50 mm/min	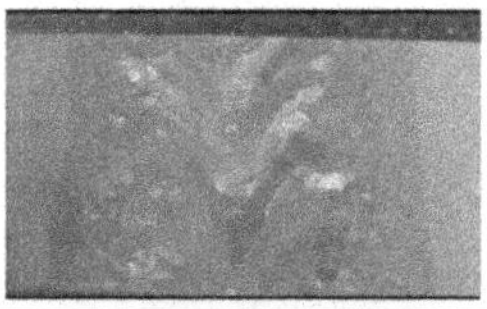	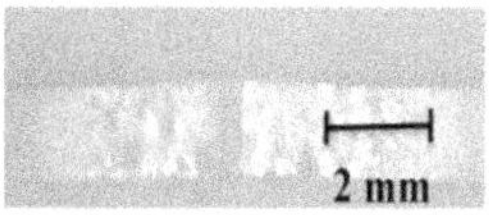 2 mm
4	P= 55 A D= 40 A F= 8 kHz S= 50 mm/min	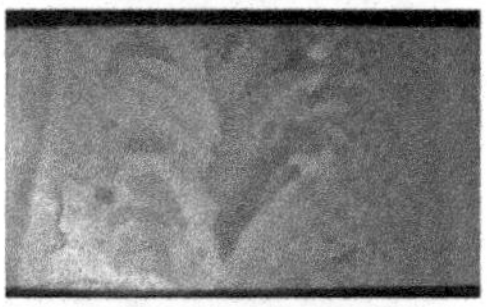	 2 mm
6	P= 55 A D= 20 A F= 16 kHz S= 50 mm/min	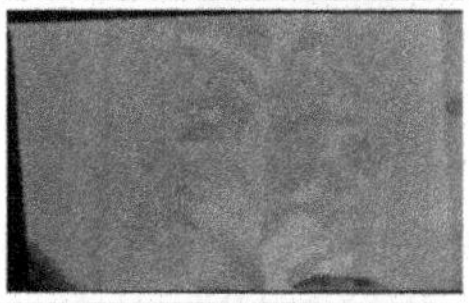	 2 mm
12	P= 55 A D= 40 A F= 8 kHz S= 70 mm/min	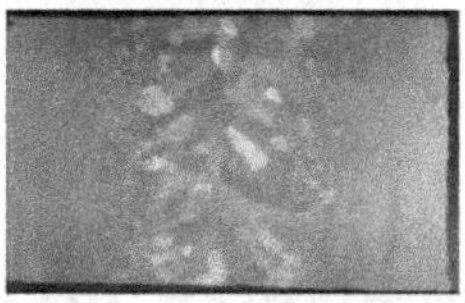	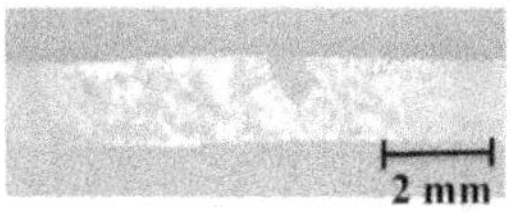 2 mm
17	P= 40 A D=30 A F= 12 kHz S= 60 mm/min	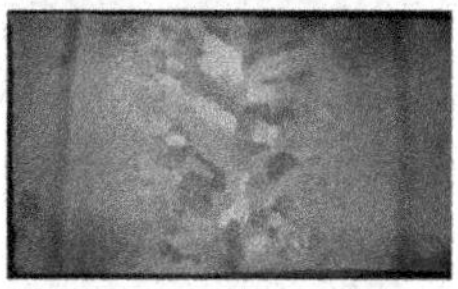	 2 mm
25	P= 50 A D= 30 A F= 12 kHz S= 60 mm/min	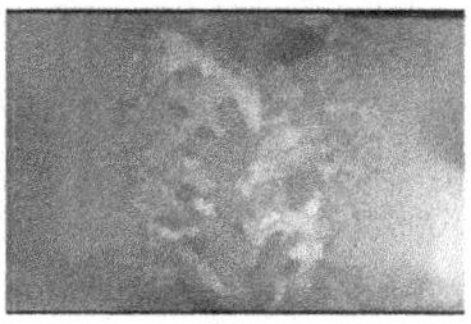	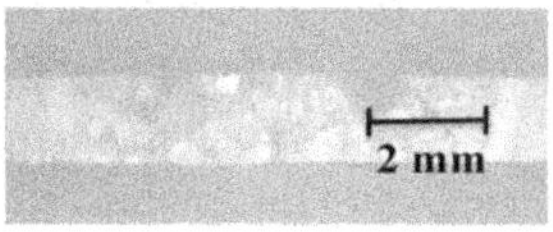 2 mm

FIGURE 1.2 Macrographs of some of the bead profiles.

TABLE 1.3 Coefficient and Their Estimated Factors

Coefficient	WB	FZA
Intercept	5.756667	6.728167
P	−0.06392	−0.09829
D	0.475083	0.597125
F	0.003917	0.001625
S	−0.19475	−0.23521
PD	−0.22538	−0.31694
PF	0.034	0.036813
PS	0.062125	0.176188

(*Continued*)

TABLE 1.3   (Continued)

Coefficient	WB	FZA
DF	−0.15963	−0.33444
DS	0.03875	−0.03031
FS	−0.05263	−0.01706
P^2	0.58575	0.658385
D^2	0.514125	0.62826
F^2	0.274	0.33376
S^2	0.404875	0.453885

The established empirical relationship is given below:

Width of bead (WB) = {82.67 − 2.315 (P) − 0.0109 (D) − 0.296 (F) − 0.563 (S) − 0.0045 (PD) + 0.0017 (PF) + 0.00124 (PS) − 0.00399 (DF) − 0.000387 (DS) − 0.00132 (FS) +0.0234 (P^2) + 0.00514 (D^2) + 0.0171 (F^2) + 0.00404 (S^2)} **mm**

Fusion zone area (FZA) = {96.46 − 2.696 (P) + 0.118 (D) − 0.315 (F) − 0.730 (S) − 0.0063 (PD) + 0.0018 (PF) + 0.00352 (PS) − 0.00836 (DF) − 0.0003 (DS) − 0.000043 (FS) + 0.02633 (P^2) +0.000628 (D^2) − 0.0208 (F^2) − 0.0045 (S^2)} **mm^2**

1.4 Checking Adequacy of the Developed Relationships (Padmanban & Balasubramanian 2011)

The acceptability of the developed relationships was confirmed using the analysis of variance (ANOVA). In this method, if the calculated F_{ratio} value of the established model is fewer than the normal F_{ratio} (from F-table) value at the preferred level of self-confidence (95%), the model is suitable within the self-confidence limit. It is implicit that the F-value for the established width of bead and fusion zone area model is 85.18 and 57.30 and proposed that the simulations are substantial.

The assessment of "r^2" for the established relationship is shown in Table 1.4, which specifies a high relationship existing between the predicted values and experimental values. All the reflection specifies admirable adequacy of the established empirical relationships to predict fusion zone characteristics.

TABLE 1.4   ANOVA Test Results for Bead Geometries

Source	F-Value (WB)	F-Value (FZA)	
Model	85.18027	57.30056	Significant
P	4.723169	5.035154	
D	260.9425	185.8274	
F	0.017735	0.001376	
S	43.84905	28.83266	
PD	39.14943	34.9008	
PF	0.890988	0.470846	
PS	2.974725	10.78545	
DF	19.63884	38.86136	
DS	1.157331	0.319251	
FS	2.13451	0.101152	
P^2	453.3377	258.1854	
D^2	349.2486	235.099	
F^2	99.19696	66.35	
S^2	216.5904	122.7054	
Lack of fit	1.390816	0.507885	Not significant

1.5 Relationship between Width of Bead and Fusion Zone Area

The experimentally measured fusion zone area and width of bead values shown in Table 1.2 are plotted in a linear graph as shown in Figure 1.3. All the points are fitted and connected by a best-fit straight line and the equation for the best-fit line is given by,

$$\text{FZA}\left(\text{mm}^2\right) = \left[-0.2434 + 1.20215\left(\text{WB}\right), \text{mm}\right] \tag{1.3}$$

The slope of the best-fit line (1.20215) is positive, and it suggests that the fusion zone area has a directly proportional relationship with width of the bead. The coefficient of determination, R^2 is found to be 96.2% of the above equation. The above-derived equation can be developed to conclude the mean value of fusion zone area for a given width of bead.

1.6 Optimization of IPTIG Welding Parameters

To find the best values of the process variables, the response surface methodology (RSM) was utilized as an optimization technique. The coded values were used to frame the empirical correlations developed in the previous section. The optimization was performed on coded values, which were then transformed to actual values. A minimal fusion zone area of 6.345 mm² was established under ideal conditions.

The response surfaces obviously designate the optimal response point. The optimum fusion zone area of IPTIG-welded Ti-6Al-4V titanium alloy was exhibited by the vertex of the response surface, as shown in Figure 1.4.

Contour plots have a characteristic circular mound shape, which indicates possible factor independence with response to depict the ideal factor settings region. The optimum is located with reasonable precision by defining the form of the surface using contour plots generated with response surface analysis software. When a circular-shaped contour appears, it tends to show that factor effects are independent, but elliptical contours may reveal factor interactions. Figure 1.4 shows the best response for IPTIG-welded Ti-6Al-4V titanium alloy.

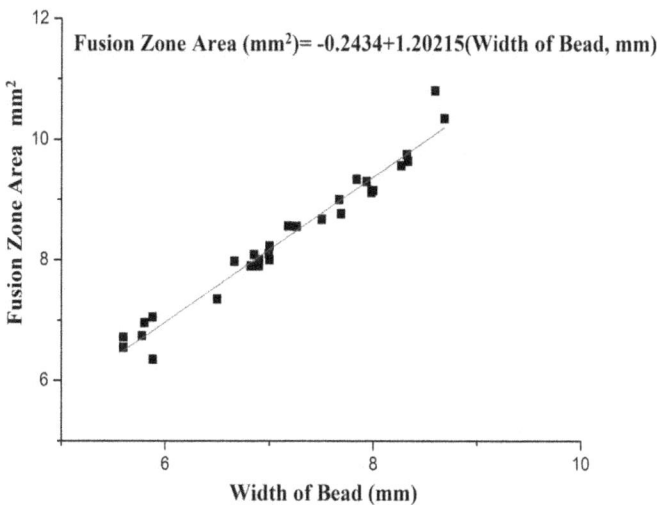

FIGURE 1.3 Relationship between width of bead and fusion zone area.

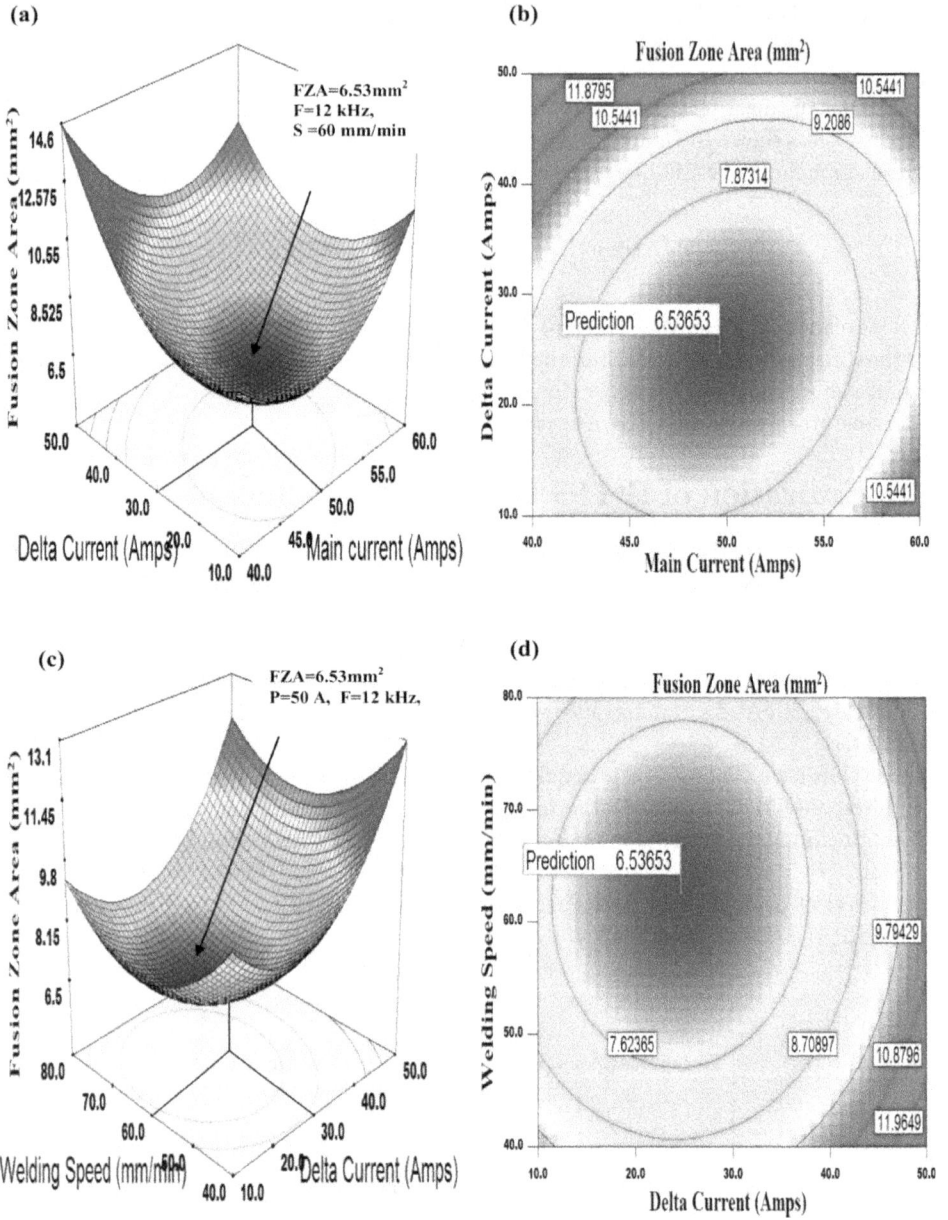

FIGURE 1.4   Interaction effects of parameters.

(*Continued*)

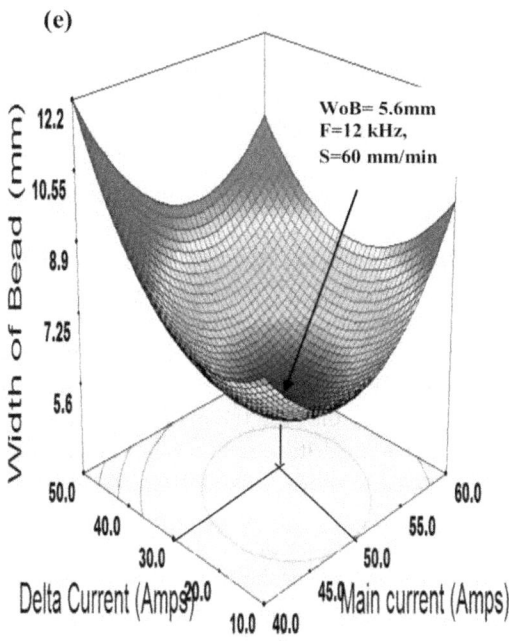

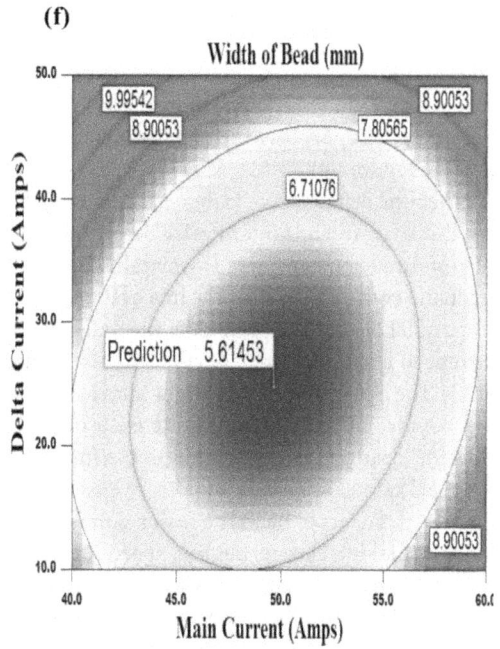

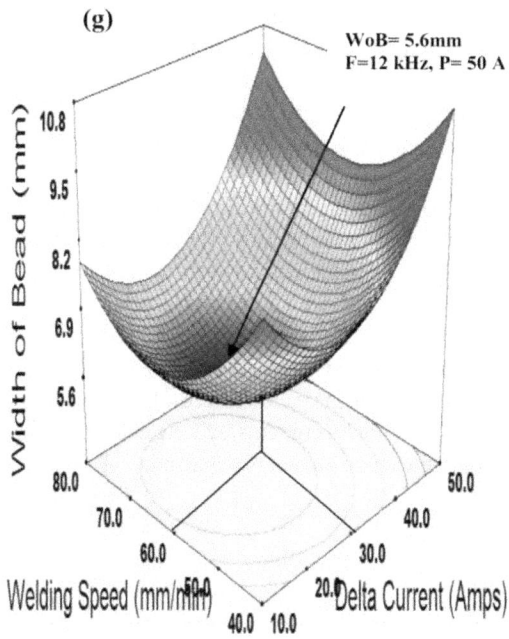

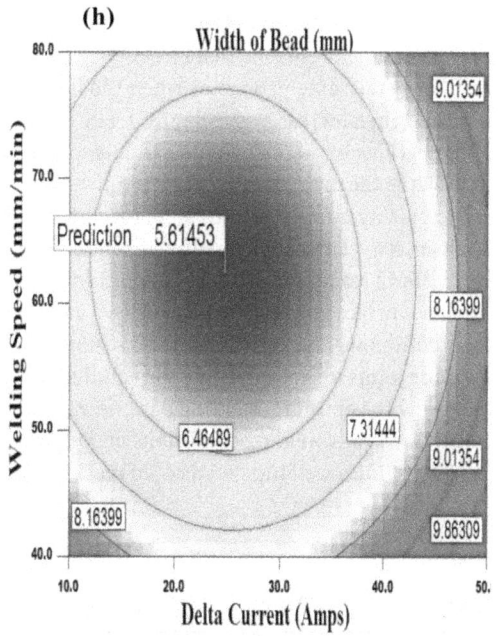

FIGURE 1.4   (Continued)   Interaction effects of parameters.

1.7 Analysis of Response Graphs and Contour Plots

Prior to optimization, it's important to understand how IPTIG welding parameters influence the bead profile. Three-dimensional response surface graphs and contour plots based on the proposed model are created for this purpose, with one parameter in the center level and the other two parameters as variables (see Figure 1.4).

When the delta current is increased for a certain main current or welding speed, the fusion zone area first decreases to a minimum value before increasing (see Figure 1.4a and c). At higher delta currents, the maximum fusion zone area is generally due to the magnetic field increasing slowly, causing the arc to gradually contract. This results in a gradual decline in arc width until the time is up. The delta current drops quickly after that, reducing the magnetic field and causing the arc to widen. Increasing the delta current to a specific range results in a slower arc compression and a faster return to the original arc width. The agitation causes a slight increase in current followed by a reduction. This works similarly to a slow pulse in that the surface tension of the pool is disturbed, allowing the material to be gently "stirred," resulting in a reduction in the fusion zone area. When the delta current reaches a high value, however, the excess heat input takes over, causing the fusion zone to widen. This is why the delta current about 30 A prefers to be in the fusion zone's minimum.

The initial increase in welding speed increases the fusion zone to a particular value at a given delta current or main current, and successive increase in welding speed makes the fusion zone almost constant (see Figure 1.4c). Lower welding speeds result in a higher heat input into the welding samples, which increases the fusion zone profile significantly. As a result, IPTIG welding joints with relatively high welding speeds (over 60 mm/min) have a lower fusion profile. When the main current is increased at a constant delta current, delta frequency, and welding speed, the development of the fusion zone profile follows the same pattern as when the delta current is increased, i.e. it first declines to a minimum value and then increases (see Figure 1.4a). The fusion zone profile, on the other hand, is less responsive to fluctuations in main current than it is to changes in delta current. The delta current has a 50% duty cycle, while the main current has a frequency of 12 kHz.

The pulsing current creates a magnetic force that constricts the arc, resulting in a stiff, narrow arc. However, if a relatively high main current is used, the increased heat input causes the fusion zone profile to grow. As a result, the primary current, which is close to 50 A, helps to reduce the fusion zone area. The response surface graphs between any two variables and the response (see Figure 1.4a, c, e, and g) show a maximum point, suggesting that the fusion area value inside the experimental zone is the smallest. A few more joints were used to test the validity of the optimization technique, using parameters just above and below the optimum welding parameters. Table 1.5 summarizes the validation results. According to the findings, only welding conditions of 50 A main current, 30 A delta currents, 12 kHz delta frequency, and 60 mm/min welding speed could reach a minimum fusion zone area of 6.345 mm².

TABLE 1.5   Validation of Optimization Procedures

Expt No	Main Current (A)	Delta Current (A)	Delta Frequency (kHz)	Welding Speed (mm/min)	Width of Bead (mm) Actual	Width of Bead (mm) Predicted	Error (%)	Fusion Zone Area (mm²) Actual	Fusion Zone Area (mm²) Predicted	Error (%)
1	52	30	12	60	5.725	5.825	−1.71	6.812	6.794	0.26
2	48	35	12	60	6.325	6.287	0.604	7.2	7.392	−2.59
3	45	30	10	65	6.5	6.476	0.37	7.56	7.498	0.826
4	50	30	12	60	5.6	5.757	−2.72	6.72	6.728	−0.11

1.8 Conclusions

1. From ANOVA test results, it has been found that the delta current has a greater effect (F = 260.67), and delta frequency has a comparatively lesser effect (F = 0.00287) on fusion zone characteristics.

2. Minimum fusion area of 6.345 mm^2 with full penetration was attained under the welding conditions of main current of 50 A, delta current of 30 A, delta frequency of 12 kHz, and welding speed of 60 mm/min. These sets of welding parameters are found to be the optimum IPTIG parameters to attain minimum fusion area with full penetration in 1.2 mm thin sheets of Ti-6Al-4V alloy used in aero-engine applications.

References

Ahmed, T., Rack, H. J. 1998. "Phase transformation during cooling in α+β titanium alloys." *Materials Science and Engineering; A* 243: 206–211.

Balasubramanian, T. S., Balakrishnan, M., Balasubramanian, V., Muthumanickam, M. 2011. "Effect of welding processes on joint characteristics of Ti–6Al–4V alloy." *Science and Technology of Welding and Joining* 16: 702–708.

Balasubramanian, M., Jayabalan, V., Balasubramaniam, V. 2008. "Developing mathematical models to predict grain size and hardness of argon tungsten pulsed current arc welded titanium alloy." *Journal of Material Processing and Technology* 196: 222–229.

Caiazzo, F., Curcio, F., Daurelio, G., Minutolo, F.M.C., 2004. "Ti-6Al-4V sheets lap and butt joints carried out by CO_2 laser: mechanical and morphological characterization". *Journal of Materials Processing Technology* 149(1–3): 546–552.

Padmanban, G., Balasubramanian, V. 2011. "Optimization of pulsed current gas tungsten arc welding process parametrs to attain maximum tensile strength in AZ31B magnesium alloy." *Transactions of Nonferrous Metals Society of China* 21: 467–476.

Yunlian, Q., Ju, D., Quan, H., and Liying, Z. 2000. "Electron beam welding, laser beam welding and gas tungsten arc welding of titanium sheets" *Materials Science and Engineering A* 280: 177–181.

Leary, R. K., Merson, E., Birmingham, K., Harvey, D., Brydson, R. 2010. "Microstructural and microtextural analysis of Inter Pulse GTCAW welds in Cp-Ti and Ti-6Al-4V" *Materials Science Engineering A* 527: 7694–7705.

Sundaresan, S., Janaki Ram, G.D. 1999. "Microstructural refinement of weld fusion zones in α-β titanium alloys using pulsed current welding" *Materials Science Engineering A* 262: 88–100.

Vaithiyanathan, V., Balasubramanian, V., Malarvizhi, S., Petlay, V., Verma, S. 2019. "High temperature tensile properties and microstructural characterization of gas tungsten constricted arc welded Ti-6Al-4V alloy." *Material Research Express* 6: 0965d6.

Wang, S.H., Wei, M.D., Tsay, L.W. 2003. "Tesnsile properties of laser beam welding in Ti-6Al-4V alloy at evaluated temperature below 450 C." *Materials Letters* 57: 1815–1823.

Gao, X. L., Zhang, L. J., Liu, J., Zhang, J. X. 2013. "A comparative study of pulsed Nd: YAG laser welding and TIG welding of thin Ti-6Al-4V titanium alloy." *Materials Science Engineering A* 539: 14–21.

2

Structural and Morphological Analysis of Drying Kinetics of Photovoltaic Thermal (PVT) Hybrid Solar Dryer for Drying of Sweet Potato Slices

E. Veeramanipriya and A. R. Umayal Sundari

Periyar Maniammai Institute of Science & Technology (Deemed to be University), Thanjavur, India

This work presents thin layer drying kinetics of photovoltaic thermal (PVT) hybrid solar dryer-assisted evacuated tube collector (ETC) for chemically untreated sweet potato slices under climatic conditions of Thanjavur, India. A comparative analysis was done with open sun drying. ETC outlet temperature was larger than the ambient temperature. This increased the rate of drying in the hybrid dryer. The initial moisture content of the dried sweet potato slices was calculated to be 93.1% (wb) for both drying methods. The equilibrium moisture content of the hybrid dryer and sun drying was estimated to be 27.1% (wb) and 29.1%, respectively. The drying time of hybrid-dried sweet potato slices was 9 hours and that of sun drying was 13 hours. To estimate the appropriate fit, nine developed mathematical models were applied to represent the thin layer drying kinetics of sweet potato slices in both drying methods. Based on the modeling results, Midilli (2001) model showed a higher equivalent of correlation coefficient (R^2) and lower equivalent of reduced chi-square and root mean square error (RMSE) for both dryers, showing the suitability of fit for drying of untreated sweet potato slices. From the XRD results, it has been observed that the hybrid-dried sweet potato starch peaks at (2θ): 14.9°, 18.1°, and 23.2° and sun-dried sweet potato starch peaks at (2θ): 15.1°, 17.2° and 23.1°. The XRD results show that both hybrid

DOI: 10.1201/9781003355755-2

and sun-dried sweet potato starch exposed 'C$_A$' type exhibited crystalline pattern and the crystallinity ranges from 24.6 to 25.4%. The granule size of the hybrid-dried sweet potato was in the range from 4.2 to 6.1 µm, and for sun-dried sample, it was from 5.3 to 8.6 µm. SEM analysis reveals that moisture was removed at periodic intervals of time and the granules showed minimal defects in the hybrid dryer. Based on the structural and morphological results, the appearance and quality of the hybrid-dried sweet potato were found to be of high grade compared to the sun-dried sample.

2.1 Introduction

Sweet potato (*Ipomoea batatas* L.) is the fourth most essential starchy tuber root crop that feeds the world and accounts for about 12% of the tuber crop production. It is grown in more than 110 countries. Fresh sweet potatoes can be prepared at home as a ready-to-eat vegetable that can be boiled, baked, or processed through industrial fermentation into food and beverages. Fresh sweet potatoes are bulky and perishable that causes difficulty in transporting and limits the distance of transportation. Developing countries face high post-harvest loss due to the high water content of the crop (Antonio et al. 2008; Latif and Müller 2015; Abong et al. 2016; Akinwande et al. 2013; Oke and Workneh 2013; Seidu et al. 2012; Sanginga 2015; Titus and Lawrence 2015).

Sweet potato contains starch, crude protein, amylopectin, oxalate, vitamin A, vitamin C, sugar, tannins, amylase, phytin, either extract, and crude fiber. Knowledge of how the tuber could be processed and stored for further use is lacking. These results in great post-harvest losses of the crop after harvest and storage process.

Drying (dehydration) is one of the oldest techniques of processing and preserving food for later use. Dehydration is the significant process of heat and mass transfer between the crop surface and its circumference medium, which leads to the removal of crop moisture and allows a safe storage level of root crops over a long period by reducing the microbiological degradation rate of the crop (Yılmaz et al. 2017; Hande et al. 2016; Sanni et al. 2012; Onu et al. 2017; Pirasteh et al. 2014).

Sun-dried products are not protected from rain, dust, wind, insects, and domestic animals during drying. Hence, it often results in poor-quality dried products. Technically, industrial dryers are operated depending on fossil fuel or electricity. These are not affordable for small-scale industries which are also not economically affordable (Sevik et al. 2019). To defeat the disadvantages, novel (PVT) hybrid solar dryer-assisted ETC is developed.

Many researchers have done a survey on mathematical modeling for thin layer drying of different crops such as banana, mango, and cassava (Kameenan et al. 2009), cocoa beans (Hii et al. 2008), apple (Kavak Akpinar 2003; Kaleta et al. 2013; Wang et al. 2007), pistachio (Adnan Midilli 2001), banana blossom (Veeramanipriya et al. 2019), cassava (Veeramanipriya and Umayal Sundari 2021), cashew (Dhanush Kodi 2017), mint (Salla et al. 2015), potato (Dusko et al. 2015; Aghbashala et al. 2009), mushroom (Reyes et al. 2014; Arumuganathan et al. 2009; Doymaz 2014), pineapple (Olanipekun et al. 2015; Reddy et al. 2017), tomato (Hussein 2016), mango slices (Wang et al. 2018), coconut coir pith (Fernando 2016), green pepper (Kavak Akpinar et al. 2008), lemongrass (Nguyen et al. 2019), pumpkin (Tunde et al. 2013), cape gooseberry (Antonio et al. 2011), etc. to study the impact of various parameters such as drying time, desired drying temperature, relative humidity, and drying air velocity. Extraction of water from agricultural products particularly tuber crops in terms of thin layer drying kinetics and appropriate drying model for different dryers has been reported.

Literature survey indicates that many PVT dryers are represented with flat plate collectors, cabinet dryers, or green house dryers. Limited studies have been disclosed on evacuated tube collector (ETC)-assisted solar dryers so far. Usually, ETCs are observed to retain higher temperature and are also found to execute better during winter and rainy days. Based on the survey, it has been observed that (PVT) hybrid solar dryer-assisted ETC has not been used for drying of tuber root crops in India until now. The major objective of the present work is to determine the thin layer drying kinetics of untreated sweet potato slices using (PVT) hybrid solar dryer-assisted ETC in comparison with sun drying. Also, this research work is done to analyze the structural and morphological characteristics of dried sweet potato using both hybrid and sun drying methods.

2.2 Methodology

2.2.1 Sample Preparation and Drying Experimentation

Two hundred grams of sweet potato slices were spread evenly on a hot air oven at a temperature of 105°C for 24 hours. From bone-dried sweet potato, the initial moisture content was determined.

Figure 2.1 shows the schematic view of a photovoltaic thermal (PVT) hybrid solar dryer-assisted evacuated tube collector (ETC). The sweet potato slices were spread uniformly on three perforated aluminium trays which were then placed inside the drying chamber. A solar PV panel was used to convert solar energy into electricity which was stored in a battery. The stored energy provided electricity to run the blower motor. The blower motor circulates the surrounding air into the ETC. In ETC, it absorbs the solar radiation and transfers the heat into the air. The fluid air acquires heat and exits through the drying chamber. The moisture gets evaporated from the sweet potato slices due to the hot air and the vaporized air is taken away by the chimney.

Different parameters were noted periodically from 9.00 a.m to 5.00 p.m until the sample hits the equilibrium moisture content.

2.2.2 Numerical Modeling

The moisture ratio of hybrid and sun-dried sweet potato was calculated using the following formula (Umayal et al. 2017; Pineda-Gomez et al. 2014; Doymaz 2010; Kalteta et al. 2013):

$$MR = \frac{\left[M - M_e \right]}{\left[M_0 - M_e \right]} \tag{2.1}$$

$$MR = \frac{M}{M_0} \tag{2.2}$$

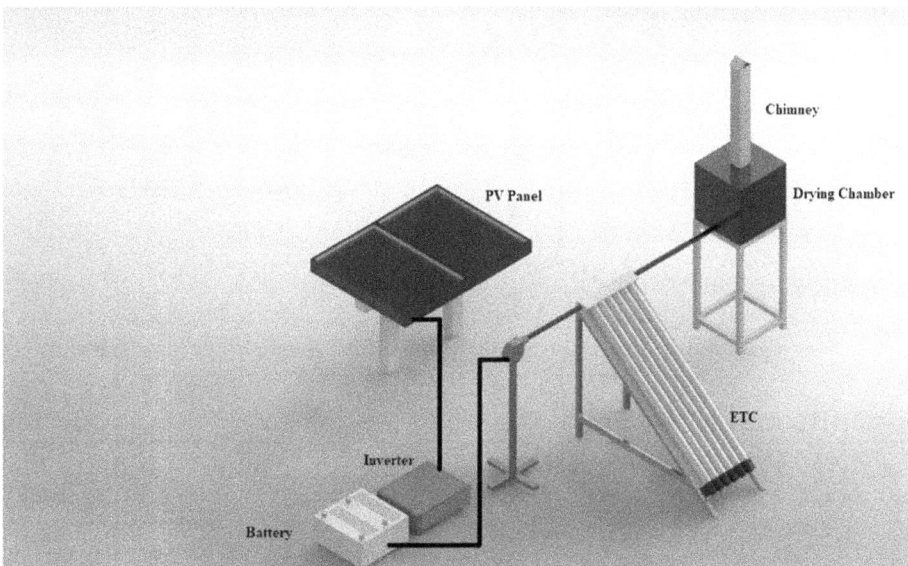

FIGURE 2.1   Schematic diagram of ETC-aided photovoltaic thermal (PVT) solar hybrid dryer.

TABLE 2.1   Various Mathematical Models Used for the Present Study

S. No.	Numerical Model	Mathematical Model Equation
1	Lewis (Newton)	MR = exp (–kt)
2	Page	MR = exp (–kt^n)
3	Henderson and Pabis	MR = a exp (–kt)
4	Logarithmic	MR = a exp (–kt) + c
5	Two term	MR = a exp (–k$_0$t) + (1–a) exp (–k$_1$t)
6	Verma et al.	MR = a exp (–kt) + (1–a) exp (–gt)
7	Wang and Singh	MR = 1 + at + bt^2
8	Midilli et al.	MR = a exp (–kt^n) + bt
9	Modified Henderson and Pabis	MR = a exp (–kt)+ b exp (–gt)+ c exp (–ht)

Numerical modeling of thin layer drying leads to the understanding of drying kinetics of sweet potato. Nine thin layer numerical models were proposed by many researchers which are appropriate for the present experimental data chosen and listed in Table 2.1 (Umayal et al. 2017; Pineda-Gomez et al. 2014; Doymaz 2010; Kalteta et al. 2013; Veeramanipriya et al. 2021). Non-linear regression analysis was performed using SPSS 23 (trial) statistical package.

The correlation coefficient (R^2) is the elementary criterion for choosing the best curve fit that is applied to define the thin layer drying kinetics of agricultural crops. The best model is also chosen based on other statistical parameters: reduced χ^2 and RMSE. The statistical parameters were estimated using the following formula (Umayal Sundari and Subramanian 2017; Pineda-Gomez et al. 2014; Doymaz 2010; Kalteta et al. 2013; Veeramanipriya and Umayal Sundari 2021),

$$R^2 = \frac{\sum_{i=1}^{n}\left(MR_{\exp,i} - \overline{MR_{\exp}}\right) \cdot \sum_{i=1}^{n}\left(MR_{pre,i} - \overline{MR_{pre}}\right)}{\sqrt{\sum_{i=1}^{n}\left(MR_{\exp,i} - \overline{MR_{\exp}}\right)^2 \cdot \sum_{i=1}^{n}\left(MR_{pre,i} - \overline{MR_{pre}}\right)^2}} \tag{2.3}$$

$$\chi^2 = \frac{\sum_{i=1}^{n}\left(MR_{\exp,i} - MR_{pre,i}\right)^2}{N - n} \tag{2.4}$$

$$RMSE = \left[\frac{1}{N}\sum_{i=1}^{n}\left(MR_{pre,i} - MR_{\exp,i}\right)^2\right]^{\frac{1}{2}} \tag{2.5}$$

These parameters were used to define the quality of exact fit. For defining the goodness of fit, R^2 should be greater (equal to 1), χ^2 and RMSE should be lesser than the correlation coefficient.

2.3 Results and Discussions

2.3.1 Parameters Determination

The various parameters of both hybrid and sun drying were noted periodically during the experimental period and are listed in Table 3.1. The performance of PVT hybrid solar dryer ETC is compared with that by natural sun drying (Table 2.2).

TABLE 2.2 Hourly Variations of Different Parameters Noted for Hybrid and Sun-Dried Sweet Potato for Day 1 and 2

Time	Solar Insolation	Wind Velocity	RH	Ambient	Chimney	Upper Tray	Middle Tray	Lower Tray	ETC Outlet	ETC Inlet
	W/m²	m/s	%	T_0	T_1	T_2	T_3	T_4	T_5	T_6
Day – 1										
9.00 am	452	1.2	59	31.4	35	36	36	37	69	55
10.00 am	789	1.42	57	31.8	44	46	44	46	84	58
11.00 am	896	1.12	54	32.3	46	48	47	51	99	61
12.00 nn	1005	0.09	52.5	33.2	43	45	45	49	98	58
01.00 pm	1120	0.05	53.7	33.7	40	41	42	42	92	51
02.00 pm	881.3	1.16	50.5	33.7	43	43	45	45	89	55
03.00 pm	549.3	1.38	53.2	33.2	43	44	44	44	84	54
04.00 pm	471.6	1.63	56	33.2	39	39	38	39	68	41
05.00 pm	456	1.14	56.8	31.6	36	37	36	37	64	39
06.00 pm	428.5	1.33	57.2	30.4	34	36	34	35	60	38
Day – 2										
9.00 am	610.9	1.61	61.5	30.8	-	-	-	-	-	-
10.00 am	582.6	1.23	62	31.2	-	-	-	-	-	-
11.00 am	544	0.82	63.6	31.6	-	-	-	-	-	-
12.00 nn	789	1.16	60.8	30.2	-	-	-	-	-	-

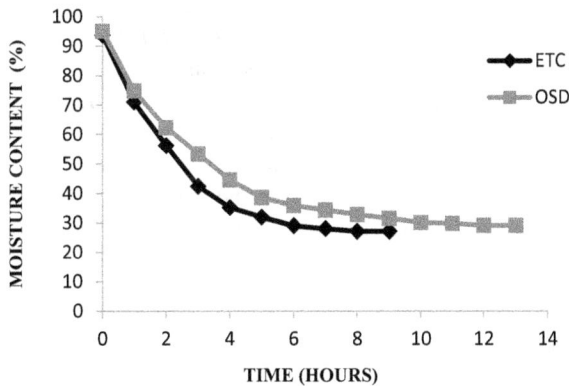

FIGURE 2.2 Moisture content vs drying time for ETC-based solar dryer and open sun drying.

From the calculated parameters, the solar insolation was found to vary from 428.5 to 1120 W/m². The performance of the ETC dryer can be determined from the average specific moisture extraction ratio (SMER) and specific extraction ratio (SER), which were 0.04375 and 58.54115, respectively. Figure 2.2 shows the moisture content versus drying time for ETC and sun drying of sweet potato. The time taken by the hybrid dryer to minimize the moisture content from 93.1 to 27.1% is 9 hours which is lower than sun drying which takes 13 hours to reach an equilibrium moisture content of 29.1%. The rate of moisture ratio of sweet potato in ETC and sun drying is shown in Figure 2.3.

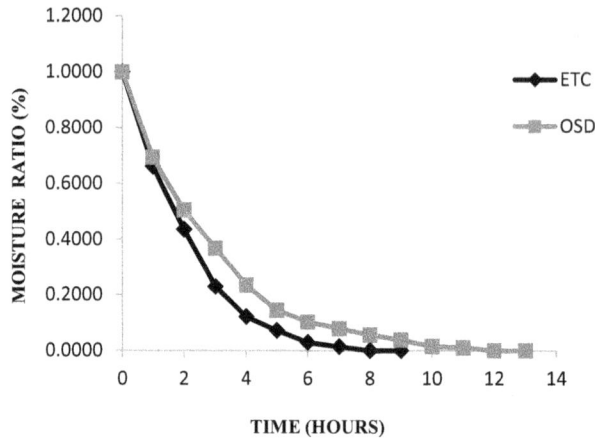

FIGURE 2.3   Moisture ratio vs drying time for ETC-based solar dryer and open sun drying.

2.3.2 Numerical Modeling

Table 2.3 shows the thin layer numerical drying models fitted for hybrid-dried sweet potatoes. In hybrid dryer, the value of R^2 varies from 0.733 to 0.998. The value of reduced χ^2 changes from 0.001869 to 0.110992 and RMSE varies from 0.000234 to 0.018499. The result shows that Midilli et al.'s model has the highest R^2 (0.998) value and the lowest value of reduced χ^2 (0.001869) and RMSE (0.000234). This shows that the predicted moisture ratio (MR) calculated from the Midilli (2001) model is accurately fitted for hybrid dryer.

In sun drying, the correlation coefficient (R^2) value changes from 0.960 to 0.999. For Midilli (2001) model, the R^2 value is close to 1 and reduced χ^2 value ranges from 0.001293 to 0.048437 and RMSE value ranges from 0.000108 to 0.004036. Midilli et al.'s model was chosen to present the thin layer drying kinetics of both ETC and sun-dried sweet potato that corresponds to the highest correlation coefficient (R^2) and lowest values of RMSE and reduced χ^2. The numerical model results are tabulated in Table 2.3.

Figure 2.4 (a) and (b) show the comparison between the predicted MR and experimental MR for untreated sweet potato slices in both the hybrid dryer and open sun drying methods of Midilli et al.'s model, respectively. The experimental MR found by performing the experiment in hybrid solar dryer

TABLE 2.3   Numerical Models of Evacuated Tube Collector-Based (PVT) Solar Dryer for Thin Layer Drying of Sweet Potato

S.No	Model	Constants	R^2	χ^2	RMSE
1	Newton	k = 0.510	0.773	0.010089	0.001121
2	Page	k = 0.405, n = 1.253	0.990	0.110992	0.018499
3	Henderson and Pabis	k = 0.521, a = 1.026	0.991	0.009280	0.001160
4	Logarithmic	k = 0.468, a = 1.056, c = −0.039	0.995	0.005300	0.000757
5	Two – Term	a = 0.942, b = 0.084, k_0 = 0.521, k_1 = 0.521	0.991	0.009280	0.001547
6	Verma et al.,	k = 0.308, a = −22.387, g = 0.315	0.996	0.003931	0.000562
7	Wang andSing	a = −0.317, b = 0.024	0.976	0.025227	0.003153
8	**Midilli et al.**	**k = 0.132, a = 0.599, b = −0.714, n = −0.00002**	**0.998**	**0.001869**	**0.000234**
9	Modified Henderson and Pabis	k = 0.361, a = 2.162, b = −0.574, g = 0.270, c = −0.575, h = 0.267	0.996	0.003784	0.000946

TABLE 2.4 Numerical models of sun drying for thin layer drying of sweet potato

S. No.	Model	Constants	R^2	χ^2	RMSE
1	Newton	k = 0.334	0.998	0.002264	0.000174
2	Page	k = 0.308, n = 1.063	0.999	0.004568	0.000245
3	Henderson and Pabis	k = 0.582, a = 1.000	0.998	0.002076	0.000173
4	Logarithmic	k = 0.447, a = 1.000, c = −0.008	0.998	0.001866	0.000170
5	Two – Term	a = 1.000, b = −1.000, k_0 = 0.985, k_1 = −0.979	0.999	0.001219	0.000122
6	Verma et al.	k = −1.000, a = 1.000, g = 1.000	0.999	0.001691	0.000154
7	Wang and Sing	a = 1.000, b = −0.968	0.960	0.048437	0.004036
8	**Midilli et al.**	**k = 0.648, a = 1.000, b = −0.127, n = −0.470**	**0.999**	**0.001293**	**0.000108**
9	Modified Henderson and Pabis	k = 0.285, a = 1.897, b = −0.446, g = 0.239, c = −0.446, h = 0.238	0.999	0.001653	0.000207

and sun drying for sweet potato slices was in good accordance with the predicted MR of Midilli et al.'s model. The results of the mathematical model were in good agreement with the results reported in the literature (Umayal et al. 2017; Pineda-Gomez et al. 2014; Doymaz 2010; Kalteta et al. 2013; Veeramanipriya et al. 2021).

2.3.3 SEM Analysis

Scanning electron microscopy analysis was applied for comparing the surface microstructure of hybrid and sun-dried sweet potato as illustrated in Figures 2.5 (a–d) at 250 × and 500 ×, respectively. This analysis was done to define the effect of drying method on the sample. The granule size of the hybrid-dried

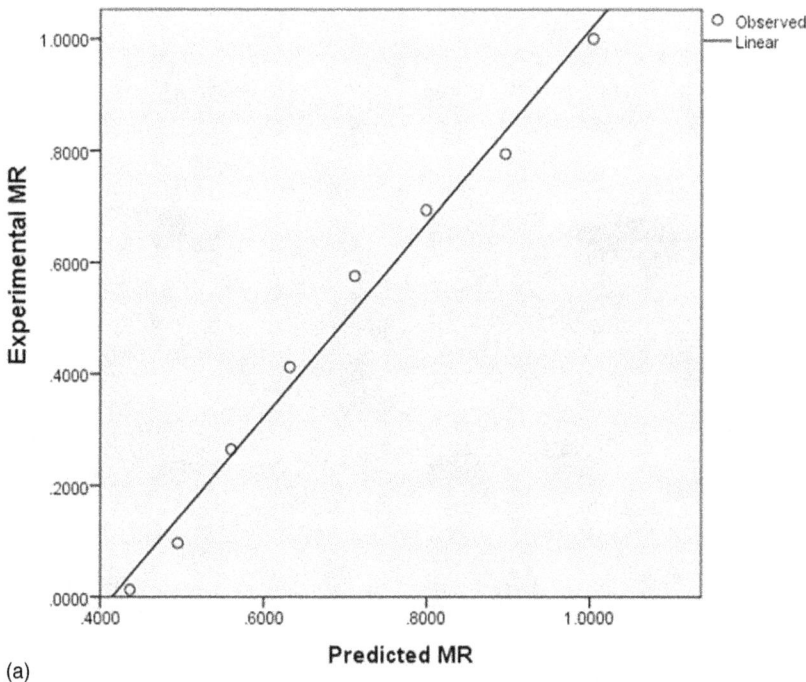

(a)

FIGURE 2.4 (a) Experimental moisture ratio vs predicted moisture ratio for sweet potato slices in ETC-aided photovoltaic thermal (PVT) solar hybrid dryer.

(Continued)

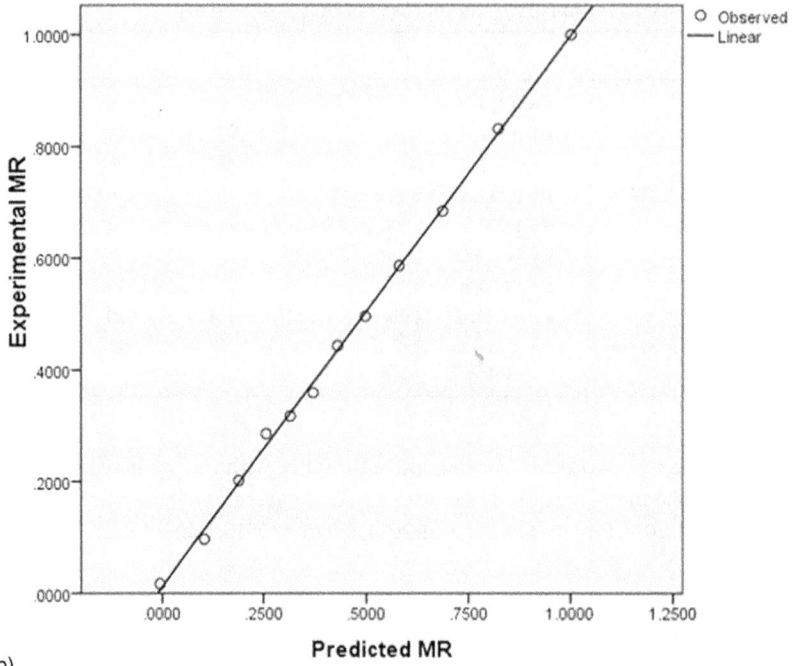

(b)

FIGURE 2.4 **(Continued)** (b) Experimental vs predicted moisture ratio for sweet potato slices in open sun drying.

(a)

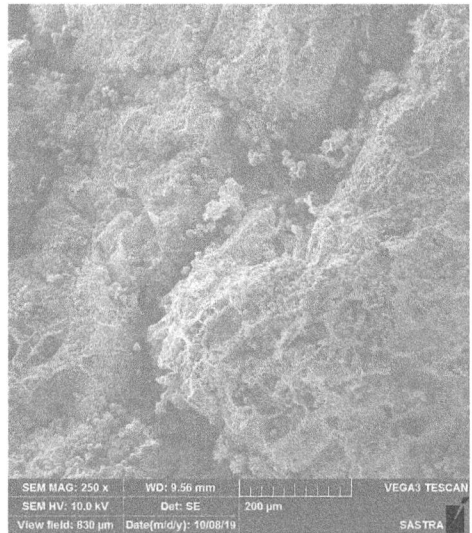

(b)

FIGURE 2.5 (a) and (b) SEM images of hybrid and sun-dried sweet potato slices (250×).

(*Continued*)

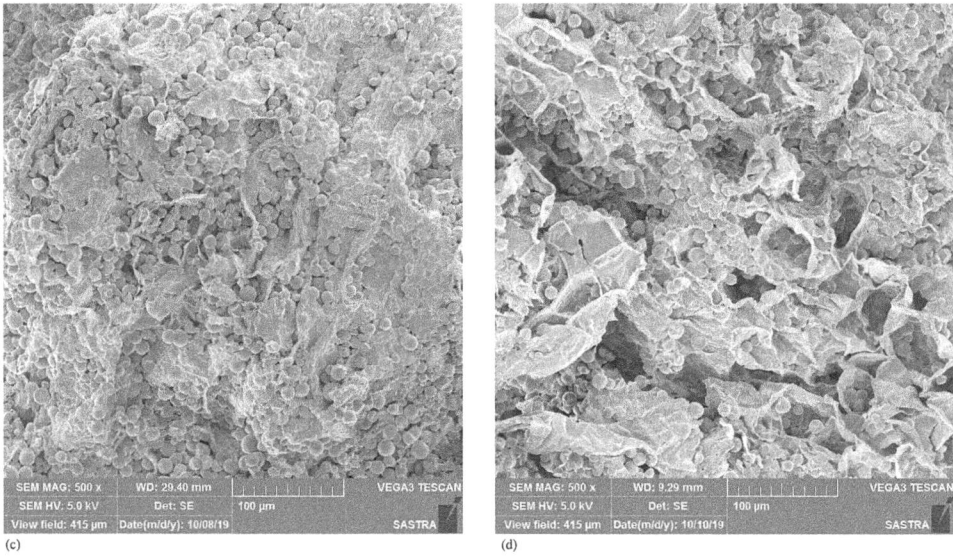

FIGURE 2.5 (Continued)   (c) and (d) SEM images of Hybrid and Sun Dried Sweet Potato Slices (500×).

sweet potato was found to vary from 4.2 to 6.1 µm, and for sun-dried sample, it is found to vary from 5.3 to 8.6 µm.

Hybrid-dried starch granules were found to have regular oval and spherical structures. The Sun-dried sample has different sizes of starch granules depending on the drying temperatures (Hartiningsih et al. 2020). Most of the starch granules were oval in shape with different sizes (Babu et al. 2015; Gasa et al. 2022; Iheagwara 2012). Based on the SEM results, it was determined that moisture removal has taken place at regular intervals of time, and the granules showed minimal defects in hybrid dryer. Similar results were accounted by many researchers in various drying methods (Guoa et al. 2019; Zhang et al. 2022; Chen et al. 2021; Pang et al. 2021; Gasa et al. 2022).

2.3.4 XRD Analysis

There are three types of starches A, B, and C classified based on the XRD patterns. C type lies on the combination of A- and B-type. Furthermore, it can be classified to C_A-, C_C- and C_B-type according to the proportion of A- and B-type crystallinity from high to low (He & Wei 2017). In this work, hybrid-dried sweet potato starch peaks at (2θ): 14.9°, 18.1°, and 23.2° and sun-dried sweet potato starch peaks at (2θ): 15.1°, 17.2°, and 23.1°. The XRD result showed that both hybrid- and sun-dried sweet potato starch exhibited 'C_A' type crystalline pattern and the crystallinity ranged from 24.6 to 25.4%, which indicates the semi-crystalline nature of starch. It also showed that the removal of moisture from hybrid-dried sweet potato was higher compared to sun-dried sample. Similar results were accounted by many researcher is illustrated in Figure 2.6 (a) and (b).

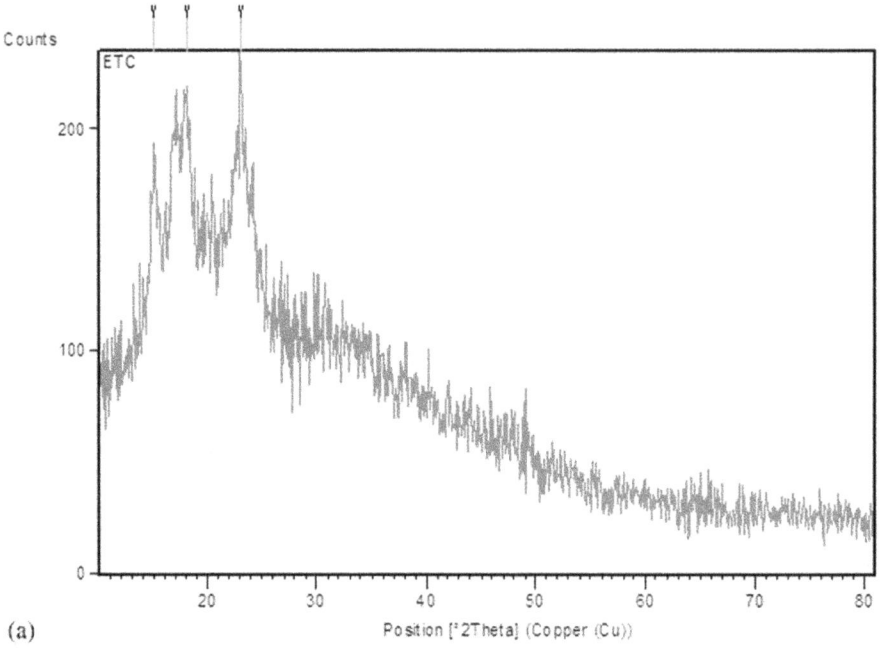

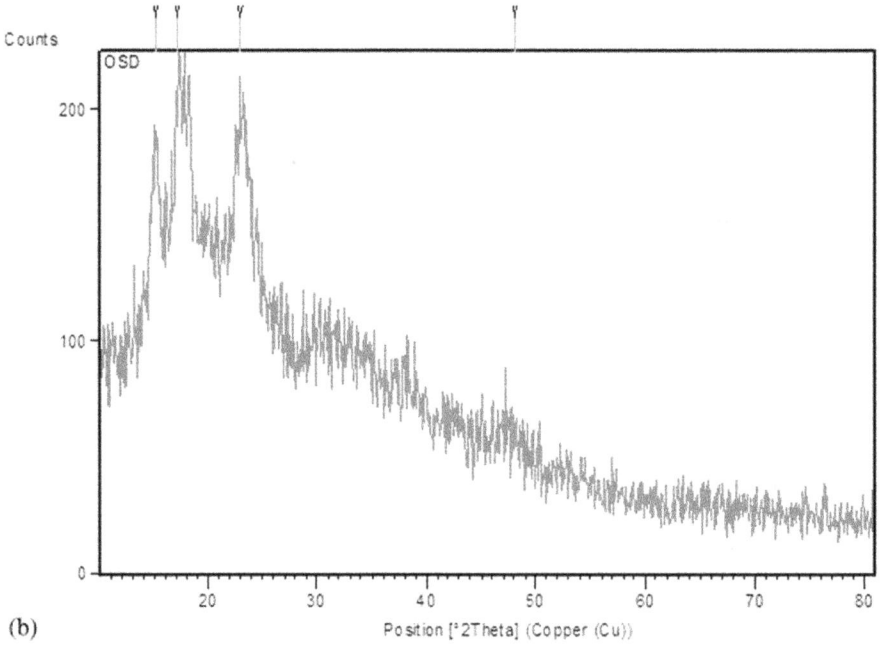

FIGURE 2.6 (a) XRD patterns of hybrid-dried sweet potato starch. (b) XRD patterns of sun-dried sweet potato starch.

2.4 Conclusion

- The present work shows the structural and morphological analysis of thin drying kinetics of PVT hybrid solar dryer-assisted ETC for chemically untreated sweet potato slices under the climatic conditions of Thanjavur, India. The comparison was done with sun drying.
- ETC outlet temperature was observed to be larger than the ambient temperature. This increased the rate of drying in the hybrid dryer.
- The initial moisture content of the dried sweet potato slices was calculated to be 93.1% (wb) for both drying methods. The equilibrium moisture content of hybrid dryer and sun drying was estimated to be 27.1% (wb) and 29.1%, respectively.
- Drying time of hybrid-dried sweet potato slices was 9 hours compared to sun drying of 13 hours.
- Nine developed mathematical models were applied to represent the thin layer drying kinetics of sweet potato slices in both drying methods to estimate the appropriate fit.
- Based on the modeling results, Midilli et al.'s model shows a higher equivalent of correlation coefficient (R^2) and lower equivalent of reduced chi-square and RMSE for both dryers, showing the suitability of fit for drying of untreated sweet potato slices.
- From the XRD results, it has been observed that the hybrid-dried sweet potato starch peaks at (2θ): 14.9°, 18.1°, and 23.2° and sun-dried sweet potato starch peaks at (2θ): 15.1°, 17.2°, and 23.1°. The XRD result shows that both hybrid- and sun-dried sweet potato starch exhibited 'C$_A$' type crystalline pattern and the crystallinity ranged from 24.6 to 25.4%.
- The granule size of the hybrid-dried sweet potato was found to be in the range from 4.2 to 6.1 μm, and for sun-dried sample, it ranges from 5.3 to 8.6 μm.
- SEM analysis reveals moisture was removed at periodic intervals of time, and the granules showed minimal defects in hybrid dryer.
- Based on the structural and morphological results, the appearance and quality of the hybrid-dried sweet potato was found to be of high grade compared to the sun-dried sample.

Nomenclature

MC	Moisture content
MR	Moisture ratio
m_i	Initial mass of cassava
m_f	Final mass of cassava
M	Moisture content at any time (% wb)
M_0	Initial moisture content (% wb)
$k, a, b, c, g, h, n, k_0, k_1$	Drying constants
$MR_{exp,i}$	Experimental moisture ratio
$MR_{pre,i}$	Predicted moisture ratio
N	Number of observations
RMSE	Root mean square error
R^2	Correlation coefficient

References

Abong, G., Ndanyi, V., Kaaya, A., Shibairo, S., Okoth, M.W., Lamuka, P., Odongo, N., Wanjekeche, E., Mulindwa, J., and Sopade, P. 2016. "A review of production, postharvest handling and marketing of sweetpotatoes in Kenya and Uganda." *Current Research in Nutrition and Food Science* 4(3): 162–181.

Aghbashala, M., Kianmehr, M.H., and Arabhosseini, A. 2009. "Modeling of thin layer drying of potato slices in length of continuous band dryer." *Energy Conversion and Management* 50: 1348–1355.

Akinwande, B., Ngoddy, P., Olajide, J., and Fakolujo, A. 2013. "Production and storage of cassava chips for reconversion into Gari." *Innovative Systems Design and Engineering* 4(9): 68–72.

Antonio, G., Azoubel, P., Murr, F.E.X., and Park, K.J. 2008. "Osmotic dehydration of sweet potato (Ipomoea batatas) in ternary solutions." *Ternary Solution OD Sweet Potato* 28(3): 696–701.

Antonio, G.C., Takeiti, C.Y., de Oliveira, R.A., and Park, K.J. 2011. "Sweet Potato: Production, morphological and physicochemical characteristics, and technological process." *Fruit, Vegetable and Cereal Science and Biotechnology* 5(2): 1–18.

Arumuganathan, T., Manikantan, M.R., Rail, R.D., Anandakumar, S., and Khare, V. 2009. "Mathematical modeling of drying kinetics of milky mushroom in a fluidized bed dryer." *International Agro Physics* 23: 1–7.

Babu, A.S., Parimalavalli, R., Jagannadham, K., and Rao, J.S., 2015. "Chemical and structural properties of sweet potato starch treated with organic and inorganic acid." *Journal of Food Science and Technology* 52(9): 5745–5753.

Cheetham, N.W.H., and Tao, L.P. 1998. "Variation in crystalline type with amylose content in maize starch granules: an X-ray powder diffraction study." *Carbohydrate Polymers* 36: 277–284, https://doi.org/10.1016/S0144-8617(98)00007-1

Chen, Lulu, Dai, Yangyong, Hou, Hanxue, Wang, Wentao, Ding, Xiuzhen, Zhang, Hui, Li, Xiangyang, and Dong, Haizhou. 2021. "Effect of high pressure microfluidization on the morphology, structure and rheology of sweet potato starch." *Food Hydrocolloids* 115: 106606.

Dhanush Kodi, S., Wilson, Vincent H., and Sudhakar, K. 2017. "Mathematical modeling of drying behaviour of cashew in a solar biomass hybrid dryer." *Resource – Efficient Technologies* 3: 359–364.

Doymaz, Ebrahim. 2014. "Drying kinetics and rehydration characteristics of convective hot-air dried white button mushroom slices." *Journal of Chemistry*, Article ID453175. http://doi.org/10.1155/2014/453175

Doymaz, Ibrahim. 2010. "Effect of citric acid and blanching pre – Treatments on drying and rehydration of amasya red apples." *Food and Bioproducts* 88: 124–132, https://doi.org/10.1016/j.fbp.2009.09.003

Salemović, D.R., Dedić, A.Đ., and Ćuprić, N.L. 2015. "A mathematical model and simulation of the drying process of thin layers of potatos in a conveyor – Belt dryer." *Thermal Science* 19(3): 1107–1118, https://doi.org/10.2298/TSCI130920020S

Fernando, J.A.K.M., and Amarasinghe, A.D.U.S. 2016. "Drying kinetics and mathematical modeling of hot air drying of coconut coir pith." *Springer Plus* 5: 807, https://doi.org/10.1186/s40064-016-2387-y

Gasa, Siyabonga, Sibanda, Sipho, Workneh, Tilahun, Laing, Mark, and Kassim, Alaika. 2022. "Thin-layer modelling of sweet potato slices drying under naturally-ventilated warm air by solar-venturi dryer." *Heliyon* 8: e08949, https://doi.org/10.1016/j.heliyon.2022.e08949

Guo, Ke, Bianc, Xiaofeng, Jia, Xiaofeng, Zhang, Long, and Wei, Cunxu. 2020. "Effects of nitrogen level on structural and functional properties of starches from different colored-fleshed root tubers of sweet potato." *International Journal of Biological Macromolecules* 164: 3235–3242, https://doi.org/10.1016/j.ijbiomac.2020.08.199

Guoa, Ke, Liua, Tianxiang, Xua, Ahui, Zhanga, Long, Bianc, Xiaofeng, and Weia, Xiaofeng. 2019. "Structural and functional properties of starches from root tubers of white, yellow, and purple sweet potatoes." *Food Hydrocolloids* 89: 829–836, https://doi.org/10.1016/j.foodhyd.2018.11.058

Hande, A.R., Swami, S.B., and Thakor, N.J. 2016. "Open-air sun drying of Kokum (Garcinia indica) rind and its quality evaluation." *Agricultural Research* 5(4): 373–383.

Hartiningsih, S., Pranoto, Y., and Supriyanto. 2020. "Structural and rheological properties of modified sago starch (Metroxylon sagu) using treatment of steam explosion followed by acid-hydrolyzed as an alternative to produce maltodextrin." *International Journal of Food Properties* 23(1): 1231–1242.

He, W., and Wei, C. 2017. "Progress in C-type starches from different plant sources." *Food Hydrocolloids* 73: 162–175, https://doi.org/10.1016/j.foodhyd.2017.07.003

Hii, C.L., Law, C.L., and Cloke, M. 2008. "Modelling of thin layer drying kinetics of cocoa bean's during artificial and natural drying." *Journal of Engineering Technology* 3(1): 1–10.

Hussein, J.B., Filli, K.B., and Oke, M.O. 2016. "Thin layer modelling of hybrid, solar and open sun drying of tomato slice." *Research Journal of Food Science and Nutrition* 1: 15–27.

Iheagwara, M. 2012. "Isolation, modification and characterization of sweet potato (Ipomoea batatas L (lam)) starch." *Journal of Food Processing & Technology* 4: 1–6.

Kalteta, Agnieszka, Gornicki, Krzysztof, Winiczenko, Radoslaw, and Chojnacka, Aneta. 2013. "Evaluation of drying models of apple (Var. Ligol) dried in a fluidized bed dryer." *Energy Conversion and Management* 67: 179–185. http://doi.org/10.1016/j.enconman.2012.11.011

Kameenan, Blaise Koua, Fassinau, Wanignon Ferdinand, Gbaha, Prosper, and Toure, Siaka 2009. Mathematical modeling of the thin layer solar drying of banana, mango and cassava. *Energy* 34: 1594–1602, https://doi.org/10.1016/j.energy.2009.07.005

Kavak Akpinar, E., and Bicer, Y. 2008. "Mathematical modeling of thin layer drying process of long green pepper in solar dryer and under open sun." *Energy Conversion and Management* 49: 1367–1375.

Kavak Akpinar, E., Bicer, Y., and Midilli, A. 2003. "Modeling and experimental study on frying of apple slices in a convective cycone dryer." *Journal of Food Process Engineering* 26: 515–541.

Latif, S., and Müller, J. 2015. "Potential of cassava leaves in human nutrition. A review." *Trends in Food Science and Technology* 44(2): 147–158.

Midilli, Adnan. 2001. "Determination of pistachio drying behavior and conditions in a solar drying systems." *International Journal of Engine Research* 25: 715–725, https://doi.org/10.1002/er.715

Nguyen, Thi Van Linh, Nguyen, My Duyen, Nguyen, Duy Chinh, Bach, Long Giang, and Lam, Tri Duc. 2019. "Model for thin layer drying of lemongrass (cymbopogon citratus) by hot air." *Processes* 7: 21, https://doi.org/10.3390/pr701002

Oke, M., and Workneh, T. 2013. "A review on sweet potato postharvest processing and preservation technology." *African Journal of Agricultural Research* 8(40): 4990–5003.

Olanipekun, B.F., Tunde-Akintunde, T.Y., Oyelade, O.J., Adebisi, M.G., and Adenaya, T.A. 2015. "Mathematical modeling of thin – layer pineapple drying." *Journal of Food Processing and Preservation* 39: 1431–1441, ISSN: 1745-4549.

Onu, C.E., Igbokwe, P.K., and Nwabanne Joseph, T. 2017. "Effective moisture diffusivity, activation energy and specific energy consumption in the thin-layer drying of potato." *International Journal of Novel Research in Engineering and Science* 3(2): 10–22.

Pineda-Gomez, Posidia, Angel-Gil, Natalia C., Valencia-Munoz, Carolina, Rosales-Rivera, Andres, and Rodriguez-Garcia, Mario E. 2014. "Thermal degradation of starch sources: Green banana, potato, cassava and corn – Kinetic study by non – isothermal procedures." *Starch*, 66, pp. 1–9, https://doi.org/10.1002/star.201300210

Pirasteh, G., Saidur, R., Rahman, S.M.A., and Rahim, N.A. 2014. "A review on development of solar drying applications." *Renewable and Sustainable Energy Reviews* 31: 133–148.

Reddy, Ravula Sudharshan, Ravula, Parabhaker Reddy, Arepally, Divyasree, Munagala, Surender Reddy, and Golla, Sheshasayana. 2017. "Drying kinetics and modelling of mass transfer in thin layer convective drying of pineapple." *Chemical Science International Journal* 19(3): 1–12. Article no. CSIJ.32746. ISSN: 2456-706X.

Reyes, Alejandro, Mahn, Andrea, and Vasquez, Fransisco. 2014. "Mushrooms dehydration in a hybrid – solar dryer using a phase change material." *Energy Conversion and Management* 83: 241–248.

Salla, Y.I., Aly, M.H., Nassar, A.F., and Mohamed, E.A. 2015. "Solar drying of whole mint plant under natural and forced convection." *Journal of Advanced Research* 6: 171–178.

Sanginga, N. 2015. *Root and Tuber Crops (Cassava, Yam, Potato and Sweet Potato)*. African Development Group, Dehar, Senegal.

Sanni, L., Onadipe-Phorbee, O., and Alenkhe, E. 2012. *Low-Cost Sustainable Cassava Drying Technologies in West Africa*. International Institute of Tropical Agriculture, Nigeria, IITA Nigeria.

Seidu, J., Kwenin, W., Tevor, W.J., Mahama, A., and Agbeven, J. 2012. "Drying of sweet potato (ipomoea batatas) (chipped and grated) for quality flour using locally constructed solar dryers." *ARPNJournal of Agricultural and Biological Science* 7(6): 466–473.

Sevik, Seyfi, Aktas, Mustafa, Dolgun, Ekin Can, Arslan, Erhan, and Tuner, Azim Dogus. 2019. "Performance analysis of solar and solar – Infrared dryer of mint apple slices using energy – Exergy methodology." *Solar Energy* 180: 537–549.

Titus, P., and Lawrence, J. 2015. Suitability of Popular Caribbean Varieties for Value Added Product Development. In *Inter-American Institute for Cooperation on Agriculture (IICA)*, Port of Spain, Trinidad and Tobago.

Tunde-Akintunde, T.Y., and Ogunlakin, G.O. 2013. "Mathematical modeling of drying of pretreated and untreated pumpkin." *International Journal of Science & Technology* 50(4): 705–713, https://doi.org/10.1007/s13197-110-0392-2

Umayal Sundari, A.R., and Subramanian, C.V. 2017. "Comparative study of solar drying characteristics and thin layer mathematical modeling of mango and cluster beans in two types of solar dryers." *International Journal of Latest Engineering Research and Applications* 2(1): 49–58.

Veeramanipriya, E., and Umayal Sundari, A.R. 2021. "Performance evaluation of hybrid photovoltaic thermal (PVT) solar dryer for drying of cassava." *Solar Energy* 215: 240–251.

Veeramanipriya, E., Umayal Sundari, A.R., and Asaithambi, R. 2019. "Numerical modelling of drying kinetics of banana flower using natural and forced convection dryers." *International Journal of Innovative Technology and Exploring Engineering* 8(10): 4193–4197.

Vega-Gálvez, Antonio, Puente-Díaz, Luis, Lemus-Mondaca, Roberto, Miranda, Margarita, and Torres, María José. 2012. "Mathematical modeling of thin-layer drying kinetics of cape gooseberry (Physalis Peruviana L.)." *Journal of Food Processing and Preservation*, ISSN 1745-4549, https://doi.org/10.1111/jfpp.12024

Wang, Wei, Li, Ming, Hassanien, Reda Hassanien Emam, Wang, Yunfeng, and Yang, Luwei. 2018. "Thermal performance of indirect forced conversion solar dryer and drying kinetics of mango." *Applied Thermal Engineering*, https://doi.org/10.1016/j.applthermaleng.2018.01.115

Wang, Z.J., Sun, X., Liao, F., Chen, G., Zhao, J., and Hu, X. 2007. "Mathematical modeling on hot air drying of thin layer apple pomace." *Journal of Food Engineering* 40: 39–46.

Xu, A., Guo, K., Liu, T., Bian, X., Zhang, L., and Wei, C. 2018. "Effects of different isolation media on structural and functional properties of starches from root tubers of purple, yellow and white sweet potatoes." *Molecules* 23: 2135, https://doi.org/10.3390/molecules23092135

Yılmaz, F.M., Yüksekkaya, S., Vardin, H., and Karaaslan, M. 2017. "The effects of drying conditions on moisture transfer and quality of pomegranate fruit leather (pestil)." *Journal of the Saudi Society of Agricultural Sciences* 16(1): 33–40.

Zhang, Min, Suo, Wenjing, Deng, Yuxin, Jiang, Lijun, Qi, Mingming, Liu, Yao, Li, Luxia, Wang, Chenjie, Zheng, Hui, and Li, Hongjun. 2022. "Effect of ultrasound-assisted dough fermentation on the quality of dough and steamed bread with 50% sweet potato pulp." *Ultrasonics Sonochemistry* 82: 105912, https://doi.org/10.1016/j.ultsonch.2022.105912

3

Armchair Carbon Nanotube Magneto Flexo Thermo Elastic Mass Sensor with Non-Linear Vibration on an Elastic Substrate

M. Mahaveer Sree
Jayan and Dr.
V. Vaithiyanathan
*Indra Ganesan College of
Engineering, Trichy, India*

M. Anusuya
*Indra Ganesan College of
Engineering, Trichy, India*

Dr. R. Selvamani
*Karunya Institute of
Technology and Sciences,
Coimbatore, India*

N. Ramya
*Saranathan College of
Engineering, Trichy, India*

The objective of this study is to research on the behaviour of non-straight ultrasonic waves in a solitary-walled carbon nanotube implanted with mass sensors on a polymer grid. Non-nearby flexibility, as proposed by Eringen, represents the impact at the microscale. The frequency equations were numerically analysed utilising non-linear foundations backed by the Winkler–Pasternak model after the proper arrangement of the numerical model comprising halfway differential conditions was done. Ultrasonic wave scattering relations were utilised for the arrangement. Quantitative investigation into the impacts of non-neighbourhood scaling, magneto-flexo-mechanical loadings, establishment boundaries, joined mass, fluctuating limit conditions, and nanotube length on the dimensionless recurrence of the structure has been conducted using the metric system. The dimensionless frequency of nano tubes is revealed to be considerably impacted by the border circumstances, non-local parameter, attached mass, and tube geometrical characteristics.

DOI: 10.1201/9781003355755-3

3.1 Introduction

The development of nano structures has been helped by the coupling of magneto flexo thermo elasticity in a chair-bound single-walled carbon nanotube (SWCNT). Recently, there has been a surge of interest in enhancing the performance of embedded nanostructures by utilising MFT-based nanomaterials in a polymer grid as an encompassing medium (Ebrahimi and Dabbagh, *2018*). The non-local Euler-Bernoulli (EBT) and Timoshenko beam theory may now analyse the scaling impact on the dispersion relations of a CNT, thanks to the development of non-local continuum theory. Multiple nanomaterials have been used to build and test non-local continuum field theories (Wang, *2005*), (Heireche et al., *2008*), (Eringen, *2002*), (Eringen and Edelen, *1972*), and (Eringen, *1983*). Wang et al. (*2006*) used non-local continuum models to examine the impact of scale on flexible locking in carbon nanotubes and indirectly addressed the effect of scale on vibration modes. Using non-local elasticity theory, Fang (*2013*) studied the non-straight free vibration of twofold walled carbon nanotubes. They found that the non-straight spread and sufficiency increment are fundamentally impacted by the encompassing versatile medium. The impacts of the limited scale boundary are generally critical in the full frequencies of the framework within the sight of liquid climate, as explored by Saadatnia and Esmailzadeh (*2017*) in their investigation of the non-straight symphonious swaying of a piezoelectric-layered nanotube conveying liquid stream. The forced displacement of a fluid carrying carbon nanotubes was read in Askari and Esmailzadeh (*2017*) while accounting for the heat impact and non-linear basis. According to Gheshlaghi and Hasheminejad (*2011*), homogeneous nano beams exhibit forced vibration due to heat and non-linear vibrational behaviour.

The vibrational behaviour of a magnetic fluid conveying CNTs in the presence of a longitudinal magnetic field was examined by Sadeghi-Goughari et al. (*2017*). The uncontrolled vibration of a viscoelastic nanotube in a longitudinal magnetic flux has been proven by Zhen et al. (*2019*), who also showed that higher natural frequencies decrease dramatically when non-local parameters rise while the first natural frequency increases modestly. Particular modes for post-buckling features of non-local nanobeams were investigated (Dai et al. *2018*) in a longitudinal attractive field. Capitalising on the leverage of a longitudinal attractive field, Ebrahimi and Barati (*2016*) explored the wave proliferation in non-neighbourhood permeable multi-stage nano glasslike nanobeams. They emphasised the idea that if inhomogeneity magnitudes are reduced, there may be an increase or fall in wave frequencies and phase velocities. Based on non-nearby strain angle hypothesis, Li et al. (*2016*) concentrated on the wave spread in viscoelastic SWCNTs with surface impact under attractive transition and concluded that damping coefficient was crucial. The impact of longitudinal attractive transition on the engendering of SWCNT waves in a fluid medium was discussed by Arani et al. (*2016*), with the Knudsen number and surface considerations utilised to illustrate the argument. To concentrate on the engendering of longitudinal pressure waves utilising a longitudinal attractive field, Zhang et al. (*2016*) led a unique examination of horn-moulded SWCNTs immersed in a viscoelastic medium. Güven (*2015*) develops a two-scale coefficient model utilising a unified non-local versatility hypothesis. Wang (*2005*) showed the non-neighbourhood versatile shell model for separating longitudinal waves in SWCNTs and found that the microstructure and the trading of the longitudinal wave and outspread movement assume a fundamental part in wave scattering. Carbon nanotubes' magneto-thermal main frequency response was studied (Azarboni, *2019*) for its effects on the material under different boundary conditions. For bigger excitation amplitudes and an assortment of limit conditions, they contemplated that an expansion in the longitudinal attractive field would cause the backward leaping to move. Non-local continuum models were utilised by Pradhan and Phadikar (*2009*) to examine the microscale impact of implanted multi-layered graphene on vibration.

Researchers (Semmah et al., *2004*) used an improved non-local model to look at the warm clasping ways of behaving of crisscross SWCNTs. In light of their discoveries, they verified that the warm clasping attributes of SWCNTs are extremely delicate to both the scale impact and the chirality of the crisscross carbon nanotube. Wave proliferation in SWCNTs has been shown in a heated environment (Naceri et al., *2011*). Baghdadi et al. (*2014*) used non-neighbourhood explanatory pillar hypothesis to concentrate on what intensity means for the vibrational properties of easy chair and crisscross SWCNTs. As well as finding out about the chirality of easy chair and crisscross carbon nanotubes and how it relates to temperature, they also found a significant correlation between natural frequencies and thermal fluctuations. Non-neighbourhood Timoshenko bar hypothesis was utilised to take a gander at the impact of power on the vibration of SWCNTs (Benzair et al., *2008*), and the outcomes showed a decline in frequencies contrasted with those got utilising the Euler bar model. Besseghier et al. (*2011*) looked at how temperature affected the movement of waves through polymer-embedded twofold walled carbon nanotubes utilising non-neighbourhood flexibility. Investigations of the impacts of temperature on the interfacial pressure move qualities of single-and multi-walled carbon nanotubes/polymer composite frameworks were directed utilising thermo-versatile hypothesis and customary fibre pullout models (Zhang et al., *2007*). Despite this, various scholars worked on distinct theories of thermoelasticity (Lata and Kaur, *2019a*, *2019b*). Lata et al. (*2016*) and Kumar et al. (*2017*) used thermal elasticity to study deformation in a transversely isotropic material.

To ascertain the non-neighbourhood scaling boundary for easy chair and crisscross SWCNTs (Narendar et al., *2011*), the creators joined atomic primary mechanics, non-nearby versatility, and wave spread. This chapter concentrates on the warm clasping qualities of an SWCNT implanted in a flexible medium as given by Bedia et al. (*2015*). The fastening assessment of chiral SWCNTs was made using the non-close by Timoshenko shaft speculation (Zidour et al., *2014*). The effects of temperature and chirality on the vibration of SWCNTs implanted in a polymeric network were explored (Bensattalah et al., *2016*) utilising non-nearby versatility models. To decide the reverberation recurrence of chiral SWCNTs, Hsu et al. (*2008*) utilised Timoshenko bar hypothesis. The pivotal vibration of SWCNTs embedded in versatile media was studied by Aydogdu (*2012*), who concluded that the conventional rod model greatly exaggerated the frequencies of SWCNT axial vibrations in elastic media by failing to account for their extremely small dimensions. Taking into account the impacts of the warm climate, Ansari et al. (*2012*) zeroed in on the powerful strength of implanted SWCNTs and presumed that the contrast between the precariousness areas anticipated by nearby and non-neighbourhood pillar hypotheses is enormous for nanotubes with lower angle proportions.

Scientists have recently examined nanoscale structures in conjunction with mass sensors. Finite element method (FEM) based on continuum mechanics was used by Wu et al. (*2006*) to explore the reverberation recurrence of a single-walled carbon nanotube (SWCNT). (Li and Chou *2006*) utilised a sub-atomic underlying mechanics demonstrating way to deal with make a CNT nano-mass sensor. Research similar to this one analysed the recurrence shift conduct of a plate-type nano-mass sensor made of functionally graded nanostructures. It was introduced by Barati and Shahverdi (*2017*). For instance, utilising continuum mechanics theory, the potential of SWCNT as a mass sensor is studied (Chowdhury et al., *2009*). The authors of this chapter examine how the nano-mass sensor is affected by factors such as the ratio of stiffness and mass, the location of the measured elastic mass, and a non-local parameter to a frequency with no dimensions (Arda and Aydogdu, *2020*). A unique nanoscale mass sensor was analysed by researchers (Liu and Lyu, *2020*). It comprised of a savvy FG-MEE centre with graphene layers on the top and base surfaces. The repeat shift and responsiveness of a carbon nanotube-based sensor with a joined mass were made utilising non-nearby versatility hypothesis (Lee et al., *2010*).

For the purpose of capturing vibrational energy, an array of flexoelectric multi-layer nanobeams was proposed by Wang et al. (*2018*). In their examination of the surface effect on the scale-subordinate vibration behaviour of flexoelectric sandwich nano beams (Ebrahimi et al., *2019*), they discovered that flexoelectricity contributes to the natural frequencies. Using the element-free Galerkin approach, they performed a static analysis of flexoelectric nanobeams that took into account surface effects (Basutkar et al., *2019*). Nematollahi et al. (*2019*) reviewed and developed a method for predicting the mass proportion and liquid speed of a piezoelectric nanotube shipping liquid. In order to explore the small-scale influence at the axial vibrations of SWCNTs with connected mass, Aydogdu and Filiz (*2011*) discovered an improved elastic rod version. To the best of the creator's information, no past work has endeavoured to apply Eringen's non-nearby versatility hypothesis to the issue of making sense of the presence of non-direct ultrasonic waves in a magneto-thermo adaptable rocker SWCNT furnished with mass sensors lying on a polymer organisation. Understanding the non-direct ultrasonic vibration lead of the arm-chair SWCNT with mass sensor lying on versatile substrate in magneto-thermoelastic environment is therefore imperative from a scientific standpoint (Selvamani et al., 2020). Single-layer graphene sheets' locking conduct in a hygro-warm climate on a versatile medium was broken down using a changed form of the two-variable plate hypothesis and the non-nearby strain slope hypothesis. Several authors, including Selvamani and Makinde (*2018*). (Ebrahami et al., *2021*) Utilising the constitutive condition from the straight hypothesis of flexibility and piezoelectricity, we investigate how rotation affects the axisymmetric waves travelling down a thin-layered piezoelectric rod.

In this investigation, we used Euler beam theory and the elasticity of linked mass sensors to investigate the non-linear magneto-flexo-thermo-versatile waves in a leaning back SWCNT upheld by a polymer lattice. Eringen's non-neighbourhood flexibility hypothesis gives the establishment for the analytical formulation, which allows for the inclusion of effects at the microscale. We utilised the relationships between the propagation of ultrasonic waves. Metric studies were used to look at how factors including magneto-electro-mechanical loadings, connected mass, non-neighbourhood boundary, and viewpoint proportion influence the avoidance attributes of nanotubes. A series of graphics and tables show the impact of each parameter.

3.2 Mathematical Formulations

3.2.1 Eringen Non-Local Theory of Elasticity

According to this hypothesis, the stress state at a certain location x in the body depends on the strain state at x, yet additionally on the strain states at any remaining areas x is in contact with. Non-local elasticity is characterised by constitutive equations that have a generic form that include an integral across the whole area of interest. The general form of the constitutive equations in the non-local form of elasticity contains an integral over the entire region of interest. The constitutive conditions of a straight, homogeneous, isotropic, non-nearby versatile strong with zero body powers are given by Eringen (2002), Eringen and Edelen (1972), and Eringen (1983).

$$\sigma_{ij} + \rho\left(f_j - \ddot{u}_j\right) = 0 \tag{3.1}$$

$$\sigma_{ij}\left(X\right) = \int_v \pi\left(\left|X - X'\right|\right), \tau)\sigma_{ij}^c\left(X'\right) dv\left(X'\right) \tag{3.2}$$

$$\sigma_{ij}^{\ c} = C_{ijkl}\varepsilon_{kl} \tag{3.3}$$

$$e_{ij}\left(X'\right) = \frac{1}{2}\left(u_{i,j} + u_{j,i}\right) \tag{3.4}$$

The equilibrium equation is given by (3.1) and it contains $\sigma_{ij,i}, \rho, f_j, u_j$ the pressure tensor, mass thickness, body force thickness, and relocation vector at a body reference point x at time t. For a traditional constitutive connection, where $\sigma^c_{ij}(X')$ is the classical stress tensor at a given location, and is the linear strain tensor $e_{ij}(X')$ at that location, see Eq. (3.3). As shown in Eq. (3.4), the old style strain relocation relationship is based on a linear connection. The non-local influence in the constitutive equations is represented by the kernel function, an $\pi(|X - X'|, \tau)$ attenuation function. The volume integral in (3.2) is defined across the area v that the body occupies. Similar equations of classical elasticity are given in Eq.

(3.2), with the only variation being the substitution of Eq. (3.2) for Eq. (3.3), where Hooke's law is used (3.2). Since the non-local modulus is defined by a set of parameters that rely on a trademark length (grid boundary, grain size, granular distance, and so forth.), these values are embedded in Eq. (3.2), where "*l*." is the system's external characteristic length (frequency, break length, size or aspects of test, and so on) Because of this, the non-neighbourhood modulus might be composed as,

$$\pi = \pi\left(\left|X - X'\right|, \tau\right), \tau = \frac{e_o a}{l} \tag{3.5}$$

where e_0 is a constant corresponding to the materials and has to be determined for each materials independently. Then, the integro-partial differential Eq. (3.2) of non-local elasticity can be simplifies to partial differential equation as follows τ

$$\left(1 - \tau^2 l^2 \nabla^2\right) \sigma_{ij}(X) = \sigma_{ij}^c(X) = C_{ijkl} e_{kl}(X) \tag{3.6}$$

where C_{ijkl} is the elastic modulus tensor of classical isotropic elasticity and e_{ij} is the strain tensor. Where ∇^2 denotes the second-order spatial gradient applied on the stress tensor σ_{ij} and $\tau = e_o a/l$. Eringen proposed $e_o = 0.39$ by the matching of the dispersion curves via non-local theory for place wave and born-Karman model of lattice dynamics at the end of the Brillouin zone ($ka = \pi$), where a is the distance between atoms is and k is the wavenumber in the phonon analysis. On the order hand, Eringen proposed $e_o = 0.31$ in his study Eringen and Edelen (1972) for Rayleigh surface wave via non-local continuum mechanics and lattice dynamics.

3.2.2 Carbon Nanotube Atomic Structure

Carbon nanotubes are considered to be tubes formed by rolling a grapheme sheet about the $\vec{T}$ vector. A vector perpendicular to $\vec{T}$ is the chiral vector denoted by $\vec{C}_h$. The chiral vector and corresponding chiral angle define the type of CNT, i.e., zigzag, armchair and chiral $\vec{C}_h$. can be expressed with respect to two base vector $\vec{a}_1$ and $\vec{a}_2$ as

$$\vec{C}_h = n\vec{a}_1 + m\vec{a}_2 \tag{3.7}$$

Where n and m are the indices of translation which decide the structure around the circumference Figure 3.1 descripts the lattice of transition (n, m) along with the base vectors $\vec{a}_1$ and $\vec{a}_2$. If the indices of translation are such that $m = 0$ and $m = n$ then the corresponding CNT are categorized as zigzag and armchair, respectively. Considering the chirality diameter and the chiral angle of the CNT is calculated by the chiral vector for each nanostructure. The diameter of armchair single-walled carbon nanotube for ($n=m$) is given by (Yamabe, 1995)

$$d = \frac{3na}{\pi} \tag{3.8}$$

The mechanical properties of *SWCNTs* were investigated utilising an energy-identical model created by Wu et al. (2006b), which considers the association between atomic mechanics and strong mechanics. A similar expression is used to express the armchair nanotube's Young's modulus:

$$E_{SWCNT} = \frac{4\sqrt{3KC}}{9Ct + 4Ka^2 t\left(\gamma^2_{21} + \gamma^2_{22}\right)} \tag{3.9}$$

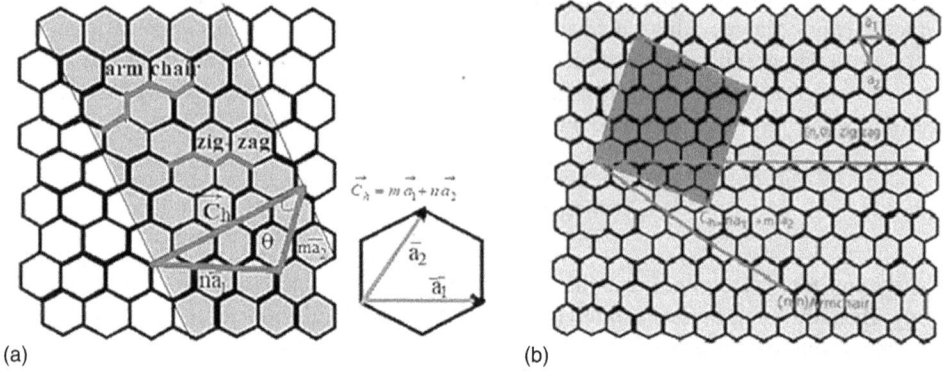

FIGURE 3.1   (a) Properties of SWCNT geometry. (b) Carbon nanotubes have a graphene structure.

where K and C are the force constants. t is thickness of the parameters γ_{21} and γ_{22} are given by:

$$\gamma_{21} = \frac{-3\sqrt{4} - 3\cos^2(\pi/2n)\cos(\pi/2n)}{8\sqrt{3} - 2\sqrt{3}\cos^2(\pi/2n)}$$

$$\gamma_{22} = \frac{12 - 9\cos^2(\pi/2n)}{16\sqrt{3} - 4\sqrt{3}\cos^2(\pi/2n)}$$

With $n \to \infty$, the expressions for a graphite sheet's Young's modulus are as follows:

$$E_g = \frac{16\sqrt{3}Kt}{18Ct + Ka^2t} \tag{3.10}$$

3.2.3 Magnetic Field Force Fundamental Equations

The carbon nanotubes are situated in an empty construction comprising of covalently fortified carbon molecules, which might be viewed as a rectangular graphite sheet moved from one side of its longest edge to make a barrel-shaped tube (displayed in Figure 3.2). A barrel-shaped coordinate framework is displayed in Figure 3.3. The surface (X, θ, Z) denoted by situated in the shell's central region, where the longitudinal direction, the circumferential direction, $Z = 0$ and the radial direction are all represented. Assuming that carbon nanotubes CNTs η and the medium surrounding them both have the same magnetic permeability, the Maxwell equation is,

$$f = \nabla \times \overline{s}, \nabla \times \varepsilon = -\eta.\frac{\partial \overline{s}}{\partial t}, div\ \overline{s} = 0, \overline{s} = \nabla \times (\overline{U} \times \overline{H}), \varepsilon = -\eta\left(\frac{\partial \overline{s}}{\partial t} \times \overline{H}\right) \tag{3.11}$$

$f, \overline{s}, \varepsilon$, and $\overline{U}$ represent the Hamilton arithmetic operators of the shell, current density, electric field strength vectors, magnetic field disturbance vectors, and displacement vectors, respectively. $\nabla = \left(\frac{\partial}{\partial X}\vec{i} + \frac{1}{R}\frac{\partial}{\partial\theta}\vec{j}\right)$. Applying Eq. (3.11) to a longitudinal magnetic field vector $\overline{H}(H_x, 0, 0)$, applied on the i layer carbon nanotube with the tube shaped coordinate (R, θ, Z) and the removal vector $\overline{U} = (W_i, V_i, Y_i)$, produces

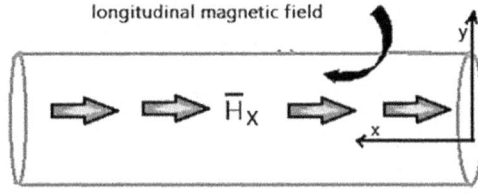

FIGURE 3.2 Axial magnetic field applied to a single-walled carbon nanotube.

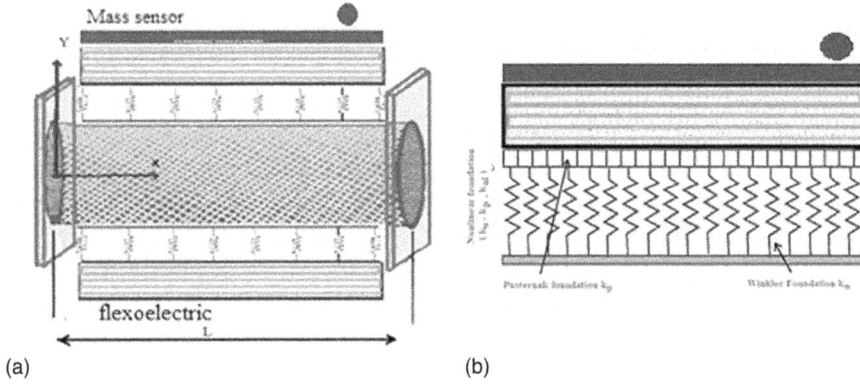

(a)   (b)

FIGURE 3.3 (a) SWCNT geometry in a flexoelectric with polymer matrix; (b) *SWCNT* geometry on a non-linear basis.

$$\bar{s} = \nabla \times \left(\bar{U} \times \bar{H} \right) = \left(-\frac{H_x}{R_i} \cdot \frac{\partial V_i}{\partial \theta}, H_x \cdot \frac{\partial V_i}{\partial X}, H_x \cdot \frac{\partial Y_i}{\partial X} \right), \tag{3.12a}$$

$$f = \nabla \times \bar{s} = \left(\frac{H_x}{R_i} \cdot \frac{\partial^2 y_i}{\partial x \partial \theta}, -H_x \cdot \frac{\partial^2 y_i}{\partial x^2}, \frac{H_x}{R_i^2} \cdot \frac{\partial^2 v_i}{\partial \theta^2} + H_x \cdot \frac{\partial^2 v_i}{\partial x^2} \right), \tag{3.12b}$$

To express the Lorentz force that a longitudinal magnetic field exerts, we have,

$$q\left(\bar{q}_x, \bar{q}_e, \bar{q}_z \right) = f \times B = f \times \eta \bar{H} = \eta \left(0, \frac{H_x^2}{R_i^2} \frac{\partial^2 V_i}{\partial X^2} + H_x^2 \cdot \frac{\partial^2 V_i}{\partial X^2}, H_x^2 \cdot \frac{\partial^2 Y_i}{\partial X^2} \right) \tag{3.12c}$$

where q_x, q_θ and q_z express the Lorentz force along the X, θ and Z axes, respectively (Figure 3.3):

$$\bar{q}_X = 0, \tag{3.12d}$$

$$\bar{q}_x = \eta \left(\frac{H_x^2}{R^2} \cdot \frac{\partial^2 v_i}{\partial x^2} + H_x^2 \cdot \frac{\partial^2 v_i}{\partial x^2} \right), \tag{3.12e}$$

$$\bar{q}_z = \eta \left(H_x^2 \cdot \frac{\partial^2 Y}{\partial X^2} \right), \tag{3.12f}$$

The external force q_{mag} is made up of the Lorentz force $\bar{q}_z$ caused by the magnetic field's longitudinal component and the surface-tension-generated F_s transverse component.

$$q_{(mag)} = \bar{q}_Z(x) + F_s \tag{3.13}$$

where the disseminated cross over force F_s may be described as, and the Lorentz force $\bar{q}_z$ is defined in Eq. (3.12f). (Li et al. 2016)

$$F_s = \left(H_s \cdot \frac{\partial^2 Y}{\partial X^2} \right), \tag{3.14}$$

Here, η denotes the magnetic permeability, H_x denotes the part of the longitudinal attractive field vector following up on the SWCNTs in the x heading, and H_s denotes a constant.

$$H_s = 2\mu(d+h) \tag{3.15}$$

where μ denotes the residual surface tension. The term $\bar{q}_z$ is the magnetic force per unit length due to Lorentz force exerted on the tube in z-direction considered form when considering the effect of surface elasticity (Yan and Jiang, 2011) the effect bending stiffness EI should be modified as (Lei et al., 2012),

$$EI^* = EI + Q_s E_s \tag{3.16}$$

where $Q_s = \frac{\pi}{8}(d+h)^3$. The surface Young's modulus is denoted by E_s, and the SWCNTs' effective thickness is denoted by d and it is already introduced in Eq. (3.8).

3.3 EBT Based on Non-Local Relations

The incomplete differential condition that portrays the free vibration of nanotubes when flexo-thermal and Lorentz forces are present may be written as,

$$\frac{\partial \pi}{\partial X} + N_i \frac{\partial^2 Y}{\partial X^2} + q_{(mag)} + \beta y + f(x) = \rho A \frac{\partial^2 Y}{\partial t^2} \tag{3.17}$$

where $f(x)$ is the nanotube's internal pressure relative to the elastic medium around it in terms of axial length. The cross-section of CNT is A.

$$\beta = \frac{f}{1 - \left(\dfrac{\alpha}{L^2}\right)f} \tag{3.18}$$

$$f = \frac{m\omega^2 L^2}{EA}, \quad \alpha = \frac{M_p}{mL} \tag{3.19}$$

The mass ratio α is the mass of the joined mass to the mass of the CNT and is the f dimensional less frequency parameter. The non-local elasticity's mass per unit of length is denoted by m. The following equilibrium equation defines the resulting shear force Π on the nanotube cross-section.

$$\Pi = \frac{\partial M}{\partial X} \tag{3.20}$$

The temperature-dependent axial force N_t is represented by the thermal expansion coefficient α. The definition of this constant force is (Narendar and Gopalakrishnan, 2010):

$$N_t = -EA\alpha T \tag{3.21}$$

where A is the cross section of nanotube and T is the temperature change. The longitudinal magnetic flux due to Lorentz force exerted on the tube in Z direction is represented by the term q_Z and is read from

$$q_z = \eta A H_x^2 \frac{\partial^2 Y}{\partial X^2} - EA\alpha T \tag{3.22}$$

H_x is the magnetic field intensity. The resulting bending moment M in Eq. (3.20) for the Euler-shaft hypothesis can be deciphered as follows:

$$M = \int_A z\sigma_{xx}\, dA, \tag{3.23}$$

where σ_{xx} is the nonlocal axial stress defined by nonlocal continuum theory. The constitutive Eq. (2.6) of a homogeneous isotropic elastic solid in non-local form for one-dimensional nanotube is taken as,

$$\sigma_{xx} - (e_0 a)^2 \frac{\partial^2 \sigma_{xx}}{\partial X^2} = E\varepsilon_{xx} \tag{3.24}$$

where E is the Young's modulus of the tube, ε_{xx} is the axial strain, $(e_0 a)$ is a nonlocal parameter which represents the impact of nonlocal scale effect on the structure. a is an internal characteristic length. The nonlocal relations in Eq. (3.8) can be written with temperature environment as follows

$$\sigma_{xx} - (e_0 a)^2 \frac{\partial^2 \sigma_{xx}}{\partial X^2} = E\varepsilon_{xx} - E\alpha T \tag{3.25}$$

The axial strain ε_{xx} for minor deflection in the Euler-Bernoulli beam model is defined as,

$$\varepsilon_{XX} = -z\frac{\partial^2 Y}{\partial X^2} \tag{3.26}$$

where z is the cross over coordinate along the twisting's positive hub. The bending moment M may be calculated using Eqs. (3.25) and (3.26), which are then plugged into Eq. (3.23):

$$M - (e_0 a)^2 \left[\frac{\partial^2 M}{\partial X^2}\right] = EI^* \frac{\partial^2 Y}{\partial X^2} \tag{3.27}$$

$$\Pi - (e_0 a)^2 \left[\frac{\partial^2 M}{\partial X^2}\right] = F^s_{11}\frac{\partial Y}{\partial X} - C^s_{11}\frac{\partial^2 Y}{\partial X^2} \tag{3.28}$$

where $I = \int_A z^2 dA$ is the moment of inertia. By substituting Eq. (3.12) and (3.11) into Eq. (3.17),

$$M - (e_0 a)^2 \left[(\rho A)\frac{\partial^2 Y}{\partial t^2} + q_{(\text{mag})} - f(x) + EA\alpha T\right] = EI^* \frac{\partial^2 Y}{\partial X^2} \tag{3.29}$$

$$\Pi - (e_0 a)^2 \left[(\rho A)\frac{\partial^3 Y}{\partial X^2 \partial t^2} + \frac{\partial^2 q_{(mag)}}{\partial X^2} - \frac{\partial f(x)}{\partial X} + EA\alpha T \right] = EI^* \frac{\partial^3 Y}{\partial X^3} + F^s_{11}\frac{\partial Y}{\partial X} - C^s_{11}\frac{\partial^2 Y}{\partial X^2} \quad (3.30)$$

Under the activity of a dispersed tension and temperature on the encompassing polymer versatile media, the condition of movement (3.17), describing the transverse vibration, may be stated as follows:

$$
\begin{aligned}
f(x) &= EI^* \frac{\partial^4 Y}{\partial X^4} + EA\alpha T \frac{\partial^2 Y}{\partial x^2} + (\rho A)\frac{\partial^2 Y}{\partial t^2} + \bar{q}_z(x) + F_s\frac{\partial^2 Y}{\partial X^2} \\
&= \left((e_0 a)^2 \left(EA\alpha T \frac{\partial^4 Y}{\partial X^4} + \bar{q}_z(x) + F_s\frac{\partial^4 Y}{\partial X^4} - \frac{\partial^2 f(x)}{\partial X^2} - \left(F^s_{11}\frac{\partial Y}{\partial X} - C^s_{11}\frac{\partial^2 Y}{\partial X^2} \right) \right) \right)
\end{aligned}
\quad (3.31)
$$

A Winkler–Pasternak type model can explain the pressure per unit pivotal length brought about by the encompassing flexible medium following up on the cylinder's surface (Barati, 2017).

$$f(x) = -\left(-K_w + K_p + K_{nl}\right) \quad (3.32)$$

where K_w, K_p, and K_{nl} include non-linear constants, Winkler–Pasternak theory, and Pasternak's non-linear constants. Substituting Eq. (3.32) for Eq. (3.31) produces.

$$
\begin{aligned}
&EI^* \frac{\partial^4 Y}{\partial X^4} + EA\alpha T \frac{\partial^2 Y}{\partial X^2} + \rho A \frac{\partial^2 Y}{\partial t^2} + \left(\eta A H_x^2 \frac{\partial^2 Y}{\partial X^2} + H_x\frac{\partial^2 Y}{\partial X^2} \right) \\
&-(e_0 a)^2 \left(\begin{aligned} &EA\alpha T \frac{\partial^4 Y}{\partial X^4} + \rho A \frac{\partial^2 Y}{\partial X^2} + \left(\eta H_x^2 \cdot \frac{\partial^4 y}{\partial X^4} - H_x\frac{\partial^4 Y}{\partial X^4} \right) \\ &-\left(-K_w + K_p + K_{nl}\right)\frac{\partial^2 Y}{\partial X^2} + \left(F^s_{11}\frac{\partial Y}{\partial X} - C^s_{11}\frac{\partial^2 Y}{\partial X^2} \right) \end{aligned} \right) = -\left(K_w + K_p + K_{nl}\right)
\end{aligned}
\quad (3.33)
$$

The cross-sectional area's surface piezoelectricity rigidities are modelled as,

$$F^s_{11} = \left(\delta^s_{11} + \frac{e_{31}e_{31}^s}{k_{33}} \right)\frac{bh^2}{4} \quad (3.34)$$

$$C^s_{11} = \left(\delta^s_{11} + \frac{e_{31}e_{31}^s}{k_{33}} \right)\frac{bh^2}{12} + \left(\frac{f_{31}e_{31}^s}{k_{33}} \right)\frac{bh^2}{4} \quad (3.35)$$

PZT-5H was used to create the flexoelectric nanotube, when the elastic stiffness constant C_{11}^s is present. The piezoelectric and dielectric coefficients are taken to be as e_{31}, k_{33}, and the elastic characteristics are taken to be c_{11} (Yang et al., 2015). The flexoelectric coefficient is also known as the f_{31} coefficient (Yang et al. 2015). Consider δ^s_{11}. and e^s_{31}. to be the surface elastic and piezoelectric constants for PZT-5H (Ebrahimi and Barati, 2017).

3.4 Ultrasonic Wave Solution

Fourier transformation may be used to convert Eq. (3.31) to the frequency domain (Narendar and Gopalakrishnan, 2010):

$$Y(x,t) = \sum_{n=1}^{N} \hat{Y}(x) e^{-j(kn-\omega_n t)} \tag{3.36}$$

where $\hat{Y}$ is the amplitude of the wave motion, $j = \sqrt{-1}$, k is the wave number ω_n the circular frequency of sampling point and N is the Nyquist frequency. The sampling rate and the number of sampling points should be sufficiently large to have relatively good resolution of both high and low frequencies respectively. Substitution of Eq. (3.36) into Eq. (3.32), we get

$$\sum_{n=1}^{N} \left[\left(EI^* \frac{\partial^4 \hat{Y}}{\partial X^2} + EA\alpha T \frac{\partial^2 \hat{Y}}{\partial X^2} + \rho A \frac{\partial^2 Y}{\partial t^2} + \left(\eta A H_x^2 \frac{\partial^2 \hat{Y}}{\partial X^2} + H_x \frac{\partial^2 \hat{Y}}{\partial X^2} \right) \right. \right.$$
$$\left. -(e_0 a)^2 \left(\begin{array}{l} EA\alpha T \dfrac{\partial^2 \hat{Y}}{\partial X^2} + \rho A \dfrac{\partial^2 \hat{Y}}{\partial X^2} - \left(\eta A H_x^2 \dfrac{\partial^2 \hat{Y}}{\partial X^2} + H_x \dfrac{\partial^2 \hat{Y}}{\partial X^2} \right) \\ -(-K_w + K_p + K_{nl}) \dfrac{\partial^2 \hat{Y}}{\partial X^2} - \left(F^s_{11} \dfrac{\partial \hat{Y}}{\partial X} - C^s_{11} \dfrac{\partial^2 \hat{Y}}{\partial X^2} \right) \end{array} \right) \right) + (-K_w + K_p + K_{nl}) \right] \tag{3.37}$$

This problem may be expressed as an ordinary differential equation in a single variable, X because it must be fulfilled for each N.

$$\sum_{n=1}^{N} \left[\left(EI^* \frac{\partial^4 \hat{Y}}{\partial X^2} + EA\alpha T \frac{\partial^4 \hat{Y}}{\partial X^2} + \rho A \frac{\partial^4 \hat{Y}}{\partial t^2} + \left(\eta A H_x^2 \frac{\partial^4 \hat{Y}}{\partial X^2} + H_x \frac{\partial^4 \hat{Y}}{\partial X^2} \right) \right. \right.$$
$$\left. -(e_0 a)^2 \left(\begin{array}{l} (\rho A) \dfrac{\partial^4 \hat{Y}}{\partial X^2 \partial t^2} - \left(\eta A H_x^2 \dfrac{\partial^4 \hat{Y}}{\partial X^2} + H_x \dfrac{\partial^4 \hat{Y}}{\partial X^2} \right) \\ -(-K_w + K_p + K_{nl}) \dfrac{\partial^4 \hat{Y}}{\partial X^2} \\ -\left(F^s_{11} \dfrac{\partial \hat{Y}}{\partial X} - C^s_{11} \dfrac{\partial^4 \hat{Y}}{\partial X^2} \right) + EA\alpha T \dfrac{\partial^4 \hat{Y}}{\partial X^2} \end{array} \right) \right) + (-K_w + K_p + K_{nl}) \right] \tag{3.38}$$

The following equation may be stated with a single variable, X, and must be fully satisfied for all values of tiny n. Eq. (3.37) may be simplified as,

$$\left[\begin{array}{l} \left[EI^* \dfrac{\partial^4 \hat{Y}}{\partial X^4} + EA\alpha T \dfrac{\partial^4 \hat{Y}}{\partial X^4} + \left(\eta A H_x^2 \dfrac{\partial^2 \hat{Y}}{\partial X^2} + H_x \dfrac{\partial^2 \hat{Y}}{\partial X^2} \right) - \dfrac{\partial^2}{\partial t^2}(\rho A) \right] \\ \left[\left(\rho A - (-K_w + K_p + K_{nl}) + EA\alpha T \right) \left(\eta A H_x^2 \dfrac{\partial^2 \hat{Y}}{\partial X^2} - H_x \dfrac{\partial^2 \hat{Y}}{\partial X^2} \right) (e_0 a)^2 \dfrac{\partial^2 \hat{Y}}{\partial X^2} \\ + \left(F^s_{11} \dfrac{\partial \hat{Y}}{\partial X} - C^s_{11} \dfrac{\partial^2 \hat{Y}}{\partial X^2} \right) \end{array} \right] + (-K_w + K_p + K_{nl}) \end{array} \right] = 0$$

$$\tag{3.39}$$

The definition of the dimensionless variables is,

$$\frac{X}{L} = x, \frac{Y}{L} = y, \alpha_i = \frac{I_i}{l}, \tau = \frac{e_0 a}{l}, K_w = \frac{k_w L^4}{EI^* D},$$

$$F_{33} = F_{11}^s \frac{L^4}{EI^* C_{11}^s}, D = \frac{1}{12} C_{11}^s bh^3, K_p = \frac{k_p L^4}{EI^* D}, \tag{3.40}$$

$$K_{nl} = \frac{k_{nl} L^4}{EI^* D}, \eta = \frac{1}{(1 + EA\alpha T)}, \bar{N}_T = \frac{N_t L^2}{EI^*}, \delta_{11}^s = \frac{L^4}{EI^*};$$

Substituting $\check{Y}(x) = \check{Y}e^{-i\omega x}$ into Eq. (3.40) employing Eq. (3.39) yields,

$$\left(1 + EA\alpha T - \eta AH_x^2\right) \frac{\partial^4 \hat{Y}}{\partial x^4} + \left[EA\alpha T - \eta AH_x^2 - \left(k_w + k_p + k_{NL}\right) + \tau^2\left(F_{33}\right) \right] \frac{\partial^2 \hat{Y}}{\partial x^2}$$

$$+ 2i\rho A \frac{\partial \hat{Y}}{\partial x} - \left[\rho A + \left(k_w + k_p + k_{nl}\right)\right] = 0 \tag{3.41}$$

The $\bar{Y}$ implication of a non-trivial solution for the amplitude of the wave is that,

$$\bar{Y}(x,t) = ye^{-ik_w x} \tag{3.42}$$

After plugging Eq. (3.42) into Eq. (3.41), we get the following non-trivial solution for the wave amplitude y

$$\left(1 + EA\alpha T - \eta AH_x^2\right)k_n^4 + \left[EA\alpha T - \eta aH_x^2 - \left(-k_w + k_p + k_{nl}\right) + \tau^2\left(F_{33}\right) \right]k_n^2$$

$$\left(2i\rho A\right)k_n - \left[\rho A + \left(-k_w + k_p + k_{nl}\right)\right] = 0 \tag{3.43}$$

It provides the characteristic equation for a SWCNT's surrounding medium and continuum structure (ECS).

3.5 Boundary Conditions

Here, the equations of motion for a vibrating nanobeam in both the simply-supported (S–S) and clamped-clamped configurations are presented. (C–C) boundary conditions are analytically solved, and the following results are given as:

3.5.1 Simply–Supported SWCNT

Its limit conditions for the just upheld issue are $(X) = (0, L)$

$$Y(x)|_{X=0} = 0$$

$$M(X) = \left(-EI^* \frac{\partial^2 Y(X)}{\partial X^2} + \left(e_0 a\right)^2 \left[\begin{array}{l} \left(\rho A\right) \frac{\partial^2 Y(X)}{\partial t^2} + q_{(mag)} \frac{\partial^2 Y(X)}{\partial X^2} \\ -f(x) \frac{\partial^2 Y(X)}{\partial X^2} + EA\alpha T \frac{\partial^2 Y(X)}{\partial X^2} \end{array} \right] \right)_{X=0} = 0,$$

$$Y(x)|_{X=L} = 0,$$

$$M(X) = \left(-EI^* \frac{\partial^2 Y(X)}{\partial X^2} + (e_0 a)^2 \left[\begin{array}{c} (\rho A) \dfrac{\partial^2 Y(X)}{\partial t^2} + q_{(\text{mag})} \dfrac{\partial^2 Y(X)}{\partial X^2} \\ -f(x) \dfrac{\partial^2 Y(X)}{\partial X^2} + EA\alpha T \dfrac{\partial^2 Y(X)}{\partial X^2} \end{array} \right] \right)_{X=L} = 0,$$

$$Y(x)|_{X=0} = 0$$

$$\Pi(X) = \left(-\left(EI^* \frac{\partial^3 Y}{\partial X^3} + F^s_{11} \frac{\partial Y}{\partial X} - C^s_{11} \frac{\partial^2 Y}{\partial X^2} \right) + (e_0 a)^2 \left[(\rho A) \frac{\partial^3 Y}{\partial X^2 \partial t^2} + \frac{\partial^2 q_{(\text{mag})}}{\partial X^2} - \frac{\partial f(x)}{\partial X} + EA\alpha T \right] \right)_{X=0} = 0$$

$$Y(x)|_{X=L} = 0,$$

$$\Pi(X) = \left(-\left(EI^* \frac{\partial^3 Y}{\partial X^3} + F^s_{11} \frac{\partial Y}{\partial X} - C^s_{11} \frac{\partial^2 Y}{\partial X^2} \right) + (e_0 a)^2 \left[(\rho A) \frac{\partial^3 Y}{\partial X^2 \partial t^2} + \frac{\partial^2 q_{(\text{mag})}}{\partial X^2} - \frac{\partial f(x)}{\partial X} + EA\alpha T \right] \right)_{X=L} = 0$$

$$(3.44)$$

3.5.2 Clamped–Clamped SWCNT

Accept that the shaft is clasped at the two finishes and is under a pivotal compressive burden. The particular conditions of this case's limits are as follows:

$$Y(x)|_{X=0} = 0 \qquad \frac{\partial Y(X)}{\partial X}|_{X=0} = 0$$
$$Y(x)|_{X=L} = 0, \qquad \frac{\partial Y(X)}{\partial X}|_{X=L} = 0.$$

$$(3.45)$$

3.6 Discussion and Numerical Findings

In this section, the ASWCNT embedded in a polymer matrix subjected to magneto -thermo elastic forces is considered as an example for the nonlinear vibration analysis. The geometrical and material parameter taken for the numerical verification is shown in Table 3.1. Table 3.2 present the comparative study between the numerical results of maximum transverse deflection of C–C CNT with and without surface effect. Results predict the reasonable agreement with the literature.

Through $L/h = 10 - 20$, $V = 0$, $K_p = 20$, $T = 20$ and $\alpha = 0.5$, Figures 3.4 and 3.5 examine the relationship between dimensionless recurrence and non-layered adequacy for a variety of non-local factors. It is discovered that a rise in amplitude led to an increase in dimensionless frequency. The amplitude of frequency reduces when non-local parameters are present, and we may infer from this that non-local values play a substantial influence in recurrence. But because of the expansion in slimness shown in Figure 3.5, a softening behaviour is seen more frequently.

Figures 3.6 and 3.7 depict the voltage-dependent change in the nanotube's dimensionless frequency in regard to the non-layered abundancy for upsides of $L/h = 10$–20, = 0.5, $Kp = Kw = 20$–30, and $T = 20$. It was discovered that an increase in non-dimensional amplitude led to a rise in the dimensionless frequency of nanotubes at varied voltage levels. Consequently, the hub malleable and compressive powers in the nanotubes were produced by the created positive and negative voltages, separately. Additionally,

TABLE 3.1   Material Properties (Lee and Chang, 2009),
(Besseghier et al., 2011), (Ebrahimi and Barati, 2017)

Materials	PZT
EI	1.1122×10^{-25} Nm^9
α^0	-1.5×10^{-6} C^{-1}
ρ	2.3 g/cm^3
e_0	0.31 nm
a	0.142 N/m
E_s	35.3 N/m
μ	$4\pi \times 10^{-7}$ N/m
H_x	2×10^8 A/m
f_{31}	10^{-7} C/(Vm)
$c11$	102 Gpa
δ^s_{11}	102 N/m
e^s_{31}	-3×10^{-8} C/m
k_{33}	1.76×10^{-8} C/(Vm)

TABLE 3.2   Most Extreme Cross Over Redirection of a C Nanotube With and
Without Surface Impacts is Looked at Basutkar et al. (2019)

	(Basutkar et al. 2019)		Author	
(L/h)	$(S.E)$	$(S.E \neq 0)$	$(S.E = 0)$	$(S.E = 0)$
10	0.6343	0.6334	0.6396	0.6363
15	0.9472	0.9459	0.9450	0.9432
20	1.2550	1.2530	1.2934	1.2914

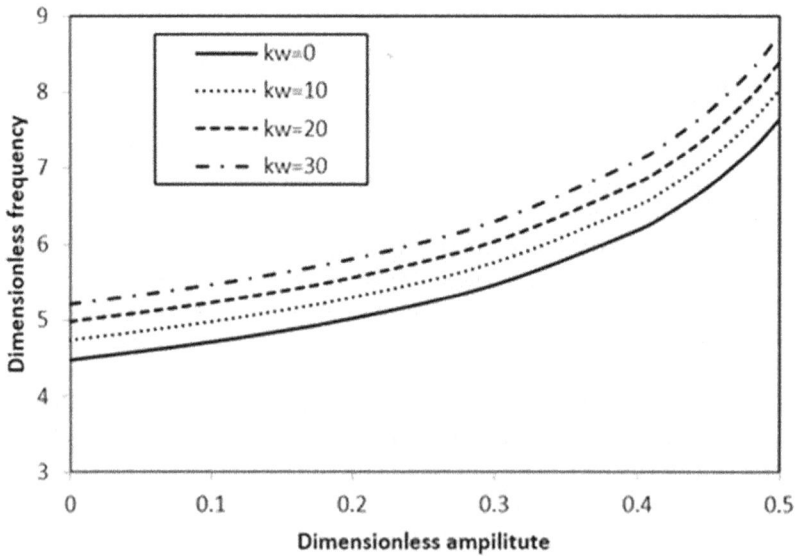

FIGURE 3.4   Chart showing the connection between recurrence and plentifulness in terms of a dimensionless scale and over a range of ($L/h = 10$, $V = 5$, $K_p = 20$, $\Delta T = 20$, $\alpha = 0.5$).

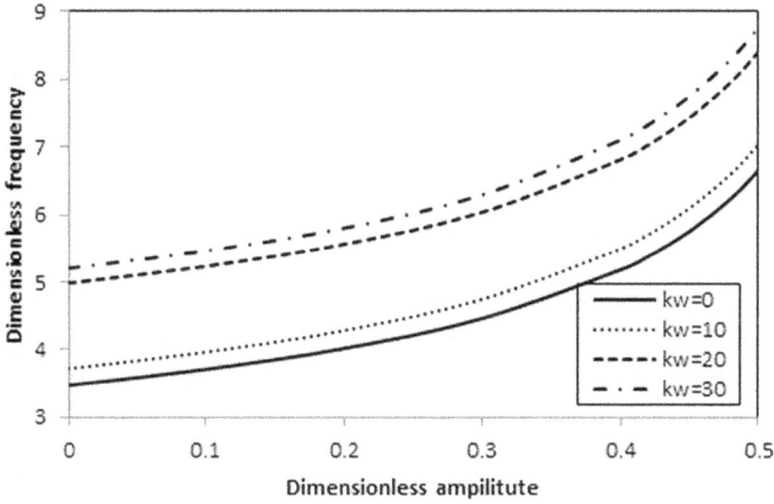

FIGURE 3.5 Graph showing the relationship between the frequency and the amplitude of a signal in the absence of any dimensions (L/h = 10, V = 5, K_p = 20, ΔT = 30, α = 0.5.).

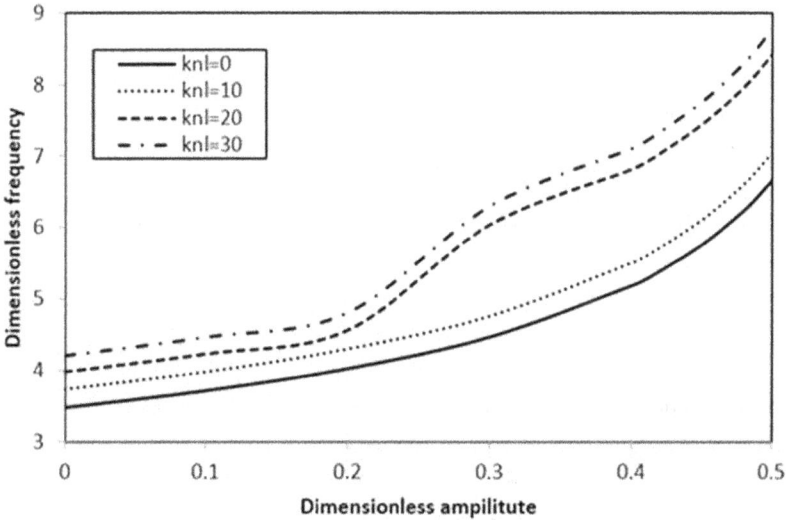

FIGURE 3.6 Diagram depicting the relationship between the non-linear basis values and the frequency/amplitude in arbitrary dimensions (L/h = 10, V = 0, K_w = K_p = 10, ΔT = 10, α = 0.5).

a little dependence between the dimensionless frequency and the establishment boundary for the consistent temperature esteem (T = 20) is noted. Figures 3.8 and 3.9 show how the bending moment varies with respect to the nanotube length under the conditions of L/h = 10, V = 0, = 0.5, Kw = Kp = 20, T = 20, and H = 10. Evidence from these calculations suggests that the bending moment develops tensile and compressive characteristics and a wider range of modes with increasing length. Additionally, the wave trend shows the effects of the magnetic field.

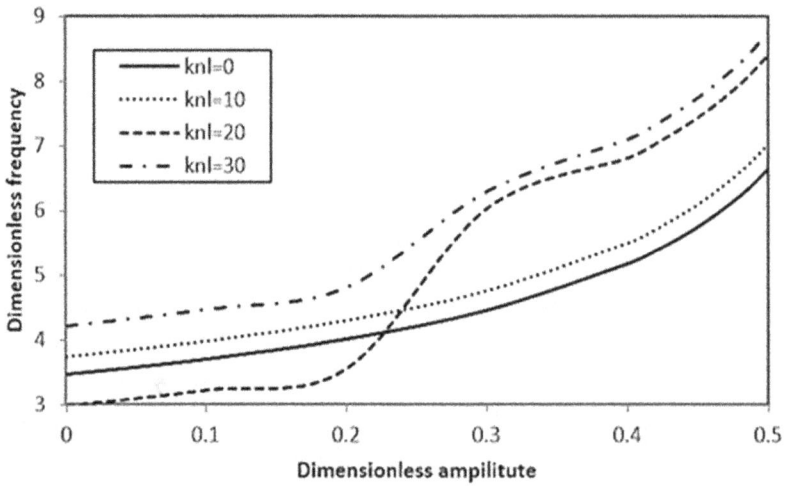

FIGURE 3.7 Diagram depicting the relationship between frequency and amplitude in arbitrary units and their non-linear bases ($L/h = 20$, $V = 0$, $K_w = K_p = 20$, $\Delta T = 20$, $\alpha = 0.5$).

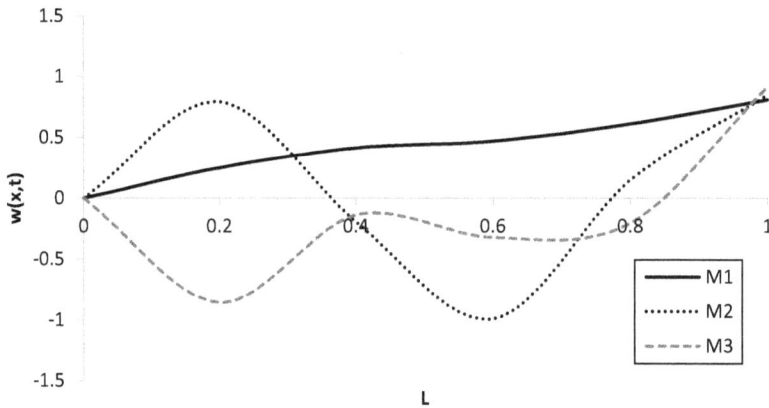

FIGURE 3.8 Graph showing the bending moment against the length for different modes. ($L/h = 10$, $V = 0$, $K_w = K_p = 20$, $\Delta T = 20$, $\Delta H = 10$, $\alpha = 0.5$).

Figures 3.10 and 3.11 show the relationship between frequency and attached mass as a function of $L/h = 10$, $V = 0.2$, $K_w = K_p = 20$, and other natural variables $(T, C)=(10, 10)$ $(20, 20)$. The recurrence boundary diminishes as the comparing mass proportion values increment. It ought to be noticed that a warm and hygro-warm climate causes the attached mass ratio to increase, which reduces plate stiffness. It has been discovered that impacts of rising temperatures and moisture concentrations are more articulated at lower mass proportion values. In general, adding a particle to the dynamics of SWCNT increases the system's overall mass while maintaining the stiffness, which causes frequency to drop. The graphic also shows how non-locality has an impact. The frequency parameter drops when the non-local parameter is increased.

The wave frequency against phase velocity in Figures 3.12 and 3.13 will now be shown to look at the effect of the Winkler and Pasternak factors. It is evident that there is sufficient potential for the Winkler

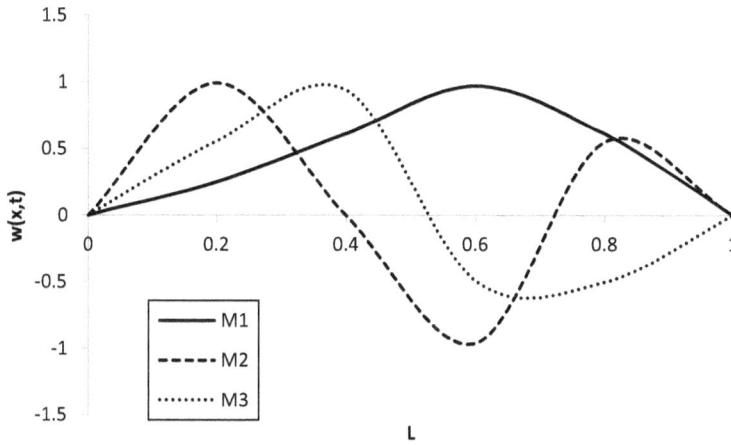

FIGURE 3.9 A diagram of bowing second versus length for various modes.

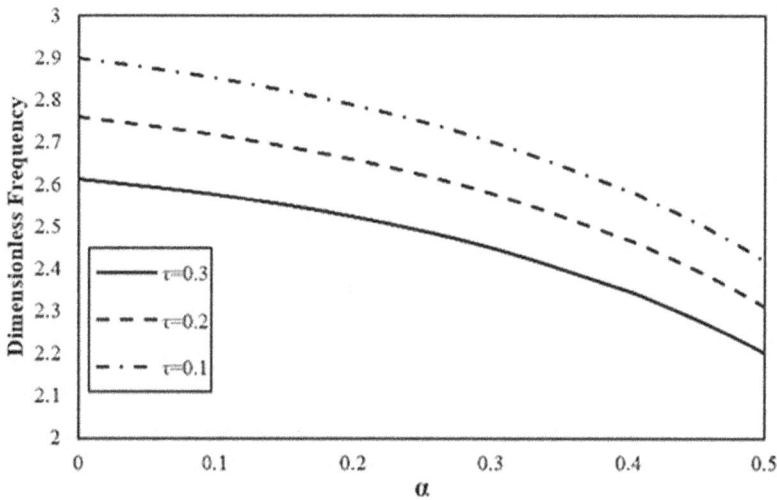

FIGURE 3.10 Frequency vs attached mass ratio as a function of dimensionless frequency.

(linear) and Pasternak (non-linear) coefficients to enhance wave frequency values. Furthermore, it's important to note that after the Winkler coefficient is adjusted from to, the most noticeable shift in the wave frequency responses may be seen. To put it simply, the most effective way to increase wave frequency is to adjust the Winkler or Pasternak coefficient's value from zero to the first non-zero number.

Figures 3.14 and 3.15, respectively, display the fluctuations of the flexoelectricity in relation to the wave number via various Pasternak and Winkler parameters. This picture demonstrates that, regardless of the size of the foundation parameter, the flexoelectric effect magnifies it at higher wave number values. This demonstrates that the solidness of the nanotube is solidifying as a result of this kind of coupling effect. It must be noted that the rise in flexo electricity with wave number has resulted in fluctuating behaviour in energy transfer at changing foundation parameters.

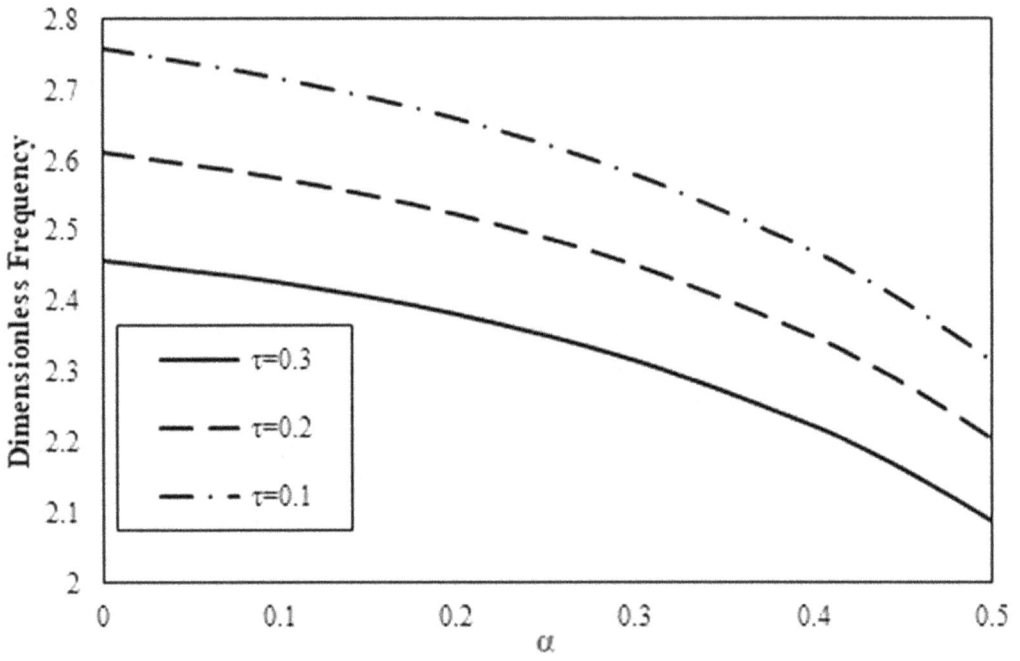

FIGURE 3.11   Frequency against attached mass in a non-dimensional graph.

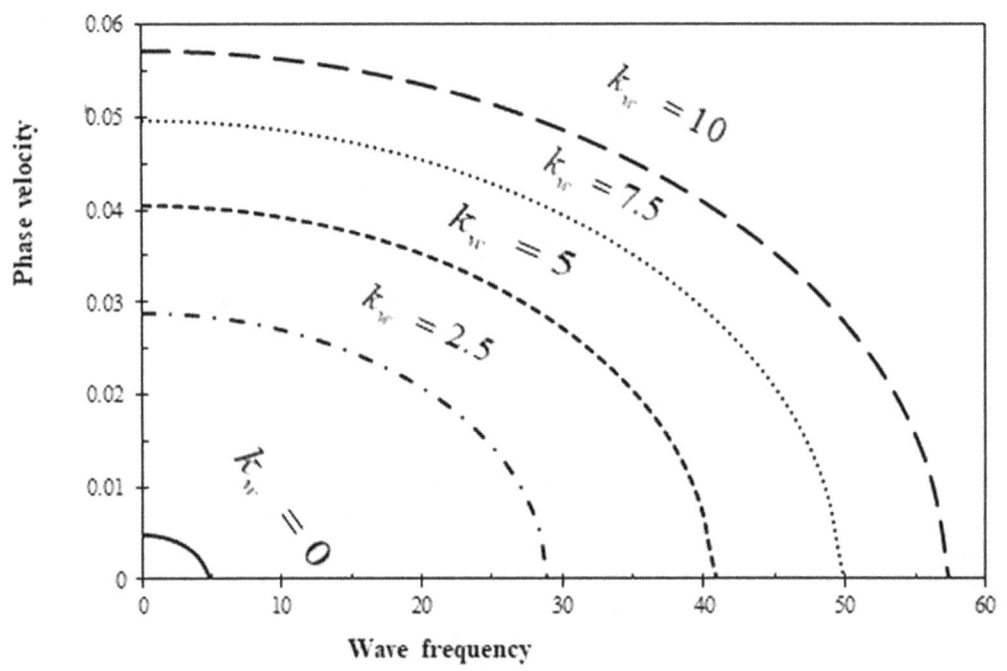

FIGURE 3.12   Phase velocity vs wave frequency as a function of varying Winkler coefficients.

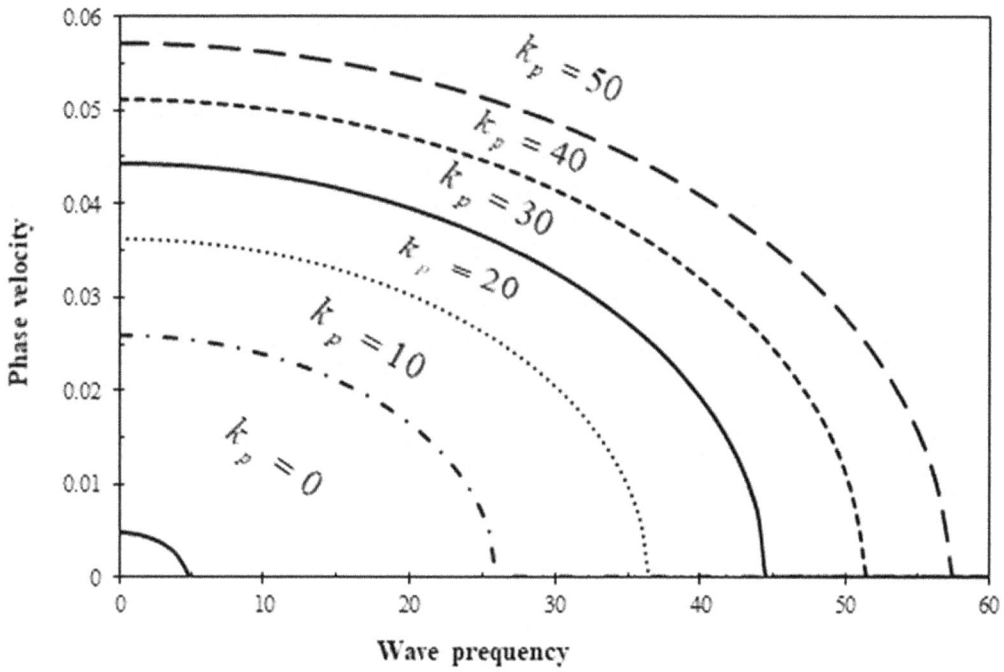

FIGURE 3.13 A graph of phase velocity vs wave frequency for a range of Pasternak coefficients.

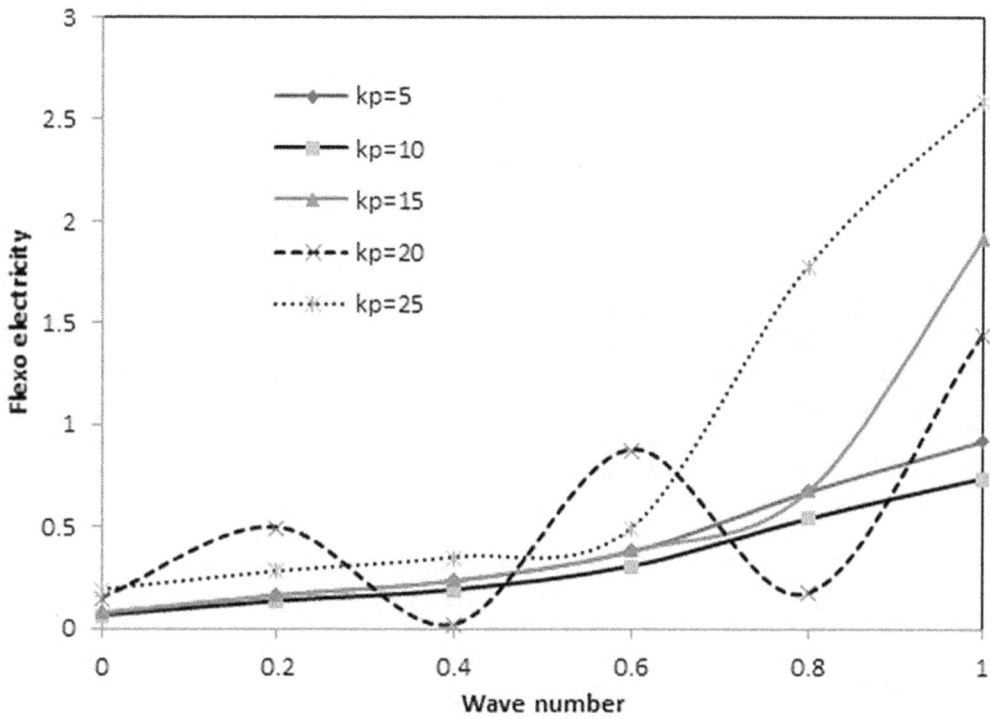

FIGURE 3.14 Flexoelectricity vs wave number as a function of Pasternak coefficients.

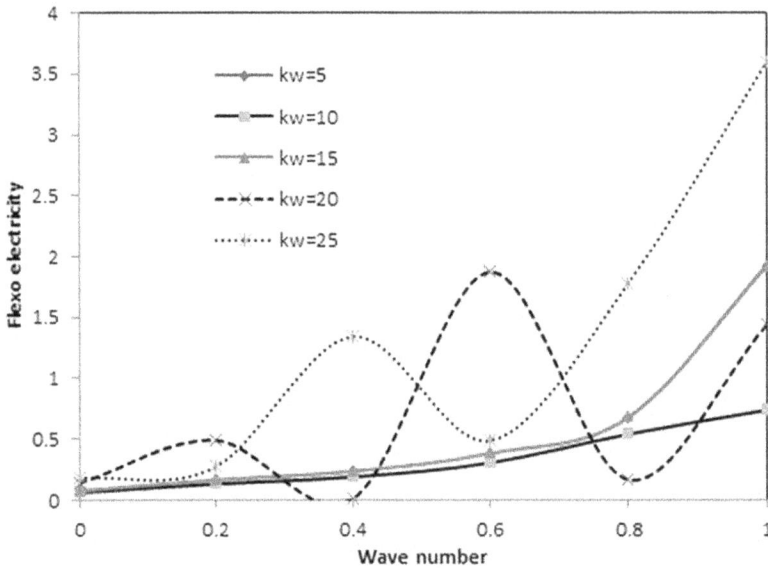

FIGURE 3.15   Differences in flexo electricity against wave number for a range of Winkler coefficients.

3.7 Conclusion

This study applies Euler beam theory to study the non-linear magneto-flexo-thermo elastic waves propagating through a solitary-walled carbon nanotube upheld by a polymer grid. Eringen's non-nearby versatility hypothesis is utilised to provide an analytical formulation that takes into consideration effects on a tiny scale. To explain the non-linear behaviour of this medium, we may turn to the Winkler–Pasternak model. The answer is found by using the dispersion relations for ultrasonic waves. Parametric analysis is presented to investigate how changes in magneto-electro-mechanical loadings, non-neighbourhood boundary, and viewpoint proportion influence the avoidance attributes of nanotubes. The outcomes are framed as follows:

- It has been observed that raising the value of the foundation constants results in a more restrictive medium.
- The outcome also demonstrates that the non-layered recurrence is upgraded by the dimensionless amplitude via the non-linear foundation parameter.
- The dynamic behaviour of the armchair nanotube was discovered to be improved by embedding the construction on an elastic base. Likewise, it was shown that the Pasternak boundary influences the design more unequivocally than the Winkler boundary.
- The results show that linked mass and very tiny length scale values dampen the vibration modes of SWCNT. In this study, it was shown that the vibration methods of SWCNT are hosed by both connected mass and low length scale values.
- Numerical analyses indicate an increase in flexo electricity via an increase in wave number values after foundation parameter is improved. The current assessment shows that raising Winkler or Pasternak parameter may be addressed by a decrease in the greatness of stage speed.

References

Ansari, R., Gholami, R., and Sahmani, S. 2012. "On the dynamic stability of embedded single-walled carbon nanotubes including thermal environment effects." *Scientia Iranica* 19, no. 3: 919–925. doi:10.1016/j.scient.2012.02.013.

Arani, A.G., Roudbari, M., and Amir, S. 2016. "Longitudinal magnetic field effect on wave propagation of fluid-conveyed SWCNT using Knudsen number and surface considerations." *Applied Mathematical Modelling* 40, no. 3: 2025–2038. doi:10.1016/j.apm.2015.09.055.

Arda, M., and Aydogdu, M. 2020. "Vibration analysis of carbon nanotube mass sensors considering both inertia and stiffness of the detected mass." *Mechanics Based Design of Structures and Machines*, 1–17. doi:10.1080/15397734.2020.1728548.

Askari, H., and Esmailzadeh, E. 2017. "Forced vibration of fluid conveying carbon nanotubes considering thermal effect and non-linear foundations." *Composites Part B: Engineering* 113: 31–43. doi:10.1016/j.compositesb.2016.12.046.

Aydogdu, M. 2012. "Axial vibration analysis of nanorods (carbon nanotubes) embedded in an elastic medium using non-local elasticity." *Mechanics Research Communications* 43: 34–40. doi:10.1016/j.mechrescom.2012.02.001.

Aydogdu, M., and Filiz, S. 2011. "Modeling carbon nanotube-based mass sensors using axial vibration and non-local elasticity." *Physica E: Low-dimensional Systems and Nanostructures* 43, no. 6: 1229–1234. doi:10.1016/j.physe.2011.02.006.

Azarboni, H.R. 2019. "Magneto-thermal primary frequency response analysis of carbon nanotube considering surface effect under different boundary conditions." *Composites Part B: Engineering* 165: 435–441. doi:10.1016/j.compositesb.2019.01.093.

Baghdadi, H., Tounsi, A., Zidour, M., and Benzair, A. 2014. "Thermal effect on vibration characteristics of armchair and zigzag single-walled carbon nanotubes using non-local parabolic beam theory." *Fullerenes, Nanotubes, and Carbon Nanostructures* 23, no. 3: 266–272. doi:10.1080/1536383x.2013.787605.

Barati, M.R. 2017. "Investigating non-linear vibration of closed circuit flexoelectric nanobeams with surface effects via Hamiltonian method." *Microsystem Technologies* 24, no. 4: 1841–1851. doi:10.1007/s00542-017-3549-8.

Barati, M. R., and Shahverdi, H. 2017. "Frequency analysis of nanoporous mass sensors based on a vibrating heterogeneous nanoplate and non-local strain gradient theory." *Microsystem Technologies* 24, no. 3: 1479–1494. doi:10.1007/s00542-017-3531-5.

Basutkar, R., Sidhardh, S., and Ray, M. 2019. "Static analysis of flexoelectric nanobeams incorporating surface effects using element free Galerkin method." *European Journal of Mechanics - A/Solids* 76: 13–24. doi:10.1016/j.euromechsol.2019.02.013.

Bedia, W. A., Benzair, A., Semmah, A., Tounsi, A., and Mahmoud, S.R. 2015. "On the thermal buckling characteristics of armchair single-walled carbon nanotube embedded in an elastic medium based on non-local continuum elasticity." *Brazilian Journal of Physics* 45, no. 2: 225–233. doi:10.1007/s13538-015-0306-2.

Bensattalah, T., Daouadji, T.H., Zidour, M., Tounsi, A., and Bedia, E.A. 2016. "Investigation of thermal and chirality effects on vibration of single-walled carbon nanotubes embedded in a polymeric matrix using non-local elasticity theories." *Mechanics of Composite Materials* 52, no. 4: 555–568. doi:10.1007/s11029-016-9606-z.

Benzair, A., Tounsi, A., Besseghier, A., Heireche, H., Moulay, N., and Boumia, L. 2008. 'The thermal effect on vibration of single-walled carbon nanotubes using non-local Timoshenko beam theory." *Journal of Physics D: Applied Physics* 41, no. 22: 1–10. doi:10.1088/0022-3727/41/22/225404.

Besseghier, A., Tounsi, A., Houari, M.S., Benzair, A., Boumia, L., and Heireche, H. 2011. "Thermal effect on wave propagation in double-walled carbon nanotubes embedded in a polymer matrix using non-local elasticity." *Physica E: Low-dimensional Systems and Nanostructures* 43, no. 7: 1379–1386. doi:10.1016/j.physe.2011.03.008.

Chowdhury, R., Adhikari, S., and Mitchell, J. 2009. "Vibrating carbon nanotube based bio-sensors." *Physica E: Low-dimensional Systems and Nanostructures* 42, no. 2: 104–109. doi:10.1016/j.physe.2009.09.007.

Dai, H., Ceballes, S, Abdelkefi, A, Hong, Y, and Wang, L. 2018. "Exact modes for post-buckling characteristics of non-local nanobeams in a longitudinal magnetic field." *Applied Mathematical Modelling* 55: 758–775. doi:10.1016/j.apm.2017.11.025.

Ebrahami, F., Selvamani, R., and Mahaveer Sree Jayan, M. 2021. "Haar wavelet method for non-linear vibration of magneto-thermo-elastic carbon nanotube-based mass sensors conveying pulsating viscous fluid." *The European Physical Journal Plus* 136: 923. 10.1140/epjp/s13360-021-01926-7

Ebrahimi, F., and Barati, M.R. 2016. "Vibration analysis of piezoelectrically actuated curved nanosize FG beams via a non-local strain-electric field gradient theory." *Mechanics of Advanced Materials and Structures* 25, no. 4: 350–359. doi:10.1080/15376494.2016.1255830.

Ebrahimi, F., and Barati, M.R. 2017. "Surface effects on the vibration behavior of flexoelectric nanobeams based on non-local elasticity theory." *The European Physical Journal Plus* 132, no. 1. doi:10.1140/epjp/i2017-11320-5.

Ebrahimi, F., and Dabbagh, A. 2018. "Magnetic field effects on thermally affected propagation of acoustical waves in rotary double-nanobeam systems." *Waves in Random and Complex Media*, 1–21. doi:10.1080/17455030.2018.1558308

Ebrahimi, F., Mahsa Karimisal, M., Civalek, O., and Vinyas, M. 2019. "Surface effect on scale –dependent vibration behaviour of flexoelectric sandwhich nanobeams." *Advances in Nano Research* 70, no. 2: 77–88. doi:10.12989/anr.2019.7.2.077.

Eringen, A. C. 1983. "On differential equations of non-local elasticity and solutions of screw dislocation and surface waves." *Journal of Applied Physics* 54, no. 9: 4703–4710. doi:10.1063/1.332803.

Eringen A. C. 2002. *Non-local continuum field theories*. Springer, Berlin.

Eringen, A. C., and Edelen, D. G. B. 1972. "On non-local elasticity." *International Journal of Engineering Science* 10, no. 3: 233–248. doi:10.1016/0020-7225(72)90039-0.

Fang, B., Zhen, Y. X., Zhang, C. P., and Tang, Y. 2013. "Non-linear vibration analysis of double –walled carbon nanotubes based on non-local elasticity theory." *Applied Mathematical Modelling* 37, no. 3: 1098–1107. doi:10.1016/j.aprm.2012.03.032.

Gheshlaghi, B., and Hasheminejad, S. M. 2011. "Surface effects on non-linear free vibration of nanobeams." *Composites Part B: Engineering* 42, no. 4: 934–937. doi:10.1016/j.compositesb.2010.12.026.

Güven, U. 2015. "General investigation for longitudinal wave propagation under magnetic field effect via non-local elasticity." *Applied Mathematics and Mechanics* 36, no. 10: 1305–1318. doi:10.1007/s10483-015-1985-9.

Heireche, H., Tounsi, A., Benzair, A., Maachou, M., and Bedia, E. A. 2008. "Sound wave propagation in single-walled carbon nanotubes using non-local elasticity." *Physica E: Low-dimensional Systems and Nanostructures* 40, no. 8: 2791–2799. doi:10.1016/j.physe.2007.12.021.

Hsu, J.C., Chang, R. P., and Chang, W.J. 2008. "Resonance frequency of chiral single walled carbon nanotubes using timoshenko beam theory." *Physics Letters A* 373: 2757–2759. doi:10.1016/j.physleta.2008.01.007.

Kumar, R., Sharma, N., Lata, P., and Abo-Dahab, S.M. 2017. "Rayleigh waves in anisotropic magneto thermoelastic medium." *Coupled System Mechanics* 6, no. 3: 317–333. doi:10.12989/csm.2017.6.3.317.

Lata, P., and Kaur, I. 2019a. "Thermomechanical interactions due to time harmonic sources in a transversely isotropic magneto thermoelastic solids with rotation." *International Journal of Microstructure and Materials Properties* 14, no. 6: 549. doi:10.1504/ijmmp.2019.103190

Lata, P., and Kaur, I. 2019b. "Transversely isotropic thick plate with two temperature and GN type-III in frequency domain." *Coupled System Mechanics* 8, no. 1: 55–70. doi:10.12989/csm.2019.8.1.055.

Lata, P., Kumar, R., and Sharma, N. 2016. "Plane waves in anisotropic thermo-elastic medium." *Steel & Composite Structures* 22, no. 3: 567–587. doi:10.12989/scs.2016.22.3.567

Lee, H., and Chang, W. 2009. "Vibration analysis of a viscous-fluid-conveying single-walled carbon nanotube embedded in an elastic medium." *Physica E: Low-dimensional Systems and Nanostructures* 41, no. 4: 529–532. doi:10.1016/j.physe.2008.10.002.

Lee, H., Hsu, J., and Chang, W. 2010. "Frequency shift of carbon-nanotube-based mass sensor using non-local elasticity theory." *Nanoscale Research Letters* 5, no. 11: 1774–1778. doi:10.1007/s11671-010-9709-8.

Lei, X., Natsuki, T., Shi, J., and Ni, Q. 2012. "Surface effects on the vibrational frequency of double-walled carbon nanotubes using the non-local timoshenko beam model." *Composites Part B: Engineering* 43, no. 1: 64–69. doi:10.1016/j.compositesb.2011.04.032.

Li, C., and Chou, T. 2006. "Atomistic modeling of carbon nanotube-based mechanical sensors." *Journal of Intelligent Material Systems and Structures* 17(3): 247–254. doi:10.1177/1045389x06058622.

Li, L., Hu, Y., and Ling, L. 2016. "Wave propagation in viscoelastic single-walled carbon nanotubes with surface effect under magnetic field based on non-local strain gradient theory." *Physica E: Low-dimensional Systems and Nanostructures* 75: 118–124. doi:10.1016/j.physe.2015.09.028.

Liu, H., and Lyu, Z. 2020. "Modeling of novel nanoscale mass sensor made of smart FG magneto-electro-elastic nanofilm integrated with graphene layers." *Thin-Walled Structures* 151: 106749. doi:10.1016/j.tws.2020.106749.

Naceri, M., Zidour, M., Semmah, A., Houari, M. S., Benzair, A., and Tounsi, A. 2011. "Sound wave propagation in armchair single walled carbon nanotubes under thermal environment." *Journal of Applied Physics* 110, no. 12: 124322. doi:10.1063/1.3671636.

Narendar, S., and Gopalakrishnan, S. 2010. "Ultrasonic wave characteristics of nanorods via non-local strain gradient models." *Journal of Applied Physics* 107, no. 8: 084312. doi:10.1063/1.3345869.

Narendar, S., Mahapatra, D.R., and Gopalakrishnan, S. 2011. "Prediction of non-local scaling parameter for armchair and zigzag single-walled carbon nanotubes based on molecular structural mechanics, non-local elasticity and wave propagation." *International Journal of Engineering Science* 49, no. 6: 509–522. doi:10.1016/j.ijengsci.2011.01.002.

Nematollahi, M.A., Jamali, B., and Hosseini, M. 2019. "Fluid velocity and mass ratio identification of piezoelectric nanotube conveying fluid using inverse analysis." *Acta Mechanica* 231, no. 2: 683–700. doi:10.1007/s00707-019-02554-0.

Pradhan, S., and Phadikar, J. 2009. "Small scale effect on vibration of embedded multilayered graphene sheets based on non-local continuum models." *Physics Letters A* 373, no. 11: 1062–1069. doi:10.1016/j.physleta.2009.01.030.

Saadatnia, Z., and Esmailzadeh, E. 2017. "Non-linear harmonic vibration analysis of fluid-conveying piezoelectric-layered nanotubes." *Composites Part B: Engineering* 123: 193–209. doi:10.1016/j.compositesb.2017.05.012.

Sadeghi-Goughari, M., Jeon, S., and Kwon, H. 2017. "Effects of magnetic-fluid flow on structural instability of a carbon nanotube conveying nanoflow under a longitudinal magnetic field." *Physics Letters A* 381, no. 35: 2898–2905. doi:10.1016/j.physleta.2017.06.054.

Selvamani, R., Mahaveer Sree Jayan, M., and Ebrahimi, F. 2020. "Static stability analysis of mass sensors consisting of hygro-thermally activated graphene sheets using a non-local strain gradient theory." *Engineering Transactions* 68, no. 3: 269–295. doi:10.24423/engtrans.1187.20200904

Selvamani, R., and Makinde, O.D. 2018. "Influence of rotation on transversely isotropic piezoelectric rod coated with a thin film." *Engineering Transactions* 66, no. 3: 211–227. doi:10.24423/engtrans.859.20180726.

Semmah, A., Beg, O. A., Mahmoud, S., Heireche, H., and Tounsi, A. 2004. "Thermal buckling properties of zigzag single-walled carbon nanotubes using a refined non-local model." *Advances in Materials Research* 3, no. 2: 77–89. doi:10.12989/amr.2014.3.2.077.

Wang, K., Wang, B., and Zeng, S. 2018. "Analysis of an array of flexoelectric layered nanobeams for vibration energy harvesting." *Composite Structures* 187: 48–57. doi:10.1016/j.compstruct.2017.12.040.

Wang, L., Hu, H., and Guo, W. 2006b. "Validation of the non-local elastic shell model for studying longitudinal waves in single-walled carbon nanotubes." *Nanotechnology* 17: 1408–1415. doi:10.1088/0957-4484/17/5/041.

Wang, Q. 2005. "Wave propagation in carbon nanotubes via non-local continuum mechanics." *Journal of Applied Physics* 98(12): 124301–124306. doi:10.1063/1.2141648.

Wang, Q., Varadan, V., and Quek, S. 2006a. "Small scale effect on elastic buckling of carbon nanotubes with non-local continuum models." *Physics Letters A* 357, no. 2: 130–135. doi:10.1016/j.physleta.2006.04.026.

Wu, D.H., Chien, W.T., Chen, C.S., and Chen, H.H. 2006a. "Resonant frequency analysis of fixed-free single-walled carbon nanotube-based mass sensor." *Sensors and Actuators A: Physical* 126, no. 1: 117–121. doi:10.1016/j.sna.2005.10.005.

Wu, Y., Zhang, X., Leung, A., and Zhong, W. 2006b. "An energy-equivalent model on studying the mechanical properties of single-walled carbon nanotubes." *Thin-Walled Structures* 44, no. 6: 667–676. doi:10.1016/j.tws.2006.05.003.

Yamabe, T. 1995. "Recent development of carbon nanotube." *Synthetic Metals* 70, no. 1–3: 1511–1518. doi:10.1016/0379-6779(94)02939-v.

Yan, Z., and Jiang, Y. 2011. "The vibrational and buckling behaviors of piezoelectric nanobeams with surface effects." *Nanotechnology* 22, no. 24: 245703. doi:10.1088/0957-4484/22/24/245703.

Yang, W., Liang, X., and Shen, S. 2015. "Electromechanical responses of piezoelectric nanoplates with flexoelectricity." *Acta Mechanica* 226, no. 9: 3097–3110. doi:10.1007/s00707-015-1373-8.

Zhang, D., Lei, Y., and Shen, Z. 2016. "Vibration analysis of horn-shaped single-walled carbon nanotubes embedded in viscoelastic medium under a longitudinal magnetic field." *International Journal of Mechanical Sciences* 118: 219–230. doi:10.1016/j.ijmecsci.2016.09.025.

Zhang, Y.Q., Liu, X., and Liu, G.R. 2007. "Thermal effect on transverse vibrations of double-walled carbon nanotubes." *Nanotechnology* 18, no. 44: 1–7. doi:10.1088/0957-4484/18/44/445701.

Zhen, Y., Wen, S., and Tang, Y. 2019. "Free vibration analysis of viscoelastic nanotubes under longitudinal magnetic field based on non-local strain gradient Timoshenko beam model." *Physica E: Low-dimensional Systems and Nanostructures* 105: 116–124. doi:10.1016/j.physe.2018.09.005.

Zidour, M., Daouadji, T. H., Benrahou, K.H., Tounsi, A., Bedia, E.A., and Hadji, L. 2014. "Buckling analysis of chiral single-walled carbon nanotubes by using the non-local timoshenko beam theory." *Mechanics of Composite Materials* 50, no. 1: 95–104. doi:10.1007/s11029-014-9396-0.

4

Atomic Layer Deposition (ALD) Utilities in Bioenergy Conversion and Energy Storage

Sugumari
Vallinayagam, Sam
Nirmala Nisha,
R. Sai Nandhini,
Azhagu Saravana
Babu, Jeyanathi
Palanivelu

*Vel Tech Rangarajan Dr.
Sagunthala R&D Institute
of Science and Technology,
Chennai, India*

S. Gayathri

*Karpaga VInayaga College of
Engineering and Technology,
Chengalpattu, India*

and

*Vel Tech Rangarajan
Dr. Sagunthala R&D
Institute of Science and
Technology, Chennai, India*

Bioenergy conversion plays an important role in energy storage in which electrodes are the storage devices based on electrochemical energy, particularly in the deciding factors, where design and fabrications decide the capacity of an electrode. One of the recent advancements in nanotechnology is atomic layer deposition (ALD), which can produce thin films, control thick films, and also change the surface of interfacial characters. As already said, ALD plays a vital role in the field modification of the surface of electrode materials. Also, deposition is a part of the chemical vapor deposition (CVD) process. These electrochemical processes provide solutions to many problems such as examining properties of the materials based on the device and surface modification (interfacial properties); and they play a critical role in a device's performance. Some devices like photo-electrochemical devices, supercapacitors, and

DOI: 10.1201/9781003355755-4

others (e.g., fuel cells and supercapacitors) have electrochemical properties. ALD can design engineered electrode materials on an atomic scale. In this chapter, we are going to study a detailed purpose of electrochemical energy generation, storage devices, ALD and its challenges, current applications, processes, etc.

4.1 Introduction

In day-to-day life, renewable resources and non-conservation energy are decreasing; to overcome this limitation, bioenergy can be used. Bioenergy is one of the mass diverse resources available to help the non-conservation/renewable energy, which decreases day by day. Bioenergy is a source of electricity and gas that is generated biologically from organic matter known as 'biomass' [1, 2]. Biomass can be any form of waste such as food, raw vegetables, fruits, and even sewage. Bioenergy is equivalent to fossil fuels (e.g., gasoline, diesel fuels, and jet fuels), which are normally in a fluid form that is converted to transportation fuels, as already mentioned. This technique involves reusable components like carbon. This reduces emission in many four-wheelers like cars and trucks and also in jets and ships. Bioenergy has played a vital role in energy security; its reliability and being available in large quantity are its advantages. Energy storage is the ability to capture the energy that is produced and to store it for subsequent use [3]. This captured energy is stored in a device called a battery or an accumulator. Different types of energy, including radiation, chemical, gravitational potential, electrical potential, electricity, raised temperature, and dormant hotness, are also dynamic. Vitality stockpiling generally includes converting vitality from structures that are challenging to store, and that's only the tip of the iceberg helpfully, or are monetarily storable. A percentage of advance gives short-term vitality storage, while at the same time others such as camwood continue to provide significantly more storage.

There is an interest and a colossal need for wellsprings of renewable vitality, as well as a need to scrutinize considerations again going as far back as decades. For example, innovations such as those that are wind and sun oriented have been broadly explored. Furthermore, in this case expressions of news persons are expository. However, prudent utilization of these advances need not be broadly expected; mostly, they should be cosseted. Moreover, the powerlessness of administrations, throughout the off-source periods, should lead to increase in these innovations and making them more competitive. The recent couple of decades have shown that there is a need to intensify examination of the under-performing vitality capacity frameworks. The clue is to devise a vitality stockpiling framework that considers the capacity of power throughout incline hours at a moderately less-expensive quality and, following this, the conveyance. Furthermore, conveyance advances the vitality capacity, for example supercapacitors such as camwood store and convey vitality at a quick rate, putting forth a current helter-skelter that was previously of a short span. Decades have gone by and we need to see work and fast development in the field of supercapacitor innovation. A few electrochemical properties of cathode material as well as electrolytes need to be included in the expositive expression of newspersons. Supercapacitor cathode materials, for example carbon and carbon-based materials, have gained growing consideration due to their helter-skelter performance on a particular surface area, beneficial electrical conductivity, previous phenomenal strength, barbarous situations, and so forth. Clinched alongside late years, there needs to be an expanding enthusiasm toward biomass-derived actuated carbons and, likewise, toward a cathode material for supercapacitor applications [5]. These improvements in an elective supercapacitor cathode material from biowaste serve two principal purposes: (1) it aids with waste disposal, converting waste into a functional product; (2) it gives an investment contention for those organizations claiming significant supercapacitor innovation. This chapter reviews late developments for carbon. Furthermore, carbon-based materials are inferred from biowaste for supercapacitor innovation. There is a correlation in the middle of those different capacity components. Also, electrochemical execution of electrodes determined, starting with biowaste, may be presented [6, 7].

Worldwide an interest in vitality may be anticipated twofold, something like two decades. Vitality assumes a paramount part of our lives, impacting us socially and with investment advancement. Advanced economies would be driven toward accessing dependable vitality sources. Customary vitality sources, for example fossil fuels (coal, gas, and oil), are being drained at a quick rate. Such decimation will claim biological communities and habitats, and lead to elimination of wildlife and contamination of the earth. The essential concern of collecting vitality fossil fuels is that it may be unsustainable in the long term; in this way, these needs determine specialists and the industry to receive maintainable, more renewable vitality advances. Over the next few decades, there is a need to increase examination of renewable vitality wellsprings, such as sun-oriented energy, geothermal energy, wind energy, biofuels, and so on, as well as electrochemical vitality stockpiling units, for example supercapacitors and rechargeable batteries [4, 5]. It will not be an exaggeration to say that there is a need for fruitful advancement in order to claim all possible renewable vitality hotspots (e.g., windmills and sun-oriented cells), mixtures, and electric vehicles. Furthermore, advanced mobile grids depend fundamentally upon the accessibility of a suitable vitality stockpiling framework.

4.2 Atomic Layer Deposition (ALD)

ALD may be additionally realized likewise with atomic layer etching (ALE; avapor-based self-terminating technique) on a place where there is a scope for conformal layered materials with regulated thickness, particularly on mind-boggling surfaces and 3D structures. It is a productive and a compelling procedure which may be produced for an affidavit claiming different metals and metal oxides/nitrides/sulfides. Since the 1970s, ALD need has been relentlessly created. Also, it has been popularized in different dainty movies. Furthermore, ALD is needed on surface coatings, for chemical, mechanical, and optical building, and also for microelectronics, an area where immaculately known cases need aid. ALD is also useful for electroluminescent shows and propelling of high-k metal oxides [5, 8]. Likewise, ALD camwood successfully covers the surface and porous structures of diverse materials (porous and surface structure). Since the 2000s, it has also been widely used for surface functionalization of materials in catalysis, fuel cells, batteries, and sensors. The ALD approach is based on successive cycles of self-terminating gas–solid surface reactions, with two people cycling alternately, with more pulses and precursors [5, 9, 10]. Because of the incomparability of the ALD process, when it is compared with different gas phase techniques such as PVD and CVD [17, 18], and based on the solution affidavit routines like hydrothermal and sol-gel techniques [15, 16], it exhibits a few advantages, for example thickness is under control (controllable) with secondary uniformity, deposition (immaculate conformal), and development in low temperature (normally following 300°C; portion materials stored during room temperature). Actually, once the mind-boggling surfaces and 3D substrates delegate a test, they gather and choose electron microscopy pictures for different sorts of nanostructures, a place ALD might have included in any event is a standout among the creation steps [13, 14]. Compared to techniques that are based on solution, ALD is based on the surface reaction which is a vapor phase. Subsequently, there is a chance indeed for conformal affidavit technique to be effectively acquired on intricate surfaces and 3D substrates.

4.3 Thin-Film Growth Mechanism

The ALD system of camwood may be utilized to both develop naturals and intensify slim movies. Of the natural films, a large portion of normal ones need aid with silicon films, which are generally utilized within the microelectronic. New precursors would ceaselessly be produced to fabricate all the complex and particular movies. This could be a chance to possibly, and eventually, attain the following methods: Tom's perusing pumping down the response space, and Tom's perusing purging the response space with a dormant gas stream in the middle of progressive pulses. In the latter method, a section of dormant gas may be framed in the conduits in the middle of those forerunner pulses. The last strategy may be only the tip of the iceberg, which broadly concerns itself with the creation scale owing to its effectiveness, and its proficiency in claiming framing and a powerful obstruction in dispersion while in the middle of progressive pulses. Regularly, the inactive purging gas may be likewise utilized as a bearer gas. Throughout, forerunner pulses, by dilution of the gas preceding it, will be nourished under the response space. Some ALD reactors utilizing purging idle gas need aid known as flow-type or voyaging wave reactors [19].

4.4 Methods for Studying ALD

ALD methodology may be developing under two primary categories: experimental and numerical. Also, considering the test, experts need to improve oath of the motion pictures by changing the states, such as pressure, temperature, cleanse time, fixation, etc. All along these lines, noticing and stockpiling of all instrumentation might be enhancing. The ALD procedure needs far-reaching materials to extend over any case, as camwood cannot be used in all materials; hence, it is essential for recreation to determine and foresee which viable response pathways need aid. Because of its nuclear-scale affidavit, which is relative to the characteristic, and the reactor scale network, those ALD numerical systems naturally include multi-scale dissection. That multi-scale strategy fuses atomic security arrangement, species chemisorption/adsorption, compound energy, and film oath. Its like manner fuses a reactor scale that incorporates material choices/associations, math impacts, and, liquid, as well as vitality transport at a perceptible level [20]. A third scale, mesoscopic scale, is introduced, which concerns providing help with the issue of delineation. At the mesoscopic scale reproduction, those continuum laws need a generous help. This scale, moreover gets its net motions from the trademark densities. The mesoscopic scale couples-decouples the center of those scale strategies by surveying a segment on the reactor scale network [21].

Handy presentation of the substrate surface is a viable option to upgrade immersion in the ALD methodology. Secondary staying likelihood might eventually help attain Tom's perusing adjustment of the surface or regulate the science of the response. To accomplish accurate ALD growth, in the long run purging the middle of the precursors' purposes of presentation ought to further bolster a chance to be sufficient. If not, there could a chance to further CVD mode development, because of gas stage precursors mixing, which will prompt poor conformity because of gas stage responses. The long run for immersion may be a great part more for heavier particles. The sub-atomic flux is conversely proportional to the square root of the sub-atomic weight [22, 23].

On analyzing, modeled and test outcomes of the ALD process are set up, along with anticipation of the concoction species, over a reactor. Likewise, a capacity for long run and space provides for an ideal ALD methodology both for the test and for the modeled outcomes. The modeled outcomes affirm that the expanded separation rates the abatements of those general centralizations in the chamber regardless of the weight. The identifier of the ideal scaling tenets over expand those uses from claiming precursors in the framework to an altered growth rate and relative consistency.

4.5 Varities of ALD

a. *Catalytic ALD*

This occurs in >32°C (Lewis base catalyst). Some precursors used are viable in ALD, such as metal oxides (i.e., titanium oxide, zirconium oxide, and SnO_2). This reaction is carried by (metal) Cl_4, H_2O. It's mainly used in the anti-reflective layers, etc. [24]

b. *Aluminum oxide ALD*

Aluminum oxide ALD is mostly used in the dielectric layers, insulating layers, etc., and solar cell surface passivation. Which metal oxide is the precursor of Al_2O_3? This reaction occurs at 30–300°C. (Metal) carbon tetrachloride, H_2O, titanium is opropoxide, (metal) $(Et)_2$—these three are involved as reactants to carry the process [25].

c. *Metal ALD using thermal chemistry*

Metal ALD occurs in 175–400°C and has different types of reactants such as hydrogen. In ALD, some metals act as precursors, such as metal fluorides and catalytic metals. Metal ALD is mainly present in conductive pathways [26].

d. *ALD on polymers*

ALD on polymers is carried out at 25–100°C, and influences the process of surface functionalization, composites creation, diffusion, and surface functionalization of polymers. These viable precursors are common polymers [27].

e. *ALD on particles*

For polymer, particles occur at the temperature 25–100°C and for metal/alloy particles 100–400°C. These applications are of protective deposition and coatings for insulators, property modification for optical and mechanical, and composite structure formation. Various gases are induced, and for individual coating fluidized bed reactors are used. These precursors are BN, ZrO_2, CNTs, and polymer particles.

f. *Plasma ALD for single element ALD materials*

Different kinds of ALD material used as reactants are organometallics, TBTDET, and ammonia (NH_3). These applications are dynamic random-access memory structures and capacitors. Pure metals and metal nitrides both act as precursors at 20–800°C [28].

4.6 ALD Precursor Requirements

It should be for the most part fluid and gas. Additionally, it ought to be unpredictable and thermally steady. At the point when it is applied, it should chemisorb onto the surface (quickly respond to the surface) and respond rapidly and quickly with one another. Additionally, it requires short immersion time, deposition rate is to be accepted, and no gas stage responses ought to take place, as well as there should be no self-decay and no influence on thickness and consistency. At that point, it ought not to carve and be licentious into the substrate or film.

4.7 ALD Process and Equipment

In ALD, at one time one layer can be made by the release of sequential precursor gas pulses. In the process chamber, a precursor gas is allowed to form a monolayer gas on the wafer surface, and this is the first step. Then again in the chamber, a second precursor gas is allowed which reacts with the precursor gas previously allowed in the first step. And this produces a monolayer film on the wafer surface.

There are two fundamental mechanisms:

- Chemisorption saturation process
- The sequential surface chemical reaction process

Example: Al$_2$O$_3$ deposition (ALD cycle)
The first and second precursor gases produce a monolayer film for one cycle, and this film thus produced can be controlled by deposition cycle (based on the number of the cycle) [29–31].

a. ***ALD cycle for Al$_2$O$_3$ deposition (Step 1a)***
In most of the surfaces, water vapor is absorbed; thus, it can form a hydroxyl group. Silicon will be in this form, (Si-O-H(S)). In the reactor, a substrate is placed at the reaction chamber, and trimethyl aluminum (TMA) is pulsed into it.

b. ***ALD cycle for aluminum oxide deposition (Step 1b)***

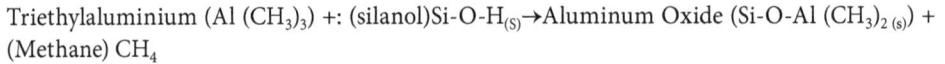

Triethylaluminium (Al (CH$_3$)$_3$) +: (silanol)Si-O-H$_{(S)}$→Aluminum Oxide (Si-O-Al (CH$_3$)$_{2\,(s)}$) + (Methane) CH$_4$

In step 1b, methane is produced as a product, where the absorbed hydroxyl reacts with the TMA.

c. ***ALD cycle for aluminum oxide deposition (Step 1c)***
Once the passivated surface is done, the absorbed hydroxyl group reacts with the TMA. In this case, TMA cannot react by itself so it reduces into one layer; this reaction can result in uniformity in thickness, and, with the product, that is, methane, excess TMA is pumped away.

d. ***ALD cycle for aluminum oxide deposition (Step 2a)***
When the reaction product methane and TMA get pumped away, water vapor is pulsed into the reaction chamber.

e. ***ALD cycle for aluminum oxide deposition (Step 2b)***
H$_2$O reacts with the dangling methyl groups to form hydroxyl surface groups and aluminum oxygen (AI-O) bridges, thus looking for a new TMA pulse. Here methane is the reaction product again.

f. ***ALD cycle for aluminum oxide deposition (Step 2c)***
Again, the methane is pumped away; thus, the remaining amount of water vapor does not react with the surface of hydroxyl groups. This reaction results from perfect passivation, to one atomic layer.

g. ***ALD cycle for aluminum oxide deposition (after three cycles)***
For one cycle, one TMA and one water vapor will form. Here three cycles are done, which is equal to one angstrom (per cycle). In each cycle, pulsing and pumping will take place (maybe 3 seconds).

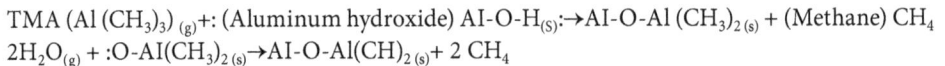

TMA (Al (CH$_3$)$_3$) $_{(g)}$+: (Aluminum hydroxide) AI-O-H$_{(S)}$:→AI-O-Al (CH$_3$)$_{2\,(s)}$ + (Methane) CH$_4$
2H$_2$O$_{(g)}$ + :O-AI(CH$_3$)$_{2\,(s)}$→AI-O-Al(CH)$_{2\,(s)}$+ 2 CH$_4$

The first and the second gas will produce a monolayer film for one cycle, and this film thus produced can be controlled by deposition cycle (based on the number of the cycle) (Figure 4.1).

4.8 Process Involved in Deposition

ALD process involves the surface of a substrate being exposed to alternating precursors, which do not overlap but instead are introduced incidentally. In each other pulse, the precursor molecule reacts with the surface in a self-limiting way, which ensures that the reaction stops once all of the reactive sites on the substrate have been used. A complete ALD cycle is determined by the nature of the precursor–surface interaction. The ALD cycle can be performed several times to increase the layers of the thin film, based on the requirement.

The process of ALD is often performed at low temperatures, which is beneficial when working with fragile substrates, and some thermally unstable precursors can still be employed with ALD as long as their decomposition rate is slow (Figure 4.2). The ALD process is widely used since it provides ultra-thin nano-layers in an exceedingly precise manner on a variety of substrates, including micron to sub-micron size particles. The nano-layers achieved with ALD are by nature conformal and pinhole-free.

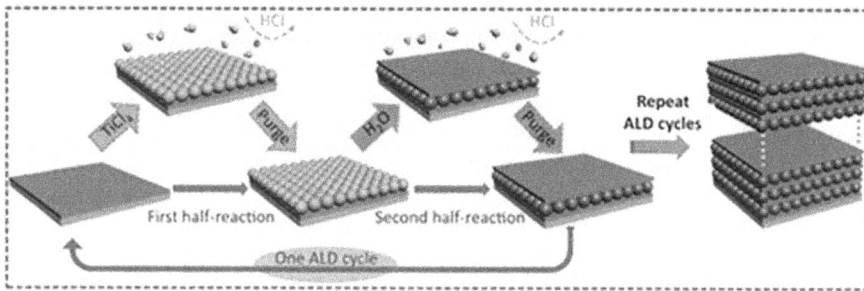

FIGURE 4.1 Deposition cycle.

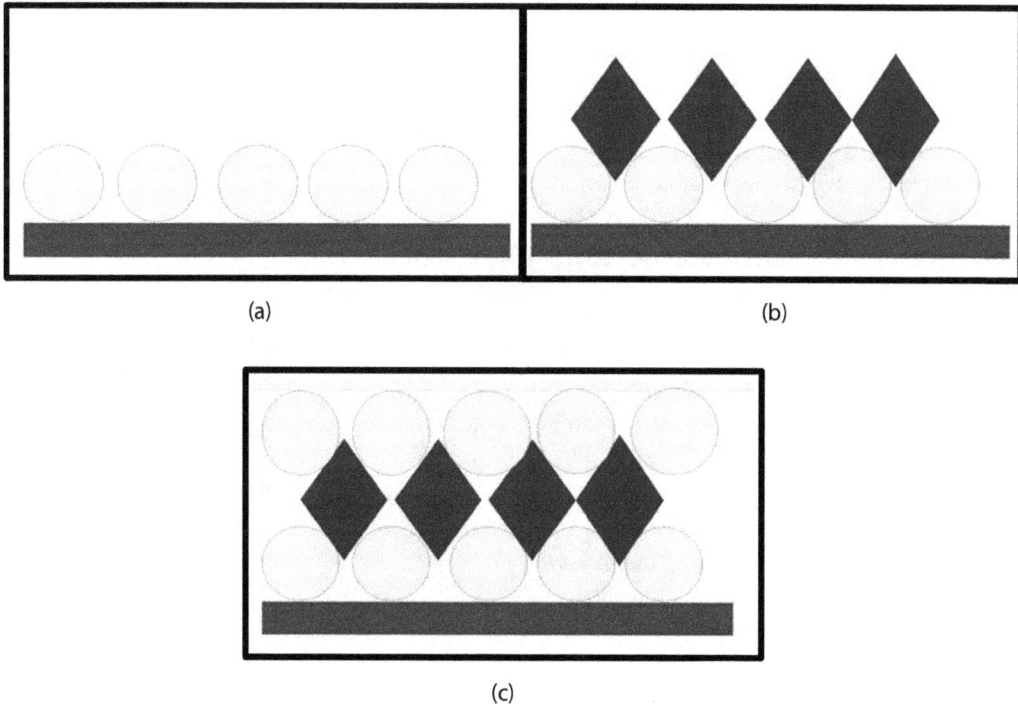

FIGURE 4.2 Main precursors coating.

Ligand precursor: To prepare the surface for the next layer and define the kind of material growth. Example: H_2O for oxides, N_2 or NH_3 for nitrides

Main precursors are

- Highly reactive
- Thermally stable
- Fulfill the requirement for self-terminating reaction
- No self-decomposition
- Enough cleanliness

ALD reactors: They are mainly for major types of reactors where there are closed-system chambers, semi-closed system chambers, open system chambers, and semi-open system chambers [32, 33].

4.9 Nanostructured Materials

Nano shaped particles (NPs) are particles in which every compound is in the form of bunches, is crystal in nature, or whose molecules range between 1 and 100 nm in size. The explosion of both scholarly and mechanical interests over these materials during the previous decade emerges from the momentous varieties in major electrical, optical, and attractive properties that happen as one advances from a "vastly broadened" strong material to a molecular dynamic comprising of a countable number of particles (Figure 4.3). New advances in the union and examination of utilitarian nanostructured materials are zeroing in on the novel size-subordinate physical science and science that results when electrons

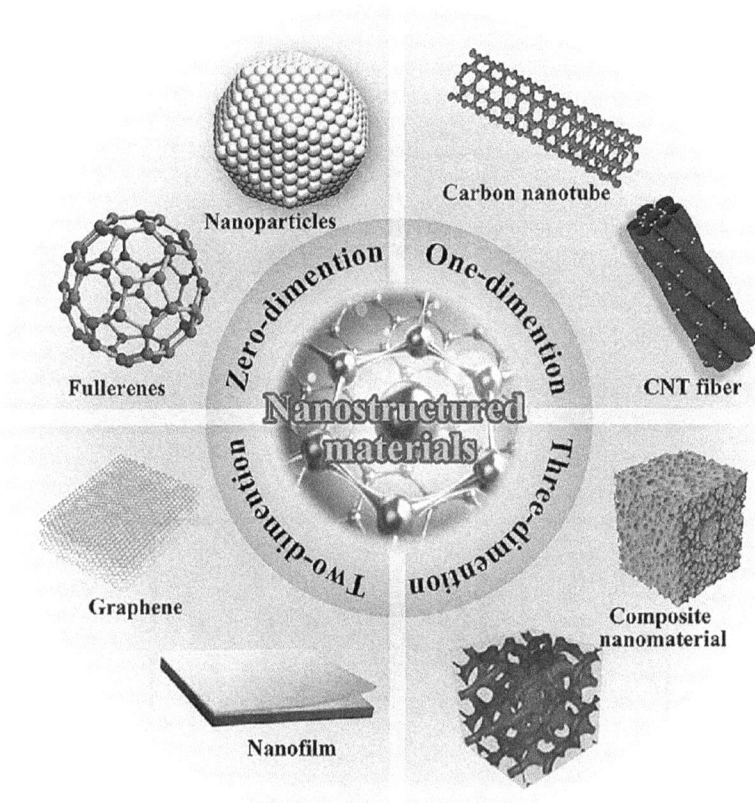

FIGURE 4.3   Various nanostructured materials.

are kept to nanoscale semiconductor and metal groups and colloids. Carbon-based nanomaterials and nanostructures, including fullerenes and nanotubes, assume an undeniable inescapable part in nanoscale science and innovation and are along these lines portrayed in some profundity. Current nanodevice manufacture strategies and the future possibilities for nanostructured materials and nanodevice is ALD.NPs can be arranged into various kinds as indicated by their shape, dimensions, and physical and substance features. Some of them are carbon-based NPs, metal NPs, semiconductor NPs, polymeric NPs, and lipid-based NPs.

4.10 Coating of Electrode

A large portion of the current transportation of lithium-ion batteries has terminal coatings of around 25–65 μm thickness, while some purchaser electronic applications may have 111 μm-thick electrodes. A vehicle's lithium-ion batteries have cathode coatings of around 22–66 μm thickness, though some shopper electronic applications may have 118 μm-thick anodes. As batteries looking for vehicles with significant all-electric reach, there is a solid motivating force to save cost by expanding the thickness of the cathode coatings. In BatPaC, a 100 μm default anode thickness limit is expected. If the determined ideal thickness is under 100 μm, the ideal determined thickness is utilized. There is an off chance that the ideal thickness is more noteworthy than 120 μm. Keeping in mind the turn of events in the field of designing, the proper thickness impediment might be more noteworthy or not exactly be default suspect. The model gauges this vulnerability by computing the complete expenses of batteries having terminal thickness limits at the 95% certainty limits, specifically 50 and 160 μm for PHEV cells and 77 and 210 μm for EV cells.

4.10.1 Nanotechnological Coating of Electrode

ALD can offer genuine nanoscale thick surface testimonies, which are profoundly conformal and pinhole-free. ALD is an ultrathin film statement strategy that is constrained by the gas stage and consecutive self-restricting substance responses of the forerunners at the material surface. Most ALD measures normally require two antecedents, which are provided in groups, each, in turn, to add to the surface covering.

ALD offers exceptionally conformal, pinhole-free, and angstrom-level surface deposition. It gives a genuine nanoscale-level covering layer, and the thickness is self-controlled for it depends on self-restricting surface responses. The thickness accomplished with each ALD cycle is measurable explicitly yet it is ordinarily in sub-nanometers. Thicker coatings are effortlessly achieved by essentially rehashing the ALD. ALD practices thinking about its exact control and high throughput, making it the most serious surface film statement procedure. It can undoubtedly deliver coatings on powder tests just as on huge region surfaces. Also, these are accomplished at low activity temperatures. In any case, it is a lethargic testimony cycle and requests ultraclean surfaces, which increases the expense. It additionally requires the end of the antecedent, while changing to the following forerunner in grouping [34–37]. The improvement in the electrochemical execution is reliant upon the kind of covering material. ALD is an arising strategy used to change the surface of cathode anodes with ultra-thin conformal assurance layers to chiefly forestall terminal/electrolyte-side responses and improve electrochemical cycling execution.

4.10.2 Electrochemical Energy Storage Device

Frameworks that convert chemical energy into electrical energy can make a significant commitment to the execution of maintainable energy. This section portrays the essential standards of conversion of chemical energy into electrical energy stockpiling and talks about three significant sorts of framework: battery-powered batteries, power modules, and stream batteries. A battery-powered battery comprises at least one electrochemical cell in an arrangement. A battery contains chemical energy and converts it into electrical energy from an outside electrical source, which is put away in the battery during charging and is then utilized to supply energy to an outer burden during releasing. Two battery-powered battery frameworks are talked about in some detail: the lead-corrosive framework, which has been in use for more than 150 years, and the significantly later lithium framework; sodium-sulfur and nickel-metal hydride frameworks are additionally momentarily examined. A power module is an electrochemical cell where the reactants providing the energy are not put away in the actual cell; rather, they are ceaselessly provided to the anodes from an outer source. A typical model is a hydrogen-oxygen power device: in this, the hydrogen and oxygen can be created by electrolyzing water; thus, the mix of the energy component and electrolyzer is adequately a capacity framework for electrochemical energy. Both high-and low-temperature energy units are portrayed, and a few models are talked about for each situation. A stream battery is like a regular battery-powered battery in that it very well may be over and again charged and released. Notwithstanding, the energy stockpiling material is broken down in the electrolyte as a fluid; thus, it can be put away in outside tanks. Different sorts of stream batteries are accessible or being worked on. Three of the most significant models are examined in some detail: the all-vanadium stream battery, the zinc-bromine mixture stream battery, and the all-iron slurry stream battery. Some different models are additionally momentarily referenced. The decision of electrochemical stockpiling framework is exceptionally subject to the particular prerequisites of the venture that is being thought of, the related forthright capital and lifetime consumption expenses, and end-of-life, nature, and wellbeing contemplations.

Chemical energy into electrical energy gadgets, for example lithium-particle batteries and supercapacitors, will assume a significant part later on for feasible energy since they have been broadly utilized in versatile hardware, electric/crossover vehicles, fixed force stations, and so forth. To fulfil the steadily developing need of superior (energy and force thickness) energy gadgets, chemical energy into electrical energy gadgets with capacitor-like rate execution and battery-like high limit are exceptionally attractive in cutting-edge energy stockpiling technology. Currently, the elite of energy stockpiling gadgets is as yet restricted by languid charge transport, in this manner motivating a lot of exploration endeavors toward tackling the innate issues of poor electronic conductivity and low cation particle diffusivity inside the terminal materials. Additionally, hypothetical demonstrating has given new bits of knowledge into the thermodynamics and energy of terminal materials, to control materials plan. As of late, constant exertion has been centered around level-headed material and terminal plans to improve the exhibition of electrochemical energy stockpiling and change gadgets [38].

Chemical energy into electrical energy stockpiling gadgets is getting progressively more significant for lessening petroleum product energy utilization in transportation and for the far-reaching organization of irregular environmentally friendly power. The uses of various energy stockpiling gadgets in explicit circumstances are largely fundamentally dependent on the anode materials, particularly carbon materials. Biomass-inferred carbon materials are accepting broad consideration as terminal materials for energy stockpiling gadgets in light of their tenable physical/substance properties, natural concern, and financial worth. In this audit, ongoing improvements in the biomass-inferred carbon materials and the properties controlling the instrument behind their activity are introduced and examined. Additionally, progress on the utilizations of biomass-inferred carbon materials as terminals for energy stockpiling gadgets is summed up, including electrochemical capacitors, lithium-sulfur batteries, lithium-particle batteries, and sodium-particle batteries. The impacts of the pore structure, surface properties, and graphitic degree on the electrochemical execution are examined in detail, which will direct further level-headed plan of the biomass-determined carbon materials for energy stockpiling gadgets.

4.10.3 Deposition and Surface Modification of Electrode

ALD has been used to deposit highly dispersed metal (or) metal oxides, particles on substrates with uniformity in several particle dimension. Although it is useful in improving the metal to utilize its energy without any lack, such that bioenergy is concerned with the catalytic reagent and used as a catalyst in gas and liquid phase reaction. Besides, it has been reported that the size of metal particles could be controlled precisely by adjusting the number of ALD cycle, and precursors dose time ALD is to prepare and optimize nanostructural catalyst for constructing 3D structural coating.

4.10.4 Selective Area Deposition

The technique developed to construct a unique nanostructure by deposition of particles/metals, or metal oxides area–selective deposition, is mainly adopted to fabricate and decorate nanostructural catalyst; much progress has been achieved for energy storage capacity.

4.10.5 Energy Storage and Conversion

Power module innovation is a promising, dependable, and clean energy innovation because of its low outflow of toxins and high energy change productivity. Among the energy component types are polymer electrolyte film power devices, direct methanol power devices, liquid carbonate power modules, strong oxide energy components, reversible energy components, phosphoric corrosive energy units, and soluble energy components. An acquaintance of ALD with this field could essentially bring down the expenses of energy units and increase the life expectancy of power devices. Impetuses as a piece of energy units are a significant commitment to both the significant expense and restricted solidness of power modules. In catalysis, a bigger surface-territory-to-volume proportion is required; accordingly, particles have inclination contrasted with films. The impetus layer covers with the chromium/gold anode around the edges of the film and gives electrical availability. The capacity to alter the outside of the stable layer empowers the improvement of better power modules.

4.11 Applications

Any strategy fit for storing slight practical layers onto organized substrates, and particularly into nano porous structures, is consulted with an immediate significance toward energy change applications. The conformal covering of non-planar examples is a property that is an interesting characterization of nuclear layer statement, which is the reason ALD s intrinsically fit to the arrangement of energy change gadgets. ALD accomplishes a meager film development by utilizing very much characterized surface science. At least two corresponding, quantitative surface responses performed accordingly and rehashed in an exchanging way bring about the testimony of a strong layer-by-layer design.

4.11.1 Applications in Photovoltaics

Aluminum oxide nanolayers, arranged by ALD, have been appearing to represent a total expansion in the transformation productivity of sunlight-based cells. This has been credited to both the fantastic synthetic and the field-impact passivation offered by the material. Nonetheless, as the interface properties firmly rely on the thickness and preparing states of the film, there is a harmony between accomplishing total surface inclusion to forestall fast decay and augmenting light conveyance through the phone. The ideal thickness of the layer is resolved to be greater than 5 nm for plasma ALD and greater than 10 nm for warm ALD. Straightforward conductive oxides by ALD are ready, with instances of indium oxide and zinc oxide for the front side and backside, respectively, of silicon heterojunction solar cells. The physical science and necessities of transparent conducting oxides (TCOs) for silicon heterojunction solar cells are examined and the benefits of ALD TCOs in high-volume fabricating (HVM) are featured. The latest pieces of the part, novel passivating and burrowing contacts and transporter particular materials are dependent on metal oxides [39].

These movies are intended to design the interfaces inside photovoltaic gadgets—for example, color-sharpened sun-based cells, quantum-spot-sharpened sun-powered cells, colloidal quantum-speck sunlight-based cells, natural sun-based cells, perovskite sun-based cells, and photoelectrochemical cells for water parting—to limit recombination misfortunes. The audit for the most part centers around the use of ALD in dye-sensitized solar cells, researching the utilization of both titanium dioxide as a minimal layer at the TCO/metal oxide interface and aluminum oxide as a hindering layer at the metal oxide/safeguard interface. It closes at 6–11 nm-thick indistinct titanium dioxide film that is both uniform and sans pinhole and is the ideal arrangement to obstruct the back move of electrons to the hole transport material. Expanding the thickness and crystallinity of the titanium dioxide film diminishes the light conveyance and productivity of impeding properties because of the presence of grain limits, and accordingly lessens the force transformation effectiveness. ALD offers the openings in the energy change field and the prerequisites for it to turn into a promising competitor in high-volume fabricating. It ought to subsequently hold any importance with general analysis in the semiconductor business just as ALD experts [40].

4.11.2 Microelectronics Applications

Dynamic random-access memory capacitors are yet another use of ALD. An individual dynamic random-access memory cell can store a solitary piece of information and comprises a solitary metal–oxide–semiconductor and a capacitor. Significant endeavors are being made to lessen the size of a capacitor, which will viably consider a more prominent memory thickness. To change the capacitor size without influencing the capacitance, distinctive cell directions are being utilized. A portion of these incorporate stacked or channel capacitors. Transition metal nitrides' potential of their utility has been discovered both as metal obstructions and as entryway metals. Metal boundaries are utilized to encase the copper interconnects utilized in present-day incorporated circuits to keep away from dispersion of copper into the encompassing materials, like covers and the Si substrate, and to forestall Cu tainting by components diffusing from the separators by encompassing each copper interconnect with a layer of metal obstructions. Testimony of the great oxides of aluminum and Hafnium (IV) oxide has been perhaps the most generally analyzed spaces of ALD [41, 42]. The inspiration for high kappa oxides comes from the high burrowing flow through the generally utilized silicon dioxide door dielectric in metal–oxide–semiconductor when it is downscaled to a thickness of 1.0 nm and below. With the high kappa oxide, a thicker entryway dielectric can be made for the necessary capacitance thickness; in this way, the burrowing flow can be diminished through the construction [43–45].

4.12 Advantages of an ALD

- Self-limiting growth process
- Precise film thickness is achieved by the number of deposition cycles
- Excellent conformity and uniformity
- Uniform dense homogeneous and pinhole-free films
- The reactant flux homogeneity need not be controlled
- Batch capable and large area
- Straightforward scale-up and good reproducibility
- Atomic-level composition control
- Why separate dozing of reactants and the surface exchange reactions are carried out

4.13 Limitations of an ALD

- Expensive equipment
- Critical assessment of the flow
    - i. Too much flow is equal to clogging of valves
    - ii. Too low floor is equal to underperformance
- Low effective deposition rate

4.14 Conclusion and Outlook

ALD has been consistently settled and popularized in different slender movies and surface coatings in compound, mechanical, and optical designing just as in microelectronics, where the ideal realized models are in electroluminescent shows and progressed high-k metal oxides. ALD is an arising procedure used to adjust the surface of cathode terminals with ultra-thin conformal insurance layers to primarily forestall anode/electrolyte side responses and improve electrochemical cycling performance. The ALD measure has a very broad material reach to look over, but not every material can be utilized; consequently, there is a requirement for recreation to decide and foresee which compelling response pathways are important. Adequate substrate openness and great cleansing of the response space are alluring for an effective ALD process. ALD offers the openings in the energy change field and the prerequisites for it to turn into a promising competitor in high-volume fabricating. It ought to subsequently hold some significance with general scientists who are in the semiconductor business, just as ALD subject matter experts.

Conflicts of Interest

The author has no conflicts of interest and of financial funding regarding publishing of this book chapter for Bentham Science.

References

1. J. Ladd, M. Amato, L.-K. Zhou, and J. Schultz, "Differential effects of rotation, plant residue and nitrogen fertilizer on microbial biomass and organic matter in an Australian alfisol," *Soil Biology and Biochemistry*, vol. 26, pp. 821–831, 1994.
2. D. Powlson, P. Prookes, and B. Christensen, "Measurement of soil microbial biomass provide an early indication of changes in total soil organic matter due to straw incorporation," *Soil Biology and Biochemistry*, vol. 19, pp. 159–164, 1987.
3. J. P. Barton and D. G. Infield, "Energy storage and its use with intermittent renewable energy," *IEEE Transactions on Energy Conversion*, vol. 19, pp. 441–448, 2004.
4. M. Fraga and R. Pessoa, "Progresses in synthesis and application of SiC films: From CVD to ALD and from MEMS to NEMS," *Micromachines*, vol. 11, p. 799, 2020.
5. E. Ahvenniemi, A. R. Akbashev, S. Ali, M. Bechelany, M. Berdova, S. Boyadjiev, D. C. Cameron, R. Chen, M. Chubarov, and V. Cremers, "Recommended reading list of early publications on atomic layer deposition—Outcome of the 'Virtual Project on the History of ALD'," *Journal of Vacuum Science & Technology A: Vacuum, Surfaces, and Films*, vol. 35, p. 010801, 2017.
6. T. Muneshwar, M. Miao, E. R. Borujeny, and K. Cadien, "Atomic layer deposition: Fundamentals, practice, and challenges," in *Handbook of thin film deposition*, ed: Elsevier, 2018, pp. 359–377.
7. S. Adhikari, S. Selvaraj, and D. H. Kim, "Progress in powder coating technology using atomic layer deposition," *Advanced Materials Interfaces*, vol. 5, p. 1800581, 2018.
8. E. Alvaro and A. Yanguas-Gil, "Characterizing the field of Atomic Layer Deposition: Authors, topics, and collaborations," *Plo S One*, vol. 13, p. e0189137, 2018.
9. M. Fraga and R. Pessoa, "Progresses in synthesis and application of SiC films: From CVD to ALD and from MEMS to NEMS," *Micromachines*, vol. 11, p. 799, 2020.
10. J. Leppäniemi, P. Sippola, M. Broas, J. Aromaa, H. Lipsanen, and J. Koskinen, "Corrosion protection of steel with multilayer coatings: improving the sealing properties of physical vapor deposition CrN coatings with Al2O3/TiO2 atomic layer deposition nanolaminates," *Thin Solid Films*, vol. 627, pp. 59–68, 2017.
11. J. H. Min, Y. A. Chen, I. T. Chen, T. Sun, D. T. Lee, C. Li, Y. Zhu, B. T. O'Connor, G. N. Parsons, and C. H. Chang, "Conformal physical vapor deposition assisted by atomic layer deposition and its application for stretchable conductors," *Advanced Materials Interfaces*, vol. 5, p. 1801379, 2018.

12. E. Ahvenniemi, A. R. Akbashev, S. Ali, M. Bechelany, M. Berdova, S. Boyadjiev, D. C. Cameron, R. Chen, M. Chubarov, and V. Cremers, "Recommended reading list of early publications on atomic layer deposition—Outcome of the 'Virtual Project on the History of ALD," *Journal of Vacuum Science & Technology A: Vacuum, Surfaces, and Films*, vol. 35, p. 010801, 2017.

13. T. Muneshwar, M. Miao, E. R. Borujeny, and K. Cadien, "Atomic layer deposition: Fundamentals, practice, and challenges," in *Handbook of thin film deposition*, ed: Elsevier, 2018, pp. 359–377.

14. S. Adhikari, S. Selvaraj, and D. H. Kim, "Progress in powder coating technology using atomic layer deposition," *Advanced Materials Interfaces*, vol. 5, p. 1800581, 2018.

15. E. A. Filatova, D. Hausmann, and S. D. Elliott, "Understanding the mechanism of SiC plasma-enhanced chemical vapor deposition (PECVD) and developing routes toward SiC atomic layer deposition (ALD) with density functional theory," *ACS Applied Materials & Interfaces*, vol. 10, pp. 15216–15225, 2018.

16. J. A. Raiford, S. T. Oyakhire, and S. F. Bent, "Applications of atomic layer deposition and chemical vapor deposition for perovskite solar cells," *Energy & Environmental Science*, vol. 13, pp. 1997–2023, 2020.

17. L. Bazli, M. Siavashi, and A. Shiravi, "A review of carbon nanotube/TiO2 composite prepared via the sol-gel method," *Journal of Composites and Compounds*, vol. 1, pp. 1–9, 2019.

18. J. Leppäniemi, P. Sippola, M. Broas, J. Aromaa, H. Lipsanen, and J. Koskinen, "Corrosion protection of steel with multilayer coatings: improving the sealing properties of physical vapor deposition CrN coatings with Al2O3/TiO2 atomic layer deposition nanolaminates," *Thin Solid Films*, vol. 627, pp. 59–68, 2017.

19. J. H. Min, Y. A. Chen, I. T. Chen, T. Sun, D. T. Lee, C. Li, Y. Zhu, B. T. O'Connor, G. N. Parsons, and C. H. Chang, "Conformal physical vapor deposition assisted by atomic layer deposition and its application for stretchable conductors," *Advanced Materials Interfaces*, vol. 5, p. 1801379, 2018.

20. E. A. Filatova, D. Hausmann, and S. D. Elliott, "Understanding the mechanism of SiC plasma-enhanced chemical vapor deposition (PECVD) and developing routes toward SiC atomic layer deposition (ALD) with density functional theory," *ACS Applied Materials & Interfaces*, vol. 10, pp. 15216–15225, 2018.

21. J. A. Raiford, S. T. Oyakhire, and S. F. Bent, "Applications of atomic layer deposition and chemical vapor deposition for perovskite solar cells," *Energy & Environmental Science*, vol. 13, pp. 1997–2023, 2020.

22. L. Bazli, M. Siavashi, and A. Shiravi, "A review of carbon nanotube/TiO$_2$ composite prepared via sol-gel method," *Journal of Composites and Compounds*, vol. 1, pp. 1–9, 2019.

23. T. Tabari, M. Ebadi, D. Singh, B. Caglar, and M. B. Yagci, "Efficient synthesis of perovskite-type oxide photocathode by nonhydrolytic sol-gel method with an enhanced photoelectrochemical activity," *Journal of Alloys and Compounds*, vol. 750, pp. 248–257, 2018.

24. T. Tabari, M. Ebadi, D. Singh, B. Caglar, and M. B. Yagci, "Efficient synthesis of perovskite-type oxide photocathode by nonhydrolytic sol-gel method with an enhanced photoelectrochemical activity," *Journal of Alloys and Compounds*, vol. 750, pp. 248–257, 2018.

25. P. Lushington, "Development of Nanostructures by Atomic and Molecular Layer Deposition," Doctoral dissertation, The University of Western Ontario, 2018.

26. H. F. Barton, A. K. Davis, and G. N. Parsons, "The effect of surface hydroxylation on MOF formation on ALD metal oxides: MOF-525 on TiO$_2$/polypropylene for catalytic hydrolysis of chemical warfare agent simulants," *ACS Applied Materials & Interfaces*, vol. 12, pp. 14690–14701, 2020.

27. H. Yu, Y. Gao, and X. Liang, "Slightly fluorination of Al2O3 ALD coating on Li1. 2Mn0. 54Co0. 13Ni0. 13O2 electrodes: Interface reaction to create the stable solid permeable interphase layer," *Journal of The Electrochemical Society*, vol. 166, p. A 2021, 2019.

28. K. J. Blakeney and C. H. Winter, "Atomic layer deposition of aluminum metal films using a thermally stable aluminum hydride reducing agent," *Chemistry of Materials*, vol. 30, pp. 1844–1848, 2018.

29. L. Astoreca Alvarez, P. S. Esbah Tabaei, D. Schaubroeck, M. Op de Beeck, R. Morent, H. De Smet, and N. De Geyter, "Visualizing the nucleation of ALD on polymers," in *20th International Conference on Atomic Layer Deposition*, 2020.

30. H. Lu, X. Chen, Y. Jia, H. Chen, Y. Wang, X. Ai, H. Yang, and Y. Cao, "Engineering Al_2O_3 atomic layer deposition: Enhanced hard carbon-electrolyte interface towards practical sodium-ion batteries," *Nano Energy*, vol. 64, p. 103903, 2019.

31. S. Seo, B. C. Yeo, S. S. Han, C. M. Yoon, J. Y. Yang, J. Yoon, C. Yoo, H.-J. Kim, Y.-B. Lee, and S. J. Lee, "Reaction mechanism of area-selective atomic layer deposition for Al2O3 nanopatterns," *ACS Applied Materials & Interfaces*, vol. 9, pp. 41607–41617, 2017.

32. B. A. Sperling, B. Kalanyan, and J. E. Maslar, "Atomic layer deposition of Al_2O_3 using trimethylaluminum and H_2O: The kinetics of the H2O half-cycle," *The Journal of Physical Chemistry C*, vol. 124, pp. 3410–3420, 2020.

33. S. Banerjee, *From radical-enhanced to pure thermal ald of aluminium and gallium nitrides*, University of Twente, 2019.

34. S. Shi, S. Qian, X. Hou, J. Mu, J. He, and X. Chou, "Structural and optical properties of amorphous Al_2O_3 thin film deposited by atomic layer deposition," *Advances in Condensed Matter Physics*, vol. 2018, 2018.

35. G. Gakis, H. Vergnes, E. Scheid, C. Vahlas, B. Caussat, and A. G. Boudouvis, "Computational fluid dynamics simulation of the ALD of alumina from TMA and H2O in a commercial reactor," *Chemical Engineering Research and Design*, vol. 132, pp. 795–811, 2018.

36. Tuoriniemi, "Coatings for ALD Reactors to Prevent Metal Contamination on Semiconductor Products," 2019.

37. P. Poodt, A. Mameli, J. Schulpen, W. Kessels, and F. Roozeboom, "Effect of reactor pressure on the conformal coating inside porous substrates by atomic layer deposition," *Journal of Vacuum Science & Technology A: Vacuum, Surfaces, and Films*, vol. 35, p. 021502, 2017.

38. D. Muñoz-Rojas, V. H. Nguyen, C. M. de la Huerta, S. Aghazadehchors, C. Jiménez, and D. Bellet, "Spatial Atomic Layer Deposition (SALD), an emerging tool for energy materials. Application to new-generation photovoltaic devices and transparent conductive materials," *Comptes Rendus Physique*, vol. 18, pp. 391–400, 2017.

39. M. Wang, G. Mi, D. Shi, N. Bassous, D. Hickey, and T. J. Webster, "Nanotechnology and nanomaterials for improving neural interfaces," *Advanced Functional Materials*, vol. 28, p. 1700905, 2018.

40. P. Mandracci, *Chemical vapor deposition for nanotechnology*, BoD–Books on Demand, 2019. Burke, "Atomic Layer Deposition Applications in Nanotechnology," 2019.

41. S. Ranjith, A. Celebioglu, H. Eren, N. Biyikli, and T. Uyar, "Nanofibrous catalysts: monodispersed, highly interactive facet (111)-oriented Pd nanograins by ALD onto free-standing and flexible electrospun polymeric nanofibrous webs for catalytic application (Adv. Mater. Interfaces 24/2017)," *Advanced Materials Interfaces*, vol. 4, p. 1770126, 2017.

42. Shin, "Application of ALD-grown metal oxide layers for energy applications: Photovoltaic and photoelectrochemical water splitting," in *ECS Meeting Abstracts*, p. 1146, 2019.

43. J. Jeong, F. Laiwalla, J. Lee, R. Ritasalo, M. Pudas, L. Larson, V. Leung, and A. Nurmikko, "Conformal hermetic sealing of wireless microelectronic implantable chiplets by multilayered atomic layer deposition (ALD)," *Advanced Functional Materials*, vol. 29, p. 1806440, 2019.

44. O. Graniel, M. Weber, S. Balme, P. Miele, and M. Bechelany, "Atomic layer deposition for biose ing applications," *Biosensors and Bioelectronics*, vol. 122, pp. 147–159, 2018.

45. Z. Song, M. Norouzi Banis, H. Liu, L. Zhang, Y. Zhao, J. Li, K. Doyle-Davis, R. Li, S. Knights, and S. Ye, "Ultralow loading and high-performing PT catalyst for a polymer electrolyte membrane fuel cell anode achieved by atomic layer deposition," *ACS Catalysis*, vol. 9, pp. 5365–5374, 2019.

5

Manufacturing of Buckypaper Composites for Energy Storage Applications: A Review

S. Roseline

MAM College of Engineering and Technology, Trichy, India

Carbon, a non-metallic element and a nature's gift to mankind, was first used as charcoal to make fire. Elemental carbon existed in two different forms, differing in structure and properties. Artificially developed elemental carbons with high aspect ratio, lengths varying from centimetres to millimetres and diameters varying from millimetres to nanometres came into existence producing major landmarks in the field of materials science. They have excellent physical, mechanical, thermal and electrical properties. Of them, the most recently developed carbon is buckypaper (BP). This is manufactured using compressed nanotubes and has excellent physical and mechanical properties and a wide range of applications. They are lighter yet stronger than steel and harder than diamonds. BP also conducts electricity and can be used as an electrode component in energy storage devices. This chapter intends to provide an extensive review of the simple and cheaper method of preparation of BP and its application in improving the performance of energy storing devices. Lithium-ion batteries have their own limits and will not be able to satisfy the growing demands of the electric vehicle markets. BP with its excellent properties of conductivity and high heat dispersion capacity can be used to meet the high demands of the electrochemical industry. Being lighter in weight, these electrodes can save space, thereby reducing the size of batteries.

5.1 Introduction

Carbon is derived from the Latin word "Carbo" meaningcharcoal (1, 2). This element forms special SP networks because of their structure and allotropes (3–5). Graphite and diamond are natural allotropes of carbon with carbon atomic layers bound together by van der Waals forces (6).

DOI: 10.1201/9781003355755-5

Graphite found abundantly in nature is a good electrical conductor, a good lubricant and thermal insulator and exhibits anisotropic behaviour (7–10). Sediments of carbonaceous material are transformed into graphite when carbon reacts with hydrothermal solutions. Synthetic graphite may also be prepared by heating petroleum coke to a temperature range of 2500°C to 3000°C (11). At this temperature, all impurities are burnt off and pure graphite is obtained in the form of a sheet-like crystalline structure (12).

Another form of carbon is graphene extracted from graphite. History says that 1 mm of graphite contains 3,000,000 layers of graphene. The main difference between the structure of graphene and graphite is that the former has a two-dimensional crystalline structure, whereas the latter has a three-dimensional crystalline structure (13–16). In graphene, the carbon atoms have a hexagonal honeycomb structure with very less space between the atoms (17–20). Many methods have been adopted by the researchers in the fabrication of graphene. The method of fabrication is mainly governed by factors such as size, quantity and purity (21). The structure and properties of graphene also vary with the fabrication techniques. A few methods used for fabrication of graphene are cleavage and exfoliation method, chemical vapour deposition (CVD) technique, pyrolysis method, unzipping of carbon nanotubes (CNTs), thermal decomposition using ruthenium crystal and thermal decomposition of silicon carbide (22–25). Graphene has high thermal and electrical conductivity, is very stiff, light in weight with high hardness and is stronger than steel (26–29). Graphene was first applied on security labels. These labels have a printed circuit on them, and if they are not disconnected properly, the circuit sounds an alarm. These labels cannot be damaged easily and the cost of these labels is very low (30). In biomedical industry, they are used as transport systems, sensors, biological agents and are also used in the field of tissue engineering (31–34). They are also used in electronics as graphene-effect transistors, field-effect transistors, light-emitting diodes, solar cells and electrodes (35–38). Its properties such as high surface area and conductivity suit them to be used in supercapacitors and batteries as energy storage devices, since they provide high storage capacity, long life and stability. They may also be used as chemical sensors in optoelectronics and nanocomposites (39–42).

Another strong material fabricated from graphite and an allotrope of carbon is CNTs. When two-dimensional graphite is rolled into the shape of a hollow cylinder, CNTs are formed (43–45). The diameter of CNTs is found to be in a few nanometres, but the length is found to be much higher than its diameter (46–48). CNTs are very stiff, stronger than steel and have a density one-fourth of steel (49). Their thermal capacity is very high and hence they do not expand on the application of heat (50). Because of this property, it is mainly used while making bridges and aircraft materials (51). It is an elastic material and a good conductor of heat and electricity and chemically neutral, which makes it corrosion resistant. It can be classified as single-walled CNTs (SWCNTs) and multi-walled CNTs (MWCNTs) (52–55). Single-walled nanotube is a one-dimensional structure with a diameter of about 2 nanometres and a length of about 2 micrometres. The diameter of multi-walled nanotubes ranges from 2 to 20 nanometres, and the length ranges from 5 to 6 micrometres (53–57). Each one has their own specific properties, and they mainly depend on the fabrication techniques. Few methods adopted for the manufacture of CNTs are laser vaporization, CVD and arc discharge. Because of its elongated shape, it can be used as a gas sensor, and because it is a good conductor, it is widely used in the electronic industries (58–60).

In this chapter, a review is made on a new material which is very light in weight, its diameter being very much smaller than a human hair but harder and stronger than diamonds (61–63). This new material is called the buckypaper (BP) fabricated using CNTs, graphene or graphite. The fabrication and application of this material are widely reviewed in this chapter.

With the extinction of mineral oils, and electric vehicles replacing the internal combustion engines, the need for energy storage devices is increasing day by day. Many low-cost electrodes are used as storage devices, but the performance of these devices is limited (64). This chapter also reviews how BP can be used to increase the capacity of energy storage devices.

5.2 Fabrication Techniques

5.2.1 Preparation of BP with CNT

The quantum mechanics calculations predict excellent mechanical and electrical properties for SWCNT papers. Experimental results show rare mechanical properties of SWCNT papers with a modulus of rigidity up to 1TPa and a tensile strength of 150–180 GPa. Through analysis, it's been found that these SWCNTs are the strongest and stiffest materials on earth. Carbon papers are supplemented with polymers to boost the mechanical properties and decrease the weight. The electrical conductivity of CNT papers is also found to be wonderful. The electrical conductivity of these papers was found to be about 5×10^7 S/m. CNTs are highly anisotropic in nature with small diameters, and they can be used as field emitters. In a research, the thermal conductivity of CNTs was found to be 3000 W/mK at room temperature. BP is a porous mat of entangled CNT, cohesively bound due to cohesive force by vander Waals interactions. Many methods are adopted in the preparation of BPs using CNTs. A few are discussed below.

A. *Vacuum filtration method*

In this method, CNT is ultrasonically dispersed in solvents like N-methylpyrrolidone (NMP) and N,N-dimethylformamide (DMF). The whole process is assisted by water-based surfactants such as sodium dodecyl sulphate (SDS) and polyoxyethyleneoctyl phenyl ether (Triton X-100). The dispersed CNT solution is then filtered in vacuum using a polytetrafluoroethylene or nylon filter with submicron-sized pores. CNT paper deposits as a thin membrane on the filter surface and can be removed after drying.

B. *Electric discharge method*

SurabhiPotnis has described the various techniques for the preparation of BP. In the electric arc discharge method, the carbon rods are placed end to end and made to vaporize (10). The enclosure is filled with inert gas at low pressure. One electrode was vaporized, and CNT paper was formed on the other. In the other method, a dual-pulsed laser is used to vaporize the carbon rods resulting in the production of fullerene and CN (12).

C. *CVD method*

In the CVD method, inert gas was made to pass through hot methane, with a metal catalyst acting as a seed for the CNTs to grow (13). The graphene oxide sheets adhere to the substrate by van der Waals forces. Strong hydrogen bonding binds the individual sheets together, making them stick on to the substrate very strongly (1). Brown et al. synthesized BP using SWCNT (16). A suspension was prepared, which consisted CNT, water and Triton-X surfactant. The suspension was sonicated, stirred for 6 hours and vacuum filtered until dry. The film edges were lifted off from the filter, and BP was removed and flattened (2).

5.2.2 Preparation of BP Using Graphene

Graphene is a material with excellent physical, chemical, electronic, electrical and mechanical properties and is used widely for research purposes. Various methods have been used to prepare papers using graphene. These papers are found to have very high electrical conductivity and tensile strength. These properties enable it to be a freestanding electrode. Introducing polymers into the graphene sheets further improves the mechanical strength and thermal and electrical conductivity of the composite. Shah et al. in their recent research prepared graphene sheets using graphite. Graphite was converted to graphene by sonification method using a Proctic solvent. These papers had a shiny lustre and were found to be very smooth throughout the cross-section. They also displayed excellent water-resisting properties and did not redisperse into water. These flexible papers with high mechanical and thermal properties can be used in energy storage devices (45). Wang et al. oxidized natural graphite to produce graphene sheets (54). The gap between the graphite layers was widened due to oxidation, enabling the monomers to fill the gap between the layers. Multi-layered composites were formed using this technique. The properties of graphene and polymers were combined to bring out a freestanding electrode. This flexible composite paper had excellent volumetric capacitance, conductivity and electrochemical activity.

Khan et al. (36) reviewed the various methods of synthesis of BP using graphene. In the micromechanical cleavage method, the graphite layers were peeled off from a silicon oxide surface using an adhesive tape to form a single graphite layer. Micromechanical exfoliation method was adopted by Liu et al. (26). They adopted a pillar method to produce graphene from graphite. Various other methods were vacuum filtration and drop-casting. Thickness of the paper and the film uniformity depended mainly on the parameters used. Vacuum filtration method produces thin bucky sheets, and it can be used in electronic devices.

In a spin coater, the substrate is rotated at a high speed causing centrifugal motion, thus spreading the materials on the surface. The solvent was volatile in nature and evaporated resulting in interaction and adhesion between the substrate and BP. The sheet thickness was controlled by graphene oxide concentration and the solvent's viscosity.

A. **Langmuir-Blodgett (L-B) assembly**

Few monolayers of GO sheets were arranged on a solid substrate to produce the GO film. The Langmuir-Blodgett (L-B) assembly utilizes the electrostatic repulsion that takes place on the edges of the sheets to form a uniform film. GO is suspended in a mixture of methanol and water and is spread over a liquid surface. Floating GO sheets can be found at the liquid/air interface. The solid substrate is now immersed into the liquid, and the GO sheets get deposited on them. The density of the GO sheet can be varied by changing the area between the liquid–air interface. Robinson et al. produced a GO film on SiO_2/Si substrates and transferred them to another substrate. GO films were treated with sodium hydroxide solution and dipped in water. Using the delamination technique, the GO film started to float on the water surface. The GO film was transferred to another substrate. An additional polymer layer was deposited to prevent the GO film disintegration.

B. **Vacuum *infiltration method to prepare polymer/GO paper***

GO suspension is obtained by dispersing the GO powder in an organic solvent. GO paper is obtained by filtering the suspension using a cellulose membrane of 0.45 mm pore size. Polymer is now filtered on the GO paper. The entire composite is then dried and removed from the porous membrane.

5.3 Properties of BP

BPs as nanofillers can be used in many applications due to their excellent electrical, thermal, electronic and mechanical properties. It is porous in nature with randomly interconnected CNTs. The density is found to be very low, but they are mechanically very strong because of high surface area and having a Young's modulus of 40 GPa and a tensile strength of125 MPa.

5.3.1 Elastic Property of BP

Cranford and Buehler explored the elastic behaviour of BP. The compression stress was increasing linearly along with strain, indicating elastic deformation. As the stress increased, the BP started deforming and underwent densification, and when the strain became 0:02, stress started increasing slowly with the increment of strain. On further loading, the inter-tube sliding occurred due to van der Waals force. From this analysis, Young's modulus and Poisson's ratio were calculated, and it was concluded that BP with short length has more Young's modulus than longer ones. The Poisson's ratio ranged between 0.18

and 0.30. It was also observed that the stiffness of the BP was enhanced by increasing the CNT content until 50 wt%. The stiffness started weakening with the increment of the content above 50 wt%. When the CNT content was below 50 wt%, the bore size of the BP was found to be small and the structure was entangled, thus increasing the stiffness value. But the pore size started increasing, and the van der Waals force weakened on further addition of CNT resulting in the decrease of the stiffness behaviour.

5.3.2 Mechanical Properties

The various methods of preparation of BPs lead to the differences in the mechanical properties of BPs. Zhang et al. conducted atensile test on the BP and observed that the BP had a very high tensile strength because it had the ability for inter-tube stress transfer. They treated the CNTs with nitric acid and found that the tensile strength was 6.3 times higher and Young's modulus was 5.8 times higher. But this treatment was found to make the paper brittle. Chen et al. exposed the CNTs containing carboxyl to ultraviolet radiation, which caused polymerization. By this method, the stress and strain of the BP were found to be 2 times higher than the normal BPs. Jakubinek et al. produced CNTs containing hydroxyl radical with six cross-linkers and found that they displayed excellent mechanical properties and the stress and strain of the BP were improved by 3.41 times than the ordinary BPs. Park et al. used suspension filtration method during synthesis of CNT, applying magnetic field thus preparing aligned BP, and the stress and strain of the ABP improved twice than the RBPs. Sakurai et al. fabricated CNTs of different lengths. BPs made out of long CNTs were found to possess good mechanical properties. Though much research is carried out on the mechanical properties of BPs, they cannot be utilized alone because of their fragile nature. They are usually fabricated with polymers to form a composite which shows excellent mechanical properties.

5.3.3 Thermal Properties

CNT is a good thermal conductor, and hence, it can be concluded that BP will also have excellent thermal conductivity. Yang found that BP is a good thermal conductor, and the thermal conductivity depends on the environment temperature. The thermal conductivity also depended on the geometry of BP. Volkov calculated the thermal conductivity of BP and analysed that it started increasing with the increase in the length and density of BP.

5.3.4 Electrical Properties

Electrical conductivity of BP differs greatly with its methods of preparation. Zhang et al. found that the thermal conductivity of BPs varies from 101 to 103 S/cm and that it can be improved by combining it with CNT. By aligning CNT, increasing the packing density or increasing the bundle length, the conductivity of BP can be increased. The following method of preparation shows an increase in the thermal conductivity of BPs.

1. Soaking BP in metal salt solution and heating the soaked BP for salt decomposition.

2. Uniform mixing of metal particles and CNT and filtering the mixture to form a composite.

3. Metal deposited on BP.

5.4 Applications

When the BP is stacked into many sheets, it looks like a carbon paper, lighter in weight but stronger than steel. It has all the properties of a conventional composite such as high strength, electrical and thermal conductivity, low optical reflectivity and high current-carrying capacity.

The applications of BP are still under research, but the following are some places where they can be applied. In the aeronautical industry, reinforcing BPs into the structure reduces the weight of the structure and saves fuel. It can be used in the protection of electronic components and also in the protection of airplanes against the effect of storms. It can also be used as electrodes in capacitors and batteries. In the medical field, this material can be used in artificial limbs. Since it reflects heat because of the dense and compact layers of CNTs, it can be reinforced with materials on the roof for fire protection. It can be used in televisions and computer screens for uniform brightness levels. It can be used in electronic equipment as a heat sink material.

BP is a highly dense and compact material, and due to this property, it can reflect heat energy. Hence, this thin layer can be used as a fire-resistant material. It can be used as a filter to trap the microparticles.

5.4.1 BP as Energy Storage Devices

With the onset of electric vehicles in every household, the need for efficient batteries plays a major role. Rechargeable batteries with high energy density are very heavy to be used in electric vehicles. BP can be used as a replacement against the metallic collectors, and by this, the weight of the batteries may be reduced to a great extent. Pushparaj et al. fabricated electrodes for lithium-ion batteries using CNTs. Hence, BP can also be used as lightweight collectors to obtain batteries for a long run. The values of voltage profiles of these collectors were found to be very near to that of the metal collectors, and it was also observed that there was no voltage drop. The cycling performance of the paper was also very close to that of metal collectors.

5.5 Conclusion

In this chapter, the allotropes of carbons are discussed in detail. Graphite is a naturally occurring allotrope with its own unique properties. With graphite as a base, other allotropes of carbon such as graphene, CNT and BP were fabricated. All these fabricated forms of carbon were found to be light in weight, stronger and harder than steel and can be used in a wide range of applications. Their size ranges in a few nanometres, and they can be easily fitted into any electronic parts. BP, a new material, also an allotrope of carbon, was discussed in detail. Many methods were adopted in its fabrication, and the structure and properties of BP were found to vary by its different manufacturing techniques. When considering the elastic properties, it was found that the stress of the paper increased linearly with the strain, indicating elastic deformation. The mechanical properties were found to be very promising with the tensile strength much greater than steel. Researchers found that the thermal properties of BP mainly depended on the environmental temperature. The electrical properties varied with the manufacturing process. When used in electric storage devices, it becomes a very lightweight collector and can be used in batteries for a long run. BP can also be used for a large variety of applications.

References

1. Arif MF, Kumar S, Shah T. Tunable morphology and its influence on electrical, thermal and mechanical properties of carbon nanostructure-buckypaper. *Materials & Design*. 2016 Jul 5; 101: 236–44.
2. Chapartegui M, Barcena J, Irastorza X, Elizetxea C, Fernandez M, Santamaria A. Analysis of the conditions to manufacture a MWCNT buckypaper/benzoxazine nanocomposite. *Composites Science and Technology*. 2012 Feb 28; 72(4): 489–97.
3. Che J, Chen P, Chan-Park MB. High-strength carbon nanotube buckypaper composites as applied to free-standing electrodes for supercapacitors. *Journal of Materials Chemistry A*. 2013; 1(12): 4057–66.
4. Chen H, Di J, Jin Y, Chen M, Tian J, Li Q. Active carbon wrapped carbon nanotube buckypaper for the electrode of electrochemical supercapacitors. *Journal of Power Sources*. 2013 Sep 1; 237: 325–31.

5. Chen X, Yin L, LvJ, Gross AJ, LeM, Gutierrez NG, Li Y, Jeerapan I, Giroud F, Berezovska A, O'Reilly RK. Stretchable and flexible buckypaper-based lactate biofuel cell for wearable electronics. *Advanced Functional Materials*. 2019 Nov; 29(46): 1905785.

6. Cooper SM, Chuang HF, Cinke M, Cruden BA, Meyyappan M. Gas permeability of a buckypaper membrane. *Nano Letters*. 2003 Feb 12; 3(2): 189–92.

7. Cottinet PJ, Souders C, Tsai SY, Liang R, Wang B, Zhang C. Electromechanical actuation of bucky-paper actuator: Material properties and performance relationships. *Physics Letters A*. 2012 Feb 27; 376(12–13): 1132–6.

8. Cranford SW, Buehler MJ. In silico assembly and nanomechanical characterization of carbon nano-tube buckypaper. *Nanotechnology*. 2010 Jun 10; 21(26): 265706.

9. DeGraffJ, Liang R, LeMQ, Capsal JF, Ganet F, Cottinet PJ. Printable low-cost and flexible carbon nanotube buckypaper motion sensors. *Materials & Design*.2017 Nov 5; 133: 47–53.

10. Elouarzaki K, Bourourou M, Holzinger M, Le Goff A, Marks RS, Cosnier S. Freestanding HRP-GOx redox buckypaper as an oxygen-reducing biocathode for biofuel cell applications. *Energy & Environmental Science*. 2015; 8(7): 2069–74.

11. Endo M, Muramatsu H, Hayashi T, Kim YA, Terrones M, Dresselhaus MS. 'Buckypaper'from coaxial nanotubes. *Nature*. 2005 Feb; 433(7025): 476.

12. Fu X, Zhang C, Liu T, Liang R, Wang B. Carbon nanotube buckypaper to improve fire retardancy of high-temperature/high-performance polymer composites. *Nanotechnology*. 2010 May 13; 21(23): 235701.

13. Gaztelumendi I, Chapartegui M, Seddon R, Flórez S, Pons F, Cinquin J. Enhancement of electri-cal conductivity of composite structures by integration of carbon nanotubes via bulk resin and/or buckypaper films. *Composites Part B: Engineering*. 2017 Aug 1; 122: 31–40.

14. Gross AJ, Holzinger M, Cosnier S. Buckypaperbioelectrodes: Emerging materials for implantable and wearable biofuel cells. *Energy & Environmental Science*. 2018; 11(7): 1670–87.

15. Han JH, Zhang H, Chen MJ, Wang D, Liu Q, WuQL, Zhang Z. The combination of carbon nanotube buckypaper and insulating adhesive for lightning strike protection of the carbon fiber/epoxy lami-nates. *Carbon*. 2015 Nov 1; 94: 101–13.

16. Han JH, Zhang H, Chen MJ, Wang GR, Zhang Z. CNT buckypaper/thermoplastic polyurethane composites with enhanced stiffness, strength and toughness. *Composites Science and Technology*. 2014 Oct 28; 103: 63–71.

17. Han JH, Zhang H, Chu PF, Imani A, Zhang Z. Friction and wear of high electrical conductive carbon nanotube buckypaper/epoxy composites. *Composites Science and Technology*.2015 Jun 19; 114: 1.

18. Hu Y, Li D, Wu L, Yang J, Jian X, Bin Y. Carbon nanotube buckypaper and buckypaper/polypro-pylene composites for high shielding effectiveness and absorption-dominated shielding material. *Composites Science and Technology*. 2019 Sep 8; 181: 107699.

19. Hussein L, Urban G, Krüger M. Fabrication and characterization of buckypaper-based nanos-tructured electrodes as a novel material for biofuel cell applications. *Physical Chemistry Chemical Physics*. 2011; 13(13): 5831–9.

20. Iurchenkova AA, Fedorovskaya EO, Asanov IP, Arkhipov VE, Popov KM, Baskakova KI, Okotrub AV. MWCNT buckypaper/polypyrrole nanocomposites for supercapacitor application. *Electrochimica Acta*. 2020 Mar 1; 335: 135700.

21. Jun LY, Mubarak NM, Yon LS, Bing CH, Khalid M, Jagadish P, Abdullah EC. Immobilization of peroxidase on functionalized MWCNTs-buckypaper/polyvinyl alcohol nanocomposite membrane. *Scientific Reports*. 2019 Feb 18; 9(1): 1–5.

22. Khan FS, Mubarak NM, Tan YH, Khalid M, Karri RR, Walvekar R, Abdullah EC, Nizamuddin S, Mazari SA. A comprehensive review on magnetic carbon nanotubes and carbon nanotube-based buckypaper for removal of heavy metals and dyes. *Journal of Hazardous Materials*. 2021 Jul 5; 413: 125375.

23. Khan ZU, Kausar A, Ullah H, Badshah A, Khan WU. A review of graphene oxide, graphene bucky-paper, and polymer/graphene composites: Properties and fabrication techniques. *Journal of Plastic Film & Sheeting*. 2016 Oct; 32(4): 336–79.

24. Kim YA, Muramatsu H, Hayashi T, Endo M, Terrones M, Dresselhaus MS. Fabrication of high-purity, double-walled carbon nanotube buckypaper. *Chemical Vapor Deposition*. 2006 Jun; 12(6): 327–30.

25. Knapp W, Schleussner D. Field-emission characteristics of carbon buckypaper. *Journal of Vacuum Science & Technology B: Microelectronics and Nanometer Structures Processing, Measurement, and Phenomena*. 2003 Jan 5; 21(1): 557–61.

26. Kumar A, Singh AP, Kumari S, Srivastava AK, Bathula S, Dhawan SK, Dutta PK, Dhar A. EM shield-ing effectiveness of Pd-CNT-Cu nanocomposite buckypaper. *Journal of Materials Chemistry A*. 2015; 3(26): 13986–93.

27. Li C, Zhang D, Deng C, Wang P, Hu Y, Bin Y, Fan Z, Pan L. High performance strain sensor based on buckypaper for full-range detection of human motions. *Nanoscale*. 2018; 10(31): 14966–75.

28. Li Y, Kröger M. A theoretical evaluation of the effects of carbon nanotube entanglement and bundling on the structural and mechanical properties of buckypaper. *Carbon*. 2012 Apr 1; 50(5): 1793–806.

29. Lopes PE, van Hattum F, Pereira CM, Nóvoa PJ, Forero S, Hepp F, Pambaguian L. High CNT con-tent composites with CNT Buckypaper and epoxy resin matrix: Impregnation behaviour composite production and characterization. *Composite Structures*. 2010 May 1; 92(6): 1291–8.

30. Lu S, Chen D, Wang X, Shao J, Ma K, Zhang L, Araby S, Meng Q. Real-time cure behaviour monitor-ing of polymer composites using a highly flexible and sensitive CNT buckypaper sensor. *Composites Science and Technology*. 2017 Nov 10; 152: 181–9.

31. Lu S, Chen D, Wang X, Xiong X, Ma K, Zhang L, Meng Q. Monitoring the manufacturing process of glass fiber reinforced composites with carbon nanotube buckypaper sensor. *Polymer Testing*. 2016 Jul 1; 52: 79–84.

32. Lv R, Tsuge S, Gui X, Takai K, Kang F, Enoki T, Wei J, Gu J, Wang K, Wu D. In situ synthesis and magnetic anisotropy of ferromagnetic buckypaper. *Carbon*. 2009 Apr 1; 47(4): 1141–5.

33. Mao R, Goutianos S, TuW, Meng N, Yang G, Berglund LA, Peijs T. Comparison of fracture proper-ties of cellulose nanopaper, printing paper and buckypaper. *Journal of Materials Science*. 2017 Aug; 52(16): 9508–19.

34. Mehmood A, Mubarak NM, Khalid M, Jagadish P, Walvekar R, Abdullah EC. Graphene/PVA buckypaper for strain sensing application. *Scientific Reports*. 2020 Nov 18; 10(1): 1–4.

35. Mirabootalebi SO. A new method for preparing buckypaper by pressing a mixture of multi-walled carbon nanotubes and amorphous carbon. *Advanced Composites and Hybrid Materials*. 2020 Sep; 3(3): 336–43.

36. Oh JY, Yang SJ, Park JY, Kim T, Lee K, Kim YS, Han HN, Park CR. Easy preparation of self-assembled high-density buckypaper with enhanced mechanical properties. *Nano Letters*. 2015 Jan 14; 15(1): 190–7.

37. Park JG, Louis J, Cheng Q, Bao J, Smithyman J, Liang R, Wang B, Zhang C, Brooks JS, Kramer L, Fanchasis P. Electromagnetic interference shielding properties of carbon nanotube buckypaper composites. *Nanotechnology*. 2009 Sep 16; 20(41): 415702.

38. Park JG, Smithyman J, Lin CY, Cooke A, Kismarahardja AW, LiS, Liang R, Brooks JS, Zhang C, Wang B. Effects of surfactants and alignment on the physical properties of single-walled carbon nanotube buckypaper. *Journal of Applied Physics* 2009 Nov 15; 106(10): 104310.

39. Patole SP, Arif MF, Kumar S. Polyvinyl alcohol incorporated buckypaper composites for improved multifunctional performance. *Composites Science and Technology*. 2018 Nov 10; 168: 429–36.

40. Pham GT, Park YB, Wang S, Liang Z, Wang B, Zhang C, Funchess P, Kramer L. Mechanical and elec-trical properties of polycarbonate nanotube buckypaper composite sheets. *Nanotechnology*. 2008 Jul 4; 19(32): 325705.

41. Rein MD, Breuer O, Wagner HD. Sensors and sensitivity: Carbon nanotube buckypaper films as strain sensing devices. *Composites Science and Technology*. 2011 Feb 7; 71(3): 373–81.

42. Ribeiro B, Botelho EC, Costa ML, Bandeira CF. Carbon nanotube buckypaper reinforced polymer composites: a review. *Polímeros*. 2017 Sep 21; 27: 247–55.

43. Rigueur JL, Hasan SA, Mahajan SV, Dickerson JH. Buckypaper fabrication by liberation of electrophoretically deposited carbon nanotubes. *Carbon*. 2010 Nov 1; 48(14): 4090–9.

44. Sakurai S, Kamada F, Futaba DN, Yumura M, Hata K. Influence of lengths of millimeter-scale single-walled carbon nanotube on electrical and mechanical properties of buckypaper. *Nanoscale Research Letters*. 2013 Dec; 8(1): 1–7.

45. Strack G, Babanova S, Farrington KE, Luckarift HR, Atanassov P, Johnson GR. Enzyme-modified buckypaper for bioelectrocatalysis. *Journal of The Electrochemical Society*. 2013 May 31; 160(7): G 3178.

46. Walgama C, Pathiranage A, Akinwale M, Montealegre R, Niroula J, Echeverria E, McIlroy DN, Harriman TA, Lucca DA, Krishnan S. Buckypaper–Bilirubin Oxidase Biointerface for Electrocatalytic Applications: Buckypaper Thickness. *ACS Applied Bio Materials*. 2019 May 2; 2(5): 2229–36.

47. Wang S, Downes R, Young C, Haldane D, Hao A, Liang R, Wang B, Zhang C, Maskell R. Carbon fiber/carbon nanotube buckypaperinterply hybrid composites: manufacturing process and tensile properties. *Advanced Engineering Materials*. 2015 Oct; 17(10): 1442–53.

48. Wang X, LuS, MaK, Xiong X, Zhang H, XuM. Tensile strain sensing of buckypaper and buckypaper composites. *Materials & Design*. 2015 Dec 25; 88: 414–9.

49. Wang Z, Liang Z, Wang B, Zhang C, Kramer L. Processing and property investigation of single-walled carbon nanotube (SWNT) buckypaper/epoxy resin matrix nanocomposites. *Composites Part A: Applied Science and Manufacturing*. 2004 Oct 1; 35(10): 1225–32.

50. Whitby RL, Fukuda T, Maekawa T, James SL, Mikhalovsky SV. Geometric control and tuneable pore size distribution of buckypaper and buckydiscs. *Carbon*. 2008 May 1; 46(6): 949–56.

51. Xia Q, Zhang Z, Liu Y, Leng J. Buckypaper and its composites for aeronautic applications. *Composites Part B: Engineering*. 2020 Oct 15; 199: 108231.

52. Yang K, HeJ, Puneet P, SuZ, Skove MJ, Gaillard J, Tritt TM, Rao AM. Tuning electrical and thermal connectivity in multiwalled carbon nanotube buckypaper. *Journal of Physics: Condensed Matter* 2010 Aug 4; 22(33): 334215.

53. Yang R, Gui X, Yao L, HuQ, Yang L, Zhang H, Yao Y, Mei H, Tang Z. Ultrathin, lightweight, and flexible CNT buckypaper enhanced using MXenes for electromagnetic interference shielding. *Nano-micro Letters*. 2021 Dec; 13(1): 1–3.

54. Yang X, Lee J, Yuan L, Chae SR, Peterson VK, Minett AI, Yin Y, Harris AT. Removal of natural organic matter in water using functionalised carbon nanotube buckypaper. *Carbon*. 2013 Aug 1; 59: 160–6.

55. Yee MJ, Mubarak NM, Khalid M, Abdullah EC, Jagadish P. Synthesis of polyvinyl alcohol (PVA) infiltrated MWCNTs buckypaper for strain sensing application. *Scientific Reports*. 2018 Nov 23; 8(1): 1–6.

56. Zaeri MM, Ziaei-Rad S, Vahedi A, Karimzadeh F. Mechanical modelling of carbon nanomaterials from nanotubes to buckypaper. *Carbon*. 2010 Nov 1; 48(13): 3916–30.

57. Zhang J, Jiang D, Peng HX, Qin F. Enhanced mechanical and electrical properties of carbon nanotube buckypaper by in situ cross-linking. *Carbon*. 2013 Nov 1; 63: 125–32.

58. Zhang J, Jiang D, Peng HX. A pressurized filtration technique for fabricating carbon nanotube buckypaper: Structure, mechanical and conductive properties. *Microporous and Mesoporous Materials* 2014 Jan 15; 184: 127–33.

59. Zhang J, Jiang D. Influence of geometries of multi-walled carbon nanotubes on the pore structures of Buckypaper. *Composites Part A: Applied Science and Manufacturing*. 2012 Mar 1; 43(3): 469–74.

60. Zhang Z, Wei H, Liu Y, Leng J. Self-sensing properties of smart composite based on embedded buckypaper layer. *Structural Health Monitoring*. 2015 Mar; 14(2): 127–36.

61. Zhao W, Tan HT, Tan LP, Fan S, Hng HH, Boey YC, Beloborodov I, Yan Q. N-type carbon nanotubes/silver telluride nanohybridbuckypaper with a high-thermoelectric figure of merit. *ACS Applied Materials & Interfaces*. 2014 Apr 9; 6(7): 4940–6.

62. Zhu T, Zhang Y, Luo L, Zhao X. Facile fabrication of NiO-decorated double-layer single-walled carbon nanotube buckypaper for glucose detection. *ACS Applied Materials & Interfaces*. 2019 Feb 25; 11(11): 10856–61.

63. Zhu W, KuD, Zheng JP, Liang Z, Wang B, Zhang C, Walsh S, AuG, Plichta EJ. Buckypaper-based catalytic electrodes for improving platinum utilization and PEMFC's performance. *Electrochimica Acta*. 2010 Feb 28; 55(7): 2555–60.

64. Zhu W, Zheng JP, Liang R, Wang B, Zhang C, AuG, Plichta EJ. Durability study on SWNT/nanofiber buckypaper catalyst support for PEMFCs. *Journal of the Electrochemical Society*. 2009 Jul 24; 156(9): B 1099.

6

Synthesis of Graphene/ Copper Oxide Nanocomposites for Supercapacitor Applications

Dr. J. Raffiea Baseri,
Dr. S. Chandra,
Dr. K.M. Govindaraju
and
Dr. J. Manikandan
PSG College of Arts and Science, Coimbatore, India

Rapid development in the field of automobiles and electronics industries has led to so many issues such as large backup devices, increased usage of electronic devices and power plants. Graphene transition metal oxides can be used to solve such issues in electronic industries. An elaborative review of recent advancements in the synthesis and applications of graphene–copper oxide composites is discussed in this chapter. Graphene, graphene oxide (GO) and reduced graphene oxide (RGO) with copper or copper oxide composites have excellent mechanical strength, conductivity and high thermal stability; hence, they find applications as supercapacitors, energy storage devices and portable electronics. The capacitance of various composite materials such as copper–graphene–GO is also discussed.

6.1 Introduction

The increased need of storage devices made researchers to focus on the developments of supercapacitors with specific properties like rapid charging, enhanced power density, long life and durability. Since graphene possesses a large surface area, high conductivity, flexibility and mechanical strength, it has attracted more attention for the fabrication of supercapacitor. However, due to the stacking effect, graphene-based supercapacitors exhibit low specific capacitance.

Graphene or graphene oxide (GO) exhibits poor electrochemical properties and therefore it is not suitable for use as supercapacitors. Transition metal oxides like Co_3O_4, RuO_2, NiO and MnO_2 exhibit good electrochemical properties and find use in supercapacitors. However, the high electrical resistivity of these transition metal oxides limits their electro catalytic applications (Liu et al., 2016). To avoid these issues, composites with enhanced qualities can be prepared by combining graphene or GO with transition metal oxides. This may improve the mechanical strength and electrical conductivity of bulk graphene–GO (Liu et al., 2019). The analysis of literature shows that graphene composites have a wide range of applications such as electro catalysts, supercapacitors, storage batteries, electrical conductors and anti-bacterial materials.

The literature survey shows that chemical exfoliation is the suitable route by which graphene or GO composites can be produced for energy storage devices. Graphene–copper oxide nanocomposites can be synthesized by solvothermal and electrochemical methods (Majeed et al., 2016). Graphene reinforced with aluminum matrix is the first graphene metal-based composite. Graphene–nano CuO composite was synthesized using Cu $[NH_3]_4(OAc)_2$ and suspension of graphene in water, and its capacitance was found to be 700 F/g which was reduced by 6% after 500 cycles. These materials were applied as electrodes in electro catalysis and electrochemical sensors (Hu et al., 2018).

Compared to noble metals and their metal oxide nanoparticles, copper oxide nano particles are known as best for synthesizing the nanocomposites with graphene. They have excellent applications as an electrode material, supercapacitor, sensor and heterogeneous catalyst. The abundant availability, potential pseudo-capacitance, environmental compatibility and reduced cost enable CuO to be considered as an eco-friendly material for the synthesis of graphene–GO nanocomposite (Weiwei Sun, Hao Li & Yong Wang, 2015). Large band gap and recombination of electron–hole pairs of nano CuO can be modified by combining with graphene, garphene oxide and reduced graphene oxide (RGO; Zhang, Cai & Ren, 2021).

6.2 Synthesis of CuO/RGO Composites

A mixture of graphite oxide and 10% N, N-dimethyl formamide (DMF) in 1:10 ratio was taken to prepare CuO or RGO composite by ultrasonic treatment. To the above composite, 0.02 g of cuprous chloride in 1 ml ammonia solution (3%) was added in nitrogen atmosphere and refluxed at 80°C for 2 h with constant agitation. The obtained product (CuO/RGO) was centrifuged at 5000 rpm and washed with anhydrous ethanol to eliminate other ions and dried in vacuum at 70°C for 5 h (Sagadevan et al., 2018).

Microwave combustion method is found to be a best route for the synthesis of $CuFe_2O_4$ (CF) and RGO integrated with $CuFe_2O_4$ (CG). The specific capacity of the fabricated CG device was found to be 360 C/g at 1 A/g current density with a resistance of 50Ω and specific energy was determined as 18.3 Wh/kg. The specific capacity of CG has been enhanced due to the low interfacial charge transfer resistance, and hence, it can be used as a suitable electrode material for hybrid supercapacitors (Mary et al., 2022).

Cu nanoparticles deposited on RGO sheets by in-situ photo reduction method and their photocatalytic applications were investigated (Aragaw and Dagnaw, 2019). The efficiency of photocatalytic degradation for GO, RGO and Cu/RGO nanocomposites was found to be 63%, 68% and 94%, respectively, at neutral pH, under light irradiation for 50 min. The catalytic effect of Cu is the main reason for the enhanced photocatalytic efficiency of Cu/RGO nanocomposite. It was suggested that graphene oxide-based composites with metal or metal oxide nanoparticles can be used to develop a cost-effective material for photocatalytic degradation.

A stable and heterogeneous graphene oxide (GO)–incorporated copper was prepared and reported, which can be used as an efficient catalyst for the organic green synthesis. A facile synthesis of GO-incorporated copper (II) complex of [Cu (II)-bis-cyclen] was prepared and studied. GO-Cu (II)-bis-cyclen nanocomposite is helpful for the bioconjugation of macromolecules via click chemistry approach (Samuel et al., 2022).

Arpita Jana et al. (2017) described the synthesis of graphene nanoparticle hybrids, where graphene was incorporated with specific nanoparticles (NPs). Such hybrid materials have been used as alternatives to TMO NPs which have some drawbacks while using in applications, such as anodes in batteries, photocatalysts and sensors. In this review, the synthesis and the various applications of graphene–TMO hybrids have been discussed in detail.

Yanrong Li et al. (2017) reported on the fabrication of novel electrode material with a large specific area by using CuO nanoparticles incorporated in the 3D network of graphene on carbon cloth for supercapacitor application. Bhusankar Talluri et al. (2019) reported spherical CuO quantum dots incorporated with graphene oxide nanocomposite with a specific capacity of 191 mA h/g at 2 mV/s and was maintained up to 63 mA h/g and at a high scan rate of 200 mV/s. Due to the high specific surface area, the CuO quantum dots play a vital role in charge storage devices which was indicated by the decreased capacitance with decreasing CuO concentration.

The capacitance retention and cycling stability of the prepared nanocomposite were examined in Na_2SO_4 solution with the help of galvanostatic charge–discharge tests for 2000 cycles. Due to double layer capacitance of graphene and pseudo-capacitance behavior of PPy and Cu_2O-$Cu(OH)_2$, the synthesized RGO/Polypyrrole/Cu_2O-$Cu(OH)_2$ composite shows a superior specific capacitance of 997 F/g at 10 A/g (Viswanathan and Shetty, 2017). Enhanced specific capacitance of 296 F/g at 1.0 A/g was reported with a retention of 96.1% even after 2000 cycles for reduced graphene oxide-wrapped CuO microspheres were synthesized using a one-step hydrothermal method (Aifeng Liu et al., 2016).

6.3 Electrochemical Analysis

With the help of a three-electrode cell set up, i.e., Ag–AgCl as the reference electrode, nanocomposite as the working electrode and platinum wire as the counter electrode, the electrochemical analysis of the nanocomposite is carried out. Usually, the electrode material is composed of 85 wt% nanocomposite, 10 wt% activated carbon and 5 wt% Teflon with ethanol. The aforementioned mixture is coated on a glassy carbon electrode and dried. Depending on the mass of the active material present in the electrode, energy density and specific capacitance may be calculated for the three-electrode system. The following equation is used to derive the specific capacitance values from cyclic voltammetry measurements.

$$C = \frac{I}{\Delta v \times m}$$

where I (mA) is the average current derived by integrating current and voltage, Δv is the voltage difference at discharge time and m (mg cm^2) is the mass of the active material. The galvanostatic discharge curves may also be used to compute specific capacitance using the following equation:

$$C = \frac{I \times \Delta t}{m \times \Delta v}$$

where I (mA) is the discharge current, Δt (s) is the discharge time, Δv is the voltage difference at discharge time and m (mg cm^2) is the mass of the active material. The following equations are used to determine the energy and power densities:

$$E = \frac{1}{2}C\left(\Delta v^2\right)$$

$$P = \frac{E}{t}$$

where E (Wh/kg) is the energy density, C (F/g) is the discharge specific capacitance, Δv (V) is the potential, P (Wk/g) is the power density and t (s) is the discharge time.

CuO/$Cu(OH)_2$ hybrid material was prepared and claimed to have an excellent specific capacitance of 278 F/g based on the cyclic voltammetry (CV) studies (Hsu et al., 2012). A nanocomposite of RGO/Cu_2O/Cu was reported to have superior electrochemical properties and a high specific capacitance of 98.5 F/g at 1 A/g (Dong et al., 2014).

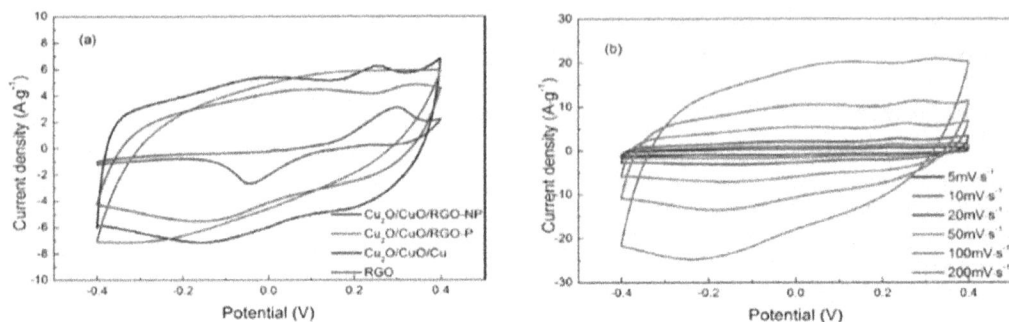

FIGURE 6.1 (a) CV curve of copper-based reduced graphene nanocomposite at a constant scan rate. (b) CV curve at different scan rates. (Reproduced with permission; Wang, Dong et al., 2015.)

Due to the rapid redox reactions that take place at the electrode/electrolyte interface, the position of the redox peaks in CV shifts to cathodic region with the increasing scan rate. Even at high scan rates, the CV curves of RGO/CuO/Cu$_2$O nanocomposite films that were recorded retain their shapes, which indicates better charge/discharge capability. In addition to that, the internal resistance of the electrode causes the specific capacitance of RGO/CuO/Cu$_2$O nanocomposites to decrease with the increase in scan rate. Figure 6.1 shows typical CV curves of bar RGO, Cu$_2$O/CuO/Cu, Cu$_2$O/CuO/RGO and its nanoparticles.

Galvanostatic charge–discharge measurements at a constant current density are used to investigate the effects of electrochemical properties, understand the charge storage mechanism and rate capacity of hybrid nanocomposite. Figure 6.2(a) shows a characteristic charge–discharge curve for the RGO–CuO nanocomposite between 0.5 and 0.7 V at a constant current of 0.5 A/g.

The excellent double layer capacitor performance is usually revealed with the symmetric charge-discharge curve. The triangular charge–discharge curve is symmetric in nature, which shows that the supercapacitor has excellent electrochemical reversibility with the linear relationship between potential and time. Therefore, the synthesis of composite using CuO/Cu$_2$O and reduced graphene is a new type of electrode material for high-performance supercapacitors. The faradaic redox reactions between the electrode and the electrolyte is the main reason for the nonlinear discharge curve of electrode.

A maximum specific capacitance value of 326 F/g at 0.5A/g was reported for G–Cu hybrid nanocomposite. This superior enhancement in the specific capacitance is mainly due to the presence of CuO in RGO. The RGO supports for the enhanced conductivity and good ion intercalation. Figure 6.2(b) shows the charge–discharge curves at various current densities. At higher current rate, ion movement in the electrode was not properly synchronized (Purushothaman et al., 2014). The specific capacitance is reduced due to this high current density. Furthermore, the power and energy density are significant parameters to select its suitability for commercial usage. One of the key factors to evaluate the supercapacitor electrodes is the retention of capacity after long-term cycling of charge–discharge processes. The percentage of stability of RGO and copper oxide are found to be 100% and 74%, respectively, up to 1500 cycles at 5 A/g, and there is no considerable degradation observed even after 1500 continuous cycles.

An improved cyclability of CuO–Cu$_2$O–RGO nanocomposite is mainly due to the redox process between copper (I) and copper (II) that occurs in the CuO-based supercapacitor, i.e., Cu$_2$O and CuO counterpart (Vidhyadharan et al., 2014). Moreover, Cu$_2$O is also successfully used as a specific semiconductor material which exhibits good electrochemical performance in supercapacitors and batteries (Goyal, Reddy & Ajayan, 2011). The supercapacitor application of CuO–Cu$_2$O–graphene were successfully studied and correlated between redox reaction of electrode materials and the charge–discharge process (Wang et al., 2015). Due to the combination of redox counterpart of CuO and Cu$_2$O into Cu$_2$O/CuO, a strong and enhanced cyclability in supercapacitors was achieved. Moreover, when utilizing this

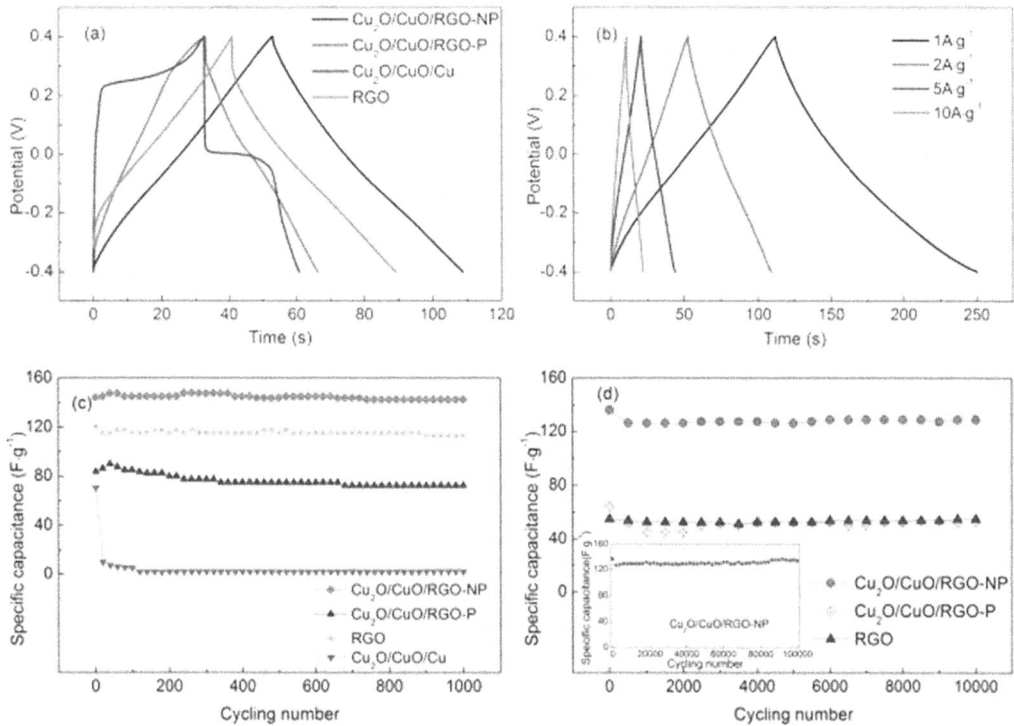

FIGURE 6.2 (a) Charge–discharge profile of nanocomposite at 500 mA/g. (b) Charge–discharge profile of CuO/ RGO nanocomposite at different current densities. (c) and (d) Cycling ability at 2 A/g and 10 A/g. (Reproduced with permission; Wang, Dong et al., 2015.)

TABLE 6.1 Specific Capacitance of Cu_2O–CuO–RGO-NP, Cu_2O–CuO–RGO-P, Cu_2O–CuO–Cu and RGO at Various Current Densities of 1, 2, 5 and 10 A/g

	Specific Capacitance (F/g)			
Samples	1 A/g	2 A/g	5 A/g	10 A/g
Cu_2O–CuO–RGO-NP	173.4	144.2	143.1	136.3
Cu_2O–CuO–RGO-P	106.3	83.6	68.3	64.1
Cu_2O–CuO–Cu	97.5	70.5	26.3	6.3
RGO	130	120.5	90.2	54.9

(Reproduced with Permission; Wang, Dong et al., 2015.)

composite as a supercapacitor, it exhibits an enhanced electrochemical performance in both rate ability and cyclability.

Table 6.1 clearly indicates that the capacitance of Cu_2O–CuO–RGO-NP is higher than the capacitance of other three materials; it may be due to the increased conductivity of RGO in the composite, which promotes ion diffusion in the host material. Also, the agglomeration of graphene sheets has been reduced efficiently and their specific area increased due to the presence of CuO–Cu_2O nanoparticles on RGO, which supports good specific capacitance (Wang et al., 2015). Compared with specific capacitance of pure RGO, the specific capacitance of Cu_2O–CuO–RGO-P is low at 2 A/g due to the aggregation of RGO in Cu_2O–CuO–RGO-P. Moreover, Table 6.1 indicates that the excellent rate capacity is retained even after increasing the discharge current density tenfold. The charge–discharge cycles of

Cu_2O–CuO–RGO-NP at a constant discharge current density of 2 and 10 A/g show superior supercapacitance due to excellent rate capability and cycling stability. The long cycle life of Cu_2O–CuO–RGO composite is due to the presence of CuO–Cu_2O redox couple and the Cu_2O–CuO–RGO nanocomposite.

6.4 Conclusions

This review describes the various methods for the synthesis of graphene-based materials and their applications for cost-effective and eco-friendly energy storage devices. Different preparation methods of composites based on graphene and copper oxide, their applications in supercapacitorand challenges and difficulties on experimental usage have been discussed. Based on the review, it is concluded that graphene–copper oxide composites are best suitable materials for super capacitance applications.

References

Aragaw, Belete Asefa, Atsedemariam Dagnaw. 2019. "Copper/reduced graphene oxide nanocomposite for high performance photocatalytic methylene blue dye degradation", *Ethiop. J. Sci. & Technol.* 12(2): 125–137.

Dong, X., K. Wang, C. Zhao, X. Qian, S. Chen, Z. Li, S. Dou. 2014. "Direct synthesis of RGO/Cu_2O composite films on Cu foil for super capacitors", *J. Alloys Compd.* 586: 745–753.

Goyal, A., A. L. M. Reddy, P. M. Ajayan. 2011. "Flexible carbon nanotube-Cu_2O hybrid electrodes for Li-ion batteries", *Small* 7: 1709.

Hsu, Y.K., Y.C. Chen, Y.G. Lin. 2012. "Characteristics and electrochemical performances of lotus-like CuO/Cu(OH)2 hybrid material electrodes", *J. Electroanal. Chem.* 673: 43–47.

Hu, Z. et al. 2018. "3D printing graphene-aluminum nanocomposites", *J. Alloy Compd.* 746: 269–276.

Jana, Arpita, Elke Scheer, Sebastian Polarz Beilstein. 2017. "Synthesis of graphene–transition metal oxide hybrid nanoparticles and their application in various fields", *J. Nanotechnol.* 8: 688–714.

Justin, P., S.K. Meher, G.R. Rao. 2010. "Tuning of capacitance behavior of NiO using anionic, cationic, and nonionic surfactants by hydrothermal synthesis", *J. Phys. Chem. C* 11: 5203.

Li, Yanrong, Xue Wang, Qi Yang, Muhammad Sufyan Javed, Qipeng Liu, Weina Xu, Chenguo Hu, Dapeng Wei. 2017. "Ultra-fine CuO Nanoparticles Embedded in Three- dimensional Graphene Network Nano-structure for High-performance Flexible Supercapacitors", *Electrochim. Acta* 234: 63–70.

Liu, Aifeng, Yongmei Bai, Yanming Liu, Mingshuo, Zhao Jingbo Mu, Chunxia Wu, Xiaoliang Zhang, Guangshuo Wang, Hongwei Che. 2016."Hierarchical dandelion-like copper oxide wrapped by reduced graphene oxide: Hydrothermal synthesis and their application in supercapacitors", *Mater. Res. Bull.* 84: 85–92.

Liu, C. et al. 2019. "Enhanced mechanical and tribological properties of graphene/bismaleimide composites by using reduced graphene oxide with non-covalent functionalization", *Compos. B Eng.*, 165: 491–499.

Liu, L., J. Lang, P. Zhang, B. Hu, X. Yan. 2016. "Facile synthesis of Fe_2O_3 nano-Dots@Nitrogen-doped graphene for supercapacitor electrode with ultra long cycle life in KOH electrolyte", *ACS Appl. Mater. Interfaces* 8(14): 9335–9344.

Majeed, A., W. Ullah, A.W. Anwar, et al. 2016. "Graphene-metal oxides/hydroxide nano composite materials: Fabrication advancements and super capacitive performance", *J Alloys Compd.* 671:1–10.

Mary, B., Carmel Jeeva, J. Judith Vijaya, Radhika R. Nair, A. Mustafa, P. Stephen Selvamani, B. Saravanakumar, M. Bououdina, L. John Kennedy. 2022. "Reduced graphene oxide-tailored CuFe2O4 nanoparticles as an electrode material for high-performance supercapacitors", *J. Nanomater.* 2022: 1–15.

Purushothaman, Kamatchi Kamaraj, Balakrishanan Saravanakumar, Inbamani Manohara Babu, Balasubramanian Sethuramana, Gopalan Muralidharan. 2014. "Nanostructured CuO/reduced graphene oxide composite for hybrid super capacitors", *RSC Adv.*, 4: 23485.

Sagadevan, Suresh, Zaira Zaman Chowdhury, Mohd Rafie Bin Johan, Fauziah Abdul Aziz, Emee Marina Salleh, Anil Hawa, Rahman F. Rafique. 2018. "A one-step facile route synthesis of copper oxide/reduced graphene oxide nanocomposite for supercapacitor applications", *J. Experim. Nanosci.*, 13(1): 284–296.

Samuel, Angel Green, Sowmya Subramanian, Vijaikanth Vijendran, Jebasingh Bhagavathsingh. 2022. "Copper (II)-bis-cyclen intercalated graphene oxide as an efficient two-dimensional nanocomposite material for copper-catalyzed azide–alkyne cycloaddition reaction", *Front. Chem.* 9: 1–11.

Sun, Weiwei, Hao Li, Yong Wang. 2015. "Microwave-assisted synthesis of graphene nanocomposites: recent developments on lithium-ion batteries", *Rep. Electrochem.* 5: 1–19.

Talluri, Bhusankar, Sourav Ghosh, G. Ranga Rao, Tiju Thomas. 2019. "Nano composites of digestively ripened copper oxide quantum dots and graphene oxide as a binder free battery-like super capacitor electrode material", *Electrochim. Acta* 321: 1–12.

Vidhyadharan, B., I.I. Misnon, R.A. Aziz, K.P. Padmasree, M.M. Yuso, R. Jose. 2014. "Superior supercapacitive performance in electrospun copper oxide nanowire electrodes", *J. Mater. Chem. A* 2: 6578.

Viswanathan, Aranganathan, Adka Nityananda Shetty. 2017. "Facile in-situ single step chemical synthesis of reduced graphene oxide-copper oxide-polyaniline nanocomposite and its electrochemical performance for supercapacitor application", *Electrochim. Acta* 257: 483–493.

Wang, Kun, Xiangmao Dong, Chongjun Zhao, Xiuzhen Qian, Yunlong Xu. 2015. "Facile synthesis of $Cu_2O/CuO/RGO$ nanocomposite and its superior cyclability in supercapacitor", *Electrochim. Acta*, 152: 433–442.

Zhang, L., W. Cai, J. Ren, Y. Tang. 2021. "Cu-Co bimetal oxide hierarchical nanostructures as high-performance electro catalyst for oxygen evolution reaction", *Mater. Today Energy* 21: 100703.

7

Nanocarbon Materials-Based Solar Cells

Dr. M. Padmavathy
and D. Meenakshi

*Shrimati Indira Gandhi
College, Trichy, India*

Solar cells are now flexible and perform well in general. because it makes it possible to combine hybridising to create a range of nanostructures. Due to the carbon–carbon bond versatility in three geometrics, the solid and molecular material propose a wide variety of features. Carbon is a versatile and necessary material used to assemble 1D, 2D, and 3D (dimensional) networks. Carbon nanoparticles are the perfect material for use as flexible solar cells because of their excellent conductivity, accurate transparency, high stability, adequate flexibility, and adjustable energy levels. The significance of carbon nanostructures like graphene for solar applications is discussed in this chapter. Graphene is perfect for use as software in flexible solar cells because of its many beneficial properties.

7.1 Introduction

According to Lian and Wu (2014), nanocarbons are conceptually distinct from "conventional" porous carbon materials like glassy carbon, graphite, and other carbon electrode materials, as well as active carbon (AC) meso-porous carbon aerogels and carbon materials, pyrolytic graphite, graphite flake, graphitic carbon, and carbon black (Desimoni & Brunetti, 2012, McCreery, 2008).

There are three distinct groups of nanocarbon materials:

The first group, which includes the three fundamentals. The different unique, morphologically simple carbon metals which can be created using three forms of nanocarbon elements (graphene, fullerene, and carbon nanotubes) include carbon quantum dots, nano-horns, and nanotubes, nano-horn-like structures, nanofibers, nanoribbons, nano capsules, and nanocages. Low dimensionality and morphology-defined features describe these nanocarbons, which also include carbon quantum dots, nano-horns, nanofibers, nanoribbons, nano capsules, and nanocages.

Heteroatoms (dopants), which are generated from the first generation's qualities, are inserted to change the electrical structure of the second group of nanocarbon materials. A new level of control concluded these micro carbons' characteristics are introduced during or after synthesis processes.

The third group of nanocarbon materials, which provide a higher level of control, was made by nano-engineering on the structure of materials on a nanoscale level to build hybrid and/or hierarchical systems (Centi & Perathoner, 2009).

There is still a dearth of knowledge of the numerous aspects that affect functional performance from the first to the third generation, materials have a greater flexibility to modify properties to fulfil specific demands (Su et al., 2012). This is in part because it is challenging to recognize the defect state in nanocarbons, which is a result of the flexible coordination and hybridization of carbon.

The presence of heteroatoms, defects, and hybridization heavily influences the properties of carbon materials, especially conductivity. Isotropic materials are diamonds with low electrical conductivity, due to their sp3 hybridization and tetrahedral structure. Graphite, on the other hand, is only a semi-metal in the basal plane and is anisotropic with sp2 hybridization. The interplanar connections between the inert, diamond-like basal plane of graphite are weak, which gives graphite its soothing and then oiling properties. The prismatic and basal edge planes of graphite react chemically in very different ways—graphite's trigonal shape results in some anisotropy in its chemical and physical characteristics. Because graphite is a homonuclear carbon compound, changes in the geometrical features of vacancies, bending, defects, and the presence of the heptagon or pentagon ring inside the network of hexagonal rings will alter these properties. Furthermore, the variety of hetero elements found in carbon compounds modifies their electrical possessions due to carbon's modest electronegativity.

The materials can be structures on nanoscale levels using nano-engineering; the third generation of nanocarbon materials, which provide a higher level of control, was developed (Centi & Perathoner, 2009) (Figure 7.1).

Under an energy scenario where fossil fuels predominate, carbon materials play a relatively minor role. Still, their significance is growing in a new energy future mainly focused on using renewable energy.

The ability to modify the electronic structure of carbon materials by adding heteroatoms, the physical possessions of carbon materials by adjusting the whole structure, and then the chemical properties of carbon materials by adding functional groups to the surface are all significant characteristics of carbon materials. TiO_2–nanocarbon (graphene) hybrids are an intriguing example, with excessive

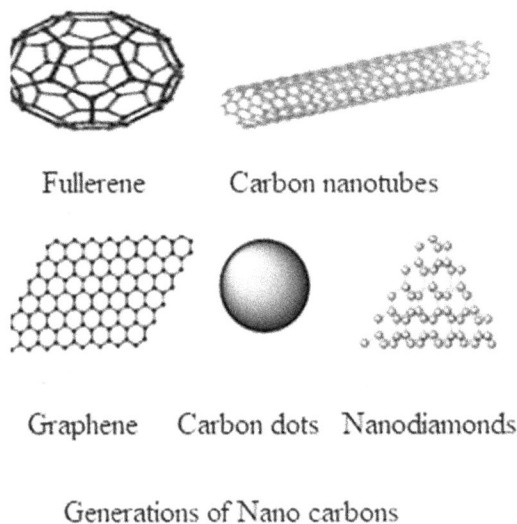

Fullerene   Carbon nanotubes

Graphene   Carbon dots   Nanodiamonds

Generations of Nano carbons

FIGURE 7.1   Generations of nanocarbons.

consideration for evolving improved structures for manufacturing solar fuels. Nanocarbons can act as electron donors or as photosensitive carbon quantum dots, mediating charge transmission between different semiconductor particles. Still, they can also alter the intrinsic aspects of TiO_2, resulting in new hybrid resources (Vilatela & Eder, 2012; Centi et al., 2014). Nanomaterials have several benefits for photochemical applications (Cheng et al., 2012).

- transfer electrons into the inorganic compound and so act as a photosensitizer in extending the absorption range;
- accept photoexcited electrons from the inorganic compound, thus separating them from photoactive holes;
- conduct excess heat away during the catalytic reaction;
- act as a template and heat sink to stabilize smaller catalyst particles whose larger specific surface area can convert more reactant molecules;
- reduce the amount of active catalyst, decrease its weight, and improve cyclability and the overall lifetime.

7.2 Materials with Nanocarbon for Energy Storage

Electrochemical energy storage is the most flexible method for nanocarbon materials, which may span a wide range of energy and time domains. For high-power, high-energy-density, and grid frequency regulation applications, batteries and electrochemical capacitors must be created. It's necessary to comprehend some of the additional benefits of nanocarbons over more traditional carbon materials before we discuss some current trends in using nanocarbons for energy storage devices. Nanocarbons offer improved access to reactants or ions because they maximize the external area. Due to their small size and distinct shape, charged particles (protons or ions) are transported to the active sites more effectively in an electrochemical process than other standard carbon materials. Generally, a hierarchical structure is necessary for carbon-based nanostructures in applications, including electrochemical energy storage, converters, and biosensors (Zhang, 2010). The hierarchical nanoarchitecture of carbon-based nanocomposites provides larger internal surface area, superior conductivity, and synergistic effects for electrochemical processes at the interface. There is now a lack of a general justification for the design of the nanoarchitectures with respect to its many applications, and this design is typically qualitative and phenomenological. Nanocarbons might also feature a cutting-edge storage system. New processes for lithium-ion storage, such as a Faradaic reaction for charge storage, are made possible by nanocarbons' small size and distinctive shape (pseudo capacity).

7.2.1 Materials with Nanocarbon for Energy Conversion

Carbon nanomaterials are being used increasingly in the stimulating field of energy conversion systems, including biofuel cells and improved solar cells in addition to conventional fuel cells (Huang et al., 2012). The need for innovative also better-quality materials that can detect and change solar energy has arisen as a result of efforts made worldwide to find practical and sustainable methods of utilizing renewable energy sources such as the progressive dye-sensitized solar cells (DSSC), which requires to increase the photovoltaic outcomes also called for a focus on employing nanocarbon for improved cell design or for developing solar fuel-producing cells (Chen et al., 2010). Creation of materials with particular nanoarchitectures, as well as effective charge transport and separation, is required for efficient collecting, conversion, or utilization of solar energy (Hasobe, 2010; Kuang et al., 2012).

Numerous qualities make nanocarbons a desirable material for such energy conversion technologies. Due to their good electric conductivity, obvious electrochemical activity, low cost, and high thermal/chemical stability, attention has been drawn to pure nanocarbon elements and nanocarbon metal (hydroxide and oxide) composites as electrodes in devices related to energy (Bensaid et al., 2012). Nanocarbon materials are an attractive option for creating high-performance electrode materials. However, due to the difficulty in discovering inexpensive materials with repeatable and consistent characteristics, their commercial usage is still in its infancy. In reality, the manufacturing process utilized to prepare nanocarbon has a considerable impact on the subsequent electrochemical performance because it affects the shape, microstructure, and composition of the manufactured material (Centi & Perathoner, 2010).

7.2.2 Graphene-Based Materials

Owing to the optimal electrical, optical, mechanical, and thermal properties, materials based of graphene show significant promise in various technological applications. Graphene materials are ideally suited for solar cells due to their exceptional transparency, conductivity, flexibility, and availability.

Many fields are interested in graphene, a two-dimensional (2D) carbon substance that is one atomic thick and sp2-bonded (Liang et al., 2011). The remarkable optical, mechanical, thermal, and electrical features of graphene have been extensively used for innovative conversion of energy as well as storage technologies such as solar cells, batteries, fuel cells, and supercapacitors.

Graphene is a monoatomic carbon film with a one-atom thickness. Graphene is a dense hexagonal form of carbon. Most people are familiar with graphene as a substance because of its toughness, which is on par with diamond.

Most graphene production occurs in the lab using chemical synthesis from graphite or another mineral representing an allotropic carbon state. As the name implies, graphene's structure typically consists of hexagonal cells; however, in cases with faults, the cells can also take on different shapes like a pentagon or a heptagon. An excellent semiconductor, graphene has outstanding optical absorbance, substantial flexibility, excellent mechanical strength, and chemical stability. Graphene is present in electrolytes translucent conducting materials, catalytic counter electrodes, transparent electrodes, and light-harvesting materials.

7.3 Results and Discussion

A single atomic layer of carbon that has been hybridized with sp2 makes up graphene in honeycomb patterns. This two-dimensional substance has developed into a valuable nanomaterial for various intriguing applications, including solar cells. Due to its unusually high crystal and electrical properties, it has grown into a promising nanomaterial for intriguing uses like solar cells. G Nanomaterials based on graphene and their use in several types of solar cells are covered in this chapter. The following explanation has been shortened for simplicity, but nanocarbon will undoubtedly be crucial to the growth of solar cells and fuels of the future.

The number and arrangement of graphene layers affect the material's electrical properties. The strongest C–C bonds are found in the plane of graphene. Van der Waal forces generate a link in the outer plane graphene and substrate or between graphene layers. Graphene conducts electricity as a result of the high electron mobility of this connection. The quantum size effect comprises charge transport structures of graphene electrical, optical, mechanical, and thermal, which range from a single to multiple atom layers.

Due to the small number of crystallographic flaws, graphene has a high level of electronic conductivity. Sp2 hybridization results in free electrons in the third dimension and significantly enhances graphene's ability to transport electrons without scattering. Due to their special mechanical, optical, and electrical properties, graphene and other graphene-based materials are used in many components of solar cells.

7.3.1 Solar Cells

Solar cells can be a tremendous green answer to the world's energy demand issues. With all its features, the DSSCs' replacement for p-n junction silicon-based solar cells, their excellent energy conversion, along with the cheaper material and manufacturing costs. DSSCs are employed in graphene nanocomposites as photo-anodes, counter electrodes, photosensitizers, and electrolytes.

Graphene's flexibility allows solar cells to bend up to 78 degrees more than standard indium tin oxides. The large surface area of graphene allows for a continuous channel and several donor and acceptor sites for better electron transport. distinct power conversion efficiency-based graphene growing techniques (PCE). Two characteristics distinguish graphene-based materials: (a) their capacity to tune with the number of graphene layers and (b) doping effects on the properties. Doping is the commonly used method for controlling graphene's bandgap, catalytic activity, work function, and charge carrier density.

7.3.2 DSSCs with Graphene/TiO₂ Active Layer

Graphene is a highly potent element with numerous uses, including solar cells made of organic materials. Additionally, it has strong optical and electrical capabilities. Because of its exceptional visual qualities, high electron mobility, and mechanical, chemical, and thermal durability, it is employed in translucent electrodes or conductive coatings.

Because its electron mobility is high, graphene is utilized as the electron transport layer in the DSSC for enhancing transfer of electron. These are DSSCs with a graphene/TiO₂ active layer. Raman scattering indicates that the electroplating method created graphene flakes, improving the properties of the material. According to Raman scattering, the electroplating technique produced graphene flakes which improve the graphene characteristics. The DSSCs with graphene scales have a low efficiency conversion of power owing to its higher series resistance due to the discarded graphene flakes. Sputtered graphene was utilized in place of the electroplated graphene flakes for enhancing the electric features of DSSCs, although it also included graphene oxide.

Radio-frequency magnetron sputtering was used to produce graphene (60 nm thick) on indium tin oxide (ITO) conductive glass substrate as an electron transport layer. A platinum layer 100 nm thick on ITO substrate served as the counter electrode. Two via-holes were left for injecting the electrolyte, and sealing films were sandwiched between the two electrodes to create the cells. Electrolytes were injected into spaces between the electrodes through "Via-Holes", which were further sealed with epoxy at a lower vapor transfer rate. The TiO₂ solution was combined to create a colloidal solution, and a thick coating was evenly daubed over the graphene electron transport layer (Figure 7.2).

Then, to investigate what we could learn, we contrasted a thick 100-nm graphene layer with 60-nm graphene electron transport layer. The absorption spectra of DSSCs without and with graphene electron transport layer are displayed in Figure 7.3. The absorbance of TiO₂ DSSCs in the visible region is shown

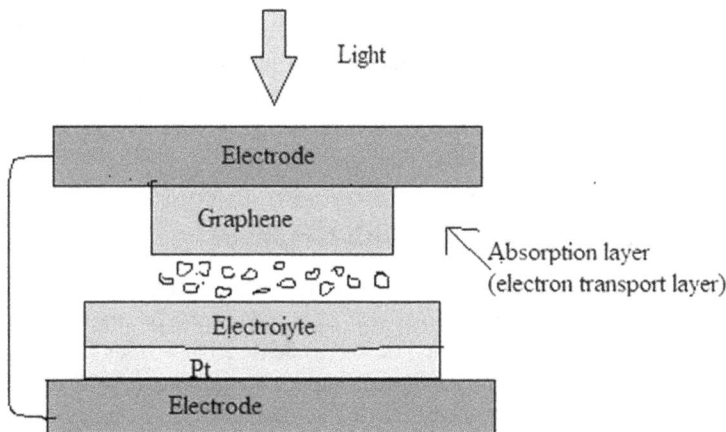

FIGURE 7.2 Electron transport layer.

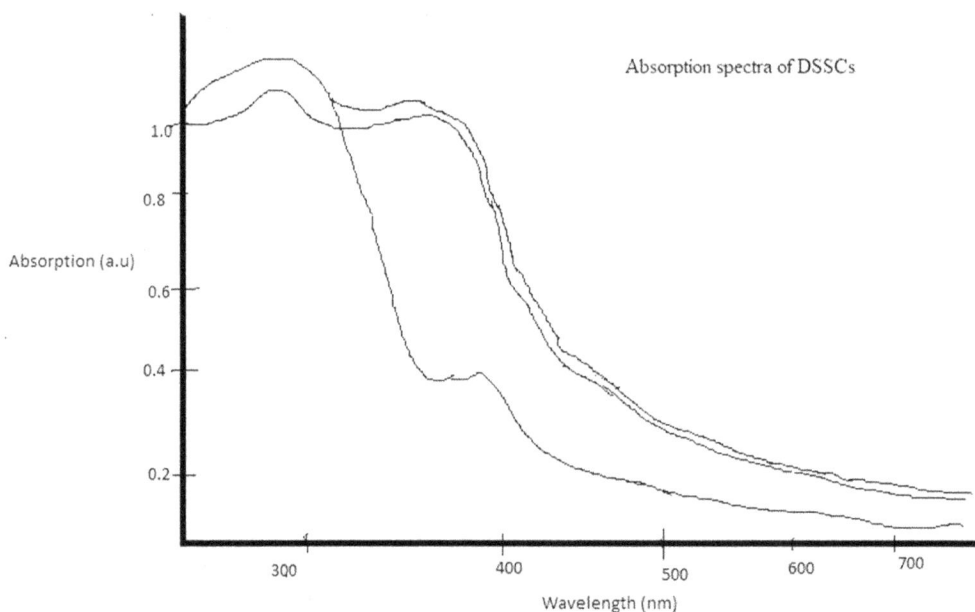

FIGURE 7.3   Absorbance spectra of DSSC's.

in Figure 7.3 with and without the graphene electron transfer layer. In the 310–400 nm region, Figure 7.3 demonstrates that the graphene electron transport layer has a more significant absorption coefficient. To boost solar cell absorption, the graphene electron transport layer also functions as an absorbent layer.

For DSSCs, the I–V curves illuminated without and with the graphene electron transport layer are shown in Figure 7.4. The DSSCs' I–V characteristics are depicted in Figure 7.3. This image displays TiO_2/graphene cell and TiO_2 DSSCs performance under AM 1.5 illumination at 25°C and a 100 mW/cm² sun intensity. The cell lacks an antireflective coating and has an active area of 33 mm². Furthermore, the outcomes were estimated by measuring the cell metrics, including *"Fill Factor (FF), Energy Conversion Efficiency (ECE), Open-Circuit Voltage (Voc) (Eff) and Short-Circuit Current (Jsc)"*. Figures 7.3 and 7.4 and Table 7.1 show that the short-circuit recent increases to 17.5 mA, the 0.45 of fill factor, and a 3.9% efficiency of energy conversion. Increased efficiency of electron transport and visible light absorption were attributed to the better performance of graphene-based DSSCs.

Improvements in visible light absorption, notably in the 310–400 nm range, and electron transport efficiency can be credited for the enhanced performance of graphene-based DSSCs. Under simulated full-sun illumination, the efficiency of solar energy conversion to electricity with graphene+TiO_2 is 1.45%–3.98%. In contrast, a sandwich of TiO_2/-graphene/-TiO_2 boosted the efficiency of solar energy conversion to electricity from 1.38% to 3.93%.

7.3.3 Schottky Junction Solar Cells with Graphene

Graphene sheets function as a transparent electrode, a way to transport holes, and a way to separate electrons from holes. Heterojunction-based solar cells have been created containing graphene in addition to metal and semiconductor components, as well as Schottky junctions (Zhu et al., 2010). First heterojunction solar cells based on graphene element reported a maximum PCE of 4.35%. An n-type silicon and graphene semiconductor formed Schottky connections. Heterojunction solar cells based

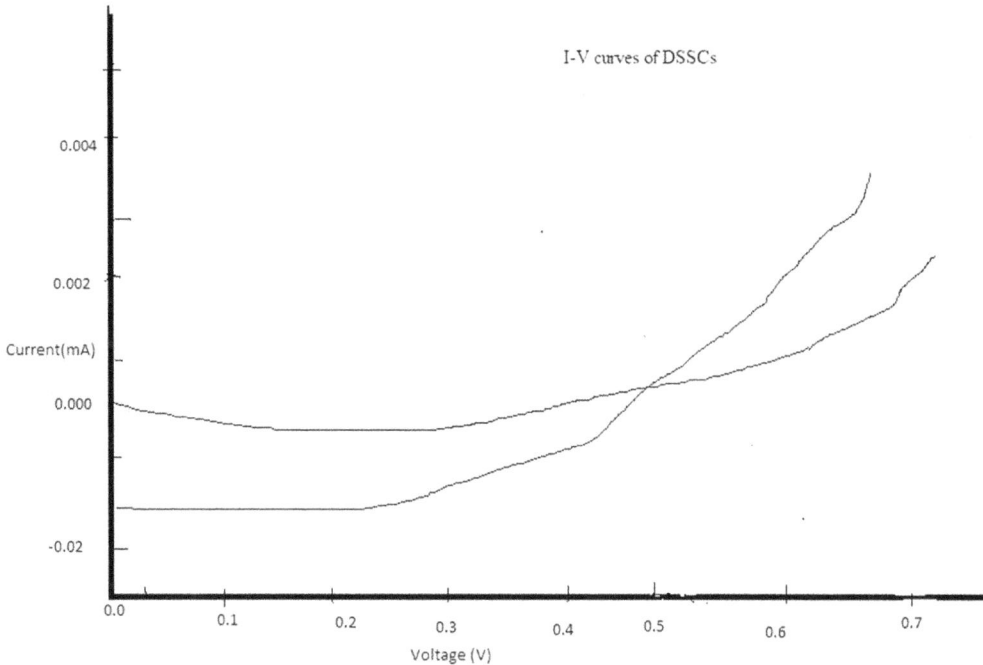

I-V curves of DSSC's

FIGURE 7.4   The I–V curves of DSSC.

TABLE 7.1   I–V Characteristics of DSSC

Parameters	TiO$_2$	Graphene+
Jsc mA/cm^2	6.9	17.5
η	1.4	3.9
FF	0.42	0.45
Voc V	0.5	0.5

on graphene-n-type-silicon have since been extensively studied and published. A brand-new flexible solar cell made of graphene and silicon was created with a PCE above 10%. These solar cells have a silicon/graphene Schottky connection, and to increase performance, a layer of graphene oxide (GRO) was placed on the graphene sheet (Huang et al., 2012).

7.3.4 Solar Cells–Organic Materials

It is possible to distinguish between inorganic and organic solar cells when analyzing the heterojunction solar cells created by the photovoltaic effect, which is initiated by p-n junctions and results in solar energy conversion into electrical energy. A cathode, a hole transport layer, an electron transport layer, and a transparent anode are the four essential components of organic solar cells. Whereas, in comparison with silicon and other types of solar cells, the organic photovoltaic (OPV) devices have a significantly lower PCE, organic heterojunction solar cells have various usage benefits over its inorganic counterparts, including lighter weight, adjustable shape, simple adjustment of physical properties, solution processability, ease of device fabrication and less manufacturing cost. Generally, the OPVs make use of poly (3,4-ethylenedioxythiophene), PEDOT: Cu2ZnSnS4, GO, ZnO, or TiO_2, as an electron collection layer and PSS as a hole collecting layer (Singh et al., 2015).

Due to their high absorption, electron mobility, and flexibility, graphene-based materials may be effectively employed to build OPV cells and enhance device performance. Numerous organizations have investigated graphene-based materials in, electron transport layers, hole transport layers, OPVs as transparent electrodes and cathodes.

7.3.5 Solar Cells Made of Graphene Polymer

A well-researched application of graphene is solar cells made of polymers. Polymeric elements have numerous benefits over inorganic-based elements because of their versatility, affordability, and simple fabrication procedures.

In polymer-based solar cells, graphene is effective as a transparent electrode substitute for ITO. After coating, stacking, reduction, and thermal annealing, the graphene in the electrode undergoes transformation into an organic–inorganic hybrid material. Because the semiconducting layer and the graphene fermi-level are closer for efficient charge injection, the hybrid material has a greater energetic link. Despite their excellent work function and conductivity, only 65% of the light can pass through transparent graphene polymer electrodes. In addition to being reduced into hybrids, chemical vapor deposition (CVD)-graphene is used as a transparent electrode. Ozone treatment of CVD-graphene causes the development on the graphene surface of carbonyl and hydroxyl functional groups.

The open-circuit voltage rises when oxygen-based functional groups are used. However, the conductivity is reduced because the sp3 bonds surrounding the functionalized carbons disrupt the sp2 hybridized covalent network. High conductivity non-covalent functionalized CVD-grown graphene is capable of 1.71 PCE, 55 fill factor, and open-circuit voltage of 0.55V. Compared with pure ITO electrodes, the solar cell may flex up to 78 degrees, because graphene is more flexible. High electron affinity is required to separate the electron-hole pairs into distinct charges in electron transporter and acceptor-based graphene polymer solar cells. When combined with conjugated polymers, graphene provides superior separation compared to other materials. The enormous surface area of graphene makes it possible to create a continuous channel and several donor/acceptor sites for efficient transport of electron. The PCE of this specific solar cell was 1.1%.

7.3.6 Bulk-Heterojunction Graphene Solar Cells

Graphene is employed in various applications in heterojunction solar cells because of its high electronic conductivity, transparency, and flexibility, including electrodes (both cathodes and anodes), donor layers, acceptor layers, active layers, and buffer. The unique tunable characteristics of graphene, such as thickness, thermal annealing temperature, doping concentration, and photovoltaic performance, are crucial for the multijunction within the solar cell.

Commonly used and studied solar cells based on graphene are graphene heterojunction solar cells. A transparent electrode, photoactive layers, and gallium arsenide (GaAs) solar cells are just a few of the ways graphene derivatives can be used in heterojunction solar cells. Consequently, graphene heterojunction solar cells aren't classified as a single type (Chen et al., 2012).

To stop current leakage and charge recombination, many solar cells need a hole transport layer. The thick film prevents electron transfer and raises electrical resistance. Hence, a 2-nm graphene sheet produces the most significant results. When graphene and a polymeric substance were mixed, materials with a high bandgap up to 3.6V were made, inhibiting the flow of electrons from the cathode to the anode. The highest PCE obtained was 9%, which is on par with or even better than other materials used as hole transit layers (Figure 7.5).

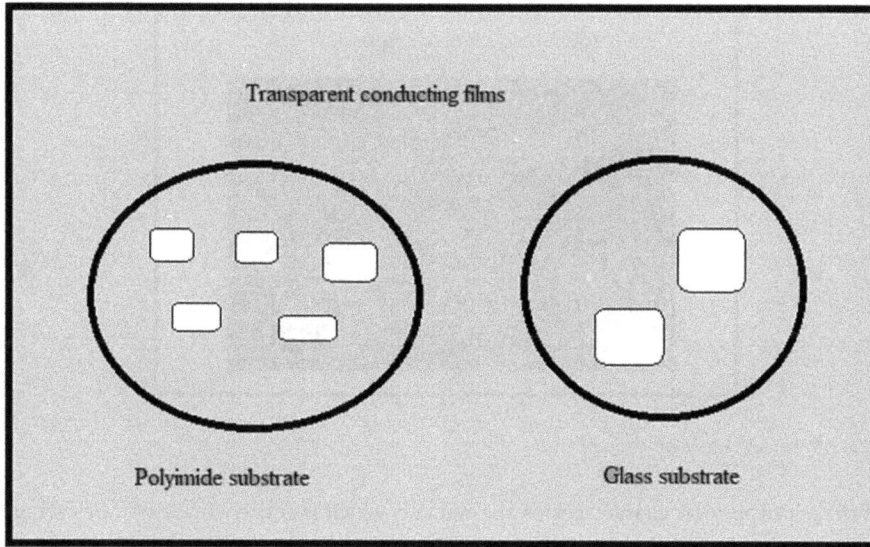

FIGURE 7.5 Transparent conducting film.

7.3.7 Solar Cells Made of Graphene Ga/As

Positive Ga/As solar cells might be improved for industrial uses. Purchase CVD-graphene or develop it on copper foil to start. Then, after immersing wafers made of Ga/As in HCl solution (10%wt., 3 minutes), remove the oxides on them by thermally evaporating gold contacts with a 60-nm thickness onto the rear surface. An 80-nm Si Nx layer is deposited to serve as an insulating layer between Ga/As and graphene on top of the Ga/As character utilizing lithography-processed mask and CVD enhanced with plasma. Dip the Ga/AS in HCl solution (10% wt., 5 minutes) and then rinse using deionized water to open a window (active area). Use NH3 plasma treatment to treat the active area (with 120W 27.5MHz radiofrequency generator for 5 minutes).

The graphene sheet is transferred to a substrate using PMMA poly (methyl methacrylate) as support. Acetone should be used to remove the PMMA. After applying silver over the six region of the graphene, bake it for five minutes at 120°C. For doping graphene, spin coat is done with (bis-trifluoro-methane-sulfonyl-amide). Include an antireflection layer, such as a 68-nm thick sheet of vaporized Al_2O_3 produced by an electron beam (Guo et al., 2015). The gate is then fabricated by applying an additional layer of graphene on top of the Al_2O_3-coated active region. Remove PMMA, connect a silver gate electrode with graphene gate, and anneal it for 5 minutes at 120°C.

7.3.8 Solar Cells Made of Graphene Could Reach 60% Efficiency

One of the many peculiar characteristics of graphene, which is made entirely of pure carbon, is that single-atom-thick sheets of it are ten times stronger than steel. Graphene may be substantially more effective at converting light to energy than silicon, with an industry standard for commercial solar cells, according to new research from "Spain's Institute of Photonic Sciences" (ICFO). Scientists found that, in contrast to silicon, graphene could produce several current-driving electrons for each photon absorbed. The study's results are excellent, even though the use of graphene in solar cells is only hypothetical at this point: graphene-based solar cells may achieve an efficiency of 60%, which is double the generally acknowledged maximum efficiency of silicon cells (Figure 7.6).

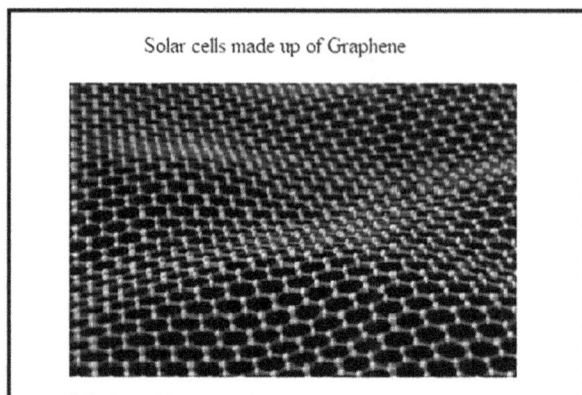

FIGURE 7.6   Solar cells made up of graphene.

One of the many peculiar characteristics of graphene, which is entirely made of pure carbon, is that single-atom thick sheets of it are ten times stronger than steel. Recently, the substance has been applied to other technologies, such as flexible batteries and water filtering. However, the Institute of Photonic Sciences examined the substance's capacity to produce electrons when it absorbed light in a study published in *Nature Physics* (Ye & Dai, 2012).

The ICFO team stated, "We employed two ultrafast light pulses". One layer of graphene received a certain amount of energy in the first experiment. The second "worked as a probe that tallied the electrons generated by the first"', according to MIT Technology Review. They first demonstrated that graphene is substantially more effective at converting light into energy and can "work with every potential wavelength you can conceive" (Feng et al., 2011).

A wide variety of light colors can be absorbed by graphene. However, we now understand that the material has a very high energy conversion efficiency once it has been taken into light. The next stage will be to find a way to extract the electrical current and enhance graphene absorption. Then, we'll be able to make graphene components better at detecting light, which might result in more effective solar cells (Jie et al., 2013).

7.4   Future Developments in Graphene Solar Cells

With developments made, solar cells are a popular study area in the industry than academia. The vast majority of solar cells currently under development are composed chiefly of silicon and inorganic materials, which will soon reach their limits. Various commercially viable solar cell topologies are expected to develop from combining organic molecules with narrow bandgap polymers, graphene derivatives, or both.

Graphene-based solar cells have made significant progress so far, and this trend will continue in the future. Due to their ability to expand in various ways, graphene-based solar cells are quite adaptable and scalable to upcoming solar examine difficulties. It doesn't matter if graphene is used to enhance currently available solar cells, enhance the features of now available non-graphene-based solar cells, or create a new line of graphene photovoltaics. It plays a part in this fascinating and quickly evolving industry.

7.5   Conclusion

The development of storage and energy conversion technologies benefited greatly from using carbon-based materials. In the energy conversion field, a pressing need is there to create novel nanocarbon-based materials. Nanocarbons for polymer solar cells appear to be one of the main application-related research drivers in the short to medium term. However, solar fuel cells are a longer-term objective for developing artificial leaf-type and photo-electrocatalytic devices. The benefits of using graphene are increased light absorption, a broader spectrum of absorption wavelengths, smaller charge transit gaps, and charge suppression recombination. Although third-generation nanocarbon materials have more complex architectures, which are needed to meet the demanding and challenging requirements for next-generation solar devices and energy storage solutions, nanocarbons still have a promising future in sustainable energy storage and use.

References

Bensaid, S., Centi, G., Garrone, E., Perathoner, S., Saracco, G., *Chem Sus Chem* 5 (2012) 500.

Centi, G., Perathoner, S., *Eur. J. Inorg. Chem.* 26 (2009) 3851–3878.

Centi, G., Perathoner, S., *Chem Sus Chem* 3 (2010) 195.

Centi, G., Perathoner, S., *Coord. Chem. Rev.* 255 (2011) 1480–1498.

Centi, G., Perathoner, S., in: D. Eder, R. Schlögl (Eds.), *Nanocarbon-inorganic Hybrids*, De Gruyter Pub., Berlin (Germany), 2014, pp. 429–454.

Chen, H.-Y., Kuang, D., Su, C.-Y., *J. Mater. Chem.* 22 (2012) 15475.

Chen, D., Tang, L., Li, J., *Chem. Soc. Rev.* 39 (2010) 3157.

Cheng, P., Yang, M., Wang, H., Cheng, W., Chen, M., Shangguan, W., Ding, G., *Int. J. Hydrogen Energy* 37 (2012) 2224.

Dai, L., Chang, D.W., Baek, J.-B., Lu, W., *Small* 8 (2012) 1130.

Desimoni, E., Brunetti, B., *Electroanalysis* 24 (2012) 1481–1500.

Feng, T., Xie, D., Lin, Y., Zang, Y., Ren, T., *App. Phys. Lett.*, 99 (2011) 233505.

Frackowiak, E., Beguin, F., *Carbon*, 39 (2001) 937–950.

Guo, X., Lu, G., Chen, J., *Frontiers in Energy Research*, 3 (2015) 50.

Hasobe, T., *Phys. Chem. Chem. Phys.* 12 (2010) 44.

Huang, X., Xiaoying, Q., Boey, F., Zhang, H., *Chem Soc. Rev.*, 41 (2012) 666–686.

Huang, C. H., Zhang, Q., Chou, T.-C., Chen, C.-M., Su, D.S., Doong, R.-A., *Chem Sus Chem* 5 (2012) 563.

Jie, W., Zheng, F., Hao, J., *App. Phys. Lett.*, 103 (2013) 233111.

Li, P., Chen, C., Zhang, J., Li, S., Sun, B., Bao, Q., *Frontiers in Materials*, 1 (2014) 26.

Liang, C., Li, Z., Dai, S., *Chem. Int. Ed.* 47 (2008) 3696–3717.

Liang, Y.T., Vijayan, B.K., Gray, K.A., Hersam, M.C., *Nano Lett.* 11 (2011) 2865.

McCreery, R. L., *Chem. Rev.* 108 (2008) 2646–2687.

Singh, E., Nalwa, H., *Journal of Nanoscience and Nanotechnology*, 15 (2015)6237–6278.

Su, D. S., Centi, G., Perathoner, S., *Catal. Today* 186 (2012) 1–6.

Vilatela, J. J., Eder, D., *Chem Sus Chem* 5 (2012) 456.

Yang, S., Bachman, R.E., Feng, X., Müllen, K., *Acc. Chem. Res.* (2012). https://doi.org/10.1021/ar3001475

Ye, Y., Dai, L., *J. Mater. Chem.*, 22 (2012) 24224.

Zhang, H.-P., Li, X.-L., Cui, Y. Lin, *J. Mater. Chem.* 20 (2010) 2801.

8

Bio-derived Nanomaterials for Energy Storage

Neelamma Gajji

Vikas College of Pharmaceutical Sciences, Suryapet, India

Santhosh Kumar Chinnaiyan

Faculty of pharmacy, Karpagam academy of higher education, Coimbatore, Tamilnadu, India

The latest developments and breakthroughs in biomeditating nanomaterials used in energy storage applications are reviewed in this chapter. According to the American Chemical Society, virus, microbes such as bacteria and fungi are used as templates or carbon resource for the fabrication of metals and carbon-based nanomaterials, as well as biomass from fauna and flora like hardwood, ground coffee, and lobster shells. In this section, we provide a summary of all of these materials as templates with a priority on the fabrication techniques and properties of nanomaterials that are manufactured utilizing these materials as templates. As a result of the surprisingly varied studies in this area, we intend that by giving a comprehensive overview of current research, we will be able to contribute to its advancement. As per our forecast, this area will continue to demonstrate extremely pertinent to the important and developing themes of renewable energy in the future.

8.1 Introduction

A revolution has occurred in the fields of nanosciences and nanotechnology, which has spread to practically every field of science and technology. Recent developments in the competence to handle and control materials on an atomic, molecular, and nanometer level, as well as the resulting understanding of principal of the processes at this scale, have paved the way for new research directions (Chen & Hu, 2018). The knowledge gained in this way can be turned into novel procedures, which can then be used to design or manufacture better products in the future. A more significant development is the emergence of new scientific developments and processes that may contribute to the resolution of the energy, ecological, and sustainable mobility problems that mankind will face in the 21st century (Al Rai & Yanilmaz, 2021). Owing to the rapidly increasing need for renewable and clean energy sources, novel material technologies are being investigated as potential answers to the world's energy concerns that are both cost-effective and environment-friendly. Supercapacitor, fuel cells, sophisticated batteries, and biofuels are among the technologies being developed to compete with petroleum-based fuels and energy storage systems (Powell et al., 2021). Despite the enormous potential of clean and renewable sources of energy such as solar and wind power, efficient electrical energy storages (EES) approaches that can supply continuous power on demand arerequired for the practical application of these sources of energy. Nanomaterials are becoming more and more important in all of these new technologies, either by making the energy storage and renovation progression more efficient or by improvingthe design and recital of the devices themselves (Kalogianni, 2021).

The Anthropocene is a new era, a time of unprecedented human impact. Never before has a sole species dominated the ecosphere in such an approach as we do nowadays on this lonely planet. Humans have an immediate and direct impact on more than 60% of the land area and 41% of the oceanic and environmental settings. Annually, people consume 40% of the flora that aregrown. The most significant factor is that we are mining and burning fossil fuels at an alarmingly fast rate, producing 160 tons of SO_2 and 36 gigatonnes of CO_2 annually, which is negatively altering global weather patterns (Zhang et al., 2019). The growing data require us to take action to slow and, eventually, turn around this troubling mode. To accomplish this, we must switch from using fossil fuels to relying on renewable energy sources. It would be necessary for technologies that would allow us to enhance acclimatize to novel, renewable-based energy approaches to receive the lead in such a drive. In this context, we are pleased to introduce this special issue, which will report on the most recent research studies on energy conversion and energy storage space conducted by renowned experts from across the world. Energy conversion and storage systems are thought to be developed for at least two primary reasons (Q. Chen et al., 2020d). Primarily, extremely efficient and economical energy conversion and energy storage are essential for tackling the challenges associated with intermittent renewable energy sources; whether they are wind, tidal, or solar. If renewable energy sources are to be more easily and sustainably integrated into our daily lives, they must first be converted into alternative types of energy, such as batteries stored a chemical energy, which can then be extracted as electricity whenever it is required. Second, addressing social needs requires an energy source that is available on demand (Jin et al., 2021).

Similarly, for the development of many other industries, the availability of suitable material is critical to the progression of energy conversion and storage technologies. Because the constraints that limit the overall performances of energy storage and conversion components are primarily material-related in nature, the reliance is perhaps predominantly strong for the subject matter covered in this chapter. The introduction of novel supplies, such as 2D materials, always serves to stimulate the development of new concepts and uses (Ma et al., 2020). Although the parameters that determine the preferred materials for energy translation and storage are numerous and diverse, they frequently share one strikingly similar property: the distinctive length scale of charge behaviors (Bhat et al., 2021). The nanoscale is a length scale that is only a few hundred nanometers long resulting in the significant effort that has been devoted to the synthesis, characterization, and application of nanomaterials. It is reasonable to predict that controlling the properties of materials next to nanoscales will be critical to making significant progress near a future powered entirely by sustainable energy resource (Figure 8.1). With all due respect, we hope that the studies collected in this special issue will act as stimuli to promote further discussion of energy storage and conversion (Q. Chen et al., 2020d).

Waste products, like coffee grounds, can be transformed into elements that are beneficial to society as a whole with waste biomass. Because biological samples are overwhelmingly abundant and have a wide range of microstructural characteristics, an extensive diversity of biomass can be taken into account. This is made possible by the large amount of biomass and the variety of microstructural characteristics of biological samples, which enable very flexible material design (Bhat et al., 2021). As opposed to viruses or microbes that bio-template precursors, plants reproduce at a slower rate than these organisms, and as a result, self-assembly processes are not characteristic of plant-derived resources. When thermally treated, macro-scale biomass has advantage in many other regions as well. It is high in carbon and

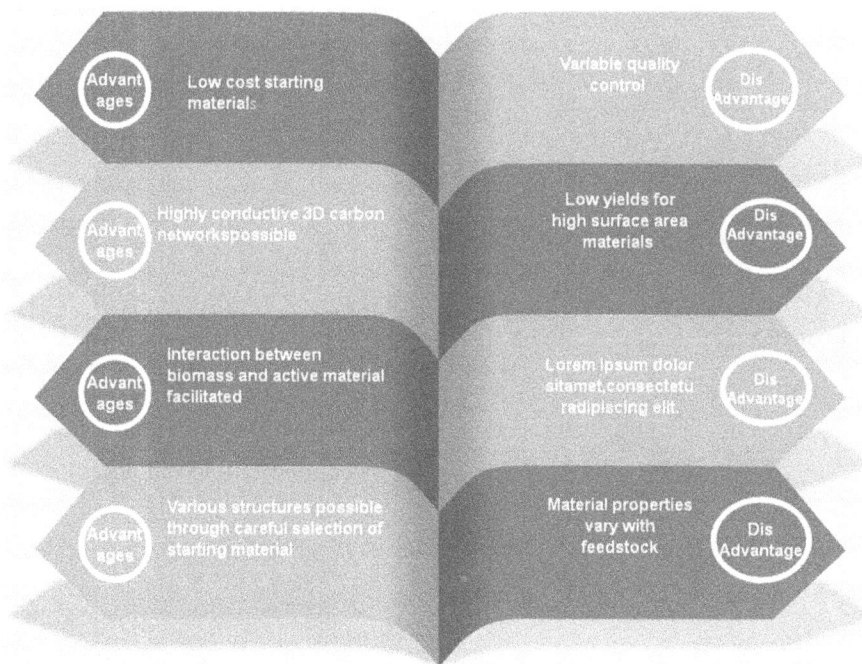

FIGURE 8.1 Advantages and disadvantages of biomass-derived materials for energy application.

heteroatoms such as nitrogen, sulfur, and phosphorus; it can provide a significant amount of intrinsic doping. When using biomass as a source of heteroatom doping, it is feasible to tailor the electrochemical properties of the resulting carbons (Cheng et al., 2020; Ma et al., 2020). The stoichiometry and precise amounts of dopants and inorganic contaminants, however, may vary among the several available options due to the nature of materials generated from biomass. Further strategies for more accurately regulating the concentrations of endogenous dopants and elements in the environment should be developed to convert biomass into a bio-derived product (Dan et al., 2020).

Nanomaterials for storing energy and transformation applications, particularly for waste products, are very intriguing. Technology that converts the already vast supply of garbage into valuable goods has the potential to be extremely beneficial to society in the areas of waste minimization, carbon sequestration, and energy-related applications. Through the use of waste resources, it is possible to realize great promise and viability for commercialization (Cai et al., 2018). The synthesis of nanocarbon can be completed in a single step rather than two steps by simply carbonizing biomass, and they do not require any template removal after they have been synthesized (Chen et al., 2019). As a result of the superior energy needs of biomass conversion to carbon, minimizing the process steps and using economical precursors are both desirable for scaling up the process (Jin et al., 2021). Moreover, because there is such a diverse range of biological starting points, there is a wide range of study on these materials. As a result, a comprehensive study is necessary to promote ongoing research progress in the field of biomaterials. Animals have been exposed to bio-derived nanoparticles in both direct and indirect ways (Li et al., 2021).

The various structures derived from many distinct templates each have their own set of characteristics that can improve the efficiency of the material being developed. The various uses' categorization and the structural properties of the material made from these resources are crucial in identifying the different types of applications that each appropriate precursor can be used (Schmitz et al., 2019). A huge range of potential nanostructures is available due to each biomass precursor's unique structural characteristics, which is a result of the wide variety of distinct starting materials available. There are various types of structures that may be found within fungi alone; molds prefer to produce branched filamentous

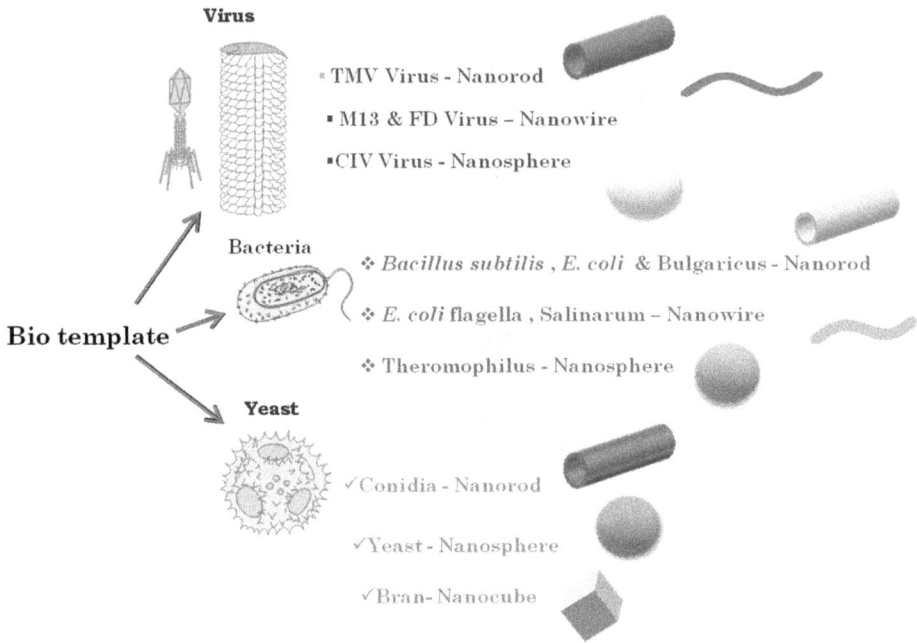

FIGURE 8.2 Bio-templates and their produced nanostructures for energy applications.

arrangement called hyphae, whereas yeasts can take on sphere-shaped, ovular, or rod-like topologies, among other shapes (Chen et al., 2019). Figure 8.2 illustrates how these structures may be effective in influencing the shape and structure of the generated biomass.

8.2 Viruses-Derived Materials for Energy Storage

According to research undertaken since the 1920s, several viral structures are well understood. Virus populations develop perfectly ordered homogeneous structures, and their structures can be modified to a certain extent using the tools that are currently accessible. It is possible to modify the shape or functionality of a specific viral structure by applying genetic techniques. The absence of a membrane and well-defined geometry of virus structures make it easier to use virus structures as bio-templates (C. Chen et al., 2020c). These architectures are frequently directly applicable for the manufacture of nanomaterials. A most typical way to generate viral structures, also called as virus-like particles (VLPs), is through bacteriophage development in *Escherichia coli* bacterial strains, which are subsequently purified. This method was chosen because of its rapidity of production and scalability (Chen et al., 2019). Additionally viruses, such as the plant-derived Tobacco mosaic virus (TMV), can be reproduced by identical expression, in which genetically engineered *E. coli* is to create virus structure, in addition to bacteriophage (Chen et al., 2018).

TMV is a rod-shaped virus that has been extensively examined as a bio-template due to its capacity to control the width of the materials produced from its template by mineralizing both its inner and outer channels. In addition, TMV is able to mineralize both its inner and outer channels. The virus is composed of 2100 unique envelope protein with dimensions of 300 nm, 18 nm on the exterior, and 4 nm on the inside (Cheng et al., 2020). They can be mineralized to produce semiconducting and insulating nanowires, which can be used to make material that is useful for the energy sector. TMV nanorods can be modified in a number of ways to enhance their functionality and performance. These are also durable and simple to purify. TMV nanorods will self-align from head to tail when placed in an acidic environment because of hydrophobic interactions at the rod extremities. This enables nanowires to form and be utilized as templates. The enormous surface area and robustness of TMV-based nanomaterials under volume expansion have led to suggestions that they be employed as battery electrodes (Zhou et al., 2012). Thiol–metal interactions make it simple for TMV1cys nanorods to attach to a metal electrode surface, which can lead to the production of nanoforest structures that enhance the electrode's surface area. This is constantly influenced by coating nanorods with metal oxide by depositing a metal oxide using electroless deposition solutions after activating the TMV1cys surface with a palladium catalyst (Li et al., 2021). According to research studies, this arrangement can improve reaction kinetics and cycling durability while reducing mechanical and electrochemical forces between both the electrode and current collector.

The M13 bacteriophage, which has a fiber-like structure and a homogeneous covering of negatively charged surface protein, is another often used virus as a bio-template. M13 is less than 10 nm in size and measures about 1 µm in length. They are homogeneous crystalline CO_3O_4 nanowires when M13-based cobalt oxide nanowires are produced by electrostatic interactions between both the viral surface and solvated metal ions. These nanowires can be used to produce other kinds of nanowires (Yu et al., 2020). As shown by Baig et al. (2021), this procedure can be changed to make discrete nanoparticles in addition to nanowires by lowering the buffering capacity of the deposition fluid. They also show that there are advantages and limitations to the various approaches to utilize viruses as bio-templates. Surface qualities are used as driving forces in the processes of bio-mineralization, deposition, and adsorption. To control pore size, pretreatment processes can be employed in conjunction with any of these methods. Deposition methods that are pretreated provide more control over the morphology of the final product. The quantity of metal that attaches to a surface can be regulated by using the right adsorption methods on the surface. Viral bio-templating of nanomaterials is an intriguing option for bio-template nanomaterials. The techniques used to generate and isolate viral structures are widely known and appropriate for the creation of VLPs in both commercial and academic sectors. Because the particle populations formed in this manner are mono-disperse, the resulting structures are typically resistant to chemical and mechanical degradation. There is a large diversity of structures and surfaces accessible, all of which can be adjusted with no causing substantial deviance from the typical manufacturing procedure (Zhou et al., 2012). The electrical and chemical properties of bio-template nanomaterials produced from virus architectures are particularly intriguing for energy applications, such as conductive material and battery electrodes, among others.

8.3 Microorganism Templates Nanostructures for Energy Applications

8.3.1 Bacteria

This section would deal with how different microbes, such as *Bacillus subtilis, E. coli, Halobacterium salinarum*, and lactobacillus bacteria strains, have been used as bio-template in the energy storage. The gram-positive bacteria with a rod shape, *B. subtilis* has a diameter of 400 nm and typical lengths ranging from 2 to 5 µm. Scanning electron microscopy and transmission electron microscopy are employed for the photographs of microbial surface structure. The peptidoglycan (PG) in the cell wall of *B. subtilis* comprises the functional group's phosphate, teichoic acids (TG), and carbonyl, among other things. Electrostatic interactions among the membrane of the cell and metal oxides are made possible by the presence of negatively charged phosphate ion group. As a result of spontaneous oxidation, metal hybrids that mimic bacteria are produced when metal oxides connect to the cell surface of bacteria through reduction on the membrane (Zhang et al., 2019). The hybrid material is subsequently calcined to eliminate the organic components, producing a permeable metal oxide hollow nanorod as a byproduct and the metal oxide result preserves the bio-templates. Adding customized polystyrene (PS) spheres–enclosed amine groups to the mixture is another way to regulate the pore size. To change the material's pore size, users can change the PS spheres' diameter. PS is utilized because it is simple to delete from the final product after it has been created. Additionally, the bacteria *B. subtilis* can be treated with glutaraldehyde and subsequently mercapto ethylamine to create thiol groups on the cell membranes of the host cell. These thiol groups then form SH–gold interactions with gold nanoparticles, forming a strong bond between them (Shaheen et al., 2020). The *E. coli* bacteria is a gram-negative microorganism with a rod-like structure as well as a typical length of 1.4 μm and width of 0.6 μm. Hydrolysis was employed by Nomura et al. (2013) to create nanomaterials from *E. coli* bio-templates. According to their findings, zirconium butoxide and alcohol are effectively introduced to a bacterial colony, resulting in the creation of nanorod hybrids. After that, the hybrids are calcined to confiscate the bacterium templates. Nomura's method is superior to the other because the nanomaterials manufactured using it retain the geometry of the bio-template, which is the most significant distinction between the two ways. In addition to the entire *E. coli* cells, flagella from *E. coli* cells can be used as bio-templates.

An archaebacterium known as *H. salinarum* is being researched as a potential bio-template for the creation of novel medications. The fact that *H. salinarum* survives in extreme conditions is helpful for bio-templating since it enables the utilization of conditions like a strongly acidic or basic milieu without damaging the structure of the template. *H. salinarum* flagella typically measure 10 nm in length and 10–15 nm in diameter and the flagella are made of proteins without any DNA. To create carboxyl groups in the flagella, it is necessary to genetically modify them with aspartate or glutamate. These groups actas the binding site for metal ions, which adhere to the flagella and result in the formation of nanowires. The *Lactobacillus*is gram-positive bacteria with cell wall made of teichuronic acid peptidoglycan (PG) and teichoic acid (W. Chen et al., 2020a). Two varieties of *Lactobacillus* employed as bio-template were *Streptococcus theromophilus* and *Lactobacillus bulgaricus*, both of which were isolated from humans. In appearance, *Streptococcus thermopiles* is globular with a diameter between 0.5 and 0.9 μm. *L. bulgaricus* is like a rod-shaped, with lengths ranging from 1 to 5 μm and dimensions varying from 0.4 to 0.8 μm (Li et al., 2021). Zhou et al. (2012) employed two alternative approaches to use *L. bulgaricus* as bio-templates to construct various nanostructures. When applied to the bacteria bio-template, nanoparticles are mono-dispersed, but nanomaterials manufactured using chemical means have a more unequal dispersion when applied to the same template. Another advantage of utilizing bacteria is the ability to control the morphology of the product. It is possible to maintain nanosheets a range of shapes from the original bio-template.

8.3.2 Fungi

Saccharomyces sp. has been utilized as a bio-template to produce Zirconia nanospheres, microspheres, or mesoporous nanostructures from zirconium (Zr), zirconium phosphate ($Zr(HPO_4)_2$), silicate (SiO_4^{4-}), iron (III) oxide (Fe_2O_3), cerium(IV) oxide (CeO_2), Cobalt (II, III), tungsten (IV), and lithium iron phosphate ($LiFePO_4$) (Zhao et al., 2018). Baker's yeast was used in the fabrication of yeast template $LiFePO_4$ particles by Chinese researchers. After being prepared by pyrolytic destruction, the yeast is then added to the mixture. Pyrolytic destruction breaks down the cell to release functional biocarbon while simultaneously limiting Fe cations from oxidizing. $FeCl_3$ is subsequently added to the template, where the Fe^{3+} cations form covalent bonds with anion biomolecules in the template. Bio-mineralization occurs when phosphate group reacts through iron cations in the presence of $(NH_4)_2HPO_4$, resulting in the formation of $FePO_4$ nanoparticles within and on the yeast cell membrane. Once the yeast cells havebeen mineralized, it is assorted with LiOH and calcinated to generate $LiFePO_4$ nucleus nanoparticles that have been biocarbon-coated. These particles have a spherical shape, a mesoporous core, and a nuclear encasing all around them, which makes them difficult to detect. The production of $LiFePO_4$ agglomeration may be inhibited by this shell (Lee et al., 2019). It has been demonstrated that the dispersed nanoparticles shorten the length of the Li^+ solid-state transport routes. Researchers were proficient to optimize the pore dimension for lithium diffusion in their studies because the pore diameter, surface area, and volume of these particles can all be changed (Zhao et al., 2018). It's been shown that coating a particle with a conductive carbon nanocomposite boosts both lithium diffusion and particle conductivity. Conidia from a variety of fungus species are used to make gold microwires. A conidium is a fungus spore that emerges because of cellular mitosis in the fungus. *Aspergillus niger, Aspergillus nidulans,* and *Neurospora crassa* live cultures were put into a solution containing gold nanoparticle and glutamates to study their growth. The disruption of the local charge configuration promotes the binding of gold nanoparticles to the fungus' growth surface when it consumes glutamate to generate conidia. This activity was not observed in samples of dead fungi or samples that did not contain a growing substrate. The chitin-rich bran of the common forest waste fungus is a useful source of nitrogen because it contains a high concentration of chitin. Afforest bran may be carbonized to provide N-doped carbon compounds, which are useful in the construction industry (C. Chen et al., 2020c). ZnO nanocubes have been created using a fungal bran-template carbon matrix, according to the findings of a recent study (Ringe, 2020). It is necessary to clean and carbonize the fungal bran before activation with potassium hydroxide in this procedure. ZnO is used to anneal the activated carbon matrix, which results in nanocubes of activated carbon being formed. In comparison to fungi templating process such as bio-mineralization and deposition, this material was found to be superior. Deposition methods make use of the surface charges of the template, but they necessitate the presence of a living template. Although bio-mineralization technologies prevent agglomeration, their effectiveness is limited by functional groups on the cell membranes (Chen & Hu, 2018).

8.3.3 Algae

Algae are abundant in nature, can be grown, are a rich source of carbon, and are economical to produce. The growing use of agrochemicals has resulted in a significant enhancement in the production of algae. The Yellow Sea, for example, has annual green tides that contain significant concentrations of *Enteromorpha algae* (EA), which are non-toxic to humans and other animals. Because of the enormous increase of algae, which is extremely detrimental to the environments, it is valuable to use EA as a bio-template (Cao et al., 2020).

Spirulina has also been investigated as a potential bio-template for energy-related applications. A blue-green alga called spirulina has a spiral coil structure and numerous solitary cells. Spirulina is a protein-rich meal that also acts as a good source of carbon and also measures 20 μm in length and 200 μm in width. Spirulina's spiral structure is incredibly resilient, and it maintains its spiral shape even after being heated and subjected to sonication. Xia et al. (2011) used spirulina as a bio-template and a source of carbon to make lithium iron phosphate/carbon. Real-time spirulina is introduced into the iron-rich solution, which has glucose, lithium, iron, and phosphate molecules, among other micronutrients. The solution is then made basic by adding ammonia, which is preferred because it helps keep the structure of the spirulina intact. The solution is mixed to make sure it is spread out evenly, and then it is moved to an autoclave lined with Teflon, where it is treated with hydrothermal heat for several hours. The product is then heated to 650°C in a diverse environment to get rid of any remaining water. According to the Brunauer, Emmett and Teller surface area calculator, template $LiFePO_4$/C has typical pore diameter of 25 nm and a bulk effective surface area of 22 m^2/g. The carbon in the spirulina joins the $LiFePO_4$ nanograins, forming junctions between them and allowing them to interact (Xia et al., 2011). These carbon linkages strengthen the stability of the structure as well as its capacity to store lithium ions. Furthermore, the carbon bridges have the potential to improve electronic contacts between grains while simultaneously lowering internal resistance between grains.

8.4 Nanomaterials Derived from Plants

Application technologies involve energy storage and conversion materials based on plant-based biomass. The transformation of biomass into carbon dioxide using thermal treatment is a process that is widely used. Although calcining biomass to produce carbon is simple, there are a great amount of distinct alterations that may be made to the process, resulting in a variety of different forms of carbon substance with dramatically varied electrochemical possessions. The conversion of biomass to carbon through calcination has the benefits of being low-cost, simple, and time-consuming regardless of the modifications that are made (Zhou et al., 2012). Scalability, on the other hand, is sometimes a challenge because of the huge amount of energy that is needed for the pyrolysis of large batches.

8.4.1 Timber Materials

Materials are a popular discussion topic because of the great number of different woods available for use in the production of soft or hard carbons. Wood is a good starting material because it has natural hierarchical structures, anisotropic, superior mechanical characteristics, and the capacity to be shaped to suit a wide variety of purposes, other advantages of wood include its ability to be shaped. Furthermore, because natural wood is abundant all over the world, being competent to tap into this naturally renewable resource is a good choice for carbon-based technology applications (Wang et al., 2018). Softwoods versus hardwoods, in particular, were taken into account when developing novel materials for a variety of diverse purposes. Softwoods are described as having a homogeneous framework with a large number of fibers. Hardwoods also significantly denser have superior bulk density as well as pore density, and as a result, they have great mechanical qualities. Softwoods are classified as having a homogeneous structure with a large number of fibers. A variety of strategy, including 3D wood along with wood paper–based electrochemical energy storage devices, can benefit from the fast ion transport provided by the unidirectional channels in the wood structure. According to the researchers, Poplar wood has been modified to serve as hierarchically permeable support for $Co(OH)_2$ type solid-state supercapacitors (Cao et al., 2019). First, the wood is acid-cleaned before being calcined at 1000°C for 2 h in an inactive environment. Following that, the active material $Co(OH)_2$ is delivered into the absorbent carbon employing electrode placement. The carbonized wood was discovered to have a synthesis pathway, as well as a $Co(OH)_2$ 3D hierarchically permeable electrode design with open channels and a high unique surface area. It is not only possible to improve a load of $Co(OH)_2$ nanoflakes by using wood-derived porous carbon networks, but they also permit for quick electron transmission and ion diffusion, which results in an acceleration of the rate performance. Using this synthesis approach, the creation of electrode materials is accomplished without the need for any binder, which results in increased active area for each unit volume of the electrode material. To create relevant technologies, it is necessary to first understand the effect of thermal treatment of wood on the structure (Mujib et al., 2020). A low-viscosity precursor solution is reported to generate hierarchical porous structures when it is infiltrated into the wood's cellular matrix and then calcined. It is additionally explained that heat treatment at various temperatures results in the production of nanosize oxide pore channels, as well as a temperature dependence that affects the connectivity and diameter of the nanopores, as well as a related temperature dependence. Because of the nature of wood, it is possible to create fascinating materials by taking advantage of the fiber arrangement and anisotropy of the wood. Recent study has shown that balsa wood can be converted to a greatly flexible substance that is ironically conducting (Schmitz et al., 2019). The hardwood is processed to create this flexible wood by first immersing it in an aqueous NaOH solution, then washing it with freshwater, frozen, and then freeze-drying it. It is believed that during this process, the wood's lignin and hemicelluloses are partially eliminated, softening the wood and increasing its elasticity, which enables the wood to maintain its shape even after being deformed. The created wood's ionic conductivity can also be adjusted by varying the amount of applied strain. This exceptionally elastic wood's characteristic makes it useful for a variety of additional applications in addition to energy storage and conversion.

As a result of its microstructure, bamboo is another intriguing possibility for the preparation of distinctively structured carbons. For example, bamboo has been utilized to generate paper-like carbonaceous nanomaterials that have been investigated for their suitability for use in supercapacitors and other applications. The nitrogen- and sulfur-doped carbon is generated under the influence of a catalyst, and this material has the potential to be used in the creation of supercapacitor materials made of bamboo. Singh's group has conducted a direct comparison of bamboo- and cassava-produced carbons, which revealed substantial degrees of graphitization in both materials. Although bamboo carbon appears to have improved stability, it appears to have lower electrochemical characteristics (Ringe, 2020). The cassava, on the other hand, revealed carbon with outstanding electrochemical characteristics, which were previously unknown. As per the assertion, synergistic effects could be evident when the biomass is mixed, which could eventually open up new research areas.

8.4.2 Materials Obtained from Latte

Owing to the fact that coffee is the widely consumed beverage on the globe, there is an excessive amount of waste coffee that needs to be controlled. Coffee's chemical composition is complicated, similar to that of wood, and it naturally contains over 27 different components (mainly Na, K, P, Li, and N). As a consequence of this, used coffee grounds have the potential to become an even more attractive use in the field of energy storage application tostore lithium, sodium, and potassium. Several research groups have been working on the preparation as well as characterizing the substance of various carbon compounds derived from discarded coffee grounds for use in energy storage along with conversion during the past decades. In their study, Liu et al., (2020) employed heat treatment of spent coffee grounds under the effect of a catalyst, followed by KOH activation, to form a hierarchically porous structure. This carbon is believed to have significant amounts of graphitization and a sizable number of pores produced by the breakdown of coffee's phytochemicals, like caffeine. The activation of hierarchically porous carbon samples with KOH results in fragmented and thinner carbon sheets, as well as a more amorphous structure. The N_2 equilibrium adsorption data further show that activating HPC samples with KOH results in the introduction of a significant number of micropores (W. Chen et al., 2020a). These findings suggest that the $FeCl_3$ catalyst was capable of generating a mesopores-dominant architecture and that the consequent KOH activation resulted in the introduction of abundant micropores, resulting in the formation of a hierarchical porous structure. Waste coffee grounds have also shown promise as a potential alternative for bipolar graphite in vanadium redox flow batteries. For example, the espresso extraction method prepares the coffee by using fine pulverizer and hot water to extract a significant proportion of the soluble chemicals from the bean. Cold brews, on the other hand, extracts at low temperature, leaving a surplus of poor solubility chemicals after extraction (Cheng et al., 2020). When heated to low temperatures, caffeine's low water solubility results in a significant amount of residue, which may contribute to an increase in the quantity of voids present, resulting in a greatly porous carbon. There should be even more research done to determinethe preprocessing (brewing) conditions that have an impact on the carbonaceous compounds that are formed from coffee. Furthermore, when coffee is pyrolyzed, a considerable amount of bio-oil is discharged into the atmosphere. Shaheen et al. (2020) used *Escherichia cognata* to develop compounds for energy storage and conversion applications. These oils might potentially be employed in the same way to create active materials for storing energy and conversion purposes (Li et al., 2021).

8.5 Materials Derived from Animals

8.5.1 Materials Derived from Crab Shells

Shellfish shells possess a high concentration of the amine-containing polysaccharide chitin, which makes them an excellent candidate for the exploitation of intrinsic nitrogen doping. Chitin can be isolated from shellfish shells using chemical or biological agents, depending on the application. As part of the chemical treatment processes, the shells are washed with acid in order to remove minerals suchas calcium. These processes are carried out in collaboration with other procedures. After that, the proteins are removed using an alkaline treatment and then the chitin is washed to separate it from other components of the material. Biological approaches entail the use of enzymes such as acylase, pepsin, papain, trypsin, and protease (Santos et al., 2020). More advanced treatments, such as alkali treatment or suitable biotherapy, can be used to deacetylate the chitin, resulting in the production of the cationic polymer chitosan (Santos et al., 2020). When carbonizing chitin or shell waste, some firms prefer to carbonize solely the chitin and leave any minerals intact. The composition can also be altered by including or eliminating particular phases from the process. The crab and shrimp show to be the majority thoroughly examined of the most often consumed shellfish. Fu et al. (2019) used crab shells as a chitosan-rich precursor for high-performance computing materials, which they are currently testing for use in supercapacitors. To get porous carbon from the shellfish, it is necessary to first wash it with an acid solution, which removes different carbonates while also forming huge pores. Once the acid-washed carbon has been dried, the carbon is next triggered with KOH at high temperature, which is followed by another drying process (Zheng et al., 2020).

In addition to its use as a supercapacitor, a $Ca(OH)_2$ carbon scaffold has been effectively made from crab shells and employed as a partition substance in Li-S batteries. As crab shells obviously have a high concentration of calcium, it is feasible to create micro- and nanostructures that are unlike anything else on the earth. It is claimed that the reported chemical is produced by calcining crab shells at varying temperatures, mixing them with gelatin films to create the separation separator, and then using the composite-based materials. The preparation of this substance is quite comparable to the preparation of the prior material, except for acid washing. This demonstrates excellent flexibility and simplicity in the manufacturing of substantially varied arrangement and chemistry from the similar starting material (Jin et al., 2021).

Crab shells have also been usedto develop carbon nanofiber aerogels that could be useful for energy storage and conversion applications. To develop the aerogels, crab shell chitin is soaked in distilled water, cross-linked, freeze-dried, and then carbonized (J. Chen et al., 2020b). Because of the numerous possible uses in a wide range of industries, synthesizing carbon as an aerogel enables the production of a conductive nanofibers matrix with a significant surface area. Crab shells–derived chitosan is employed as a template for the preparation of mesoporous $CoFeO_4/C$ materials with helical hierarchical structures, which are then converted to nanoflakes under pyrolysis conditions. This serves to further demonstrate the utility and adaptability of this material (C. Chen et al., 2020c).

8.5.2 Materials Derived from Shrimp Shells

Because of the quantity of chitin and proteins in shrimp shells, they are also an excellent source of porous carbon compounds. Recent research has revealed that shrimp shells may be used to create mesoporous carbon networks that can be used in a variety of applications. The use of non-platinum group metals in catalyst research and design is of interest as a means of lowering the cost of catalyst development and design. The final catalyst does not rely on the use of any extra metals, precious or otherwise, indicating that it has considerable potential for usage in electro-catalysts for other applications. Furthermore, the similarities between shrimp and crab shells allow for the use of identical processing procedures and, as a result, the preparation of equivalent materials. Because of the high quantities of nitrogen and phosphorous doping in shrimp shell-based aerogels, these materials may be of interest for modifying the characteristics of the final material.

8.5.3 Shell-less Fish

Fish lacking shells also have a significant deal of promise for use as energy storage along with conversion materials because of their high energy density. It has been demonstrated that several parts of fish waste (skin and bones) are excellent candidates for recycling trash from the fishing industry. Because of the presence of collagen and hydroxyapatite crystals in fish skin and bones, these materials are particularly suitable for the production of biomass-derived permeable carbon materials. Collagen and hydroxyapatite crystals have an impact on the shape and porosity of the resultant carbons. A considerable opportunity for the creation of novel clean energy materials is also provided by the volume of this collagen-rich biomass (Santos et al., 2020). Using a combination of proteins and inorganic materials, it is possible to create porous carbon compounds. Fish also contain amino acids, which can be used as a precursor to the creation of N/S co-doped hierarchically porous materials (Li et al., 2021). These amino acids can be obtained from a variety of biomass waste proteins, including fish. For instance, pseudocapacitive storage mechanisms have been found in the lithium and sodium-ion batteries chemistry, among others, and fish collagen has been employed as an anode material in a range of battery applications. Baig et al. (2021) prepared porous carbons with governed doping of heteroatoms and well-defined porosity, so that they could be used as high-performance electrodes for resilient Na-ion capacitors. Another study analyzes the use of fish collagen–based materials in magnesium, sodium, and lithium-ion battery, and the results show that they provide satisfactory performance. This type of method has great assurance for the improvement of innovative anode materials derived from animal biomass waste. Protein containing biomass such as collagen, with inorganic substance like hydroxyapatite observed in fish scales and animal bones, haspiqued the scientific community's attention.

8.5.4 Materials Obtained from Terrestrial Animals

Animal bone, nails, and hair are examples of biomass sources that have the potential to be used as electronic predecessors. Liu et al. (2017) investigated solid-state supercapacitors crafted of fiber that were made from human hair, with the hair helping as a source of fibrous carbon in the capacitors. It is considered that the characteristics of hair mentioned above account for the extraordinary performance of solid-state supercapacitors created by depositing nanostructure MnO_2 on hair. It is critical to the area because the majority of people have hair and require haircuts; therefore it's a great way to advance the field to be able to use this waste product to create something valuable for emerging technology (Dan et al., 2020). Additionally, it may be helpful to look into the association between hair structure and the carbon qualities that follow because various breeds have distinct types of hair. Bone marrow is another potential byproduct of the meat industry that could be employed in storing energy and conversion devices, in addition to animal bones. Carbon produced from waste bones has a large surface area and a distinct hierarchically porous structure, which is due to the bone's composition (Chen et al., 2019). As demonstrated in this work, when carbon is used as a cathode in Li-Se batteries, the consequential carbon performs admirably, enabling the development of a method for reducing the quantity of animal waste deposited in landfills. In recent years, egg yolk has also been studied for its possible application in energy storage and conversion materials, with promising findings. Zhang et al. (2019) reported the fabrication of egg yolk–based, high-performance supercapacitor materials. Egg yolk and an iron precursor are mixed, hydrothermally reacted, and then annealed to produce nanoparticles of iron enclosed in carbon nanotubes. The significance of these findings lies in the fact that carbon nanotubes appear to spontaneously form during the technique without the requirement for any template in a simple and inexpensive method.

8.6 Conclusion

Nanomaterials developed from bacterial and viral templates frequently have properties of materials that are advantageous to battery electrodes. In addition to providing a carbon source, algae can serve as a template for one-step synthesis. There are numerous species of plants and animals, each possesses a structurally distinct heteroatom composition and abundance. Despite extensive study on biomass-derived carbon compounds, there is no discernible pattern or principle that can be used to develop the next generation of carbon-based materials. For example, having a database that contains carbon characteristics (e.g. doping of carbon atoms, graphitization level, geometries, and porosity) created from diverse biomass resources and under various preparation conditions would be quite beneficial. Due to the abundance of data available, deep learning may be used to recognize and draw conclusions about the relationship between the properties of biomass treatment and performance. This will greatly aid lab work in producing the needed carbon materials in the least period of time. To summarize, in the pursuit of a sustainable future on the planet, the efficient and effective use of the millions of tons of biomass waste produced worldwide for energy storage and conversion will considerably cut the costs and environmental implications.

References

Al Rai, A., & Yanilmaz, M. (2021). High-performance nanostructured bio-based carbon electrodes for energy storage applications. *Cellulose*, *28*(9), 5169–5218. https://doi.org/10.1007/s10570-021-03881-z

Baig, N., Kammakakam, I., & Falath, W. (2021). Nanomaterials: A review of synthesis methods, properties, recent progress, and challenges. *Materials Advances*, *2*(6), 1821–1871. https://doi.org/10.1039/D0MA00807A

Bhat, V. S., Jayeoye, T. J., Rujiralai, T., Sirimahachai, U., Chong, K. F., & Hegde, G. (2021). Acacia auriculiformis-derived bimodal porous nanocarbons via self-activation for high-performance supercapacitors. *Frontiers in Energy Research*, *9*. https://doi.org/10.3389/fenrg.2021.744133

Cai, T., Wang, H., Jin, C., Sun, Q., & Nie, Y. (2018). Fabrication of nitrogen-doped porous electrically conductive carbon aerogel from waste cabbage for supercapacitors and oil/water separation. *Journal of Materials Science: Materials in Electronics*, *29*(5), 4334–4344. https://doi.org/10.1007/s10854-017-8381-5

Cao, M., Cheng, W., Ni, X., Hu, Y., & Han, G. (2020). Lignin-based multi-channels carbon nanofibers @ SnO_2 nanocomposites for high-performance supercapacitors. *Electrochimica Acta*, *345*, 136172. https://doi.org/10.1016/j.electacta.2020.136172

Cao, W., Zhang, E., Wang, J., Liu, Z., Ge, J., Yu, X., Yang, H., & Lu, B. (2019). Potato derived biomass porous carbon as anode for potassium ion batteries. *Electrochimica Acta*, *293*, 364–370. https://doi.org/10.1016/j.electacta.2018.10.036

Chen, C., & Hu, L. (2018). Nanocellulose toward advanced energy storage devices: Structure and electrochemistry. *Accounts of Chemical Research*, *51*(12), 3154–3165. https://doi.org/10.1021/acs.accounts.8b00391

Chen, C., Song, J., Cheng, J., Pang, Z., Gan, W., Chen, G., Kuang, Y., Huang, H., Ray, U., Li, T., & Hu, L. (2020c). Highly elastic hydrated cellulosic materials with durable compressibility and tunable conductivity. *ACS Nano*, *14*(12), 16723–16734. https://doi.org/10.1021/acsnano.0c04298

Chen, H., Liu, T., Mou, J., Zhang, W., Jiang, Z., Liu, J., Huang, J., & Liu, M. (2019). Free-standing N-self-doped carbon nanofiber aerogels for high-performance all-solid-state supercapacitors. *Nano Energy*, *63*, 103836. https://doi.org/10.1016/j.nanoen.2019.06.032

Chen, J., Liu, Y., Liu, Z., Chen, Y., Zhang, C., Yin, Y., Yang, Q., Shi, Z., & Xiong, C. (2020b). Carbon nanofibril composites with high sulfur loading fabricated from nanocellulose for high-performance lithium-sulfur batteries. *Colloids and Surfaces A: Physicochemical and Engineering Aspects*, *603*, 125249. https://doi.org/10.1016/j.colsurfa.2020.125249

Chen, Q., Tan, X., Liu, Y., Liu, S., Li, M., Gu, Y., Zhang, P., Ye, S., Yang, Z., & Yang, Y. (2020d). Biomass-derived porous graphitic carbon materials for energy and environmental applications. *Journal of Materials Chemistry A*, *8*(12), 5773–5811. https://doi.org/10.1039/C9TA11618D

Chen, S., Koshy, D. M., Tsao, Y., Pfattner, R., Yan, X., Feng, D., & Bao, Z. (2018). Highly tunable and facile synthesis of uniform carbon flower particles. *Journal of the American Chemical Society*, *140*(32), 10297–10304. https://doi.org/10.1021/jacs.8b05825

Chen, W., Gong, M., Li, K., Xia, M., Chen, Z., Xiao, H., Fang, Y., Chen, Y., Yang, H., & Chen, H. (2020a). Insight into KOH activation mechanism during biomass pyrolysis: Chemical reactions between O-containing groups and KOH. *Applied Energy*, *278*, 115730. https://doi.org/10.1016/j.apenergy.2020.115730

Cheng, D., Tian, M., Wang, B., Zhang, J., Chen, J., Feng, X., He, Z., Dai, L., & Wang, L. (2020). One-step activation of high-graphitization N-doped porous biomass carbon as advanced catalyst for vanadium redox flow battery. *Journal of Colloid and Interface Science, 572*, 216–226. https://doi. org/10.1016/j.jcis.2020.03.069

Dan, R., Chen, W., Xiao, Z., Li, P., Liu, M., Chen, Z., & Yu, F. (2020). N-doped biomass carbon/reduced graphene oxide as a high-performance anode for sodium-ion batteries. *Energy & Fuels, 34*(3), 3923–3930. https://doi.org/10.1021/acs.energyfuels.0c00058

Jin, C., Nai, J., Sheng, O., Yuan, H., Zhang, W., Tao, X., & Lou, X. W. (David). (2021). Biomass-based materials for green lithium secondary batteries. *Energy & Environmental Science, 14*(3), 1326–1379. https://doi.org/10.1039/D0EE02848G

Kalogianni, D. P. (2021). Nanotechnology in emerging liquid biopsy applications. *Nano Convergence, 8*(1), 13. https://doi.org/10.1186/s40580-021-00263-w

Lee, D. W., Jang, J., Jang, I., Kang, Y. S., Jang, S., Lee, K. Y., Jang, J. H., Kim, H., & Yoo, S. J. (2019). Bio-derived Co 2 P nanoparticles supported on nitrogen-doped carbon as promising oxygen reduction reaction electrocatalyst for anion exchange membrane fuel cells. *Small, 15*(36), 1902090. https://doi. org/10.1002/smll.201902090

Li, L., Hu, X., Guo, N., Chen, S., Yu, Y., & Yang, C. (2021). Synthesis O/S/N doped hierarchical porous carbons from kelp via two-step carbonization for high rate performance supercapacitor. *Journal of Materials Research and Technology, 15*, 6918–6928. https://doi.org/10.1016/j.jmrt.2021.11.076

Ma, L., Bi, Z., Xue, Y., Zhang, W., Huang, Q., Zhang, L., & Huang, Y. (2020). Bacterial cellulose: An encouraging eco-friendly nano-candidate for energy storage and energy conversion. *Journal of Materials Chemistry A, 8*(12), 5812–5842. https://doi.org/10.1039/C9TA12536A

Mujib, S. Bin, Vessalli, B., Bizzo, W. A., Mazon, T., & Singh, G. (2020). Cassava- and bamboo-derived carbons with higher degree of graphitization for energy storage. *Nanomaterials and Energy, 9*(1), 54–65. https://doi.org/10.1680/jnaen.19.00040

Nomura, T., Tanii, S., Ishikawa, M., Tokumoto, H., & Konishi, Y. (2013). Synthesis of hollow zirconia particles using wet bacterial templates. *Advanced Powder Technology, 24*(6), 1013–1016. https://doi. org/10.1016/j.apt.2013.02.009

Powell, M. D., LaCoste, J. D., Fetrow, C. J., Fei, L., & Wei, S. (2021). Bio-derived nanomaterials for energy storage and conversion. *Nano Select, 2*(9), 1682–1706. https://doi.org/10.1002/nano.202100001

Ringe, E. (2020). Shapes, plasmonic properties, and reactivity of magnesium nanoparticles. *The Journal of Physical Chemistry C, 124*(29), 15665–15679. https://doi.org/10.1021/acs.jpcc.0c03871

Santos, V. P., Marques, N. S. S., Maia, P. C. S. V., Lima, M. A. B. de Franco, L. D. O., & de Campos-Takaki, G. M. (2020). Seafood waste as attractive source of chitin and chitosan production and their applications. *International Journal of Molecular Sciences, 21*(12), 4290. https://doi.org/10.3390/ijms21124290

Schmitz, Auza, Koberidze, Rasche, Fischer, Rainer, & Bortesi, Luisa. (2019). Conversion of chitin to defined chitosan oligomers: Current status and future prospects. *Marine Drugs, 17*(8), 452. https://doi.org/10.3390/md17080452

Shaheen, I., Ahmad, K. S., Zequine, C., Gupta, R. K., Thomas, A., & Malik, M. A. (2020). Organic template-assisted green synthesis of CoMoO 4 nanomaterials for the investigation of energy storage properties. *RSC Advances, 10*(14), 8115–8129. https://doi.org/10.1039/C9RA09477F

Wang, Y., Lin, X., Liu, T., Chen, H., Chen, S., Jiang, Z., Liu, J., Huang, J., & Liu, M. (2018). Wood-derived hierarchically porous electrodes for high-performance all-solid-state supercapacitors. *Advanced Functional Materials, 28*(52), 1806207. https://doi.org/10.1002/adfm.201806207

Xia, Y., Zhang, W., Huang, H., Gan, Y., Xiao, Z., Qian, L., & Tao, X. (2011). Biotemplating of phosphate hierarchical rechargeable LiFePO4/C spirulina microstructures. *Journal of Materials Chemistry, 21*(18), 6498. https://doi.org/10.1039/c1jm10481k

Yu, K., Wang, J., Wang, X., Liang, J., & Liang, C. (2020). Sustainable application of biomass by-products: Corn straw-derived porous carbon nanospheres using as anode materials for lithium ion batteries. *Materials Chemistry and Physics, 243*, 122644. https://doi.org/10.1016/j.matchemphys.2020.122644

Zhang, S., Li, H., Wang, S., Liu, Y., Chen, H., & Lu, Z.-X. (2019). Bacteria-assisted synthesis of nanosheet-assembled TiO_2 hierarchical architectures for constructing TiO_2-based composites for photocatalytic and electrocatalytic applications. *ACS Applied Materials & Interfaces*, *11*(40), 37004–37012. https://doi.org/10.1021/acsami.9b15282

Zhao, B., Shao, Q., Hao, L., Zhang, L., Liu, Z., Zhang, B., Ge, S., & Guo, Z. (2018). Yeast-template synthesized Fe-doped cerium oxide hollow microspheres for visible photodegradation of acid orange 7. *Journal of Colloid and Interface Science*, *511*, 39–47. https://doi.org/10.1016/j.jcis.2017.09.077

Zheng, F.-Y., Li, R., Ge, S., Xu, W.-R., & Zhang, Y. (2020). Nitrogen and phosphorus co-doped carbon networks derived from shrimp shells as an efficient oxygen reduction catalyst for microbial fuel cells. *Journal of Power Sources*, *446*, 227356. https://doi.org/10.1016/j.jpowsour.2019.227356

Zhou, J. C., Soto, C. M., Chen, M.-S., Bruckman, M. A., Moore, M. H., Barry, E., Ratna, B. R., Pehrsson, P. E., Spies, B. R., & Confer, T. S. (2012). Biotemplating rod-like viruses for the synthesis of copper nanorods and nanowires. *Journal of Nanobiotechnology*, *10*(1), 18. https://doi.org/10.1186/1477-3155-10-18

9

A Conceptual Approach to Analyse the Behaviour of Nano Materials for Hydrogen Storage

Dr. J. Vidhya

M.Kumarasamy College of Engineering (Autonomous), Karur, India

R. Gayathri

Cauvery College for Women (Autonomous), Tiruchirappalli, India

Efficient and high-capacity storage materials are pivotal for a hydrogen-based economy. Nickel, palladium, and platinum are used most significantly, with uses in a wide range of industrial applications for hydrogen storage in industry. Following a two-pronged strategy, we have found density functional approximations in our study that perform well for hydrogen-binding applications. First, a dataset accurately depicts the hydrogen-binding problem and captures the variety of chemical and mechanistic binding sites found in materials that are used to store hydrogen. The coupled cluster theory–based reference interaction energies are calculated for this dataset. Second, two hybrid density functionals have been established based on the effectiveness of density functional approximations in forecasting H_2 interaction energies. Consistently underbinding density functionals such asrevPBE, BLYP, and B3LYP with the inclusion of empirical dispersion corrections for minor performance gains. To enable high-throughput screening of promising materials, density functional theory is therefore ideally positioned to become a very helpful technique.

9.1 Introduction

As the threat of global warming increases as a result of the use of fossil fuels, our world needs to develop new ways to use unlimited energy sources [1, 2]. A novel form of clean and sustainable energy carrier with various applications is hydrogen [3]. Hydrogen is a fuel that can be instantly transformed into

DOI: 10.1201/9781003355755-9

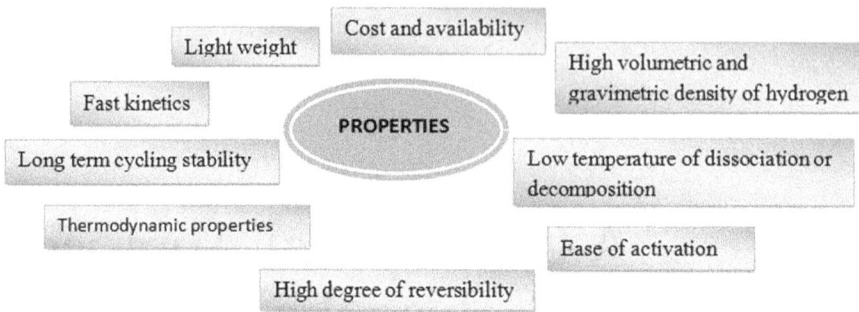

FIGURE 9.1   Properties of hydrogen storage materials.

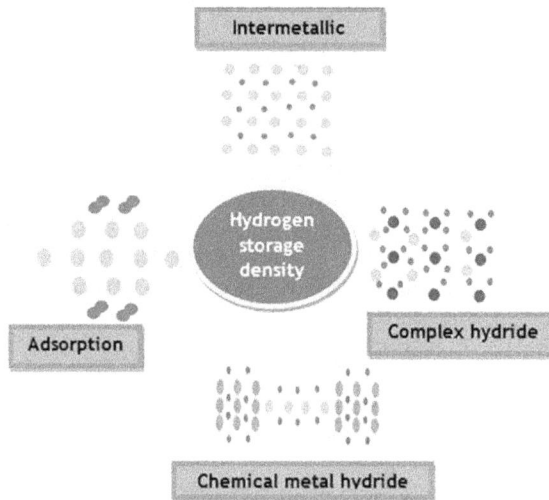

FIGURE 9.2   Hydrogen storage density.

any type of energy without emitting harmful pollutants. It is practical, secure, and versatile. Because it reduces greenhouse gas emissions, lessens the world's reliance on fossil fuels, and increases the energy conversion efficiency of both internal combustion engines and proton exchange membrane fuel cells, hydrogen is the fuel of choice for the future [4, 5]. Fuel cells use hydrogen to instantly convert energy. Figure 9.1 shows the important properties of hydrogen storage materials that need to be studied in automotive applications.

All of the aforementioned characteristics call for a thorough understanding of the underlying mechanistic behaviour of catalyst-containing compounds as well as their atomic-or molecular-scale physicochemical reactivity with hydrogen. Applications for onboard hydrogen storage [6–11] are shown in Figure 9.2.

9.2 Hydrogen Storage

Hydrogen storage is a challenging task. Hydrogen gas has an extremely low density because it is the lightest molecule; 1 kg of hydrogen gas occupies over 11 m^3 at room temperature and atmospheric pressure [12]. As a result, increasing the storage density of hydrogen is required to make it economically viable. There are several methods for storing hydrogen at a higher density. All of these approaches, however, necessitate some type of energy input, whether in the form of work, heat, or, in some circumstances, hydrogen-binding materials.

9.3 Importance of the Hydrides in Hydrogen Storage

9.3.1 Metal Hydrides

Hydrogen is chemically bound to metal hydrides. Compared to the physical ones involved in hydrogen adsorption, these connections are significantly more potent. The chemically bound hydrogen must therefore be released with additional energy. On the other hand, the stronger bonds make it possible for hydrogen to be maintained at a high density even at room temperature [13]. There are two ways to release hydrogen from metal hydrides: by heating (thermolysis) or by interacting with water (hydrolysis). Thermolysis is endothermic, whereas hydrolysis is exothermic; it may sometimes be reversed, but hydrolysis cannot; it occurs in the solid phase, whereas hydrolysis occurs in solution; it also requires high temperatures, whereas hydrolysis can happen spontaneously at normal temperatures [14].

9.3.2 Elemental Metal Hydrides

Most of the metallic elements, known as elemental hydrides, can combine with hydrogen to form binary compounds. However, the majority of devices are unsuitable for hydrogen storage due to thermodynamics, hydrogen storage capacity, or both [14]. The elemental metal hydrides are deemed most promising for large-scale hydrogen storage (Figure 9.3).

9.3.3 Intermetallic Hydrides

A few elemental metal hydrides possess properties that make them suitable for storing hydrogen. Intermetallic hydrides are used because they have hydrogen storage capabilities that are midway between those of their component components [15]. This alloy, AxByHz, is made up of one element A that binds hydrogen strongly and one element B that binds hydrogen weakly. Because both the choice of elements A and B, as well as their ratio in the alloy (x/y), can vary, numerous unique intermetallic hydrides are feasible. In actuality, intermetallic metal hydride complexes with the crystal shapes AB_5, AB_2, and AB that are used most frequently applied in hydrogen storage [16].

9.3.4 Complex Metal Hydrides

In complex metal hydrides, hydrogen is incorporated as a component of a complex anion that is bonded to a metallic cation. The primary kinds of complex hydrides of interest for hydrogen storage are alanates (based on the $[AlH_4]$ anion), borohydrides (based on the $[BH_4]$ anion), and amides (based on the $[NH_2]$ anion) [17]. Contrary to intermetallic hydrides, complex metal hydrides are primarily composed of light

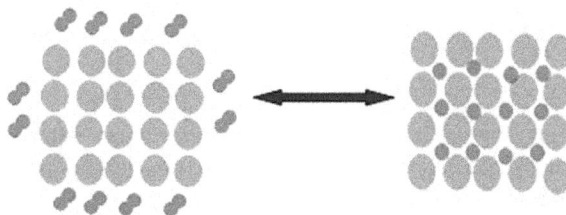

FIGURE 9.3 Hydrogen occupied interstitial sites of metal.

elements. This makes it possible for the complex metal hydrides to have astonishingly high gravimetric hydrogen storage capacities, which has prompted a considerable amount of interest in their potential usage in FCV applications [18]. Unfortunately, only a small number of the more complex hydrides can be dehydrogenated by thermolysis since it requires high temperatures.

9.3.5 Chemical Hydrides

Hydrogen is chemically bonded by chemical hydrides, as with metal hydrides. The properties of chemical hydrides, in contrast, differ greatly from those of metal hydrides since they are composed of lighter elements. The most noteworthy distinction is that chemical hydrides are often liquids under normal circumstances, which makes heat and mass transfer during dehydrogenation and hydrogenation processes more simpler as well as transportation and storage. The chemical hydrides proposed for hydrogen storage include methanol, ammonia, and formic acid. These are alreadyexisting bulk compounds that are routinely made from natural gas. To put it another way, these chemicals' utility extends beyond hydrogen storage. The necessary infrastructure for the manufacture, handling, and transportation of these compounds is already in place in large part due to the widespread production of these compounds. Producing these bulk chemicals with hydrogen obtained from water electrolysis rather than natural gas reforming is thus not only useful for storing hydrogen but also minimises the use of fossil fuels in bulk chemical synthesis. Furthermore, it is important to note that different chemical hydrides have been suggested as hydrogen substitutes rather than hydrogen storage materials [19].

9.4 Role of Nickel, Platinum, and Palladium Nanoparticles in Hydrogen Storage Devices

9.4.1 Nickel

The most important metal, nickel, is used in numerous sectors. Nickel metal hydride (NiMH) and nickel cadmium (NiCd) batteries are used in the majority of plug-in hybrid vehicles (PHEVs), and it is predicted that the rise in demand for these batteries will drive down prices by 40% in the upcoming year [20]. By 2025, it's anticipated that the nickel content of lithium-ion batteries would increase to 58% from its current level of roughly 39%. Additionally, nickel is used in the absorber component of carbon capture systems to withstand the corrosiveness of wet flue gas, which is now more frequently produced [21]. Furthermore, stainless steel, one of the most often used building materials, contains more than 60% nickel. Nanocrystals [22], nanochains [23], flower-like nanostructures [24], nanobeads [25], nanowires [26], and microspheres [27] are a few of the different types of nickel superfine particles that have been developed over time. In addition to pure nickel particles, Ni/CNT nanocomposites [28] and Fe-Ni alloy powders [29] have also been created. These nickel superfine particles have been produced using a variety of reduction techniques, including hydrothermal [30], solvothermal [31], water quenching [32], ethanol water system [33], gas phase [34], liquid phase [35], ultrasonic [36], microwave-assisted [37], and selective leaching methods [38–43]. The superfine particle was created using nickel salts such nickel acetate, nickel chloride, nickel acetylacetonate, and nickel nitrate, among others.

9.4.2 Palladium

Pd-based nanoparticles play crucial roles in the purification, storage, detection, and fuel cell applications of hydrogen. A hydrogen economy is anticipated to be enabled and developed in large part because ofPd-based nanomaterials. Cost and availability will be the main determining factors in the scope of use for these nanoparticles. To maximise the catalytic and hydrogen absorbing efficiencies of Pd-based nanomaterials, the following three goals should be prioritised in research on these materials: (i) design new synthesis methods that provide precise control over product morphology and composition; (ii) design and fabricate robust nanoscale palladium entities that can endure under the challenging circumstances; and (iii) enhance our understanding to a point where it will be able to successfully tailor high-performance Pd-based nanomaterials by design. Metal hydrides have recently been the subject of extensive research aiming at improving their adsorption/desorption properties for hydrogen storage. To better understand bimetallic or trimetallic Pd-based hydride systems, it may be helpful to develop more sophisticated computational methods for the detailed analysis of hydride synthesis and hydrogen desorption. To summarise, cost, hydrogen embrittlement, poisoning species, hydrogen solubility, and hydrogen adsorption strength are critical areas for exploratory research in order to develop functional Pd-based nanomaterials that will facilitate technologies that will drive the hydrogen economy.

9.4.3 Platinum

Platinum (Pt) has sparked a lot of attention because it is the best catalyst for a variety of applications. As a noble metal, Pt exhibits the highest catalytic activity of all the pure metals, particularly in fuel cells [44]. The size, distribution, and manufacturing procedure of Pt particles all have an effect on their electrocatalytic activity [45, 46]. The catalytic activity of Pt is enhanced by the generation of nanosized Pt particles. High electrocatalytic activity can be obtained by utilising a nanosized Pt catalyst that has been properly disseminated [44, 47, 48]. Colloidal systems [49], reduction [50], leveraging bacterial cellulose (BC) as a hydrophilic matrix [51], microemulsion [52], sol–gel [53], sonochemical technique [54], and electrodeposition have all been utilised to manufacture nanosized Pt particles.

9.5 Carbon with Metal Hydrides

As a result, an essential topic is whether a material is best for use as an additive or support for metal hydrides. Because a structural additive or a support/scaffold constitutes such a large portion of the composite material, only light and low-cost materials are appropriate. Carbon materials excel in this regard because they are light, inert, and abundant. Carbon's capacity to bond with itself (and other atoms) in various hybridisations, resulting in allotropes with radically distinct characteristics, is a fascinating phenomenon. Aside from allotropes, there are a variety of carbon materials that have a variety of intriguing features. In addition, carbon is a good heat conductor (Figure 9.4).

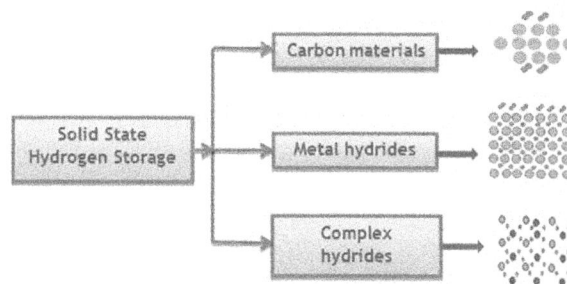

FIGURE 9.4 Carbon with metal hydrides.

9.6 Gaussian Program Implementation

Carbon-based nanomaterials such asgraphene, fullerene, and coronene have proven to be promising materials for solid-state hydrogen storage. Hydrogen storage on carbon-based materials provides a safer alternative to current methods that require extreme temperatures or pressures. The low hydrogen storage capacity in carbon materials, such as graphene, is due to hydrogen being stored through weak van der Waals forces. Several studies have shown that one possible method for increasing hydrogen storage capacity on carbon-based materials is by metal atom adsorption. In metal-doped carbon systems, the metal atom is relaxed on the surface as an intercalated system. The charge is exchanged between the hydrogen and the metal atom during the bond interaction, which typically absorbs the hydrogen.

The GAUSSIAN 09 software is used for the energy calculations in this study about coronene exfoliation rates and hydrogen storage applications. The GAUSSIAN09 software package is used here to relax the structure geometries, calculate system energies, and obtain the necessary energy barriers for exfoliation. The GAUSSIAN09 program provides standard molecular modelling methods such asab initio, density functional, semi-empirical, and empirical methods. Molecular mechanics is an example of the empirical method. Molecular mechanics force fields available within GAUSSIAN include AMBER, UFF, and DRYING. On the basis of direct ab initio molecular dynamics (AIMD) calculations, we can theoretically design a hydrogen storage device with reversible H_2 adsorption–desorption properties using carbon materials, along with nickel, palladium, and platinum systems.

9.7 Conclusion

Metal hydrides are a type of material that can be employed in numerous clean energy applications, including enabling grid and renewable-energy technologies and enhancing building and vehicle energy efficiency. It is noteworthy that metal hydrides can take part in the full "energy cycle," from harvesting renewable energy (solar, wind, and water) to storing it for use in power production. A specific application's requirements are satisfied by altering the metal hydride's composition, which controls the working temperature and pressure. To fill in the gaps in the energy cycle, many metal hydride–based technologies are emerging. With more governmental and corporate investments, a faster shift to a renewable-energy economy is anticipated. This dataset allows for the evaluation of density functionals to determine whether they are appropriate for applications involving hydrogen storage or to test the functionality of non-covalent interactions.

References

[1] S. Satyapal, J. Petrovic, and G. Thomas, "Gassing up with hydrogen," *Scientific American*, vol. 296, no. 4, pp. 80–87, 2007.

[2] M. S. Dresselhaus and I. L. Thomas, "Alternative energy technologies," *Nature*, vol. 414, no. 6861, pp. 332–337, 2001.

[3] B. Sakintuna, F. Lamari-Darkrim, and M. Hirscher, "Metal hydride materials for solid hydrogen storage: A review," *International Journal of Hydrogen Energy*, vol. 32, no. 9, pp. 1121–1140, 2007.

[4] E. K. Stefanakos, D. Y. Goswami, S. S. Srinivasan, and J. T. Wolan, "Hydrogen energy," in *Environmentally Conscious Alternative Energy Production*, M. Kutz, Ed., vol. 4, chapter 7, pp. 165–206, John Wiley & Sons, New York, NY, 2007.

[5] S. A. Sherif, F. Barbir, T. N. Vieziroglu, M. Mahishi, and S. S. Srinivasan, "Hydrogen energy technologies," in *Handbook of Energy Efficiency and Renewable Energy*, F. Kreith and D. Y. Goswami, Eds., chapter 27, CRC Press, Boca Raton, FL, 2007.

[6] A. M. Seayad and D. M. Antonell, "Recent advances in hydrogen storage in metal-containing inorganic nanostructures and related materials," *Advanced Materials*, vol. 16, no. 9–10, pp. 765–777, 2004.

[7] F. E. Pinkerton and B. G. Wicke, "Bottling the hydrogen genie," *The Industrial Physicist*, vol. 10, no. 1, pp. 20–23, 2004.

[8] F. Schüth, "Technology: Hydrogen and hydrates," *Nature*, vol. 434, no. 7034, pp. 712–713, 2005. F. Schüth, B. Bogdanović, and M. Felderhoff, "Light metal hydrides and complex hydrides for hydrogen storage," *Chemical Communications*, vol. 10, no. 20, pp. 2249–2258, 2004.

[9] N. B. McKeown, S. Makhseed, K. J. Msayib, L.-L. Ooi, M. Helliwell, and J. E. Warren, "A phthalocyanine clathrate of cubic symmetry containing interconnected solvent-filled voids of nanometer dimensions," *Angewandte Chemie International Edition*, vol. 44, no. 46, pp. 7546–7549, 2005.

[10] M. Fichtner, "Nanotechnological aspects in materials for hydrogen storage," *Advanced Engineering Materials*, vol. 7, no. 6, pp. 443–455, 2005.

[11] A. G. Wong-Foy, A. J. Matzger, and O. M. Yaghi, "Exceptional H2 saturation uptake in microporous metal-organic frameworks," *Journal of the American Chemical Society*, vol. 128, no. 11, pp. 3494–3495, 2006.

[12] L. Schlapbach and A. Zuttel, "Hydrogen-storage materials for mobile applications," *Nature*, vol. 414, no. 6861, pp. 353–8, 2001.

[13] J.B. Von Colbe, J.R. Ares, J. Barale, M. Baricco, C. Buckley, G. Capurso, et al., "Application of hydrides in hydrogen storage and compression: Achievements, outlook and perspectives," *International Journal of Hydrogen Energy*, vol. 44, no. 15, pp. 7780–808, 2019.

[14] L. E. Klebanoff, K. C. Ott, L. J. Simpson, K. O'Malley, and N. T. Stetson, "Accelerating the understanding and development of hydrogen storage materials: A review of the five-year efforts of the three DOE hydrogen storage materials centers of excellence," *Metallurgical and Materials Transactions*, vol. 1, no. 2, pp. 81–117, 2014.

[15] G. Sandrock, "A panoramic overview of hydrogen storage alloys from a gas reaction point of view," *Journal of Alloys and Compounds*, vol. 293, no. Supplement C, pp. 877–88, 1999.

[16] D. Chandra, "Intermetallics for hydrogen storage," in *Solidstate Hydrogen Storage*, Woodhead Publishing, UK, pp. 315–56, 2008.

[17] S. I. Orimo, Y. Nakamori, J. R. Eliseo, A. Züttel, and C. M. Jensen, "Complex hydrides for hydrogen storage," *Chemical Reviews*, vol. 107, no. 10, pp. 4111–32, 2007.

[18] J.M. Pasini, C. Corgnale, B. A. van Hassel, T. Motyka, S. Kumar, and K. L. Simmons, et al., "Metal hydride material requirements for automotive hydrogen storage systems," *International Journal of Hydrogen Energy*, vol. 38, no. 23, pp. 9755–65, 2013.

[19] A. Goeppert, M. Czaun, J. P. Jones, G. S. Prakash, and G. A. Olah, "Recycling of carbon dioxide to methanol and derived products closing the loop," *Chemical Society Reviews*, vol. 43, no. 23, pp. 7995–8048, 2014.

[20] J.C. Clare Richardson, P. Chhabra, G. Coates, P. Kelly-Detwiler, C. Leonida, G. Moe, et al., "Powering future," *Nickel Magazine*, vol. 2017, p. 3, 2017.

[21] C. Richardson, "Five years of carbon capture and storage," *Nickel Magazine*, vol. 2019, p. 2, 2019.

[22] H.T. Zhang, G. Wu, X.H. Chen, and X.G. Qiu, "Synthesis and magnetic properties of nickel nanocrystals," *Materials Research Bulletin*, vol. 41, no. 3, pp. 495–501, 2006.

[23] H. Lin, Z. Wangzhi, Z. Wei, D. Honglin, C. Chinping, and G. Lin, "Size-dependent magnetic properties of nickel nanochains," *Journal of Physics: Condensed Matter*, vol. 19, no. 3, p. 036216, 2007.

[24] L. Qiang and D. Weimin, "Large-scale synthesis and catalytic properties of nearly monodispersive nickel 3D nanostructures," *Rare Metals Materials And Engineering*, vol. 38, no. 12, pp. 2080–2084, 2009.

[25] M. Benelmekki, A. Montras, A.J. Martins, P.J.G. Coutinho, and L.M. Martinez, "Magnetophoresis behaviour at low gradient magnetic field and size control of nickel single core nanobeads," *Journal of Magnetism and Magnetic Materials*, vol. 323, no. 15, pp. 1945–1949, 2011.

[26] X. Li, H. Wang, K. Xie, Q. Long, X. Lai, and L. Liao, "Self-assembly mechanism of Ni nanowires prepared with an external magnetic field," *Beilstein Journal of Nanotechnology*, vol. 6, pp. 2123–2128, 2015.

[27] Y. Teng, L.X. Song, W. Liu, L. Zhao, J. Xia, and Q.S. Wang, et al., "Facile one-pot synthesis of highly mono disperse nickel microspheres with raised nickel dots and their adsorption performance for heavy metal ions," *Dalton Transactions*, vol. 45, no. 23, pp. 9704–9711, 2016.

[28] X. Zhang, W. Jiang, D. Song, J. Liu, and F. Li, "Preparation and catalytic activity of Ni/CNTs nanocomposites using microwave irradiation heating method," *Materials Letters*, vol. 62, no. 15, pp. 2343–2346, 2008.

[29] Z. Xu, C. Jin, A. Xia, J. Zhang, and G. Zhu, "Structural and magnetic properties of nanocrystalline nickel-rich Fe–Ni alloy powders prepared via hydrazine reduction," *Journal of Magnetism and Magnetic Materials*, vol. 336, pp. 14–19, 2013.

[30] F. Zhang, Y. Chen, J. Zhao, and H. Li, "Preparation of nanosized nickel particles by hydrothermal method," *Chemistry Letters*, vol. 33, no. 2, pp. 146–147, 2004.

[31] M. Khizar Shafique, T. Muhmood, S. Lin, H. Xiaobin, and S. Gull, "Synthesis of superfine paramagnetic nickel particles using anionic and cationic surfactants," *Materials Research Express*, vol. 6, p. 108001, 2019.

[32] A.S. Bolokang and M.J. Phasha, "Novel synthesis of metastable HCP nickel by water quenching," *Materials Letters*, vol. 65, no. 1, pp. 59–60, 2011.

[33] H.-G. Zheng, J.-H. Liang, J.-H. Zeng, and Y.-T. Qian, "Preparation of nickel nanopowders in ethanol-water system (EWS)," *Materials Research Bulletin*, vol. 36, no. 5, pp. 947–952, 2001.

[34] W. Gao, S. Shen, and Y. Cheng, "Synthesis of dispersed superfine fcc nickel single crystals in gas phase," *Journal of Physical Chemistry C*, vol. 117, no. 18, pp. 9223–9228, 2013.

[35] L. Lei, D. Jinghong, G. Guoyou, Y. Jikang, Z. Jiamin, L. Yichun, et al., "Study on preparation technology of nickel powder with liquid phase reduction method," *Rare Metals Materials And Engineering*, vol. 44, no. 1, pp. 36–40, 2015.

[36] Y. Zhu, Y. Liu, Y. Gao, Q. Cheng, L. Zhao, and Z. Yang, "Magnetic properties of aristate spherical Ni nanoparticles synthesized through ultrasound reduction method," *Materials Research Bulletin*, vol. 87, pp. 135–139, 2017.

[37] S.B. Kashid, R.W. Raut, and Y.S. Malghe, "Microwave assisted synthesis of nickel nanostructures by hydrazine reduction route: Effect of solvent and capping agent on morphology and magnetic properties," *Materials Chemistry and Physics*, vol. 170, pp. 24–31, 2016.

[38] A. Michalcova, P. Svobodova, R. Novakova, A. Len, O. Heczko, D. Vojtěch, et al., "Structure and magnetic properties of nickel nanoparticles prepared by selective leaching," *Materials Letters*, vol. 137, pp. 221–224, 2014.

[39] Z. Libor and Q. Zhang, "The synthesis of nickel nanoparticles with controlled morphology and SiO_2/Ni core-shell structures," *Materials Chemistry and Physics*, vol. 114, no. 2, pp. 902–907, 2009.

[40] B. Ebinand S. Gurmen, "Synthesis and characterization of nickel particles by hydrogen reduction assisted ultrasonic spray pyrolysis (USP-HR) method," *Kona: Powder Science and Technology in Japan*, vol. 29, pp. 134–140, 2011.

[41] Y. Pan, R. Jia, J. Zhao, J. Liang, Y. Liu, and C. Liu, "Size-controlled synthesis of monodisperse nickel nanoparticles and investigation of their magnetic and catalytic properties," *Applied Surface Science*, vol. 316, pp. 276–285, 2014.

[42] B. D. Adams and A. Chen, "The role of palladium in a hydrogen economy," *Materials Today*, vol. 14, pp. 282–289, 2011.

[43] D.-H. Lim, W.-D. Lee, and H.-I. Lee, "Highly dispersed and nano-sized Pt-based electrocatalysts for low-temperature fuel cells," *Catalysis Surveys from Asia*, vol. 12, no. 4, pp. 310–325, 2008.

[44] Y. Takasu, N. Ohashi, X.-G. Zhang, et al., "Size effects of platinum particles on the electroreduction of oxygen," *Electrochimica Acta*, vol. 41, no. 16, pp. 2595–2600, 1996.

[45] K. Bergamaski, A. L. N. Pinheiro, E. Teixeira-Neto, and F. C. Nart, "Nanoparticle size effects on methanol electrochemical oxidation on carbon supported platinum catalysts," *The Journal of Physical Chemistry B*, vol. 110, no. 39, pp. 19271–19279, 2006.

[46] Z. He, J. Chen, D. Liu, H. Tang, W. Deng, and Y. Kuang, "Deposition and electrocatalytic properties of platinum nanoparticals on carbon nanotubes for methanol electrooxidation," *Materials Chemistry and Physics*, vol. 85, no. 2–3, pp. 396–401, 2004.

[47] C. V. Rao and B. Viswanathan, "Monodispersed platinum nanoparticle supported carbon electrodes for hydrogen oxidation and oxygen reduction in proton exchange membrane fuel cells," *Journal of Physical Chemistry C*, vol. 114, no. 18, pp. 8661–8667, 2010.

[48] M. P. Pileni, "Nanosized particles made in colloidal assemblies," *Langmuir*, vol. 13, no. 13, pp. 3266–3276, 1997.

[49] S.-H. Wu and D.-H. Chen, "Synthesis and characterization of nickel nanoparticles by hydrazine reduction in ethylene glycol," *Journal of Colloid and Interface Science*, vol. 259, no. 2, pp. 282–286, 2003.

[50] B. R. Evans, H. M. O'Neill, V. P. Malyvanh, I. Lee, and J. Woodward, "Palladium-bacterial cellulose membranes for fuel cells," *Biosensors and Bioelectronics*, vol. 18, no. 7, pp. 917–923, 2003.

[51] M. Sanchez-Dominguez, M. Boutonnet, and C. Solans, "A novel approach to metal and metal oxide nanoparticle synthesis: The oil-in-water microemulsion reaction method," *Journal of Nanoparticle Research*, vol. 11, no. 7, pp. 1823–1829, 2009.

[52] H. B. Suffredini, G. R. Salazar-Banda, and L. A. Avaca, "Carbon supported electrocatalysts prepared by the sol-gel method and their utilization for the oxidation of methanol in acid media," *Journal of Sol-Gel Science and Technology*, vol. 49, no. 2, pp. 131–136, 2009.

[53] C. A. Angelucci, M. D'Villa Silva, and F. C. Nart, "Preparation of platinum-ruthenium alloys supported on carbon by a sonochemical method," *Electrochimica Acta*, vol. 52, no. 25, pp. 7293–7299, 2007.

[54] J. M. Sieben, M. M. E. Duarte, and C. E. Mayer, "Supported Pt and Pt-Ru catalysts prepared by potentiostatic electrodeposition for methanol electrooxidation," *Journal of Applied Electrochemistry*, vol. 38, no. 4, pp. 483–490, 2008.

10

Investigation of Nanomaterials: An Energy Storage and Conversion Device

Dr. V. Koushick

Vel Tech Rangarajan Dr. Sagunthala R&D Institute of Science and Technology, Chennai, India

J. Eindhumathy

Saranathan College of Engineering, Tiruchirappalli, India

K. S. Vinod

Vel Tech Rangarajan Dr. Sagunthala R&D Institute of Science and Technology, Chennai, India

C. Divya

Centre for Information Technology and Engineering, Manonmaniam Sundaranar University, Tirunelveli, India

Nanoscience and nanotechnology have infiltrated substantially all realms of science and technology. Knowledge of the underlying processes at the nanoscale, as well as the capacity to govern and switch resources at the fissionable and molecular level (nanometre series), has opened up new options in recent decapods. The knowledge gathered can then be used to develop innovative techniques that result in enhanced product design or production. More crucially, new-fangled systematic events and progressions have evolved that have the potential to give moreover radical or creative explanations to humanity's vigour, ecological, and bearable transportation concerns in the twenty-first century. Lithium-ion batteries are one of the most astonishing capabilities of contemporary supplies electrochemistry. Supercapacitors are vital for maintaining a system's voltage with escalating lots in all from transferrable apparatus to rechargeable cars. Fuel cell technologies are on the verge of becoming commercially viable, particularly in the arenas of transferrable supremacy bases –the creation of fragmented and remote electrical energy.

DOI: 10.1201/9781003355755-10

10.1 Introduction

Nanotechnology is the learning of matter edifices with sizes of a billionth of a metre or smaller. People learned that different materials had different qualities depending on their size and form when nanotechnology was introduced. It all started that 'There is adequately of area at the bottom'. Nanomaterials combine macroscopical solids, atomic systems, and molecular systems. Materials with as a minimum one exterior measurement of 100 nmor inner edifices of 100 nmor less are called nanomaterials. Nanomaterials having the similar chemical maquillage as recognised unpackaged materials can have different physicochemical properties.

Materials lowered to the nanoscale can have quite unlike possessions than macroscale materials. For instance, dense materials can convert clear (copper), inert materials can develop catalysts (platinum), stable materials can develop flammable (aluminium), solids at room temperature can convert liquids (gold), and insulators can develop conductors (silicon). Nanomaterials are more than just the next stage in the shrinking of materials or particles. They usually demand a variety of manufacturing methods. 'Top-down' and 'bottom-up' are two approaches for making nanomaterials of various sizes. Top-down techniques for manufacturing extremely trivial edifices from bigger amounts of material, such as engraving to produce circuits on the shallow of a silicon microchip, can be used to create nanomaterials. They can also be built from the ground up, particle by particle, or bit by bit. Self-assembly is one method in which particles or bits organise themselves into an edifice based on their inherent possessions. Self-assembly can be seen in semiconductor crystals as well as in the chemical production of massive molecules. While 'positional assembly' gives you greater control over the construction process, it's still time-consuming and unsuitable for industrial application.

Nanomaterials vary from substance materials in that they have a great fraction of superficial particles, a high superficial vigour, three-dimensional (3D) confinement, and fewer flaws. As the global energy crisis intensifies, an increasing number of governments and individuals are focusing on the group and use of renewable energy sources like gale, aquatic, and astral power. In the interim, scientists are getting more interest in developing a trustworthy new energy storage technology. To find low-cost, spotless, lithe, clear, and high-performing materials or to create novel structures towards harness this novel energy. Supercapacitors [1–5] and lithium-ion batteries [6–11] both offer advantages as green energy sources, such as high vigour mass, high power density, extended service life, no contamination, and so on. As a result, a lot of study has been done on lithium-ion batteries and supercapacitors.

This chapter will describe the present state of lithium-ion battery and supercapacitor research and development. Because electrode materials are easily removed from the collector, most earlier energy storage systems are stiff and cannot fulfil today's strict criteria. Then it has an impact on electrochemical performance, can cause a short circuit, and poses serious safety concerns as well as a significant amount of waste while twisting and folding. Lithe and clear vigour stowing systems have thus become more important in applications such as transferrable vesture devices, LED [12], transistors [13], vigour storing smart windows, gas sensors [14], and others.

10.2 Nanomaterials

If 50% or additional of the component atoms in the numerical extent supply have one or more exterior sizes in the extent range 1–100 nm, the substance is a nanomaterial. In slightly additional widely used extent supply material [15], such as shallow area, capacity, physique, or dispersed light concentration, a proportion of 50% in a mathematical extent delivery with one or more exterior sizes among 1 and 100 nm is always less than 50%. It could be a minuscule percentage of the substance's overall mass in reality. Even if a product contains nanomaterials or emits nanoparticles as a result of use or ageing, it is not a nanomaterial except it is a particulate matter substantial that meets the atom extent and segment criteria.

Under certain conditions, the capacity precise shallow zone (CPSZ) can be rummage-sale to determine if a substance is a nanomaterial. The total surface area of all atoms is alienated by the total volume

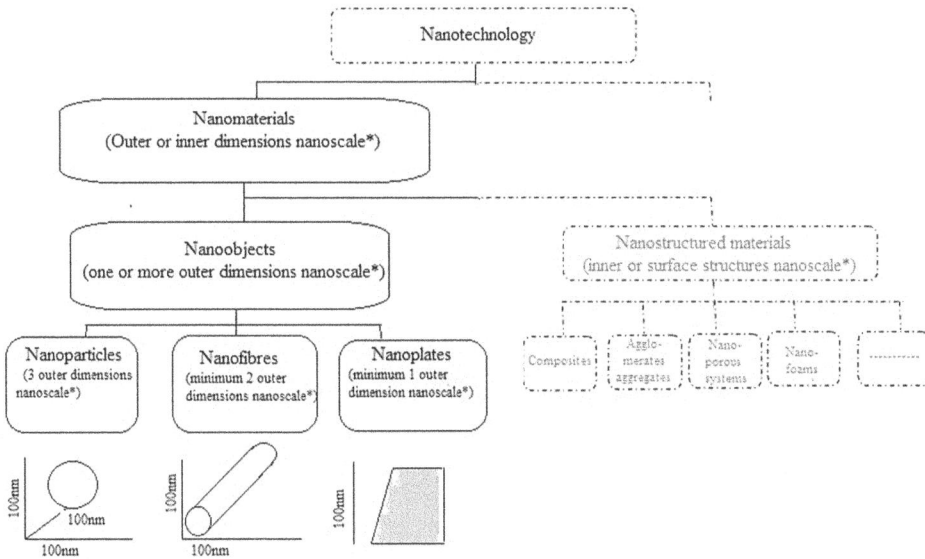

FIGURE 10.1 Structure of nanomaterials.

of all atoms to get CPSZ. Unless the atoms are absorbent or have uneven exteriors, CPSZ > 60 m^2/cm^3 is expected to be a strong gauge that a material is a nanomaterial; nonetheless, many nanomaterials (rendering to the fundamental size-based principle) will have a CPSZ of lesserthan 60 m^2/cm^3. As a result, the CPSZ > 60 m^2/cm^3 criteria may only be rummage-sale to demonstrate that a material is a nanomaterial, not vice versa. If the atom size supply and atom shap (s) of a sample are recognised in detail, the CPSZ of the sample can be approximated [16]. The size distribution cannot be calculated using the CPSZ value. Figure 10.1 depicts the basic structure of nanomaterials.

10.2.1 Dimensions of Nanomaterials

The quantity of larger-than-nanoscale properties in a material impacts how nanomaterials are classed (100 nm). The nanoscale dimension classification is depicted in Figure 10.2.

As an outcome, all sizes of zero-dimensional (0D) nanomaterials remain leisurely on the nanoscale (no extents are greater than 100 nm). Nanoparticles are the maximum recurrent 0D nanomaterials. One dimension of one-dimensional (1D) nanomaterials is external on the nanoscale. Nanotubes, nanorods, and nanowires all drop inside this group.

Two dimensions are outside the nanoscale in two-dimensional (2D) nanomaterials. Graphene, nanofilms, nanolayers, and nanocoatings are all plate-like materials that fall into this category.

Nanomaterials that are 3D do not have any dimensions that are limited to the nanoscale. This includes bulk powders, nanoparticle dispersions, nanowire and nanotube bundles, and multi-nanolayers.

The dimensionality distribution of nanomaterials in marketed goods is depicted in the graph of Figure 10.3. Nanoparticles account for 78% of all nanoproducts, while 3D nanomaterials account for 85% of all materials [17].

10.2.2 Properties of Nanomaterials

We'll look at a variety of nanomaterials to get a better understanding of their properties. Some nanoparticles' behaviour is well understood, while others pose significant obstacles, as we'll see. For decades, 1D nanomaterials like reedy flicks and bespoke exteriors have been produced and employed in industries

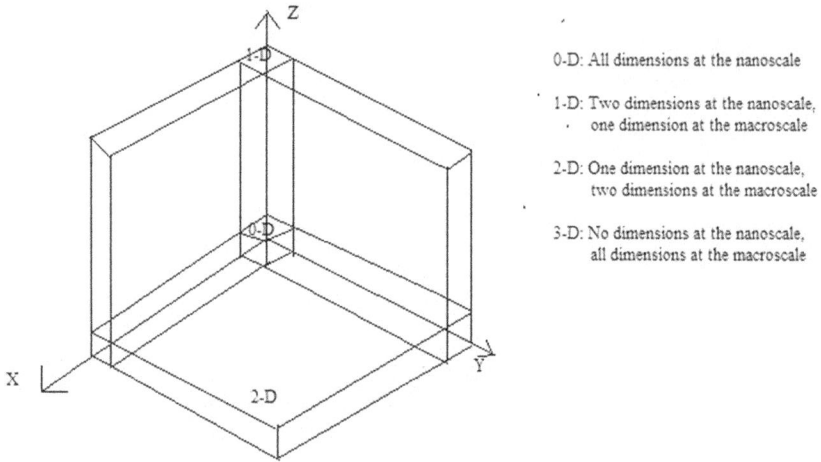

FIGURE 10.2 Classification of nanoscale dimension.

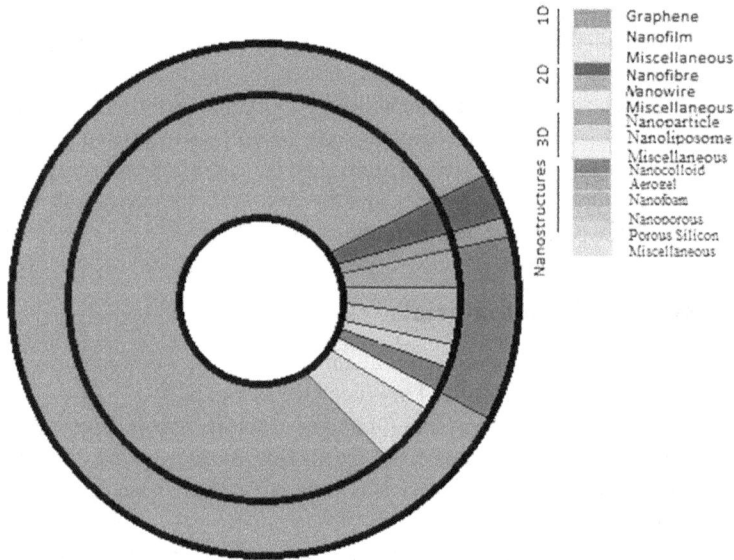

FIGURE 10.3 Distribution of nanomaterials.

like electrical device production, chemistry, and engineering. In the silicon-integratedcircuit sector, for instance, many products depend on reedy films to function, and atomic-level switch of film thickness is typical. Monolayers (layers one particle or bit thick) are also commonly complete and rummage-sale in chemistry. Graphene [18] is the maximum well-known member of this novel class of materials. Even with complex layers (such as lubricants) and nanocoatings, the growth and behaviour of these layers may be predicted from the atomic level upwards. The development of coatings, as well as the control of surface composition and smoothness, is progressing. Engineered surfaces with unique properties like huge shallow part or high reactivity are widely rummage-sale in a variety of applications, including fuel cells and substances. The high shallow part of nanoparticles and their capacity to self-assemble on a provision exterior should assist all of these applications [19]. Surfaces with better properties, although being minor improvements, should find use in the chemical and energy industries. Increased motion and selectivity in reactors and separation processes could have advantages beyond obvious cost and resource savings, allowing for small-scale distributed processing (making chemicals as near as likely to the opinion of use). This is previously taking place in the chemical trade.

10.3 Types of Nanomaterials

Many distinct types of nanomaterials have been created for specific purposes, and many more are predicted to emerge in the future. Intended for the chapter, maximum modern nanomaterials can be divided into four categories:

- Carbon-based materials
- Metal-based materials
- Dendrimers
- Composites

10.3.1 Carbon-Based Materials

Carbon nanostructures in the shape of resonating spheres, ellipsoids, or pipes make up the majority of these nanostructures. Fullerenes are sphere-shaped and spheroidal carbon nanoparticles, although nanotubes are cylinder-shaped carbon nanomaterials. These atoms have possible applications in better films and coverings, solider and igniter materials, and electromagnetic applications [20].

10.3.2 Metal-Based Materials

Nanomaterials comprise major dots, nanogold, nanosilver, and metal oxides, for instance, titanium dioxide. A quantum dot is a small, densely crowded semiconductor crystal covering hundreds or thousands of particles that range in extent from scarce nanometres to scarce hundred nanometres. The optical characteristics of quantum dots alter as their size changes.

10.3.3 Dendrimers

These nanomaterials are made up of nanoscale polymer branching components. Many chains that end on the surface of a dendrimer can be customised to behaviour-specific chemical tasks. This characteristic could be useful in catalysis as well. Besides, since 3D dendrimers have inner spaces into which other particles can be injected, they may be positive for medicine administration.

10.3.4 Composites

Nanoparticles are combined with other nanoparticles or more, bulk-type resources to create composites. Nanoparticles, for instance, nanosized clays, are previously being employed to improve mechanical, thermal, barrier, and flame-retardant qualities in goods ranging from vehicle components to packaging materials.

These nanomaterials feature unique electrical, catalytic, magnetic, mechanical, thermal, or imaging capabilities that remain exceedingly wanted for commercial, medicinal, military, and ecological applications. These materials could be employed in increasingly complicated nanostructures and systems [21]. As new-fangled applications for resources with these particular qualities are discovered, the quantity of goods that use such nanoparticles and their prospective applications increases.

10.4 Types of Nanoparticles

Nanoparticles are categorised into several classes based on their extent, figure, corporeal qualities, and chemical makeup. Carbon nanoparticles, ceramic nanoparticles, metal nanoparticles, semiconductor nanoparticles, polymeric nanoparticles, and lipid nanoparticles are just a scarce instance.

10.4.1 Carbon-Based Nanoparticles

Carbon nanoparticles are ended up mostly of carbon nanotubes (CNTs) and fullerenes. CNTs are graphene pieces folded into tubes. Because these resources are 100 times sturdier than steel, they are largely employed for structural reinforcement. CNTs are alienated into two types: single-walled carbon nanotubes (SWCNTs) and multi-walled carbon nanotubes (MWCNTs). Heat is conducted down the length of CNTs but not across their width. Fullerenes are carbon allotropes with a resonating crate structure of 60 or more carbon particles. The structure of C-60 is buckminsterfullerene, which looks like a resonating football [22]. In these formations, the carbon units are organised in pentagonal and hexagonal patterns. They are employed for commercial purposes for their strong electrical conductivity, edifice, forte, and electron kinship.

10.4.2 Ceramic Nanoparticles

Ceramic nanoparticles are made up of solid oxides, carbides, carbonates, and phosphates. These nanoparticles are chemically inert and exceedingly heat-resistant. Applications include photocatalysis, dye photodegradation, medication administration, and imaging. Ceramic nanoparticles can be used as an excellent drug distribution agent by adjusting various properties such as extent, superficial part, sponginess, superficial-to-capacity ratio, and so on. These nanoparticles have been rummage-sale to treat bacterial infections, glaucoma, cancer, and other diseases with great effectiveness.

10.4.3 Metal Nanoparticles

Metal ancestors are rummage-sale to create metallic nanoparticles. Chemical, electrochemical, and photochemical processes can all be rummage-sale to create these nanoparticles. Metal-ion precursors are reduced in solution using chemical reducing agents, resulting in metal nanoparticles. These have a lot of surface energy and can adsorb very small molecules [23]. Research, biomolecule recognition and imaging, environmental and bioanalytical applications, and ecological and bioanalytical applications are all possible uses for these nanoparticles. Prior to SEM investigation, gold nanoparticles, for example, are employed to coat the material. To boost the electrical current and obtain high-quality SEM images, this is routinely done.

10.4.4 Semiconductor Nanoparticles

Semiconductor nanoparticles are alike to alloys and nonmetals in terms of their characteristics. They are classified as II-VI, III-V, or IV-VI on the periodic table. These atoms have broad bandgaps that change their characteristics depending on how they are modified [24]. Only a few of the applications include photocatalysis, electronics, photo-optics, and water splitting. GaP, GaN, InAs, and InP belong to group III-V semiconductor nanoparticles, whereas ZnS, CdS, ZnO, CdTe, and CdSe fit to group II-VI semiconductor nanoparticles, and Si and Ge fit to group IV semiconductor nanoparticles.

10.4.5 Polymeric Nanoparticles

Organic nanoparticles, also known as polymeric nanoparticles, are nanoparticles comprised of organic compounds. Depending on the manufacturing procedure, these have forms that resemble nanocapsules or nanospheres. The nanosphere atom has a matrix-like construction, whereas the nanocapsular particle has a core-shell shape. In the former, the lively chemicals and polymer are equally disseminated, while in the latter, the lively compounds are contained and then encased by a polymer shell. Polymeric nanoparticles offer a variety of pros, including skilful release, drug molecule defence, the capability to combine treatment with imaging, selective targeting, and then many others. They can be employed in the delivery of drugs as well as diagnostics [25]. Recyclable and biocompatible polymeric nanoparticles are employed in medicine delivery.

10.4.6 Lipid-Based Nanoparticles

Lipid nanoparticles have a diameter of 10–100 nm and are spherical. It is complete with hard phospholipid core and a matrix of solvable lipotropic particles. Surfactants and emulsifiers stabilise the outer essential of these nanoparticles. These nanoparticles are employed in biomedicine as pharmaceutical carriers and distribution systems, and incancer treatment for RNA release. As a result, nanotechnology is far from saturated, with statistics indicating that it is on the verge of an exponential development trend. It roughly corresponds to the 1960s in terms of information technology and the 1980s in terms of biotechnology. As a result, it's easy to forecast that this field will increase at the same rate as the previous two [26].

10.5 Energy Storage Techniques

One of the most remarkable achievements of contemporary materials electrochemistry is lithium-ion batteries [27]. Their science and technology have attracted a lot of interest. A lithium-ion battery is complete with a negative (typically graphite) and a positive lithium-ion intercalation electrode (usually lithium metal oxide, $LiCoO_2$) detached by a lithium-ion conducting electrolyte like $LiPF_6$ in ethylene carbonatediethylcarbonate. Despite the fact that such batteries are economically practical, current electrode and electrolyte materials are approaching their performance limits. More material discoveries, such as the usage of nanomaterials devices, will be required in future peers of rechargeable lithium batteries, not onlyaimed at use in user electronics, but also for long-term energy storage and application in hybrid electric cars. Supercapacitors [28] are rummage-sale the lot from transferrable apparatus to electric cars to keep a system's voltage stable during high loads. The two types of electrochemical supercapacitors are electric double-layer capacitors (EDLC) and redox supercapacitors. Unlike batteries, which consume a finite cycle life owing to periodic reduction and growth of the electrode through cycling, EDLC has an indefinite lifespan due to electrostatic surface charge build-up. Fast faradic charge transport occurs in redox supercapacitors, resulting in a large pseudocapacitance [29]. Switching from conventional to nanostructured electrodes can develop supercapacitor technology.

Because power surpasses energy density, supercapacitors' electrode supplies not as much of stringent than batteries', at any rate in relationto electrode compaction. As a result, the benefits of nanopowders with a large superficial part (main nanoparticles) may be greater, which explains nanopowders' extensive interest and growing use in supercapacitor-based storage devices.

Fuel cell technologies [30] remain on the verge of being commercially viable, particularly trendy in the areas of transferrable influence bases and dispersed and distant electrical energy cohort. Nanostructured materials are previously consuming an influence on processing approaches for low-temperature fuel cells (T200), precious metallic catalyst dispersion, nonprecious catalyst development and dispersion, fuel reformation and hydrogen storage, then membrane-electrode assembly (MEA) fabrication.

Molten carbonates fuel cells (MCFCs), phosphoric acid fuel cells (PAFCs), and solid oxide fuel cells (SOFCs) continue to proposecompensations for immobile applications, particularly cogeneration. Platinum-based catalysts are the most lively resources for hydrogen, reformate, or methanol-fuelled low-temperature fuel cells. To save money, platinum-loading necessity be lowered (although MEA recital is maintained or improved), and continuous techniques for mass-producing MEAs necessity be devised. Scarce approaches are being explored to increase the electrocatalytic activity of Pt-based catalysts. Pt is usually alloyed with changeover alloys or has its atom extent altered.

TiO_2 photochemistry is a rapidly emerging area of study and commercial activity due to its potential uses in environmental protection [31]. The amount of TiO_2 research activity during the last decade is demonstrated by the exponential upsurge in pertinent investigate prose produced (11500 papers among 1993 and 2003) and the quantity of copyrights addressing, for example, photocatalysis (3000 among 1996 and 2001, and additional lately 2500 in Japan then 500 in the United States). These characteristics are responsible for the scientific community's massive basic research effort.

Ecological fortification done heterogeneous catalysis (aquatic purification, air cleaning, and self-cleaning ingredients) and renewable vigour production in photoelectrochemical solar cells and dye-sensitised solar cells (DSSC) are two major environmental priorities that TiO_2's remarkable photoactive properties and, as a result, its frequent applications are linked to [32]. These materials' photo electrochemical characteristics have also been improved by the inclusion of TiO_2 nanostructures.

10.6 Applications

The relations in the correct pilaster of this side provide information on a variety of applications for nanotechnology and nanomaterials. As a result, we won't go over it here. To summarise, nanotechnology is already prevalent in everyday life through commodity objects, and its adoption is accelerating.

The distribution of objects using nanomaterials and nanotechnology across industries is depicted in Figure 10.4.

When total revenue from nanotechnology is split down by sector, it's clear that materials and manufacturing contributed the most, as shown in Figure 10.5. Given that fundamental multi-disciplinary investigation, which in the circumstance of nanotechnology interprets the detection of substantial possessions and the fusion of nanoscale components, is required at the earliest stages of development for any general-purpose technology, such a tendency is to be expected. In the coming years, however, if nanodevices are combined with existing technologies, this tendency should shift more towards application areas [33].

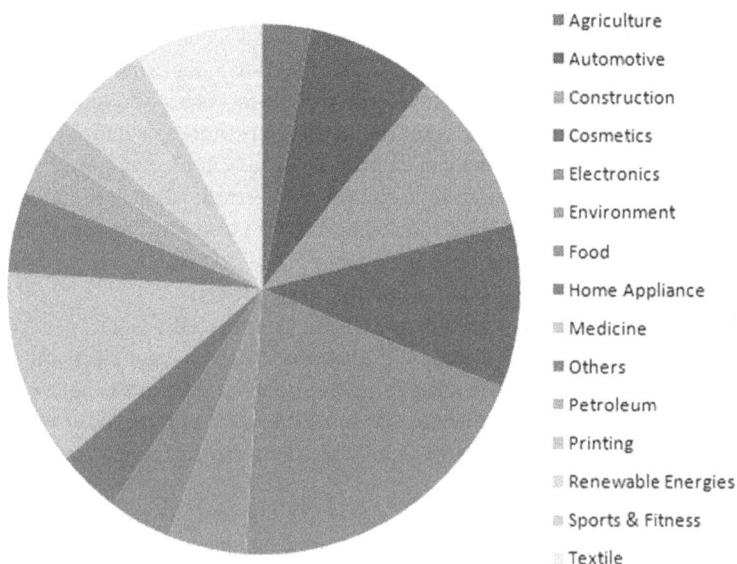

FIGURE 10.4   Development of nanotechnology in industries.

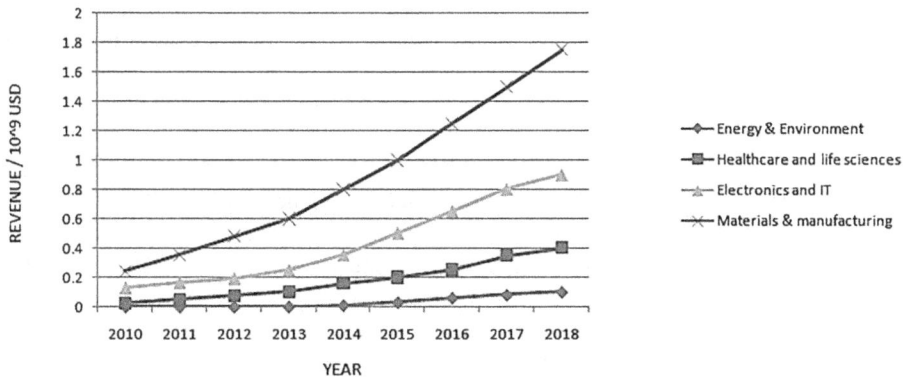

FIGURE 10.5 Revenue generated by nanotechnology and its products (in USD).

10.7 Conclusion

Through the swift advancement of electronic technology, it appears that numerous professionals have a strong need for wearable technology that is both comfortable and multi-functional. One of the most pressing challenges that must be addressed is energy supply equipment. Vigour storage systems require lithium-ion batteries and supercapacitors; however, fit in clear and stretchy possessions in avigour storage device, which has slowed the progress of modern electronic devices, is a significant problem. Materials, physical plan, trade systems, and unified muster of lithe and clear vigour storage devices have all advanced dramatically in recent years. Energy stowage devices that are lithe and translucent can be made in two ways: through revolutionary structural design and the discovery of novel materials. This chapter explains how to make lithe and translucent lithium-ion batteries and supercapacitors. Many unique materials and technologies are employed in lithium-ion batteries and supercapacitors as a consequence of continuing efforts. Carbon nanomaterials such as CNTs, carbon nanofibers, and graphene, as well as their innovative designs, have been used to create energy storage devices. To improve electronic conductivity, CNTs are typically employed as a supporting component. Graphene, on the other hand, has a better electrical property than CNTs. Graphene with a huge precise superficial part can also efficiently control capacity growth during the lithiation procedure, making it an auspicious choice for forthcoming lithe and translucent stowage systems. To achieve this, researchers in the disciplines of electrode physical research, electrode edifice design, and production method development will have to work harder. On the other hand, we anticipate that, as a result of rapid technical breakthroughs and ongoing research efforts, a high-performance energy storage device with unique features will be realised in the near future, considerably improving our quality of life.

References

1. Ge, J.; Cheng, G.; Chen, L. Transparent and flexible electrodes and supercapacitors using polyaniline/single-walled carbon nanotube composite thin films. *Nanoscale* 2011, 3, 3084–3088.
2. Huang, J.Q.; Xu, Z.L.; Abouali, S.; Garakani, M.A.; Kim, J.K. Porous graphene oxide/carbon nanotube hybridfilms as interlayer for lithium-sulfur batteries. *Carbon* 2016, 99, 624–632.
3. Jung, S.; Lee, S.; Song, M.; Kim, D.-G.; You, D.S.; Kim, J.-K.; Kim, C.S.; Kim, T.-M.; Kim, K.-H.; Kim, J.-J.; et al. Extremely flexible transparent conducting electrodes for organic devices. *Adv. Energy Mater.* 2014, 4, 1300474.
4. Jung, H.Y.; Karimi, M.B.; Hahm, M.G.; Ajayan, P.M.; Jung, Y.J. Transparent, flexible supercapacitors fromnano-engineered carbon films. *Sci. Rep.* 2012, 2, 773–778.

5. Kanninen, P.; Luong, N.D.; Anoshkin, I.V.; Tsapenko, A.; Seppala, J.; Nasibulin, A.G.; Kallio, T. Transparentand flexible high-performance supercapacitors based on single-walled carbon nanotube films. *Nanotechnology* 2016, 27, 235403–235409.

6. Cong, H.P.; Ren, X.C.; Wang, P.; Yu, S.H. Flexible graphene–polyaniline composite paper forhigh-performance supercapacitor. *Energy Environ. Sci.* 2013, 6, 1185–1191.

7. Cao, S.; Feng, X.; Song, Y.; Xue, X.; Liu, H.; Miao, M.; Fang, J.; Shi, L. Integrated fast assembly of free-standinglithium titanate/carbon nanotube/cellulose nanofiber hybrid network film as flexible paper-electrode forlithium-ion batteries. *ACS Appl. Mater. Interfaces* 2015, 7, 10695–10718.

8. Chen, T.; Dai, L. Flexible supercapacitors based on carbon nanomaterials. *J. Mater. Chem. A* 2014, 2, 10756–10775.

9. Cai, G.; Darmawan, P.; Cui, M.; Wang, J.; Chen, J.; Magdassi, S.; Lee, P.S. Highly stable transparent conductivesilver grid/PEDOT: PSS electrodes for integrated bifunctional flexible electrochromic supercapacitors. *Adv. Energy Mater.* 2016, 6, 1501882–1501889.

10. Ghosh, A.; Manjunatha, R.; Kumar, R.; Mitra, S. A facile bottom-up approach to construct hybrid flexiblecathode scaffold for high-performance lithium-sulfur batteries. *ACS Appl. Mater. Interfaces* 2016, 8, 33775–33785.

11. Gittleson, F.S.; Hwang, D.; Ryu, W.H.; Hashmi, S.M.; Hwang, J.; Goh, T.; Taylor, A.D. Ultra-thinnanotube/nanowire electrodes by spin–spray layer-by-layer assembly: A concept for transparent energystorage. *ACS Nano* 2015, 9, 10005–10017.

12. Kim, J.-S.; Lee, Y.-H.; Lee, I.; Kim, T.-S.; Ryou, M.-H.; Choi, J.W. Large area multi-stacked lithium-ion batteriesfor flexible and rollable applications. *J. Mater. Chem. A* 2014, 2, 10862–10868.

13. K. Gao; Shao, Z.; Wu, X.; Wang, X.; Zhang, Y.; Wang, W.; Wang, F. Paper-based transparent flexible thin filmsupercapacitors. *Nanoscale* 2013, 5, 5307–5311.

14. Kang, Y.J.; Chun, S.J.; Lee, S.S.; Kim, B.Y.; Kim, J.H.; Chung, H.; Lee, S.Y.; Kim, W. All-solid-state flexiblesupercapacitors fabricated with bacterial nanocellulose papers, carbon nanotubes, and triblock-copolymerion gels. *ACS Nano* 2012, 6, 6400–6406.

15. Liu, L.; Niu, Z.; Zhang, L.; Zhou, W.; Chen, X.; Xie, S. Nanostructured graphene composite papers for highlyflexible and foldable supercapacitors. *Adv. Mater.* 2014, 26, 4855–4862.

16. Lv, T.; Yao, Y.; Li, N.; Chen, T. Highly stretchable supercapacitors based on aligned carbon nanotube/molybdenum disulfide composites. *Angew. Chem.* 2016, 128, 9191–9195.

17. Liu, Y.; Zhou, J.; Chen, L.; Zhang, P.; Fu, W.; Zhao, H.; Ma, Y.; Pan, X.; Zhang, Z.; Han, W. Highlyflexible freestanding porous carbon nanofibers for electrodes materials of high-performance all-carbonsupercapacitors. *ACS Appl. Mater. Interfaces* 2015, 7, 23515–23520.

18. Li, X.; Zhang, C.; Xin, S.; Yang, Z.; Li, Y.; Zhang, D.; Yao, P. A facile synthesis of MoS_2/reduced grapheneoxide@polyaniline for high-performance supercapacitors. *ACS Appl. Mater. Interfaces* 2016, 8, 21373–21380

19. Mukesh, C.; Gupta, R.; Srivastava, D.N.; Nataraj, S.K.; Prasad, K. Preparation of a natural deep eutecticsolvent mediated self-polymerized highly flexible transparent gel having super capacitive behaviour. *RSC Adv.* 2016, 6, 28586–28592.

20. Mendoza-Sanchez, B.; Gogotsi, Y. Synthesis of two-dimensional materials for capacitive energy storage. *Adv. Mater.* 2016, 28, 6104–6135.

21. Niu, Z.; Liu, L.; Zhang, L.; Zhou, W.; Chen, X.; Xie, S. Programmable nanocarbon-based architectures forflexible supercapacitors. *Adv. Energy Mater.* 2015, 5, 1500677–1500696.

22. Niu, Z.; Du, J.; Cao, X.; Sun, Y.; Zhou, W.; Hng, H.H.; Ma, J.; Chen, X.; Xie, S. Electrophoretic build-up ofalternately multilayered films and micropatterns based on graphene sheets and nanoparticles and theirapplications in flexible supercapacitors. *Small* 2012, 8, 3201–3208.

23. Niu, Z.; Luan, P.; Shao, Q.; Dong, H.; Li, J.; Chen, J.; Zhao, D.; Cai, L.; Zhou, W.; Chen, X. A "skeleton/skin" strategy for preparing ultrathin free-standing single-walled carbon nanotube/polyaniline films for high-performance supercapacitor electrodes. *Energy Environ. Sci.* 2012, 5, 8726–8733.

24. Peng, L.; Feng, Y.; Lv, P.; Lei, D.; Shen, Y.; Li, Y.; Feng, W. Transparent, conductive, and flexible mul-tiwalledcarbon nanotube/graphene hybrid electrodes with two three-dimensional microstructures. *J. Phys. Chem. C* 2012, 116, 4970–4978.

25. Peng, L.; Xu, P.; Liu, B.; Wu, C.; Xie, Y.; Yu, G. Ultrathin two-dimensional MnO_2/graphene hybridna-nostructures for high-performance, flexible planar supercapacitors. *Nano Lett.* 2013, 13, 2151–2157.

26. Qiu, T.; Luo, B.; Giersig, M.; Akinoglu, E.M.; Hao, L.; Wang, X.; Shi, L.; Jin, M.; Zhi, L. Au@MnO_2 core-shellnanomesh electrodes for transparent flexible supercapacitors. *Small* 2014, 10, 4136–4141.

27. Qi, D.; Liu, Z.; Liu, Y.; Leow, W.R.; Zhu, B.; Yang, H.; Yu, J.; Wang, W.; Wang, H.; Yin, S. Suspendedwavy graphene microribbons for highly stretchable microsupercapacitors. *Adv. Mater.* 2015, 27, 5559–5566.

28. Rodríguez, J.; Navarrete, E.; Dalchiele, E.A.; Sánchez, L.; Ramos-Barrado, J.R.; Martín, F. Polyvinylpyrrolidone–LiClO4 solid polymer electrolyte and its application in transparent thin film supercapacitors. *J. Power Sources* 2013, 237, 270–276.

29. Wang, K.; Wu, H.; Meng, Y.; Zhang, Y.; Wei, Z. Integrated energy storage and electrochromic func-tion in oneflexible device: An energy storage smart window. *Energy Environ. Sci.* 2012, 5, 8384–8389.

30. Wang, X.; Li, G.; Seo, M.H.; Lui, G.; Hassan, F.M.; Feng, K.; Xiao, X.; Chen, Z. Carbon-coated silicon nanowireson carbon fabric as self-supported electrodes for flexible lithium-ion batteries. *ACS Appl. Mater. Interfaces* 2017, 9, 9551–9558.

31. Wu, Y.; Wu, H.; Luo, S.; Wang, K.; Zhao, F.; Wei, Y.; Liu, P.; Jiang, K.; Wang, J.; Fan, S. Entrapping electrodematerials within ultrathin carbon nanotube network for flexible thin film lithium ion bat-teries. *RSC Adv.* 2014, 4, 20010–20016.

32. Xu, J.; Yang, C.; Xue, Y.; Wang, C.; Cao, J.; Chen, Z. Facile synthesis of novel metal-organic nickel hydroxidenanorods for high performance supercapacitor. *Electrochim. Acta* 2016, 211, 595–602.

33. Yang, Y.; Jeong, S.; Hu, L.; Wu, H.; Lee, S.W.; Cui, Y. Transparent lithium-ion batteries. *Proc. Natl. Acad. Sci. USA* 2011, 108, 13013–13018.

11

Nanomaterials for Supercapacitors

Dr. S. Vijayalakshmi,
Dr. K. Rajkumar,
C. Pearline Kamalini
and R. Sridhar
*Saranathan College of
Engineering, Trichy, India*

The supercapacitor is a kind of energy storage system. Basically, the fast rise in worldwide energy consumption and the conservational impression of conventional energy sources have led to extremely improved research events and clean and renewable energy sources all through the past eras. Till now, researchers are working on batteries, but batteries have a problem that sometimes during the chemical reactions, they can generate some kind of toxic gases which can corrode the cathode and anode, and they also do not have a longer life. Maybe if one uses solid materials for the batteries, it is maybe one-time use sometimes, can recharge the battery, but that is also having some limitations. So, an alternative solution is supercapacitor. In this chapter, we discuss about supercapacitor, how it replaces the conventional batteries, types of supercapacitors, history of supercapacitor, and applications of supercapacitors, especially those made from nanomaterials.

11.1 Introduction

In the daytime, the solar or sunlight generates a huge amount of solar energy. Nevertheless, it is not that each time needs that specific energy into the day; It is used to avail that specific energy in the night time. So, that is why whatsoever the energy producing hooked on the day time are storing that the particular energy either perhaps batteries or perhaps around kind of storage devices or perhaps like Supercapacitors, and as and once essential in the evening, just taking out that particular energy. It is to put up the predictable extraordinary penetration of PV and wind in upcoming grids with lower grid denial loss.

Now, what is the characteristics of energy storage systems? So, first, one is called the storage capacity; storage capacity is nothing but the amount of energy available in the storage systems after charging; in other words, it's like a tank. The second is known as available power, which is represented as an average or peak value and is frequently used to express the highest charge or discharge power. Hence, "available power" refers to the amount of power or energy that is present at any one moment. This is most obvious during the day, particularly in the early morning or late evening when the sun is not yet at its highest point. Our energy generation is less at that particular time. But at 12 o'clock in the noon, one can get the maximum sunlight. So, automatically getting the maximum energy at that particular point. The power transfer rate is the time required to extract the stored energy. That is dependent on the discharge capacity or perhaps capability of the medium in which we have stored the energy.

The second one is efficiency, which is the most crucial factor because it always pertains to energy generation or, possibly, energy storage in terms of efficiency. The fifth one, which is also known as cycling capacity or durability, refers to the maximum number of cycles that an energy storage device may release the energy it has stored after being recharged. For normal battery; it is having already some stored energy; it for certain times, and after certain times it can be called as the battery is dead; that means what? It has released all the energy; that it has already stored inside it. But, now talks about the rechargeable battery; so we are utilizing that battery for a certain time when the battery is getting fully drained. So, what we are doing? Again charging that battery; means, keeping the charge inside it and then after certain time, again utilizing it for a longer time. So, that is known as the cycle capacity, or maybe sometimes we are calling; it as durability. The sixth one is called autonomy: The maximum period of time for which a system may continuously release energy is equal to Pd, where Pd is nothing but the maximum discharge power. That is, energy for use; practicality and flexibility to the producing source; high-efficiency storage systems must be carefully fitted to the types of applications and production [1].

Self-discharge: When a battery is not being used, some of the energy it has stored dissipates. What will happen if you keep a battery for a long time? If it is not used, the charge will automatically be discharged. because every material has a life of its own.; so after a certain time the material will automatically degrade, and automatically it will lose its energy. And next one is the mass and volume density; which refers to the maximum amount of energy stored per unit mass of the storage system or maybe the volume of that particular storage system. So, this is the idea that how to utilize the energy.

The oldest battery type is a lead-acid battery, which is 1 kiloWatt to 100 kiloWatt in size and has high-energy supercapacitors as well as other nickel-cadmium batteries. Nowadays, people are working on the lithium-ion battery, then some nickel-metal hybrid battery, or maybe some kind of flywheels for the tidal energy or high power supercapacitors. When higher energy storage capacity, say suppose 1 Mega Watt to 1 Giga Watt; moving towards the hydrogen fuel cells, cryogenic energy storage, supercapacitors, magnetic energy storage. So, these all are the kind of things; that means, simply increasing its mass and volume density so that one can store the maximum energy inside it.

Now, there are several types of energy storage systems; so basically that depends upon the electrochemical, electrical, chemical, thermal, and mechanical energy systems. Electrochemical energy systems are based on the batteries, flow batteries, and fuel cells. The Electricals; it is about the capacitors, supercapacitors, superconducting magnetic energy storage or maybe, in short, calling it a SMEs. The last one is mechanical; it is about the flywheel, compressed air energy storage; in short, it is CAES and the pumped hydroelectric storage system.

Batteries store energy chemically and produce electricity at a fixed voltage through electrochemical reactions. The batteries are best suited for applications that demand relatively significant amounts of energy storage, such as more than 1 megawatt hour, for lengthy periods of time (15 minutes or more), when rapid recharge is not required and maintenance can be conducted satisfactorily. When used as a rechargeable battery, they are not ideal for environmentally sensitive places, distant locations, or applications that demand quick discharge and adsorption of energy [2].

The next one is electrochemical which is known as supercapacitors, ultracapacitors or electrical double-layer capacitors in short basically we are calling it an EDLC. So, in this particular case, using two-electrode over there, both are made by the activated carbon electrode and inside that, we are putting the electrolytes. So, when giving the charge over there, the electrolyte is divided into two parts: the plus and the negative one and simple capturing over there. They store energy in the electrical double layer at an electrode or maybe the electrolyte interface. The energy and power densities of electrochemical capacitors fall between those of batteries and conventional capacitors; their advantages include having a high power density, high cycle life, and quick recharge. That means when the whole energy will be drained out; simply one can charge it and then again can use it for a longer time. Certain disadvantages are there; it is having the low energy density; that means, the volume or maybe the mass is very less over here and it is also made by very expensive material.

Energy storage is achieved by technologies or physical media that store energy in order to perform beneficial actions at a later time. Because one can directly consume the energy generated as it is generated, but this is not always practical. So, how was that energy going to be utilized? That surplus energy will be saved and used as needed in the future. To boost the reliability and flexibility, energy storage is essential, combined with renewable energy generators or possibly photovoltaics. Because renewable energy sources like solar and wind are intermittent, storage is required to produce the proper amount of power at the right quality [3].

What kind of nanomaterials is basically used for energy storage? Smaller size offers a great deal of advantages because it is having the maximum surface area to the advancement of existing technologies and to the explorations and development of new technologies. Nanostructures benefit this device by providing a large surface area. In order to speed up electrochemical reactions or molecular adsorption at the solid-liquid or solid-gas interface, a material's surface area naturally increases when it is reduced in size. The controllable fabrication of nanostructures with the fully desired morphology structure, facets, and surface chemistry remains a challenge for nanomaterials synthesis technology; a more insightful understanding of the relationship between device performance and material structure, including chemical properties.

It is necessary to continue developing the benefits of nanostructured materials in order to improve device performance in terms of reaction activity, optical absorption, electron or ion transport, and so on. New processes based on nanostructures are expected to improve both the energy and power densities of lithium-ion batteries by increasing optical absorption and reducing charge recombination, as well as other energy resources related to electron transport in solar cells.

In general, supercapacitors can store electrical energy and should be considered a green energy source because they do not emit any poisonous gases and are not dangerous to humans. So, in general, because these new energy systems or energy forms are intermittent or regionally constrained, there is a pressing need to develop superior energy storage devices, such as supercapacitors, for efficient storage. The structure and qualities of the component materials determine the efficiency of an energy storage system. Nanotechnology's recent advancements have opened up new vistas by generating novel materials and architectures for various energy storage applications. In the current global market climate, the need for supercapacitors will continue to rise.

What are capacitors in general? The term "capacitor" refers to an electronic component that stores charge in the form of electric potential. A capacitor is made up of two parallel plates that are separated by a few millimetres. Capacitance of capacitors is the amount of charge stored in the capacitor and is determined by the capacitor's dimensions.

Supercapacitors are capacitors that have been deliberately developed to have a large capacitance value and high energy density as compared to ordinary capacitors. The energy storage principle of a supercapacitor can be either pure charge storage on an electrode or electrochemical double layer capacitance at the electrolyte interface. By combining high power output with a small design, supercapacitor cells can bridge the gap between traditional capacitors and batteries.

The energy capacitor, the supercapacitor, plays a bridge role in between the capacitors and the batteries. Working of the supercapacitor consists of two porous electrodes, a membrane which separates positive and negative plated is called the separator. Electrodes are electrically connected with ionic liquid called the electrolyte, and this area is totally filled by this electrolyte. When the voltage is applied to positive electrode, it attracts negative ions from the electrolyte, so negative ions are coming over here. When the voltage is applied to negative electrode, it attracts the positive ions from the electrolyte. These ions are stored near the surface of the electrode, and these ions decrease the distance between the electrodes because it is making one layer, then another layer, then another layer same thing is these sides also, so automatically it is reducing the distance in between the anode and cathode. Due to decrease of distance of electrodes, the capacitance become very huge that is C directly proportional to capital A by small d. The following sections explain about the history of supercapacitors, types of supercapacitors and applications of nanomaterial for supercapacitors.

11.2 History of Supercapacitors

In 1876, Fitzgerald developed a **Wax-impregnated dielectric** paper **capacitor** with foil electrodes [4]. For power supply filtering, the foil type and wax-soaked capacitors were utilized in RF receivers (Figure 11.1), then dielectric mica capacitors for radiofrequency circuits (Figure 11.2). Since foil type and wax-soaked paper capacitors are bulky, these radios' power supplies typically used filter chokes, in conjunction with the rectifier unit and capacitive filters to minimize the ripple content in supply.

Charles Pollak discovered the electrolytic capacitor theory in 1886 as part of his study into anodizing aluminum and other metals. Pollack noticed that the capacitance between the aluminum and the electrolyte solution was extremely high because of the thinness of the aluminum oxide layer formed. When the power was removed, most electrolytes tended to dissolve the oxide layer anew, but he eventually discovered that sodium borate (borax) allowed the layer to form and did not attack it afterward. In 1897, he received a patent for the aluminum electrolytic capacitor with borax solution [5].

Howard Becker was granted a patent for a "low-voltage electrolytic capacitor" in 1957 [6]. Two porous electrodes made of carbon served as the plates. The output voltage of the capacitor is 0.5 V and its capacity is 0.8 Farad. Donald Boos constructed the first production of a supercapacitor sample termed "Electrolytic capacitor with carbon paste electrodes" in October 1970 [7]. Two symmetrical tablet-shaped plates made up the capacitor. A conductive carbon paste was used to make each plate. The suspension was dried to make the paste. Then a paste of 0.25 grams was kept in a die and soaked. Because the plate wasn't strong enough, they were held together with a ring. In today's supercapacitors, similar designs can be seen Figure 11.3 picturizes the industrial design.

The plates are spherical in shape and consist of a carbon suspension with a viscous binder substance. Between the plates, a cellulose fiber membrane is inserted. The capacity of the sample displayed is 2F, and the voltage is 1.25V.

Bert Hart and Richard Pikema, IBM workers, were granted a patent for an "Electrochemical capacitor having a double layer" [8] in 1972. A two-layer capacitor with lower value series resistance and large capacitance was used in the experiment. The plate utilized activated carbon was separated by a gasket with a thickness of 0.00127 cm which was exceedingly porous. The seal is impregnated with a high concentration of the electrolyte. KOH or H_2SO_4 served as electrolytes. The possibility of using solid electrolytes allows operating temperatures for a wider range, elasticity and longer life period (25 years)., and outstanding resistance to a variety of other variables. Currently, a significant amount of research is being done in this area.

FIGURE 11.1 Antique radio capacitors made of waxed paper and foil. When restoring a radio, such capacitors are frequently replaced by polymer film capacitors.

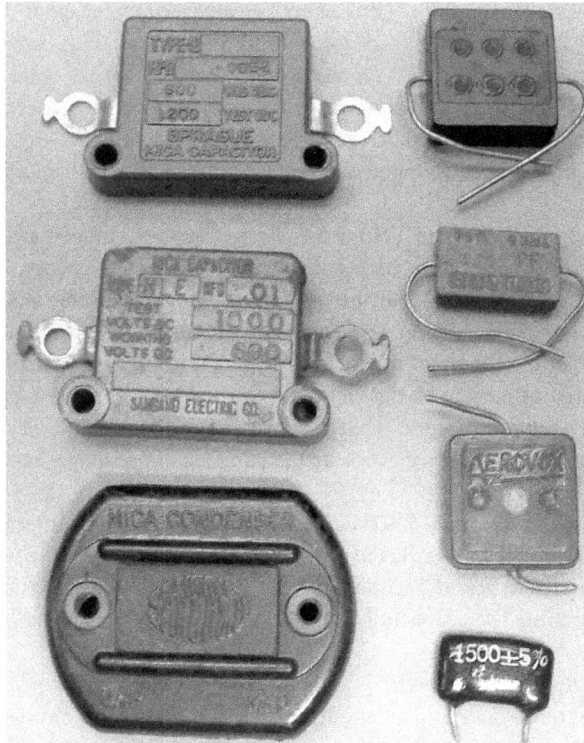

FIGURE 11.2 Mica capacitors in their infancy. Mica capacitors have always been, and continue to be, extremely dependable. Although they are still accessible, their high cost prevents broad use today. (Used with permission from William Harris, nbcblue@hotmail.com.)

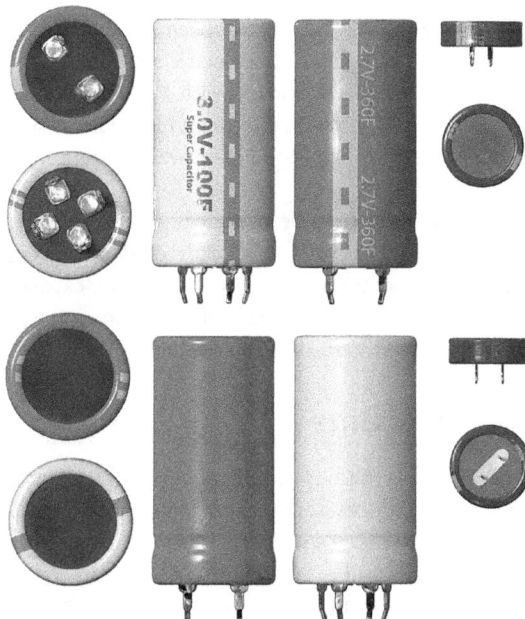

FIGURE 11.3 The industrial design of internal components of a supercapacitor.

In 1975, the Soviet Union (Laboratory "Gillicondo") created a supercapacitor (ionizer) based on solid electrolytes RbAg4I5 [9, 10] In 1975, the word "ionization" was coined. This term will be used later in the article to refer to supercapacitors. The KI1-1 monitors were rated by a 0.5 V voltage and capacity ranges from 0.1 to 50 F. In this structure, a solid electrolyte was used in the two-electrode cell. The Electric Double Layer Capacitor effect was re-discovered by Standard Oil of Ohio in 1966. In 1978, Standard Oil of Ohio gave the licensing to NEC (Nippon Electric Corporation), which marketed the product as a "supercapacitor".

Somewhere in the range of 1975 and 1980, Brian Evans Conway chipped away at ruthenium oxide electrochemical capacitors broadly, both concerning principal innovative work. In 1991, he recognized the way of behaving of "supercapacitor" and "battery" in electrochemical energy stockpiling. In 1999, he instituted the expression "supercapacitor" to portray the expansion in capacitance saw because of surface redox processes including faradaic charge move among terminals and particles. His "supercapacitor" held electrical charge somewhat in the Helmholtz twofold layer and somewhat through faradaic responses including "pseudocapacitance" charge move among anode and electrolyte. Redox cycles, intercalation, and electrosorption are the working instruments of pseudocapacitors (adsorption onto a surface)[11, 12].

In 1982, Pinnacle Research Institute (PRI) developed the first low-value internal-resistance supercapacitor for military uses, which was marketed as the "PRI Ultracapacitor". The work was taken over by Maxwell Laboratories (later Maxwell Technologies) in 1992. To emphasize its use for power applications, Maxwell took the name Ultracapacitor from PRI and referred to them as "Boost Caps" [David A. Evans et al.].

David A. Evans invented an "Electrolytic-Hybrid Electrochemical Capacitor" in 1994, employing the electrolytic capacitor as the cathode and an anode of 200V [13]. Electrolytic and electrochemical capacitor characteristics are combined in these capacitors and they create a hybrid electrochemical capacitor by means of combining the excessive dielectric power of an electrolytic capacitor's anode with the excessive capacitance of an electrochemical capacitor's pseudocapacitive steel oxide (ruthenium (IV) oxide) cathode. Evans'Capattery capacitors have a 5 times greater electricity content material than a tantalum electrolytic capacitor of equal size [14]. Because of their high cost, they were limited to military applications only.

Since 2005, electric buses in Shanghai (China) have been equipped with supercapacitors that are promptly recharged at each bus stop using so-called electric umbrellas. In Nurnberg (Germany), the public transportation company VAG tested a hybrid bus with a diesel-electric battery propulsion system and supercapacitors in 2001–2002. In Mannheim (Germany) and Heidelberg (Germany), regenerative hybrid systems based on Bombardier MITRAC equipment were used in light-rail trains without overhead wires from 2003 to 2008 [15].

By 2007, lithium-ion capacitors are a current development. Fujitsu's FDK turned into the primary to expand hybrid capacitors [16]. They utilize an anode made of electrostatic carbon and a cathode of lithium-particle electrochemical type that has been pre-doped. The capacitance value increased as a result of this combination. Besides the pre-doping method lowers down the potential of anode and produces a high cell voltage, improvingenergy considerably further.

11.3 Types of Supercapacitors

Supercapacitors store electrical energy and are categorized into three groups based on their functioning principles. Double-layer Capacitors are the first group, which is further split into Activated Carbons, Carbon Aerogels, and Carbon Nanotubes (CNT). Pseudocapacitors are split into Conducting Polymers and Metal Oxides in the second group. Hybrid Capacitors are split into three categories: Asymmetric Pseudo/EDLC, Composite, and Rechargeable batterie-type. Figure 11.4 depicts the classification chart.

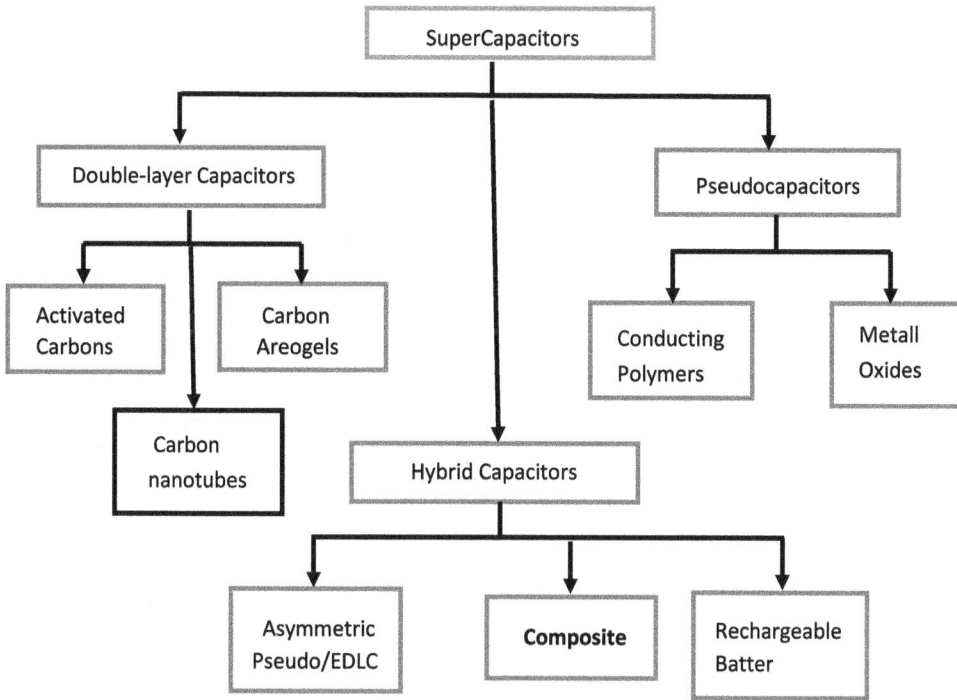

FIGURE 11.4   Types of supercapacitors.

11.3.1 Double-Layer Capacitors

Electrical Double-Layered Capacitors is another name for them (EDLC). Charge separation in a double layer, as seen in Figure 11.5, stores electrical energy. It is made up of two electrodes made of carbon-based materials, an electrolyte, and a separator. When a voltage is applied across the electrodes, charge builds up on the electrode surface, and ions from the electrolyte solution diffuse across the separator and into the pores of the electrode with the opposite charge. The electrodes have been designed to prevent ion recombination, resulting in a double-layer of charge at each electrode. As a result, the surface area of the electrodes increases while the space between them decreases. EDLCs can reach higher energy densities than ordinary capacitors as a result of this. The electrodes are made of to enhance the area of the surface, carbon electrodes or derivatives such as charcoal powder rods are utilized. This allows for more charge to be stored electrostatically. Figure 11.6 shows a picture of EDLC.

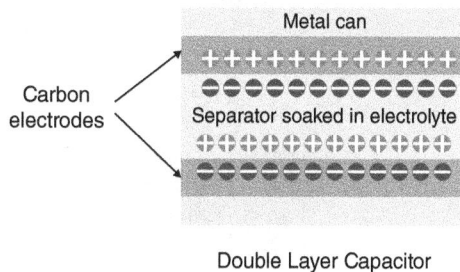

FIGURE 11.5   Double-layer capacitor.

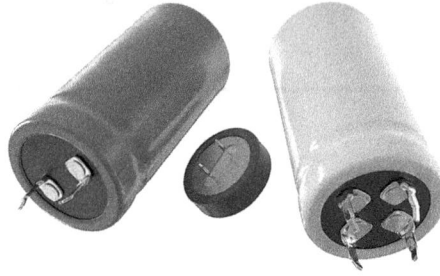

FIGURE 11.6 EDLC.

i) **Activated Carbon (AC):**
   The traditional approach of EDLCs is porous AC with a large specific surface area. Their particular surface area increases the charge accumulation capability at the electrode/electrolyte contact. Because of the large pore size range of activated carbon materials, which have random pore connection as illustrated in Figure 11.4, the specific capacitance does not necessarily rise linearly. As a result, ion transport through pore channels will significantly slow down, limiting the energy and power density of EDLCs. Researchers are working on a new form of narrow distribution ordered mesoporous carbons (OMCs) with a size of 2–50nm. The picture of an activated carbon supercapacitor is shown in Figure 11.5.

ii) **Carbon nanotubes (CNTs):**
   Due to innate high electrical conductivity and chemical stability, carbon nanotubes have been considered as promising electrodes for supercapacitors. These tubes are produced by catalytic decomposition of some hydrocarbons and by carefully manipulating different parameters as shown in Figure 11.9. CNTs have interconnected mesopores that contribute toa continuous charge distribution which most of the surface area are utilized. CNTs are chemically activated with potassium hydroxide to improve its specific capacitance.

iii) **Graphene:**
   It's a visually appealing electrode with good electrical conductivity and charge transport mobility. The great power performance of charge nano sheets grown on a metal current collector was demonstrated. Many studies are being conducted on the synthesis of porous graphene oxide using KOH activation andon graphene synthesis for supercapacitor applications as well.
   The 2.3V, 50F Graphene Supercapacitor for Electric Vehicles with a 7.5 mm pitch and a 2.3V 50F Graphene Supercapacitor for Electric Vehicles (Figures 11.7 and 11.8).

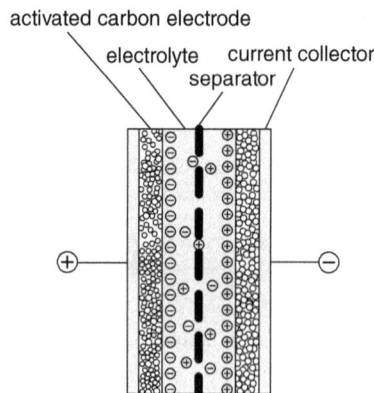

FIGURE 11.7 Process of AC Supercapacitor.

FIGURE 11.8   Image of AC Supercapacitor.

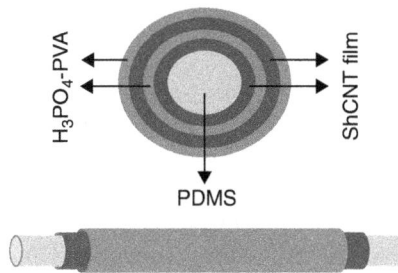

FIGURE 11.9   Cross-section view of carbon nanotubes mdpi.com.

Pseudocapacitors: Energy is stored electrochemically in this form due to charge transfer between the electrode and the electrolyte. This charge transfer is accomplished through a redox reaction or by ions intercalating on the electrode surface (see Figure 11.9). Figure 11.9 depicts a typical picture of 2.3 V, 300 F Pseudocapacitors.

Hybrid Capacitors: It's a hybrid of Pseudocapacitors and Double-layer capacitors. The electrodes are asymmetric, with one having electrostatic properties and the other having electrochemical capacitance. As seen the charge is stored in the electrostatic and electrochemical processes.

Criteria for choosing nanomaterials for supercapacitors: The criterion for selecting a supercapacitor nanomaterial and the materials' qualities are listed below:

i. Electrode materials:

The properties of electrodes materials used in supercapacitor are:
1. High cyclability
2. Long-term stability
3. High specific surface area
4. High electric conductivity
5. Resistance to electrochemical oxidation reduction
6. High temperature

ii. Electrolytic materials:

The properties of electrolytic materials used in superconductors are:
1. High ionic conductivity
2. Low viscosity to access small pores
3. Chemical stability
4. Easy availability
5. Cost effectiveness

11.4   Applications of Supercapacitors Using Nanomaterials

Generally, the supercapacitors are highly useful in compensating the high power and lifetime deficiencies of batteries. Hence, they occupy a strong role in between the conventional storage devices such as batteries and capacitors. In many cases, the supercapacitors are used to supply high energy pulse for very short duration, probably for milliseconds, where compensation is required against the power disturbances and/or energy supply to extend the period of operation. The application range varies from cameras to energy generation system. While used in movable energy supply systems, the battery size can be reduced and the power pulses occur in the system are managed by the supercapacitors and hence increases the lifetime of the battery. In recent years, remarkable researchstudies are going around the world on the commercial and industrial usage of supercapacitors made of electrochemical and nanomaterials [17].

11.4.1 Low Power Appliances

Supercapacitors can help in stabilizing the fluctuations in low power consuming appliances such as hand-held equipment, computers, photovoltaic systems, etc. In recent past, the drastic developments in consumer electronics can be noticed. Particularly, laptop computers, mobile phones and other communication devices are flooding in to the market with the lithium-ion batteries because of their higher energy density characteristics. However, the multi-functionalities of those devices challenge the capabilities of those batteries. Hence, intensive research studies for alternative energy storage techniques point out that supercapacitors area reliable candidate to help the batteries during their under-performing times [18]. Use of nanocomposites ($NiMoO_4$@$NiWO_4$) in the electrodes of supercapacitor exhibits very good cyclic stability electrochemical performance [19].

11.4.2 Energy Buffering

The power density per unit weight of supercapacitors is approximately 10 times greater than the batteries [20]. This superior characteristic makes the supercapacitor as an indispensable device in high power applications in industries, like regenerative braking in hybrid mobility, elevators, etc. It has been noticed that the power drawn from the distribution network fed to an elevator and trolley buses were being constant because of supercapacitor incorporated in the system. The voltage profile has revealed that nearly 20V compensation was made by the supercapacitors at the time of demand [21]. A review of supercapacitor management system architectures implemented in various automobile applications shows that more than one type of supercapacitors need to be incorporated based on the primary objectives, such as control strategy, safety, and thermal characteristic boundaries, with a good trade-off. Among the active and passive balancing methods, the simplicity of the passive balancing method attracts the majority of manufacturers, whereas active balancing is preferred in racing cars that require fast charging and discharging [22]. Performance of the systems can be improved by energy buffering to mitigate the shortfall occurs by power interruptions and the current peaks. Replacement of electrolytic capacitors by supercapacitors in uninterrupted power supplies lengthens the lifetime of batteries and reduces the cost per cycle [23].

11.4.3 Voltage Stabilizers

Supercapacitors are highly useful in stabilizing the voltage in power systems. The role of supercapacitors found very good while gusting of winds experienced by wind energy conversion systems [24] and shading in solar photovoltaic systems [25]. The stresses due to unsafe level of current spikes in wind energy conversion system can be avoided with the use of supercapacitors and thus increases the lifetime of the same. Performance analysis of regenerative braking of an electric rail system shows that the supercapacitor not only improves the voltage stabilization of the system but also the response of the system [26]. Supercapacitor helps to improve the voltage profile, frequency and also meets the power fluctuations.

11.4.4 Energy Harvesting

Better charging and discharging cycles and less maintenance put the supercapacitor forward as a candidature of reliability in off-grid or on-grid renewable energy conversion systems. Wide spreading of distributed generation needs it more [27]. Proper selection of supercapacitors maximizes the energy efficiency of batteries and the driving efficiency of the sub-watt scale energy conversion system [28]. Energy harvesting from electromagnetic radio frequency waves uses printable supercapacitors to store the harvested energy [29].

11.4.5 Supercapacitor-Battery Applications

Over a period of decade, the automobile industries concentrate on the hybrid energy storage system (HESS) to escalate the performance of electric vehicles [30]. Combining the better energy density battery and excellent energy density of the supercapacitors, the electric drive can offer very good performance in dynamic conditions [31]. Commonwealth Scientific and Industrial Research Organisation (CSIRO), and Australian Government Agency, invented "Ultrabattery" as a hybrid combination of lead-acid battery known for better energy storage and a supercapacitor named for higher charging and discharging cycles [32]. The tests conducted in rural parts of India by the Institute of Transformative Technologies proved that the Ultrabattery improves energy security [33]. The Ultrabattery installed in large scale power applications offers better frequency regulation and smoothing operations.

11.4.6 Solar/Wind Powered Street Lighting

Solar photovoltaic/wind powered street lighting system uses supercapacitor as energy storage device [34, 35]. In both the cases, the stand-alone architecture energizes the LED street lighting system. During the fluctuations, the supercapacitor reduces the charging and discharging of batteries and reduce the discharge depth by which the lifetime of the batteries can be prolonged. Because of the ability of supercapacitors to perform better in various atmospheric conditions (+40°C to −20°C), their usage in street lighting is increasing.

11.4.7 Railways

It was observed that the diesel-electric locomotive saved 25% of fuel in which 20% by theoptimized operation of internal combustion engine and 5% by the regenerative braking using supercapacitors [36]. Operational cost of a high-speed railway with a perfect controller for energy storage system containing supercapacitor has been significantly reduced. Apart from the economical point of view, the efficient and safest operation of the railway were obtained by peak shaving and power quality improvement [37]. In 2018, Ducati Energia Spa installed a supercapacitor-based uninterrupted power supply for the Italian railways aiming to save energy by recuperation systems and to stabilize the voltage in the network [38]. In China, a light metro train has been introduced in 2013 with the supercapacitor as energy source without any external power source [24].

11.4.8 Biomedical Engineering

Generally, battery-operatedimplantable medical sensors are used in medical sciences. Eventhough batteries are named for high energy density, they need more time tocharge. Quick chargingsupercapacitors are replacing the batteries now a day in this regard [39]. Soft and stretchable materials have made possible to get stretchable supercapacitors for health monitoring systems, wearable biomedical devices, implantable biomedical devices and consumer electronics as energy storage unit. In some applications, low energy densities of these stretchable supercapacitors can be compensated by piezoelectric and triboelectric devices, that is these devices convert the mechanical energy obtained from heart beat to electric energy and stores in supercapacitors for continuous usage. These implantable medical devices need to perform robustly and should be controllably degradable when immersed in physiological fluids to avoid further surgical removal [40]. Various hierarchical nanostructures are possible by combining graphene oxide sheets and carbon spheres at different weight ratios and help to create flexible fibrous electrodes of supercapacitors [41]. Flexible supercapacitors constructed using biopolymer, conducting polyene and ionic liquids have given excellent result in bioelectronics with nearly cent percent efficiency and long cyclic life [42]. A self-chargeable nano-biosupercapacitor with a volume of 1 nanoliter in tubular geometry was tested in the vascular system to measure the local pH level in blood and to monitor the formation of cancer cells. On-chip manufacturing techniques ensure promising self-chargeable microsystems in medical electronic applications [43].

11.4.9 Power Quality Improvement

Improvement of voltage regulation has been achieved by supercapacitor energy storage system (SESS) in Aircraft Distributed Power Systems as per MIL-STD-00704E (AS), a military standard. SESS used in the substation of distribution system for trolley buses compensates the voltage drop and maintains the same within the limits prescribed by standards. It is also recorded that the SESS maintain the reliability of supply by active and reactive power compensation in the distribution system in all modes such as islanding, transient switching and grid connection [44]. Energy storage system using supercapacitors in microgrid applications have given very good power quality in transient conditions. The supercapacitors quickly responded to the faults and maintained the stability through the voltage regulation [45]. Faster charging and discharging characteristic of supercapacitors used with STATCOMs help to improve the voltage quality of the power system by injecting real power [46]. Grid Codes restricts grid connection of fixed speed wind generation systems by several constraints. Low voltage ride through (LVRT) is one among them. LVRT is the ability of the system to overcome the voltage dip in various situations. The supercapacitor-incorporated STATCOM successfully injects the maximum amount of reactive current into the system during fault without any dead band by which the grid codes can fairly be satisfied [47].

11.5 Conclusion

In this chapter, the types of energy sources and how supercapacitor replaces a battery, types of supercapacitors, and the history of supercapacitors are explained. How the supercapacitor supports biomedical engineering, railways, power quality improvement, energy harvesting, energy buffering, and solar/wind-powered street lighting systems is also explained. The supercapacitor is not only useful for the above-said application, it also plays a vital role in EV applications as mentioned in this chapter.

References

1. P. Y. Burchenko et al., "A Storage Device that Stores Information when the Power is Turned Off," *Journal of Electronics*, 1984, pp. 34–45.
2. https://teemp.ru/en/news/prospects-for-the-application-of-supercapacitor-based-solutions-in-railway-transport/
3. V. P. Kuznetsov, "Results of Research in the Field of Creating Ionistorsfor Surface Mounting and Planar Ionistors Based on the Superionic Conductor AgI-RbITechnologies and Materials for

Extreme Conditions (Creation and Development of Technologies for Manufacturing Electroactive Materials for Converters and Energy Storage Devices)," 2015b, pp. 20–29.

4. D. G. Fitzgerald, "Improvements in Electrical Condensers or Accumulators," British Patent No. 3466/1876, September 2, 1876.

5. J. Ho, R. Jowand S. Boggs, "Historical Introduction to Capacitor Technology," *IEEE Electrical Insulation Magazine*, vol. 26, no. 1, 20–25, 2010.

6. H. Becker, "Low Voltage Electrolytic Capacitor. U.S. Patent 2, 800, 616A," July 23, 1957.

7. D. I. Boosand G. Heights, "Electrolytic Capacitor Having Carbon Paste Electrodes," 1970, United State Patents.

8. Y. K. Sun and P. V. Kamat, "Advances in Solid-State Batteries, a Virtual Issue," 2021.

9. E. A. Gailish, M. N. Diakonov, V. P. Kuznetsovand E. V. Kharitonov, "Ionistors-Electrochemical Solid-state Elements Electronic Industry No. 8," 1975.

10. V. Kuznetsov, "Solid-State Ionistors: New Series, Parameters and Characteristics," 2015a.

11. B. E. Conway, *Electrochemical Supercapacitors: Scientific Fundamentals and Technological Applications (in German)*. Berlin, Germany: Springer, 1999, pp. 1–8, ISBN 978-0306457364

12. B. E. Conway, "Transition from 'Supercapacitor' to 'Battery' Behavior in Electrochemical Energy Storage," *Journal of the Electrochemical Society*, vol. 138, no. 6, pp. 1539–1548, May 1991. Bibcode: 1991JElS.138.1539C. doi:10.1149/1.2085829.

13. N. A. Marcus, "A Survey of Electrochemical Supercapacitor Technology (PDF)," *Journal of Electrical Engineering*, 2015, pp. 203–208.

14. D. A. Evans, Evans Company, East Providence, RI 02914 (401) 434–5600, High Energy Density Electrolytic-Electrochemical Hybrid Capacitor, 2016.

15. P. Kurzweil, *Electrochemical Energy Storage for Renewable Sources and Grid Balancing*, 2015, pp. 345–407.

16. FDK, Corporate Information, FDK History 2000s. FDK. Retrieved February 21, 2015.

17. The Wayback Machine Technical Paper, Evans Capacitor Company, 2007.

18. https://www.csiro.au/en/research/production/optimisation/ultrabattery

19. A. E. Reddy and H. Kim, "Supercapacitor Applications in Electric Home Appliances and Smartphones," *2018 International Conference on Information and Communication Technology Robotics (ICT-ROBOT)*, 2018, pp. 1–3, doi:10.1109/ICT-ROBOT.2018.8549915.

20. N. Gekakisand A. Nadeau, *Modeling of Supercapacitors as an Energy Buffer for Cyber-Physical Systems*, CRC Press, 2015.

21. P. Barrade and A. Rufer, "Supercapacitors as Energy Buffers: A Solution for Elevators and for Electric Busses Supply," *Proceedings of the Power Conversion Conference-Osaka 2002 (Cat. No. 02TH8579)*, vol. 3, 2002.

22. F. Naseriand S. Karimi, "Supercapacitor Management System: A Comprehensive Review of Modeling, Estimation, Balancing, and Protection Techniques," *Renewable and Sustainable Energy Reviews*, vol. 155, pp. 111913, 2021.

23. S. Ekanayake, P. Wijesinghe, C. Dassanayake, V. Sooriarachchi, N. Gurusinghe and N. Kularatna, "UPS Capability and End-to-End Efficiency Improvement Technique for DC Microgrids with Supercapacitor Energy Storage," *2021 IEEE 30th International Symposium on Industrial Electronics (ISIE)*, 2021, pp. 01–06, doi:10.1109/ISIE45552.2021.9576327.

24. I. H. Panhwar et al., "Mitigating Power Fluctuations for Energy Storage in Wind Energy Conversion System Using Supercapacitors," *IEEE Access*, vol. 8, pp. 189747–189760, 2020, doi:10.1109/ACCESS.2020.3031446.

25. L. Wang, Q.-S. Vo and A. V. Prokhorov, "Stability Improvement of a Multimachine Power System Connected With a Large-Scale Hybrid Wind-Photovoltaic Farm Using a Supercapacitor," *IEEE Transactions on Industry Applications*, vol. 54, no. 1, pp. 50–60, January-February 2018, doi:10.1109/TIA.2017.2751004.

26. M. Khodaparastan and A. Mohamed, "Supercapacitors for Electric Rail Transit Systems," *2017 IEEE 6th International Conference on Renewable Energy Research and Applications (ICRERA)*, 2017, doi:10.1109/icrera.2017.8191189

27. M. Habibzadeh, M. Hassanalieragh, A. Ishikawa, T. Soyataand G. Sharma, Hybrid Solar-Wind Energy Harvesting for Embedded Applications: Supercapacitor-Based System Architectures and Design Tradeoffs. *IEEE Circuits and Systems Magazine*, vol. 17, no. 4, pp. 29–63, 2017.

28. S. Kim and P. H. Chou, "Size and Topology Optimization for Supercapacitor-Based Sub-Watt Energy Harvesters," *IEEE Transactions on Power Electronics*, vol. 28, no. 4, pp. 2068–2080, April 2013, doi:10.1109/TPEL.2012.2203147.

29. S. Lehtimäki, M. Li, J. Salomaa, J. Pörhönen, A. Kalanti, S. Tuukkanen, P. Heljo, K. Halonenand D. Lupo, "Performance of Printable Supercapacitors in an RF Energy Harvesting Circuit," *International Journal of Electrical Power & Energy Systems*, vol. 58, pp. 42–46, 2014, doi:10.1016/j.ijepes.2014.01.004.

30. S. R. Soni, C. D. Upadhyay and H. Chandwani, "Analysis of Battery-Super Capacitor Based Storage for Electrical Vehicle," *2015 International Conference on Energy Economics and Environment (ICEEE)*, 2015, doi:10.1109/energyeconomics.2015.

31. G. Subramanianand J. Peter, "Integrated Li-Ion Battery and Super Capacitor Based Hybrid Energy Storage System for Electric Vehicles," *2020 IEEE International Conference on Electronics, Computing and Communication Technologies (CONECCT)*, 2020, doi:10.1109/conecct50063.2020.9198317.

32. N.P.H. Duraman, K.L. Lim and S.L.I. Chan, "Chapter 16- Batteries for Remote Area Power (RAP) Supply Systems," in *Advances in Batteries for Medium and Large-Scale Energy Storage*, Woodhead Publishing, 2015, pp. 563–586.

33. https://www.idtechex.com/en/research-report/batteries-and-supercapacitors-in-consumer-electronics-2013-2023-forecasts-opportunities-innovation/336

34. G. Tian, X. Dingand J. Liu, "Study of Control Strategy for Hybrid Energy Storage in Wind-Photovoltaic Hybrid Streetlight System," *2011 IEEE Xplore*, 2012, doi:10.1109/OSSC.2011.6184698.

35. D. Jayananda, N. Kularatnaand D. A. Steyn-Ross, "Design approach for Supercapacitor Assisted LED Lighting (SCALED) Technique for DC-Microgrids," *2018 IEEE International Conference on Industrial Electronics for Sustainable Energy Systems (IESES)*, 2018, doi:10.1109/ieses.2018.8349845

36. C. Mayet, J. Pouget, A. Bouscayrol and W. Lhomme, "Influence of an Energy Storage System on the Energy Consumption of a Diesel-Electric Locomotive," *IEEE Transactions on Vehicular Technology*, vol. 63, no. 3, pp. 1032–1040, March 2014, doi:10.1109/TVT.2013.2284634.

37. J. Chen, H. Hu, Y. Ge, K. Wang, W. Huang and Z. He, "An Energy Storage System for Recycling Regenerative Braking Energy in High-Speed Railway," *IEEE Transactions on Power Delivery*, vol. 36, no. 1, pp. 320–330, February 2021, doi:10.1109/TPWRD.2020.2980018.

38. https://www.railwaygazette.com/supercapacitor-light-metro-train-unveiled/37218.article

39. A. Pandey, F. Allos, A. P. Hu and D. Budgett, "Integration of Supercapacitors into Wirelessly Charged Biomedical Sensors," *2011 6th IEEE Conference on Industrial Electronics and Applications*, 2011, pp. 56–61, doi:10.1109/ICIEA.2011.5975550.

40. X. Chen, N. S. Villa, Y. Zhuang, L. Chen, T. Wang, Z. Li and T. Kong, "Stretchable Supercapacitors as Emergent Energy Storage Units for Health Monitoring Bioelectronics," *Advanced Energy Materials*, vol. 10, 1902769, 2019, doi:10.1002/aenm.201902769. Components and Technologies, No. 9, pp. 52–54.

41. X. Zhang, Y. Zheng and K.-Q. Zhang, "Fibrous and Flexible Supercapacitors with a Hierarchical Nanostructure Comprised of Carbon Spheres and Graphene," *2013 IEEE MTT-S International Microwave Workshop Series on RF and Wireless Technologies for Biomedical and Healthcare Applications (IMWS-BIO)*, 2013, doi:10.1109/imws-bio.2013.6756249

42. G. Srinivasanand M. Lorenzo Fernandez, "Flexible Supercapacitors for Biomedical Applications," Poster session presented at Faraday Discussion 154, Cambridge, United Kingdom, 2017.

43. Y. Lee, V.K. Bandari, Z. Li et al., "Nano-Biosupercapacitors Enable Autarkic Sensor Operation in Blood," *Nature Communications*, vol. 12, pp. 4967, 2021, doi:10.1038/s41467-021-24863-6

44. I. M. Syed, "Power Quality Improvement by Supercapacitor Energy Storage," Ryerson University, Thesis, 2010, doi:10.32920/ryerson.14662359.v1

45. D. A. Evans, The Littlest Big Capacitor - an Evans Hybrid Archived 3 March 2016 at Dehua Zheng, Dan Wei, Wei Zhang and Zhaojun Meng, "The Study of Supercapacitor' Transient Power Quality Improvement on Microgrid," 2015 IEEE Eindhoven Power Tech, 2015, pp. 1–5, doi:10.1109/PTC.2015.7232600.

46. P. Srithorn, M. Sumner, L. Yao and R. Parashar, "Power System Stabilisation Using STATCOM with Supercapacitors," *2008 IEEE Industry Applications Society Annual Meeting*, 2008, pp. 1–8, doi:10.1109/08IAS.2008.327.

47. A.F. Obando-Montaño, C. Carrillo, J. Cidrás and E. Díaz-Dorado, "A STATCOM with Supercapacitors for Low-Voltage Ride-Through in Fixed-Speed Wind Turbines," *Energies*, vol. 7, pp. 5922–5952, 2014, doi:10.3390/en7095922

12

Aspects of Nanotechnology Applied in the Energy Sector: A Review

Dr. N. Sathammai Priya

Cauvery College for women (A), Affiliated to Bharathidasan University, Tiruchirappalli, India

P. Archana

Institute of Organic and Polymeric Materials, Research and Development Centre of Smart Textile Technology, Taiwan

N. Srinivasan alias Arunsankar

Sri Sai Ram Engineering College, Chennai, India

Human life would not be possible without energy because of its benefits as a primary supply for human activities. Global energy demands are expected to increase until 2030, according to the International Energy Agency (IEA). By implementing new technologies, including nanotechnology, and optimizing manufacturing processes, it is possible to enhance energy efficiency across a wide array of industries and value renewable energy generation to meet economic needs. It will be possible to fully contribute to the global climate protection policy and the sustainable energy supply in the long run. Here, the application of nanotechnology is brought to bear on each element of the value creation chain in the energy sector. Nearly all branches of economics consider nanotechnology to be a key technology for innovation and technological advancement.

In many scientific researchstudies, application of nanotechnology or nanomaterials in the area of energy has become a hot topic, which includes lithium-ion batteries, fuel cells, light-emitting diodes (LEDs), and solar cells. Unfortunately, its current development is hampered by the expensive cost of production compared to conventional technologies. Therefore, priority should be given to nanotechnology in the energy sector in order to achievehigher efficiency, lower production cost, and easy application. With this view, the review attempts to explain how nanotechnology can help address both present and future sustainable energy needs.

DOI: 10.1201/9781003355755-12

12.1 Introduction

Energy has a great influence on human life, since it is used for various purposes, such as transportation, appliances, services, and industry. Energy demand is expected to rise each year due to the dependence of many tools and instruments on external energy sources. Consequently, energy issues are an important topic of discussion internationally (Christian et al. 2013). As a result of serious problems related to energy production and consumption, developing renewable energy technologies is one of the great technological challenges of the 21st century. Nanotechnology has become an increasingly popular area of research over the past few years and is regarded now as one of the best options to address this issue (Hussein 2015). In addition to improved performance potentials for conventional energy sources (such as fossil fuels and nuclear fuels), nanotechnology also offers significant improvement potentials for renewable energy sources such as geothermal energy, solar, wind, water, tides, and biomass (Wolfgang Luther 2008). At present, its growth has been hampered by the high manufacturing costs, which are comparable to those of traditional technologies. In order to increase the efficiency of energy production by lowering production costs and simplifying its use, nanotechnology should receive priority in the energy sector.

In order to achieve sustainable power supply in the long run, it is not only necessary to maximize the efficiency and environmental friendliness of existing energy sources, but it is also necessary to minimize loss of energy during transport from the energy source to the end user, to provide and distribute energy as efficiently and effectively as possible for each application purpose, as well as to reduce demand for energy in industry and private households (Deshmuk et al. 2013). There is a growing importance attached to nanotechnology in improving energy efficiency and leveraging renewable energy production through new technological solutions and optimizing production technologies to enhance economic benefits in every field of industry.

12.2 Energy Applications of Nanotechnology

Energy storage based on nanotechnology will likely be used primarily in decentralized energy production and electromobility in the near future, so the storage of energy generated decentrally (e.g. using a combination of roof-top solar panels and storage in private homes and businesses) will become a significant factor in reducing carbon emissions. Among the key technologies for the future are lithium-ion batteries and supercapacitors (Seitz et al. 2013). With the increase in wearable electronics and functionality, it is becoming more and more important to have lightweight electric power for wearable electronics and devices. A number of innovations have been made in this area, such as flexible solar cells for charging batteries, fuel cells, and supercapacitors. There can be no doubt that energy storage is one of the greatest challenges of the 21st century. Today, we need to find new, low-cost, and environment-friendly technologies for energy conversion and storage in response to the needs of modern society and environmental concerns. This has spurred research in this field to develop rapidly. An examination of opportunities and barriers involved in preparing next-generation battery and supercapacitor products based on nanotechnology is of particular importance (Balakrishnan et al. 2014).

12.2.1 Lithium-Ion Batteries

Several studies have shown that lithium-ion cells have a high voltage, high energy, and power density, and are regarded as a promising alternative to traditional batteries for energy storage (Wolfgang Dubbert et al. 2014). As energy is released from the battery when it is discharged, lithium is oxidized to lithium oxide, freeing oxygen from the atmosphere. Regenerating the lithium electrode produces oxygen when the battery is charged. In this case, oxygen is used as the cathode, which simplifies the process since it will not add to the weight or volume of the battery. This type of battery involves nanostructured materials as a structural component of the anode (usually carbon) and separator membrane.

Nanostructures allow oxygen to pass while preventing moisture from entering (Kraytsberg and Ein-Eli 2011). Nanotechnology or new materials capable of offering better performance are currently being utilized to develop lithium-ion batteries. It has been discovered that there is no electrode material that can execute the requirements for high aspects of performance, such as high capacity, high operating voltage, and long life cycle, for increasing the performance of Li-ion batteries (Bullies et al. 2010).

By utilizing nanotechnology, battery manufacturing is more cost-effective and environment-friendly than using conventional Li-ion batteries. Compared to a conventional Li-ion battery, nanotechnology offers the following advantages:

The electrodes in the battery have a high heat resistance, so there is less risk of a battery catching fire. Reducing the time required to recharge the battery increasesthe battery's power. A coating is applied to the electrodes to give them a higher power output. The electrode surface is coated with nanoparticles to create this excess. In order to improve the current flow between electrodes and chemical cells, the surface area of the electrodes would be increased. By reducing the weight of the battery, this technique can also improve the efficiency of hybrid vehicles. When not in use, liquid electrodes can be separated by nanoscale materials, thereby increasing the battery life. As a result, conventional batteries are no longer subjected to the low-level discharges that occur in conventional batteries (Christian et al. 2013).

12.2.2 Solar Cells

Solar power, as the world's most abundant and inexpensive source of energy, has a special place in the energy mix because it is a renewable energy source that emits no emissions. A person can consume all the resources available to them in a year with the sunlight that the Sun reflects on the planet every hour (Panchabikesan et al. 2019). In today's world, solar energy has produced more than 178 GW of electric power (Esmaeili and Najafi 2019). The resource assigned a solar power potential of more than 500 GW to build power plants by 2020 once barriers were removed (Murray et al. 2019). Due to the fact that these green energy resources, such as solar energy, are inherently local and distributed, it makes sense that their consumption will be centralized. This is because these resources, such as solar energy, are all about local and localized output; to avoid the use of global grids, distributed output would be necessary (Ghasemzadeh et al. 2020). Solar cells generally perform better when nanotechnology is used from a variety of perspectives. With the development of new materials and solar cell types, in addition to simplified production processes, nanotechnologies are considered a key factor for photovoltaics to be widely accepted as an economic power source. Significant cost savings and improved efficiency are attributed to nanotechnology. It is possible to adjust semiconductor band gaps to the incident radiation spectrum by using nanostructures, including quantum dots (QDs), or to emit multiple charge carriers per photon to enhance conversion efficiency.

12.2.2.1 Silicon-Based Solar Cells

Based on a single-crystal or multiple-crystal structure, silicon wafers of the first generation have a thickness between 300 and 400 microns. In order to find enough electron holes in silicon, it is polluted with different elements. Solar cells such as these are made up of layers of perforated silicon coated in electron-contaminated material, which emits light through electron cavities (Esmaeili Shayan et al. 2020). The external circuit generates an electrical current when the load is connected to it. The high performance of these solar cells has made them commercially successful, but they have two major drawbacks: the high cost of processing silicon raw materials and the high energy intake (Mahmoudi et al. 2018).

12.2.2.2 Thin-Film Solar Cells

The growth of thin-film solar cells enhanced by nanotechnology has recently caught the attention of many researchers (Akimov et al. 2010). The development of thin-film solar cells is also related to nanotechnology, since these cells are devoted to achieving thinner solar cells. As part of thin-film solar cell technology, nanoparticles or nanocrystalline structures can be embedded into thin films to reduce the size or thickness. Moreover, carbon nanotubes (CNTs) have also shown promise in nanotechnology, as they have high carrier mobility, reliable continuous wave illumination, and good light absorption properties (Gorji et al. 2013). As another potential technology, nanowires or nanorods can be used to form a P–N junction, allowing the surface-to-volume ratio of the junction to be increased, thereby increasing conversion efficiency. Nanowires or nanorods have also been developed to form nanoscaled solar cells structures with a high efficiency (Dutta et al. 2012). It is important to note that nanowire solar cells are also capable of improving band gap, reducing reflection, increasing defect tolerance, trapping good amounts of light, and reducing the amount of strain required (Garnett et al. 2011). Thin-film solar cells are developing nanotechnology beyond the concentration on nanoparticles or nanoscale-based structures. Other new technologies are being considered for incorporation into the solar cells such as nanowires, nanorods, and CNTs that could improve cell performance (Md Yunus et al. 2015).

12.2.2.3 Dye-Sensitized Solar Cells

DSSCs (dye-sensitized solar cells) have long been under study, as they have low production costs, simple preparation methods, low toxicity, and are simple to manufacture. Nevertheless, given their low abundance, high cost, and long-term stability, there is considerable scope for DSSC materials to be replaced. As a result of optimizing the material and structural characteristics, existing DSSCs can be improved up to 12% efficiency, which is still lower than the efficiency of other thin-film solar cells and Si-based solar cells, which provide 20%–30% efficiency (Sharma et al. 2018).

12.2.3 Fuel Cells

In an electrochemical reaction, a fuel cell converts fuel directly into electricity. It is more common to generate electrical energy by mechanical means rather than burning fuel and capturing the heat. Fuel cells could be a very efficient form of power generation. The fuel cells can theoretically operate on a wide variety of fuels, and the technology can scale from a battery in a laptop all the way up to a huge power station for data centres. Platinum is the most common material for the electrodes of fuel cells. Using platinum nanoparticles instead of a solid surface of platinum has been known for some time to improve efficiency and use of metal. In order to improve current technology, platinum nanoparticles could be mounted on surfaces such as activated carbon, CNTs, or nanowalls to create porous surfaces. In addition to this, the platinum surfaces are more accessible, so less of the expensive metal is required for producing an effective electrode (https://www.azonano.com/authors/will-soutter).

The fuel cell can be applied from supplying power to mobile phones or laptop computers to supplying power to electric vehicles or providing heat and electricity to houses up to small power plants. Electrocatalysts and fuel electrodes are specific applications of nanotechnology. Platinum alloys and platinum are known to be effective electrocatalysts, and platinum-ruthenium is regarded as the most effective catalyst in oxidizing methanol (Chan et al. 2004). In the case of platinum electrocatalysts in internal combustion engines for all automobiles, this would have the unfortunate consequence of depleting the global supply of platinum several times over.

Nanotechnology can be used to develop materials that are either equally effective or even more so than existing electrocatalysts if nanotechnology has been applied. By manipulating materials on a nanoscale, the platinum usage is reduced while the efficiency in desired applications is increased (Gondal et al. 2018).

12.2.4 Wind Power

In recent years, interest in wind resources has grown rapidly all over the world. A process called wind energy is responsible for generating electricity from the wind. Converting kinetic energy from the wind into electricitycan be achieved using turbines. These turbines will become more affordable and more efficient, and demand for green manufacturing methods will grow as they are developed. Utilizing the unique properties of nanomaterials to develop wind technology will result in greater potential for the technology. A variety of materials can be used for this technology, such as polymer nanocomposites, conductive nanomaterials, and nanostructured surfaces. By combining polymer nanocomposites with windmills, electricity can be generated more efficiently. Using an epoxy with CNTs will provide these properties, as CNTs have unique properties, making it stronger and lighter. To increase the durability and the efficiency of turbines, conductive and strong nanomaterials can be used. The nanoparticles are made up of several different types of materials, including metallic nanoparticles, rare earth nanoparticles, and carbon nanomaterials, such as CNTs and graphene. In wind turbines, nanostructured surfaces protect metals from corrosion, enhancing durability. They are used to protect metals from the weather when they are exposed to wind (https://medium.com/@nanografi/development-of-wind-energy-technology-with-nanomaterials-360c278fb22d).

Nanotechnology is being applied to wind energy applications in a broad range of ways that will address some of the challenges that wind energy scientists currently face. In order to make this a success, it is essential to stimulate the combination of scientific research and technological development for the next generation of wind turbines. By using nanoscale technology for energy system components, critical components will be more durable, have better performance during production, distribution, and even transport with lower maintenance costs as well as dramatically lower greenhouse gas (GHG) emissions. Furthermore, nanomaterials and nanosensors can help enhance the integration of renewable energy into smart grids and decentralize energy production. Researchers already implemented nanoscale technologies and materials into the wind industry, creating novel standards for wind turbine performance, availability, operability, and reliability. Wind turbines can be lengthened, and structural fatigue failures are minimized by developing nanoscale models. This would reduce the complete cost of energy generation (http://www.windengineering.eu/wind_power/nanotechnologies.html#:~:text=Nanoscale%20 models%20are%20being%20developed,lighter%2C%20safe%2C%20and%20sustainable).

12.2.5 Supercapacitors

It is the mechanism of energy storage at the electrochemical interface between the electrode and electrolyte within supercapacitors that allows them to store and deliver energy at relatively high rates. They also have the advantage of being able to store small amounts of electricity (Sivaraman et al. 2006). In addition to having a higher energy density than conventional capacitors, supercapacitors possess other advantages. In addition to their high cycle life and high power, supercapacitors are used in a wide range of other products including electronic vehicles, mobile phones, and quick charge products, e.g. wireless power tools and medical equipment (Yu et al. 2013; Halper et al. 2006; El-Kady et al. 2015). The use of supercapacitors today would solve a problem of injecting power for a short period (Fawzy et al. 2017). For the supercapacitors to function correctly, an important aspect of nanotechnology is achieving a balanced size distribution of the surface area and the pore size distribution (so the electrolyte has easy access to the surface) (Aricò et al. 2005). The majority of electrode materials used in commercial supercapacitors currently are activated carbons, so there is an ongoing effort to determine what factors influence each material's capacitance and series resistance. A recent study clearly demonstrated that electrodes based on verticallyaligned CNTs with open tips, graphene sheets with tunable through-thickness stacking interactions and/or edge functionalities, and 3D pillared graphene-CNT networks can be prepared for high-performance supercapacitors. By coating with conductive polymers and/or metal oxide, pseudocapacitance may be introduced to further enhance their performance. Despite the progress made on supercapacitors based on carbon nanomaterials, further improvements are still necessary. CNTs or graphene aggregation reduces the surface area, therefore affecting device performance for one reason or another. There are two kinds of nanomaterials designed to prevent aggregation: carbon nanomaterials with various 3D structures (such as CNT arrays, graphene foams, and pillared vertically aligned [VA]-CNT/

graphene networks), along with 2D graphene sheets attached to crumpled structures. Another challenge for carbon nanomaterials to be scaled up for practical applications in supercapacitors is their high cost relative to commercial mesoporous or activated carbon. A low-cost way to develop carbon nanomaterials with high charge capacity (e.g. through ball milling) is therefore highly desirable (Tao Chen et al. 2013).

12.2.6 Quantum Dots

Among the many optical properties of QDs, which are affected by quantum physical effects, the size of the clusters determines their optical and electronic properties. Solar cells could exploit QDs because on one hand, multiple electron-hole pairs can be produced per photon, and on the other hand, the absorption bands can be tailored to match the wavelengths of the incident light. Essentially, QDs are single nanoparticles (nanocrystalline crystals) that range in size between 2 and 10 nm (https://www.azom.com/article.aspx?ArticleID=20875). The unique property of QDs is their optical and electrical properties. Exposed to light or electrified, they emit their own monochromatic light. Based on particle size and shape, colour is determined by light. On-screen displays are continuously made brighter, higherresolution, and more vibrant through the development of new technologies. With the advent of QDs, the industry has adopted the technology quickly. Quantum dot enhancement films (QDEFs) are commonly used to enhance images on light-emitting diode (LED) backlight screens. By using QDs, blue light from a blue LED backlight becomes pure red and green light, which is combined with blue to create red, green, and blue (RGB) pixels. The optical and electronic properties of QDs can be exploited in multiple ways through display applications (Fang et al. 2012).

The bioimaging industry has been using organic dyes for decades. The quantum yield and photostability of organic dyes are low, however. Despite this, QDs have been considered superior to organic dyes in many ways with the progress of nanotechnology. There has been research demonstrating that QD reporters exhibit 20 times more brightness and 100 times greater stability than conventional fluorescent reporters. Today, QDs with high brightness can be generated using well-established inorganic synthetic techniques. In order to successfully perform bioimaging, stable fluorescent probes that remain well-dispersed and stable require water with a wide range of pH and ionic strengths. QDs can be water-dispersible by a number of approaches. The use of QDs for in vitro and in vivo imaging has received a great deal of attention in the past, and this is expected to have significant ramifications for diagnosis of many diseases, the understanding of embryogenesis, and the study of lymphocyte immunity.

A QD beam–emitting diode may lead to improved LEDs, thereby allowing LEDs to perform more efficiently. Considering QDs' unique optical properties, they can be of tremendous value when it comes to display devices. Their uniquely accurate and outstanding colour-rendering capabilities make them a very useful display medium. Technically, QD displays have been successfully demonstrated in a proof-of-concept form years ago, showing good performance and good emissions in the visible and near-infrared regions (https://www.cd-bioparticles.com/t/Properties-and-Applications-of-Quantum-Dots_56.html).

12.2.7 Hydrogen Storage

As an energy storage medium, hydrogen shows great promise. There is no naturally occurring fuel like methane or coal, nor is it a renewable energy source. It produces only pure water as exhaust gases, since the process by which it releases energy is extremely efficient. At this time, steam reformation of natural gas is the predominant method of generating hydrogen. Steam reformation is highly energy intensive and requires a source of natural gas, which isn't renewable, even though it is a well-established, relatively efficient process. Electrolysis of water is another method of producing hydrogen. As well as heat, electrical power is often needed for this process. In short, hydrogen is not an extremely "green" way of storing energy because most of it is generated from fossil fuels. Several large projects have been underway in recent years to use renewable energy – primarily solar energy – to directly produce hydrogen, thereby transforming the fuel into what it should be: a clean and efficient alternative to batteries for storing energy in large quantities. As a result, some nanoparticles (like titanium dioxide, which in its bulk form is a common white pigment) exhibit extremely high photocatalytic activity – the ability to decompose molecules using sunlight as an energy source. Research has focused predominantly on self-cleaning surfaces, but there has been a great deal of interest in tailoring the photocatalytic properties of nanomaterials to split water into hydrogen and oxygen (https://www.azonano.com/article.aspx?ArticleID=3067). Hydrogen is a potent fuel, but storing it safely and effectively is a major obstacle. Gases and liquids that are compressed into hydrogen can only be stored under extremely high pressures, which can result in expensive tanks and explosion risks. Hydrogen is often seen as a crucial source of energy by encapsulating it in solid materials. These materials can store hydrogen securely at greater densities than those available through other methods, by absorbing hydrogen and forming hydrogen molecules. Standard materials, on the other hand, cannot store very large amounts of hydrogen and require very high temperatures to be able to capture and release hydrogen efficiently enough for commercial use. Hydrogen storage application is better suited to nanostructured materials due to their unique and tunable properties. Nanomaterials capable of storing hydrogen at high densities have been discovered by researchers around the world. It is important to create a material that can release its full fuel capacity in the shortest amount of time and has a controllable affinity for hydrogen. Lawrence Berkeley from National Laboratory developed a magnesium nanoparticle composite encased in a polymer matrix composed of a flexible organic polymer in 2011. Using this material, hydrogen can be selectively absorbed, stored safely as magnesium hydride at high densities, and released rapidly when needed. A number of nanotechnology-based solutions could solve the two biggest problems preventing hydrogen's economic development: producing hydrogen with renewable energy resources and safely storing and distributing hydrogen. It is desirable that novel nanomaterials be used in the production of hydrogen from water and subsequently to store hydrogen, so that solar-fuelled cars, combined heat and power (CHP) systems, and other devices can run on clean, efficient hydrogen fuel (Soutter, 2019).

12.3 Conclusion

A sustainable economy and society are built around energy. Although energy supply must be unrestricted as much as possible, this causes conflicting feedback loops on the ecological environment. In order to decrease negative impact on the environment from energy consumption, research is growing worldwide. Such efforts involve the generation, harvesting, converting, and storing of energy. In recent advances in energy-related applications, nanomaterials and nanostructures have played an important role due to their unique mechanical, electrical, and optical properties. Materials science and engineering are opening a new frontier for deal with this challenge thanks to nanotechnology, which allows for the development of new materials and technologies for energy conversion and storage. Global energy consumption and GHG emissions continue to increase exponentially, necessitating us to develop clean and renewable energy systems and advanced energy storage devices.

Nanotechnology will likely have the greatest impact on the renewable energy sector, where it may break through cost and efficiency barriers. As well as providing an efficient and environment-friendly way to exploit traditional energy sources, nanotechnology may also be used to develop a sustainable method to produce energy sources. Managing power grid and load varying and decentralizing feed in stations are also equally important to reduce losses. Nanotechnology can provide viable solutions to reduce power grid losses through load varying and decentralizing feed in stations.

References

Akimov Yu, A., Koh, W. S., Sian, S.Y., and S. Ren. 2010. Nanoparticle-enhanced thin film solar cells: Metallic or dielectric nanoparticles. *Applied Physics Letters* 96(7). doi:10.1063/1.3315942.

Aricò, A. S., Bruce, P., Scrosati, B., Tarascon, J. M., and W. Van Schalkwijk. 2005. Nanostructured materials for advanced energy conversion and storage devices. *Nature Materials* 4: 366–377.

Balakrishnan, A., and K.R.V. Subramanian. 2014. *Nanostructured Ceramic Oxides for Supercapacitor Applications* (1st ed.). Boca Raton, Florida: CRC Press.

Bullis, K. 2010. Higher-capacity lithium-ion batteries technology review. https://www.technologyreview.com/2006/06/22/228932/higher-capacity-lithium-ion-batteries/

Chan, K.Y., Ding, J., Ren, J., Cheng, S., K.Y. Tsang. 2004. Supported mixed metal nanoparticles as electro-catalysts in low temperature fuel cells. *Journal of Materials Chemistry* 14: 505–516.

Chen, Tao, and Liming Dai. 2013. Carbon nanomaterials for high-performance supercapacitors. *Materials Today* 16 (7–8): 272–280.

Christian, Ferric, Edith, Selly, Adityawarman, Dendy, and Antonius Indarto. 2013. Application of nano-technologies in the energy sector: A brief and short review. *Frontiers in Energy* 7: 6–18.

Deshmuk, P, and S. Kataria. 2013. Nanotechnology applications in the energy sector. *International Journal of Advancements in Research & Technology* 2: 1–8.

Dutta, Achyut K. 2012. Prospects of nanotechnology for high-efficiency solar cells. In *Electrical & Computer Engineering (ICECE)*.

El-Kady, M. F., Ihns, M., Li, M., Hwang, J. Y., Mousavi, M. F., Chaney, L., and R.B. Kaner. 2015. Engineering three-dimensional hybrid supercapacitors and micro supercapacitors for high-performance inte-grated energy storage. *Proceedings of the National Academy of Sciences* 112 (14): 4233–4238.

Esmaeili, M. S., and G. Najafi. 2019. Energy-economic optimization of thin layer photovoltaic on domes and cylindrical towers. *International Journal of Smart Grid* 3: 84–91.

Esmaeili Shayan, M., Najafi, G., and S. Gorjian. 2020. *Design Principles and Applications of Solar Power Systems (In Persian)*(1st ed.). Tehran: ACECR Publication-Amirkabir University of Technology Branch.

Fang, M. 2012. Quantum dots for cancer research: current status, remaining issues and future perspec-tives. *Cancer Biology & Medicine* 9(3): 151–163.

Fawzy, Ahmed, and Ashraf Eessaa. 2017. Review on Supercapacitor Based on Nonomaterials for Energy Storage. https://www.researchgate.net/profile/Ahmed-Fawzy-6/publication/321010336_Review_on_Supercapacitor_Based_on_Nonomaterials_for_Energy_Storage/links/5a104c61aca27287ce288f57/Review-on-Supercapacitor-Based-on-Nonomaterials-for-Energy-Storage.pdf

Garnett, Erik C., Mark, L., Brongersma, Yi Cui, and Michael D. McGehee. 2011. Nanowire solar cells. *Annual Review of Materials Research* 41: 269–295.

Ghasemzadeh, F., and M. E. Shayan. 2020. Nanotechnology in the Service of Solar Energy Systems. In Jesus Alberto Pulido Arcas (Ed.), *Nanotechnology and the Environment*. London, United Kingdom: Intech Open.

Gondal, Irfan. 2018. The role of nanotechnology in fuel cells and renewable energy. https://www.researchgate.net/publication/329140196_The_role_of_nanotechnology_in_fuel_cells_and_renewable_energy

Gorji, Nima E., and Mohammad Houshmand. 2013. Carbon nanotubes application as buffer layer in Cu (In, Ga) Se2 based thin film solar cells. *Physica E: Low-dimensional Systems and Nanostructures* 50: 122–125.

Halper, M. S., and J.C. Ellenbogen. 2006. *Supercapacitors: A Brief Overview*. McLean, Virginia, USA: The MITRE Corporation, 1–34.

Hussein, Ahmed Kadhim. 2015. Applications of nanotechnology in renewable energies-A comprehensive overview and understanding, *Renewable and Sustainable Energy Reviews* 42: 460–476.

Kraytsberg, A., and Y. Ein-Eli. 2011. Review on Li-air batteries – Opportunities, limitations and perspec-tive. *Journal of Power Sources* 196: 886–893.

Mahmoudi, T., Wang, Y., and Y.B. Hahn. 2018. Graphene and its derivatives for solar cells application. *Nano Energy* 47: 51–65.

Murray, R.L., and K.E. Holbert. 2019. *Nuclear Energy: An Introduction to the Concepts, Systems, and Applications of Nuclear Processes*. Amsterdam, Netherlands: Elsevier, 460–476.

Panchabikesan, K., Swami, M.V., and V. Ramalingam. 2019. Influence of PCM thermal conductivity and HTF velocity during solidification of PCM through the free cooling concept—A parametric study. *Journal of Energy Storage* 21: 48–57.

Radiant Vision Systems. 2021. Shining a light on nanotechnology & quantum dots. AZoM. https://www.azom.com/article.aspx?ArticleID=20875

Seitz, R., Moller, B. P., Thielman, A., Sauer, A., Meister, M., Pero, M., Kleine, O., and V. Kayser. 2013. Nanotechnology in the sectors of solar energy and energy storage: Technology report.

Sharma, K., Sharma, V., and S.S. Sharma. 2018. Dye-sensitized solar cells: Fundamentals and Current status. *Nanoscale Research Letters* 13, 381.

Sivaraman, P., Thakur, A., Kushwaha, R. K., Ratna, D., and A.B. Samui. 2006. Poly (3-methyl thiophene)-activated carbon hybrid supercapacitor based on gel polymer electrolyte. *Electrochemical and Solid-State Letters* 9(9): A435–A438.

Soutter, Will. 2019. Production and storage of hydrogen using nanotechnology. AZoNano. https://www.azonano.com/article.aspx?ArticleID=3067

Dubbert, Wolfgang. 2014. Use of nanomaterials in energy storage. https://www.umweltbundesamt.de/sites/default/files/medien/376/publikationen/use_of_nanomaterials_in_energy_storage.pdf

Wolfgang Luther. 2008. Application of Nanotechnologies in the Energy Sector Volume 9 of the series Aktionslinie Hessen Nanotech of the Hessian Ministry of Economy, Transport, Urban and Regional Development.

Yu, A., Chabot, V., and J. Zhang. 2013. *Electrochemical Supercapacitors for Energy Storage and Delivery: Fundamentals and Applications*. Boca Raton: CRC Press.

Yunus, Nurul Amziah Md, Abdul, Halin Izhal, Sulaiman, Nasri, Ismail, Noor, and Nik Aman. 2015. A compilation of nanotechnology in thin film solar cell devices. *World Academy of Science, Engineering and Technology, International Journal of Electrical, Computer, Energetic, Electronic and Communication Engineering* 9: 724–728.

13

Synthesis of Graphene-Based Nanomaterials from Biomass for Energy Storage

Gitanjali
Jothiprakash,
Sriramajayam
Srinivasan, Ramesh
Desikan, Deepa
Jaganathan and
Karthikeyan
Subburamu

Agricultural Engineering
College and Research
Institute, Tamil Nadu
Agricultural University,
Coimbatore, India

Graphene is a promising nanomaterial with a honeycomb carbon lattice. Biomass is a rich source of carbon and can be used to make graphene in a sustainable way. To combat the problem of energy and sustainable pollution, there is a great need for sustainable development that will be used in efficient energy storage systems and environmental applications. However, it is a huge undertaking to keep up with energy density in energy storage systems using nanomaterials embedded in biomass. The increase in demand for graphene-based nanomaterials has improved as it has higher mechanical strength, electrical conductivity, chemical durability, thermal durability, and a higher surface area with lower production costs. This chapter discusses various techniques and characteristics of graphene-based nanomaterials.

13.1 Introduction

The Hummers' technique is one of the most common methods for producing graphene materials with the risk of explosion, high production costs, and emissions (Yu et al. 2016). Energy-saving devices, namely, batteries, fuel cells, and capacitors require something with significant electrical, chemical, thermal, and mechanical properties. Graphene nanomaterials synthesized from biomass will be a suitable material that meets all the requirements for use as an energy reservoir (Thirupathi et al. 2020). The various syntheses or production of graphene-based nanomaterials from biomass are solvothermal, pyrolysis, vapor deposition, oxidation, exfoliation, microwave, and plasma. These nanomaterials are characterized for application in electrical power devices such as electrical capacitors, supercapacitors, and batteries based on surface, structural, thermal, optical, and electrochemical properties.

13.2 Raw Material and Its Properties

Lignocellulosic biomass is a promising raw material for graphene-based nanomaterials (Poorna et al. 2021). Biomass is divided into four categories namely field-level residues, crop residues, agricultural industry residues, and biomass from uncultivated land. The remnants of the field are grass, nuts, sugarcane, corn cobs, cotton stalks, red gram stalks, tapioca stalks, sunflower heads, bean stalks, etc. Crop residues are casuarina, peanut shell, coconut shell, arecanut shell, jatropha, etc. Some of the remnants of the agricultural industry are sawdust, logs, rice husks, and peanut shells. *Prosopis juliflora*, *Melia dubia*, and other wild woods fall under the biomass from uncultivated land. Mostly the biomass produced is burned or disposed of before it is processed, and a very small amount is processed as fuel. Efforts are needed to convert this massive resource into a more sustainable resource, thus reducing environmental problems. Biomass with high-flexibility content produces a large number of hydrocarbons during the energy conversion process. These hydrocarbons can be used as a carbon source for the synthesis of graphene-based nanomaterials (Safian et al. 2020). The choice of biomass in graphene-based nanomaterials depends on their physicochemical properties, and therefore it is important to place them in the right energy storage medium (Yan et al. 2020). Product integration function is governed by physical, chemical, and thermal biomass factors. Visual factors that have helped to understand the dynamic behavior of biomass-responsive systems and chemical structures are important in studying the reactions that occur within the reactor. Basic trace elements essential to the analysis and design of a biomass conversion unit include proximate analysis (ash content, moisture content, volatile matter, and fixed carbon), ultimate analysis or elemental composition (carbon, hydrogen, oxygen, sulfur, calcium, nitrogen, etc.), and chemical composition (cellulose, hemicellulose, and lignin).

13.3 Fabrication/Synthesis of Graphene-Based Nanomaterials from Biomass

13.3.1 Methods of Synthesis

Graphene-based nanomaterials can be synthesized in a top-down and bottom-up manner. Electrochemical oxidation and electrochemical exfoliation come from the top-down method and the vapor depositor, pyrolysis, solvothermal, microwave, and plasma-assisted synthesis fall intobottom-up method (Xu et al. 2021).

13.3.1.1 Electrochemical Oxidation

Electrochemical oxidation of graphite to produce a combination of graphite intercalation is an ancient method that emerged in the 1970s using high voltage or current. The same method can be used to assemble graphene nanomaterials using a current or voltage higher than that required to form a graphite intercalation composite (Lowe et al. 2019). These high-energy forces lead to the formation of oxidation and the expansion of carbon materials into layers of graphene. The classification of composite graphene nanomaterials is now attracting more researchers as it overcomes many of the barriers caused by the chemical oxidation process (Komoda and Nishina 2021). This method avoids the use of chemicals, oxidation can be controlled by various electrical forces, the structure can be altered by providing the right reaction gases, and in particular, this is a non-explosive process. In general, this method has an amazing advantage compared to other routes, meeting the industry in terms of mass production.

13.3.1.2 Electrochemical Exfoliation

Electrochemical exfoliation is a useful and increasingly effective way of converting carbon material into graphene. This compounding method is a liquid chemical method that minimizes energy-intensive processes and thus costs that lead to a higher level of graphene production performance (Xu et al. 2018). This can be achieved by the oxidation of carbon electrodes into mineral acids or alkaline solutions. Charged ions cause graphite content in carbon electrodes to be expanded by graphene by expansion or extraction. Chemical spraying can be accompanied by increased filtration in the composition of nano-materials based on graphene.

13.3.1.3 Physical Vapor Deposition

Graphene carbon films are produced from biochar materials using the filtered cathodic vacuum arc (FCVA) front-facing system. Biomass is converted to biochar using any of the methods of pyrolysis, i.e., slow pyrolysis and rapid pyrolysis. The biochar produced is crushed into microparticles. These biochar microparticles are packaged as carbon films by the FCVA system. The system contains a curved magnetic filter to reduce the deposition of macroparticles (Oldfield et al. 2015).

13.3.1.4 Chemical Vapor Deposition

Through the chemical vapor deposition (CVD) process, graphene-based nanomaterials can be repaired by decomposition and binding of the binding agent (any carbon source) in the carbon pore structure to produce the smallest structure (Santhiran et al. 2021). Biomass with higher volatile content yields a substantial quantity of hydrocarbons upon pyrolysis. These hydrocarbons can be utilized as a carbon source for the synthesis of graphene-based nanomaterials. The CVD technique is of greater importance in preparing carbon molecular sieves with better adsorption ability and higher micropore volume. It is possible to accumulate carbon by using various organic chemicals such as benzene or methane at higher temperatures. During the process, the pore diameters became narrow with vapor. Thus, the process creates more micropore structures (Saeed et al. 2020). Biomass with high-flexibility content produces a large number of hydrocarbons during pyrolysis. These hydrocarbons can be used as a carbon source for the synthesis of graphene-based nanomaterials. The CVD method is very important in preparing carbon molecular filters with better advertising ability and higher micropore volume. It is possible to accumulate carbon using various natural chemicals such as benzene or methane at high temperatures. During the process, the pore diameters become smaller with evaporation. Therefore, the process creates many micropore structures. The activated carbon samples taken from the biomass are placed in a fixed bed reactor and heated to a set temperature of 800–1000°C under the nitrogen environment. Once the deposition was stabilized, the gas stream was converted to benzene at different flow rates and held at different deposit times. Once the installation was complete, the reactor was allowed to cool by flowing nitrogen gas (Ahlatc 1999).

13.3.1.5 Solvothermal

Solvothermal is one of the thermo-chemical modifiers with high pressure and low temperature. It incorporates hydrolysis, pyrolysis, dehydration, polymerization, and organic aromatization and produces a product composed of the content of the oxygen group and the aromatic content (Piao 2016). In the solvothermal method, biomass is converted to carbon materials, i.e., char and graphene by the process of carbonization and reduction. Char is based on the polymerization of aromatics, polysaccharides, ketone, aldehydes, furan, and graphene on water-soluble acid and low-molecular fatty acids through pyrolysis, rearrangement, and dehydration (Chin et al. 2019).

13.3.1.6 Pyrolysis

Graphene nanomaterials are synthesized by biomass carbonization at 600°C under anaerobic conditions and carbon material decomposition at a high temperature of 1000°C under natural nitrogen (Kong et al. 2020). These materials must be cooled without external disturbances to produce several layers of graphene-based nanomaterials and pyro products. The gases released during this process can be shortened to produce biocrude.

13.3.1.7 Microwave-Assisted Synthesis Methods

Microwave technology is widely used in the food processing industry as it provides uniform heating of food products from within. Due to its energy-saving features, this technology protects against high usage in the field of material science. Microwave radiation in graphite or carbon material provides homogeneity in dispersing particle size and low crystalline duration. Heat is formed during the interaction of radiation with the active groups of carbon materials. Polar solvent and oxidation in surface areas cause similarity of graphene sheets (Hong et al., 2012). Graphene properties can be altered by changing the reaction temperature and concentration of the polar solvent (Al-Hazmi et al. 2015). This combination of graphene-based nanomaterials is an easy-to-use technology with the added advantage of low reaction time and low-cost production technology.

13.3.1.8 Plasma-Assisted Synthesis Methods

Plasma processes have received increasing attention in the development of a renewable energy pathway to overcoming pollution barriers. A high plasma temperature of about 6000°C, according to the low power input, will facilitate the production of graphene compared to other existing production techniques (Ouyang et al. 2021). This is due to the very low graphene yield obtained due to its cost of raw materials and the high cost of graphite energy. Surprisingly, synthetic graphene-based nanomaterials will become an eco-friendly way from biomass-derived carbon through a low-cost plasma assisting the process (Jothiprakash et al. 2015). High efficiency based on graphene and nanomaterials should be maintained by plasma-assisted systems, using dissolving carbon material feeders (Dey et al. 2016). This leads to the development of this production method to satisfy the commercial prices of graphene-based nanomaterials.

13.3.2 Problems That Apply to Synthesis Methods

Graphene-based nanomaterials are synthesized using a step-by-step method using acids and oxyl-groups. These routes have economic and environmental impacts associated with the real-world potential for product development. The constant supply of electrical energy of the electrochemical process causes abnormal formation. However, a redesigned system for amplification may allow for a much higher process of assembling nanomaterials. In the middle of the road to assembling graphene-based nanomaterials, evaporation techniques have many problems. This process requires expensive equipment operating at high temperatures. The vapor suspension depends on substrates with geometrics with many characteristics and is slow compared to other processes. The installation of chemical vapors depends on the precursor characteristics and the efficiency of the vaporizer and the flow rate. This leads to inefficient product integration. Pyrolysis cannot stand alone in a combination of things and should be integrated with other integration routes (Huang et al. 2021). However, it is more effective in producing graphite-rich carbon dioxide, which can be used as a raw material for graphene production. The solvo-thermal process is the best in lab-scale production, while the scale requires a modified system and uses additional power to maintain reactor pressure. Microwave- and plasma-assisted mixing routes require proper management of energy, time, and temperature to maintain the solids to integrate well withthe different types of carbon dioxide found in biomass.

13.4 Benefits of Graphene Synthesis from Biomass

Biomass-based fuels mainly biochar, hydro-char, activated carbon, and syngas are used in a variety of industrial and domestic applications. Rapid economic development in recent years has allowed for the rapid production and use of biomass due to strict environmental laws and economic considerations. Coal and wood are the most common precursors for the production of graphene-based nanomaterials (Abbas et al. 2021). The path to such a biodiversity-based economy will only be successful if the feedstock is adequately available not only for energy production but also for other biomass-based chemicals and products (Yuan et al. 2021). However, biomass materials in the form of powder, granule, and pellet can be an alternative to coal and wood for graphene synthesis.

13.5 Characterization of Graphene-Based Nanomaterials

13.5.1 Surface Characterization

The surface characterization is performed using a scanning electron microscope (SEM) and atomic force microscope (AFM). These microscopes provide detailed information on the topography of objects and how they are constructed under the modification and distribution of pores in the graphene product (Myung et al. 2019).

13.5.2 Structural Characterization

The structural characterization can be performed using Fourier-transform infrared spectroscopy (FT-IR), Raman spectroscopy, transmission electron microscopy (TEM), and X-ray powder diffraction (XRD). These techniques are used to identify structural compounds and changes in morphology and ultimately validate compounds and structures in advanced graphene materials (Liu et al. 2021).

13.5.3 Thermal Characterization

Thermal stability is studied in thermogravimetric analysis (TGA) by changing reaction temperature and time (Yakovlev et al. 2019).

13.5.4 Optical Characterization

Ultraviolet-visible spectroscopy (UV-Vis) is used to study the detection of graphene-based nanomaterials by absorbing electromagnetic radiation for use in solar cells (Mohandoss et al. 2018).

13.5.5 Electrical Characterization

Electrical conductivity is one of the important indicators of the use of graphene-based nanomaterials in energy storage applications (Zhang et al. 2019).

13.5.6 Microwave Characterization

The dispersing structure of graphene-based nanomaterials should be studied by measuring complex clearance, bright light, and dielectric power with different wavelengths (Savi et al. 2017).

13.6 The Use of Graphene-Based Nanomaterials for Energy Storage

Graphene-based nanomaterials have a variety of applications such as fuel cells and solar power cells, capacitors, energy storage batteries, wastewater, gas purification membranes, and biomedical equipment (Zhang et al. 2019). Having amazing properties, these nanomaterials play a major role in energy storage devices.

13.6.1 Electric Capacitors

Graphene compounds embedded in biomass can be used as electrode materials in capacitors. Graphene nanomaterials have high performance to withstand high pressures, high chemical resistance, and temperature, which increase the overall life of the capacitor (Kovalchuk et al. 2020).

13.6.2 Supercapacitors

The electrodes in the supercapacitor require high-strength building materials, mechanical stability, electrical stability, and especially electrochemical stability (Yeleuov et al. 2021). This can be achieved with graphene derived from biomass. The structural morphology of graphene materials enhances the efficiency and health of the supercapacitor cycle. Film-shaped, fiber-based, substrate-supported supercapacitors can be developed using these nanomaterials (Purkait et al. 2017). Long cyclic life and high-energy density are also achievable while using these materials. Supercapacitor energy storage occurs at the anode and cathode without chemical reactions. Electrical double layer and pseudocapacitors are different types of supercapacitors in energy storage systems (Agudosi et al. 2020).

13.6.3 Batteries

Batteries play a vital role in electrical activity as they are an important component. Electronic devices (mobile phones and power backups), portable devices, and renewable energy storage devices require long-lasting batteries with low weight and size. Graphene materials appear to be the best choice of electrodes for lithium-ion, lithium-sulfur, and lithium-air batteries (Sui et al. 2021). Graphene nanomaterials will replace these graphene materials as they are excellent for electrochemical performance, regenerative property, specific density, durability, pore formation, and surface area.

13.7 Challenges and a Vision for the Future

Graphene-based nanomaterials from biomass endure a major challenge that must be overcome in order to be marketed and widely used. The proper mixing method must be combined with a combination of two or more composite methods for producing these high-quality and quantitative nanomaterials in an environment-friendly manner. Production parameters control is required to obtain the complete or desired product from a scientific point of view (Saha et al. 2022). In a well-rounded view of the future, a piece of better knowledge of these material products is needed to match the other relevant assets for energy storage applications.

13.8 Conclusion

The production of graphene-based nanomaterials from carbonaceous materials by many efficient technologies provides a separate energy supply by storing improper materials. Future global economic development may result in the rapid growth of electricity demand, shortages, and problems such as global warming and climate change. Therefore, there is a need to identify sustainable energy storage equipment in order to conserve energy without polluting the environment. These graphene-based nanomaterials will play a key role in balancing energy demand by storing them. These materials act as building blocks with their superconductivity, electron mobility, stability, and durability. This chapter addresses various integration approaches and their limitations in production in order to meet the need for the development of an energy storage device.

References

Abbas, A., S. Abbas, T.A. Tabish, S.J. Bull, A.N. Phan, and T.M. Lim. 2021. "Role of Precursor Microstructure in the Development of Graphene Quantum Dots from Biomass." *Journal of Environmental Chemical Engineering* 9, no. 5: 106154.

Agudosi, E.S., E.C. Abdullah, A. Numan, N.M. Mubarak, M. Khalid, and N. Omar. 2020. "A Review of the Graphene Synthesis Routes and Its Applications in Electrochemical Energy Storage." *Critical Reviews in Solid State and Materials Sciences* 45, no. 5: 339–77.

Ahlatc, H. 1999. "Production, Modification and Usage Areas of Diamonds Thin Films." *Meta* 23, no. 122: 24.

Al-Hazmi, F.S., G.H. Al-Harbi, G.W. Beall, A.A. Al-Ghamdi, A.Y. Obaid, and W.E. Mahmoud. 2015. "One-Pot Synthesis of Graphene-Based on Microwave-Assisted Solvothermal Technique." *Synthetic Metals* 200: 54–57.

Chin, S.J., M. Doherty, S. Vempati, P. Dawson, C. Byrne, B.J. Meenan, V. Guerra, and T. McNally. 2019. "Solvothermal Synthesis of Graphene Oxide and Its Composites with Poly (ε-Caprolactone)." *Nanoscale* 11, no. 40: 18672–82.

Dey, A., A. Chroneos, N. St, J. Braithwaite, R.P. Gandhiraman, and S. Krishnamurthy. 2016. "Plasma Engineering of Graphene." *Applied Physics Reviews* 3, no. 2: 021301-1.

Hong, Y.C., S.J. Lee, D.H. Shin, Y.J. Kim, B.J. Lee, S.Y. Cho, and H.S. Chang. 2012. "Syngas Production from Gasification of Brown Coal in a Microwave Torch Plasma." *Energy* 47, no. 1: 36–40.

Huang, H., N.G. Reddy, X. Huang, P. Chen, P. Wang, Y. Zhang, Y. Huang, P. Lin, and A. Garg. 2021. "Effects of Pyrolysis Temperature, Feedstock Type and Compaction on Water Retention of Biochar Amended Soil." *Scientific Reports* 11, no. 1: 1–19.

Jothiprakash, G., S. Pugalendhi, S. Kamaraj, S. Karthikeyan, V.J.F. Kumar. 2015. "Feasibility test of agricultural residues through characterization for utilization in plasma gasification." *Indian Journal of Agricultural Sciences* 85, no. 12: 1534–39.

Komoda, M., and Y. Nishina. 2021. "Electrochemical Production of Graphene Analogs from Various Graphite Materials." *Chemistry Letters* 50, no. 3: 503–9.

Kong, X., Y. Zhu, H. Lei, C. Wang, Y. Zhao, E. Huo, X. Lin, Q. Zhang, M. Qian, W. Mateo, R. Zou, Z. Fang, and R. Ruan. 2020. "Synthesis of Graphene-like Carbon from Biomass Pyrolysis and Its Applications." *Chemical Engineering Journal* 399: 125808.

Kovalchuk, O., S. Uddin, S. Lee, and Y.W. Song. 2020. "Graphene Capacitor-Based Electrical Switching of Mode-Locking in All-Fiberized Femtosecond Lasers." *ACS Applied Materials & Interfaces* 12: 54005–54011.

Liu, H., W. Chen, R. Zhang, C. Xu, X. Huang, H. Peng, C. Huo, M. Xu, and Z. Miao. 2021. "Bioinspired in Situ Self-Catalyzing Strategy towards Graphene Nanosheets with Hierarchical Structure Derived from Biomass for Advanced Supercapacitors." *Applied Surface Science* 566: 150692.

Lowe, S.E., G. Shi, Y. Zhang, J. Qin, L. Jiang, S. Jiang, M. Al-Mamun, P. Liu, Y.L. Zhong, and H. Zhao. 2019. "The Role of Electrolyte Acid Concentration in the Electrochemical Exfoliation of Graphite: Mechanism and Synthesis of Electrochemical Graphene Oxide." *Nano Materials Science* 1, no. 3: 215–23.

Mohandoss, M., and A. Nelleri. 2018. "Optical Properties of Sunlight Reduced Graphene Oxide Using Spectroscopic Ellipsometry." *Optical Materials* 86: 126–32.

Myung, Y., S. Jung, T.T. Tung, K.M. Tripathi, and T. Kim. 2019. "Graphene-Based Aerogels Derived from Biomass for Energy Storage and Environmental Remediation." *ACS Sustainable Chemistry & Engineering* 7, no. 4: 3772–82.

Oldfield, D.T., D.G. McCulloch, C.P. Huynh, K. Sears, and S.C. Hawkins. 2015. "Multilayered Graphene Films Prepared at Moderate Temperatures Using Energetic Physical Vapour Deposition." *Carbon* 94: 378–85.

Ouyang, D.D., L.B. Hu, G. Wang, B. Dai, F. Yu, and L.L. Zhang. 2021. "A Review of Biomass-Derived Graphene and Graphene-like Carbons for Electrochemical Energy Storage and Conversion." *New Carbon Materials* 36, no. 2: 350–72.

Piao, Y. 2016. *Preparation of Porous Graphene-Based Nanomaterials for Electrochemical Energy Storage Devices*, pp. 229–52. Springer, Dordrecht.

Poorna, A.R., R. Saravanathamizhan, and N. Balasubramanian. 2021. "Graphene and Graphene-like Structure from Biomass for Electrochemical Energy Storage Application- A Review." *Electrochemical Science Advances* 1, no. 3: 1–16.

Purkait, T., G. Singh, M. Singh, D. Kumar, and R.S. Dey. 2017. "Large Area Few-Layer Graphene with Scalable Preparation from Waste Biomass for High-Performance Supercapacitor." *Scientific Reports* 7, no. 1: 1–14.

Saeed, M., Y. Alshammari, S.A. Majeed, E.A. Nasrallah. 2020. "Chemical Vapour Deposition of Graphene Synthesis, Characterisation, And Application: A Review." *Molecules* 25, no. 3856: 2–62.

Safian, M.T., U.S. Haron, and M.N.M. Ibrahim. 2020. "A Review on Bio-Based Graphene Derived from Biomass Wastes." *Bio Resources* 15(4), 9756–9785.

Saha, J.K., and A. Dutta. 2022. "A Review of Graphene: Material Synthesis from Biomass Sources." *Waste and Biomass Valorization* 13, no. 3: 1385–1429.

Santhiran, A., P. Iyngaran, P. Abiman, and N. Kuganathan. 2021. "Graphene Synthesis and Its Recent Advances in Applications—A Review." *C* 7, no. 4: 76.

Savi, P., K. Naishadham, S. Quaranta, M. Giorcelli, and A. Bayat, 2017. "Microwave characterization of graphene films for sensor applications." *2017 IEEE International Instrumentation and Measurement Technology Conference (I2MTC)*.

Sui, D., M. Chang, Z. Peng, C. Li, X. He, Y. Yang, Y. Liu, and Y. Lu. 2021. "Graphene-Based Cathode Materials for Lithium-Ion Capacitors: A Review." *Nano* 11, no. 10: 1–28.

Thiruppathi, A.R., B. Sidhureddy, E. Boateng, D.V. Soldatov, and A. Chen. 2020. "Synthesis and Electrochemical Study of Three-Dimensional Graphene-Based Nanomaterials for Energy Applications." *Nano* 10, no. 7: 1–25.

Xu, M., A. Wang, Y. Xiang, and J. Niu. 2021. "Biomass-Based Porous Carbon/Graphene Self-Assembled Composite Aerogels for High-Rate Performance Supercapacitor." *Journal of Cleaner Production* 315: 128110.

Xu, Y., H. Cao, Y. Xue, B. Li, and W. Cai. 2018. "Liquid-Phase Exfoliation of Graphene: An Overview on Exfoliation Media, Techniques, and Challenges." *Nano* 8, no. 11: 942.

Yakovlev, A.V., E.V. Yakovleva, V.N. Tseluikin, V.V. Krasnov, A.S. Mostovoy, L.A. Rakhmetulina, and I.N. Frolov. 2019. "Electrochemical Synthesis of Multilayer Graphene Oxide by Anodic Oxidation of Disperse Graphite." *Russian Journal of Electrochemistry* 55, no. 12: 1196–1202.

Yan, Y., F.Z. Nashath, S. Chen, S. Manickam, S.S. Lim, H. Zhao, E. Lester, T. Wu, and C.H. Pang. 2020. "Synthesis of Graphene: Potential Carbon Precursors and Approaches." *Nanotechnology Reviews* 9, no. 1: 1284–1314.

Yeleuov, M., C. Daulbayev, A. Taurbekov, A. Abdisattar, R. Ebrahim, S. Kumekov, N. Prikhodko, B. Lesbayev, and K. Batyrzhan. 2021. "Synthesis of Graphene-like Porous Carbon from Biomass for Electrochemical Energy Storage Applications." *Diamond and Related Materials* 119: 108560.

Yu, H., B. Zhang, C. Bulin, R. Li, and R. Xing. 2016. "High-Efficient Synthesis of Graphene Oxide Based on Improved Hummers Method." *Scientific Reports* 6: 1–7.

Yuan, S.J., B. Dong, and X.H. Dai. 2021. "Facile and Scalable Synthesis of High-Quality Few-Layer Graphene from Biomass by a Universal Solvent-Free Approach." *Applied Surface Science* 562: 150203.

Zhang, H., Z. Zhang, J.D. Luo, X.T. Qi, J. Yu, J.X. Cai, J.C. Wei, and Z.Y. Yang. 2019. "A Chemical Blowing Strategy to Fabricate Biomass-Derived Carbon-Aerogels with Graphene-Like Nanosheet Structures for High-Performance Supercapacitors." *Chem Sus Chem* 12, no. 11: 2462–70.

14

Distributed Optical Fiber Sensing System for Leakage Detection in Underground Energy Storage Pipelines Using Machine-Learning Techniques

Dr. T. Kavitha

Veltech Rangarajan
Dr. Sagunthala R&D
Institute of Science and
Technology, Chennai, India

Dr. P. Nagarajan

Asso.Professor/ECE
SRMIST, Vadapalni Campus,
Chennai, Tamililnadu, India

A. Arulmary

Anna University, Chennai,
India

Dr. A. Adaikalam

Anna University BIT
Campus, Trichy, India

Pipeline transmission is the most supportable and harmless method of transporting energy sources from the production facilities to the many end-users. Pipeline integrity is critical for a safe operation in this environment, and it must be prioritized while passing metropolitan areas. Despite the system operators' best efforts, there is no such thing as zero risk when it comes to energy transmission and industrial activity, so extra caution is required to avoid damage to the pipeline. This is especially true when we take into account that most problems with natural gas transmission infrastructure are brought on by intervention from outside sources, mostly because of third-party pipeline construction, some of which tragically result in human lives. In addition to personal losses, incidents resulting in energy supply disruptions and gasoline leaks result in significant economic losses and environmental damage.

DOI: 10.1201/9781003355755-14

Pipeline security has improved tremendously, still a considerable demand is there for high-performance and economic systems for continuous monitoring of possible threats to pipeline integrity. Distributed acoustic sensing technology is particularly well adapted to the length of the given pipeline. Researchers and industries are becoming increasingly interested in combining distributed acoustic sensing with a pattern recognition system. It's used to identify and categorize possible dangerous events that happen above fiber-optic cables installed along pipelines, with the goal of constructing pipeline surveillance system. This chapter aims to implement the optical fiber–based pipeline leak detection and then suggest a machine-learning technique for monitoring the pipeline. This technique uses the time-domain characteristics to identify the leaks and the frequency-domain data to identify the location. Signal to noise ratio (SNR) value and correlation coefficient will be increased by using this approach. The accuracy of detecting pipeline leaks and locating the leakage spots will be significantly improved.

14.1 Introduction

Pipeline integrity is critical for a harmless process in this environment, and it should be prioritized while passing metropolitan areas. Despite the system operators' best efforts, there is no such thing as zero risk when it comes to energy transmission and industrial activity, so extra caution is required to avoid damage to the pipeline. Along with personal losses, incidents due to energy supply disruptions and gasoline leaks result in significant economic losses and environmental damage [1].

Fiber-optic technology research and development has encouraged innovative changes in today's life. Technological developments in fiber-optics have covered the way for the widespread networks of fiber-optic communication, commonly recognized as the super highway data. Today, we depend strongly on fiber-optic networks to transmit and receive online information, voice, and video signals. Fiber-optic sensor technology can be regarded a fiber-optic communication technology extension. In applications such as industrial automation, healthcare, aerospace, and aviation, it is widely used. For health monitoring applications, many optical fiber sensor technologies have been created with commercially available products. Large-scale engineering structure specifications now frequently include instrumentation to address surveillance demands, not only during the building period but also to enable structural health monitoring (SHM) throughout the lifetime. Controlling the state of health of a structure requires a more number of sensors, commonly referred to as SHM. Optical fiber sensors are discovered to be outstanding instruments, particularly as they allow distributed measurements. It provides the information over the entire structure rather than being limited to point data at sensor location. Single fiber monitoring can thus provide data on the general behavior of the structure and thus overcome the constraints of traditional sensors whose data are limited to local effects [2]. To overcome the initial disappointments and make full use of the specifications of these sensors, whose implementation has become start of the art, some 20 years of advances have been essential. Pipelines are widely utilized around the world to transport flammable and explosive gas over millions of kilometers [3]. To reduce fiber usage, the fiber is suspended freely outside the pipeline.

The proposed work objectives are

- Pipelines are widely utilized around the world to transport flammable and explosive gas over millions of kilometers.
- Pipeline structures are designed to withstand a variety of environmental loading conditions to enable safe and dependable distribution from the point of production to the coast or distribution station.
- Leaks in pipeline networks are one of the leading causes of losses (environmental disasters, human deaths, and financial loss) for pipeline operators and nature.
- Pipeline leak detection and localization using various approaches have been a major focus of study to avoid this issue and maintain a secure and dependable pipeline system.

- The main objective of the proposal is to use Rayleigh scattering–based DAS system for monitoring gas pipeline leakage. To reduce fiber usage, the fiber is suspended freely outside the pipeline.
- The next objective is to apply the machine-learning technique to pipeline surveillance system.

14.2 Review of Status

Pipeline safety planes and monitoring requirements are increasing as the existing gas pipeline infrastructure ages. The development of a stable and dependable leak monitoring system has piqued the interest of both domestic and international researchers. Stress wave method is the traditional method for monitoring the leakage in the pipeline [4]. Due to short-distance monitoring, this approach is not suitable for complex pipe networks. Although the negative pressure wave approach overcomes the pipeline length constraint, it gives poor result on monitoring effect on minor leaks. Ultrasonic testing needs costly equipment and stringent pipeline standards, which cannot be checked in real-time [5]. The pressure gradient method and mass flow balance method are two software-based detection methods that provide real-time monitoring, but they have a high false alarm rate, hence they are rarely used. Online pipeline leakage monitoring is possible with a Supervisory Control and Data Acquisition (SCADA) system; however, the system is huge, and the reliability is low [6].

Impulse response method was investigated by Kim [7] to identify the oscillatory flows in pipeline. He employed a genetic algorithm–based impulse response method to locate the leak. Hou et al. [8, 9] presented a negative pressure wave approach utilizing a Bragg grating sensor with slight pipeline width distortion. Two sensors were connected on the exterior and interior surfaces of the Bragg grating pressure sensor that Huang et al. created and used to detect leaks in prestress concrete pipelines. Ma et al. [10] developed a flow balance–based negative wave pressure solution to deal with the issue of poor positioning accuracy and a high false alarm rate. To tackle the problem of false leak detection, Mandal et al. [11] suggested a novel leak detection approach based on rough set theory and support vector machine (SVM). They also used an SVM classifier to inspect the examples that were missed by the imposed criteria. Da Silva [12] described an approach for detecting pipeline leaks that included clustering and classification algorithms for fault identification. They classified the running mode and identified operational and process transients using a fuzzy method. Qiong and Shidong [13] created a pipeline leak detection system using LabVIEW as a virtual instrument. The proposed system continuously measured pressure and flow at both ends of the pipeline, analyzed the data using LabVIEW's wavelet analysis tools, recognized leaks through flow variations, and located leaks using the negative pressure wave approach. A group of expert systems were proposed by Laurentys et al. [14] to track and identify pipeline breaches in real-time. Fuzzy logic, neural networks, genetic algorithms, and statistical analysis are among the approaches they used to enable the system to perform its function. Lu et al. [15] presented the structure of leakage finding system, they provided a full description of the properties of the wavelet transform and the denoizing concept. Verde [16] suggested a method for detecting leak locations using flow and pressure sensors at the duct's ends, assuming a nonlinear fluid model. Lee et al. [17] proposed a method for single-pipe leak detection in which the system frequency response diagram (FRD) behavior was employed as an indicator of pipe integrity. The presence of a leakage in a pipe executes a pattern on the FRD resonance peaks that can be utilized to identify leakage. Dos Santos et al. [18] introduced a new method for detecting gas leaks in high-pressure distribution networks, in which two leak detectors were treated as a linear parameter varying (LPV).

Fiber-optic sensing (FOS) approaches have been studied as a superior alternative to electrical sensing for a number of reasons over the last decade: (a) immune to Electro magnetic interference and radio frequency interference; (b) inherently safe due to the employment of nonelectrical sensor components; (c) low weight and compact; (d) appropriate for real-time automated data acquisition; and (e) more sensing points. To detect leaks in oil and gas pipelines, Raman scattering–distributed optical fiber sensing technology was used by Vogel et al. [19]. They were able to detect leaks by monitoring temperature changes around the pipeline, the system was easily affected by the surrounding environment. Brillouin-scattered distributed optical fiber System technology was discussed by Jia et al. [20] used to monitor the pipeline leakage. Stress changes were monitored during leakage, it removes the effect of environmental temperature but simulated the corresponding experiment. Rayleigh scattering–based distributed optical fiber sensing technology is used by Monica et al. [21] to simulate pipeline leakage by measuring fiber bending loss, but they overlooked the effect of pipeline vibration on the experiment. Brillouin optical time-domain analyzer–based online monitoring system was complex to design. Stajanca et al. [22] proposed the system based on the vibration, in their system optical fiber is wound spirally on the pipe wall.

Indian researchers and structural engineers are interested in health monitoring and evaluation systems because of their ability to provide spatial and quantitative information about structural deterioration and performance throughout a structure's life cycle. Pipeline inspection systems based on sensor networks have received a lot of attention in natural gas pipeline checking and monitoring applications. The latest research and development and the use of various methods of health monitoring and assessment of underground oil and gas pipelines are presented by Arun et al. [23] The use of distributed fiber sensing, macro fiber ceramics, piezo sensors, and wireless sensors for pipeline health monitoring was considered. They addressed that monitoring and assessing the health of oil and gas pipelines is a hot research topic. Mishra et al. [24] studied and examined the idea of using distributed fiber-optics sensing systems to actively and frequently monitor leakages and also their article focuses on the leakage detection and monitoring of 20 petroleum product pipelines with lengths ranging from 7 to 10 km. In their paper, they reviewed that, this is one of the first initiatives in India to use distributed temperature sensing to identify leaks. Ramji and Prasant [25] examined the current state of fiber-optic–distributed temperature sensing systems, as well as their principles of operation, problems, and possible applications in their research paper.

Sampath and Bhattacharya's [26] paper highlighted the use of optical sensing to find anomalies in pipes using an inline crawler robot (ICR). There are two types of optical sensors used to analyze abnormalities and the health of a pipeline (light-dependent resistor (LDR) sensors and light-to-voltage sensors). The experiment takes place in a lab test bed with optical sensor modules moving across a pipeline with simulated physical abnormalities on the inner surface. Himanshu and Sahu [27] studied an overview of the current developments in distributed fiber-optic sensors as well as their performance. The performance characteristics of several fiber-optic–distributed sensors are investigated.

The objective of Ajay et al. [28] research's was to monitor and control pipelines using wireless sensors and LabVIEW software. Wireless sensors are linked to pipelines in their system, which continuously monitor and transmit their data to a control room, where LabVIEW software was used to monitor these values and manage the pipes automatically based on reference values. Sudeep et al. [29] presented a three-layered architecture for monitoring and safeguarding over ground oil, gas, and water pipelines using wireless sensor networks.

14.3 Importance of the Proposed Work

Machine-learning technology provides a standard framework for processing data and allowing strong and high-level knowledge to aid decision-making processes as the amount of data in improved storage facilities grows. Many fields take the advantages of machine-learning technology such as speech processing, biometric, and text processing. The field of DAS is very new, and it can offer high-quality information that can illustrate physical impacts over great distances. Based on the literature, it is understood that real-world deployment of a PRS based on DAS technology is still a tough area of research. Improved performance rates and accurate results in field positioning should be established to permit an industry-extensive implementation of the DAS+PRS approach for pipeline surveillance.

14.4 Methodology

Coherent Rayleigh scattering–based DAS system is used in this proposal to monitor leakage in gas pipeline. Direct-modulated laser source emits the coherent light pulse which is sentto the sensing fiber via a modulator. When the interruption occurs, the fiber refractive index at the invader point varies, this will lead to the changes in phase of the backscattered Rayleigh light. The intensity of scattering light will alter as a result of the interference effect. The detector notices the Rayleigh-scattered light reflected from various fiber points and extracts the faded disturbance signal. When the optical fiber is affected by the pipeline leaking sound, the refractive index of the interruption point varies because of elastic effect. As a result, the phase difference between two scattered waves changes, therefore the intensity of backscatter Rayleigh light changes.

The linear pipe is positioned in the muffler chamber, and test will be conducted with a background noise of 40 dB and a frequency of less than 1000 Hz. Pipeline is made up of three steel pipe segments that are joined together by flanges. The diameter and length of the pipeline will be 0.005 and 15 m, respectively. In the leak detection zone, every 10 cm simulated leak hole is there, which is situated directly above the pipe and both the ends of the pipe areclosed, and pipe has a leak hole with different diameters. The hose connects the controllable valve to the inside of the pipe, which is then coupled to air compressor via the pressure gauge. The air compressor pressurizes the pipe's internal pressure after the controlled valve is opened as shown in Figure 14.1.

The 1550 nm wavelength laser diode (LD) is used to emit the light signal with a linewidth of 1.6MHz. The LD coherence length is 50 m, and the pulse length in the fiber will be 10 ns, thus minimizes the coherence noise between the consecutive pulses. The size of the leak hole and the pipe's internal pressure were combined and tested. The pressure inside the pipe is initially increased to specific value for each combination test, and the leaking hole is then opened after a small voltage stabilization. In the meantime, the DAS system will be tested for 5 minutes, and the pressure is slowly decreased when the gas leaks are discovered.

The proposed architecture consists of distributed acoustic system with PRS. The acoustic signal is generated by the acquisition equipment which is connected to DAS system, and that signal will be used for classification and training stage. The modules that contain the training stage are responsible for creating models that appropriately describe the input data attributes for each of the classes that are being considered. To create a final decision, the classification stage modules choose trained models which closely represent the input acoustic signals. The learned signal is fed into a feature extraction module, which extracts significant patterns. The relevant feature vectors are then given into a pattern classification algorithm, which either classifies the feature vector as a certain kind or develops the right models.

The goal of extraction is to enhance useful, discriminative data from acoustic signals captured by DAS system, and then the pattern classification module can identify each individual activity occurring at the top of the cable. Discrete or continuous wavelet transform–based features and short-time fast Fourier transform–based features are examples of time-frequency-domain features that strive to take advantage of both the time and frequency domains at the same time.

A pattern classification system's purpose is to categorize each input feature vector x as belonging to a specific class from a previously learned set of models. Maximum a Posteriori (MAP) criterion is commonly employed for this, with the assigned class label being chosen as the one that maximizes the posterior probability of the class given the input feature vector, which may be determined using the Bayes method. Two processes are required to develop a pattern classification system: training and classification (testing), both of which need the use of counterpart data. The training stage contains in learning about (usually statistical) models from the training data's associated set of feature vectors. Classification stage will use these statistical models. The classification stage uses the models created in the training stage to run the system and make predictions.

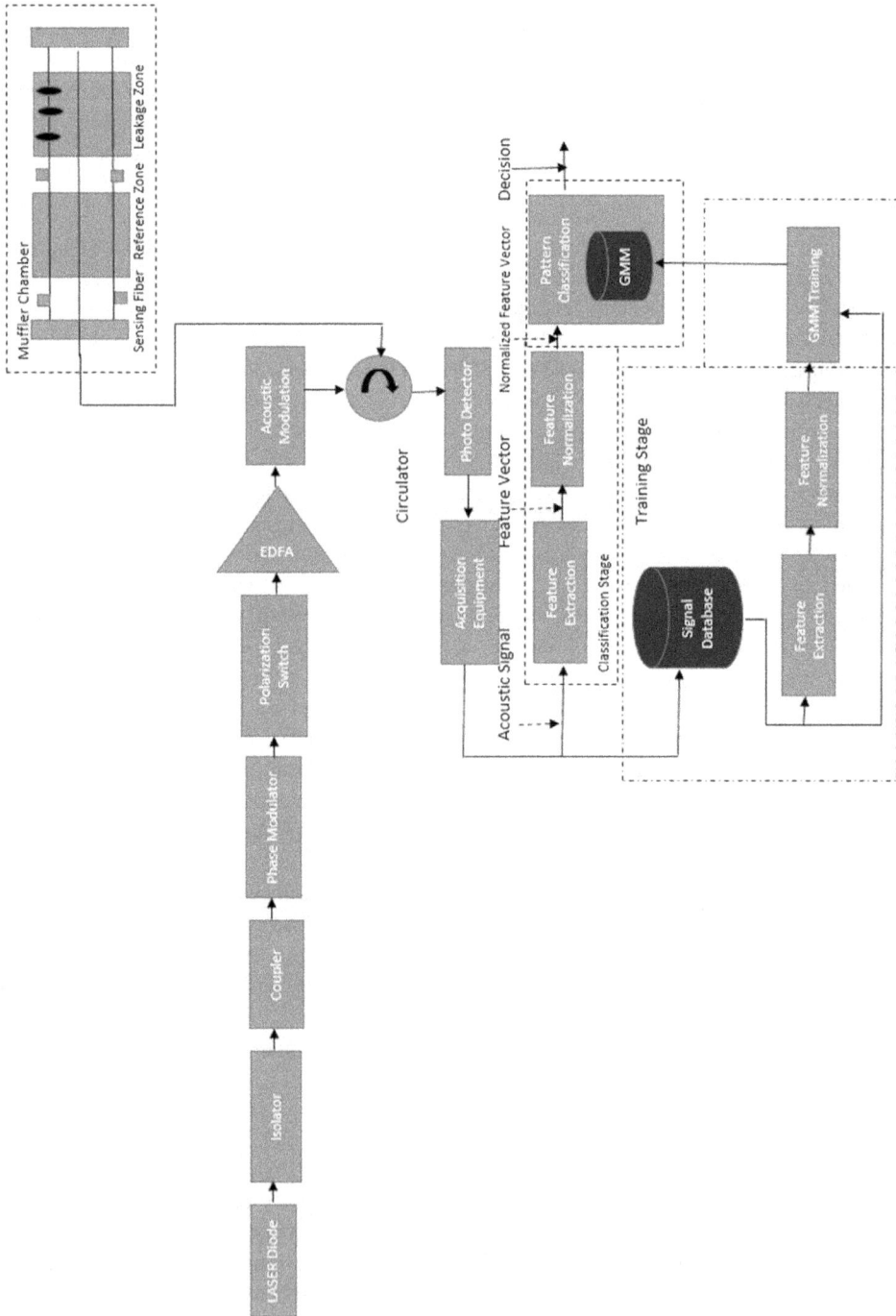

FIGURE 14.1 Proposed system.

The incident laser in a single-mode fiber is a rectangular pulse, according to the one-dimensional impulse response model of fiber back-Rayleigh scattering, and the amplitude of the back-Rayleigh scattering wave can be derived by injecting the fiber at time $t = 0$.

Rectangular pulse is applied to the single-mode fiber, at a time $t = 0$, back-Rayleigh scattering wave amplitude will be calculated

$$e(t) = \sum_{i=1}^{N} a_i \exp\left(-\frac{a}{2}\frac{c\tau_i}{n}\right)\exp[j2\pi v(t-\tau_i)]\,rect\left(\frac{t-\tau_i}{w}\right) \tag{14.1}$$

where $\tau_i = \dfrac{2nl_i}{c}$, c = number of scattering centers, l_i = length of the fiber, τ_i = time delay of the ith scattering wave, a_i = amplitude of the ith scattering wave, a = attenuation coefficient, c = speed of light, n = refractive index, v = frequency, and w = pulse width.

The detector's input will acquire a continuous backscatter Rayleigh wave with a period T = $1/f$ after continually injecting m pulses, and its amplitude is calculated as

$$e(t') = \sum_{i=1}^{N} a_i \exp\left(-\frac{a}{2}\frac{c\tau_i}{n}\right)\exp\left[j2\pi v\left(t'-\frac{k}{f}-\tau_i\right)\right]rect\left(\frac{t'-\dfrac{k}{f}-\tau_i}{w}\right) \tag{14.2}$$

Next the power of the light will be calculated by using

$$p(t') = |e(t')|^2 = p_a(t') + p_b(t') \tag{14.3}$$

where

$p_a(t')$ = sum of optical powers of each independent backscattering center
$P_b(t')$ = sum of the light power generated by the interference of the backscatter Rayleigh

The phase difference between the twoscattering waves is φ_{ij}

$$\Phi_{ij} = \cos 2\pi v(\tau_i - \tau_j) = 4\pi vn(li - lj) \tag{14.4}$$

The refractive index of the incursion point varies when the optical fiber is affected by the pipeline leaking sound due to the elastic light effect. As a result, the phase difference between the two scattered waves of the interference changes, and the backscatter Rayleigh light intensity changes.

14.5 Results and Discussion

The outcome of the proposed work is to identify pipeline failures caused by either vandalism or device/material failure and corrosion damages. The proposed system avoids the financial losses, injuries, and extreme environmental pollution and reduces the impacts of gas/oil spillage on society. All the electronic components of the DAS system such as LD, detector, modulators, and optical fiber are of lightweight. Design is validated for structural integrity, and also endurance through analysis, simulation, and testing. Risk associated with their in-service failure and consequential harm to the stack holders are substantially mitigated.

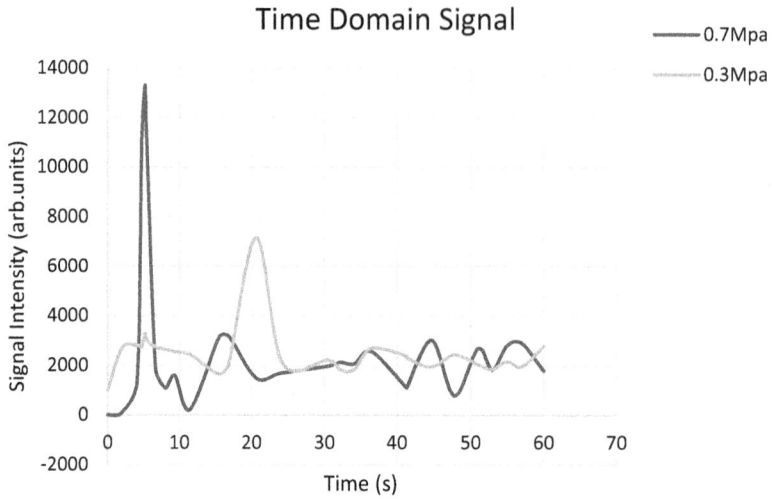

FIGURE 14.2   Time-domain signal characteristics of pipeline leakage.

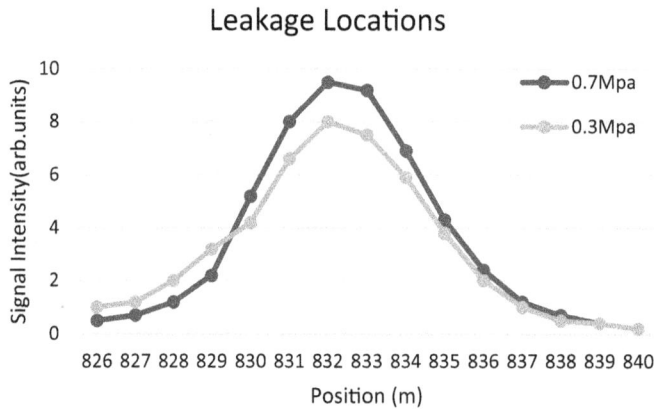

FIGURE 14.3   Leakage location with different pressure.

The monitoring signals are processed using a leakage detection technique based on wavelet transform. When the leaking aperture is 1 mm and the internal pressure is 0.3 and 0.7Mpa, the original time-domain signal is shown in Figure 14.2. Pipe leaking occurs in the 5th, 11th, and 19th seconds, making it difficult to distinguish between the effective pipeline leakage signals.

Figure 14.3 shows the positioning diagram when the pressure is 0.3 and 0.7Mpa, and leakage aperture is 0.1 mm. The maximum DAS signal intensity is around 835 m, which is nearly the same as the actual pipeline leakage hole. On both sides of the leak hole, the DAS signal gradually declines and tends to zero. This implies that the DAS system is capable of locating the pipeline leakage spot and that the leak-finding technique based on the time-domain accumulation average can help increase positioning accuracy.

14.6 Conclusion

Output: Real-time monitoring of gas pipeline is proposed based on DAS system using machine-learning concept. The sensing fiber is placed outside the pipeline to sense acoustic vibrations produced by leakage. Based on the previously obtained data sets, the machine-learning algorithm predicts the unknown data sets for similar conditions. Typically, this kind of data is used to create statistical models that may then be utilized to generate predictions. This algorithm can extract the time-domain and frequency-domain characteristics for pipeline leakage and location and hence increases the signal-to-noise ratio greatly boosting leak detection accuracy. The field trials will find the efficiency of the technology, which places the groundwork for DAS system to monitor gas pipeline leakage in real-time and accurately detect the leakage spot in real-world engineering applications.

Outcome: Pipeline failures are caused by either vandalism or device/material failure and corrosion damages. The proposed system avoids the financial losses, injuries, and extreme environmental pollution and reduces the impacts of gas/oil spillage on society. Environmental and structural parameters are continuously recorded 24/7 in SHM system. The obtained information is used to plan and design maintenance activities and also enhances the knowledge about the structure being monitored. Therefore, economic losses are reduced, safety is increased, and maintenance cost is optimized.

References

1. Huang, J., Zhou, Z., Zhang, D., Wei, Q., "A fiber bragg grating pressure sensor and its application to pipeline leakage detection," *Advances in Mechanical Engineering*, vol. 2013, pp. 1–6, 2013.
2. Liou, J.C.P., "Leak detection by mass balance effective for Norman wells line," *Oil & Gas Journal*, vol. 94, no. 17, pp. 69–74, Apr. 1996.
3. Wang, L.K., *Research on Some Key Issues in Leak Detection of Crude Oil Pipeline*. Tianjin, China: Tianjin University, pp. 9–11, 2002.
4. Cheng, Z.B., Wang, Z.H., Ma, H.W., "A brief review on damage detection in pipes using stress wave factor technique," *Journal of Taiyuan University of Technology*, vol. 34, no. 4, pp. 426–430, Apr. 2003.
5. Chen, J.Z., Yu, N.T., *Ultrasonic Testing New Technology*. Beijing, China: Science Press, 1991, pp. 43–45.
6. Cao, Y.L., Wang, W.M., Ge, L., Shi, J.J., Li, W.H., "Numerical simulation of the pressure gradient method to position pipeline leak," *Journal of Liaoning University of Petroleum & Chemical Technology*, vol. 32, no. 2, pp. 45–48, Feb. 2014.
7. Kim, S.H., "Extensive development of leak detection algorithm by impulse response method," *Hydraulic Engineering*, vol. 131, no. 3, pp. 201–208, 2005.
8. Hou, Q., Wenling, J., Shuhui, Z., Liang, R., Ziguang, J., "Natural Gas Pipeline Leakage Detection Based on FBG Strain Sensor," *2013 Fifth International Conference on Measuring Technology and Mechatronics Automation*, pp. 712–715, 2013.
9. Hou, Q., Ren, L., Jiao, W., Zou, P., Song, G., "An improved negative pressure wave method for natural gas pipeline leak location using FBG based strain sensor and wavelet transform," *Mathematical Problems in Engineering*, vol. 2013, pp. 1–9, 2013.
10. Ma, C., Yu, S., Huo, J., "Negative Pressure Wave-Flow Testing Gas Pipeline Leak Based on Wavelet Transform," *2010 International Conference on Computer, Mechatronics, Control and Electronic Engineering, CMCE 2010*, pp. 306–308, 2010.
11. Mandal, S.K., Chan, F.T.S., Tiwari, M.K., "Leak detection of pipeline: An integrated approach of rough set theory and artificial bee colony trained SVM," *Expert Systems with Applications*, vol. 39, pp. 3071–3080, 2012.
12. Da Silva, H.V., Morooka, C.K., Guilherme, I.R., da Fonseca, T.C., Mendes, J.R.P., "Leak detection in petroleum pipelines using a fuzzy system," *Journal of Petroleum Science and Engineering*, vol. 49, pp. 223–238, 2005.

13. Qiong, H., Shidong, F., "Development of Pipeline Leak Detection System Based on Lab VIEW," *2008 IEEE International Symposium on Knowledge Acquisition and Modeling Workshop*, pp. 671–674, 2008.

14. Laurentys, C.A., Bomfim, C.H.M., Menezes, B.R., Caminhas, W.M., "Design of a pipeline leakage detection using expert system: A novel approach," *Applied Soft Computing*, vol. 11, pp. 1057–1066, 2011.

15. Lu, X., Sang, Y., Zhang, J., Fan, Y., "A Pipeline Leakage Detection Technology Based on Wavelet Transform Theory," *2006 IEEE International Conference on Information Acquisition*, pp. 1432–1437, 2006.

16. Verde, C., "Accommodation of multi-leak location in a pipeline," *Control Engineering Practice*, vol. 13, pp. 1071–1078, 2005.

17. Lee, P.J., Vítkovský, J.P., Lambert, M.F., Simpson, A.R., Liggett, J.A., "Leak location using the pattern of the frequency response diagram in pipelines: A numerical study," *Journal of Sound and Vibration*, vol. 284, pp. 1051–1073, 2005.

18. Lopes dos Santos, P., Azevedo-Perdicoúlis, T.P., Jank, G., Ramos, J.A., Martins de Carvalho, J.L., "Leakage detection and location in gas pipelines through an LPV identification 58 approach," *Communications in Nonlinear Science and Numerical Simulation*, vol. 16, pp. 4657–4665, 2011.

19. Vogel, B., "Leakage Detection Systems by Using Distributed Fiber Optical Temperature Measurement," *Proceeding SPIE's 8th Annual International Symposium Smart Structure Materials*, Newport Beach, CA, pp. 23–34, 2001.

20. Jia, Z.A., Wang, H., Qiao, X.G., "A study for monitoring strain of oil and gas pipeline based on distributed optical fiber Brillouin scattering," *Journal of Optoelectronics Laser*, vol. 23, no. 3, pp. 534–537, Mar. 2012.

21. Mendoza, M., Carrillo, A., Márquez, A., "New distributed optical sensor for detection and localization of liquid hydrocarbons: Part II: Optimization of the elastomer performance," *Sensors and Actuators, A: Physical*, vol. 111, no. 2–3, pp. 154–165, Mar. 2004.

22. Stajanca, P., Chruscicki, S., Homann, T., Seifert, S., Schmidt, D., Habib, A., "Detection of leak-induced pipeline vibrations using fiber— Optic distributed acoustic sensing," *Sensors*, vol. 18, no. 9, p. 2841, Aug. 2018.

23. Arun Sundaram, B., Kesavan, K., Parrivallal, S., "Recent advances in health monitoring and assessment of in-service oil and gas buried pipelines," *Journal of The Institution of Engineers*, vol. 99, pp. 729–740, 2018.

24. Mishra, A., Soni, A., "Leakage Detection Using Fibre Optics Distributed Temperature Sensing," *6th Pipeline Technology Conference*, 2011.

25. Tangudu, R., Sahu, P., "Review on the developments and potential applications of the fiber optic distributed temperature sensing system," *IETE Technical Review*, vol. 39, pp. 553–567, 2021.

26. Sampath, S., Bhattacharya, B., "Optical sensors for inspection and monitoring of pipeline using inline crawler robot, ISSS national conference on MEMS, smart materials," *Structures and Systems*, vol. 8, pp. 1–6, 2016.

27. Pradhan, H.S., Sahu, P.K., "A Survey on the Performances of Distributed Fiber Optic Sensors," *International Conference on Microwave, Optical and Communication Engineering*, IIT, Bhubaneswar, India, Dec. 18–20, 2015.

28. More, A., Parab, N., Zinjad, P., Patil, S., "Pipeline monitoring using Lab VIEW," *International Journal of Engineering Research & Technology (IJERT)*, vol. 3, no. 6, pp. 1–3, 2015.

29. Varshneya, S., Kumara, C., Swaroop, A., "Leach Based Hierarchical Routing Protocol for Monitoring of Over ground Pipelines Using Linear Wireless Sensor Networks," *6th International Conference on Smart Computing and Communications, ICSCC 2017*, pp. 7–8, Dec. 2017.

15

Influence of Nanomaterials on the Ionic Conductivity and Thermal Properties of Polymer Electrolytes for Li⁺-Ion Battery Application

D. Ravindran

Velammal College of Engineering and Technology, Madurai, India

P. Vickraman

Gandhigram Rural University, Dindigul, India

Polymer electrolytes find a wide variety of applications in solid-state rechargeable batteries, fuel cells, supercapacitors, etc. The incorporation of nano-/micron-sized particles in polymer electrolyte system yields better favorable electrical and thermal properties. In this work, we have synthesized CdO, CuO, SnO$_2$ and ZnO nano-sized particles through cost-effective co-precipitation and solid-state milling method. The synthesized particles were subjected to XRD and SEM analysis to confirm the purity of sample and observe its morphological properties. These nanomaterials were dispersed in PVC-PVdF (poly(vinyl chloride)-poly(vinylidene fluoride)) blend–based electrolytes with lithium perchlorate (LiClO$_4$) as dopant salt, propylene carbonate and poly(ethylene glycol) as plasticizers. XRD study shows a reduction in crystallinity of the electrolyte membrane due to the incorporation of fillers that favor ion transport mechanism. The influence of various fillers on the ionic conductivity and thermal properties is studied in interest. Almost all the system showed two maxima behavior that could be ascribed to filler effect at lower concentration and formation of conducting interfacial space-charge region between

DOI: 10.1201/9781003355755-15

the filler particle grain boundaries and polymer host at higher content. Among the fillers dispersed, CdO exhibits better conductivity, which could be attributed to its particle size which might have facilitated uniform dispersal in reducing the agglomeration effect. The SEM analysis reveals the formation of larger number of micropores due to the incorporation of fillers. TGA studies show an increase in thermal stability in the normal operating temperature range.

15.1 Introduction

The demand for solid polymer electrolytes (SPEs) with high ionic conductivity, good transport number, electrochemical stability and sufficient mechanical integrity is increasing as it is an integral part of an effort to fabricate a new generation of rechargeable solid-state batteries. SPE has emerged as an attractive electrolyte/separator material for application in all solid-state batteries, supercapacitors, fuel cells, dye-sensitized solar cells, etc., due to their highly desirable properties such as ease in process ability, versatility and dimensional stability (Sankaran et al. 2018 nano-/micro-sizedactive/inert). But their lower ionic conductivity at ambient temperature minimized their device application. To enhance the conductivity, various techniques have been carried out, such as blending of compatible polymers, plasticization and dispersion of inert/active nano/microfillers (Aziz et al. 2018). The inclusion of lower molecular weight and higher dielectric constant plasticizers such as ethylene carbonate (EC), propylene carbonate (PC), dimethyl carbonate (DMC) to the polymer-salt system has been regarded as Fruitful technique. Plasticized polymer electrolytes are more amorphous than unplasticized electrolytes and improve segmental motion of longer polymer chain molecules that enhances the conductivity of the system, but with poor mechanical properties. The issue of poor mechanical stability can be rectified with a polymer like poly(vinyl chloride) (PVC) as one of the component in the blend system. It has unique feature of phase-separated morphology due to the lower miscibility with plasticizers and solvent resulting in distinct plasticizer-rich region and polymer-rich region. The former provides sufficient pathways for mobile ions, and the latter act as a substrate to aid mechanical strength. Polymer blending is widely used method for synthesizing polymer-based materials with improved properties, which is unachievable by single constituent. But demonstration of desired characteristic properties depends on the better compatibility and miscibility of the blend components (Rocco et al. 2001). Due to blending of different structured polymers, molecular-level interactions occur across the blend constituents that yield relatively lower degree of crystalline phase for the resulting polymer blend system. Another feasible approach as mentioned earlier, dispersion of nano/micro –sized active/inert fillers to the polymer matrix improves mechanical integrity and also enhances the lithium cation transport properties such as electrochemical stability and interfacial stability with electrodes. Earlier studies reveal that the composite polymer electrolytes (CPEs) possess enhanced electrolyte/electrode compatibilities and safety hazards (Pradeep and Shikha 2011). Composite electrolytes are comprised of three main basic components, polymer matrix acting as a substrate to hold the ingredients, dopant salt the charge carriers and filler. The role of filler is to maneuver polymer-ion and ion-ion interactions to facilitate efficient pathways for ion migration (Ciosek et al. 2007). The formation of interfacial region between inorganic fillers and polymer host matrix may provide the necessary channel for ion migration.

Several polymeric materials such as poly(vinylidene fluoride) (PVdF; Mohamed and Arof 2004), poly (methyl methacrylate) (PMMA; Sun et al. 2019), polyacrylonitrile (PAN; Biying Huang et al. 1996), PVC (Subramaniam et al. 2002), poly(vinyl alcohol) (PVA; Sunitha et al. 2020) and poly(vinyl pyrrolidine) (PVP; Venkata Subba Rao et al. 2012) have been used as host polymers for synthesizing polymer electrolytes. Among these, PVdF ($[-CH_2-CF_2-]_n$) is a semi crystalline polymer, and PVdF-based polymer electrolytes are attractive due to their salient features such as high anodic stability (Tsutsumi et al. 1991), higher permittivity, relatively low dissipation factor and higher dielectric constant ($\varepsilon \approx 8.4$) facilitating ionization of lithium salts and thereby providing larger number of charge carriers. PVdF matrix acts as inert substrate, retains gel constituent part within the porous framework, and PVC provides necessary mechanical integrity and has good compatibility with additives and plasticizers (Koh Singh Ngai et al. 2016).

TABLE 15.1 Composition and Conductivity of CdO-, SnO$_2$-, CuO- and ZnO-Dispersed Polymer Electrolytes at Room Temperature

S. No.	PVdF (wt%)	PVC (wt%)	PEG (wt%)	PC (wt%)	LiClO$_4$ (wt%)	Filler (wt%)	$\sigma \times 10^{-5}$ S/cm			
							CdO	CuO	SnO$_2$	ZnO
1	5	20.0	15	50	10	0	0.181	0.181	0.181	0.181
2	5	19.5	15	50	10	0.5	1.921	0.259	0.326	0.696
3	5	19.0	15	50	10	1.0	2.035	0.129	2.305	0.654
4	5	18.0	15	50	10	2.0	3.485	0.576	1.502	1.748
5	5	16.0	15	50	10	4.0	6.214	0.101	3.566	5.124

In this work, we have synthesized PVC-PVdF blend polymer electrolyte with lithium per chlorate (LiClO$_4$) as salt and CdO, CuO, SnO$_2$ and ZnO as nano additives. PC is chosen as a plasticizer due to its higher dielectric constant ($\varepsilon' = 64.4$ at 25°C) and low molecular weight, which will improve the detachment of ion pairs of the salt (Ravindran 2016). Poly (ethylene glycol) (PEG) is added as an additional component as it can serve as an additional plasticizer.

The polymer blend composite electrolytes are synthesized by widely practiced casting procedure by systematically changing (PVC: Filler) weight ratio as 20.0: 0 wt%, 19.5: 0.5 wt%, 19.0:1 wt%, 18.0: 2 wt% and 16.0:4 wt%, respectively, while PVdF, PEG and PC are fixed at 5, 15 and 50 wt% (Table 15.1). The membranes without any trace of filler are labeled as FX, and coded as FD2–FD5 with CdO, FU2–FU5 for CuO dispersed, FS2–FS5 for SnO$_2$ dispersed and FZ2–FZ5 for ZnO incorporated.

15.2 Experimental Details

15.2.1 Synthesis of CdO Nanoparticles

Cadmium oxide nanoparticles were synthesized by cost-effective co-precipitation method using cadmium acetate and ammonium hydroxide. Cadmium acetate (0.5 M) was dissolved in a cleaned glass beaker containing 100 mL deionized water, and ammonia solution was introduced dropwise until pH value of 8 was attained with continuous stirring. The milky white color precipitate formed was left to settle for 6–7 h and then filtered using high quality filter paper. The resulting crystalline powdery materials were collected by centrifugation and washed repeatedly with distilled water and ethanol 6–7 times. The resulting powder was dried at 100°C in air and grinded. The powder was collected and calcined at 400°C for 2 h. The powder after calcinations turns yellowish, which confirms the formation of CdO.

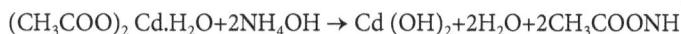

$$(CH_3COO)_2 \, Cd.H_2O + 2NH_4OH \rightarrow Cd \, (OH)_2 + 2H_2O + 2CH_3COONH$$

During calcination, the prepared powder loses the water content, and CdO was formed.

$$Cd \, (OH)_2 \rightarrow CdO + H_2O$$

15.2.2 Synthesis of CuO Nanoparticles

Copper oxide (CuO) nanoparticles were prepared by co-precipitation method (Narsinga Rao et al. 2009). Aqueous solution of 0.1M Cu (NO$_3$)$_2$.2H$_2$O was prepared in a clean glass container. To this solution, 0.9 M NaOH solution was introduced dropwise under continuous stirring at ambient temperature. The NaOH solution was added until the pH of the system reaches 13. The blue-colored gel formed was washed with deionized water several times, until it is free of nitrate ions. Then it was centrifuged and dried in air at 60°C for 12h. The resulting final material was annealed in air at 500°C for 3 h to obtain CuO particles.

15.2.3 Synthesis of Tin Oxide (SnO₂) Particles

Crystalline tin oxide (SnO_2) was synthesized by adding ammonia solution $NH_4.OH$ dropwise to 0.05 M aqueous solution of $SnCl_4.5H_2O$ at 40°C. This dropwise addition was continued till pH value attains 8 to obtain a milky white precipitate of tin hydroxide. The settled product was detached using a centrifuge operating at 2500 rpm for 25 min. The resulting whitish gel was washed with deionized water several times and again with ethanol 4–5 times to remove excess chlorine ions. The final resulting mass was dried for 24 h at 100°C in a hot oven to remove the moisture present. The dried powder was gently ground in a mortar and finally calcined at 600°C in a muffle furnace for 2h (Patil et al. 2012).

15.2.4 Synthesis of Zinc Oxide (ZnO) Nanoparticles

Zinc oxide nanoparticles were synthesized by a non-aqueous solid-state grinding method. Zinc acetate dehydrate $[Zn (CH_3COO)_2.2H_2O)]$ and citric acid ($C_6H_8O_7$) were used in this process. High-purity zinc acetate and citric acid powders were weighed separately and gently mixed in 1: 1 molar ratio. The powders were slowly ground for 2h at room temperature using a well-cleaned mortar and pestle. The final resulting product was collected in heat-resistant crucible and calcined in air at 600°C for 2 h.

15.2.5 Preparation of Polymer Electrolyte

High-purity PVC with average M_w ~62,000 and PVdF with an average molecular weight of 5.3×10^5 were procured from Sigma-Aldrich, and PEG with average M_w ~200 from S.D Fine chemicals, India. The dopant salt lithium perchlorate ($LiClO_4$) purchased from Sigma-Aldrich was dried at 110°C under vacuum for at least 24 h prior to use to remove the moisture present if any at the time of usage. The appropriate wt% of salt was first dissolved in anhydrous tetrahydrofuran (THF), and after complete dissolution, other ingredients such as polymers, plasticizer and filler were added and continuously stirred for 18–20 h at ambient temperature. The resulting highly viscous slurry was cast on cleaned glass petri dish. Prior to casting the obtained gel solution was degassed to remove air bubbles. The residual solvent was removed by evaporation at ambient temperature and further dried in a temperature-controlled oven at 50°C for 12 h to eliminate any traces of THF present in the sample. The polymer free–standing flexible thin films thus obtained were preserved in desiccators. The same procedure was repeated with 0, 0.5, 1, 2 and 4 wt% of filler, by varying the (PVC: filler) ratio without altering other constituents.

15.3 Characterization Techniques

15.3.1 X-ray Diffraction Analysis

The degree of crystalline features of the electrolyte membrane was examined using a JEOL, JDX 8030 X-ray diffractometer with Cu Kα radiation of wavelength λ = 1.5418 A° at a scan rate of 10° per minute. The samples were kept on glass slides to make their insertion into the machine easier. The area of the films on the glass slides was maintained the same for all the samples, so that comparison of intensities would be possible.

15.3.2 Conductivity Measurements

AC conductivity studies were performed using a computer-enabled micro-autolab type III impedance analyzer in the frequency range of 50–100 KHz with an oscillation potential of 10 mV. The samples for analysis were cut into smaller circular disks and sandwiched between two stainless electrodes specifically fabricated with a small spring system to confirm good firm contact between the electrode and the electrolyte membrane. The measurements were carried out in a dry inert atmosphere to prevent absorption of moisture.

15.3.3 SEM Analysis

The surface morphological features of the prepared samples were observed by a JEOL, JSM-840A scanning electron microscope. All the samples were gold coated with a sputter coater prior to imaging, and all the samples' images were recorded at three different magnifications.

15.3.4 Thermal Analysis

Thermal stability properties were analyzed using a thermogravimetric analyzer (Perkin Elmer; Pyres diamond thermogravimetry/differential thermal analysis, TG/DTA) under N_2 atmosphere at a heating rate of 10°C per minute from 30°C to 600°C.

15.4 Results and Discussion

15.4.1 XRD Studies

15.4.1.1 XRD Pattern of Cadmium Oxide Particles

Figure 15.1 depicts the XRD profile of the synthesized cadmium oxide (CdO) particles. The prominent fingerprint diffraction peaks at 2θ values of 32.9°, 38.2°, 55.2°, 65.8° and 69.2° corresponding to the reflections of planes (111), (200), (220), (311) and (222) are clearly seen and match standard reference patterns for the face-centered cubic (fcc) phase of CdO (Joint Committee for Powder Diffraction Studies (JCPDS) File No. 05-0640). The breadth of the Bragg peak is due to the combination of instrument and sample-dependent effects. To separate these contributions, the diffraction pattern of silicon is obtained under identical conditions to determine the instrumental broadening. The instrumental corrected broadening (β) corresponding to the diffraction peak of CdO was determined using the relation,

$$\beta = [\beta_o^2 - \beta_i^2]^{1/2}$$

$$D = K\lambda / (\beta \cos\theta),$$

where β_o is the observed line width at half-maximum height (FWHM) in radians, β_i is the FWHM for silicon powder, D is the crystallite size, λ is the wavelength of the X-ray radiation ($\lambda = 0.15406$ nm) for

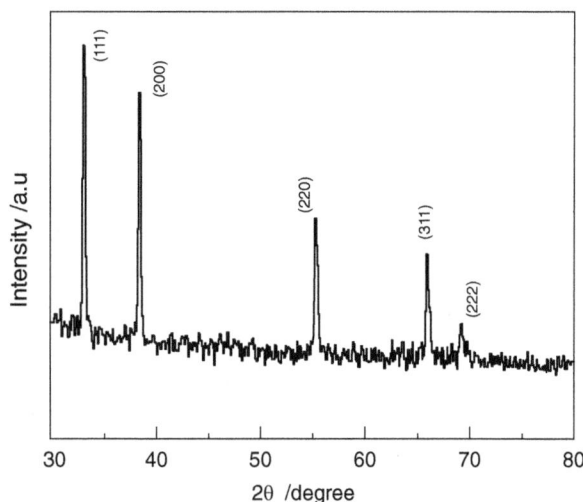

FIGURE 15.1 XRD pattern of synthesized cadmium oxide (CdO).

CuKα; K is usually taken as 0.89 and θ is the angle of reflection [19]. The crystallite size of CdO particles was found to be 64.2 nm. The same method was followed in estimating the crystallite size of other filler particles.

15.4.1.2 XRD Pattern of Copper Oxide Particles

Crystalline features of the synthesized CuO nanoparticles were known from their corresponding powder XRD patterns (Figure 15.2). The observed prominent diffraction peaks were well matched with monoclinic phase of CuO, space group C2/c (standard JCPDS File No: 05-661). Diffraction prominent peaks with $2\theta = 35.45°$ and $38.73°$, respectively, were indexed to (002) and (111), planes. The peaks are broader due to the smaller size of the synthesized particles, and the average particle size was 12 nm.

15.4.1.3 XRD Pattern of Tin Oxide Particles

The XRD profile of the SnO_2 is shown in Figure 15.3 and the peaks observed corresponding to 2θ values of 26.6°, 33.8°, 37.9°, 51.8°, 54.7°, 61.9° and 65.9° can be associated with (110), (101), (200), (211), (220), (310) and (301), respectively. It can be matched with standard (hkl) planes confirming SnO_2 having

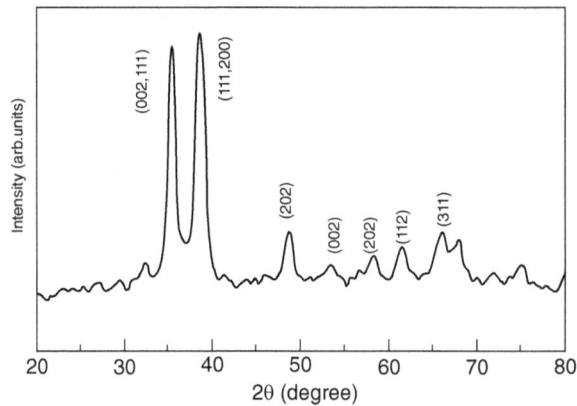

FIGURE 15.2   XRD pattern of synthesized copper oxide (CuO).

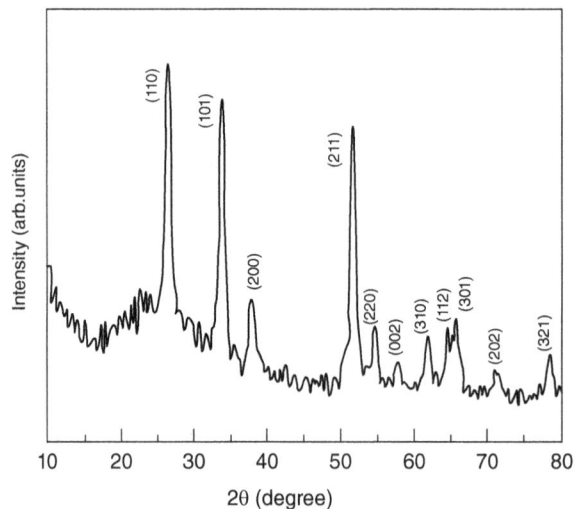

FIGURE 15.3   XRD pattern of synthesized tin oxide (SnO_2).

a tetragonal rutile structure which is in good agreement with the standard values (JCPDS card no. 41–1445) and earlier reported studies (Ganesh et al. 2012 and Patil et al. 2012). The average crystallite size (D) was determined to be 12.4 nm using the Scherrer equation.

15.4.1.4 XRD Pattern of Zinc Oxide Particles

The XRD profile of the ZnO powder shown in Figure 15.4 is having a hexagonal phase (wurtzite structure). No impurity peaks related to Zn, Zn (OH$_2$), or other ZnO phases are detected indicating the purity of the sample. The characteristic fingerprint peaks intensity is sharper and narrow, confirming that the sample is of good quality with good crystallinity. The peaks that originated from (100), (002), (101), (102), (110), (103), (200), (112) and (201) planes correspond to the hexagonal ZnO (JCPDS card no: 36-1451) was observed. The pattern coincides well with the result reported in a previous study (Yang et al. 2005). The mean crystallite size of the particles estimated by modified Debye Scherrer's formula is 21.3 nm (Figure 15.4).

15.4.2 XRD Studies of Polymer Electrolyte

The XRD pattern of the pure polymers PVC, PVdF, lithium salt, lithium per chlorate (LiClO$_4$) and fillers used in this system such as cadmium oxide (CdO), copper oxide (CuO), tin oxide (SnO$_2$) and Zinc oxide (ZnO) are given in Figure 15.5(a). The peaks observed at 2θ = 18.2°, 20°, 26.6° and 38° due to the reflections corresponding to the planes (100), (020), (110) and (020) pertaining to pure PVdF confirm its semi crystalline nature.

The XRD pattern of the composite film with PVC-rich phase: filler zero phase without filler (FX) and PVC poor phase: filler-rich phase, with CdO (FD), CuO (FU), SnO$_2$ (FS) and ZnO (FZ) is shown in Figure 15.5(b). The film without filler exhibits a broad hump indicating the amorphous property of the composite film due to the inclusion of the plasticizer and salt. The absence of the crystalline peaks of PVdF in the blend system shows the amorphous nature of the resulting composite film. The polymer segmental mobility of polymer molecular chains is much higher in amorphous region than the crystalline domains. The amorphous structure facilitates greater ionic diffusivity, and high ionic conductivity can be observed in amorphous membranes that have entirely flexible backbone (Frech and Chintapalli 1996). The peaks related to the lithium salt are absent in the film signifying that dopant salt is completely

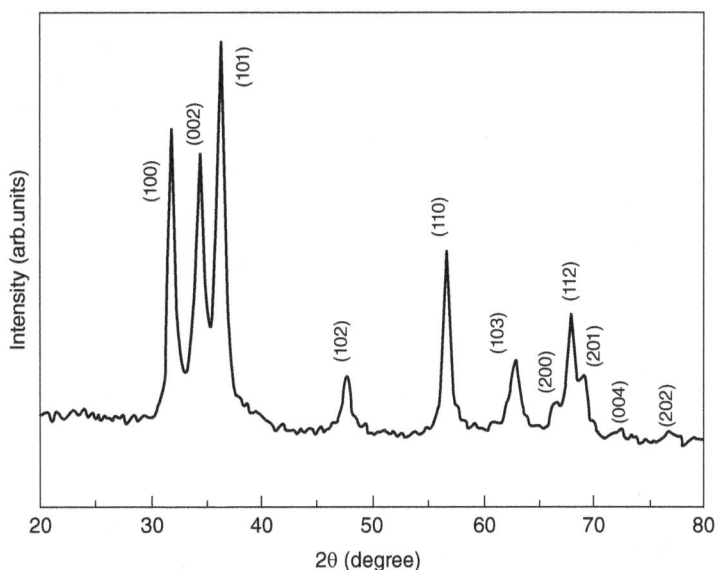

FIGURE 15.4 XRD pattern of synthesized zinc oxide (ZnO).

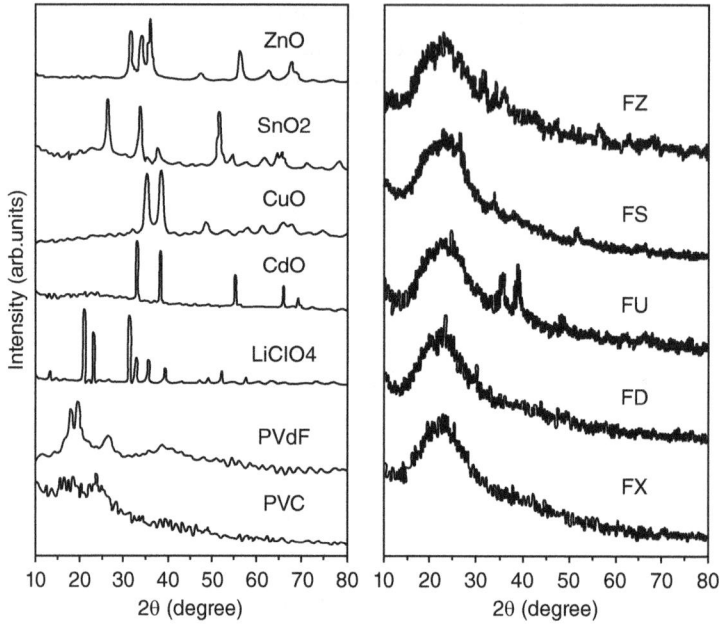

FIGURE 15.5 (a) XRD pattern of pure polymer and (b) XRD pattern of film without *and nanofillers filler and with filler.*

dissolved by the host polymer matrix and confirms good miscibility (Bashir Abubakar Abdulkadir et al. 2021). Some of the peaks related to the fillers (CuO and ZnO) are distinctly seen in the XRD pattern of the composite film, suggesting that the filler exists as a separate phase in the composite electrolyte films.

15.4.3 Ionic Conductivity Studies

The magnitude of ionic conductivity of electrolyte film is related to density of conducting species and their mobility features. The ionic conductivity value was estimated using the relation, $\sigma = l / R_b (\pi r^2)$, where l is the average thickness of the electrolyte film, r is the radius of the membrane and R_b is the bulk resistance which can be observed from the plots of real impedance Z_r against imaginary impedance Z_i. The complex impedance plot of filler-free and filler-dispersed films is shown in Figure 15.6. The plot depicts a depressed semicircle at the higher frequency region that relates to the effective bulk resistance, and a spur in the lower frequency region could be related to the interfacial impedance (Subba Reddy et al. 2007). If the electrode/electrolyte interface is ideal, vertical spikes parallel to imaginary impedance axis at low frequency region might have been observed. The linear spikes inclined at an angle less than 90° to the Z_r axis are generally related to the roughness of the electrode/electrolyte interface.

Table 15.1 depicts the composition and conductivity values of without filler and filler-dispersed electrolyte films. The content (wt%) of the fillers is kept at same value to compare and analyze its effect on the conductivity properties of the electrolytes.

The ionic conductivity of a polymer electrolyte can also be expressed by the mathematical equation $\sigma = \sum \mu_i n_i q_i$; where μ represents the mobility, n the concentration and q the charge of the conducting species. The profiles of ionic conductivity at ambient temperature with respect to nanofiller concentration are shown in Figure 15.7. Non-linearity in variation of conductivity with respect to filler content is observed. The conductivity of filler-dispersed electrolyte films shows two maxima behavior. This type of two maxima behavior has been reported in ion conducting–plasticized polymer electrolyte composite systems (Pandey et al. 2009). The first conductivity maximum is related to the creation of free charge

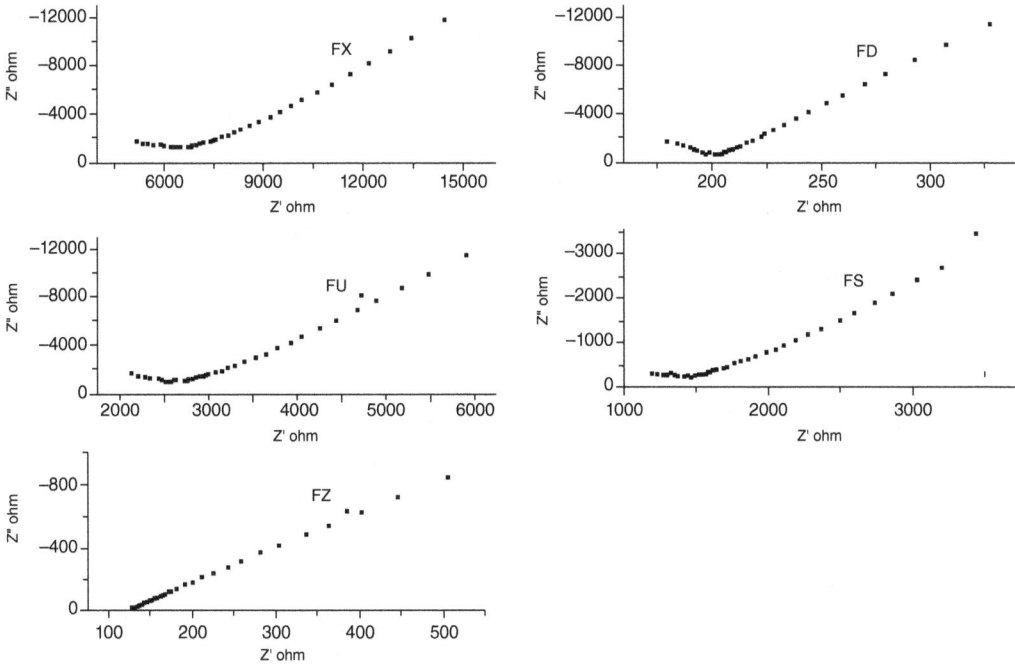

FIGURE 15.6 Impedance plot of PVdF/PEG/PVC– polymer film: without filler (FX), with CdO (FD), with CuO (FU) and with ZnO (FZ).

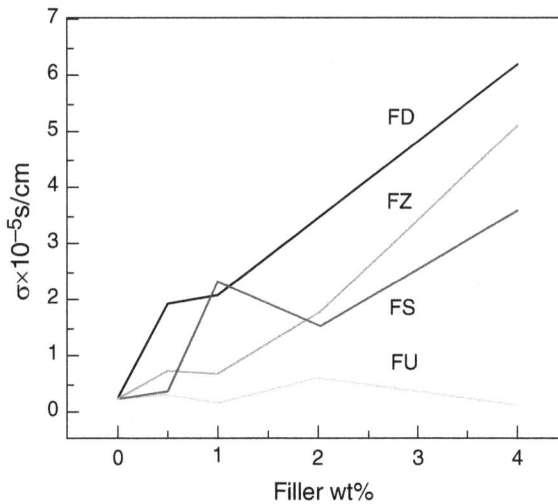

FIGURE 15.7 Variation conductivity with filler content.

carriers due to the dispersion of filler particles in the blend polymer host. The second maximum is attributed to the composite effect and can be elucidated due to the development of a conducting interfacial space-charge double region between the filler particle grain boundaries and polymer host (Maier 1994). The decline in the conductivity value after the second conductivity maximum is due to the obstructive effect of filler particles, which hampers the motion of mobile ions (Kumar et al. 2006).

The CdO- and ZnO-added films exhibit first maxima at 0.5 wt% and second at 4 wt% of filler, whereas SnO_2-dispersed films show at 1 wt% and 4 wt% of filler, and CuO-incorporated films exhibit at 0.5 and 2 wt% with filler content. Yap et al. (2013) have studied the impact of filler particle size on the conductivity profile of PEO-based polymer electrolytes and observed a waning in the conductivity value when smaller size particles are employed. In this work, the average particle size of CdO (64.2 nm) and ZnO particles (21 nm) is much higher than that of SnO_2 (12.4 nm) and CuO (12 nm), which is in good agreement. This could be attributed to the inadequate uniform dispersal of filler particles leading to aggregation even at lower concentration. The interaction of nanoparticles among themselves is higher than with polymer host or cation and anion of lithium salt.

15.4.4 Thermal Studies

In order to compare thermal properties of the electrolytes, synthesized samples were exposed to TG/DTA analysis in the temperature range of 30–600°C. The TGA curves for the samples without filler (FX), with CdO (FD), with SnO_2 (FS), with CuO (FU) and with ZnO (FZ) are shown in Figure 15.8.

The film without filler (FX) is almost stable upto 100°C without losing any significant weight loss. At 150°C, it undergoes 5% weight loss and at 200°C, it suffers 12% loss in weight, and this could be attributed to the vaporization of moisture absorbed or remaining solvent present if any in the specimen. The second weight loss starts at 175°C and ends at 280°C, where the specimen suffers nearly 22% weight loss, and this may be due to the decomposition of PVdF in the specimen. The third drastic decrease in weight occurs in the temperature range 270–299°C in which 34% loss in weight is observed. The TGA curve for the sample with CdO filler (FD) is stable up to 200°C without losing any weight, and this could be due to the dryness of the film and moisture-free environment during measurement. There is sharp weight loss at 200°C, where the film loses nearly 8% weight followed by a gradual slow loss in weight in the temperature range 210–230°C. After this, the film suffers a drastic fall in its weight losing nearly 59% up to 300°C. From the graph, we can observe that the film is thermally stable up to 220°C, which is much above the normal working temperature.

The TGA curve for SnO_2-added film (FS) shows a marginal weight loss of 5% at 100°C and around 8% at 150°C, and this may be ascribed to the residual plasticizer/solvent present in the specimen at the time of loading or absorption of moisture by the sample during measurement. A gradual weight loss starts at 160°C and ends at 280°C, where the film undergoes 23% weight loss.

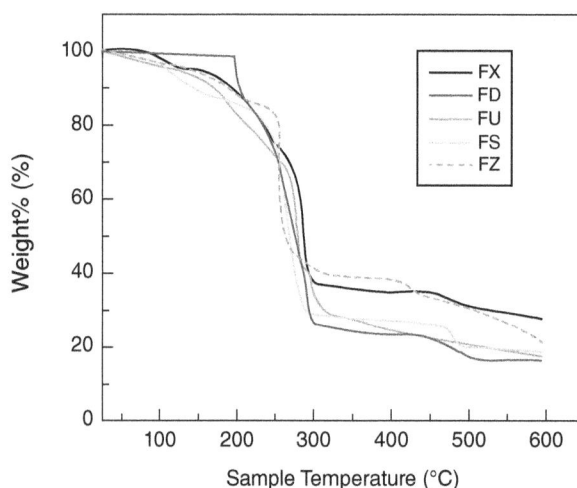

FIGURE 15.8   Variation conductivity with filler content.

The TGA curve for CuO particles-doped electrolytes film (FU) is almost similar to that of the SnO_2-added film. At 100°C, the film loses only 2–3% weight loss and at 150°C, around 10% weight loss, which is 2–3% higher than CuO-added film (FU). From 150 to 250°C, the film starts losing weight gradually, suffering 13% weight loss. A steeper loss in weight occurs in the temperature region 250–285°C, where the film loses nearly 45% weight, and this could be the decomposition temperature of the PVC-PVdF blend.

The film with ZnO filler (FZ) exhibits a better thermal stability losing nearly 3% and 5% weight at 100°C and 150°C, respectively. At 200°C, it loses only 12% weight, which is much higher value than filler-free (FX), CuO (FU)- and SnO_2 (FS)-added polymer electrolytes. In the temperature range 150–251°C, there is a gradual weight loss of 13%, which is also lower value than the abovementioned films. There is steeper weight reduction in the narrow temperature range 251–275°C in which the film loses nearly 40% weight loss. This is followed by a horizontal line indicating a stable region in the temperature range 320–410°C.

Comparing the TGA curves of the various fillers-incorporated films, we can observe that the CdO-doped film (FD) exhibits better thermal stability followed by ZnO (FZ)-, SnO_2 (FS)- and CuO (FU)-added films. The residual weight present in the samples at 150°C and 300°C is given in Table 15.2.

From the thermogravimetric analyses of the filler-free and filler-dispersed electrolyte films, the initial slow weight loss of about 10–15% from the ambient temperature to 100°C can be attributed to the vaporization of moisture in the film and due to the evaporation of fluorine. From the analysis, we found that the films were thermally stable up to 150°C losing only marginal weight, which is above the normal working temperature. The TGA curve resembles previous-reported studies in PVC-PEG-TiO_2 system, where the electrolyte suffers degradation in the temperature region 280–300°C (Rajendran et al. 2009).

15.4.5 Morphological Studies

The morphology of filler particles, PVC-rich phase: filler zero phases (without filler) (FX) and with PVC poor phase: filler-rich phase exhibiting higher conductivity are shown in Figure 15.9(a) and (b). The pores in the SEM images confirm the development of polymer-rich phase and electrolyte-rich phase (phase separation) in the electrolyte during the casting process, which is the characteristic feature of PVC-based polymer electrolytes (Hee-Tak Kim et al. 2000 and Subramaniam et al. 2002). The presence of PVdF seems to decrease the driving force for phase separation and minimize the dominance of PVC features.

A drastic change is found due to the inclusion of various fillers. Larger variations in the size and number of pores are observed in the filler-dispersed film. It seems that the function of nanoparticles is to modify the porosity and hence solution retention ability, which in turn affects the conductivity. The variation in the porous structure is correlated to the variation in driving force for phase separation, and it is attributed to influence filler-dispersed film. The development of the porous structure is a multifarious process that is influenced by the interaction of the filler with host polymer, plasticizer and solvent molecules (Ulaganathan and Rajendran 2011) (Figure 15.9).

TABLE 15.2 Residual Weight (%) Present in the Sample at Different Temperatures

Sample Code	At 150°C	At 300°C
FX	95	38
FD	99	27
FS	93	35
FU	89	29
FZ	95	40

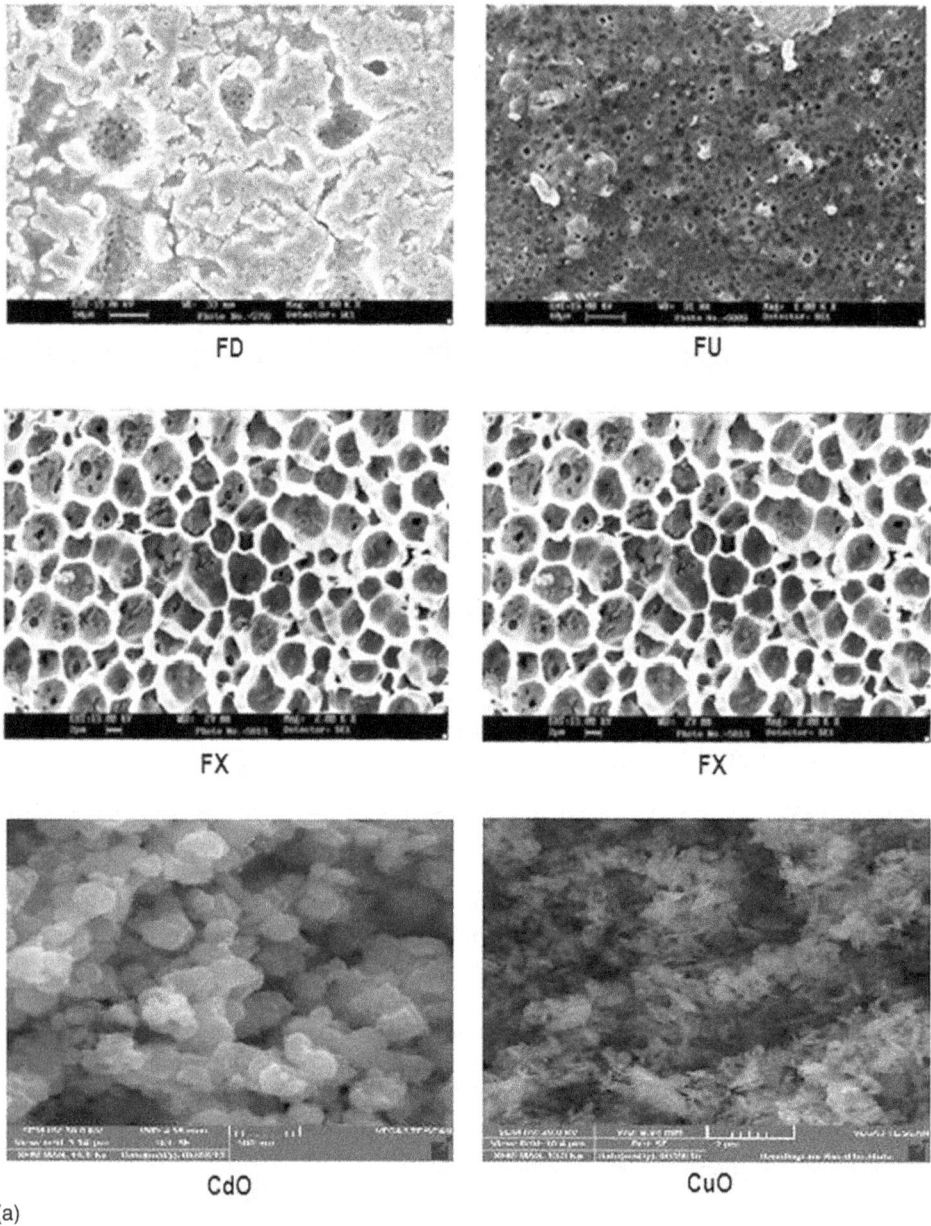

(a)

FIGURE 15.9   (a): SEM images of the pure fillers, filler-free and filler-added polymer electrolytes: FX – without filler, FD – with CdO and FU – with CuO.

(Continued)

The addition of CdO particles results in larger pores within which numerous micropores are seen. These larger-sized porous region and microporous domains are capable of trapping adequate liquid electrolyte, which facilitate better mobility for the free ions, and this is reflected in the conductivity profile data (Vickraman et al. 2000).

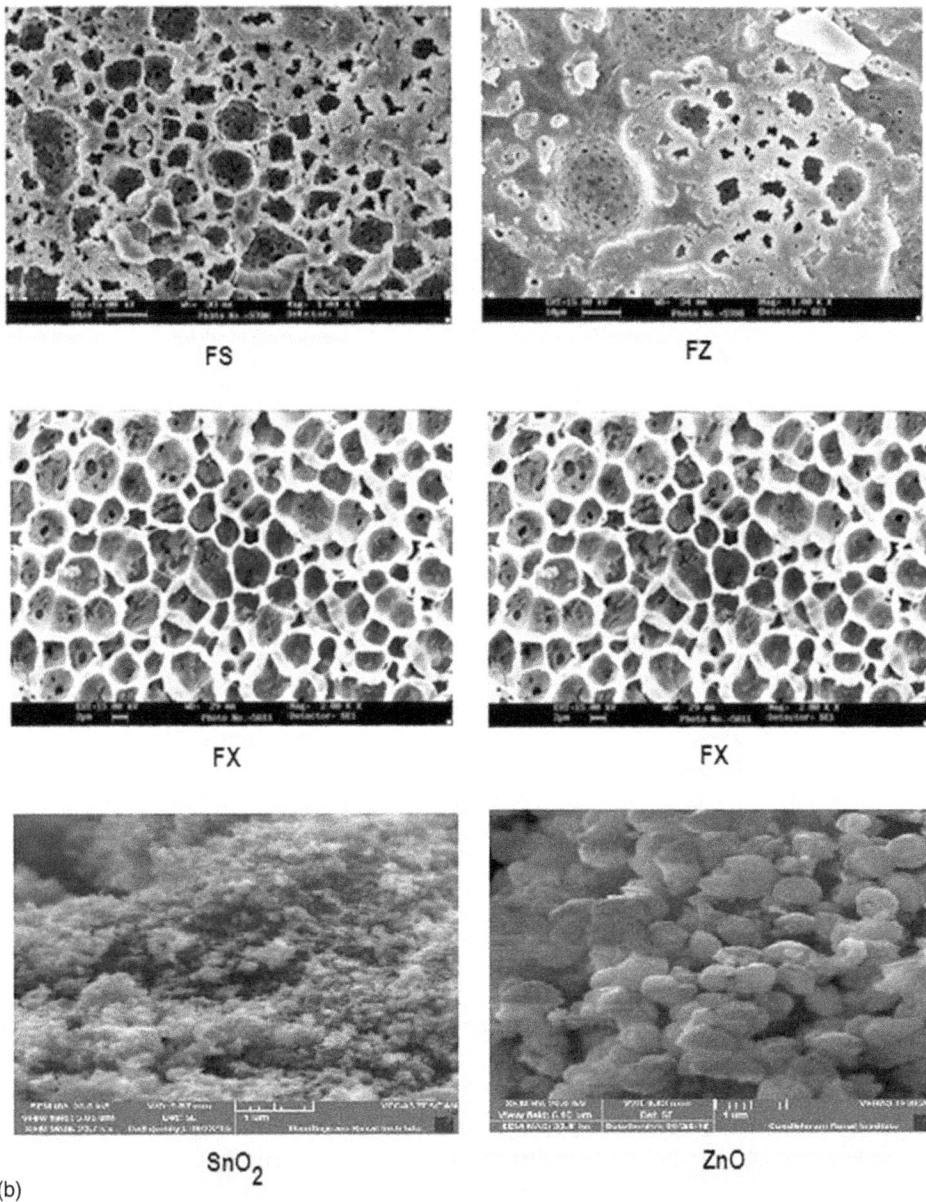

(b)

FIGURE 15.9 **(Continued)** (b): SEM images of the pure fillers, filler-free and filler-added polymer electrolytes: FX – without filler, FS – with SnO2 and FZ – with ZnO.

The dispersion of smaller sized CuO particles restricts the development of the porous region, and a larger number of tiny pores are seen in the polymer electrolyte. This reduces the plasticizer-rich region and decreases the effective channel for ionic conduction.

The agglomeration of SnO_2 filler particles is found in the filler-dispersed (FS) film. There is an increase in the number of micropores within the larger-sized pores due to the incorporation of the filler.

The inclusion of a relatively larger-sized filler (ZnO) creates numerous tiny pores within the larger pores similar to a CdO-added film. This helps higher mobility, and hence higher conductivity is observed.

Comparing the SEM images of all the films relatively, smooth morphology was developed with CdO filler (FD), showing PVdF-characteristic features with numerous micropores and pores. A smooth morphology suggests that the higher amorphous region and the ion conduction are more favored in this region rather than the crystalline phase. Similar morphology is developed in ZnO-added film (FD), which aids better mobility of the ions.

15.5 Conclusion

- A significant increase in conductivity was observed due to the addition of fillers in PVC-PVdF-PEG ternary blend polymer electrolytes, and two maxima behavior is observed. This is correlated with the generation of charge carriers and formation of interfacial region accounting for conducting channels for transport of ions.
- The XRD studies show slight enlargement of broad hump at $2\theta = 22°$ due to the incorporation of fillers, indicating the reduction in the crystalline nature of the electrolyte, which is a desired property. The peaks pertaining to fillers are seen in electrolyte films with decreased intensity.
- Among the four fillers namely CdO, CuO, SnO_2 and ZnO used in the present PVC-PVdF-PEG ternary blend electrolyte system, CdO-dispersed electrolyte exhibits higher conductivity, followed by ZnO, SnO_2 and CuO. The trend in the conductivity profile of various fillers-incorporated electrolytes can be correlated to their particle size, which facilitates effective uniform dispersal hindering the agglomeration which is prevalent in nano-sized materials.
- The morphological studies show the decrease in PVC dominance features found in CdO- and ZnO-dispersed films that facilitate better ion mobility.
- The thermal analyses of the films reveal improvement in the thermal stability of filler-added film, and CdO-added film exhibits better thermal performance in the normal operating temperature range.

References

Abdulkadir BA, Dennis JO, Abd MFB, Shukar MM, Nasef E, and Usman F. 2021. "Study on dielectric properities of gel polymer electrolytes based on PVA-K_2CO_3 composites." *Int. J. Electrochem. Sci.* 16: 1–15.

Aziz SB, Faraj M, and Abdullah OG. 2018. "Impedance spectroscopy as a novel approach to probe the phase transition and microstructure existing in CS: PEO based blend electrolytes." *Sci. Rep.* 8: 14308. DOI:10.1038/s41598-018-32662-1.

Ciosek M, Sannier L, Siekierski M, Golodnitsky D, Peled E, Scrosati B, Glowinkowski S, and Wieczorek W. 2007. "Ion transport phenomena in polymer electrolytes." *Electrochim. Acta* 53: 1409–1416.

Frech R, and Chintapalli S. 1996. "Effect of propylene carbonate as plasticizer in high molecular PEO-LiCF3SO3 electrolytes." *Solid State Ion* 85: 61–66.

Ganesh E Patil, Dnyaneshwar D Kajale, Vishwas B Gaikwad and Gotan H Jain, 2012. "Preparation and characterization of SnO_2 nanoparticles by hydrothermal route." *International Nano Letters*, 2(17): 1–5.

Huang B, Wang Z, Chen L, Xue R, Wang F. 1996. "The mechanism of lithium ion transport in polyacrylonitrile-based polymer electrolytes." *Solid State Ion* 91: 279–284.

Kim H-T, Kim K-B, Kim S-W, and Park J-K. 2000. "Li-ion polymer battery based on phase-separated gel polymer electrolyte." *Electrochim. Acta* 45: 4001–4007.

Kumar B, Nellutla S, Thokchom JS, and Chen C. 2006. "Ionic conduction through heterogeneous solids: Delineation of the blocking and space charge effects." *J. Power Sources* 160:1329–1335.

Maier J. 1994. "Defect chemistry at interfaces." *Solid State Ion* 70–71: 43–51.

Mohamed S, and Arof AK. 2004. "Investigation of electrical and electrochemical properties of PVDF-based polymer electrolytes." *J. Power Sources* 132: 229–234.

Narsinga Rao G, Yao YD, and Chen JW. 2009. "Evolution of size, morphology, and magnetic properties of CuO nanoparticles by thermal annealing." *J. Appl. Phys.* 105: 093901–093906.

Ngai KS, Ramesh S, Ramesh K, and Juan JC. 2016. "A review of polymer electrolytes: fundanmental, approaches and applications." *Ionics* 22: 1259–1279.

Pandey GP, Agarwal RC, and Hashmi SA. 2009. "Magnesium ion-conducting gel polymer electrolytes dispersed with nanosized magnesium oxide." *J. Power Sources* 190: 563–572.

Patil GE, Kajale DD, Gaikwad VB, and Jain GH. 2012. "Preparation and characterization of Sno$_2$ nanoparticles by hydrothermal route." *Int. Nano Lett.* 2: 1–5.

Pradeep K Varshney and Shikha Gupta. 2011. "Natural polymer based electrolytes for electrochemical devies: a review." *Ionics* 17: 479–483.

Rajendran S, Babu RS, and Renuka Devi K. 2009. "Ionic conduction behavior in PVC-PEG blend polymer electrolytes upon the addition of TiO$_2$." *Ionics* 15: 61–66.

Ravindran D. 2016. "A study on ceramic and as synthesized semiconducting oxides in ternary/binary polymer electrolytes." Ph. D thesis, Gandhigram Rural University, Dindigul, India.

Rocco AM, Pereora RP, and Felisberti MI. 2001. "Miscibility, crystallinity and morphological behavior of binary blends of poly (ethylene oxide) and (polymethyl vinyl ether- maleic acid)." *Polymer* 42: 5199–5205.

Sankaran S, Deshmukh K, Ahamed MB, Pasha SKK, Sadasivuni KK, and Ponnama D. 2018. "Investigation on the electrical properties of lithium ion conducting polymer electrolyte films based on biodegradable polymer blends". *Adv. Sci. Lett.* 24(8): 5496–5502.

Subba Reddy CV, Zhu Q-Y, Mai L-Q, and Chen W. 2007. "Electrochemical studies in PVC/PVdF blend based polymer electrolytes." *J. Solid State Electrochem.* 11: 543–548.

Subramaniam RT, Yahaya AH, and Arof AK. 2002. "Dielectric behavior of PVC-based polymer electrolytes." *Solid State Ion.* 152: 291–294.

Sun CC, Ah Heng Y, and Teo LL. 2019. "Characterizations of PMMA-based polymer electrolyte membranes with Al$_2$O$_3$." *J. Polym. Eng.* 7: 612–619.

Sunitha VR, Suraj KM, Kabbur PGS, Sandesh N, Suhas MR, Lalithnarayan C, Laxman N, and Radhariskhnan S. 2020. "Lithium ion conduction in PVA-based polymer electrolyte system modified with combination of nanofillers." *Ionics* 26: 823–829.

Tsutsumi N, Davis GT, and Dereggi AS. 1991. "Measurement of the internal electric field in a ferroelectric copolymer of vinylidene fluoride and trifluoroethylene using electrochromic dyes." *Macromolecules* 24: 6392–6398.

Ulaganathan M, and Rajendran S. 2011. "Novel Li-ion conduction on poly (vinyl acetate)- based hybrid polymer electrolytes with double plasricizers." *J. Appl. Electrochem.* 41: 83–88.

Venkata Subba Rao C, Ravi M, Raja V, Balaji Bhargav P, and Sharma AK. 2012. "Preparation and characterization of PVP-based polymer electrolytes for solid-state battery applications." *Iran. Polym. J.* 21: 531–536.

Vickraman P, Aravindan V, and Lee Y-S. 2000. "Lithium ion transport in PVC/PEG 2000 blend polymer electrolytes complexed with LiX (X=ClO–4, BF–4, and CF3SO–3)." *Ionics* 16: 263–267.

Yang L, Guozhong W, Chujuan T, Hongqiang W, and Lide Z. 2005. "Synthesis and photoluminescence of corn-like ZnO nanostructures under solvothermal-assisted heat treatment." *Chem. Phys. Letts* 409: 337–341.

Yap YL, You AH, Teo LL, and Hanapei H. 2013. Inorganic fillers size effect on Ionic conductivity in polyethylene oxide (PEO) composite polymer electrolyte. *Int. J. Electrochem. Sci.* 8: 2154–2163.

16

Prospective Materials for Potential Applications in Energy Storage Devices

Dr. S. Vidhya and
Dr. K. Sujatha

Shrimati Indira Gandhi College, Tiruchirappalli, India

In recent years, the world is witnessing a global increase in the demand for efficient energy storage devices. These storage devices power a wide range of everyday electronic systems such as electric vehicles, portable electronic gadgets, and so on. The need for cleaner, more sustainable, and efficient energy sources has resulted in the exploration and research of new material technologies which can provide cost-effective and environmentally friendly alternative solutions. Extensive research is being carried out to analyze nanomaterials as an alternative for high-efficiency energy storage. The high surface-to-volume ratio and short diffusion pathways of nanomaterials help in achieving high power and energy density. In this chapter, an attempt is made to discuss the latest developments and the application of nanomaterials in the field of energy storage.

16.1 Introduction

A good energy storage material not only contributes to efficient energy usage but is also critical in exploiting renewable energy. Since there is a demand for lightweight, flexible, and versatile storage material skyrockets, researchers are looking into new materials that come with extraordinary properties and net-zero emissions. Advanced materials like nanomaterials have drawn the attention of researchers due to their dynamic applications in batteries, catalysts, supercapacitors, and similar energy storage devices. The latest advancements in batteries, fuel cells, and biofuels are also becoming a strong competition to petroleum-based sources owing to their clean, eco-friendly nature.

DOI: 10.1201/9781003355755-16

Nanomaterials are a smart alternative because they:

- Boost the efficiency of energy storage devices.
- Improve the design and performance of the storage device.

16.2 Classification of Storage Systems

Energy storage technologies include a wide range of systems, which can further be broadly divided into five major categories, thermal, mechanical, electrical, electrochemical (or batteries), and hydrogen storage technologies. This chapter focuses mainly on materials for electrochemical storage devices.

16.2.1 Mechanical Storage

Mechanical energy storage devices are some of the very efficient and sustainable available energy storage systems. Below are the main types: flywheel, pumped hydro, and compressed air. This technology stores energy as gravitational potential energy, the potential energy of compression, or the kinetic energy of motion. Some examples are gravity energy storage, flywheel energy storage systems (FESS), and pumped storage hydropower.

16.2.2 Electrochemical Storage

Electrochemical energy storage comprises all kinds of secondary storage batteries. These batteries convert chemical energy into electrical energy through electrochemical processes. Examples include lead acid, Ag-Zn, Li-ion, and Na-S. Based on their nature, these batteries find application for short-duration usage (ranging from a few minutes) or long-duration usage (over 8 hours).

16.2.3 Thermal Storage

Thermal energy storage devices store energy by heating or cooling a medium. The energy stored can be utilized on demand for heating, cooling, or generating power. Thermal energy systems are economic and reduce carbon emissions when compared with other storage systems.

Thermal energy storage materials are widely used in solar heaters, air-conditioning, etc.

16.2.4 Electrical Storage

Electrical energy storage systems are essential to develop energy technologies where the energy can never be depleted. Transformational changes are needed to permit maximum energy storage economically and with a longer lifetime in both battery and capacitor technologies. A lot of these changes need materials with higher redox capacities that react rapidly and reversibly. The need for highly efficient electrical energy storage materials fascinated the researchers to invent materials with high dielectric constant and high breakdown fields. Polystyrene, polyethylene, ferroelectric polymers, and paraffin are some of the organic materials used in capacitors as dielectrics due to their high breakdown fields. In order to improve the dielectric constant of such organic materials, nanoparticles of metal oxide can be added. Commonly used metal oxide particles with high dielectric constants are $BaTiO_3$, $SrTiO_3$, TiO_2, and $BaSrTiO_3$.

16.2.5 Hydrogen Storage Technologies

Hydrogen storage technology employs electricity to electrolyze water to produce oxygen and hydrogen. High-pressure containers are used to store the hydrogen produced and it can then be used as a fuel for direct combustion such as cooking and heating applications or electricity generation via proton exchange membrane (PEM) fuel cells.

16.3 Batteries

Primary batteries cannot be recharged due to which they are one-time use batteries. The electrodes are permanently changed during the discharge process. Common examples of primary batteries are dry cells (zinc-carbon) and alkaline batteries. AA and AAA batteries used in clocks, remotes, and other electrical gadgets use these disposable batteries.

Secondary Batteries: Secondary batteries are rechargeable and can be used over and over again. The advantage of secondary batteries is that they can be used while simultaneously being recharged. Common examples of secondary batteries include lead-acid and lithium-ion batteries. They are used in everyday gadgets like cell phones, automobiles, etc.

16.4 Fuel Cells

Fuel cells have the ability to transform chemical energy into electrical energy. Fuel cells are alike batteries as both of them are capable of generating electricity, but in a way, unlike typical batteries, because of their dependency on a continuous supply of fuel. As long as there is a consistent fuel supply, fuel cells will continue to produce electricity. Hydrogen is mostly used as fuel. Common applications of hydrogen fuel cells include a power supply for satellites, automobiles, boats, and submarines.

16.5 Supercapacitors

A supercapacitor is a special kind of energy storage device with an extremely huge capacitance. This happens by combining the capacitor and battery properties into a single device. These capacitors can store more energy as compared with other conventional capacitors and provide high output power than batteries. Supercapacitors are known as ultracapacitors, which are easy and safe to use. Supercapacitors are used to provide an uninterrupted power supply for trains, cranes, elevators, and more. They can be classified into:

  i. Electrochemical double-layer capacitor (EDLC)
 ii. Pseudocapacitor
iii. Hybrid capacitor.

A graphical classification of supercapacitors is depicted in Figure 16.1.

16.5.1 Electrochemical Double-Layer Capacitors

Electrochemical energy storage systems are called supercapacitors or ultracapacitors. They are classified into EDLCs and pseudocapacitors. Electrochemical double layers are formed at the interface of an electrode and electrolyte. EDLCs save charge electrostatically. This process is reversible. They are used as memory backup devices due to their long life cycle. The electrode material and the type of electrolyte ensure the quality and performance of electrochemical capacitors. The electrode materials used in electrochemical capacitors are transition metal oxides, carbon-based materials, and conductive polymers.

The major classes of electrolyte materials used for electrical energy storage devices are organic electrolytes, aqueous electrolytes, and ionic liquids. Innovative electrode materials like $LiNixCoyMnzO_2$ and electrolytes like fluoroethylene carbonate, methyl-(2,2,2-trifluoroethyl) carbonate, etc., are

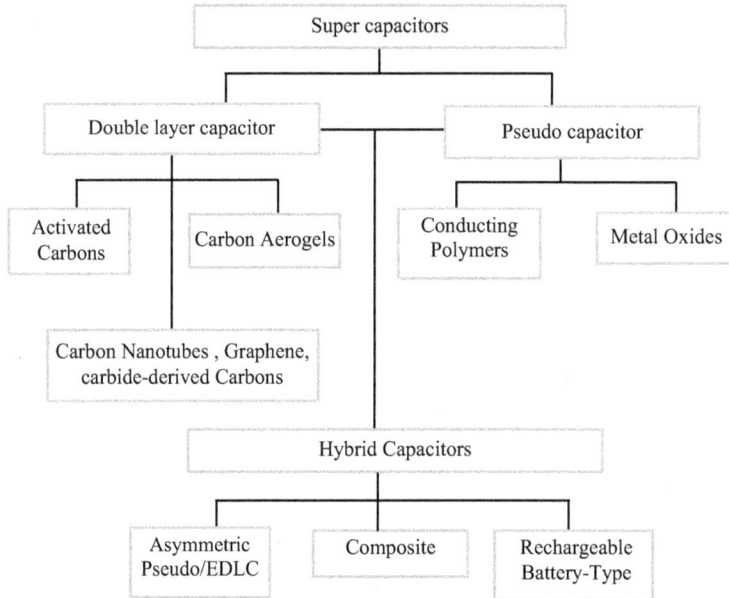

FIGURE 16.1 Classification of supercapacitors.

analyzed for the improved performance with respect to the cost, energy and power density, life cycle, and battery safety.

EDLCs are suitable for applications where the availability of user service is scanty, for example, in high altitudes and underwater environments (Conway 1999, Burke 2000, Kotz and Carlen 2000). Introducing changes in the property of the electrolyte modifies the performance of electrochemical double layers. Organic electrolytes or water-based electrolytes are used in EDLCs. It is found that the water-based electrolytes such as H_2SO_4, KOH, NaOH, Na_2SO_4, etc., have smaller equivalent series resistance values, lower minimum pore size criterion, and low breakdown voltage when compared with organic electrolytes. Carbon aerogels, activated charcoal, and carbon nanotubes (CNTs) (Simon and Burke 2008) are some of the commonly used electrode materials to store charges.

Using activated charcoal is economical, as it is extremely porous, and a greater surface area is available for adsorption and chemical reactions when compared with other carbon-based materials. The complex porous structure of activated charcoal has different size micropores, mesopores, and macropores, which provides a greater surface area. By a continuous network of conducting carbon nanoparticles with distributed mesopores, the carbon aerogel is created. Using the surface area more efficiently, the CNT-based electrode can provide a higher value of capacitance compared with the activated carbon electrode for supercapacitors.

CNT-based electrodes have a reasonable surface area when compared with activated carbon-based electrodes (Frackowiak and Beguin 2002, Frackowiak and Metenier 2000). CNT-based electrodes also have smaller equivalent series resistance values compared with activated carbon (Niu and Sichel 1997, An and Kim 2001a, 2001b). By increasing the applied potential window and the electrostatic charge related to the capacitance of the electrode material, the energy stored in EDLCs can be increased.

16.5.2 Pseudocapacitors

16.5.2.1 Types of Pseudocapacitor

Pseudocapacitors store energy through Faradaic reactions. They store charge electrostatically by transferring charges between the electrode and the electrolyte. When a voltage is applied to a pseudocapacitor, both oxidation and reduction take place on the electrode material. The Faradic process used in pseudocapacitors improves the electrochemical reactions resulting in high specific capacitance and energy densities when compared with EDLCs. Pseudocapacitors are classified into two types based on the electrode materials used to store charge within pseudocapacitors.

16.5.2.2 Metal Oxide

Metal oxide is a pseudocapacitive material that demonstrates reversible as well as fast redox reactions at the outside of the electrode materials. It displays low resistance along with high specific capacitance. Because of this, constructing supercapacitors with high power is easy. MnO_2, RuO_2, NiO, SnO_2, IrO_2, Fe_3O_4, V_2O_5, Co_2O_3, and MoO are the commonly used metal oxides as electrodes. Among these metal oxide electrode materials, RuO_2 is considered the most capable electrode material in supercapacitor applications due to its high electrical conductivity and specific capacitance.

16.5.2.3 Conducting Polymers

Conducting polymers are utilized as redox pseudocapacitors because of their fast, quick, and reversible oxidation or reduction processes, affordability, and stable electrical conductivity. Commonly used conducting polymers are polypyrrole (PPy), polyaniline (PANI), polythiophene (PTh), p-poly p-phenylene vinylene (PPV), and p-polyethylene dioxythiophene (PEDOIT). Conducting polymers are usually produced by electrochemical oxidation and rendered conductive using a unified bond system with polymer as the backbone. As compared with carbon-based electrode materials, conducting polymer materials provide capacitance, enhanced conductivity, and decreased equal series resistance. When the oxidation and reduction process takes place, the ions are simply absorbed or drifted away from the electrolyte toward the conducting polymers and are eventually released into the electrolyte. Three types of electrochemical processes involved to develop pseudocapacitors are:

  i. surface adsorption of ions from the electrolytes
  ii. redox reactions involving the ions from the electrolytes
  iii. doping and undoping of active ions into the electrode materials.

The electrode materials used in pseudocapacitors are conducting polymers or transition metal oxides. They have relatively high capacitance, high conductivity, and low ESR values. The main advantage of conducting polymers is that they can be easily processed. As transition metal oxides possess high conductivity, they are used as electrode materials for pseudocapacitors, (Burke 2000, Kotz and Carlen 2000, Kim and Kim 2001, Zheng and Jow 1995, Zheng and Cygan 1995). RuO_2-based pseudocapacitors can produce maximum energy and power density compared to EDLCs and conducting polymer-based pseudocapacitors, but their high cost and toxicity limit their use.

16.5.3 Hybrid Capacitors

In order to bridge the gap between traditional batteries with high energy density and supercapacitors with high power output and long life, hybrid capacitors were introduced. They have dual electrodes with both a battery type and a capacitor type. Recently, potassium-ion batteries and potassium-ion hybrid capacitors have drawn great attention. Generally, Faradaic and non-Faradaic processes are used for storing charges in hybrid capacitors. These capacitors are fabricated with one electrode made of double-layer carbon materials, and the other electrode with pseudocapacitance materials. They have greater energy and power density in comparison to EDLCs.

In the recent past, nickel oxide was used as the pseudocapacitance material in the positive electrode for most hybrid supercapacitors. Current research is focused on the composite, asymmetric, and battery types of hybrid supercapacitors, which are differentiated by electrode configuration. Composite electrodes blend carbon-based materials with a metal oxide or conducting polymer materials. Asymmetric cells employ both Faradaic and non-Faradaic mechanisms through coupling an EDLC electrode with a pseudo capacitor electrode (Jacob 2009). Particularly, the combination of an activated charcoal negative electrode with a conducting polymer positive electrode drives today's research (Kim and Kim 2001, Mastragostino and Arbizzani 2002, Stoller and Ruoff 2010).

As in the case of asymmetric hybrid supercapacitors, battery-type hybrid supercapacitors combine two different electrodes. This configuration implies the need for higher-energy supercapacitors and higher-power batteries. Hybrid supercapacitors include the advantages of both EDLCs and pseudocapacitors such as high energy and power density. Copper oxide is used as an electrode material for lithium-ion batteries and electrochemical capacitors, heterogeneous catalysts, and selective gas sensors (Shaikh et al. 2011 and Xiang et al. 2010). In nanocomposite electrode systems, it has been chosen as one of the components as it has enhanced electrochemical properties.

16.6 Ferroelectric Materials

Different types of ferroelectric materials are ferroelectric single crystals, ceramics, polymers, ceramic-polymer composites, and films. Quartz, lithium tantalite ($LiTaO_3$), and lithium niobate ($LiNbO_3$) are the widely used piezo and ferroelectric materials. Ferroelectric single crystals are anisotropic in nature. They exhibit different material properties depending on the cut of the material and the direction of the wave propagation.

In recent years, some ferroelectrics, called relaxor ferroelectrics, have received special attention because of their intriguing and extraordinary dielectric, ferroelectric, and piezoelectric properties (Vidhya et al. 2013, 2018a). Currently, these properties are explained by a model with polar nano regions in a non-polar matrix. Relaxor-based ferroelectric single crystals, such as PIN-PMN-PT (lead indium niobate-lead magnesium niobate-lead titanate) crystals, find application as explosives in ferroelectric generators. These ferroelectric crystals are widely used in various sensors, and actuator applications, in the development of acoustic projectors, motors based on ultrasonic technique underwater transducers for Navy sonars, and microfluidic and microbeam transducers (Vidhya et al. 2018b).

16.7 Nanomaterials in Lithium-Ion Batteries

Lithium is mostly preferred for the production of efficient batteries. By using enlarged electrode surfaces and optimized separators, the energy density and stability of the batteries are improved, allowing novel applications in electric vehicles and stationary power storage.

Lithium-ion battery has several advantages:

- High specific charge
- Highly negative standard potential
- Stability in most of the organic electrolytes.

Lithium-ion batteries are currently well-known, rechargeable mobile energy storage devices. It is an amazing fact that consumers are buying billions of lithium-ion batteries every year for use in cameras, mobile phones, laptops, and electric vehicles. Lithium-ion batteries have high voltage and capacity, a longer lifespan, and are highly safe and reliable to be used for such purposes. Although lithium-ion batteries have higher voltages, they come with good stability. Nanomaterials are used as anodes and cathodes in lithium-ion batteries. The anode nanomaterials are oxides of titanium, nickel, iron, cobalt, or copper, lithium titanate, nano dispersions of carbon, silicon, metals like aluminum and tin, red phosphorus nanoparticles, transition metal oxides containing graphene, carbon nanotubes, nanospheres and Co_3O_4 nanomaterials with various microstructures.

The cathode nanomaterials are lithium cobalt oxide, lithium nickel oxide, oxides of manganese, lithium cobalt phosphates, lithium iron phosphate, lithium nickel phosphate, lithium manganese phosphate, and oxides of vanadium (V_2O_5, LiV_3O_8). But high power density and efficient working capacity of lithium-ion batteries are obtained, when nanomaterials are used as opposed to their bulk counterparts.

16.8 Nanomaterials in Electrochemical Storage Devices

The two-dimensional nanomaterial graphene is useful for lithium-ion batteries, storage of hydrogen, and supercapacitors. On combustion, it produces non-polluting water as the by-product. Though hydrogen gas is a good fuel, it is difficult to store and transport hydrogen in pressurized gas or liquid form. This can be achieved by using graphene as it is cheap, lightweight, chemically inert with high mechanical strength, and large surface area. The dispersion forces enable hydrogen to be absorbed on the graphene surfaces. The storage capacity is improved with the addition of hydrogen molecules between assembled graphene sheet layers.

In mobile applications, hydrogen storage with graphene sheet used as a substrate is found to be very much safer. When hydrogen is stored in its atomic form, it is known as hydrogen spillover. There is a dissociation of hydrogen molecule over the nanometal (Ni, Pt, or Pd) surface on a graphene sheet. In this, one hydrogen atom is attached to the metal and the other diffuses into the graphene substrate.

The high surface area of graphene is the main factor that enables it application in supercapacitors. The alternation in the chemical constitution or structure of graphene shows higher capacitance values, in aqueous and organic electrolytes. Carbon nanotubes are also used in energy storage devices. They are better anode materials compared with graphite as they exhibit higher capacity. This is because of their nano-dimension and significant specific area. They can be employed as additives for anode materials such as silicon and metal oxides (e.g., TiO_2, MnO_2, SnO_2, and Fe_3O_4) due to their high conductivity.

16.9 Conclusion

The discovery and development of innovative materials cut across the entire energy technology scenario, from energy generation and storage to delivery and end-use. They are the building blocks of every clean energy innovation, such as advanced batteries, solar cells, thermal storage devices, and ultracapacitors. New materials such as graphene, CNT, and others on nanoscale concepts offer the prospect of a higher level of efficiency in supercapacitors. These new materials constitute one of the milestones for the global transition to storage in the zero-carbon system of the future.

References

An K H, Kim W S 2001a. "Electrochemical properties of high-power supercapacitors using single-walled carbon nanotube electrodes", *Advanced Functional Materials* 11(5): 387–392.

An K H, Kim W S 2001b. "Supercapacitors using single-walled carbon nanotube electrodes", *Advanced Materials* 13(7): 497–500.

Burke A 2000 "Ultracapacitors: why, how, and where is the technology", *Journal of Power Sources* 91(1): 37–50.

Frackowiak E, Beguin F 2002. "Electrochemical storage of energy in carbon nanotubes and nanostructured carbons", *Carbon* 40(10): 1775–1787.

Frackowiak E, Metenier K 2000. "Supercapacitor electrodes from multiwalled carbon nanotubes", *Applied Physics Letters* 77(15): 2421–2423.

Jacob G M 2009. "Nanocomposite electrodes for electrochemical supercapacitors", PhD thesis, McMaster University.

Kim I H, Kim K B2001. "Electrochem. Ruthenium oxide thin film electrodes for supercapacitors", *Electrochemical and Solid State Letters* 4(5): A62–A64.

Kotz R, Carlen M 2000. "Principles and applications of electrochemical capacitors", *Electrochimica Acta* 45(15–16): 2483–2498.

Mastragostino M, Arbizzani C 2002. "Conducting polymers as electrode materials in supercapacitors", *Solid State Ionics* 148(3–4): 493–498.

Niu C M, Sichel E K 1997. "High power electrochemical capacitors based on carbon nanotube electrodes", *Applied Physics Letters* 70(11): 1480–1482.

Shaikh J S, Pawar R C, Moholkar A V, Kim J H, Patil P S 2011. "CuO-PAA hybrid films: chemical synthesis and supercapacitor behaviour", *Applied Surface Science* 257: 4389–4397.

Simon P, Burke A F 2008. "Nanostructured carbons: Double layer capacitance and more", *Journal of The Electrochemical Society Interface* 17(1): 38.

Stoller M D, Ruoff R S 2010. "Best practice methods for determining an electrodematerials performance for ultracapacitors", *Energy & Environmental Science* 3(9): 1294–1301.

Vidhya R G, Ramasamy R, Vijayalakshmi L 2013. "Optical characterization of gel grown zinc tartrate crystal", *International Journal of Physics and Research* 3(3): 97–100.

Vidhya R G, Ramasamy R, Vijayalakshmi L 2018a. "Comparative analysis on optical, thermal and structural characterization of barium tartrate crystal grown in the presence and absense of magnetic field", *International Journal of Recent Scientific Research* 9(12(A)): 29832–29837.

Vidhya R G, Ramasamy R, Vijayalakshmi L 2018b. "Effect of magnetic field on the structural and thermal properties of manganese tartrate dihydrate crystal", *International Journal of Scientific Research and Reviews* 7(4): 1944–1953.

Xiang J Y, Tu J P, Zhang L, Zhou Y, Wang X L, Shi S J 2010. "Self-assembled synthesis of hierarchical nanostructured CuO with various morphologies and their 194 application as anodes for lithium ion batteries", *Journal of Power Sources* 195: 313–319.

Zheng J P, Cygan P J 1995. "Hydrous ruthenium oxide as an electrode material for electrochemical capacitors", *Journal of the Electrochemical Society* 142(8): 2699–2703.

Zheng J P, Jow T R 1995. "A new charge storage mechanism for electrochemical capacitors", *Journal of the Electrochemical Society* 142(1): L6.

17

Food Waste Mixed with Carbon Nanotechnology for Energy Storage

Deepa Jaganathan,
Rajkumar
Perumal, Gitanjali
Jothiprakash
and Arulmari
Rajachidambaram

*Agricultural Engineering
College and Research
Institute, Tamil Nadu
Agricultural University,
Coimbatore, India*

Solid waste has already become a global problem because of its impact on ecosystems. Continued population growth and demand are the main causes of excessive production of solid waste such as plastic, agro-residues, organic waste, inorganic waste, e-waste, and processing waste from industry. It has been identified as a major threat to human health and the environment at the same time. The best way to live is to convert solid waste into essential carbon nanomaterials. Waste materials can be converted into various nanomaterials and may be used in construction, energy, pharmaceutical, and aeronautical industry. This chapter summarizes the production of carbon nanotubes from solid waste especially food waste and its potential use in many fields.

17.1 Introduction

There is a need for a storage device to store solar and wind energy. Although water storage is widely used, batteries are receiving a lot of attention as a promising solution (Mukhtar et al. 2021). Electricity generation in renewable energy has increased steadily in recent years, and this increase is expected to continue. Increased battery consumption and flywheel power storage have occurred over the past five years, according to the Environmental Impact Assessment (Yusik et al. 2019). This helps to harness the power of chemical and mechanical equipment. But energy-saving like the heat is also possible.

Population growth has led to the development of the universe and its related needs are increasing day by day. There is a dramatic increase in solid waste production due to these developments, which include

DOI: 10.1201/9781003355755-17

several negative effects on our environment and health. The global problem of solid waste management with distressing effects on living system on the biosphere is the major tasks threatening our future now. These extremes are the result of global population growth and dramatic global economic growth and globalization. To overcome this situation, scientific progress requires a new approach that uses advanced scientific technology. This approach can solve and control the use of such waste through the use of nanoscience and nanotechnology, to function as sustainable solid waste management and other global obligations that include energy problems everywhere (Chen et al. 2016). Waste management technology needs to be enhanced in order for this solution to be fully resolved.

Two major tasks for scientists include the development of a new way to produce cost-effective, environmentally friendly energy tools to address the global energy crisis and the need for more scientific reports on waste management and the production of less expensive carbon nanomaterials (carbon nanotubes), graphene, and carbon quantum dots (CQDs). Carbon nanomaterials from solid waste provide a wide range of potential applications (Choi et al. 2021). Thus, a large amount of waste produced can be used. Carbon nanomaterials, especially carbon nanotubes, CQDs, and graphene, have many distinctive features that include high electrical conductivity and mechanical strength as well as heat resistance (Packiyalakshmi et al. 2019).

With all the high carbon content found in solid waste, it is possible to convert other carbon nanomaterials such as graphene, carbon nanotubes, and CQDs. These are gradually adjusted and packaged for use as energy resources (Rai and Yanilmaz 2021). With an eco-friendly disposal option for solid waste, you can build the last resort of working equipment as the best practice is encouraged. By converting solid waste into carbon nanomaterials, it can be transformed into energy materials.

17.1.1 Food Waste as a Feedstock

Waste recycling is an ongoing and important indicator of the management of the reserve. The entire biosphere is capable of recycling household waste and vegetables. Recycling of discarded or leftover food does not occur as it can be eaten by animals or humans. Global food loss and waste are one-third and one-half of all food production (Abdelbasir et al. 2020). Redefining the purpose of waste products is becoming increasingly popular, as food waste management creates a serious problem nowadays.

The fruit and berry/vegetable products can be converted into raw materials for producing compounds of nanomaterials through chemical engineering. The food industry is a combination of agriculture and industry such as agro commodity processing (fruits, vegetables, millets, pulses, oil seeds, etc. The waste produced from these kinds of industries is from the processing of fruits and vegetables. In addition to being eaten fresh, they are grown for sale and processed into juices, jams, jars, and nutritious ingredients (Saravanakumar et al. 2007)). Similarly, pomace makes up about 30 percent of the total amount of waste in the juice industry. After squeezing the juice, the dirty pomegranate leaves take up 12 percent of the fruit in terms of mass. For example, grape juice production leads to about 50 percent of waste such as seed and pomace. This pomace contains antioxidant activity which has more health benefits. In the grape juice industry, about 45 percent of grapes are used for juice and some are made from grape pomace. In the same way, pomegranate peels contain compounds viz., phenolic, flavonoids, and acids which have the same antioxidant property. This antioxidant property paves the way to use as raw materials for carbon nanomaterial synthesis. Other food waste namely potato peels, pomace of rapeseed, pulp from sugar beet, and papaya can also be used for the production of nanomaterials (Xiaoxia et al. 2020). Green nanomaterials are a very actual way to treat industrial waste in the development of an environmentally friendly manner.

17.2 Production of Carbon Nanomaterials from Food Waste

There are various ways to obtain carbonic waste conversion into carbon nanomaterials, including deposition using physical and chemical vapor, the modified Hummers method, pyrolytic methods and Organic redox methods. Each method has its own specific characteristics for the production of different carbon nanomaterials, which are highly dependent on local reactions and a variety of reagents. In the pyrolytic process, temperature, residence time, pressure, and inert condition are very important, which change the layer structure and diversity of carbon nanomaterials (Vinay et al. 2020).

In the combination of graphene with carbon nanotubes, frequency, pyrolysis, and chemical vapor deposition processes are followed, while quantum dots are heated in a hot way. Comparison in the structure and properties of carbon nanomaterials found in solid debris is due to debris caused by other components present in raw material, which cause certain defects, such as edge flexibility and displacement of carbon nanomaterials (Reddy and Rhim 2014). Low pyrolysis levels, the number of catalysts, and the absorption temperature are important factors in the continuous conversion of solid carbon dioxide into carbon nanomaterials. The role of catalyst decomposition of carbonic layers and the role of high-temperature reduction are critical to the formation of graphene (Hui et al. 2018). In methods based on chemical vapor permeability, the quality of carbon nanomaterials is strongly associated with processing temperatures as well as the gas flow rate. In the development stage of chemical vapor infiltration, a high flow rate of gas per m^3/min over a period of time is required to form graphene nanosheets. In general, the graphene nanomaterial high-satisfactory relies upon the technique condition viz., catalyst concentration, temperature, pressure, time, and gas flow rate. Chemical vapor deposition is a result of the emission of carbonic gases to create the van der Waals gravitational force among the layers of carbon substances that pave the manner to create nanolayers easily. This purity of graphene composite can be stricken by fluidic ions or metallic ions.

Carbon nanotubes are made of a compound similar to graphite. These materials derived from coal are used to make pencils and diamonds. These can be modified to form different materials by the combination of nanoparticle sheets. Nanosheets containing nitrogen-doped carbon nanosheets can be produced using fruit waste, especially as a precursor and simultaneous activation and carbonization. The carbon nanomaterial produced from this waste was very efficient and its application in supercapacitor devices showed improved performance (Dhivyashree et al. 2016).

Food waste also can be transformed into carbon nanotubes with the aid of using temperature, pressure, and time of the pyrolytic chamber. The carbon nanotubes can be single-walled or multi-walled. The single-walled carbon nanotubes are portrayed with the aid of using structural, performance, and aesthetic foundations for metal, metalloid, and metalloid types. The parallel combination of single-walled carbon nanotubes makes the carbon nanotube as multi-walled. Strong carbon–carbon sigma bonds offer sub-atomic energy and stabilization of carbon nanotubes.

Food waste can be converted to carbon nanotubes using nickel/molybdenum/magnesium oxide as a catalyst. The above three categories depend on the specific function of objects. However, the effect of these catalysts was found to be very effective. A good composition of catalysts will produce good quality carbon nanotube.

17.3 Structure and Properties of Carbon Nanotube

In 1991, Ijima's discovery of carbon nanotubes is another example of a carbon allotrope with a cylinder form. Strong energy, electrical resistance, and thermal conductivity are only a few of the unusual features resulting from the peculiar structure of carbon nanotubes. Since the discovery of carbon nanotubes, scientists have taken a remarkable step in testing the concept of their utility, electrical, and thermal properties. Although scarce, this material is more desirable for light applications because it has a lower density than aluminum (2.7 g cm^{-3}).

It is a tube-shaped object, made of carbon, measuring nanometer scales. Carbon nanotubes are different because the interaction between atoms is very strong and the tubes can have extreme aspect ratios. It can be as small as a few nanometers but can be hundreds of microns long. These tubes can be automatic or multi-colored by wrapping layers of various diameters around one another. The way graphene sheets and carbon nanotubes are formed in the production process includes chemical alchemy and heat treatment but is similar to the way carbon-fiber sheets are produced (Yanyan et al. 2018).

It works both as part of a building and as a conductor. A single-walled carbon nanotube can be as small as one carbon atom. That means the whole tube is about one nanometer in diameter, about 1/10,000 the size of a human hair. When acting as a semiconductor, carbon nanotubes can be remarkably similar to transistors in a chip of the same size as modern silicon versions.

Many transistors equate to additional power, low-temperature production, and high reliability throughout the life of the product service. Carbon nanotube chips are not only smaller (or more powerful in comparable size) but also faster and more efficient while producing much less heat than modern silicon chips. It is widely used as a stabilizer for carbon-fiber products.

High power, small size, low temperature, and fast carbon nanotube technology also promise greater improvements than modern technology. X-ray machines can be more accurate with greater processing power, digital cameras can be incredibly fast and incredibly advanced, and smartphones can easily deliver the equivalent of modern laptops and computers (Tailong et al. 2018).

The small size of nanotubes, the flexibility, and the chemical composition of food waste provide promising high-tech technologies. There are still some technological barriers that need to be overcome before carbon nanotubes become commonplace in consumer electronics and other public applications, but it is a matter of time before these tiny graphite tubes become a major part of our daily lives.

Many applications are already being tested or used at limited prices before major industrial and consumer products are ready for normal use (Wei et al. 2019). Many more are coming up for nanotubes to see widespread production attainments for potential interruptions and for prices low enough to integrate equipment into existing designs (Wild et al. 2012).

Nanotechnology is widely used in many industries, including health and medical, electronics, energy, and the environment. Nanotechnology has the ability to remove bacteria and contaminants from water in an efficient manner. Due to its numerous benefits, adsorption has emerged as the finest method for cleaning water. Nanotubes have been used in the transport of proteins, drugs, and cancer-related peptides. There are several kinds of nanoparticles viz., carbon dioxide nanoparticles, and fiber-based nanosheets available to treat cancer (Rai and Yanilmaz 2021).

17.4 Production of Carbon Nanotube from Food Waste

Sugary alcohol is a major commodity in the food industry and is mixed with carbon nanotubes. Carbon nanotube of various sizes is mixed with two types of sugary alcohol namely erythritol and xylitol, both compounds that occur naturally in food but the heat transfer between the mixture decreases as the nanotube diameter decreased. Also, high cramming has led to better heat transfer. It can significantly improve the thermal conductivity of heat exchangers, such as sugary alcohol. Some enhancements depend on the rate of heat transfer throughout the carbon structure (Dangsheng and Centi 2011).

There are three main methods, namely arc discharge, laser vaporization, and chemical vapor deposition to synthesize carbon nanotubes, each with its own benefits and difficulties in terms of the quality and size of the nanotube synthesized (Tatrari et al. 2021).

17.4.1 Arc Discharge Method

By releasing arc discharge, carbon is evaporated by plasma induced by helium gas and cobalt metal catalyst, which contains higher energy than energy caused by anode and cathode (Andrea et al. 2010). It enhances a strong van der Waals interaction and forms strong bonds in a carbon nanotube (Yoshinori and Zhao 2006).

17.4.2 Laser Vaporization

Laser vaporization uses a constant or sharp laser beam to burn the carbon at a high temperature of around 1200°C in the furnace. The laser beam evaporates a graphite sample along with a metal catalyst.

This metal is converted into a carbon-saturated nanomaterial where carbon is deposited on it, allowing for the production of carbon nanomaterials (Justyna et al. 2015). This produces single-walled nanomaterial. The relative quantity of nanotubes as products and the emissions produced by this method depends on the specific conditions of the process. Fullerene vapors, metal nanoparticles, and carbon nanoparticles are major emissions. The impurities in the carbon nanomaterials are removed by washing in acid solutions (Elich 2004).

17.4.3 Chemical Vapors Deposition

The deposition of chemical vapors through the transfer of hydrocarbon gases over the catalyst within the reactor. The catalyst is, in particular, used as a transition alumina nanomaterial. Catalytically grown nanomaterials are collected after the reactor has cooled to ordinary room temperature. The most important technical parameters are the sort of hydrocarbon and catalyst, and reactor temperature. Normally used hydrocarbon gases are ethylene and acetylene for multi-walled nanomaterial, methane, and ethane for single-walled nanomaterial, metallic catalysts are steel, nickel, and cobalt for multi-walled nanomaterial and alumina for single-walled nanomaterial, and reactor temperature around 500 to 800°C for multi-walled nanomaterial and 800 to 1200°C for single-walled carbon nanomaterial (Notarianni et al. 2016).

Alternative methods are used that produce carbon nanotubes with abridged temperature growth and bigger bulk yield. These methods are: plasma-enhanced chemical vapor deposition, in which gases such as hydrocarbon gases are supplied to the chamber, and high pressure and temperature are required; laser-assisted thermal chemical vapor deposition, in which a medium-strength CO_2 laser is applied perpendicular to the substrate, and then pyrolyze sensitive acetylene mixtures in the reactor under high catalytic pressure of hydrocarbon gas, where gas and catalyst particles produced by the decomposition metals when higher gases flow at 10 atm pressure and around 1200°C temperature in the reactor (Hussein and Firas 2019).

17.5 Advantages of Food Waste-Based Carbon Nanomaterials Synthesis

The worldwide position of solid waste has stretched to a point where the waste management is necessary to use it efficiently (Sabah et al. 2020). The food waste contains higher carbon content, so it can be converted into carbon nanomaterials with different additional properties that allow it to use in different fields. The features are mainly electrical conductivity and a high degree of precision, which makes them easy to use as building materials for the energy storage device (Suan et al. 2020). Energy storage devices are the ultimate electrochemical device that stores and delivers energy and provide the highest power in a short span. Nowadays, it is used in electric vehicles, uninterrupted power supply, memory backups in Artificial intelligence systems, and in various other areas. The green approach to nanoparticle production has recently provoked much attention because of its benefits such as low cost, ease, eco-friendly, biogenesis, and prevalent usage in the conventional chemical and physical methods.

17.6 Disadvantages of Food Waste-Based Carbon Nanomaterials Synthesis

The disadvantages of the synthesis of carbon nanomaterials from food waste are low purity, high material damage, and higher remains at the end of the process due to the presence of damp bonds to eliminate these damp bonds, a higher temperature should be involved in the synthesis process.

17.7 Future Aspects of Food Waste-Based Carbon Nanomaterials Synthesis

Dispersion of carbon nanotube structures plays an important role in enhancing electrochemical, elasticity, and mechanical properties while mass production and good thermal stability are a major challenge. Economic productivity is an important aspect of the feasibility study. Substantial growth is required to produce the nanomaterial with good quality and quantity. This nanomaterial usage in large industries leads to higher growth in global economic conditions. However, before considering the industrial scale, a short study on economic analysis, material balance, and energy balance should be carried out.

17.8 Conclusion

The global level of solid waste has reached a point where waste has to be managed by using it as a raw material for generating useful products from it. The handful properties of carbon nanomaterials viz., carbon content plays a major role in wide applications. The main bottleneck is the production of carbon nanomaterials in a large-scale system in a cost-effective manner. However, before approving the performance scale, a really short and insightful assessment should be done on shocking and maintenance assets, provider type, advantages, mitigation, and related disadvantages.

References

Abdelbasir, S.M., K. M. McCourt, C. M. Lee and D.C. Vanegas. 2020. Waste-Derived Nanoparticles: Synthesis Approaches, Environmental Applications, and Sustainability Considerations. *Frontiers in Chemistry* 8: 1–18. doi:10.3389/fchem.2020.00782.

Andrea, S., C. Perri, A. Csató, G. Giordano, D. Vuono and J. B. Nagy. 2010. Synthesis Methods of Carbon Nanotubes and Related Materials. *Materials* 3(5): 3092–3140. doi:10.3390/ma3053092.

Chen, C., D. Yu, G. Zhao, B. Du, W. Tang, L. Sun, Y. Sun, F. Besenbacher and M. Yu. 2016. Three-Dimensional Scaffolding Framework of Porous Carbon Nanosheets Derived from Plant Wastes for High-Performance Supercapacitors. *Nano Energy* 27: 377–3789. doi:10.1016/j.nanoen.2016.07.020.

Choi, J., T. Wixson, A. Worsley, S. Dhungana, S. R. Mishra, F. Perez and R. K. Gupta. 2021. Pomegranate: An Eco-Friendly Source for Energy Storage Devices. *Surface and Coatings Technology* 421: 127405. doi:10.1016/j.surfcoat.2021.127405.

Dangsheng, S. and G. Centi. 2011. Carbon Nanotubes for Energy Applications. *Nanoporous Materials for Energy and the Environment* 173–202. doi:10.4032/9789814303125.

Dhivyashree, A., S.A.B.A. Manaf, S. Yallappa, K. Chaitra, N. Kathyayini, G. Hegde. 2016. Low Cost, High Performance Supercapacitor Electrode Using Coconut Wastes: Eco-Friendly Approach. *Journal of Energy Chemistry* 25(5): 880–887.

Elich, J. M. 2004. Laser Vaporization Synthesis of Single Wall Carbon Nanotubes. Thesis, Rochester Institute of Technology, 117.

Hui, C., G. Wang, L. Chen, B. Dai and F. Yu. 2018. Three-Dimensional Honeycomb-like Porous Carbon with Both Interconnected Hierarchical Porosity and Nitrogen Self-Doping from Cotton Seed Husk for Supercapacitor Electrode. *Nanomaterials* 8(6) doi:10.3390/nano8060412.

Hussein, F. H. and H.A. Firas. 2019. Synthesis of Carbon Nanotubes by Chemical Vapor Deposition. *Nanomaterials: Biomedical, Environmental, and Engineering Applications* 105–132. doi:10.1002/9781119370383.ch4.

Justyna, C., J. Hoffman, A. Małolepszy, M. Mazurkiewicz, T. A. Kowalewski, Z.B. Szymanski and L. Stobinski. 2015. Synthesis of Carbon Nanotubes by the Laser Ablation Method: Effect of Laser Wavelength. *Physica Status Solidi (B) Basic Research* 252(8): 1860–1867. doi:10.1002/pssb.201451614.

Mukhtar, Y., C. Daulbayev, A. Taurbekov, A. Abdisattar, R. Ebrahim, S. Kumekov, N. Prikhodko, B. Lesbayev and K. Batyrzhan. 2021. Synthesis of Graphene-like Porous Carbon from Biomass for Electrochemical Energy Storage Applications. *Diamond and Related Materials* 119: 108560. doi:10.1016/j.diamond.2021.108560

Notarianni, M., J. Liu, K. Vernon and N. Motta. 2016. Synthesis and Applications of Carbon Nanomaterials for Energy Generation and Storage. *Beilstein Journal of Nanotechnology* 7(1): 149–196. doi:10.3762/bjnano.7.17.

Packiyalakshmi, P., B. Chandrasekhar and N. Kalaiselvi. 2019. Domestic Food Waste Derived Porous Carbon for Energy Storage Applications. *Chemistry Select* 4(27): 8007–8014. doi:10.1002/slct.201900818.

Rai, A. A. and C. Yanilmaz. 2021. High-Performance Nanostructured Bio-Based Carbon Electrodes for Energy Storage Applications. *Cellulose* 28: 5169–521810.1007/s10570-021-03881-z

Reddy, J.R. and J.W. Rhim. 2014. Isolation and Characterization of Cellulose Nanocrystals from Garlic Skin. *Material Letters* 129: 20–23.

Sabah, M. A., K. M. McCourt, C. M. Lee and D.C. Vanegas. 2020. Waste-Derived Nanoparticles: Synthesis Approaches, Environmental Applications, and Sustainability Considerations. *Frontiers in Chemistry* 782(8): 1–18. doi:10.3389/fchem.2020.00782

Saravanakumar, A., T. M. Haridasan, T. B. Reed and R.K. Bai. 2007. Experimental Investigation and Modelling Study of Long Stick Wood Gasification in a Top Lit Updraft Fixed Bed Gasifier. *Fuel* 86(17–18): 2846–2856. doi:10.1016/j.fuel.2007.03.028.

Suan, N.H., P.E. Kee, H.S. Yim, P. T. Chen, Y.H. Wei and J.C.W. Lan. 2020. Recent Advances on the Sustainable Approaches for Conversion and Reutilization of Food Wastes to Valuable Bioproducts. *Bioresource Technology* 302(135): 122889. doi:10.1016/j.biortech.2020.122889.

Tailong, C., H. Wang, C. Jin, Q. Sun and Y. Nie. 2018. Fabrication of Nitrogen-Doped Porous Electrically Conductive Carbon Aerogel from Waste Cabbage for Supercapacitors and Oil/Water Separation. *Journal of Materials Science: Materials in Electronics* 29(5): 4334–4344. doi:10.1007/s10854-017-8381-5.

Tatrari, G., M. Karakoti, C. Tewari, S. Pandey, B.S. Bohra, A. Dandapat and N. G. Sahoo. 2021. Solid Waste-Derived Carbon Nanomaterials for Supercapacitor Applications: A Recent Overview. *Material Advances* 2: 1454–1484.

Vinay S.B., P. Kanagavalli, G. Sriram, B.R. Prabhu, N. S. John, M. Veerapandian, M. Kurkuri. and G. Hegde. 2020. Low Cost, Catalyst Free, High Performance Supercapacitors Based on Porous Nano Carbon Derived from Agriculture Waste. *Journal of Energy Storage* 32: 101829. doi:10.1016/j.est.2020.101829.

Wei, C., E. Zhang, J. Wang, Z. Liu, J. Ge, X. Yu, H. Yang and B. Lu. 2019. Potato Derived Biomass Porous Carbon as Anode for Potassium Ion Batteries. *Electrochimica Acta* 293: 364–370. doi:10.1016/j.electacta.2018.10.036.

Wild, J., P. Kudrna, M. Tichý, V. Nevrlý, M. Střižík, P. Bitala, B. Filipi and Z. Zelinger. 2012. Electron Temperature Measurement in a Premixed Flat Flame Using the Double Probe Method. *Contributions to Plasma Physics* 52(8): 692–698. doi:10.1002/ctpp.201200005.

Xiaoxia, B., Z. Wang, J. Luo, W. Wu, Y. Liang, X. Tong and Z. Zhao. 2020. Hierarchical Porous Carbon with Interconnected Ordered Pores from Biowaste for High-Performance Supercapacitor Electrodes. *Nanoscale Research Letters* 15(1). doi:10.1186/s11671-020-03305-0.

Yanyan, X., H. Cao, Y. Xue, B. Li and W. Cai. 2018. Liquid-Phase Exfoliation of Graphene: An Overview on Exfoliation Media, Techniques, and Challenges. *Nanomaterials* 8(11). doi:10.3390/nano8110942.

Yoshinori, A. and X. Zhao. 2006. Synthesis of Carbon Nanotubes by Arc-Discharge Method. *New Diamond and Frontier Carbon Technology* 16(3): 123–137.

Yusik, M.S.J., T.T. Tung, K.M. Tripathi and T. Kim. 2019. Graphene-Based Aerogels Derived from Biomass for Energy Storage and Environmental Remediation. *ACS Sustainable Chemistry and Engineering* 7(4): 3772–3782. doi:10.1021/acssuschemeng.8b04202.

18

A Facile Microwave-Assisted Synthesis of Nanoparticles in Aspect of Energy Storage Applications

P. Anitha

Roever Engineering College,
Perambalur, India

N. Muruganantham
and R. Govindharaju

Thanthai Hans Roever
College, Perambalur, India

P. Sakthivel

Urumu Dhanalakshmi
College, Trichy, India

M.M. Senthamilselvi

Directorate of Collegiate
Education, Trichy, India

Green synthesis is done in a single step; it is possible to synthesise metal ions into nanoparticles using biomolecules found in herbal extracts. They are easily scaled up and easily conduct at ambient pressure and temperature. This bioactive conversion of metallic ions to base metal takes place very quickly. According to reports, silver nanoparticles exhibit antiinflammatory, magnetic, catalytic, and optical properties. By reducing a solution of zinc sulphate with an aqueous solution of *Ocimum basilicum* root extract, zinc nanoparticles were successfully produced in this study. This approach was straightforward, naturally flexible, non-lethal, and reasonably priced. Physical colour alterations, UV-visible spectroscopy, and Fourier-transform infrared spectroscopy (FTIR) were used for the identification of the zinc nanoparticles that were produced. Zinc nanoparticle creation was confirmed by the combination of reaction of zinc solution and *O. basilicum* root extract, which displayed a black hue. The zinc nanoparticles exhibit an absorption band in the UV-vis spectra between 300 and 450 nm. The average particle size of the nanoparticles was determined to be 25 nm based on the spectra of the biosynthesised zinc nanoparticles at various magnifications from X-ray diffraction (XRD) and scanning electron microscopy

DOI: 10.1201/9781003355755-18

(SEM). The OH stretching of the phenolic group can be recognised by the absorption band at 3887.9 cm^{-1}, which is visible in the FTIR spectra. ZnO nanocrystals demonstrated increased lithium storage capacity and superior cycling properties, and after 100 cycles at 200 mA g^{-1}, they revealed a reversible discharge capacity of 500 mAh g^{-1}. The absorption bands at 1173.8 cm^{-1} and 1605 cm^{-1} corresponds to the carbonyl groups present in the extract.

18.1 Introduction

The development of novel medications benefits greatly from the use of therapeutic herbs. It is now well acknowledged that Indian medicinal herbs have a huge potential of producing therapeutically useful medications that might potentially be employed by allopathic physicians. Ancient civilisations, including Egypt, China, India, and Greece, have long used terrestrial plants as medicine, and an astounding number of contemporary pharmaceuticals have been created from them. Around 2600 BC, the Sumerians and Akkadians began to record the medical benefits of herbs in written form (Sartorelli et al. 2009).

The practice of conventional medicine is the culmination of generations of doctors who have treated patients using their own systems of medicine. Plant-based medications have a significant role in the treatments that are used. Researchers have discovered that members of the Liliaceae, Asteraceae, Solanaceae, Caesalpinaceae, Rutaceae, Piperaceae, Sapotaceae, and Apocynaceae families can be utilised as healing plants (Samy and Gopalakrishnakone). One such plant in this group is *Ocimum basilicum*, which belongs to the Caesalpinaceae family. The Leguminosae plant *O. basilicum* Linn. is quite ubiquitous and wellknown for its therapeutic benefits.

Known in English as "Indian Laburnum," *Cassia fistula* Linn., family Caesalpiniaceae, is also known by the common names Fistula, Amaltas, Laburnum, Purging Fistula, and Golden Shower. It can be found in evergreen and mixed tropical forests throughout most of India, up to an elevation of 1300 m in the outer Himalaya. It grows as a sporadic tree in Konkan regions and Maharashtra's Deccan. For its lovely bunches of yellow roots, it is widely utilised as an ornamental tree in many nations, including South Africa, Mauritius, India, Mexico, the West Indies, Brazil, East Africa, and China. It is found in larger areas of India in deciduous and mixed-monsoon forests, rising to a height of 1300 metre in the outer Himalaya and is extensively utilised in traditional medical systems (Nirmal et al. 2008; Bahorun et al. 2005). This plant is beneficialinskin diseases, liver problems, tuberculous glands, diabetes, hematemesis, rheumatism, leucoderma, pruritus and the use of its roots and pods as purgatives for cough, biliousness, retained excretions have all been mentioned (Alam et al. 1990 and Asolkar 1992).

It is a perennial shrub that is 6–9 m tall, with a straight trunk and smooth, light grey bark that becomes hard and darker brown with age. Compound leaves with a greenish grey bark have pairs of leaves that are 5–12 cm long and have a total length of 23–40 cm. The pods are abundant with (40–100) horizontal seeds submerged in a brown coloured, sweet-tasting meat. They are pendulous, cylindrical, virtually straight, smooth, shiny, and dark brown. The majority of seeds are ovulated 8-mm long, 5-mm thick, and slightly less wide. This tree is prized for its beautiful clustering with yellow roots that are utilised in traditional medicine for a number of purposes (Kirtikar and Basu 2006; Danish et al. 2011).

18.2 Phytochemical of *O. basilicum*

The majority of the biological effects reported to *O. basilicum* extracts have been related to the composition of their primary and secondary metabolites. The seeds, pollen, fruit, leaves, and pod have essentially been the focus of primary metabolite studies. Two known bioactive substances, 6-dimethoxybenzoate, benzyl 2-hydroxy-3 and its dimer dibenzyl 2,2'-dihydroxy-3,6,3'',1'-dicarboxylate, 6''-tetramethoxy-biphenyl-1, which showed an innovative structural formation, have been produced by fractionation depending on the bioactivity of the n-hexane phase from the MeOH extract of the seeds of *O. basilicum* L (Rastogi 2004). The principal fatty acids in the seeds include linoleic, oleic, stearic, and palmitic acids, with levels of caprylic and myristic acids. They also were significant in glycerides (Senthil et al. 2006). According to studies, *O. basilicum*'s stem bark may also include lupeol, ß-sitosterol, and hexaconazole (Bahorun et al. 2005). In order to investigate alternate and efficient sources of protein, it is well known that 31% of raw proteins, primarily albumin, and globulin are present in wild berries (Shankar and Mathew 2012). Galactomannan is an abundant carbohydrate found in seeds that is composed of eight various types of sugar moieties. Similar research showed that seeds had 11.8% carbs and were efficient suppliers of cephalin and lecithin phospholipids (Sharma et al. 2010).

A thorough biochemical investigation of the pollen from the root, which is thought to have a major antiallergenic effect, revealed that it contains 12% protein and considerable levels of unattached amino acids like methionine, phenylalanine, and glutamic acid (Panda 2011).

An analysis of the phytochemicals, carbohydrates, alkaloids, glycosides, protein, flavonoids, saponins, amino acids, and tri-terpenoids showed that polar extracts such as methanol, ethanol, and aqueousextracts contained the majority of the constituents against non-polar extracts (petroleum-ether and chloroform). However, it was observed that every Laughton et al. (1991) mixture contains proteins, tannins, amino acids, and phenols. The herb *C. fistula* Linn. (Leguminosae) is readily available and well-known for its therapeutic benefits. This plant's application in treating hematemesis, rheumatism, pruritus, leucoderma, and diabetes has been mentioned in the Indian literature as being therapeutic against skin illnesses, tuberculous glands, and liver problems (Alam et al. 1990). Additionally, it contains hypoglycaemic and anti-inflammatory properties and is frequently employed as a traditional remedy that is safe for both pregnant women and children (Kumar and Cauhan 2006).

There are numerous publications on the antioxidant, antifertility, and hepatoprotective activities of *C. fistula* (Siddhuraju et al. 2002). Along with other Indian medicinal plants, *C. fistula* root and seed have been the subject of a few investigations on their antibacterial activities (Duraipandiyan and Ignacimuthu 2007, Vimalraj et al. 2009). These studies provide scant information about the antibacterial capability of this plant's leaves. The initial analysis of this plant, which was gathered from the Similipal Biosphere Reserve (SBR), revealed in reports by Thatoi et al. (2008) that it exhibited antibacterial activity against specific bacterial strains. For three days, the bark paste is administered externally to the mouthful location two to three periods daily at regular intervals. A half scoop of the extracts is taken orally three times a day to treat diarrhoea. All kinds of skin infections are treated externally using the leaf paste and neem. There are several publications on the antibacterial properties of the bark of *C. fistula*, roots, seeds, roots, and fruits, but there are only a few on its leafy parts. *O. basilicum* is a medium-sized deciduous tree that has become a popular ornamental species due to its stunning, flamboyant yellow roots.

The Leguminosae family includes the golden shower plant *O. basilicum* (Perumal Samy et al. 1998). In albino rats, the root extract suppresses ovarian activity while stimulating uterine function. Fruits are used to cure a variety of conditions, including antipyretic, diabetes, demulcent, abortifacient, and lessening inflammation in the chest, throat, liver, and eye illnesses (Daniel and Astruc 2004). Numerous natural antioxidant substances, including reducing sugars, phenolic acids, proteins, amino acids, anthraquinones, flavonoids, saponins, alkaloids, carbohydrates, and tannins, which have stronger antioxidant activity are found in *O. basilicum* plants (Wong et al. 2004). Tribal people frequently use the *O. basilicum* plant to cure a variety of illnesses, especially ringworm and other fungus skin infections. The Malayalis tribe of India utilises it to cure nose infections (Kumaret al. 2006). In Indian literature, the *O. basilicum* herb has been utilised to treat diabetes, haematemesis, leucoderm, liver disorders, and tuberculous glands. It can also be used as a laxative, analgesic, hypoglycaemic, antipyretic, and antiinflammatory drug (Rajeswari et al. 2006). ZnO particles have a theoretical capacity of 978 mAh gl when used as a lithium-ion battery anode materialby Pan et al. (2010). However, its practical application is hampered by weak electronic conductivity, a significant volume shift during the delithium/lithium process, and the accompanying significant capacity fading (Huang 2011). There have been some attempts up to this time to enhance its fatigue resistance, such as the creation of nanostructures (Huang et al. 2012), creating composite with metal by doping. For instance, dandelion-like ZnO nanorod arrays showed higher lithium capacity and improved cycling properties when compared with ZnO in powder form (Yuan et al. 2015). According to Wang et al. (2011), mesoporous ZnO nanostructures exhibit better capacitance and cycling performance than traditional solid ZnO particles, with a 50th charge capacity of 420 mAh g^{-1}. The performance of ZnO lithium-ion batteries needs to be improved even more, which will require extensive research.

18.3 Materials and Methods

18.3.1 Preparation of Root Extract

O. basilicum root was harvested when it was still young and fresh. It was completely cleaned with sterile double-distilled water (DDW). Small chunks of 20 g of sterilised root samples were obtained. Hundred millilitres of sterilised DDW and finely chopped root were placed in a 500-ml Erlenmeyer flask. The mixture was then boiled for 5 minutes and after that, it was filtered and the extract was stored at 40°C (Figure 18.1).

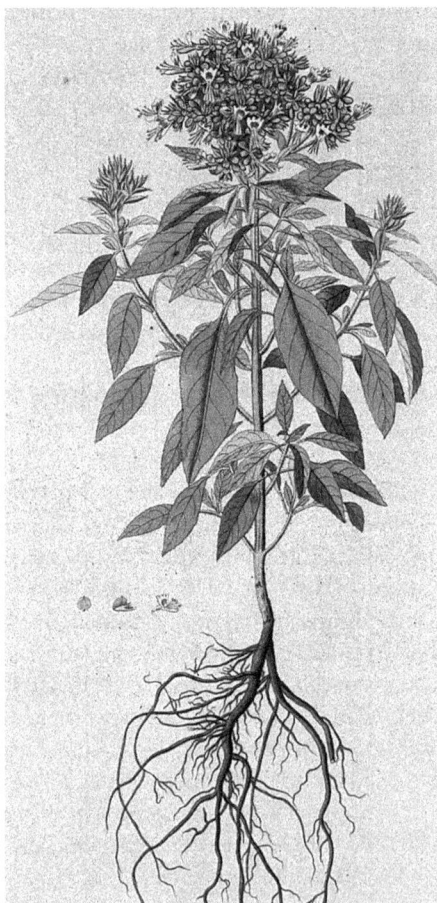

FIGURE 18.1 *Ocimum basilicum* root.

18.3.2 Synthesis of Silver Nanoparticles

The precursor used in the creation of zinc nanoparticles was zinc sulphate. Using a conical flask with a 250-ml capacity, 100 ml of the root extract was mixed with 100 ml of 0.1N $ZnSO_4$ aqueous solution at room temperature. The reaction was then carried out in the flask for 12 hours while being shaken at 100 rpm and 500°C. The mixture was then kept in the microwave to expose it to heat. After 20 minutes, the mixture was dried entirely, yielding nanoparticles in the form of granules.

18.3.3 UV-Visible Spectroscopy Analysis

Visual observation was used to document the colour effect on reaction mixture (metal ion solution + root extract). The UV-visible spectra of the solid sample were then measured in order to track the bio reduction of Ag^+ ions in aqueous solution. The UV-visible spectra of the sample were examined using a PERKIN ELMER (Lambda 35 model) spectrometer in the wavelength range of 190–1100 nm as a function of the reaction's duration on UV-visible spectroscopy.

18.3.4 FTIR Measurement

A Perkin Elmer (Spectrum RXI) instrument was used to conduct a Fourier-transform infrared (FTIR) study in the 400-4000 cm^{-1} range. The peak assignments were used to identify the functional groups.

18.3.5 XRD Measurement

By repeatedly dropping a little bit of the sample onto the nickel plate, allowing it to dry, and then preparing a thick layer of the sample, the sample was drop-coated onto the plate. X-ray diffraction was used to determine the silver nanoparticle's size and kind (X-ray diffraction [XRD]). Using a Rigaku miniflex-3 model with 30 kv, 30mA, and Cuk radians at a 2° angle, this was accomplished.

18.3.6 Scanning Electron Microscopy Analysis of AgNPs

A Quanta 200 FEG scanning electron microscope was used to conduct the scanning electron microscopic (SEM) study. By simply removing a very tiny portion of the specimen onto a copper grid that has been coated with carbon, thin films of the sample were created. The films on the SEM grid were then dried for 5 minutes under a mercury lamp after any excess solution had been blotted away with blotting paper.

18.3.7 Electrochemical Performance

To evaluate how well the manufactured electrodes operate electro chemically, coin cells (CR2025) were put together. In order to create a homogeneous slurry, prepared samples were combined with carboxymethyl cellulose and acetylene black in an aqueous solution at a weight ratio of 60: 20: 20. An electrode sheet with a coating thickness of 9–10 m and a loading level of roughly 1.4 mg cm^2 was made by applying the sludge to a sheet of copper foil that was 10 m thick, allowing it to dry for 12 hours at 60°C in a vacuum oven, and then pressing it. The electrolyte was purchased from Zhangjiagang Guotai-Huarong New Chemical Materials Company (China) and composed of 1 M LiPF6 and 5% fluoroethylene carbonate (FEC) in ethylene carbonate (EC, >99.9%), diethylene carbonate (DEC, >99.9%), and dimethyl carbonate (DMC, >99.9%) (v:v:v = 1:1:1, water content 20 ppm).

In a glove box filled with argon, the cells were put together. On a Shenzhen Neware cell cycler (China), the battery was galvanostatically charged and discharged in the fixed voltage range from 0.01 Volt to 3 Volt. The cycler was operated at 25°C. EIS was measured by applying a 5-mV alternating voltage at rates between 10 and 105 Hz. Unless otherwise stated, all impedance measurements in this work were performed after the constructed electrode had cycled through once.

18.4 Results and Discussion

18.4.1 Characterisation of Zinc Nanoparticles from *O. basilicum* Root Extracts

18.4.1.1 UV-Visible Spectroscopy Analysis

After UV-visible spectrophotometer examination, colour variations in the zinc nanoparticles proved their synthesis. The ratio of light that passes through a sample over those that are incident, or the transmittance of a sample or T, is defined as T = I/I0. We estimate the photons in such a standard UV-vis spectroscopy experiment that are not destroyed or scattered by the specimen. It is customary to report the sample's absorbance (A), which is correlated with the transmittance by the formula A = −log10 (T). The correlation between transmittance and wavelength is shown in Figure 18.2.

For the examination of various metal nanoparticles, a UV-visible spectrophotometer has proven to be quite helpful. It is also a remarkable method for confirming the creation and durability of ZnNPs in an aqueous phase. ZnNPs are well known for having dark brown colours that vary based on the particle's size and intensity. The excitation of the ZnNPs plasmon resonance surface (SPR) is what causes the colours to appear.

It is well accepted that size- and shape-controlled nanomaterials in aqueous solution can be examined using UV-vis spectroscopy. It is one of the techniques most frequently employed to characterise the structural properties of zinc nanoparticles. UV-vis spectra of aqueous plant leaf extracts were taken. The range of the absorption peaks was 268.70–453.30 nm. Nanoparticles in UV-vis spectrometers can be as large as 300–500 nm. The results from the UV-vis spectrometer ranged from 268.70 to 453.30 nm in this investigation. The produced particles identity as zinc nanoparticles is thus confirmed.

18.4.1.2 FTIR Measurement

Using the peak assignments, the relevant functional groupings of *O. basilicum* was found. Phytochemical groups, such as phenolic groups, amines, ether, carboxylic acid, and a hydroxyl group, that served as a reducing agent for the creation and stability of NPs were identified using FTIR spectroscopy analysis. The spectrum was captured in the 400 cm^{-1} to 4000 cm^{-1} wavelength range. FTIR measurements were

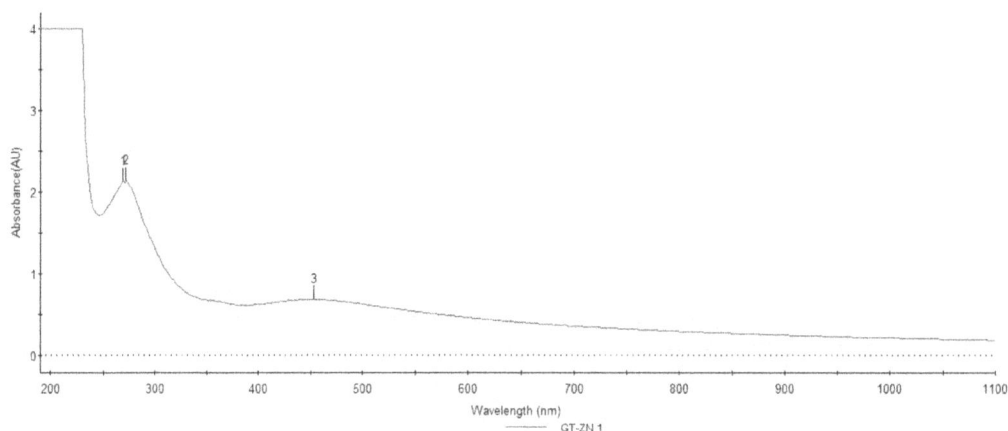

FIGURE 18.2 UV-visible spectra of synthesised zinc nanoparticles using root extract of *Ocimum basilicum*.

TABLE 18.1 Functional Groupings Using FTIR Spectrum

Frequency Range	Intensity	Functional Group	Compounds
3560.66 cm^{-1}	strong, broad	O-H stretching	alcohol
2161.18 cm^{-1}	medium	C=C=C stretching	allene
1617.67 cm^{-1}	strong	C=C stretching	α,β-unsaturated ketone
1384.80 cm^{-1}	medium	O-H bending	phenol
1150.70 cm^{-1}	strong	C-F stretching	fluoro compound
1107.79 cm^{-1}	strong	C-O stretching	primary alcohol
1008.31 cm^{-1}	strong	C-O stretching	primary alcohol
992.01 cm^{-1}	strong	C=C bending	alkene
740.63 cm^{-1}	strong	C=C bending	alkene
617.80 cm^{-1}	strong	C-Br stretching	halo compound

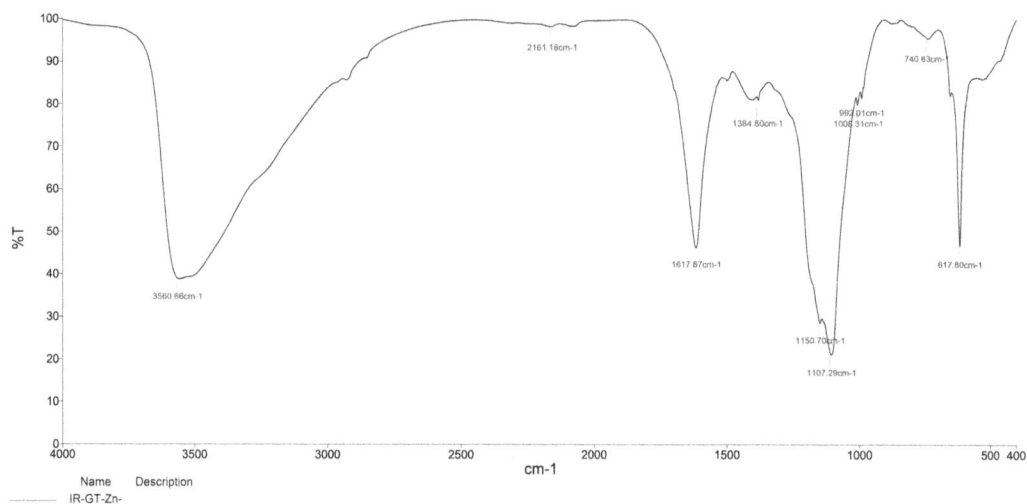

FIGURE 18.3 FTIR spectrum of synthesised zinc nanoparticles from Ocimum basilicum root extracts.

used to identify the possible molecules in *O. basilicum* that are in charge of capping and reducing the bio reduced zinc nanoparticles (Table 18.1).

Three bands were present at approximately 3560.66 cm^{-1}, 2161.18 cm^{-1}, 1617.67 cm^{-1}, 1384.80 cm^{-1}, 1150.70 cm^{-1}, 1107.79 cm^{-1}, 1 cm^{-1}, 108.31 cm^{-1}, 992.01 cm^{-1} and 617.80 cm^{-1} (Figure 18.3). The infrared spectroscopy of phytocapped ZnNPs showed the same absorption bands that can be found in the aqueous extract's IR spectrum. This demonstrates how the phytoconstituents, primarily tannins, prevent ZnNP aggregation.

18.4.1.3 XRD Measurement

X-ray scattering (XRD) is a quick analytical method that primarily identifies the phase of crystalline materials and can reveal details about unit cell dimensions. Therefore, by utilising XRD to look at the plant's diffraction peaks, it is possible to detect the presence of metal nanoparticles in the tissues. It is also frequently used to determine the chemical content and crystalline structure of a substance.

The XRD study of the dehydrated nanoparticles derived from colloid samples further demonstrated the crystalline character of Zn nanoparticles. The detected peak dispersion and noise were most likely caused by macromolecules found in the plant extract, which may have reduced zinc ions. The XRD

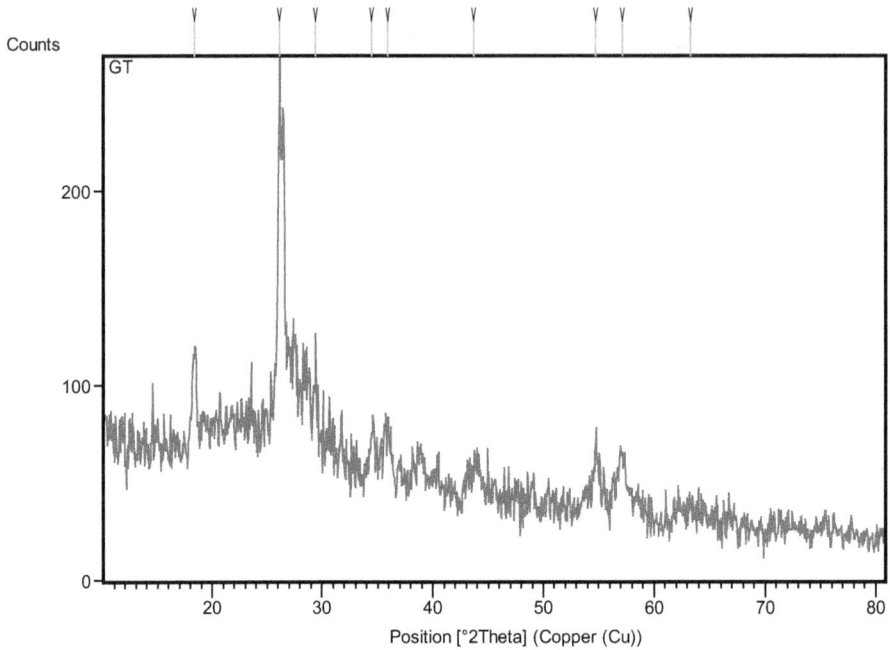

FIGURE 18.4 XRD spectrum of synthesised zinc nanoparticles *from* Ocimum *basilicum root* extracts.

TABLE 18.2 XRD Spectra Peak of Synthesised Silver Nanoparticles from Ocimum Basilicum Root Extracts

Peak List:

Position [°2Th.]	Height [cts]	FWHM Left [°2Th.]	d-spacing [Å]	Rel. Int. [%]
18.3825	47.52	0.3936	4.82648	24.49
26.1613	194.06	0.1968	3.40637	100.00
29.4448	42.47	0.2952	3.03356	21.88
34.4985	18.79	0.5904	2.59987	9.68
35.9515	22.00	0.5904	2.49806	11.33
43.6544	14.72	1.1808	2.07348	7.58
54.6986	33.45	0.3936	1.67809	17.24
57.1017	29.94	0.5904	1.61305	15.43
63.2458	5.94	2.3616	1.47034	3.06

pattern so amply demonstrated the crystallinity of the zinc nanoparticles produced in this current synthesis. Small particle size is the main cause of the peaks' lines becoming more pronounced.

The size of the synthesised ZnNPs is approximately 19.40 nm (Figure 18.4 and Table 18.2).

Scherrer Formula:

$$D = (0.94 \times \lambda) / (\beta \times Cos\theta)$$

where

Dp = Average crystallite size,

β = Line broadening in radians,

θ = Bragg angle,

λ = X-ray wavelength

18.4.1.4 SEM Analysis

SEM images were also used to determine the size and shape of the produced nanoparticles. Figure 18.5 displays pictures of ZnNPs produced by *O. basilicum*. The ZnNPs are mostly spherical in appearance and do not touch each other, as shown in the images. An image at a lower magnification shows that the nanoparticles are immersed in a thick matrix that may be one of the *O. basilicum*'s organic stabilising components. By looking at the strong Scherrer reflection in the XRD spectrum, it is possible to further demonstrate the existence of organic content linked to ZnNPs. ZnNPs that have been produced typically have diameters between 15 and 25 nm.

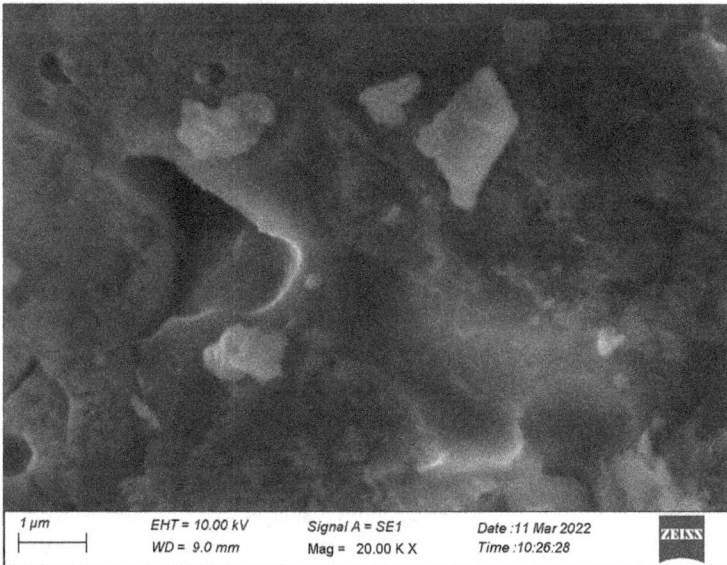

(a)

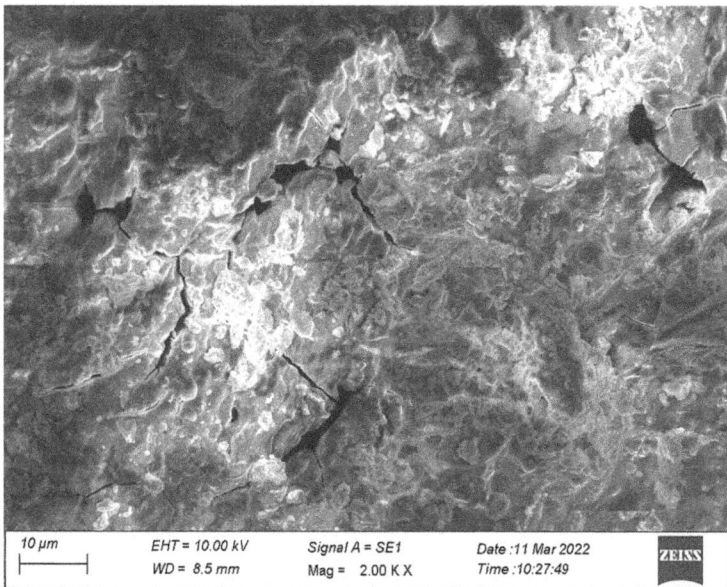

(b)

FIGURE 18.5   (a, b) SEM images of synthesised ZnNPs by root extract of Ocimum basilicum.

18.4.1.5 Electrochemical Performance

Investigations were made on the electrochemical characteristics of commercial ZnO nanoparticle electrodes and ZnO nanocrystal electrodes for lithium-ion batteries. Figure 18.7 displays the electrode CV in an operating voltage of 0.01–3 V at a scanning rate of 0.5 mV s^{-1}. Broadened peaks around 0.5–0.7 V and 1.2–1.0 V and substantial lowering peak below 0.3 V were visible during the first release process of zinc oxide nanoparticles and commercial ZnO electrodes, which were indicative of the irreversible processes. When Li-ion is inserted into ZnO, FEC breaks down, and a primordial solid electrolyte (SEI) coating forms on the composite interface.

The comparative peak at 0.5 Volt resulted from the reduction of ZnO into Zn and the production of amorphous Li_2O, whereas a significant peak at 0.25V was attributed to the dissolution of the electrolyte and the creation of Li-Zn alloy. These peaks vanished in subsequent rounds, signifying that ZnO with a significant irreversible capability underwent an irreversible decline in the preceding cycle. After the primary cycle, fresh reduction peaks at roughly 0.80 volt and 0.30 volt developed and migrated to a low voltage level over subsequent cycles. Six-week oxidation peaks with centres at 0.27 V, 0.52 V, 0.63 V, 1.48 V, 1.78 V, and 2.20 V can be clearly distinguished in the following delithium processes for ZnO nanocrystals (Figure 18.7(a)), which could be a result of the Li-Zn alloy undergoing many dealloying steps. However, only two oxidation peaks of 0.65 and 1.52 V could be seen in the CV curves for the commercial ZnO electrode (Figure 18.7(b)). Additionally, the ZnO nanocrystal electrode displayed a higher peak current in both oxidation–reduction reaction pathways, demonstrating a better and even more active electrode reaction.

Figure 18.6 shows the repeated results of a zinc oxide nanocrystal and a conventional zinc oxide nanoparticle electrode at current densities of 200 and 400 mAg^{-1}. While commercial ZnO nanoparticle electrode produced a lower initial charge/discharge capacity of 489/1273 mAhg^{-1} and showed a lower average discharge capacity of 112 mAh g^{-1} over 100 cycles, ZnO nanocrystal electrode delivered a higher initial discharge/charge capacity of 709/1563 mAhg^{-1} and demonstrated a high average discharge capacity of 500 mAhg^{-1} more than 100 cycles. The electrodes displayed comparable results at 400 mA g^{-1}. The ZnO nanocrystal electrode displayed excellent cycling performance with initial charge/discharge rates of 676/1475 mAhg^{-1} and a coulombic efficiency of 45.8%, as well as a discharge capacity of 428 mAh g^{-1}

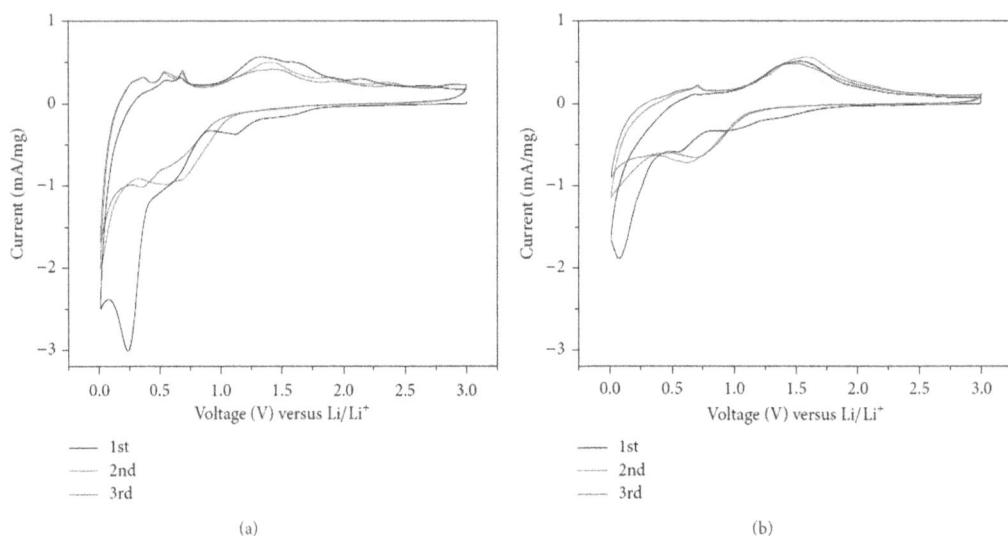

FIGURE 18.6 Cyclic voltammograms of (a) ZnO nanocrystal and (b) commercial ZnO nanoparticle electrode tested at 0.5mVs^{-1} in 0.01–3V.

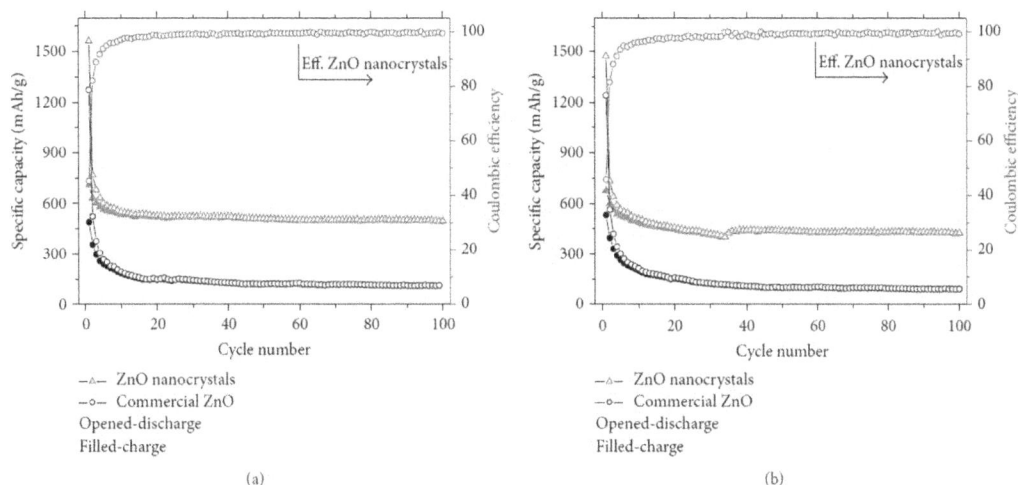

FIGURE 18.7 Cycle performances of ZnO nanocrystal and commercial ZnO nanoparticle electrode at a current density of (a) 200 mAg⁻¹ and (b) 400 mAg⁻¹.

following 100 cycles. Figure 18.9 displays the discharge capacities of a ZnO nanocrystal electrode. For the first two cycles, a ZnO nanocrystal electrode was operated at a current density of 200 mA g^{-1}.

The current density was again steadily raised to 4000 mA g^{-1} before being reduced to 200 mA g^{-1} again. With a coulombic efficiency of 44.8%, the ZnO nanocrystal electrode had an initial discharge capacity of 570/1272 mAh g^1. The capacity decreased to roughly 309, 187, 114, 75, and 51 mAh g^{-1} at increasing current densities of 500, 1000, 2000, 3000, and 4000 mA g^{-1}, respectively.

EIS was commissioned to describe the impedance characteristics of electrodes in order to evaluate the change in electrochemical performance. Figure 18.7 displays the electrodes' Nyquist complex plane conductivity charts following a first two charge/discharge cycle procedures. Both Nyquist plots featured a compressed semi-circle in the increased frequency band that was connected to the resistance to electron transfer at the electrode/electrolyte contact. The analogous circuit (inset) was used to fit these Nyquist plots, as illustrated in Figure 18.10. This equivalent circuit comprises three constant-phase elements (CPE), a Warburg diffusion element, and a series of four resistor elements. The analogous circuit's R1 consisted of the electrolyte resistance (Rs) and the electrode resistance, and the charge-transfer resistance across the electrode/electrolyte interfaces was R4 (Re). In the SEI film and pore stream of the electrode materials, respectively, CPE1 and CPE2 contributed to the Li+ diffusion. CPE3 denoted a double-layer electric capacitive of the electrode/solution interface, in contrast to ZW, which stood for Warburg impedance. In the high-volume electrode, it was linked to the semi-infinite dispersion of lithium ions.

The normal parameters were determined by applying the significant results listed in Table 18.1. It was discovered that the smaller R3, which is typically favourable for the quick transport of lithium ions and electrons across the contact, was also found in ZnO nanocrystal electrodes. Additionally, it showed that the carbon layer provided quick charge-transfer channels at the ZnO nanocrystal interfaces. The ZnO nanocrystal electrode's Y0,1, Y0,2, and Y0,3 values were higher than those of commercially ZnO electrodes. Charge transfer for the electrode reaction was promoted by the rise in Y0,3 reflecting the electric double-layer capacitance. Additionally, an increase in Y0,1 and Y0,2 encouraged Li+ transport in the SEI films and in the electrode's pore channels, respectively.

The ZnO nanocrystal electrode demonstrated a lower charge-transfer resistance and a higher electric double-layer capacitance in comparison to the commercial ZnO electrode, indicating an enhanced kinetic character of the electrochemical process (i.e., charge transfer and polarisation), which could be attributed to the improved availability of electron density and perhaps also Li+. The faster channel for

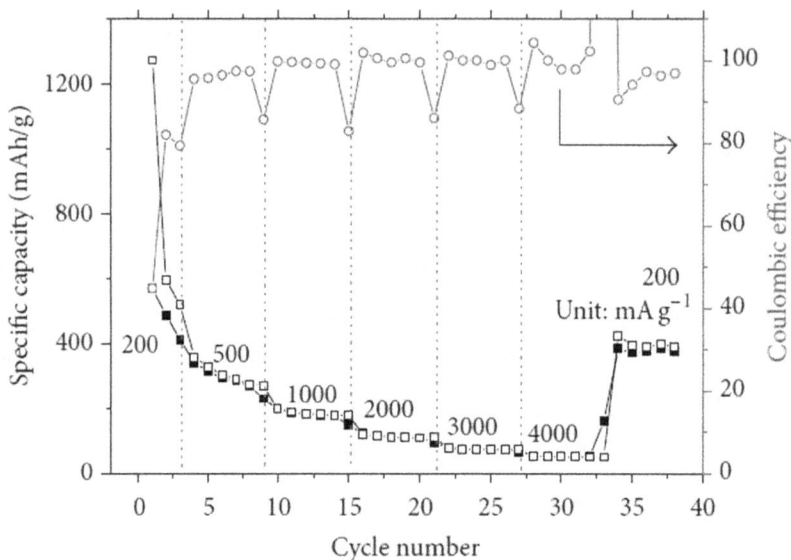

FIGURE 18.8   Rate performance of ZnO nanocrystal electrode.

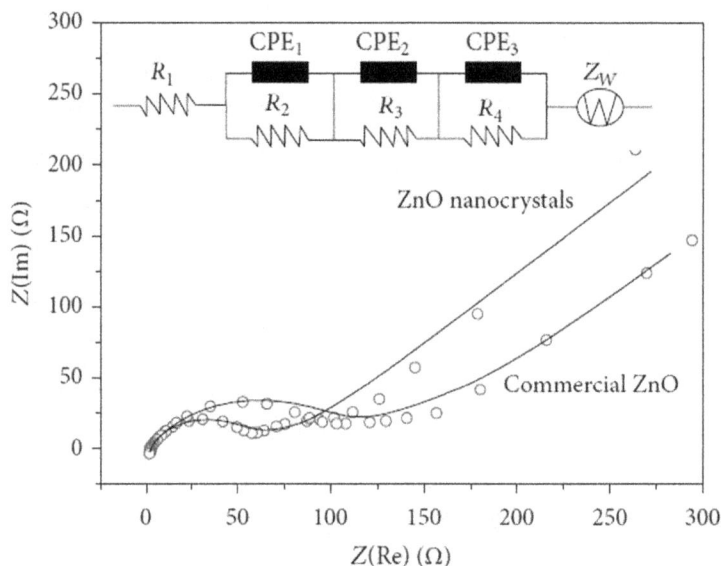

FIGURE 18.9   Nyquist plots for ZnO nanocrystal and commercial ZnO nanoparticle electrodes. The spots correspond to the experimental data, and the solid lines stand for the calculated data from the equivalent circuits of the *inset.*

mass movement and electron transfer created by the smaller size of particles and superior crystal type increased Li storage capacity. The ZnO nanocrystals are quite attractive when compared to other known ZnO-based anode materials due to their high capacity (Figure 18.8 and Tables 18.3–18.4).

TABLE 18.3 Comparison of Commercial ZnO and ZnO Nanocrystal

Sample ID	R1	R2	R3	R4	CPE$_1$		CPE$_2$		CPE$_3$		W
	(Ω)	(Ω)	(Ω)	(Ω)	Y0,1 (μF)	n	Y0,2 (μF)	n	Y0,3 (μF)	n	DW
Commercial ZnO	0.3	1.7	107.1	43.1	0.55	0.4	4.7	0.8	103.5	0.7	69.2
ZnO nanocrystals	0.3	1.3	53.4	22.2	0.91	0.4	6.2	0.6	280.5	0.6	47.6

TABLE 18.4 Electrochemical Performance Comparison of ZnO-based Anodes for Lithium-ion Batteries

The structure of the material	Specific discharge capacity (mAh g^{-1})	Cycle number	Current density
ZnO radial hollow microparticles	320	100	200 mA g^{-1}
ZnO nanowires	252	30	120 mA g^{-1}
ZnO root-like microaggregates	179	200	1 C
ZnO dandelion-like nanorod arrays	310	40	250 mA g^{-1}
ZnO root-like arrays	238	50	0.5 C
ZnO microrod arrays	150	50	500 mA g^{-1}
ZnO nanowire arrays	200	40	120 mA g^{-1}
ZnO nanocrystals 500	428	100	400 mA g^{-1}

18.5 Summary and Conclusion

In the fields of energy conversion, catalysis, medicine, and water treatment, nanomaterials may offer answers to environmental and technological problems. Environmentally friendly synthesis techniques must be used in tandem with this expanding need. Numerous significant applications exist for nano-silver. It functions as a disinfectant. It is used in textiles, household appliances, medical equipment, cosmetics, and home water filtration systems. Metal nanoparticles created chemically are expensive, bad for the environment, and energy intensive. As a valuable complement to chemical methods, biological methods using phytoconstituents for metal nanoparticle manufacturing have been proposed.

The villages near Trichy are where the samples were gathered. The plant extract in aqueous form was made. To ascertain the existence of secondary metabolites, a qualitative examination of plant extracts was conducted. The production of zinc nanoparticles was assisted by plants. The development of ZnNPs was indicated by the hue changing from green to brown to black. AgNPs that had been produced were characterised using a variety of techniques, including UV-vis spectra, XRD, and FTIR. ZnNPs have a UV wavelength range of 268.70–453.30 nm.

The FTIR data demonstrate that the functional groups are present in AgNPs. XRD pattern amply demonstrated the crystalline character of the produced silver nanoparticles. According to SEM findings, the form of the produced silver nanoparticles is spherical. By looking at the sharp Braggs reflections in the XRD spectrum, it is possible to further demonstrate the existence of organic content linked to AgNPs. The generated AgNPs are about 25 nm in size.

The green production of nanoparticles demonstrates that they are economical, eco-friendly, and suitable for therapeutic usage in people. After 100 cycles at 200 mAg^{-1}, ZnO nanocrystals showed an increased reversed discharge capacity of 400 mAh g^{-1}. The lithium storage capacity of ZnO nanocrystals was higher than that of commercial ZnO nanoparticles. These results are explained by the structural diversity of ZnO nanocrystals, which affects Li-ion diffusion and results in varied cell impedance.

References

Alam MM, Siddiqui MB, and Hussian W. Treatment of diabetes through herbal drugs in rural India. *Fitoterapia* 1990; 61: 240–2.

Asolkar LV, Kakkar KK, and Chakre OJ. *Second supplement to glossary of Indian medicinal plant with active principles*. New Delhi: Publication and Information Directorate, CSIR; 1992: 414.

Bahorun T, Neergheen VS, Aruoma OI. Phytochemical constituents of Ocimum Basilicum. *Afr J Biotechnol* 2005; 4(13): 1530–40.

Daniel MC, and Astruc D. Gold nanoparticles: Assembly, supramolecular chemistry, quantum-size-related properties, and applications toward biology, catalysis, and nanotechnology. *Chem Rev* 2004; 104(1): 293–346.

Danish M, Singh P, Mishra G, Srivastava Jha SKK, Khosa RL. Ocimum basilicum Linn. (Amulthus)- An Important Medicinal Plant: A Review of Its Traditional Uses, Phytochemistry and Pharmacological Properties. *J Nat Prod Plant Resour* 2011; 1(1): 101–18.

Duraipandiyan V, and Ignacimuthu S. Antibacterial and antifungal activity of Ocimum basilicum L.: An ethnomedicinal plant. *J Ethnopharmacol* 2007; 112: 590–4.

Huang XH, WuJB, Lin Y, and Guo RQ. ZnO microrod arrays grown on copper substrates as anode materials for lithium-ion batteries. *Int J Electrochem Sci* 2012; 7(8): 6611–6621.

Huang XH, Xia XH, Yuan YF, and Zhou F. Porous ZnO nanosheets grown on copper substrates as anodes for lithium-ion batteries. *Electrochim Acta* 2011; 56(14): 4960–5.

Kirtikar KR, and Basu BD. Indian Medicinal Plants. In *International Book Distributors* 2006; Lalit Mohan Basu, Allahabad, vol. 2: 856–60.

Kumar VP, and Cauhan SN. padh H and antifungal agents from selected Indian medicinal plants. *J Ethnopharmacol* 2006; 107(2): 182–8.

Kumar VP, Chauhan NS, Padhi H, and Rajani M. Search for antibacterial and antifungal agents from selected Indian medicinal plants. *J Ethnopharmacol* 2006; 67: 241–45.

Laughton MJ, Evans PJ, Moroney MA, Hoult JRS, and Halliwell B. Inhibition of mammalian 5-lipoxygenase and cyclo-oxygenase by flavonoids and phenolic dietary additives-Relation to antioxidantactivity and to iron reducing ability. *Biochem Pharmacol* 1991; 42: 1673–81.

Nirmal A, Eliza J, Rajlakshmi M, Priya E, and Daisy P. Effect of Hexan extract of Ocimum basilicum barks on blood glucose and profile in Straptocoin. *Int J Pharm* 2008; 4(4): 292–6.

Pan Q, Qin L, Liu J, and Wang H. Root-like ZnO-NiO-C films with high reversible capacity and rate capability for lithium-ion batteries. *Electrochim Acta* 2010; 55(20): 5780–5.

Panda SK, Padhi LP, and Mohanty G. Antibacterial activities and phytochemical analysis of Ocimum basilicum (Linn.) leaf. *J Adv Pharm Technol Res* 2011: 2(1): 62–7.

Perumal Samy R, Ignacimuthu S, and Sen A. Screening of 34 medicinal plants for antibacterial properties. *J Ethnopharmacol* 1998; 62: 173–82.

Rajeswari R, Thejomoorthy P, Mathuram LN, and Narayana Raju KVS. Anti-inflamatory activity of Ocimum basilicum Linn. bark extracts I sub-acute modules of inflammation in rats. *Tamilnadu J Vet Anim Sci* 2006; 2(5): 193–9.

Rastogi RP, and Mehrotra BN. Compendium of Indian Medicinal plants, Central Drug Research Institute, Lucknow and National Institute of Science Communication and Information. *Resources* 2004; 4: 155–6.

Sartorelli P, Carvalho CS, Reimao JQ, Ferreira MJP, Tempone AG. Antiparasitic activity of biochanin A, an isolated isoflavone from fruits of Ocimum basilicum (Leguminosae). *Parasitol Res* 2009: 104(2); 311–4.

Senthil KM, Sripriya R, Vijaya RH, and Sehgal PK. Wound healing potential of Ocimum basilicum on infected albino rat model. *J Surg Res* 2006; 131(2): 283–9.

Shankar S, and Mathew L. Chemopreventive potential of methanol extract of stem bark of Ocimum basilicum L. In Mice. *IJPI'S J Pharmacognosy Herbal Formul* 2012; 2: 8.

Sharma A, Laxmi V, Goel A, Sharma V, and Bhatia AK. Anti-viral activity of Ocimum basilicum against IBR virus. *J Immunol Immunopathol* 2010; 12(2): 114–9.

Siddhuraju P, Mohan PS, and Becker K. Studies on the antioxidant activity of Indian laburnum (Ocimum basilicum L.): A preliminary assessment of crude extracts from stem bark, leaves, fl owers and fruit pulp. *J Agric Food Chem* 2002; 79: 61–7.

Thatoi HN, Panda SK, Rath SK, and Dutta SK. Antimicrobial activity and ethnomedicinal uses of some medicinal plants from Similipal Biosphere Reserve, Orissa. *Asian J Plant Sci* 2008; 7: 260–7.

Vimalraj TR, Saravanakumar S, Vadivel S, Ramesh S, and Thejomoorthy P. Antibcaterial effects of Cassia fistula extracts on pathogenic bacteria of veterinary importance. *Tamilnadu J Veter Anim Sci* 2009; 5: 109–111.

Wang J, DuN, Zhang H, YuJ, and Yang D. Layer-by-layer assembly synthesis of ZnO/SnO2 composite nanowire arrays as high-performance anode for lithium-ion batteries. *Mater Res Bull* 2011; 46(12): 2378–84.

Wong TS, and Schwaneberg U. Protein Engineering in Bioelectrocatalysis. *Curr Opin Biotechnol* 2004; 14(6): 590–6.

Yuan G, Wang G, Wang H, and Bai J. Synthesis and electro- chemical investigation of radial ZnO microparticles as anode materials for lithium-ion batteries. *Ionics* 2015; 21(2): 365–71.

19

A Critical Review on Role of Nanoparticles in Bioenergy Production

N. Jeenathunisa,
S. Jeyabharathi and
V. Aruna

*Cauvery College for Women
(Autonomous), Affiliated to
Bharathidasan University,
Tiruchirappalli, India*

Scarcity of fossil fuels, high energy demand, and global warming have made countries worldwide search for alternative fuel resources. Biofuel that emits lower greenhouse gas dragged the attention to serve as an alternative fuel source. The limitations in currently available technologies correlated with the high production expense of petroleum-derived fuels, so the interest towards nanotechnology tremendously increased in the current perspective. Due to their energy efficiency, selectivity, and time management with lower cost investment, nanoparticles such as carbon tubes, metal, magnetic, and metal oxide can be employed for biomass use. The current review study will investigate a thorough overview of the nano-materials employed in the manufacturing of biofuels and their potential future applications.

19.1 Introduction

An increase in population growth and urbanisation leads to excessive dependence on petroleum-derived fuels and the need for fossil fuels is a global issue in search for alternative forms of bioenergy [1–6].

In response to the rising need for energy and excessive reliance on fossil fuels, experts are interested in developing bioenergy, a type of sustainable energy made from organic materials called biomass.

Cost-effectiveness and sustainable development are bridged by the generation of biofuel from organic waste [7, 8]. Due to their unique qualities, biofuels have drawn attention from all over the world [9]. Plant sources utilised as feedstock for the manufacture of biofuels include vegetables, corn, soybeans, sugar-cane, palm oil, and *Jatropha* (used in Africa) on almost every continent [10, 11]. The promising, efficient methods of nanotechnology can be used to raise the calibre of biofuel production.

With their special qualities including a high surface area to volume ratio, a large amount of catalytic activity, adsorption capacity, crystallinity, and stability, nanoparticles (NPs) as a source for biofuel synthesis provide a number of benefits [12–14].

In general, nanoparticles like metal oxide and carbon nanotubes (CNTs) help with high-potential recovery and are used as nanocatalysts for the generation of biofuel [15]. Biofuels were created as a result of combining nanotechnology with other processes like pyrolysis, gasification, hydrogenation, and anaerobic digestion [16, 17]. These biofuels are further divided into primary and secondary biofuels, which lessen the effects of these environmental problems. While secondary biofuels are produced by combining nanomaterials or microorganisms, primary biofuels can be acquired directly from animals, woods, plants, and crop wastes.

Depending on the type of biomass/feedstock and technique, secondary biofuels are divided into three categories. First-generation biofuels include biomass generated from food plants such as wheat, barley, sunflower, corn, sugarcane, and sorghum. While the third-generation fuels use algal biomass and species of microalgal to produce biodiesel, biogas, and biohydrogen under various kinetics conditions, the second-generation biofuels are primarily made from biomass of lignocellulosic, *Jatropha*, cob, grass, straw, miscanthus, and cassava (Figure 19.1).

Third-generation biofuels, on the other hand, also use carbon dioxide as a feedstock during biofuel synthesis. Large-scale interest in the display is sparked by second- and third-generation biofuels, which causes deforestation, a water scarcity, and problems with food security [18, 19]. However, using carbon and maximising energy were related to the efficiency of the biofuel production process.

Nanoparticles play a significant role in energy optimisation and increase overall product output. Non-metal and metallic substances, bioactive molecules, biocatalysts or enzymes, and carbon-based nanomaterials are some of the categories for nanomaterials that are successfully used in a variety of science and technology fields for fluid flow, water purification, biopharmaceuticals, cancer treatment, biological sensing, and biofuel production.

Nanomaterials in the perspective of biofuels improve the chemical kinetics, or the catalyst's and electron transfer's activities, which reduces the production of inhibitors and boosts the activity of impoverished microbial populations. The stability of a nanomaterial is dependent on its size, structure, and shape, which is crucial for catalytic activities during the synthetic process. The use of nanomaterials to maximise energy production remains in its development.

Due to their greater interfacial surface area, which enhances the chemical kinetics of biological reactions, nanoparticle applications in the manufacture of biofuel have drawn significant attention from researchers worldwide. For this, a variety of nanocrystals, including cornerstone, nanotubes, nanocage, composite, nanoring, nanowire, and magnetic nanosystems, have been studied. This section examines the various biofuel production improvements made with the assistance of nanomaterials.

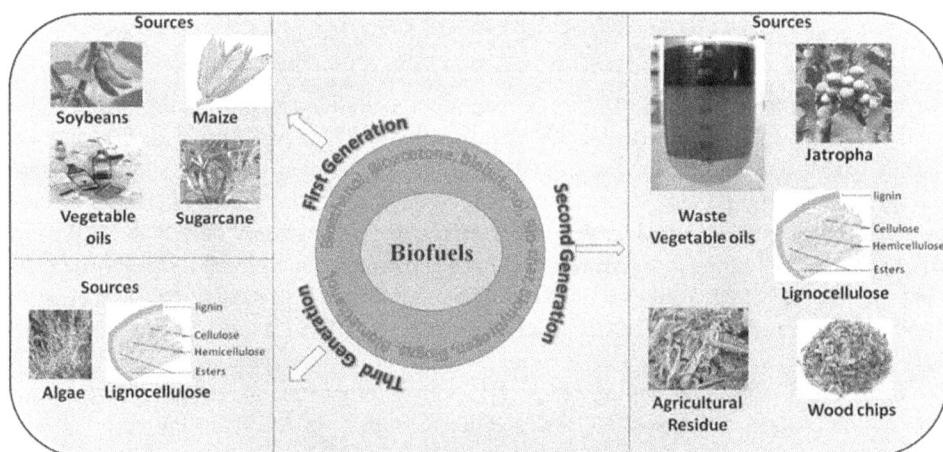

FIGURE 19.1 Representation of biofuel types and their sources. (Taken from Arya, I.; Poona, A.; Dikshit, P.K.; Pandit, S.; Kumar, J.; Singh, H.N.; Jha, N.K.; Rudayni, H.A.; Chaudhary, A.A.; Kumar, S. Current Trends and Future Prospects of Nanotechnology in Biofuel Production. Catalysts 2021, 11, 1308. https://doi.org/10.3390/catal11111308.)

19.2 Carbon Nanotubes

The fourth-most prevalent element in the cosmos, graphite exists in a variety of shapes known as polymorphs based on the configurations of its atoms. Dielectric strength and toughness are two distinct characteristics allotropes of carbon [20, 21]. At room temperature, solid graphite has two known forms—gemstone and graphene. A new area of carbon nanochemistry has emerged as a result of the discovery of a novel graphite crystalline forms with 60 carbon atoms aligned in perfect symmetry. One of the key developments in carbon nanochemistry is the fullerene.

CNTs are tubular molecules that contain single-layer carbon atoms in the form of rolled-up sheets (graphene). They could be single-walled nanotubes (SWCNTs) with a diameter of just 1 nm or multi-walled nanotubes (MWCNTs), which have diameters of over 100 nm and are made up of many concentrically interconnected nanotubes. They can reach many micrometres or millimetres in length [22]. With the help of sp2 bonds, a potent sort of molecular contact, CNTs were chemically joined. Using this characteristic along with the CNTs' innate propensity to stick together through van der Waals forces, it is possible to create extremely strong, lightweight materials with high conductivity thermal and electrical capabilities.

Researchers from the National Institute of Advanced Industrial Science and Technology (AIST) in Japan, the University of Vienna in Austria, and the Leibniz Institute for Solid State and Materials Research (IFW) Dresden in Germany came together to publish a paper in the journal Advanced Materials [23, 24]. The investigation researchers discovered that saturating a nanotube resulted in the tube being bombarded with electrons.

These findings reflect this limited chemical process' utility in fabricating carbon nanomaterials dielectric strength. The optical and electrical characteristics of CNTs can be adjusted using the appropriate filler. This research would lay a solid foundation for the use of CNTs in diverse energy transformation procedures.

The distinct thermo-mechanical characteristics of CNTs plays an essential role in current advances [25]. Nanotubes are mechanically 400 times more durable than steel, and their superior thermal conductivity outperforms diamond. Unless subjected to high temperatures at the same time, nanotubes are resistant to chemicals and can withstand almost any chemical effect. Additionally, oxygen offers nanotubes exceptional hardenability [26]. The cylindrical core, which is supplied with nanostructured materials, is beneficial for the conversion of bioenergy.

Organic matter is a clean alternative hydrocarbons commodity that can be used to make activated carbons, greener pharmaceuticals, and hydropower [27]. An active nanomaterial-based nanocellulose biomass can regulate the production of nanoparticles in a radiofrequency [28]. Another research found a proportional link between CNTs and the simple sugars produced during biomass pyrolysis. The expulsion of volatiles has purportedly been pushed by hot-spot fabrication and selective heating. Thermal treatment enhanced the development and future direction for producing CNTs from lignocellulosic materials by combining the gradient of biomass particles peculiar to microwave heating and the existence of inherent inorganic compounds in biomass.

Co-pyrolysis of polymers to produce CNTs is a potential and ecologically sound method of producing energy or other materials from debris [29]. Different metallic mediators (Fe/cordierite, Ni/cordierite, and Ni-Mg/cordierite) were used in the chemical torrefaction. This was investigated how the reacting tension (0.5–1.25 MPa) affected the production of CNTs [30]. According to their findings, materials were studied by temperature programme oxidation (TPO), scanning electron microscopy (SEM), high propagation microscopy (TEM), and Raman spectroscopy in order to determine the fullerene synthesis during the thermal decomposition of plastics in a fixed bed reactor.

The production of helical carbon produced by the Ni-based accelerator is around 93 mass percentage. The Ni-Mg-based trap's potent metal-support interaction restricted CNT expansion, leading to shorter and crooked allotropes of carbon tubes as a result. The output of more homogenous and dense CNTs improved with the optimum response pressure, notably at 1.0 MPa (198 mg/gPP).

In a brief study, Liu and his colleagues investigated CNTs, carbon allotropes made by rolling up carbon sheets, and they postulated that their rod carried redox processes and electrocatalytic dynamics [31]. According to Liu et al., the inclusion of 100 mg/L CNTs to microbial sludge blanket processors boosted the production of biohydrogen at a rate of 5.55 L/L/d. In comparison to certain other activated carbon nanoparticles, the anaerobic fermentation procedure led to a shorter start-up time and increased performance.

The quantities of metal ions in rivers close to industrial sites were succinctly examined by Anh Tuan Hoang and his colleagues [32]. They discovered that the quantity frequently surpasses the permitted thresholds, continuing to be hazardous to both the environment and to population lives. Therefore, the authors acknowledge that there has been a larger focus on developing effective methods for sequestering ions from heavy metals. This literature review focuses on the production of CNTs from bioenergy and how to use them to filter out water that contains heavy metals. Toxic metals were eliminated energetically, electrostatically, topically, and by interactions between organic compounds.

It has been demonstrated that gas produced during the combustion of biomass can be used to create CNTs [33]. Because the gas produced by the gasification of biomass contains a lot of charcoal, it is a rich carbon feedstock gas. The above research focuses on the thermogravimetric and flocculated bed processes utilised to produce CNTs from a combination of CO and CH4, which were the main components of the gas produced during the thermochemical of biomass. Reported to the authorities were the response heat and vapour pressure of the mixture gas' critical roles. Outcomes of the decomposition of mixed gas (CO/CH4) on a fluidised bed were contrasted with those of pure gas.

Each of the findings demonstrated that the mixture gaseous disintegration component's temperature range for carbon excess weight was lower than that of the refined CH4 digestion process in the thermogravimetric reactor. The combination gas breakdown in the liquid increased the transformation of CO and CH4. Due to the medium's role in the CO/CH4 breakdown process, CO and CH4 worked in conjunction with H2. The production and quality of CNTs dramatically improved with the elevation in the CO/CH4 pressure gradients ratio.

19.3 Magnetic Nanoparticles

Nanomaterials have been employed continuously for energy production over the years, including nanocomposites (oxidised iron), silver and gold nanoparticles, nanoshells, and nanocages. In the biofuel trade, catalysts like lipases and cellulases are frequently used. A lot of studies on magneto strictive nanofillers propose as they play a vital purpose in incapacitating enzymes for bioenergy synthesis. After being immobilised to a durable template covered in certain nanomaterials, enzymes can be reused, and this technique is suitable for the breakdown of cellulosic biomass [34].

By altering various enzyme characteristics, the immobilisation of biocatalysts utilised for the breakdown of lignocellulosic materials can be improved. The magnetic nanoparticles' strong spellbinding characteristic is useful in the partition of excipients, increasing utilisation [35]. A number of initiatives have to encapsulate cellulose on magnetic particles for biomass degradation [36]. Combining calcium and iron magnetic nano will indicate how *Jatropha curcus*is used to make biodiesel [37].

Nanomaterials have pores with a size of 90 nm, a volume of 0.55 cm^3/g, and a significant surface area of 391 m^2/g. In all but most rare cases, the yield of biodiesel from *Jatropha* oil is 94%. Nevertheless, after cycles, it plummeted to 85% and then further dropped gradually as a result of the reactivation of the nanostructures.

The reason for the deactivation of NPs was investigated further to determine the cause. The presence of response-relevant properties clogging the pores was demonstrated by the data after the fourth and sixth cycles. Furthermore, the exterior space was reduced to 252 m^2/g, which was a smaller amount than ever before. In a different work, reduced graphene oxide and iron were mixed with aniline to create a polyaniline matrix, a nanocomposite. The nanocomposite is likely to work better at glucose oxidase's bio-electro processing.

Incapacitated glucose oxidase on iron oxide or polyaniline was demonstrated to have an exceptionally high enzymatic activity. Compressed graphene oxide that has been coated in polyaniline also has a large external surface region and a high conductivity. The outcomes proved that the composite is capable of transporting electrons. The highest amount of energy generated when applied to an enzymatic biocatalyst was 31.9 mA cm^2 at a glucose concentration of 49 mM [38]. Magnetic micro ferrites doped with calcium have been demonstrated to significantly improve the yield of biodiesel produced from soybean cooking oils, increasing it by almost 85% [39].

It was determined that the use of manganese dioxide nanoparticles and sugarcane leaves boosted transesterification. This process converts sugarcane leaves to bioethanol. Manganese dioxide nanoparticles' extremely large surface area allows them to effectively attach catalysts to their adsorption sites, improving ethanol generation [40]. This was determined that the accelerated generation of ethanol was caused by the entrapment of yeast cells on the magneto strictive nanocrystals. A previous study determined the possibility of using cellulase anchored on manganese in conjunction with lipid extraction to emulsify the microalgae cell wall.

Using water hyacinth as the substrate, the inclusion of iron nanoparticles improves microbial degradation and hydrogen production [41]. The results of this investigation demonstrated how precisely focusing on iron nanoparticles increased the hydrogen yield, reaching 56 mL/g of the dry weight of plants. Research has also shown that when glucose is used as the substrate, H2 generation improves. Ferric oxide nanoparticles, such as iron oxides, were investigated for biorefinery using sugar, sewage, and bagasse in addition to zero-valent nanoparticles [42]. Iron oxide and nano zero-valent iron (nZVI) have also been investigated to improve biogas production by employing waste-activated sludge.

19.4 Metallic Nanoparticles

Despite substantial research into metallic nanoparticles, numerous studies have been conducted to confirm their suitability for biofuel production. Due to their superior external area and nano-size, metallic nanoparticles are known to attract numerous enzymes, including oxidoreductase, allowing for the recovery of electron transport [43].

For an enhanced ion transfer rate and oxygen decrease frequency functionality, numerous reactive nanoparticles are created. When used in a layer-by-layer assembly with the proper polymers and enzymes, it has been predicted that metallic nanoparticles may be added in a planned method to advance electro-reactive exercise and create a biofuel cell with the highest absorption ability and acceptable electron shifting rate [44]. By putting them in multi-walled nanomaterials that have been acid-functionalised, metallic gold and platinum nanoparticles from hybrid nano reactants are intended to be used.

Another technique, however, used polydendrimer-structured gold nanoparticles. Analyses using high-performance electron microscopy have revealed how straight and capable polymer hybrid nanoparticles are. Gold, platinum, and Pt0.75-Tin0.25 supported transition metal capacity has been constructed in biofuel cells. Gold and platinum particles together demonstrated significant alcoholic oxidation capability [45].

In a different investigation, an aqueous solution was used to create gold nanoparticles, establishing high catalytic activity even on the ninth cycle and enormous electromagnetic responsiveness. For the synthesis of biohydrogen, several nanomaterials have been used. Gold nanoparticles (4 nm) boost biohydrogen production rate by 47% and substrate utilisation capability by 59% [46]. Gold nanoparticles allow the production of biohydrogen by clinging microbes to dynamic regions because of their smaller size and bigger outer region. Such nanomaterials also increase the enzymatic activity in the system that produces biohydrogen, which is essential for its generation. Silver nanoparticles have also been successfully used to reduce substrate degradation and promote the generation of bioenergy.

The principal biohydrogen generating mechanism, the acid effect, is activated by these nanoparticles, which also shortens the lag period of bacterial and algal development. Nanoparticles promote biohydrogen production in photosynthetic microorganisms. The growth of microorganisms, physiological functions, photosynthetic capacity, protein synthesis, and nitrogen metabolism are all enhanced by the addition of nanoparticles to the growth medium. The ideal amounts of silver and gold nanoparticles were found to boost the photosynthetic activity of *Chlorella vulgaris*. Zero-valent iron nanoparticles have been shown to improve the production of biogas from waste materials. Nickel nanoparticles have also been widely used to convert glucose into sorbitol during the hydrogenation process [47].

19.5 Conclusion

Use of fossil fuels caused the ecosystem irreversible harm. As a result, biofuels are able to supply the world's energy needs, and recycling the available cellulose as feedstocks is environmentally friendly. The potential cannot be used with current technologies. Therefore, one of the main objectives in the current situation is the idea of employing nanomaterials to speed up the manufacturing of new alternatives, such as biofuels [48].

Nanotechnology-based biofuel production is the way of the future for sectors dependent on petroleum because it is extremely efficient and environmentally, inexhaustible, healthier, and extremely safe to use. Moreover, the price increases of petroleum-derived fuels have been caused by constraint and desire, making biofuels a technologically and scientifically viable alternative. The importance and potential of biofuel as a different source of sustainable energy will soon be implemented globally.

References

1. Farrell, A.E., Plevin, R.J., Turner, B.T., Jones, A.D. and M.D. O'Hare Kammen. 2006. "Ethanol can contribute to energy and environmental goals." *Science* 311, 506–508.
2. Cherubini, F. 2010. "The biorefinery concept: Using biomass instead of oil for producing energy and chemicals." *Energ. Conver. Manage.* 51, 1412–1421.
3. Silva, S.S. and A.K. Chandel. 2014. *Biofuels in Brazil. Fundamental Aspects, Recent Developments and Future Perspectives*. IIT BHU, Varanasi, Springer International Publishing.
4. Kumar, M. and M.P. Sharma. 2014. "Potential assessment of microalgal oils for biodiesel production: A review." *J. Mater. Environ. Sci.* 3, 757–766.
5. Lee, A.F., Bennett, J.A., Manayil, J.C. and K. Wilson. 2014. "Heterogeneous catalysis for sustainable biodiesel production via esterification and transesterification." *Chem. Soc. Rev.* 43, 7887–7916.
6. Rai, M., Dos Santos, J.C., Soler, M.F., Franco Marcelino, P.R., Brumano, L.P., Ingle, A.P., Gaikwad, S., Gade, A. and S.S. Da Silva. 2016. "Strategic role of nanotechnology for production of bioethanol and biodiesel." *Nanotechnol. Rev.* 5, 231–250.
7. Naik, S.N., Goud, V.V., Rout, P.K. and A.K. Dalai. 2010. "Production of first and second generation biofuels: A comprehensive review." *Renew. Sustain. Energy Rev.* 14, no. 2: 578–597.
8. Ho, D.P., Ngo, H.H. and W. Guo. 2014. "A mini review on renewable sources for biofuel." *Bioresour. Technol.* 169, 742–749.
9. Bhattarai, K., Stalick, W.M., McKay, S., Geme, G. and N. Bhattarai. 2011. "Biofuel: An alternative to fossil fuel for alleviating world energy and economic crises." *J. Environ. Sci. Health Part A Toxic* 46, 1424–1442.
10. Shalaby, E.A. 2013. "Biofuel: Sources, Extraction and Determination." In *Liquid, Gaseous and Solid Biofuels*. Fang, Z., Ed.; Intech Open, Cario University, UK.
11. Folaranmi, J. 2013. "Production of biodiesel (B100) from Jatropha oil using sodium hydroxide as catalyst." *J. Pet. Eng.* 2013, 1–6.
12. Do Nascimento, R.O., Rebelo, L.M. and E. Sacher. 2017. "Physicochemical Characterizations of Nanoparticles Used for Bioenergy and Biofuel Production." In *Nanotechnology for Bioenergy and Biofuel Production*. Rai, M., da Silva, S.S., Eds.; Springer, Cambridge, MA, 173–191.

13. Saoud, K. 2018. "Nanocatalyst for Biofuel Production: A Review." In *Green Nanotechnology for Biofuel Production*. Srivastava, N., Srivastava, M., Pandey, H., Mishra, P.K., Ramteke, P.W., Eds.; Springer, Doha, Qatar, 39–62.

14. Dikshit, P.K., Kumar, J., Das, A.K., Sadhu, S., Sharma, S., Singh, S., Gupta, P.K. and B.S. Kim. 2021. "Green synthesis of metallic nanoparticles: Applications and limitations." *Catalysts* 11, 902.

15. Singh, N., Dhanya, B.S. and M.L. Verma. 2020. "Nano-immobilized biocatalysts and their potential biotechnological applications in bioenergy production." *Mater. Sci. Energy Technol.* 3, 808–824.

16. Hussain, S.T., Ali, S.A., Bano, A. and T. Mahmood. 2011. "Use of nanotechnology for the production of biofuels from butchery waste." *Int. J. Phys. Sci.* 6, 7271–7279.

17. Ali, S., Shafique, O., Mahmood, S., Mahmood, T., Khan, B.A. and I. Ahmad. 2020. "Biofuels production from weed biomass using nanocatalysttechnology." *Biomass Bioenergy* 139, 105–595.

18. Ahmed, W. and B. Sarkar. 2018. "Impact of carbon emissions in a sustainable supply chain management for a second-generation biofuel." *J. Clean. Prod.* 186, 807–820.

19. Belayin, V.V., Bulusheya, L.G. and A.V. Okotrub. 2003. "Modifications to the electronic structure of carbon nanotubes with symmetric and random vacancies." *Int. J. Quantum Chem.* 96, no 3: 239–246.

20. Wang, Changlong, Han, Honggui, Wu, Yufeng and Didier Astruc. 2022. "Nano catalyzed up cycling of the plastic wastes for a circular economy." *Coord. Chem. Rev.* 458, no. 1: 214422.

21. Wang, Jianqiao, Shen, Boxiong, Lan, Meichen, Kang, Dongrui and Chunfei Wu. 2020. "Carbon nanotubes (CNTs) production from catalytic pyrolysis of waste plastics: The influence of catalyst and reaction pressure." *J. Catal. Today* 351, no 1: 50–57.

22. Zhang, Baiqiang, Piao, Guilin, Zhang, Jubing, Bu, Chan Sheng, Hao Xie, Bo Wu and Nobusuke Kobayashi. 2018. "Synthesis of carbon nanotubes from conventional biomass-based gasification gas." *Fuel Process. Technol.* 180: 105–113.

23. Dai, Hongjie. 2002. "Carbon nanotubes: Opportunities and challenges." *Surf. Sci.* 500, no. 1–3: 218–241.

24. Shiozawa, Hidetsugu, Pichler, Thomas, Gruneis, Alexander, Hans, Pfeiffer, Liu, Zheng, Suenaga, Kazu and Hiromichi Kataura. 2008. "A catalytic reaction inside a single-walled carbon nanotube." *Adv. Mater.* 20, no. 8: 1443–1449.

25. Hong, J., Park, D.W. and S.E. Shim. 2010. "A review on thermal conductivity of polymer composites using carbon-based fillers: Carbon nanotubes and carbon fibers." *Carbon Lett.* 11: 347–356.

26. Baskar, Chinnappan, Baskar, Shikha and Ranjit S. Dillon. 2012. *Biomass Conversion: The Interface of Biotechnology, Chemistry and Materials*. Springer, Heidelberg, New york.

27. Ma, Jie, Wang, Jian-Nong, Tsai, Chung-Jung, Nussinov, Ruth and Buyong Ma. 2010. "Diameters of single-walled carbon nanotubes (SWCNTs) and related Nano chemistry and nanobiology." *Front. Mater. Sci. in China* 4: 17–28.

28. Jin, Fan-Long and Soo-Jin Park. 2011. "A review of the preparation and properties of carbon nanotubes-reinforced polymer composites." *Carbon Lett.* 12, no. 2: 57–69.

29. Omoriyekomwan, Joy Esohe, Tahmasebi, Arash, Dou, Jinxiao, Wang, Rou and Jianglong Yu. 2021. "A review on the recent advances in the production of carbon nanotubes and carbon nanofibers via microwave-assisted pyrolysis of biomass." *Fuel Process. Technol.* 214: 106686.

30. Peng, F., Zhang, L., Wang, H., Lv, P. and H. Yu. 2005. "Sulfonated carbon nanotubes as a strong protonic acid catalyst." *Carbon* 43: 2405–2408.

31. Hong, Anh Tuan, Sandro Nizetic, Chin Kui, Cheng, Rafael Luque, Thomas, Sabu, Banh, Tien Long, Pham, Van Viet and Xuan Phuong Nguyen. 2022. "Heavy metal removal by biomass-derived carbon nanotubes as a greener environmental remediation: A comprehensive review." *Chemosphere* 287, no. 1: 131959.

32. Liu, Z., Lv, F., Zheng, H., Zhang, C., Wei, F. and X.H. Xing. 2012. "Enhanced hydrogen production in a UASB reactor by retaining microbial consortium onto carbon nanotubes (CNTs)." *Int. J. Hydrogen Energy* 37: 10619–10626.

33. Tran, D.T., Chen, C.L. and J.S. Chang. 2012. "Immobilization of Burkholderia Sp. Lipase on a Ferric Silica Nanocomposite for biodiesel production." *J. Biotechnol.* 158, no. 4: 112–119.

34. Alftrén, J. and T.J. Hobley. 2015. "Covalent immobilization of β-glucosidase on magnetic particles for lignocellulose hydrolysis." *Appl. Biochem. Biotechnol.* 169, no. 2: 2076–2087.

35. Huang, P.J., Chang, K.L., Hsieh, J.F. and S.T. Chen. "Catalysis of rice straw hydrolysis by the combination of immobilized cellulose from aspergillus niger on β -cyclodextrin-fenanoparticles and ionic liquid." *Biomed. Res. Int.* 54, no. 8: 1–9.

36. Teo, S.H., Islam, A., Chan, E.S., Thomas Choong, S.Y., Alharthi, N.H., Taufiq-Yap, Y.H. and M.R. Awual. 2019. "Efficient biodiesel production from jatrophacurcus using $CaSO_4/Fe_2O_3$-SiO_2 core-shell magnetic nanoparticles." *J. Clean. Prod.* 208, no. 11: 816–826.

37. Shakeel, N.M.I. Ahamed, A. Ahmed, M.M. Rahman and A.M. Asiri. 2019. "Functionalized magnetic nanoparticle-reduced graphene oxide nanocomposite for enzymatic biofuel cell applications." *Int. J. Hydrogen Energy* 44, no. 5: 28294–28304.

38. Dantas, J.E.A., Leal, B., Mapossa, D.R. and A.C.F.M. Cornejo 2017. "Magnetic nanocatalysts of Ni0.5Zn0.5Fe2O4 doped with Cu and performance evaluation in transesterification reaction for biodiesel production." *Fuel* 191, no. 3: 463–471.

39. Cherian, E., Dharmendirakumar, M. and G. Baskar. 2015. "Immobilization of cellulase onto MnO_2 nanoparticles for bioethanol production by enhanced hydrolysis of agricultural waste." *Chin. J. Catal.* 36, no. 6: 1223–1229.

40. Mahmood, T., Zada, B. and S.A. Malik. 2013. "Effect of iron nanoparticles on Hyacinthâ€™sfermentation." *Int. J. Sci.* 21, no. 2: 106–121.

41. Engliman, N.S., Abdul, P.M. and J.M. Jahim. 2017. "Influence of iron (II) oxide nanoparticle on biohydrogen production in thermophilic mixed fermentation." *Int. J. Hydrogen Energy* 42, no. 4: 27482–27493.

42. Vincent, K.A., Blanford, C.F., Belsey, N.A., Weiner, J.H. and F.A. Armstrong. 2007. "Enzymatic catalysis on conducting graphite particles." *Nat. Chem. Biol.* 31, no. 5: 761–762.

43. Kwon, C.H., Shin Kwon, D., Park, J., Bae, W.K., Lee, S.W. and J. Cho, 2018. "High-power hybrid biofuel cells using layer-bylayer assembled glucose oxidase-coated metallic cotton fibers." *Nat. Commun.* 9, no. 2: 1–11.

44. Aquino Neto, S., Almeida, T.S., Palma, L.M., Minteer, S.D. and A.R. De Andrade, 2014. "Hybrid nanocatalysts containing enzymes and metallic nanoparticles for ethanol/O_2 biofuel cell." *J. Power Sources* 259, no. 15: 25–32.

45. Zhang, Y. and J. Shen 2007. "Enhancement effect of gold nanoparticles on biohydrogen production from artificial wastewater." *Int. J. Hydrogen Energy*, 32, no. 12: 17–23.

46. Yigezu, Z.D. and K. Muthukumar, 2014. "Catalytic cracking of vegetable oil with metal oxides for biofuel production." *Energy Convers. Manag. Green Chem* 84, no. 3: 326–333.

47. Kumar, Yogendra, Yogeshwar, Prerna, Bajpai, Sushant, Jaiswal, Pooja, Yadav, Shalu, Pathak, Diksha Praveen, Sonker, Muskan and Saurabh Kr Tiwary. 2021. "Nanomaterials: Stimulants for biofuels and renewables, yield and energy optimization." *Mater. Adv.* 2, 5318–5343.

48. Demirbas, A. 2008. "Biofuels sources, biofuel policy, biofuel economy and global biofuel projections." *Energ. Conver. Manage.* 49, 2106–2116.

20

Copper Oxide Nanoparticles for Energy Storage Applications

Dr. A. Arun kumar

Methodist College of Engineering & Technology, Abids, Hyderabad, India

Dr. R. Subramaniyan@ Raja

KPR Institute of Engineering & Technology, Coimbatore, India

Dr. G. Padmasree

Stanley College of Engineering and Technology for Women, Hyderabad, India

Kodumuri Veerabhadra Rao and Dr. K. Anuradha

Methodist College of Engineering and Technology, Hyderabad, India

Dr. A. Rathika

Stanley College of Engineering and Technology for Women, Hyderabad, India

Copper oxide nanoparticles are synthesized using chemical precipitation method. The synthesized nanoparticles are confirmed by X-ray diffraction technique. The average grain size of the synthesized nanoparticles is calculated by the Scherrer formula. The optical absorption studies are studied using UV–visible spectral analysis. Scanning electron microscopy is used to analyze the surface morphology of the synthesized nanoparticles. Photoluminescence studies are calculated carried out at an excitation wavelength of 330 nm.

DOI: 10.1201/9781003355755-20

20.1 Introduction

A fundamental stepping-stone in the development of functional nanomaterials is provided by metal oxides, in particular. When it comes to scientific and technological applications, metal oxides have a very broad bandgap. While there appear to be many different metal oxides in nature, some are more beneficial than others in terms of scientific and technological applications [1–3]. A large number of transition metals may be found on the periodic table, and they have several applications in a wide range of industries. In that they are metallic in bulk, CuO nanoparticles exhibit semiconductor behavior when reduced to nanoscale, which is extraordinary. Because of their practical utility in electrical and opto-electronic devices, semiconducting materials have piqued the curiosity of researchers.

In the field of nanotechnology, copper oxide (CuO) nanoparticles are a form of transition metal oxide nanoparticle that has found a wide range of applications. CuO nanoparticles are a p-type oxide semi-conductor with a bandgap between 1.2 and 1.5 eV [4, 5]. Nanoparticles are used in solar cells, which is one application [6]. Long-term energy sources such as solar electricity, often known as solar cells, are a realistic possibility. However, due to the intermittent nature of solar resources, the self-conversion of solar energy, like many other renewable energy systems, suffers from a scarcity of resources [7]. Molten salt has the potential to be employed in solar power plants as a heat transfer fluid (HTF) and as a thermal storage medium, according to researchers. In contrast, organic HTF becomes exceedingly unstable when exposed to high temperatures. These techniques are crucial for the practical evolvement of minimal feature size semiconducting integrated circuits. The optical conductivity of metal oxides is one of the essential properties of metal oxides, and it may be measured experimentally using reflectance and absorption tests to determine its value [8, 9]. Recognizing that absorption features are responsible for the bulk of absorption behavior observed in solids, reflectivity features are responsible for the majority of absorption behavior noticed in liquids. Light absorption becomes discontinuous and size-dependent with quantum-size confinement. The microwave-assisted hydrothermal method [10] was successful in the preparation of C/CuO hollow spheres and CuO nanocrystals. A practical technique for the synthesis of CuO with specific electrochemical performance is provided by this microwave-assisted hydrothermal method, and the performance can be further improved by increasing the dispersivity of the resulting particles [11].

Methods, solvents, surfactants, beginning precursors, temperature, and copper (II) acetate as substances can regulate the size and shape of desired nanostructure materials. CuO nanostructures by a sonochemical method using copper (II) acetate as precursors, urea and sodium hydroxide as reducing agents, and polyvinyl pyrrolidone (PVP) as copper oxide nanoparticles are extremely attractive as heat transfer nanofluids, which makes them a promising candidate for further research [12]. Flower-shaped CuO nanostructures using the reflux condensation process will be investigated as a potential biomaterial in the near future [13].

20.2 Applications

CuO is finding applications for use as antioxidants, antibacterial agents, thermal conductive materials, catalytic agents, batteries, and solar cells. Copper oxide is extensively used in marine paints as a pigment, fungicide, and antifouling agent, among other things. The use of rectifier diodes made of this material dates back to the time when silicon became the standard. Copper is used as a pigment in ceramics to create blue, red, green, gray, pink, and black glazes.

20.3 Synthesis of Pure CuO Nanoparticles

The co-precipitation process was used to create pure CuO nanoparticles, which were then characterized. A conical flask was filled with approximately 39.93 g of copper acetate, and a known quantity of deionized water was added to it. A solution of 8 g of NaOH was dissolved in a known quantity of water, and this solution was slowly added to the aforesaid solution. The solution was stirred every 4 hours, and the temperature was maintained at a steady level. When NaOH was added to copper acetate solution, the solution turns black, and a substantial amount of precipitate was created at the bottom of the flask while the solution is being added. After that, it was rinsed with methanol to eliminate any remaining organic matter. After that, the final product was oven dried at 150°C for 4 hours in a closed atmosphere before being calcined at 400°C for 4 hours in a muffle furnace. After that, a black powder with a high purity is obtained [14–16].

20.4 Characterization Details

The calcinated nanoparticles were ground to a fine powder in order to conduct characterization experiments. The crystalline phases and grain size of nanoparticles were determined using an X-ray diffractometer, a Rigaku Model RAD II A, by the researchers.

A Shimadzu 8400S infrared spectrophotometer was used to analyze the functional groups present in the samples, which were measured in the range of 4000–400 cm^{-1} utilizing the samples.

The optical measurements were carried out by Systronics, an Indian company that specialized in UV–Vis spectrometers. A scanning electron microscope (SEM) was used to examine the particle size and surface morphology of the nanoparticles that were generated, and the images were captured with a Tescan vega3 SBU camera.

20.5 Results and Discussion

20.5.1 XRD and Surface Morphology Studies

Figure 20.1 depicts the XRD pattern of CuO nanoparticles as they were synthesized. It results in a single phase with a monoclinic structural arrangement. Its cell parameters are as follows:

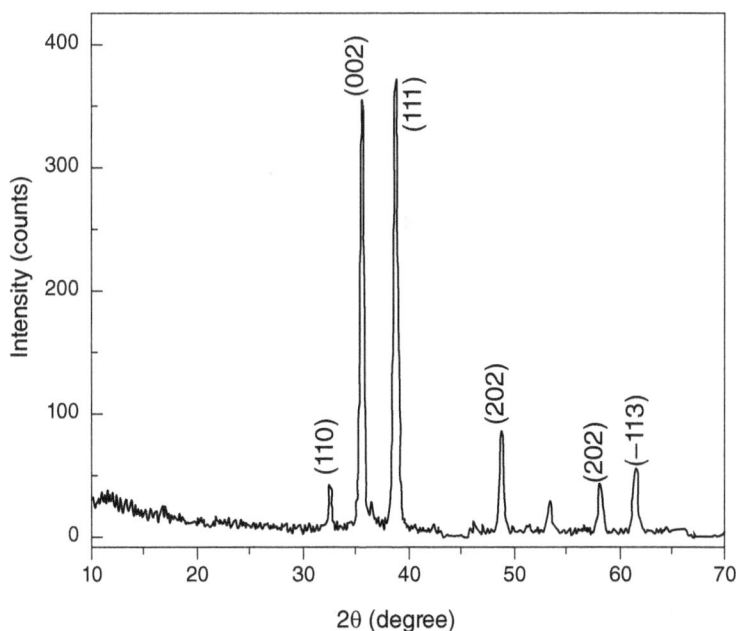

FIGURE 20.1 X-ray diffraction pattern of copper oxide nanoparticles.

a = 4.84(2) Å, b = 3.45 (2) Å, c = 5.35(4) Å. There is good agreement between the reported values and the intensities and positions of the peaks (JCPDS file No. 05-661). Using a computer program, we computed the average grain size of the synthesized nanoparticles, and the results revealed that the values were 15 nm. Smaller grains have a high value of strain, while larger grains have a lower value of strain. This clearly demonstrates that smaller particles experience greater strain, whereas larger particles experience less strain [17, 18]. In the instance of pure CuO spectrum, they were able to get diffraction peaks that were much narrower and peaks that were clearly visible.

20.5.2 FTIR Studies

The FTIR peaks obtained in the present study are presented in Figure 20.2. Table 20.1 shows the observed frequency and their corresponding assignments. Metal oxide generally gives absorption band below 1000 cm^{-1} that arising from interatomic vibration in the present study it is observed at 995 cm^{-1}, 578 cm^{-1}, and 424 cm^{-1}. The absorption peaks at 2299 cm^{-1} ,1558 cm^{-1} ,1373 cm^{-1} ,1064 cm^{-1}, and 586 cm^{-1}

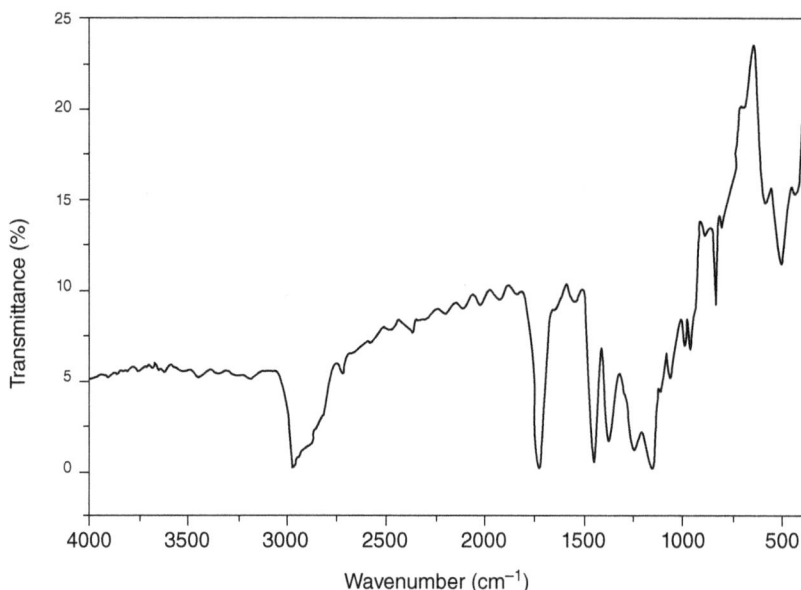

FIGURE 20.2   FTIR spectrum of copper oxide nanoparticles.

TABLE 20.1   Observed Vibrational Wavenumber and Their Corresponding Assignments of Copper Oxide of Nanoparticles

Pure CuO	Assignments
2299.15	Cu absorption band
1728.22	O-H bending
1651.07	O-H bending
1558.48	CuO peak
1458.18	O-H bending
1373.32	CuO peak
1064.71	Cu absorption band
586.36	Cu absorption band

show the presence of CuO which supports the presence of monoclinic phase [19–22]. The absorbed band at 1458 cm^{-1}, 1728 cm^{-1}, and 1651 cm^{-1} represents O-H bending vibration [23].

20.5.3 Optical Studies

The UV–Vis absorption spectra were acquired using Systronics, India, in the wavelength range of 200–1100 nm. The absorption spectrum of copper oxide nanoparticles in the UV–visible range is depicted in Figure 20.3. The absorption peak of pure CuO is determined to be 320 nm in wavelength. When the graph between photon energy and (hv)^2 is shown, the direct band gap of the sample may be computed, as illustrated in Figure 20.4. On the basis of these results, the band gap energy values are found to be

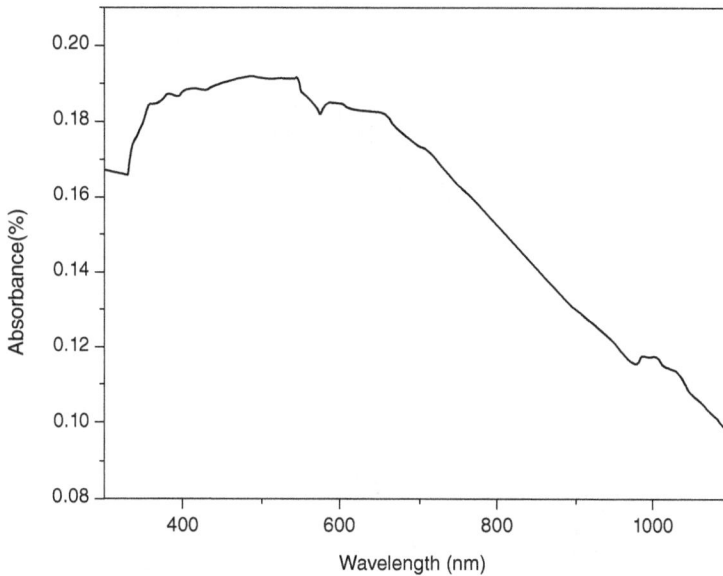

FIGURE 20.3 UV-Vis spectrum for pure CuO.

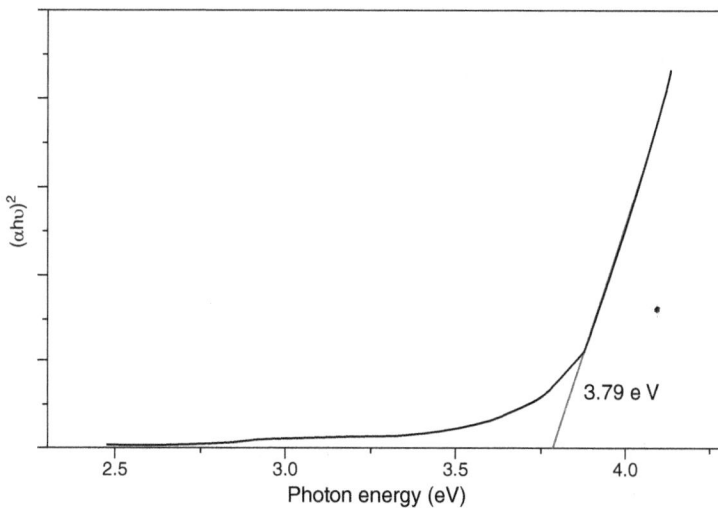

FIGURE 20.4 Band gap energy estimation graph for pure CuO.

3.79 eV. Higher levels of the absorption coefficient result in a power load dependence on photon energy in the case of optical absorption [24–26].

20.5.4 Scanning Electron Microscopy

Figure 20.5 shows a scanning electron micrograph of CuO nanoparticles that were synthesized. SEM examination of pure CuO reveals that the sample comprises agglomerated CuO nanoparticles and that the particles are of more uniform size and morphology [27, 28].

20.5.5 Photoluminescence Studies

The photoluminescence spectra of CuO nanoparticles as synthesized following excitation at 330 nm at room temperature are depicted in Figure 20.6. Infrared wavelengths of 398 nm (violet), 470 nm (blue),

FIGURE 20.5   SEM image of copper oxide nanoparticles.

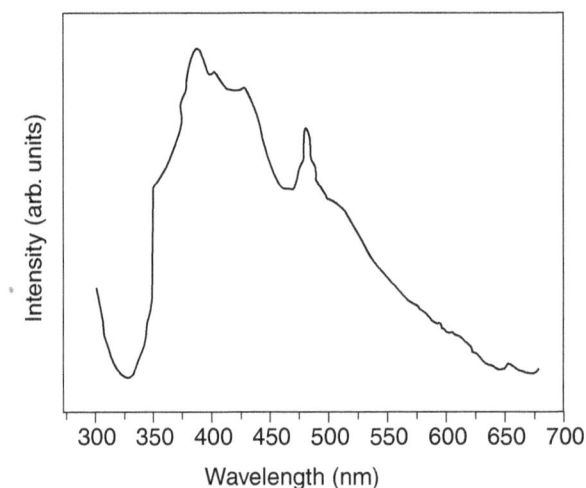

FIGURE 20.6   Emission spectra of CuO nanoparticles.

and 527 nm (green) are emitted by copper oxide (green). The first of them is connected with band-edge emission [29]. The second is due to the presence of an artifact. The third form of emission occurs when a photogenerated hole combines with a valence band electron [30, 31], which results in the emission of green light from CuO materials.

20.6 Conclusion

The X-ray diffraction technique is castoff to confirm the presence of the synthesized nanoparticles. The Debye–Scherrer formula is used to calculate the average grain size of the produced nanoparticles, which is a measure of their size distribution. The particle diameters range from 15 to 30 nm. The nanoparticles that were produced had a monoclinic structure. Fourier transform infrared spectrum (FTIR) studies revealed the presence of CuO, which supported the presence of the monoclinic phase. Pure CuO has an optical absorption peak at 320 nm, and the matching band gap energy values for CuO are found to be 3.79 eV. In order to examine the surface morphology of the CuO nanoparticles that were produced, scanning electron microscopy was used. The SEM picture of pure CuO revealed a higher tendency for agglomeration, indicating that the sample comprises CuO nanoparticles that had clumped together. It also demonstrates that it has a more homogeneous size and morphology than the other. The particle size found in the powder X-ray diffraction investigations was consistent with the results of the SEM examination as well.

References

1. Nikhil J, Zhong LW, Tapan KS, Tarasankar P, Seed-mediated growth method to prepare cubic copper nanoparticles, *Current Science*, 2000, 79, 1367–1370.
2. Zhu H, Zhang C, Yin Y, Novel synthesis of copper nanoparticles: Influence of the synthesis conditions on the particle size, *Nanotechnology*, 2005, 16, 3070–3083.
3. Wang H, Qiao X, Chen J, Ding S, Preparation of silver nanoparticles by chemical reduction method, *Colloids and Surface A: Physicochemical and Engineering Aspects*, 2005, 256, 111–115.
4. Zhao M, Wang X, Ning L, Jia J, Li X, Cao L, Electrospun Cu-doped ZnO nano fibers for H_2S sensing, *Sensors and Actuators B*, 2011, 156, 588–592.
5. Qader A, Faisal Zainab D, Jameel N, Raid A. Ismail, synthesis of colloidal copper oxide nano particles using pulsed Nd: YAG laser ablation in liquid, *Journal of Engineering and Technology*, 2013, 32, 14–23.
6. Poizot P, Laruelle S, Grugeon S, Dupont L, Tarascon JM, Nano-sized transition-metal oxides as negative-electrode materials for lithium-ion batteries, *Nature*, 2000, 407, 496–499.
7. Berndt C, Fischer TE, Ovid'ko I, Skandan G, Tsakalakos T, Nanomaterials for structural applications materials, *Research Society*, 2003, 740, 15–20.
8. Kim JB, Byun D, Ie SY, Park DH, Choi WK, Choi JW, Angadi B, Cu-doped ZnO-based p-n heterojunction light emitting diode, *Semiconductor Science and Technology*, 2008, 23, 4–25.
9. Milenova K, Stambolova I, Blaskov V, Eliyas A, Vassilev S, Shipochka M, The effect of introducing copper dopant on the photocatalytic activity of ZnO nanoparticles, *Journal of Chemical Technology and Metallurgy*, 2013, 48, 259–264.
10. Fernandes D, Silva R, Winkler Hechenleitner A, Radovanovic E, Custydio Melo M, Pineda E, Synthesis and characterization of ZnO, CuO and a mixed Zn and Cu oxide, *Materials Chemistry and Physics*, 2009, 115, 110–115.
11. Renjith SG, Krishnan C, Synthesis of cadmium-doped copper oxide nanoparticles: Optical and structural characterizations, *Advances in Applied Science Research*, 2013, 4, 103–109.
12. Sankara Reddy B, Venkatramana Reddy S, Koteeswara Reddy N, Pramoda Kumari J, Synthesis, structural, optical properties and antibacterial activity of Co doped (Ag, Co) ZnO nanoparticles, *Research Journal of Material Science*, 2013, 1, 11–20.

13. Visvanatha R, Naik YA, Venkatesh TG, Synthesis, characterization and optical properties of Sn-ZnO nanoparticle, *Nanoscience and Nanotechnology*, 2013, 3, 16–20.

14. Habibi HM, Fatemeh F, Sol-gel combustion synthesis and characterization of nanostructure copper chromite spinel, *Journal of Thermal Analysis and Calorimetry*, 2014, 115, 1329–1333.

15. Suleiman M, Mousa IM, Hussein A, Hammouti B, Hadda TB, Warad I, Copper (II)-oxide nanostructures: Synthesis, characterizations and their applications, *Journal of Materials and Environmental Science*, 2013, 4, 792–797.

16. Ma MG, Qing SJ, Li SM, Zhu JF, Fu LH, Sun RC, Microwave synthesis of cellulose/CuO nanocomposites in ionic liquid and its thermal transformation to CuO, *Materials Science and Technology*, 2013, 91, 162–170.

17. Zeng S, Zhang W, Sliwa M, Su H, Comparative study of CeO$_2$/CuO and CuO/CeO$_2$ catalysts on catalytic performance for preferential CO oxidation, *International Journal of Hydrogen Energy*, 2013, 38, 3597–3605.

18. Tamayo J, Garcia R, Deformation, contact time, and phase contrast in tapping mode scanning force microscopy, *Langmuir*, 1996, 12, 4430–4435.

19. Nia KZ, Montazer M, Synthesis of nano copper/nylon composite using ascorbic acid and CTAB, *Physicochemical and Engineering Aspects*, 2013, 439, 167–175.

20. Rahman MM, Bahadar Khan S, Asiri AM, Marwani HM, Qusti AH, Selective detection of toxic Pb (II) ions based on wet-chemically prepared nanosheets integrated CuO–ZnO nanocomposites, *Composites Part B: Engineering*, 2013, 54, 215–223.

21. Mageshwari K, Sathyamoorthy R, Flower-shaped CuO nanostructures: Synthesis, characterization and antimicrobial activity, *Journal of Material Science and Technology*, 2013, 29, 909–914.

22. Phiwdang K, Suphankij S, Mekprasart W, Pecharapa W, Synthesis of CuO nanoparticles by precipitation method using different precursors, *Energy Procedia*, 2013, 34, 740–745.

23. Safarifard V, Morsali A, Sonochemical syntheses of a nano-sized copper (II) supramolecule as a precursor for the synthesis of copper (II) oxide nanoparticles, *Ultrasonics Sonochemistry*, 2012, 19, 823–829.

24. Aparna Y, Enkateswara Rao KV, Srinivasa Subbarao P, Synthesis and characterization of CuO nano particles by novel sol-gel method, *International Conference on Environment Science and Biotechnology*, 2012, 48, 30–36.

25. You MaY, He ZQ, Xiao ZB, Huang KL, Xiong LZ, Wu XM, Synthesis and electrochemical properties of SnO$_2$-CuO nanocomposite powders, *Transactions of Nonferrous Metals Society of China*, 2012, 16, 791–795.

26. Bai Y, Yang T, Gu Q, Cheng G, Zheng R, Shape control mechanism of cuprous oxide nanoparticles in aqueous colloidal solutions, *Journal of Powder Technology*, 2012, 227, 35–42.

27. Meshram SP, Adhyapak PV, Mulik UP, Amalnerkar DP, Facile synthesis of CuO nanomorphs and their morphology dependent sunlight driven photocatalytic properties, *Journal of Chemical Engineering*, 2012, 204–206, 158–168

28. Mallick P, Sahu S, Structure, microstructure and optical absorption analysis of CuO nanoparticles synthesized by sol-gel route, *Nanoscience and Nanotechnology*, 2012, 3, 71–74.

29. Hao Q, Maa H, Jua Z, Li G, Li X, Xua L, Qiana Y, Nano-CuO coated LiCoO$_2$: Synthesis, improved cycling stability and good performance at high rates, *Electrochimica Acta*, 2011, 56, 9027–9030.

30. Winnubst L, de Veen PJ, Ran S, Blank DHA, Synthesis and characteristics of nanocrystalline 3Y-TZP and CuO powders for ceramic composites, *Ceramics International*, 2010, 36, 847–853.

31. Zhang DW, Chen C, Zhang J, Ren F, Fabrication of nanosized metallic copper by electrochemical milling process, *Journal of Material Science*, 2008, 43, 1492–1496.

21

Enhanced Thermal Energy Effectiveness in Storage, Conversion, and Heat Transfer Utilizing Graphene-Based Devices

J Femila Roseline

Saveetha School of Engineering, Saveetha Institute of Medical and Technical Sciences, Saveetha University, Chennai, India

N.M. Nandhitha

Sathyabama Institute of Science and Technology, Chennai, India

R. Rekha

Saranathan Engineering College, Trichy, India

Graphene is a carbon-based hexagonal lattice on an atomic scale. Graphene's use in energy conversion and management has advantages based on energy transmission and storage growth and its impact in diverse, innovative ways. Numerous study findings on energy transformation, memory, and transmission improvements demonstrate various graphene components, comprising graphene foam, film, fiber, and paper. A recent study of energy conversion in GP has recently proven that adding graphene to nano-triboelectric generators improves output performance above standard generators, which can withstand a maximum strain percent without losing electrical qualities. The introduction of graphene fibers on the battery might substantially enhance charge and discharge speeds while also increasing storage capacity in terms of energy transmission. According to studies on energy storage, graphene foam can increase the efficiency of solar thermal energy retention for longer storage. Nanographene and graphene coatings enable energy storage devices to be reduced in size, resulting in portable and flexible electronics development. The research presented in this study has the potential to boost the use of graphene in the energy area.

DOI: 10.1201/9781003355755-21

21.1 Introduction

The conversion and usage of thermal energy in integrated circuits is an important research issue. However, there are two basic factors to managing and utilizing thermal energy. On the one hand, appropriate materials and structures must be used to extract heat energy for the circuits to operate normally. The conversion and utilization of thermal energy in embedded electronics is a major research topic. Graphene is an innovative two-dimensional material with improved mechanical, electrical, optical, and thermoelectric properties expected to extend Moore's law. Due to their excellent conductivity and temperature rectification, graphene-based synthesized semiconductors are employed in various application domains, notably thermal breakthroughs. Nonlinear thermal products, thermoelectric bonding devices, thermal linking equipment, and thermo-optical coupling devices are the four kinds of graphene-based thermal devices based on coupling distinct physical variables. The construction, operating mechanism, and effectiveness of such devices and the ways of linking physical quantities are all examined. Molecule mechanics is a popular simulation tool for studying material thermodynamic characteristics. We may use simulation research to forecast the qualities of new materials and, therefore excellent, enhance our research. Recent molecular mechanisms modeling experiments have revealed that graphene framework thickness, flaw, doped kind, external stain, and heterogeneous structure can affect thermal rectifier performance. The findings of earlier graphene thermal rectification characteristics' molecular dynamics simulations are described in this part, which will aid in the development of novel graphene-based thermal rectifiers. Figure 21.1 shows the microscopy image of graphene (G) and the structural image of graphene (G).

Graphene is a stable two-dimensional carbon chemical structure with a unique structure that has been created by removing carbon atoms from graphite sheets using a highly sophisticated process (Guo et al., 2020; Kashyap et al., 2020). Graphene's impact on the present materials world has been linked to various benefits, depending on its outstanding physical and chemical capabilities. The graphene conductor of the sensor can detect at least low glucose levels, which are beyond the range of traditional sensors (Ji et al., 2015). In the biological field, graphene possesses several unique properties. The material science movement, which featured enhanced ceramics, carbon fiber, and unique metals, became connected with graphene (Kant et al., 2017a). It was decided to throw out the old atomic formulas and develop a new technique to represent a clear structure directly tied to both vertical and horizontal approaches due to its incredible layout.

The most incredible qualities of graphene lead to the most major advancements in numerous industries, notably for its domain with energy (Kashyap et al. 2020). Graphene foam, film, fiber, and paper are

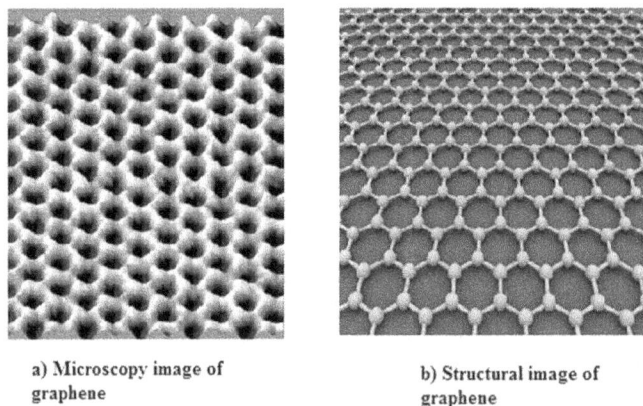

a) Microscopy image of
graphene

b) Structural image of
graphene

FIGURE 21.1   (a) Microscopy image of graphene (G), (b) structural image of graphene (G).

the four different graphene materials considered during this work. Their applications are categorized according to their unique qualities to reflect their differences and benefits (Li et al., 2020; Lin et al., 2020). Due to its small weight and flexibility, GP is commonly utilized in flexible electronics such as capacitors, storage, and associated sensor devices.

Graphene sheets are frequently used in electron transmission, photo electronics, and electron field effects due to their high mechanical strength. The porous features of graphene foam have aided its evident applicability in solar energy and biosensors. Because of their foldability, graphene fibers are particularly popular in wearable information expression goods (Li et al., 2017). Furthermore, nanographene and graphene coatings show great promise in energy transformation, memory, and heat conduction improvement. The research presented in this study has the potential to assist boost the use of graphene in the energy area (Luo and Lloyd, 2012; Ma et al., 2022). This will be a new research guide in using multidisciplinary thermodynamic and thermal energy transformation device optimization.

21.2 Related Work

Energy security has become one of the most pressing challenges facing the world due to fast-growing fuel consumption and the degradation of existing energy sources. Electricity, the most frequently utilized kind of energy, is generated primarily by the combustion of petroleum-based fuels, coal, and oil and gas. Fossil fuel combustion releases massive volumes of harmful gases and climate-altering carbon dioxide into seas and food systems. More importantly, natural fuels are finite resources, with supplies diminishing significantly more quickly than newer sources are identified or exploited. As a result, human beings face tremendous hurdles in creating, storing, and consuming alternative energy sources today. Owing to its special framework and outstanding properties, including high temperature and electrical conductivities, mechanical reliability, charge carrier movement, large substrate area, molecular stability, and optical permittivity, graphene – a single or few layers of two different (2D) adhesively carbon plates have piqued interest in recent years. Like an emerging legend in molecular science and technology, graphene is thought to have the ability to address several critical difficulties in energy technology or to help the creation of a sustainable energy-generating system that makes optimal use of energy sources. Photovoltaic cells, fueling cells, rechargeable batteries, and supercapacitors have recently seen significant attempts to leverage graphene's uses in fuel technologies to improve efficiencies, ease of processing, durability, and pricing. Given its fast-expanding interest in it, it is critical to emphasize current new inventions and successes in this slashing scientific topic. This research report examines the most recent developments in the use of graphene in an emerging field.

Scientific study on graphene is undoubtedly vital and valuable to society, and the existence of monolayer carbon atoms has puzzled scientists for some years (Messina and Ben-Abdallah 2013). One of the most often cited reasons for graphene's broad use in the energy engineering field is its adaptability, which allows it to be used in various applications and installation scenarios (Olabi et al., 2021). Lithium-ion batteries, nanogenerators, fuel cells, and supercapacitors are among the sectors developing, and more adaptable graphene components for energy transmission and conservation are being deployed. Along with the trendy energy sector, graphene's use in energy conversion and management has certain advantages in heat transfer and mass transfer improvement and its impact on diverse, innovative ways (Ren et al. 2017). In graphene, the carbon atom distribution is both axisymmetric and centrosymmetric (Min et al., 2018). Because the amount of molecules in carbon equals six, and the proportion of particles in the outermost phase is four, covalent bond structures are particularly simple to construct. Thermal graphene conductivity range was determined by using the direct thermal conductivity (Shamsaei et al., 2022).

Graphene film, fiber, foam, and paper are the four kinds of graphene discussed in this work (Yang et al., 2016a). Their applications are classified and categorized based on their distinguishing characteristics to highlight their distinctions and benefits (Sun et al., 2021). Graphene material is widely used in stretchable electronics such as capacitors, storage, and sensor devices due to its lightweight and

GRAPHENE (G)   GRAPHENE OXIDE(GO)

reduced GRAPHENE OXIDE(rGO)

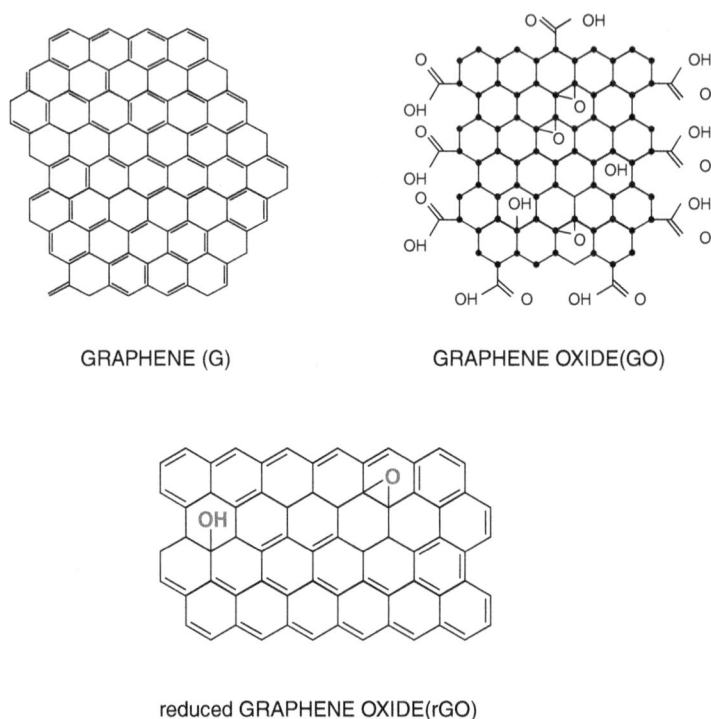

FIGURE 21.2 Crystal structure of (a) graphene (G), (b) graphene oxide (GO), and (c) reduced graphene oxide (rGO).

flexibility (Wu et al., 2022; Wu et al., 2019). Graphene sheets are commonly used in electron transmission, photo electronics, and electron field effects due to their strong tensile qualities (Wang et al., 2022). Figure 21.2 depicts the crystal structure of (a) graphene (G), (b) graphene oxide (GO), and (c) reduced graphene oxide (rGO).

Research is intended to effectively identify graphene materials (graphene film, fiber, foam, and paper) in future research findings.

21.3 Demonstrating the Most Common Graphene Components

One of the most often significant purposes for graphene's widespread use in energy engineering is its flexibility, which allows it to be adapted to various applications and installation circumstances. Lithium-ion batteries, nanogenerators, fueling cells, and supercapacitors are among the sectors developing and deploying highly adaptable graphene technologies for energy transformation and memory capacity. Graphene is available in four different forms, as illustrated in Figure 21.3, including extensible GP, thin-film, foam, and fibers; almost all are widely employed in various products, including systems. The graphene sheet is extremely lightweight and bendable, making it convenient to carry everywhere. Compared to GP, graphene film may be employed as a particular thickness if larger strength is needed. Because graphene foam possesses the features of a porous media, its heat transfer ability is enhanced. The graphene fibers' structural properties are flexible in form and simple to install. The rise of stretchable graphene appeared to encapsulate the growing excitement of researchers and noncovalent functionalized graphene's distinct excellent advantage.

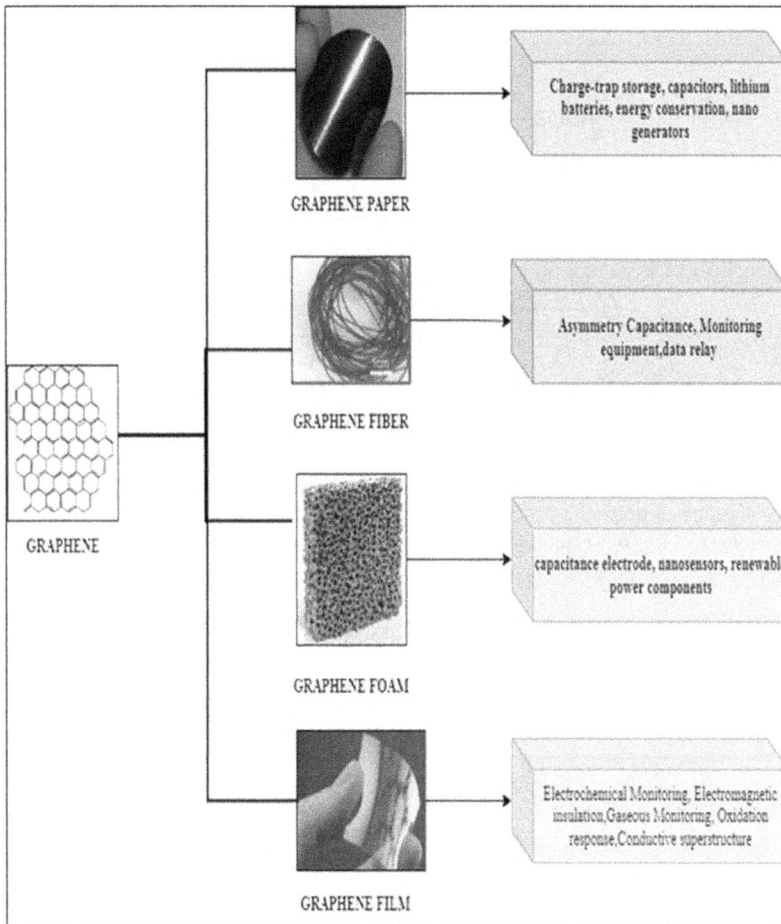

FIGURE 21.3   Energy conversion and storage components made of graphene materials.

21.4 Storage of Electric Energy

GP is a one-of-a-kind material paper with graphene that is thin enough to be lightweight and porta-ble, providing it desirable for application in energy storage technologies. The present graphene sheet is a compound component sheet to get additional properties. Polymer-based combination graphene sheets and inorganic nanoparticle-based composite graphene sheets are the most common compound graphene sheets. GP, also known as graphite paper, is a graphite oxide-based substance. Graphite oxide membranes and GO membranes (from more recently) are micrometer-thick GO paper sheets. Slow evaporating of graphene solution or screening are two common methods for producing cell walls. Because of the essential nature of the multi graphene backbone and the interwoven layer structure that distributes stresses, the material exhibits extraordinary strength and rigidity. Figure 21.4 shows the structure of GP with functional groups for electric energy storage.

21.4.1 Performance Measures

The use of graphene sheet with graphene polyaniline compound sheet in the construction of low-cost electrodes for future energy storage devices has a lot of promise. Applying composite metal components

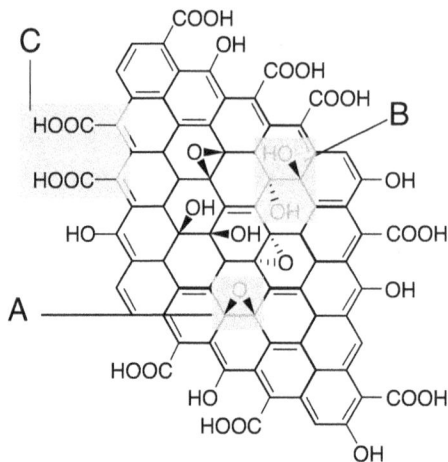

FIGURE 21.4 Structure of graphene paper with functional groups for electric energy storage. e – MnO$_2$, and graphene – polyaniline in terms of weight (mg^{-3}), specific capacitance (Fg^{-1})

TABLE 21.1 Comparison of Performance Measures of Pillared Graphene Paper, Graphen, and Current Density (Amg^{-1})

Performance Measures	Pillared Graphene Paper	Graphene — MnO$_2$	Graphene Polyaniline
Weight (mg^{-3})	160	240	200
Specific capacitance (Fg^{-1})	265	256	763
Current density (Amg^{-1})	500	500	1000

to graphene material can increase heat conduction, capacitance, and density of the current flow, all of which can enhance the quality of products.

Compared to graphene – MnO$_2$ at the same density of the current, pillared GP holds a lot of promise for future flexible and ultralight electrochemical energy storage supercapacitors. Table 21.1 shows the comparison of performance measures graphene–MnO$_2$, pillared graphene sheet, and graphene–polyaniline in terms of weight (mg^{-3}), specific capacitance (Fg^{-1}), and current density (Amg^{-1}).

Another intriguing discovery is that graphene MnO$_2$, pillared graphene sheet, and graphene polyaniline have the same output performance. To acquire less weight, utilize pillared GP (i.e., lightweight of material requirements). The performance measures of graphene – MnO$_2$, pillared graphene sheet, and graphene–polyaniline are shown in Table 21.1 and Figure 21.3. A Simple line graph of the performance measures of pillared GP, graphene – MnO$_2$, and graphene polyaniline in terms of weight (mg^{-3}), specific capacitance (Fg^{-1}), and current density (Amg^{-1}) is shown in Figure 21.5.

21.4.2 Graphene Paper-Based Materials

In recent years, the technology for employing GP as a substrate has evolved dramatically, and adaptable graphene-based sheet materials can now be employed in electronic equipment. Theoretical graphene energy research has accelerated the consumption use of graphene sensors. A graphene monolayer was used to conduct a strain energy theoretical investigation of graphene transfer-based sensors. The graphene energy of absorption and energy of strain is shown in Figure 21.6.

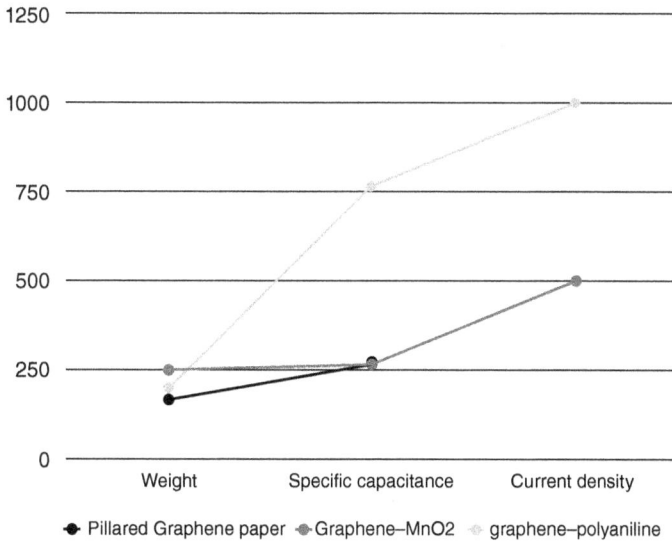

FIGURE 21.5 Simple line graph of performance measures ofpillared graphene paper, graphene–MnO$_2$, and graphene–polyaniline in terms of weight (mg^{-3}), specific capacitance (Fg^{-1}), and current density (Amg^{-1}).

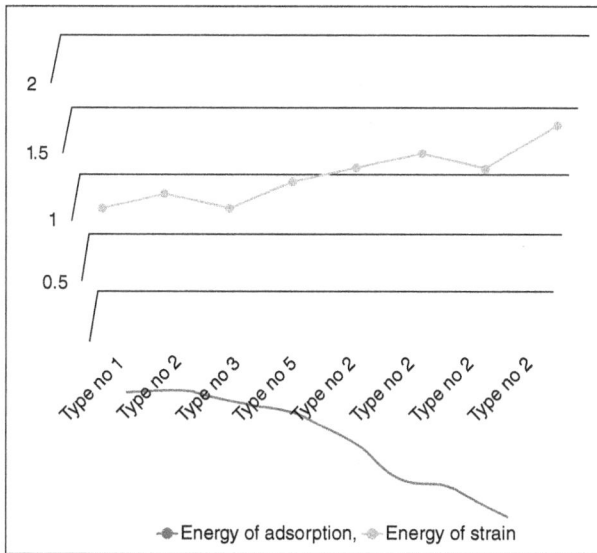

FIGURE 21.6 Simple line graph showing performance metrics versus absorption and strain energy.

The energy of absorption can be represented as

$$The\ energy\ of\ absorption_N = \frac{1}{N}\left(ES_{total} - E_{graphene\ monolayer} + E_{error\ correction}\right)$$

The energy of strain can be denoted as

$$Energy\ of\ strain_N = \left(ES_{total} - E_{graphene\ monolayer}\right)$$

The total energy system is$_{total}$, with $E_{graphene\_monolayer}$ representing pure graphene monolayer energy, N representing the number of functionalized type groups, *Energy of absorption* representing the absorption energy, and *Energy of strain*$_N$ representing the strain energy.

In Figure 21.6, a simple line graph of performance measures compares *the Energy of absorption*$_N$ representing the absorption energy and the *Energy of strain*$_N$ representing the strain energy for different types.

21.4.3 Application of Thin Graphene Films toward Energy Transfer

Graphene thin films of interacting and overlaying graphene or GO platelets have different electrical, photonic, and heating properties than single coating graphite. The optical transparency of a single graphene sheet is roughly 98 percent, although this reduces as the quantity of sheets grows. To increase the electrical permeability of graphene films of steel nanowires, thin graphene films in the random substrates are produced by chemical vapor transfer methods using large graphene filters. In addition, rGO films can decrease hydrohalic acids, and monolayer graphene film can be used as a nanorod array for Schottky junction UV photodetectors, which have been developed in multiple research studies. Figure 21.7 represents the application of thin graphene films toward energy transfer.

21.4.4 Electrochemical Catalyst's Oxygen Reduction Process

The demand for an extensible graphene sheet as an electrode component of supercapacitor electrodes should not be the only option; graphene film may also significantly improve the capacitor's performance. It created an extensible graphene carbon nanotube sheet from complicated dispersion that retained a high percentage of its specific capacitance after continual load or discharge cycles. Contrastive research on various metals and graphene is carried out for the oxygen reduction process. According to experts, phosphorous, boron-doped graphene, and nitrogen can also significantly speed up the process due to graphene's unique and active capabilities.

21.4.5 Electrical and Optoelectronics

The ultra-hydrophilic moisture control in graphene films and the wettability and surface clean energy of graphene films make them an attractive material for biology and identification, with a recent upsurge in attention as a microcosmic diagnosis solution. Graphene can be used for supersensitive multimodal

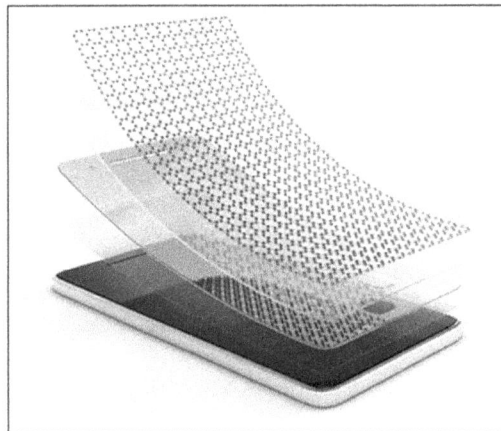

FIGURE 21.7 Application of thin graphene films toward energy transfer.

TABLE 21.2 Performance Measures of Decreased Graphene Oxide, Graphene Oxide, and Simultaneous Waveform Compared to the Peak Intensity of Distinct Processes

Performance Measures	Peak_Intensity (Process1)	Peak_Intensity (Process2)	Peak_Intensity (Process3)
Graphene oxide	0.7	0.6	0.3
Reduced graphene	0.2	0.1	0.3
Continuous-wave	2.3	2.7	1.0

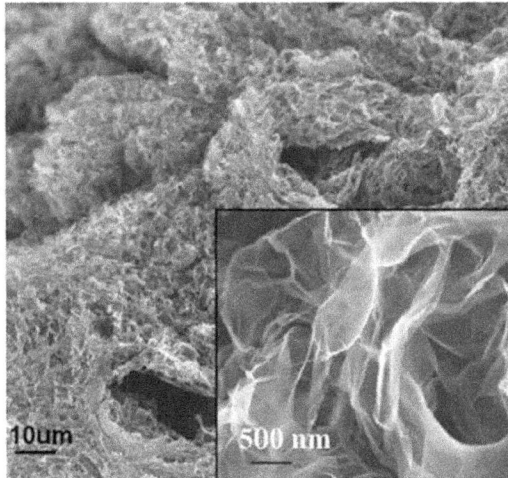

FIGURE 21.8 Electron photomicrograph exhibiting graphene foam made with a Ni pattern. (Taken from Paronyan et al. (2017).)

identification of tumor indicators and nonenzymatic hydrogen peroxide prediction due to its superior electrical properties and biocompatibility. By modifying the immune component substrate with graphene, the obtained waveform enhancement can accelerate the transfer of electrons as a marker label for designating the waveform peptides. Table 21.2 compares the peak intensity of several procedures using performance indicators such as decreased GO, GO, and simultaneous waveform. We employed a simple line graph to compare peak intensity in different procedures and GO, decreased GO, and continuous-wave performance assessments in Figure 21.5.

21.4.6 Translation and Utilization of Energy

A regulated chemical formation and a flavoring process may be utilized to create graphene foam, a specific morphological derivation of graphene with holey and exceptional electrical attributes effectiveness. Graphene foam is an open-cell solid foam constructed of single-layer graphene sheets. It is a potential electrode substrate for lithium batteries. Figure 21.8 portrays the electron photomicrograph exhibiting graphene foam made with a Ni pattern. The foam may be made by coating a metal foam, a multi mesh of metal filaments, with vapor deposition. After that, the metal is removed. Carbon nanotubes were employed to strengthen a foam in 2017. When unweighted, the latter substance can sustain 3,000 times its weight and return to its former shape. Sugar, nanotubes, and a dusted nickel catalyst were combined. The substance's dried pellets were then squeezed into screw-shaped steel die. A propeller portion from foam was left when the nickel was removed. The outer layers of the nanotubes broke and merged with the graphene.

21.4.7 Material for Supercapacitor Electrodes

Supercapacitor electrodes are typically thin films coated on a conducting, metal power source and electronically linked. The quality produced high heat durability, good corrosive environment friction, and greater substrate areas per square mass and volume. Minimal expense and environmental compatibility are two further factors to consider.

The electrode substrate area governs the number of twofold pseudocapacitors preserved in each standard voltage in a supercapacitor. As a reason, activated carbon, porosity, and sponge material with a large effective contact area are frequently used in supercapacitor electrodes. The overall capacitance is also increased by the electrode material's capacity to undertake faradaic charge transfers. Evaluating graphite foam with lithium carbonate additional storage or otherwise, supercapacitor substrates require the tremendous capacitive performance that can produce significant percentage potential having an excellent duration for high energy management devices, as per the experts, as well as exceptional cyclic stability performance. Table 21.3 shows the maximum capacitance, scan rate, current density, and cycle periods of supercapacitors for energy conversion and management.

The porous structure of graphene froth generated by various procedures varies greatly, resulting in a wide range of sensor sensitivity; however, the graphene detectors are highly sensitive to ordinary sensors. The simple line graph comparing peak intensity in different processes, performance measurements of GO, rGO, and continuous-wave were used in Figure 21.9. Table 21.3 represents the supercapacitor performance metrics for energy conversion and management, including maximum capacitance, scan rate, current density, and cycle times. Figure 21.10 depicts, on a light microscope, a micrograph of triggered carbon beneath intense area lighting of graphene foam. Figure 21.11 represents the structure of graphene foam after removing Ni particles.

TABLE 21.3 Supercapacitor Performance Metrics for Energy Conversion and Management Include Maximum Capacitance, Scan Rate, Current Density, and Cycle Times

Performance Measures	Max_capacitance	Scan Rate	Current Density	Cycle Times
Supercapacitors	100–400 Fg^{-1}	10–20mVs^{-1}	2.0–7.0Ag^{-1}	1000–2000

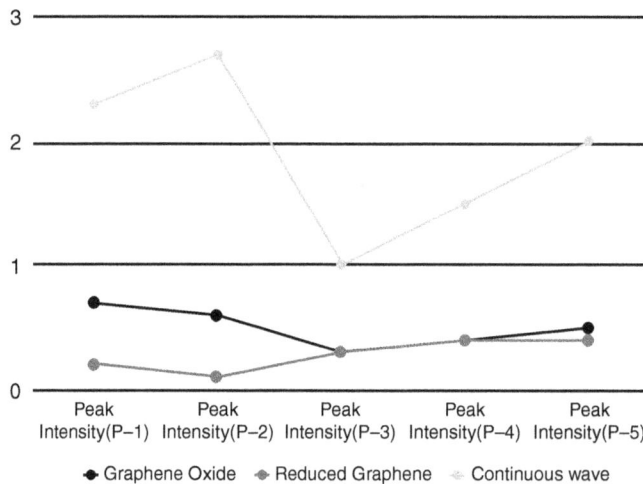

FIGURE 21.9 Simple line graph comparing peak intensity in different processes, performance measurements of graphene oxide, reduced graphene oxide, and continuous-wave were used.

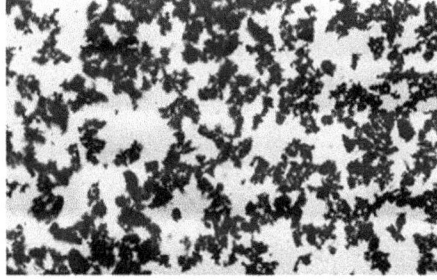

FIGURE 21.10 A micrograph of triggered carbon beneath intense area lighting of graphene foam. (Courtesy of Zephyris https://commons.wikimedia.org/wiki/File:ActivatedCharcoalPowder_BrightField.jpg [accessed 07/12/2022].)

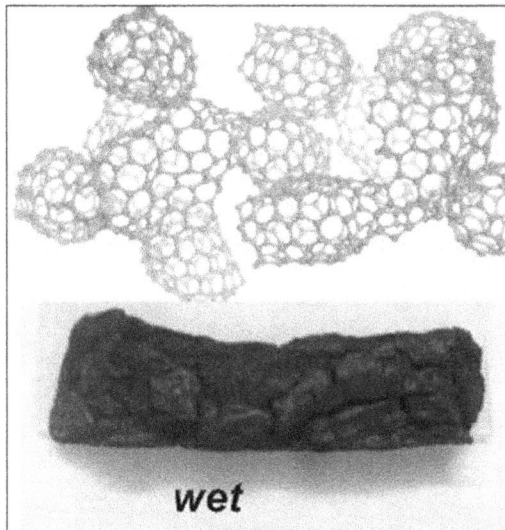

FIGURE 21.11 Structure of graphene foam after removing Ni particles.

21.4.8 Solar Thermodynamic Energy

Solar thermal energy is a sort of resource and a method of capturing solar radiation to produce heat energy for use in commerce and residential and commercial settings. The US Department of Energy divides solar collectors into poor, medium, and large collectors. Acrylic-reduced collectors are typically used to heat pools or exhaust air. Moderate-temperature collectors, such as smooth sheets, are employed to warm water or air for household and industrial purposes. Solar thermal energy is still being preserved at the nanoscale, namely nanographene. The simple diagram of solar thermal energy using graphene substances is shown in Figure 21.12.

Graphene seems never merely restricted with perspective as energy acquisition; solar thermodynamic energy is useful, yet it could be used to make electronic components. Experts seem fascinated with using holey graphite in solar thermal energy transition systems in its preferable electric permeability and mechanical durability. As the prevalence incidence increases, multilayered graphene foam outperforms in terms of solar thermodynamic energy conversion durability.

The simple line graph shows the performance measures of graphene in terms of sensitivity and detection limit concerning other sensors for energy conversion and management in Figure 21.13. Table 21.4 represents the performance measures of graphene in terms of sensitivity and detection limit concerning other sensors for energy conversion and management.

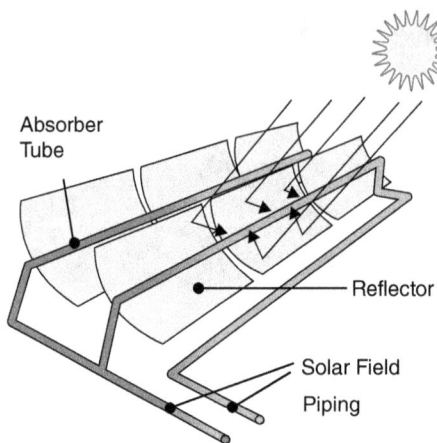

FIGURE 21.12   Simple diagram of solar thermal energy utilizing graphene substances.

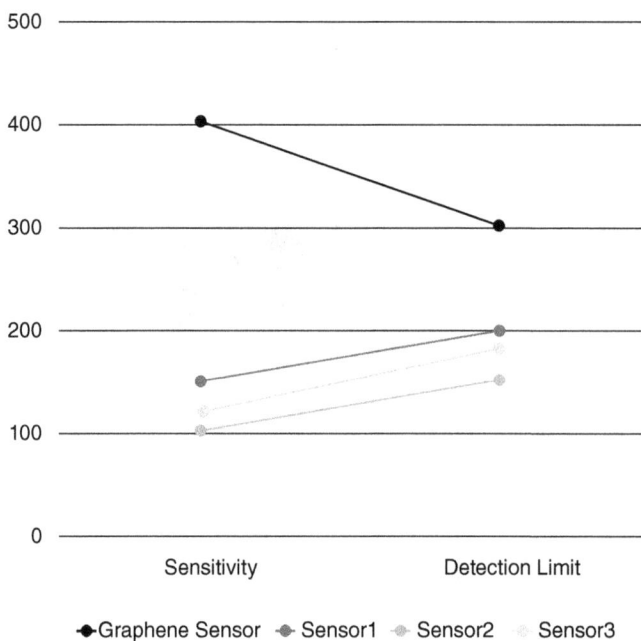

FIGURE 21.13   Simple line graph shows the performance measures of graphene in terms of sensitivity and detection limit concerning other sensors for energy conversion and management.

TABLE 21.4   Graphene's Performance Measures in Terms of Sensitivity and Detection Limit Concerning Other Energy Conversion and Management Sensors

Performance Measures	Graphene Sensor	Sensor1	Sensor2	Sensor3
Sensitivity	400	150	100	120
Detection limit	300	200	150	180

21.4.9 Usage of Graphene in Heat Transfer

Scientists are seeking to analyze the stimulatory impacts between temperature distribution and mass transfer advantageous function of graphene in composites, and scientists have classed graphene as a unique energy material. Coulomb force effects have been shown to generate energy transfer drag, which significantly influences lateral electrical conduction but does not impede thermal conductivity enhancement.

21.4.10 Nanographene Heat Transfer

Nanographene affects the performance of heat transfer thermal energy produced using the phase transition material's thermal conductivity. Nano-superior graphene's heat transmission can assist finned heat transfers, shifting working fluids, among similar thermal transfer components. Therefore, the heat transfer system for the new nanoparticles based on nanographene is constructed along with fundamental processing and characteristics.

21.4.11 Coated Graphene Heat Transfer

Many other graphene multilayers coating research directions are bringing better and more efficient thermal performance facilities across the usages of graphene monolayer or multilayers coating, varying from unstable natural mixture analysis to composite component, graphene, and graphene enclosed metamaterials effective influence of near-field heat transfer.

Table 21.5 depicts, in terms of temperature exchange factor and contentious thermal flow, performance measurements in comparison of coated graphene and uncoated graphene. Figure 21.14 shows

TABLE 21.5 Comparison of Performance Measures of Coated Graphene and Uncoated Graphene in Terms of Temperature Exchange Factor and Contentious Thermal Flow

Performance Measures	Coated Grapheme	Uncoated Graphene
Critical heat flux	184 W/cm	268W/cm
Heat transfer coefficient	98w/m	145w/m

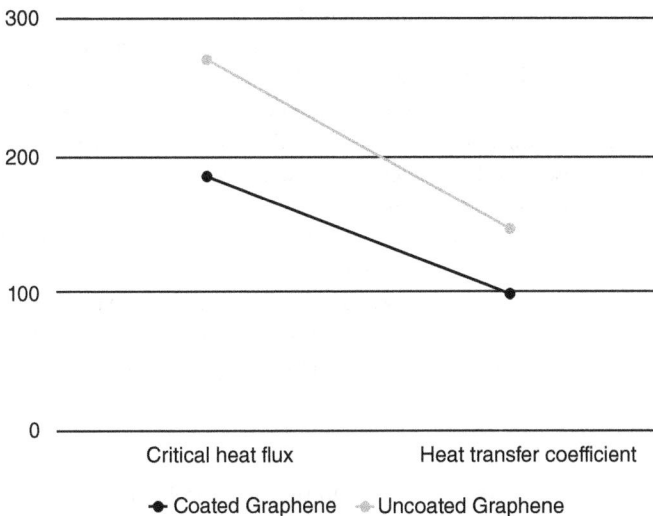

FIGURE 21.14 Simple line graph shows the performance measures compared to coated graphene and uncoated graphene in terms of crucial thermal flux and heat transmission coefficient.

performance measurements compared to coated graphene and uncoated graphene in terms of temperature exchange factor and contentious thermal flow.

As research has progressed, authors have become more fascinated with the workable benefits of graphene functionality in heat permeability. The development of graphene conductivity measurement and related investigations confirmed that graphene might increase thermal distribution. The cause for another thermal permeability mechanism is that current transmission performs an increasingly important function in energy thermoelectric permeability.

21.5 Discussion

Along with being the latest trend in the energy business, graphene's use in energy conversion and management has several advantages regarding energy distribution and preservation development and is the latest trend in the energy industry. Influence is exerted in a variety of innovative ways. In current and prior research on energy transformation, energy retention, and energy optimization, several study findings indicate that all graphene products are developing regarding storage and energy transfer. GP, film, foam, and fiber, among others, play an important role in boosting effectiveness. In the current energy transformation study on graphene material, it was currently proved that incorporating graphene into triboelectric nanogenerators improves output performance over standard generators with a maximum strain as big as without affecting electrical qualities.

Regarding fuel transmission, the use of graphene fibers in batteries can dramatically improve charge and discharge speeds while also increasing storage capacity. In energy preservation, graphene foam has been shown to improve the long-term density of solar thermal storage. Nanographene and graphene coatings enable energy storage tools to be reduced in size, resulting in transportable and flexible electronics development. The findings of this study can increase the use of graphene in the energy sector. Researchers will be able to pick graphene goods (GP, graphene film, graphene foam, or graphene fiber) effectively in the future based on our findings.

21.6 Conclusion

The energy problem is rapidly expanding, and worldwide financial growth and manufacturing production exacerbate it. The work is dedicated to identifying optimal methods to enhance energy volume fraction and preserve energy, possibly explicitly and sometimes implicitly. Because graphene is yet another compound, it founded new scientific topics and methodologies in energy investigation and certain novel prospects. Graphene is obtainable in numerous distinct dimensions, every one offering its unique array of applications and impacts. This research assessed graphite foam, fiber, film, and paper, considering its perspective on graphene's progress in energy storage, conversion, and transport. The author must choose graphite foam, fiber, film, and paper depending upon effectiveness, load, extruding, and use conditions if graphene is utilized in energy-related materials. Finally, this study will aid in the marketing and usage of graphene materials in the energy industry and provide a simpler to consumers that analyze the graphene component used. Finally, nanographene and graphene coatings results showed tremendous potential within the growth of energy storage, conversion, and transfer and recent research direction of interdisciplinary design enhancement, including the thermal exchanger field efficiency assessment on combustion compartment with coated graphene adhesives and field efficiency analysis on graphene implementation in energy temperature distribution.

Acknowledgment

The authors would like to express their gratitude to the Saveetha Institute of Medical and Technical Sciences for their contributions to this study.

References

Guo, Qiang, Nan, and Liang Qu. "Two-dimensional materials of group-IVA boosting the development of energy storage and conversion." *Carbon Energy* 2, no. 1 (2020): 54–71.

Ji, Yuanchun, Lujiang Huang, Jun Hu, Carsten Streb, and Yu-Fei Song. "Polyoxometalate-functionalized nanocarbon materials for energy conversion, energy storage and sensor systems." *Energy & Environmental Science* 8, no. 3 (2015): 776–789.

Kant, Karunesh, Amritanshu Shukla, Atul Sharma, and Pascal Henry Biwole. "Heat transfer study of phase change materials with graphene nanoparticle for thermal energy storage." *Solar Energy* 146 (2017a): 453–463.

Kant, Karunesh, Amritanshu Shukla, Atul Sharma, and Pascal Henry Biwole. "Heat transfer study of phase change materials with graphene nanoparticle for thermal energy storage." *Solar Energy* 146 (2017b): 453–463.

Kashyap, Shaswat, Shruti Kabra, and Balasubramanian Kandasubramanian. "Graphene aerogel-based phase changing composites for thermal energy storage systems." *Journal of Materials Science* 55, no. 10 (2020): 4127–4156.

Li, Chuanchang, Baoshan Xie, and Jian Chen. "Graphene-decorated silica stabilized stearic acid as a thermal energy storage material." *RSC Advances* 7, no. 48 (2017): 30142–30151.

Zhou, Yang, Chunhai Li, Hong Wu, and Shaoyun Guo. "Construction of hybrid graphene oxide/graphene nanoplates shell in paraffin microencapsulated phase change materials to improve thermal conductivity for thermal energy storage." *Colloids and Surfaces A: Physicochemical and Engineering Aspects* 597 (2020b): 124780.

Li, Yu-Tao, Ye Tian, Meng-Xing Sun, Tao Tu, Zhen-YiJu, Guang-Yang Gou, Yun-Fei Zhao, et al. "Graphene-based devices for thermal energy conversion and utilization." *Advanced Functional Materials* 30, no. 8 (2020a): 1903888.

Lin, Keng-Te, Han Lin, Tieshan Yang, and Baohua Jia. "Structured graphene metamaterial selective absorbers for high efficiency and omnidirectional solar thermal energy conversion." *Nature Communications* 11, no. 1 (2020): 1–10.

Luo, Tengfei, and John R. Lloyd. "Enhancement of thermal energy transport across graphene/graphite and polymer interfaces: A molecular dynamics study." *Advanced Functional Materials* 22, no. 12 (2012): 2495–2502.

Ma, Fukun, Liqiang Liu, Liangqing Ma, Qian Zhang, Jianing Li, Min Jing, and Wenjie Tan. "Enhanced thermal energy storage performance of hydrous salt phase change material via defective graphene." *Journal of Energy Storage* 48 (2022): 104064.

Messina, Riccardo, and Philippe Ben-Abdallah. "Graphene-based photovoltaic cells for near-field thermal energy conversion." *Scientific Reports* 3, no. 1 (2013): 1–5.

Min, Peng, Jie Liu, Xiaofeng Li, Fei An, Pengfei Liu, Yuxia Shen, Nikhil Koratkar, and Zhong-Zhen Yu. "Thermally conductive phase change composites featuring anisotropic graphene aerogels for real-time and fast-charging solar-thermal energy conversion." *Advanced Functional Materials* 28, no. 51 (2018): 1805365.

Olabi, A. G., Mohammad Ali Abdelkareem, Tabbi Wilberforce, and Enas Taha Sayed. "Application of graphene in an energy storage device – A review." *Renewable and Sustainable Energy Reviews* 135 (2021): 110026.

Paronyan, T., Thapa, A., Sherehiy, A. et al. Incommensurate Graphene Foam as a High Capacity Lithium Intercalation Anode. *Scientific Reports* 7, 39944 (2017). doi:10.1038/srep39944.

Ren, Huaying, Miao Tang, Baolu Guan, Kexin Wang, Jiawei Yang, Feifan Wang, Mingzhan Wang, et al. "Hierarchical graphene foam for efficient omnidirectional solar–thermal energy conversion." *Advanced Materials* 29, no. 38 (2017): 1702590.

Shamsaei, Ezzatollah, Felipe Basquiroto de Souza, Amirsina Fouladi, Kwesi Sagoe-Crentsil, and Wenhui Duan. "Graphene oxide-based mesoporous calcium silicate hydrate sandwich-like structure: Synthesis and application for thermal energy storage." *ACS Applied Energy Materials* 5 (2022): 958–969.

Sun, Keyan, Hongsheng Dong, Yan Kou, Huning Yang, Hanqing Liu, Yangeng Li, and Quan Shi. "Flexible graphene aerogel-based phase change film for solar-thermal energy conversion and storage in personal thermal management applications." *Chemical Engineering Journal* 419 (2021): 129637.

Wang, Chen, Wenjun Dong, Ang Li, Dimberu G. Atinafu, Ge Wang, and Yunfeng Lu. "The reinforced photothermal effect of conjugated dye/graphene oxide-based phase change materials: Fluorescence resonance energy transfer and applications in solar-thermal energy storage." *Chemical Engineering Journal* 428 (2022): 130605.

Wu, Guanzheng, Nanci Bing, Yifan Li, Huaqing Xie, and Wei Yu. "Three-dimensional directional cellulose-based carbon aerogels composite phase change materials with enhanced broadband absorption for light-thermal-electric conversion." *Energy Conversion and Management* 256 (2022): 115361.

Wu, Haiyan, Sha Deng, Yaowen Shao, Jinghui Yang, Xiaodong Qi, and Yong Wang. "Multiresponsive shape-adaptable phase change materials with cellulose nanofiber/graphene nanoplatelet hybrid-coated melamine foam for light/electro-to-thermal energy storage and utilization." *ACS Applied Materials & Interfaces* 11, no. 50 (2019): 46851–46863.

Yang, Jie, Guo-Qiang Qi, Yang Liu, Rui-Ying Bao, Zheng-Ying Liu, Wei Yang, Bang-Hu Xie, and Ming-Bo Yang. "Hybrid graphene aerogels/phase change material composites: thermal conductivity, shape-stabilization and light-to-thermal energy storage." *Carbon* 100 (2016b): 693–702.

Yang, Yingkui, Cuiping Han, Beibei Jiang, James Iocozzia, Chengen He, Dean Shi, Tao Jiang, and Zhiqun Lin. "Graphene-based materials with tailored nanostructures for energy conversion and storage." *Materials Science and Engineering: R: Reports* 102 (2016a): 1–72.

Yu, Zepei, Daili Feng, Yanhui Feng, and Xinxin Zhang. "Thermal conductivity and energy storage capacity enhancement and bottleneck of shape-stabilized phase change composites with graphene foam and carbon nanotubes." *Composites Part A: Applied Science and Manufacturing* 152 (2022): 106703.

Zabek, Daniel, Kris Seunarine, Chris Spacie, and Chris Bowen. "Graphene ink laminate structures on poly (vinylidene difluoride)(PVDF) for pyroelectric thermal energy harvesting and waste heat recovery." *ACS Applied Materials & Interfaces* 9, no. 10 (2017): 9161–9167.

Zahir, Noura, Pierre Magri, Wen Luo, Jean Jacques Gaumet, and Philippe Pierrat. "Recent advances on graphene quantum dots for electrochemical energy storage devices." *Energy & Environmental Materials* 5, no. 1 (2022): 201–214.

Zhang, Lianbin, Renyuan Li, Bo Tang, and Peng Wang. "Solar-thermal conversion and thermal energy storage of graphene foam-based composites." *Nanoscale* 8, no. 30 (2016): 14600–14607.

Zhang, Linlin, Yijing Wang, Zhiqiang Niu, and Jun Chen. "Single atoms on graphene for energy storage and conversion." *Small Methods* 3, no. 9 (2019): 1800443.

22

Nanomaterials in Energy Storage: Groundbreaking Developments

Dr. P.V. Premalatha
M.I.E.T. Engineering College, Trichy, India

K. Jayachitra
Oxford Engineering College, Trichy, India

L.K. Rex
M.I.E.T. Engineering College, Trichy, India

Nanomaterials and nanotechnology are widely growing technology in all fields of engineering, science, and medicine. The use of these materials is growing in every sector and further research in this field is considered very essential. The solar cell is currently very useful for electrical production. Solar cells made with graphene nanomaterials can improve efficiency and are of low cost. To avoid environmental issues, nanomaterial-based fuel cells are used to produce electricity by chemical energy. The use of nanotechnology has greatly helped in creating more energy-efficient and sustainable materials such as waterproofing layers and self-repairing concrete and windows. Nanotechnology coatings improve fire resistance, waterproof layer for concrete, footpath generator, corrosion protection, energy storage in concrete buildings, insulation, and countless other applications. This chapter features the various applications of nanomaterials in groundbreaking developments and ongoing research activities in this field.

22.1 Introduction

Nanotechnology is an interdisciplinary branch of engineering and research that deals with materials that range in size from 1 to 100 nanometers. Materials that don't have or have low levels of certain properties in bulk can show up as new properties when measured down to the nanometer scale. Nanotechnology applications in a wide range of fields, such as biology, materials science, electronics, and energy science, have been demonstrated to be useful. To counteract global warming's increasing greenhouse gas emissions and depleting fossil fuel supply chains, today's scientists and researchers are tasked with finding a new clean, efficient, and biocompatible energy source that is both renewable and environmentally benign (Fan, 2020). Industrial and scientific communities are increasingly interested in renewable energy sources such as biomass from fuel cells, tidal power, solar energy, hydropower, wind energy, and biofuels. With solar power being the world's most abundant energy source, it has a unique

DOI: 10.1201/9781003355755-22

place in the energy supply. A single hour's worth of sunlight is more than what the entire world's population consumes in a year's food supply. Therefore, the production of this energy is of great importance. Different types of solar power production systems have been established in many nations with the right capacity for solar radiation as a result of the requirement to harvest solar energy, so that their electricity may be sent to the national grid (Otanicar and DeJarnette, 2016).

Over the last few years, there has been a significant increase in graphene (GE) research. Since graphene's discovery in 2004, it has profoundly impacted a wide range of industries. The essential graphene material is graphene oxide (GO). However, there are significant drawbacks to the conventional methods that have been reported for removing contaminants from water and wastewater. Photocatalytic degradation for water treatment shows great promise. However, traditional methods have several disadvantages. Due to their application in water purification, semiconductor-based photocatalysis has drawn more attention. In addition, the optoelectronic properties of semiconductors and other material scan often be studied using the photoluminescence technique.

Nanoparticles can help reduce the use of natural resources by improving the performance of construction materials and reducing energy consumption. In construction, the use of nanoparticles offers a more cost-effective, faster, and more environmentally friendly method of producing construction materials. The principal applications of nanotechnology in the building industry are lighter and more durable structural composites, low-maintenance coatings, improved cementitious material characteristics, and decreased thermal conductivity of insulation and fire-retardant materials.

22.2 Literature Survey and Theoretical Concepts

The application of nanomaterials in engineering scenarios has tremendous performance in today's life. This chapter covers topics such as the process of photocatalysis, solar cell operation, fuel cell function, as well as emerging applications in Civil Engineering. Since oxygenated moieties can be manipulated on the surface of two-dimensional carbonaceous nanocomposites, they can be used as photocatalysts to degrade pollutants (Arumugasamy et al. 2022). The photocatalytic performance of metal-semiconductor-based nanocomposites has been better than pure nanoparticles in the degradation of organic pollutants and has attracted considerable attention (Li et al. 2021). The photosensitive material absorbs a photon to degrade the toxic aromatic dye. This sets off a photochemical or photophysical reaction (Balasurya et al. 2021). Nanocomposites based on graphene and GO are more effective at photocatalysis because of GO's unique electronic band structure, which has a strong acceptor, conductive channel, and reservoir of photogenerated electrons (Imran et al. 2021 & Farooq et al. 2021).

Photocatalytic efficiency is hampered by the high recombination rate of electron–hole pairs generated during photosynthesis (Shoreh et al. 2021). The development of renewable energy sources, such as solar, hydro, wind, tidal, and oceanic energy, is crucial to addressing the issues of global warming and greenhouse gas emissions (Kumar et al. 2017). Due to their lightweight, ability to withstand complex deformations, ability to be integrated into curved surfaces, compatibility with roll-to-roll manufacturing, ease of storage and transportation, and compatibility with roll-to-roll manufacturing, flexible solar cells have recently emerged as a promising photovoltaics direction (Xuemei Fu et al. 2018). Due to their low production costs, robustness, environmental friendliness, and simplicity, dye-sensitized solar cells (DSSC) have gained much attention (Kanmani et al. 2012).

Carbon dots, graphene, reduced GO, and grapheme oxide have all been extensively studied as transparent conducting electrodes, counter electrodes, and active electron donors or acceptors in emerging solar cells, among other materials (Rodrigo Szostak et al. 2018). Perovskites with organic–inorganic hybrid structures and excellent optical and electrical properties have been discovered. One of the most fascinating groups of semi conductive organ lead halide perovskites is the AMX3 structure, where lead cations occupy the M sites, halide anions occupy the X sites, and an organic component occupies the A site within the crystal structure (Moore and Wei, 2021). Proton exchange membrane fuel cells (PEMFCs) research and development has been an important part of the alternative, renewable, and sustainable energy sector for several decades (Yan-Jie Wang et al. 2018, Das et al. 2015, Dutta et al. 2016, and You et al. 2016).

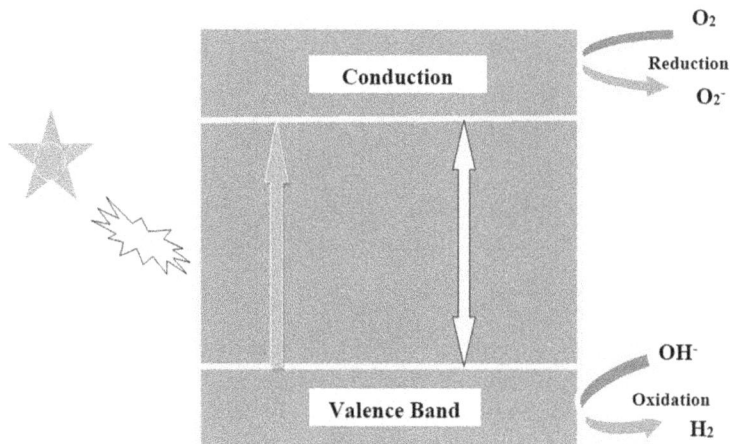

FIGURE 22.1 Photocatalytic reaction.

Advancements in building materials have been observed from a physical and chemical standpoint due to the growing demand for engineering constructions. The speed and aesthetics of construction were greatly influenced by these materials have had a significant impact on the speed and visual appeal of construction, as well as offering improved properties at a low cost. This type of material is known as a nanotechnology product (Adil Hatem Nawar 2021). The use of nanotechnology in civil engineering has increased in recent years. Nanomaterials can improve structural materials like steel, concrete, and others. Nanotechnology has created new materials to improve the mechanical properties and durability of concrete, cement paste, and other structural materials. Numerous experimental programs have recently been conducted to study the mechanical properties of modified concrete. Researchers (Selective laser-induced preparation of metal-semiconductor nanocomposites and application for enhanced photo-catalytic performance in the degradation of organic pollutants, 2021; Hawreen & Bogas, 2019; Shekari & Razzaghi, 2011) used a variety of nanomaterials, such as nano-silica (NS), nano clay (NC), carbon nanotubes (CNTs), and nano aluminum (NA), to improve the mechanical properties of concrete (A. Hawreen et al. 2019, Mohamed O et al. 2019, Shekari and Razzaghi, 2011). Figure 22.1 shows the photocatalytic reaction.

22.3 Applications

22.3.1 Photocatalysis

In photocatalysis, a photo-activated chemical reaction, free radical mechanisms are activated when the compound and photons with sufficient energy levels come into contact. In the presence of a catalyst, photocatalysis speeds up a photoreaction. It is in this process that irradiated materials are transformed into radicals that have a direct effect on the surrounding environment. During photocatalysis, the adsorption substrates absorb light. The bandgap allows electrons to jump from the conduction band (CB) to the valence band (VB) by leaving positive holes (Hussain M 2010). Reduction is the term for this. Oxygen species (ROS) like O_2 and OH are formed when electrons and holes in a molecule are reduced (oxidation). A number of ROS are formed due to the sequential reduction of oxygen through the addition of electrons. The impact of the photocatalysis process on the environment, such as degradation, adoption, reduction, or antibacterial activity, can be used to assess its efficiency (Fajrina & Tahir 2019). Photocatalytic processes can be evaluated by comparing the concentration of unwanted substances before and after photocatalytic reactions, which is the most common method. Photocatalysis relies heavily on GO (Dasgupta 2017).

22.3.1.1 Calculation of Percent Degradation of Dye

The decolorization and photocatalytic degradation efficiency have been calculated as,

$$\text{Efficiency}\left(\%\right) = \frac{C_o - C_e}{C_o o} \times 100$$

C_o and C_e are the dye concentrations before and after photoirradiation, respectively. Degradation percentage of the dye photocatalyst is shown by E percent in this equation.

22.3.1.2 Role of Nanomaterials in Photocatalysis

We must increase the efficiency of photocatalytic reaction seven though they help us in the energy industry and the fight against pollution. We can use nanotechnology as a catalyst rather than bulk material. Particularly for catalysts, the surface-to-volume ratio characteristic of nanoparticles outperforms bulk materials. The use of nanomaterials as a catalyst in the water-splitting process for hydrogen production in the energy sector can increase the efficiency of H_2 production. The use of nanotechnology for water purification is transforming industries in both developed and developing countries. In addition to their photocatalytic properties, nanomaterials have a wide range of hydrophobic and hydrophilic properties, a short intra-particle diffusion distance, and the ability to compress without changing their surface areas. Furthermore, high surface-to-volume ratio of nanomaterials regulate bacterial-pollutant interactions. The conventional water purification method has a problem, and nanotechnology can help us solve it.

Photocatalytic Degradation of Various Dyes

- Alizarin S
- Crocein orange G
- Methyl red
- Congo red
- Methylene blue

22.3.1.3 Environmental Protection of Photocatalysis

22.3.1.3.1   Water Splitting

Using the photocatalytic process for hydrogen production is a more efficient, long-lasting, and environmentally friendly option. Solar light and photocatalyst are required for its operation. It is the photocatalyst that plays a major role in the process of producing hydrogen through photocatalysis. TiO_2, ZnO, Ag_2O, and other semiconductor metal oxides are used as photocatalysts. A photon striking the catalyst initiates the chemical reaction. An electron–hole pair is created, which is the catalyst for the reaction, when an electric field is applied. In order to form an electron–hole pair, the photocatalyst must have specific properties. But EgBand's Gap is the most important one. The valance band of a semiconductor contains electrons and holes that are excited to move into the CB when exposed to photons with energy greater than or equal to the gap between the bands. The most common response is (Figure 22.2),

$$\text{Catalyst} + \text{Photon} \rightarrow \text{e}- + \text{h} +$$

22.3.1.4 Waste Water Treatment

Light and an appropriate catalyst can degrade pollutants in water through photocatalytic reactions, which are green waste treatment methods. A change in reaction rate caused by a catalyst band under the appropriate ultraviolet, visible, or infrared light energy is what this term refers to. A highly reactive reducing or oxidizing radical is generated on the CB by this energy's generation of an electron–hole

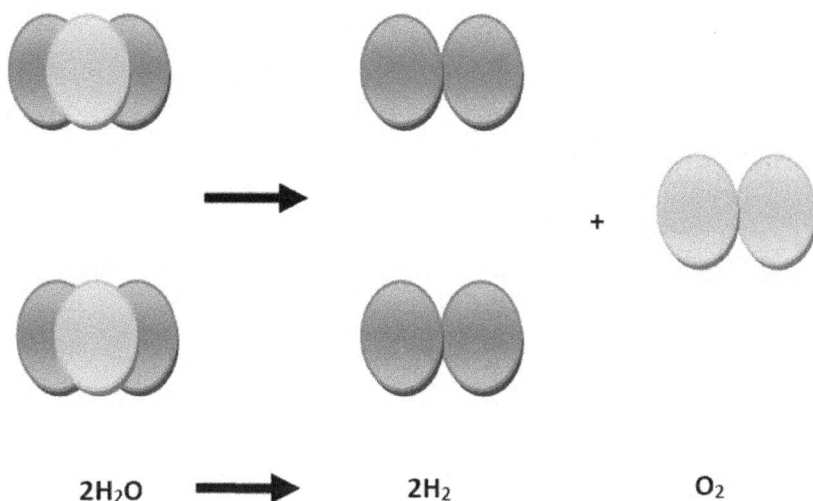

FIGURE 22.2 Splitting of water.

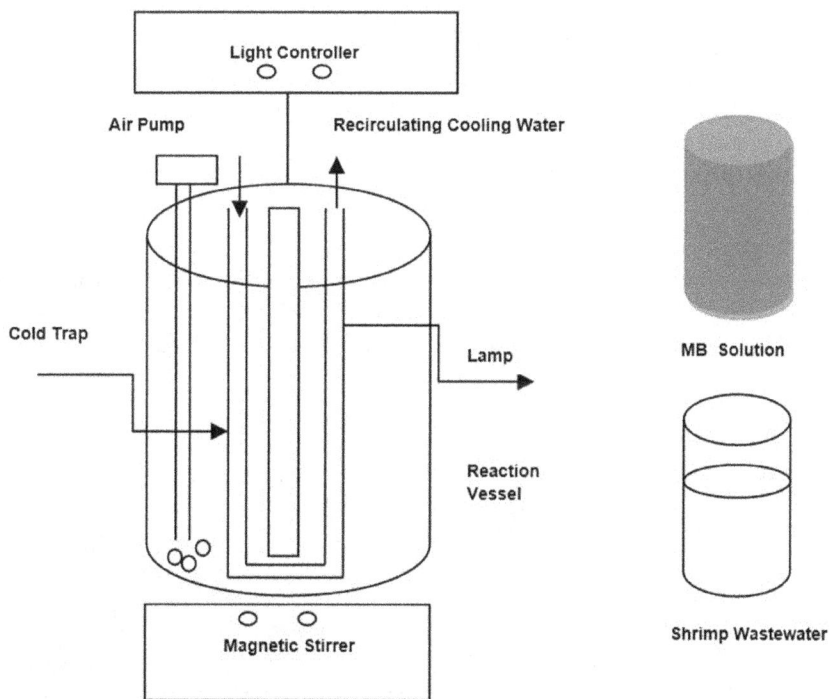

FIGURE 22.3 Degradation of pollutants in water.

pair (eh). In polluted water, these radicals react with organic or inorganic pollutants; they degrade them through some secondary reaction (Hussain M. Synthesis, 2010). Because semiconductor nanoparticles have a wide band gap and can be seen in the visible spectrum, they are ideal catalysts for wastewater treatment processes as shown in Figure 22.3.

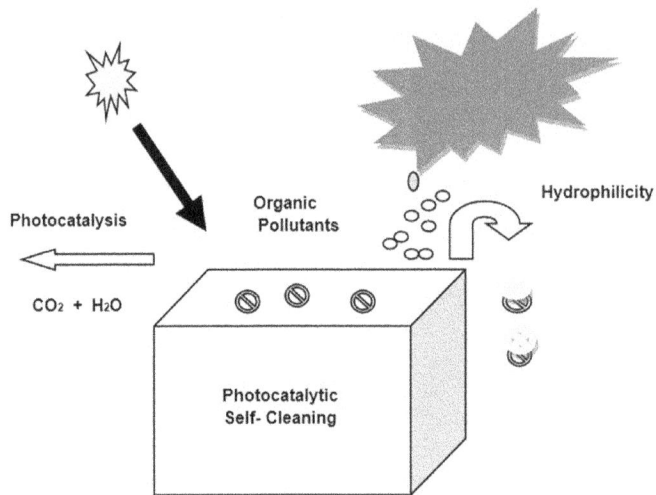

FIGURE 22.4   Photocatalytic self-cleaning process.

22.3.1.5 Self-Cleaning Surface

The method of photocatalytic self-cleaning, as shown in Figure 22.4, is low-maintenance and trouble-free. Glazing and translucent membranes benefit from improved light transmission because surface dirt and grime reduces the amount of daylight obscured. Lighting can save a lot of money on energy costs this way. Building materials coated with titanium dioxide (TiO_2) play an essential role in self-cleaning building materials (Banerjee et al. 2015).

22.3.1.6 Photoelectrochemical Conversion

Solar energy conversion processes via homogeneous and/or heterogeneous photocatalysis have been described with mechanistic insights. These include photocatalytic CO_2 reduction to fuels, light-driven nitrogen fixation, water splitting for hydrogen or oxygen evolution, etc. Figure 22.5 shows the photo-electrochemical conversion process.

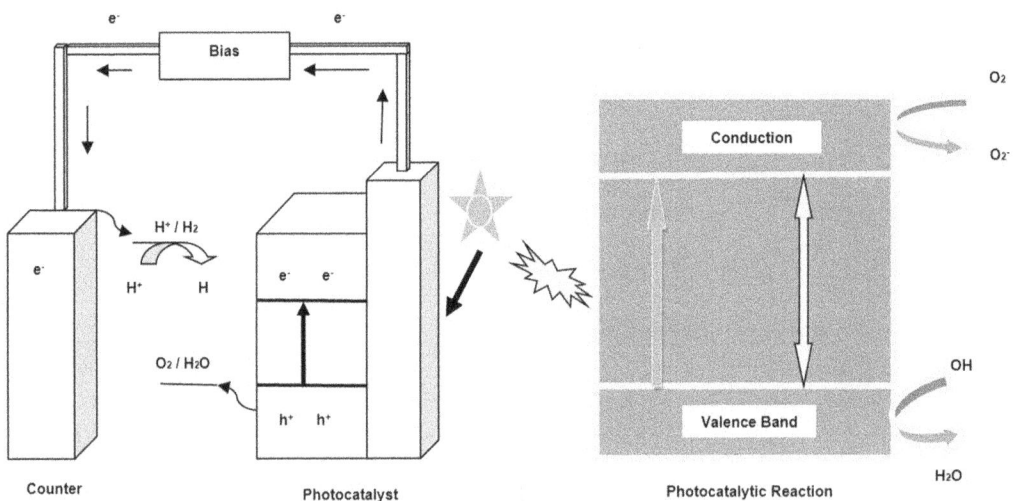

FIGURE 22.5   Photoelectrochemical conversion process.

22.3.1.7 Air Treatment

There are a plethora of impurities in the air we breathe. There are various methods for removing pollutants, including electrostatic air purification, gas-adsorption filtration, ventilation, and air filtration. These are just a few examples. The disadvantages of these physical methods include the need for frequent replacement of materials and the time and expense associated with their removal and disposal. As a result, there is a need to use environmentally friendly, safe, and low-energy decomposition methods to remove air pollutants from the atmosphere. Nanomaterials can be used as catalysts in photocatalytic processes that generate reactive oxygen species which react with air pollutants to decompose them and thus disinfect the air. This process is environmentally friendly, inexpensive, and timesaving.

22.3.2 Solar Cell

Solar energy extraction will be greatly aided by nanotechnology. Both power generation and heat production will be enhanced thanks to this new technology. The light spectrum of the sun has a wide range of hair lengths. The wavelength of these photons determines their intensity. Sunlight energy can be directly converted into electricity by a photovoltaic effect in a solar cell (Figure 22.6). As long as the semiconductor is exposed to sunlight, electrons will move from the capacitance band to the semiconductor CB, where they will form electron–hole pairs that can participate in the load transfer cycle of the semiconductor and generate user-initiated variations.

22.3.3 Types of Solar Cells

- Silicon solar cells (first generation)
- Thinfilm of solar cells (second generation)
- Dye-sensitized solar cell (third generation)

22.3.4 Aspects of Nanomaterials in Solar Cells

Magnetic nanoparticle-based solar cells have been synthesized in studies. They are extremely efficient. Nanomaterials like these have the potential to convert solar energy into electricity. Because of this, it will be more efficient than currently available solar cells. As a result, nanostructures can improve efficiency and lower the costs of photovoltaic (PV) cells.

In this graphene, a sheet will give a compelling performance. In rainwater, these materials could separate positively charged ions like sodium, calcium, and ammonium.

FIGURE 22.6 Photovoltaic cell.

22.3.5 Applications of Nanotechnology in Solar Cells

- Self-cleaning
- Anti-reflecting nanocoatings
- Nanocatalysts are a new window

Quantum dots, fluorescent nanofibers, gold (or silver) nanoparticles, and other fluorescent nanomaterials are frequently employed to increase the efficiency of solar cells. Even though they increase the spectrum of sunlight absorption beyond visible light due to their photoelectric capabilities, quantum dots will be employed to replace pigments in solar cells because of their exceptional optoelectronic properties. One of the most recent innovations in solar cell technology is using silver sulfide (Ag_2S) quantum dots to enhance their light response and increase their absorption. Such quantum dots are resistant to a wavelength in the solar spectrum between 400 and 1000 nm.

22.3.6 Recent Trends in Nanomaterials for Fuel Cell Applications

In fuel cells, as shown in Figure 22.7, platinum is frequently used to make the catalytic electrodes. Since platinum nanoparticles have long been known to be more efficient than solid platinum surfaces, this has

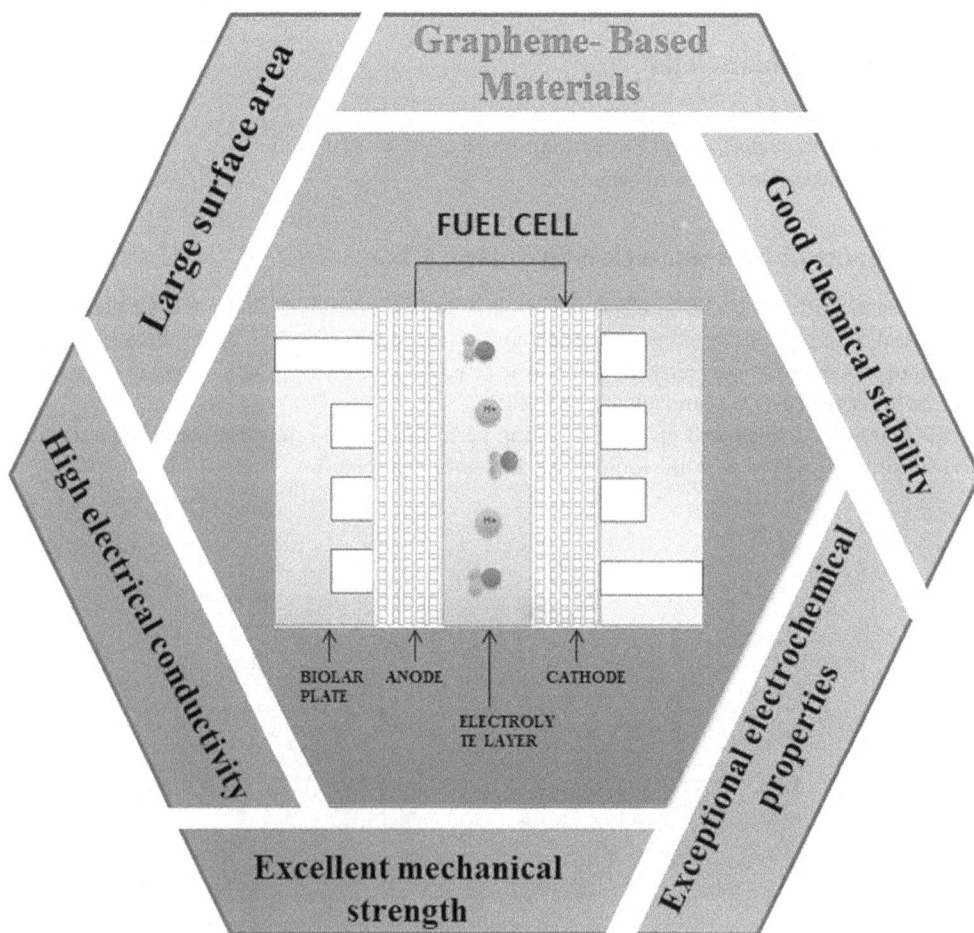

FIGURE 22.7   Characteristics of fuel cells.

been a well-known fact. It is possible to generate electricity from fuel cells for as long as fuels are available. Chemical energy from fuels is directly converted into electricity, making fuel cells more efficient than combustion engines and releasing fewer pollutants into the environment during operation. This is why fuel cells are seen as an attractive technology for dealing with global energy and environmental issues.

Fuel cells typically contain an electrolyte layer between two nanomaterial electrodes. To reduce O_2, electrons released from the anode surface flow through an external circuit to the cathode. To complete the circuit, the mobile charge carriers (H+, OH, or O_2) transfer through electrolytes simultaneously. An electrolyte cell can be classified as Phosphoric Acid Fuel Cell (PAFC), Polymer Electrolyte Membrane Fuel Cell (PEMFC), Solid-Oxide Fuel Cell (SOFC) and Molten Carbonate Fuel Cell (MCFC). The choice of an electrolyte cell is influenced by a number of variables, including performance requirements, cost, and operating conditions. This is because PAFC and PEMFC are better suited for mobile and vehicular uses. The MCFC and SOFC, on the other hand, can use a variety of fuels and are better suitable for stationary applications due to their high working temperatures (500°C).

Selecting the right material for a fuel cell component must take into account electrochemical performance, efficiency, and durability. For use in alternative fuel cell applications, graphene and its derivatives have exceptional chemical, electrical, and mechanical characteristics. Thus, graphene-based fuel cell materials have drawn a lot of interest in recent years. Graphene-based materials offer excellent electrocatalyst supports for oxygen reduction and fuel oxidation because they increase active sites and facilitate electron transport (ORR). Additionally, metal-free graphene materials are interesting ORR candidates due to their strong electrocatalytic activity, high poisoning tolerance, and low cost.

22.3.7 Energy Generation Using Genostep and Graphene Cement Battery

Energy is mostly generated by consuming natural resources such as fossil fuels, crude oil, forest wood, etc. These are also generated using renewable energy sources such as solar power, wind energy, and water in the form of potential and kinetic energy. Unknowingly, we ignore many energy systems generated through human being, which are also sustainable. In other words, this is also called as green energy.

Pedestrians' biomechanical activity can be harnessed to generate electricity, say, researchers. Genostep is a footpath generator that converts biomechanical activity into electrical energy by converting the foot's movement. Through footfalls on the ground, a person's weight is transferred to the road surface, causing impact, vibrations, and other forms of energy to be transferred to the road. An electrical energy source capable of supporting oneself can be derived from this kinetic energy. We're working on a footstep electricity-generating device to figure out how to connect the energy from footsteps. Incorporating this device into a sidewalk can convert the energy of a person's footsteps into electricity. Because of the weight of the pedestrian, the upper plate of the device will dip down when they step on it. That's how it works. Because the plate descends, the piezoelectric materials used in the device to generate electricity are more durable. As long as pedestrians keep moving over the device, much power is generated. A prototype of the power-generating tiles is built and put through a series of stress tests (Mohanapriya et al. 2021).

Researchers used graphene cement batteries for the energy storage generated through Genostep. This energy stored through human biomechanical activity is converted into electrical energy and stored in a nearby convenient area. Additionally, graphene serves as the primary energy conductor in both the Genostep and graphene cement battery, enabling the highest possible energy efficiency.

Since the sources of generation and storage are integrated into the infrastructure, there will be less strain on the grid, fewer power outages, and less money lost due to transmission loss fromdistant power plants (nuclear, wind, and solar). Transmission of power is another major benefit of this collaboration.

A cement matrix composite is used to create graphite cement batteries. Both the anode and cathode areas of the matrix benefit from the conductive addition of carbon black that is spread throughout the matrix (Meng & Chung 2010).

The Genostep and graphene cement battery will enable a direct energy source, reducing power anxiety while also aiding the circular economy by reducing deforestation and environmental effect from nuclear, solar, wind, and hydroelectric power plants. The combination of Genostep and Graphene Cement Battery can be utilized in various public places like sidewalks, train stations, airports, schools, business establishments, and box stores. This combination can play a crucial role in providing zero-emission energy, which is imperative in reducing carbon emissions and enhancing the overall quality of life. By implementing this solution, we can ensure a sustainable and eco-friendly environment for everyone.

However, despite population growth and rapid urbanization, the cement industry faces significant challenges, including the need to reduce its carbon footprint (CO_2 contribution).

Concrete additives made from graphene have the potential to improve durability, reduce the clinker factor in cements, and increase strength.

The technology has the potential to produce concrete structures that are more durable and less porous, opening the door to new concrete design possibilities. A new generation of environmentally friendly building and infrastructure approaches may be enabled by graphene-based concrete, mortar, and cement additives. The compressive strength of cement mortar has increased by 34%, while the tensile strength has increased by 27% when tested using international standard methods. As a result, carbon dioxide emissions are reduced.

The increased durability of graphene-induced concrete can also be attributed to its ability to reduce the corrosion of steel reinforcing rods.

22.3.8 Energy Storage in Concrete Building

Researchers from the civil and architecture field are exploring the possibilities of using concrete structures for energy storage. Each upcoming high-story building can act as a giant battery for energy storage in future. Because of the integration of nanotechnology with current electronic and information technology systems, innovative concrete now can self-monitor and even generate energy through the thermoelectric or piezoelectric effects (Cwirzen, 2021).

When added to cement mortar, the use of Graphene oxide nanosheet (GONS) additives shows increased flexural strength of concrete. This results in an ultra-high strength concrete (UHSC) of notable strength (Lu & Ouyang 2017).

Workability and setting times were both sped up when using concrete that contained nanomaterials. In addition, researchers found that adding nanomaterials to cement concrete increased the material's mechanical strength by a significant margin. Aside from reducing porosity, nanomaterials improved the concrete's durability by decreasing chloride and carbonation depths and increasing its resistance to chemical attacks (Huseien et al. 2021).

Engineers from the Department of Architecture and Civil Engineering at Sweden's University of Technology (UT) recently published an article outlining an innovative new concept for rechargeable batteries – made from cement Chalmers. Emma Zhang and Luping Tang's researchers have developed a world-first concept for a cement-based battery that can be recharged.

To achieve high strength and flexural toughness, this concept uses a mixture of nanomaterials and a small amount of short carbon fibers. A metal-coated carbon fiber mesh is incorporated into the mixture. In these metals, iron and nickel are used. Iron acts as an anode and nickel as cathode. This research is successful and available in the laboratory as a scale down model and will soon be implemented realistically in buildings. For example, a concrete battery could power sensors on highways and bridges to detect corrosion or cracking. In addition, it could be used in conjunction with solar cell panels to generate electricity.

A rechargeable cement-based battery developed by Luping Tang and Emma Zhang has an average energy density of seven watthours per square meter (or 0.8 Watthours per liter). The battery's capacity is measured in terms of energy density, and one conservative estimate suggests that the new Chalmers battery could perform ten times better than previous attempts at concrete batteries. Commercial batteries have a higher energy density, so this shortcoming could be overcome. Buildings can be equipped with a large-scale battery because of its large volume.

22.3.9 Nanomaterials as Waterproofing Layer in Construction

As a result of their inherent porosity and micro-cracks, building materials are known to have leaks and seepage. Adding waterproofing to a material makes it water-resistant. Waterproofing products, particularly those with polymeric backbones and other materials, have undergone a great deal of technological and product development over the last 50 years. Nanotechnology has ensured that this approach's service life will lead to life cycles of more than 20 to 30 years at a meager cost.

Nanotechnology-based waterproofing coatings have the potential to improve waterproofing in a number of ways, including speed, safety, cost, and variety, leading to smart building. Nanotechnology-based waterproofing employs monomeric water-repellent compounds that can pierce several millimeters into the core of the substrate. The majority of the substrate structure is protected by nanotechnology foundation waterproofing chemicals since they are extremely stable and do not change chemically after 2000 hours of UV testing. Many cutting-edge waterproofing materials are now available, thanks to nanotechnology, which can help prevent concrete structures from deteriorating from environmental assaults (such as surface scaling, spalling, internal frost damage, and penetration of water and waterborne harmful agents like chlorides and sulfates) and chemical/corrosion damage (caused by chemicals such as $NaCl$, $CaCl_2$, and $MgCl_2$). Some of the commercial nano waterproofing products available in the market include REINSTE, ZYCOSIL+, and NsANEX coatings.

Reactive nano-based waterproofing material can generate an impermeable monolithic membrane by penetrating concrete pores and cementitious surfaces as deep as 2 mm. Monomeric water-repellent chemicals are used in nanotechnology-based waterproofing to reach deep into the substrate's core. Since the development of nanotechnology, almost any material may now be waterproofed. As the name implies, a nanocoating and waterproofing fills every pore on the surface of a substance, making it waterproof. The nanocoating provides a second, water- and dirt-repellent skin (Arumugasamy et al., 2022). Ceramic coating for waterproofing is one of the most widely used nanotech solutions in everyday life. This type of coating can be applied to a non-wettable surface. Natural examples include the lotus leaf and some insect wing feathers (Figure 22.8).

Nanotechnology-based waterproofing compounds can be applied to a wide range of construction materials, including concrete, brick, sandstone, granite, lime stone, marble, plaster, cement sheet, and natural stones. A monomeric compound is the norm when it comes to these water-repellent substances. The molecule is smaller than 6–8 nm in diameter. They have no problem penetrating the substrates' pores. Due to the molecules' small size, they can flow through small pores and branches within the

FIGURE 22.8 Bricks after applied with nanocoating spray.

substrate. They are often applied as a water solution, but the manufacturer may specify another solvent. These waterproofing compounds are highly effective at covering the surface and penetrating deeply.

The advantages of using these waterproofing nanomaterials include their ability to withstand the hydraulic pressure brought on by heavy rainfall and strong winds, protect structures against water damage brought on by abrasion from heavy traffic or natural surface deterioration well, and prevent rusting of the reinforcement. The sealer should not be applied below 500°F or exposed to freezing conditions during the first 24 hours of drying; however, they are generally not advised for frozen/frosted surfaces. After mixing, these substances must be utilized within 24 hours. They should not be applied to previously sealed or painted surfaces, windy areas, or places where rain is expected within the next two hours. They are not applied to concrete until they have attained 80% design strength and are dry.

Nanotechnology expands on already existing technologies to a new scale where conventional methods and ideas no longer hold true. As such, it is not truly new technology. Although disruptive, nanotechnology has the potential to create enormous advancements, in contrast to traditional approaches, which can only achieve minimal advancements. New generations of nanomaterials will appear, posing novel and possible unexpected problems. Future enhanced advancement will be enabled through nanotechnology.

22.4 Future Scope of Nanomaterials

One of the most promising research areas in the rapidly expanding field of nanotechnology is its potential to improve the service life and life-cycle costs of infrastructure construction, potentially paving the way for a completely new world, for example, building insulation, coatings, and solar technology use of nanotechnology-based materials and products. To preserve energy and improve material properties and functions, micro-nano materials (MNMs) should be considered in the construction industry.

This is a major achievement because a large amount of the energy used in buildings is used in commercial and residential ones (including heating, lighting, and air conditioning). Energy savings options include solar-powered self-cleaning nanoTiO_2-coated surfaces and improved thermal management employing silica nanoparticles in insulating ceramics, paint, and coatings (aside from using MNMs to harvest solar energy or other renewable energy). Energy transmission, lighting, and heating devices as well as energy storage technologies like batteries and capacitors that harness energy from sporadic renewable sources can all benefit from the utilization of carbon nanotubes (CNTs), fullerenes, and grapheme (e.g., solar and wind).

Bibliography

Arumugasamy, S. K., Ramakrishnan, S., Yoo, D. J., Govindaraju, S., & Yun, K. (2022). Tuning the interfacial electronic transitions of bi-dimensional nanocomposites (pGO/ZnO) towards photocatalytic degradation and energy application. *Environmental Research*, 204, 112050. doi:10.1016/j.envres.2021.112050

Balasurya, S., Das, A., Alyousef, A. A., Alqasim, A., Almutairi, N., & Sudheer Khan, S. (2021). Facile synthesis of Bi2MoO6-Ag2MoO4 nanocomposite for the enhanced visible light photo-catalytic removal of methylene blue and its antimicrobial application. *Journal of Molecular Liquids*, 337, 116350. doi:10.1016/j.molliq.2021.116350

Banerjee, S., Dionysiou, D. D., & Pillai, S. C. (2015). Self-cleaning applications of TiO2 by photo-induced hydrophilicity and photo-catalysis. *Applied Catalysis B: Environmental*, 176–177, 396–428. doi:10.1016/j.apcatb.2015.03.058

Cwirzen, A. (2021). Introduction to concrete and nanomaterials in concrete applications. In *Carbon Nanotubes and Carbon Nanofibers in Concrete-Advantages and Potential Risks*, 1–58. doi:10.1016/b978-0-323-85856-4.00003-0

Das, S., Dutta, K., Hazra, S., & Kundu, P.P. (2015). Partially sulfonated poly (vinylidene fluoride) induced enhancements of properties and DMFC performance of Nafion electrolyte membrane. *Fuel Cells*, 15, 505–515. doi:10.1002/fuce.201500018.

Dasgupta, A. (2017). Covalent three-dimensional networks of graphene and carbon nanotubes: Synthesis and environmental applications. *Nano Today*, 12, 116–135.

Directorate Works Guidelines/Specification & standardization/Reports Guidelines (S. No. 8). (n.d.) http://www.rdso.indianrailways.gov

Dutta, K., Das, S., & Kundu, P.P. (2016a). Effect of the presence of partially sulfonated polyaniline on the proton and methanol transport behavior of partially sulfonated PVdF membrane. *Polymer Journal*, 48, 301–309. doi:10.1038/pj.2015.106.

Dutta, K., Das, S., & Kundu, P.P. (2016b). Highly methanol resistant and selective ternary blend membrane composed of sulfonated PVdF-co-HFP, sulfonated polyaniline and nafion. *Journal of Applied Polymer Science*, 133, 43294. doi:10.1002/app.43294.

Fajrina, N., & Tahir, M. (2019). A critical review in strategies to improve photo-catalytic water splitting towards hydrogen production. *International Journal of Hydrogen Energy*, 44(2),540–577.

Fan, Z., Wang, D., and Yuan, Y. (2020). A lightweight and conductive MXene/graphene hybrid foam for superior electromagnetic interference shielding. *Chemical Engineering Journal*, 381, 122696.

Farooq, N., Rehman, A., Qureshi, A. M., Rehman, Z., Ahmad, A., Aslam, M. K., ... Alomar, T. S. (2021). Au@GO@g-C3N4 and Fe2O3 nanocomposite for efficient photo-catalytic and electrochemical applications. *Surfaces and Interfaces*, 26, 101399. doi:10.1016/j.surfin.2021.101399

Hawreen, A., & Bogas, J.A. (2019). Creep, shrinkage and mechanical properties of concrete reinforced with different types of carbon nanotubes. *Constr. Build. Mater.*, 198, 70–81. doi:10.1016/j.conbuildmat.2018.11.253.

https://firstgraphene.net/applications/composites/

https://news.cision.com/chalmers/r/world-first-concept-for-rechargeable-cement-based-batteries,c3342255

https://smartcoatings.in/waterproofing/

https://www.constrofacilitator.com/innovation-in-waterproofing-technologies/

https://www.constructionplacements.com/nanotechnology-in-waterproofing

https://www.einnews.com/pr_news/549000830/genostep-and-graphene-cement-battery-collaboration

https://zydexindustries.com/blog-post/nanotechnology-for-waterproofing/

Huseien, G. F., Khalid, N. H. A., & Mirza, J. (2021). Nanomaterial-Based Cement Concrete. In *Nanotechnology for Smart Concrete*, 21–34. doi:10.1201/9781003196143-3

Hussain, M. (2010). Synthesis, characterization, and photo-catalytic application of novel TiO_2 nanoparticles. *Chemical Engineering Journal*, 157(1), 45–51.

Imran, M., Alam, M. M., Hussain, S., Ali, M. A., Shkir, M., Mohammad, A., ... Irshad, K. (2021). Highly photo-catalytic active r-GO/Fe3O4 nanocomposites development for enhanced photo-catalysis application: A facile low-cost preparation and characterization. *Ceramics International*, 47(22), 31973–31982. doi:10.1016/j.ceramint.2021.08.083

Ioelovich, M., Figovsky, O., and Leykin, A. (n.d.). Polymate Ltd and Nanotech Industries Inc 2012 Biodegradable nano-composition for application of protective coatings onto natural materials US Patent 8,268,391.

Kanmani, S. S., & Ramachandran, K. (2012). Synthesis and characterization of TiO2/ZnO core/shell nanomaterials for solar cell applications. *Renewable Energy*, 43, 149–156. doi:10.1016/j.renene.2011.12.0149.

Kumar, S., Nehra, M., Deep, A., Kedia, D., Dilbaghi, N., & Kim, K.-H. (2017). Quantum-sized nanomaterials for solar cell applications. *Renewable and Sustainable Energy Reviews*, 73, 821–839. doi:10.1016/j.rser.2017.01.172

Li, Y., Li, S., He, C., Zhu, C., Li, Q., Li, X., ... Zeng, X. (2021). Selective laser-induced preparation of metal-semiconductor nanocomposites and application for enhanced photo-catalytic performance in the degradation of organic pollutants. *Journal of Alloys and Compounds*, 867, 159062. doi:10.1016/j.jallcom.2021.159062

Lu, L., & Ouyang, D. (2017). Properties of Cement Mortar and Ultra-High Strength Concrete Incorporating Graphene Oxide Nanosheets. *Nanomaterials*, 7(7), 187. doi:10.3390/nano7070187

Meng, Q., & Chung, D. D. L. (2010). Battery in the form of a cement-matrix composite. *Cement and Concrete Composites*, 32(10), 829–839. doi:10.1016/j.cemconcomp.2010.08.009

Mohanapriya, R. (2021). Generating electricity by footpath power generator floor tiles. *Information Technology in Industry*, 9(2), 576–582. doi:10.17762/itii.v9i2.389

Mohsen, M.O., Al Ansari, M.S., Taha, R., Al Nuaimi, N., Taqa, A.A., & Barron, A.R. (2019). Carbon nanotube effect on the ductility, flexural strength, and permeability of concrete. *J. Nanomater.* doi:10.1155/2019/6490984.

Moore, Katherine, & Wei, Wei (2021). Applications of carbon nanomaterials in perovskite solar cells for solar energy conversion. *Nano Materials Science*, 3(3), 276–290. doi:10.1016/j.nanoms.2021.03.005

Nawar, Adil Hatem (2021). Nanotechnologies and Nano-materials for civil engineering construction works applications. *Materials Today: Proceedings*. doi:10.1016/j.matpr.2021.01.49717.

Otanicar, T., and DeJarnette, D. (2016). *Society of Photo-optical Instrumentation Engineers. Solar energy harvesting: How to generate thermal and electric power simultaneously.* Bellingham, Washington, USA: SPIE Publications.

Pradesh, H. (2012). Application of nanotechnology in building materials. *International Journal of Engineering Research and Applications*, 2, 1077–82.

Shekari, A.H., & Razzaghi, M.S., Influence of nano particles on durability and mechanical properties of high performance concrete. *Procedia Eng, Elsevier*, 2011, 3036–3041. doi:10.1016/j.proeng.2011.07.382.

Shoreh, S. K. H., Ahmadyari-Sharamin, M., Ghayour, H., Hassanzadeh-Tabrizi, S. A., Pournajaf, R., & Tayebi, M. (2021). Two- stage synthesis of SnO2-Ag/MgFe2O4 nanocomposite for photo-catalytic application. *Surfaces and Interfaces*, 26, 101326. doi:10.1016/j.surfin.2021.101326

Szostak, Rodrigo, Morais, Andreia, Carminati, Saulo A., Costa, Saionara V., Marchezi, Paulo E., & Nogueira, Ana F. (2018). Application of Graphene and Graphene Derivatives/Oxide Nanomaterials for Solar Cells. In *The Future of Semiconductor Oxides in Next-Generation Solar Cells*, 395–437. doi:10.1016/b978-0-12-811165-9.00010-7

Wang, Y.-J., Long, W., Wang, L., Yuan, R., Ignaszak, A., Fang, B., Wilkinson, D.P. (2018). Unlocking the door to highly active ORR catalysts for PEMFC applications: polyhedron-engineered Pt-based nanocrystals. *Energy and Environmental Science*, 11, 258e275. doi:10.1039/C7EE02444D.

Wangidjaja, W. (2022). Waterproofing using nano technology as an alternative solution. *5th International Conference on Eco Engineering Development-IOP Conf. Series: Earth and Environmental Science*, 998, 012024.

WASHINGTON, DC, UNITED STATES, August 18, 2021. /EINPresswire.com/

Xuemei, Fu, Limin, Xu, Jiaxin, Li, Xuemei, Sun, & Huisheng, Peng (2018). Flexible solar cells based on carbon nanomaterials. *Carbon*, 139, 1063–1073. doi:10.1016/j.carbon.2018.08.017

You, S., Gong, X., Wang, W., Qi, D., Wang, X., Chen, X., & Ren, N. (2016). Enhanced cathodic oxygen reduction and power production of microbial fuel cell based on noble-metalfree electrocatalyst derived from metal-organic frameworks. *Advanced Energy Materials*, 6, 1501497. doi:10.1002/aenm.201501497.

23

Prospects of Graphene and MXene in Flexible Electronics and Energy Storage Systems: A Review

S Jana

*Vel Tech Rangarajan
Dr. Sagunthala R&D
Institute of Science and
Technology, Chennai, India*

S Selvaganesan

*JNN Institute of Engineering,
Tiruvallur, India*

S Sanjay
Sethuganesh

*Visvesvaraya National
Institute of Technology,
Nagpur, India*

Depletion of energy sources and need for efficient, cost-effective materials for flexible electronics is on demand and has led to the search for alternative materials that could be used in energy storage systems and devices for flexible electronics. The discovery of two-dimensional nanomaterials graphene and MXene with their flexible properties has led to the development of composite materials which could be used in energy storage devices and flexible electronic products like wearable devices, displays, and touch panels. This article reviews the process of synthesis of MXene, properties of graphene and MXene. Also, a review of the applications of graphene and MXene in flexible and transparent electronics and how these emerging popular nanomaterials could be used in energy storage systems has been presented.

23.1 Introduction

In the current scenario of gradually reducing energy resources and need for more suitable material for flexible electronics, research is being carried out in developing more energy-efficient, environmentally friendly energy storage devices and devices for flexible electronics. Though devices like batteries, capacitors, and supercapacitors (SC) are available, its use as energy storage devices is based on the efficiency, cost, and stability of the electrodes used [1]. Less availability of noble metals and poor performance of noble non-metals as electrodes [2–5] has led to the research in search of new electrode materials that

DOI: 10.1201/9781003355755-23

could be used in next-generation energy storage devices and excellent conducting materials that could be used in flexible electronics.

The widespread adoption of portable, smart electronics has accelerated the development of energy storage devices as well as other cutting-edge items like screens and touch panels. Mechanically durable transparent conductive electrodes are required for interactive gadgets such as smartphones, tablets, and other touchable devices. Materials with good conductivity, good optoelectronic properties, and excellent storage properties that act as current collectors and storage materials are best suited for transparent flexible energy storage systems [6]. The next generation of transparent electronics will benefit from the development of transparent supercapacitors as a power source. SC respond to the growing need for energy storage by providing extraordinary performance attributes namely faster rate of charging and discharging, extended cyclic life, and increased power density. The usage of improved electrode materials is a major aspect in improving SC performance. Exorbitant aspect ratios and a few atomic layer thicknesses characterise two-dimensional (2D) materials and this has ignited a lot of interest on the materials research horizon in the recent past. Because of their qualities, structure, and quantity, as well as the fact that they are environmentally friendly, carbon-based materials have been the subject of several recent investigations aiming towards electrochemical applications. Reports state that carbon derivatives and conducting polymers, such as activated carbon, carbon nanotubes, graphene, and polypropylene, are best suited as SC electrodes [1].

The most recent allotrope form of carbon, graphene, is among these materials. Graphene and MXene, two members of the 2D family, have recently demonstrated exceptional electronic conductivity and sparked a lot of interest in the energy storage industry.

Graphene, which is superior to plastic and silicon, is the most revolutionary substance of today, having evolved from graphite [7]. It is a 2D allotrope of carbon whose structure resembles a planar honeycomb lattice that makes it the strongest material. The basic structure of graphene is shown in Figure 23.1.

Graphene is the fundamental component of carbon nanotubes and big fullerenes. Carbon nanotubes get their characteristics from its component graphene which is usually in the form of a sheet. In a perfect graphene whose surface plane extends infinitely, where the valence and conduction bands meet, there is an almost negligible electronic band gap in which the electrons have negligible effective mass. As a result, graphene is a unique substance which neither possesses the characteristics of a metal nor a semiconductor. Due to their high aspect ratio (AR), these 2D nanomaterials provide improved fatigue resistance and fracture toughness through crack deflection and bridging mechanics [8]. Due to its extraordinary electric, mechanical, and chemical properties, graphene has attracted a huge number of researchers to carry out research with graphene and its derivative, for example, graphene oxide, porous graphene, reduced graphene oxide, and graphene quantum dots. Many efforts have been made to change the graphene structure since its discovery in order to integrate this promising material into a wide range of applications mainly in energy storage devices and flexible electron devices. The structure of the various types of graphene derivatives is given in Figure 23.2.

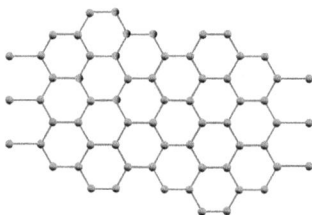

FIGURE 23.1 Basic structure of graphene [6].

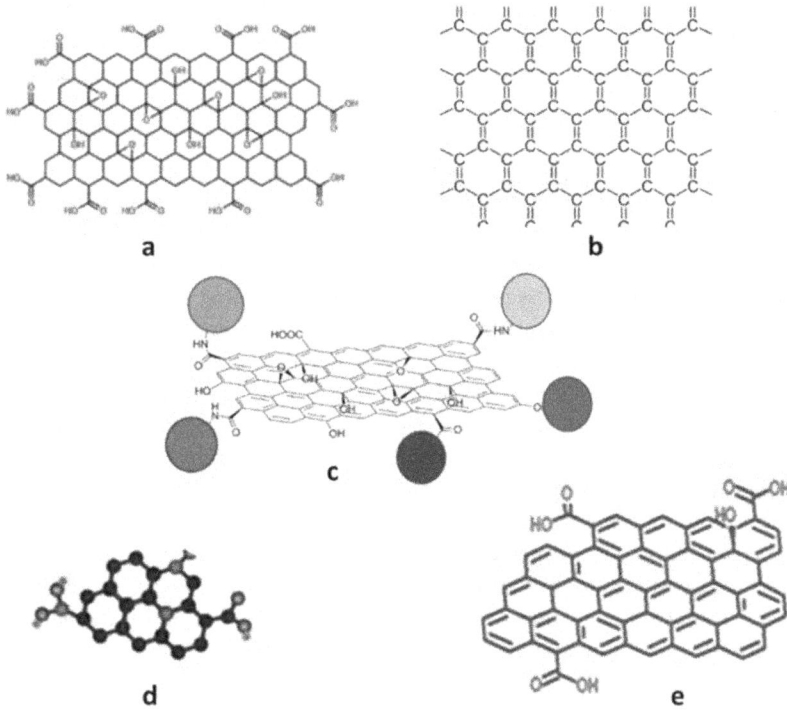

FIGURE 23.2 Different forms of graphene: (a) graphene oxide, (b) pristine graphene, (c) functionalised graphene, (d) graphene quantum dot, and (e) reduced graphene oxide [9].

MXene, has emerged as a novel material in the 2D family which is usually represented as $M_{n+1}X_n$ where M refers to an early transition metal, and X refers to carbon or nitrogen, has received significant attention in the recent past due to its unique features. The suffix 'ene' indicates that these 2D-layered MXenes are akin to graphene. The structure of MXene is given in Figure 23.3. New innovative electrode

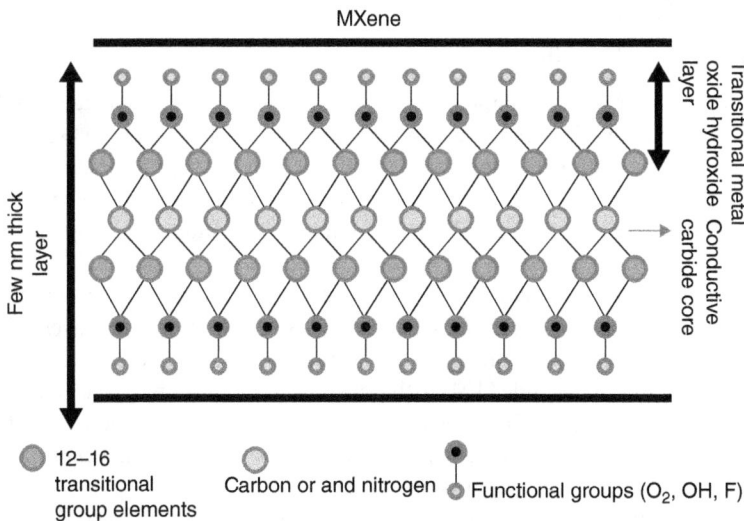

FIGURE 23.3 Structure of MXene [1].

shapes of MXene enable batteries to get charges at a faster rate comparatively to conventional battery shapes. The design of MXene enables them to be used in supplying quick bursts of energy essential for camera flashes. MXenes are materials with a lot of application potential and are a great addition to the 2D materials family.

23.2 Peculiarity of MXene and Graphene

Graphene has a number of unusual physical features that have never been seen before at the nanoscale such as the detection of quantum phenomena at ambient temperature. In the category of 2D nanomaterials, graphene scores higher as lightweight material with outstanding thermal and electrical conductivity, transparency, and mechanical resistance. Graphene is the thinnest and strongest substance [10]. At normal temperature, its charge carriers have a high degree of inherent mobility, negligible effective mass, and can traverse for micrometres without dispersing. Graphene is impermeable to gases and can endure current densities whose magnitude is six times higher than that of copper. The electronic structure of graphene is unique and the behaviour of electrons is more similar to that of relativistic massless particles rather than charge carriers, leading to increased device speeds and scope of being used in flexible electronics when compared to current existing device speeds. Graphene is practically transparent, as it can absorb only 2.3 percent of light that is incident on it, making it ideal for photovoltaic applications such as screens and solar cells in optoelectronics [11]. Graphene's notable features that make it desirable in electronics are the Hall Effect, ultrahigh electron mobility and ballistic transport, long electron mean-free pathways, outstanding thermal conductivity, great mechanical strength, and incredible flexibility.

Graphene sheet, interestingly, is impenetrable to molecules as tiny as helium. Graphene has a conductivity of up to 2,104 S/cm and an electronic mobility of 2,105 cm^2/Vs, making it 100 times more efficient than silicon [7]. Its thermal conductivity at room temperature can approach 5,000 W/mK (vs. 400 W/mK for copper) [7], implying that it could be used in thermal applications. It has a substantially bigger surface area (2,600 m^2/g) than graphite (10 m^2/g) and carbon nanotubes (1,300 m^2/g)[12].

Because of the presence of surface terminations (–O, –F, and –OH) as well as transition metal carbide/nitride, MXenes have a unique combination of high hydrophilicity and metallic conductivity [13]. MXenes also have a number of unusual properties, including high strength and stiffness.

23.3 Synthesis of MXene and Graphene

23.3.1 Synthesis of MXene

The MAX phases which are orderly aligned carbides and nitrides of ternary metals are the precursors of MXenes. The formula for MAX phase is represented as $M_{n+1}AX_n$, (MAX), where n can take up a value between 1 and 3, M represents a transition metal (such as Mo, Ti, Zr, Cr, etc.), A can be a group 13–16 element in the periodic table. The layered hexagonal MAX phases are made up of almost densely packed M layers and octahedral sites inhabited by X atoms, with A atoms bonding the layers of M and X together [13].

In the MAX phase, the M-X bond possesses a combination of ionic, metallic, and covalent properties, while the M-A bond possesses a metallic character in most cases. The strong interlayer bonding in MAX phase inhibits the mechanical exfoliation or in other words the shearing of these bonds, while in graphite and transition metal dichalcogenide (TMD), the layers are kept intact by a weak vander Waals force. MAX phases feature an unusual combination of properties that aid in the production of ceramics and metals. They are stiff, brittle, thermally stable, and strong in the same way that ceramics are, and they carry heat and electricity in the same way that metals do.

MXenes are created from MAX phases by selectively etching the A layers alone [14]. The steps involved in MXene synthesis is shown in Figure 23.4. The characteristics of MXenes differ from those of their

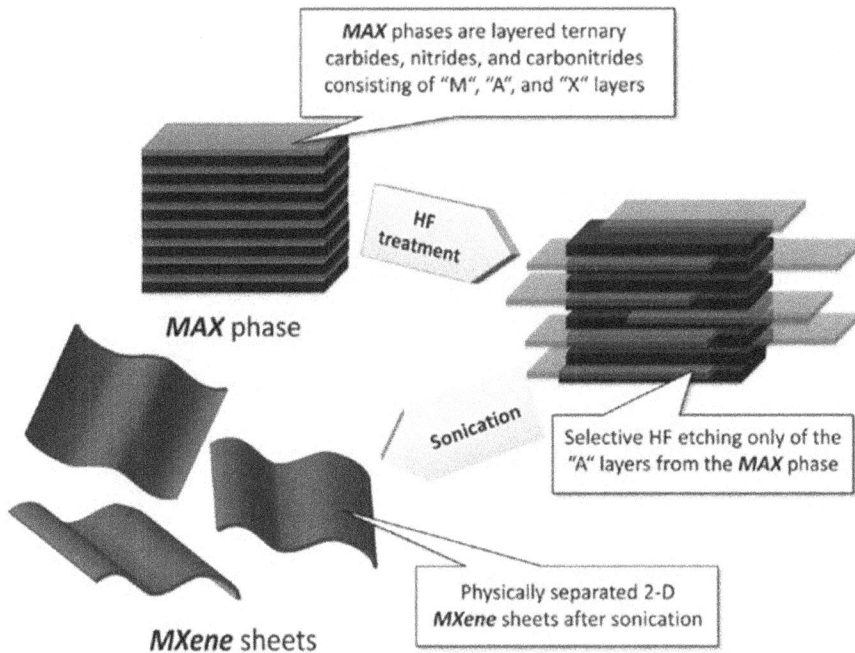

FIGURE 23.4 MXene synthesis process. (Reproduced with permission [13] https://doi.org/10.1016/j.mtchem.2019.08.010.)

predecessors. MXenes are described as $M_{n+1}X_nT_x$ in general, where n ranges between 1 and 3 and the functional groups (–O, –F, and/or –OH) generated during the etching stage are denoted by the letter T. As the MAX phases' n values range from 1 to 3, the resultant MXene can have one of three lattice structures: M_2X, M_3X_2, or M_4X_3 with 3, 5, or 7 atomic layers, respectively [13, 14]. The first MXene to be synthesised was $Ti_3C_2T_x$ and there are also MXenes, which include multiple M elements. Ordered phases and solid solutions are the two techniques to obtain this structure. The structure of the synthesised MXene is checked using a scanning electron micrograph (SEM).

23.3.2 Synthesis of Graphene

Graphene was created via mechanical exfoliation of graphite [15]. For basic research, this approach generated a small number of high-quality samples. Later, numerous ways for producing graphene sheets were developed, which can be divided into bottom-up and top-down procedures [16], as shown in Figure 23.5, employing molecules containing carbon and graphite as beginning materials, respectively. When choosing a graphene synthesis method, sheet quality, cost, sheet size, chemical composition, throughput, and compatibility with the conventional chip fabrication processes are some of the most critical factors that need to be considered.

Chemical vapour deposition (CVD)-based processes combined with the conventional lithographic methods is more suited for chip fabrication. The alternative technologies, on the other hand, can create enormous amounts of graphene at substantially lower costs.

Graphene-based materials with exceptional characteristics were recently successfully produced using the electrophoresis deposition process [10]. Films, non-metallic graphene composites, metal-graphene-based nanoparticle composites, and graphene-polymer composites are examples of graphene-based products generated via electrophoresis [17].

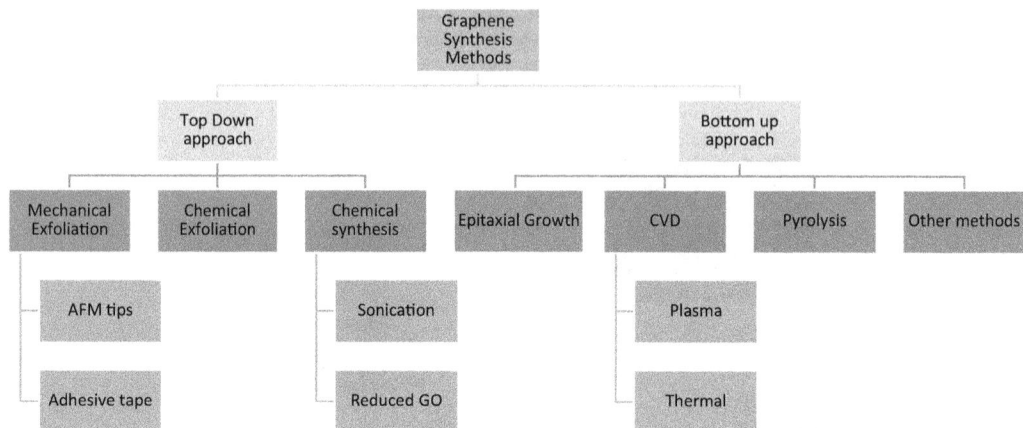

FIGURE 23.5 Common methods of graphene synthesis.

The physical approach is the easiest way to make composite materials since it does not create new substances and can be mixed uniformly with varied materials components, avoiding the need for complex processes and parameters in the chemical reaction process. Mechanical mixing and self-assembly methods are the two types of synthetic approaches for $Ti_3C_2T_x$/rGO composites that have been used presently [17].

23.4 Application of MXene and Graphene

MXene and graphene have recently been used in a number of applications of which energy storage and flexible electronics are important. Figure 23.6 shows the distribution of graphene and MXene applications namely supercapacitor, lithium sulphur, and lithium-ion batteries.

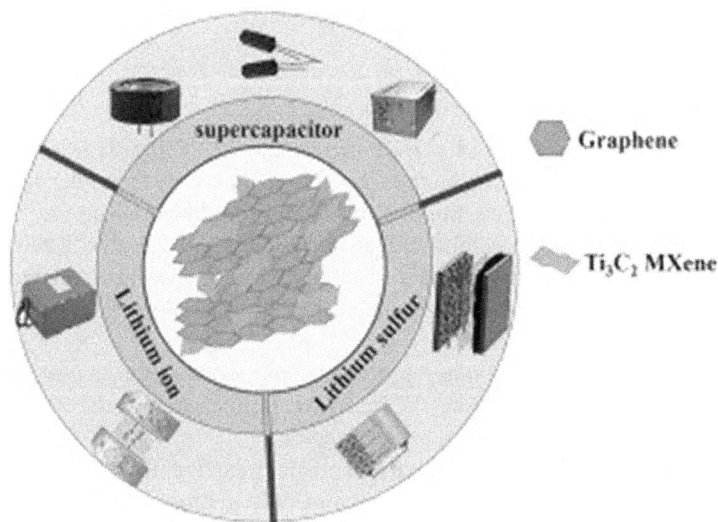

FIGURE 23.6 Graphene and MXene in energy storage application [17].

23.4.1 Applications of MXene

MXene could be used as Li-ion battery electrodes, pseudocapacitors, and other energy storage devices. According to researchers, MXene might potentially be utilised as a composite reinforcement, in a way similar to clays or graphene. This enriches the mechanical properties in polymers while at the same time it lowers the gas permeability. MXene's range of surface chemistries, high surface area, and presence of transition metal oxides make it potentially appealing for catalytic applications. MXenes have shown promise as Li-ion and Na-ion battery electrodes, as well as supercapacitor electrodes [16, 17]. MXenes are thought to have better super capacitive performance than other electrode materials because of their surface-terminated functional groups and layered architectures.

According to computational studies, the process of exfoliating or delaminating some MXenes would enhance the charging capacity of the layers, making it suitable for anode electrode in batteries. Scientists successfully intercalated MXenes with a variety of organic chemicals, one of which is dimethyl sulfoxide (DMSO), allowing them to completely peel stacked layers into MXene sheets and manufacture MXene 'paper' obtained by filtering flakes from the solution. Intercalation of cations in $Ti_3C_2T_x$MXenes is shown in Figure 23.7.

This flexible, electrically conductive paper has four times the lithium-ion capacity of typical MXene material and the rate of charging and life cycle duration are more compared to those of graphite used in the lithium-ion batteries existing in market. The supercapacitance of MXenes can be changed by adjusting the interlayer spacing.

With its 2D structure and unique features, MXenes have found uses in almost every sector. MXenes offer a wide range of uses as catalysts for hydrogen storage material dehydrogenation, oxygen reduction and evolution reactions, and hydrogen evolution reactions. They've also found uses in sensor technologies, such as electrochemical biosensors, gas sensors, macromolecule and cell detection, and so on. MXenes can be used in a variety of applications, including energy storage.

23.4.2 Applications of Graphene

After the silicon era, graphene has earned a special focus in the field of electronics. Graphene is a material with exemplary potential for electronics due to its unique 2D networks of sp2 hybrids of carbon atoms organised in a honeycomb structure. Applications include sensors, photovoltaic cells, organic thin film transistors, super capacitors, photo-detectors, catalytic applications, and diodes. Because of its unique properties, graphene could be used in a variety of applications, including electrochemical energy systems, electromechanical systems, hydrogen storage, field-effect transistors, and polymer composite materials.

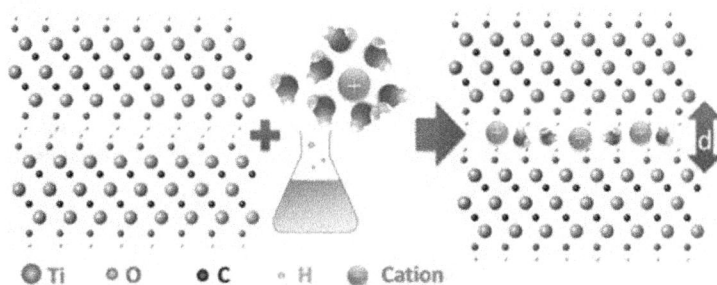

● Ti   ○ O   ● C   · H   ● Cation

FIGURE 23.7 Intercalation of cations in $Ti_3C_2T_x$MXenes. (Reproduced with permission [18] https://doi.org/10.1016/j.pnsc.2018.03.003.)

In the last two decades, graphene has emerged as the perfect photovoltaic material and has played an important role in cell devices as a transparent electrode, hole/electron transport medium, and interfacial buffer layer. Physiochemical qualities, great mechanical performance, and high thermal and electric conductivity distinguish this one-thick carbon crystal, making graphene a revolutionary alternative to many traditional materials in different applications.

23.4.3 MXene and Graphene: Energy Storage Devices

Energy conservation is one of the century's most pressing issues, and it is intertwined with the global economy's long-term viability. The increasing global demand for energy necessitates the development of high-performance systems and technologies that allow for more efficient usage while minimising environmental damage and long-term resource depletion. As a result, research and development of novel materials for application in energy storage devices is being undertaken.

$Ti_3C_2T_x$ is the most researched MXene variety that is most commonly used in electrochemical applications due to its large surface area characteristic, super conductivity behaviour, and improved chemical stability [16]. Many researchers have demonstrated that $Ti_3C_2T_x$ is a potential candidate material for high-performance supercapacitor electrode development. The advantages of $Ti_3C_2T_x$ include excellent charge storage and extended cycle life qualities, which makes them suitable for MXene-based electrodes are extremely widely employed as cathode materials for SCs or anodes for Li-ion batteries [19]. MXene films have a high capacitance, and a number of studies have documented advanced fabrication processes. The creation of MXene-based composites is of great importance. MXenes have been combined with polymers, carbon nanotubes, graphene, metal oxides, and hydroxides to create advanced composites. A number of researchers have evaluated the performance electrodes made using MXene in various electrolytes [8, 13, 14, 17].

MXene–polypyrrole (PPy) composites [17] have been developed as a result of efforts to enhance the capacitance of MXene-based electrodes. This method made it easier to make flexible electrodes and gadgets. PPy is a pseudocapacitive electrochemically active conducting polymer. In MXene–PPy composites, the synergistic effect of the separate layers of MXene and PPy has been observed to improve charge–discharge rates and electron transport. The polymerisation of PPy has been carried out in the spaces between the MXene layers in the absence of an oxidant. The chemical properties of MXene and PPy enable amazing interactions at their interfaces, such as the formation of MXene surfaces with PPy layers on them and increased PPy cyclic and capacitive performance.

Homogeneous PPy nanoparticles when subjected to intercalation can further increase the interlayer spacing of $Ti_3C_2T_x$, allowing for more electrolyte transport channels [20]. Literature reports that in H_2SO_4, KOH, and Na_2SO_4 electrolytes, $Ti_3C_2T_x$-PPy electrodes were examined. In a number of previous research, MXene–PPy electrodes were developed and tested by applying a low range of positive potential. Symmetric devices for operation in the 0.5–1 V range of voltage, were created in the H_2SO_4 electrolyte, whereas symmetric devices for operation in the 0.5–0.6 V range, were developed in the H_2SO_4/PVA electrolyte.

Despite advancements in the manufacturing of MXene–PPy composites, the demand for SC electrodes having areal capacitance still exists. The lowering of specific capacitance with increased active mass loading (AML) is a fundamental issue in the development of such electrodes. Because neutral electrolytes are environmentally friendly, non-corrosive, stable, economical, and allow the use of a wide range of current collectors, they provide a substantial advantage in the creation of asymmetric SC with a wide voltage range. It is commonly agreed that the asymmetric SCs provide a way to expand the device voltage window while also boosting the device's specific power and energy.

Graphene-based perovskite solar cells have an energy conversion efficiency of 20.3 percent, while organic solar cells with bulk heterojunction (BHJ) have 10 percent efficiency [20]. Apart from its involvement in charge extraction and transmission between electrodes, graphene also plays a unique role in protecting photovoltaic (PV) devices from environmental deterioration due to its dense 2D network structure, which ensures long-term environmental stability. Graphene electrodes can boost the supercapacitor's capacitance by 20% to 30% [20]. When compared to graphite electrodes, several articles highlight the advantages of employing graphene or its nanocomposites as electrode material in Li-ion batteries, with significantly improved cycle performance and higher reversible capacity.

23.4.4 MXene and Graphene: Flexible Electron Devices

Graphene is very flexible due to its fineness modulus and acts as the best heat conductor, which enables it to be used in the development of flexible electronics. The physical properties of graphene make it excellent for electrical applications. The unusually high carrier mobility, in particular, has gotten a lot of attention. There have been reports of mobilities in excess of 100,000 cm²/Vs. The mechanical strength, and flexibility of the material, and its excessive current carrying capacity (up to 109A/cm²) and heat conductivity (up to 5,000 W/mK) [20], all contribute to its appeal.

While it is observed that the electrical properties of the graphene channel are excellent, what happens in other portions of the graphene device might have a significant impact on, and even dominate, transport. Carriers are initially introduced into the graphene channel and subsequently collected via metal contacts. Contacts create possible energy blocks for carriers which should be overcome, affecting device performance significantly. Charge transfer (CT) occurs between graphene and metal because their work functions are different. The resulting dipole layer produces local band bending and graphene doping beneath the metal.

The graphene band structure is also significantly altered with more reactive metals. The metal-doped to -undoped channel graphene (p-n junction-like) barrier and the dipole barrier created by CT must both be passed by a carrier injected into graphene. The gate-dependent resistance of metal-graphene contacts varies from a few hundred Wm to several kWm. The creation of asymmetry between electron and hole transport is another key consequence of CT at contacts.

For example, electrical transport is susceptible to a variety of scattering interactions. During the process of scattering, long-range interactions on graphene take place with charged impurities or at the junction of supporting substrate which is an insulator. The short-range interactions in the scattering process are obtained with neutral defects or adsorbates, roughness, and phonons. Based on the quality of graphene and the conditions that exist, there is a variation in the mechanism that is dominant. When graphene is in contact with polar substrates like SiO_2 or Al_2O_3, for example, Coulomb scattering from charged impurities often prevails at low temperatures. When the substrate is removed and the graphene is suspended, dispersion occurs due to the presence of adsorbates. When the adsorbates are volatilised and the graphene is heated, phonon scattering becomes dominant, allowing for large mobilities. When graphene is faulty, carrier transport is dominated by scattering from neutral point defects. The carrier mobility amplitude and its relationship with temperature (T) and carrier density (n) could be utilised to determine the type of scatterer that dominates in a graphene sample.

Graphene has been a promising material over the conventional indium tin oxide (ITO) [21] material in applications which require transparent conductive electrodes (TEC) in a stretchable and flexible format. They are suitable in applications involving wearable devices and biomedical applications. The unique characteristics of graphene such as ease of processability, higher conductivity have enabled it to be used in printed flexible electronics, as shown in Figure 23.8. Graphene composite inks which is a combination of graphene oxide, graphene nanoplatelets, and silica fillers are used in 3D printing [22].

In the recent past, MXene has found its significance as biosensors with greater antiviral activity in the fabrication of face masks, personal protective equipment (PPE) kits, face shields, and biomedical instruments. MXenes have been found to provide viable solutions in Covid-19 patient's for their comorbidities during drug loading and in controlling the drug release. The biocompatibility nature of MXene has enabled it to be used in diverse biomedical applications [23].

MXene's advantage of flexibility and bonding nature with fibre because of surface functional groups have enabled it to be used in skin-mountable devices, leading to a tremendous increase in smart textiles.

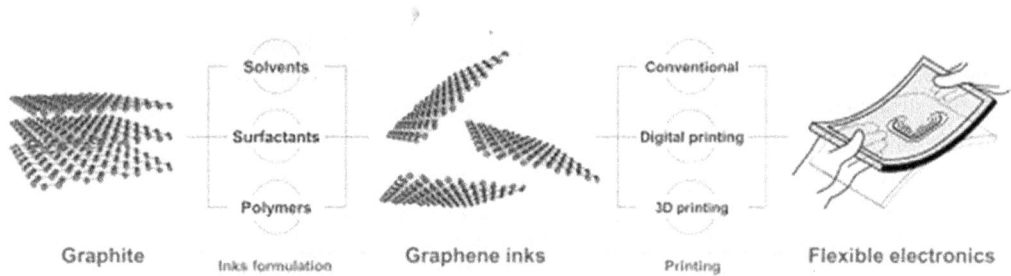

FIGURE 23.8   Graphene inks in flexible electronics [22].

Real-time healthcare monitoring, flexible displays, biomedical therapy, flexible sensing, soft-robotics, and many more applications have all benefited from textile-based electronic systems [20]. MXenes have also found its application in electromagnetic interference (EMI) shielding and sensors and imitates nacre-like nanocomposite structures in bioinspired materials [20]. The MXene-based sensor could be employed in applications which are based on sensitivity to pressure and physiological signal detection. MXene-based hydrogel is used in acquiring biomedical signals like ECG, EOG, and EMG so as to improve the polymerisation and cross-linking of polymer chains and their conductivity. MXene has a broad range of applications in the design and development of wearable and flexible electronic devices suitable for energy storage platforms that are portable, medical surveillance, human–machine interfaces, and electromagnetic interference shielding.

23.5 Summary

In this article, the 2D nanomaterials graphene and MXene are studied focussing on its unique characteristics, the various synthesis methods and its application in the area of energy storage and flexible electronics. Applications of graphene and Ti_3C_2 MXene can be found in a number of diverse disciplines, one of which is electrochemical energy storage. The electrochemical energy storage feature of graphene is due to its electric double-layer effect and that of MXene is due to its pseudo capacitance property. The MXene/rGO composite, in particular, enhances the material characteristics such as increased surface area and faster ion diffusion in electrolyte. This feature is attributed to the lack of self-stacking between graphene and MXene sheets. Improvement in MXene nanosheet's carrier mobility feature can be achieved with further research on the methods of etching mechanism, ion insertion, and ion exchange.

Ti_3C_2 MXene/rGO has shown a significant increase in volume capacitance, more energy/power density, and extended cycle life when used in high-performance energy storage devices. The future of Ti_3C_2 MXene/rGO composites looks quite promising right now. In the coming years, MXene together with graphene, nanotechnology, and advanced chemical synthesis technology is projected to result in a rise in nanostructured materials for a wide range of applications including high-performance electrochemical energy storage devices and in the area of flexible electronics as sensors, in EMI shielding, in 3D printing, as efficient hydrogels in biomedical applications, as wearable devices in textiles, etc. This promising research discovery is a result of continuing research in exploring the properties of the 2D nanomaterials graphene and MXene by researchers.

References

1. Garg, Ruby; Agarwal, Alpana; Agarwal, Mohit. A review on MXene for energy storage application: effect of interlayer distance. *Materials Research Express* 2020, 7, 022001.
2. Yan, Y.; Xia, B. Y.; Zhao, B.; Wang, X. A review on noble-metal-free bifunctional heterogeneous catalysts for overall electrochemical water splitting. *Journal of Materials Chemistry A* 2016, 4, 17587–17603.

3. Burke, M. S.; Enman, L. J.; Batchellor, A. S.; Zou, S.; Boettcher, S. W. Oxygen evolution reaction electrocatalysis on transition metal oxides and (Oxy)hydroxides: activity trends and design principles. *Chemistry of Materials* 2015, 27, 7549–7558.

4. Hunter, B. M.; Gray, H. B.; Müller, A. M. Earth-abundant heterogeneous water oxidation catalysts. *Chemical Reviews* 2016, 116, 14120–14136.

5. Wang, J.; Xu, F.; Jin, H.; Chen, Y.; Wang, Y. Non-noble metal-based carbon composites in hydrogen evolution reaction: fundamentals to applications. *Advanced Materials* 2017, 29, 1605838.

6. Rudrapati, Ramesh. Graphene: Fabrication Methods, Properties, and Applications in Modern Industries. In: *Graphene Production and Application* Chapter 2, Intechopen. doi:10.5772/intechopen. 83309.

7. Mbayachi, Vestine B.; Ndayiragije, Euphrem; Sammani, Thirasara; Taj, Sunaina; Mbuta, Elice R.; Khan, Atta Ullah. Graphene synthesis, characterization and its applications: A review. *Results in Chemistry* 2021, 3, 100163.

8. Kilikevičius, Sigitas; Kvietkaitė, Saulė; Mishnaevsky, Leon; Omastová, Mária, Jr.; Aniskevich, Andrey; Zeleniakienė, Daiva. Novel hybrid polymer composites with graphene and mxene nanoreinforcements: Computational analysis. *Polymers (Basel)* 2021, 13, no. 7, 1013.

9. Tiwari, Santosh; Sahoo, Sumanta; Wang, Nannan; Huczko, Andrzej. Graphene research and their outputs: Status and prospect. *Journal of Science: Advanced Materials and Devices* 2020, 5. doi:10.1016/j.jsamd.2020.01.006.

10. Taghioskoui, Mazdak. Trends in graphene research. *Materials Today* 2009, 12, no. 10, 34–37.

11. Mohan, Velram Balaji; Lau, Kin-tak; Hui, David; Bhattacharyya, Debes. Graphene-based materials and their composites: A review on production, applications and product limitations. *Composites Part B: Engineering* 2018, 142, 200–220.

12. Stoller, Meryl D.; Park, Sungjin; Zhu, Yanwu, An, Jinho; Ruoff, Rodney. Graphene-based ultracapacitors. *Nano Letters* 2008, 8, no. 10, 3498–3502.

13. Salim, O.; Mahmoud, K.A.; Pant, K.K.; Joshi, R.K. Introduction to MXenes: synthesis and characteristics. *Materials Today Chemistry* 2019, 14, 100191.

14. Giménez, R.; Serrano, B.; San-Miguel, V.; Cabanelas, JC. Recent advances in MXene/epoxy composites: Trends and prospects. *Polymers (Basel)* 2022, 14, no. 6, 1170.

15. Papageorgiou, Dimitrios G.; Kinloch, Ian A.; Young, Robert J. Mechanical properties of graphene and graphene-based nanocomposites. *Progress in Materials Science* 2017, 90, 75–127.

16. Avouris, Phaedon; Dimitrakopoulos, Christos. Graphene: Synthesis and applications. *Materials Today* 2012, 15, no. 3, 86–97.

17. Liu, Yanyue; Yu, J.; Guo, Dongfang; Li, Zijiong; Su, Yanjie. Ti3C2Tx MXene/graphene nanocomposites: Synthesis and application in electrochemical energy storage. *Journal of Alloys and Compounds* 2020, 815, 152403.

18. Tang, Hao; Hu, Qin; Zheng, Mingbo; Chi, Yao; Qin, Xinyu; Pang, Huan; Xu, Qiang. MXene–2D layered electrode materials for energy storage. *Progress in Natural Science: Materials International* 2018, 28, no. 2, 133–147.

19. Xu, S.; Wei, G.; Li, J.; Han, W.; Gogotsi, Y. Flexible MXene–graphene electrodes with high volumetric capacitance for integrated co-cathode energy conversion/storage devices. *Journal of Materials Chemistry A* 2017, 5, no. 33, 17442–17451.

20. Wu, Wenling; Wei, Dan; Zhu, Jianfeng; Niu, Dongjuan; Wang, Fen; Wang, Lei; Yang, Liuqing; Yang, Panpan; Wang, Chengwei. Enhanced electrochemical performances of organ-like Ti3C2 MXenes/polypyrrole composites as supercapacitors electrode materials. *Ceramics International* 2019, 45, 7328–7337.

21. Das, Tanmoy; Sharma, Bhupendra K.; Katiyar, Ajit K.; Ahn, Jong-Hyun. Graphene-based flexible and wearable electronics. *Journal of Semiconductors* 2018, 39, 1.

22. Tran, Tuan Sang, Dutta, Naba Kumar, Choudhury, Namita Roy. Graphene inks for printed flexible electronics: Graphene dispersions, ink formulations, printing techniques and applications. *Advances in Colloid and Interface Science* 2018, 261, 41–61.

23. Panda, Subhasree; Deshmukh, Kalim; Hussain, Chaudhery Mustansar; Pasha, S.K. Khadheer. 2D MXenes for combatting COVID-19 pandemic: A perspective on latest developments and innovations. *Flatchem* 2022, doi:10.1016/j.flatc.2022.100377.

24. Lukatskaya, Maria; Mashtalir, Olha; Ren, Chang; Gogotsi, Yury, et al. Cation intercalation and high volumetric capacitance of two-dimensional titanium carbide. *Science* 2013, 341, no. 6153, 1502–1505, doi:10.1126/science.1241488

24

PAN-Based Composite Gel Electrolyte for Lithium-Ion Batteries

Dr. M. Malathi,
Dr. S. Kiruthika and
Dr. K. V. Gunavathy
*Kongu Engineering College,
Perundurai, India*

Composite polymer electrolytes were prepared from PAN (polyacrylonytrile), lithium perchlorate (LiClO4), and with three different dielectric reinforcements such as lead zirconium titanate (PZT)-12000, barium titanate (BT)-1000, and alumina (Al_2O_3)-6. Differential scanning calorimetry and X-ray diffractometry, impedance spectroscopy, FTIR, NMR spectroscopy, and scanning electron microscop were employed to reveal the crystalline nature of the electrolytes. The conductivity of the composite polymer electrolytes was measured by impedance spectrometry. Among the three systems, PZT-reinforced composite exhibits maximum ionic conductivity of 3.42×10^{-3} S/cm at room temperature. The ionic conductivity of the polymer composites increases with an increase in dielectric constant of the reinforcement. The composite with alumina reinforcement displayed strongly modified properties with very weak temperature dependence of conductivity. The morphology of PAN gel with ceramic reinforcement depicts the uniform distribution of reinforcement in the polymer matrix. FTIR studies show that the Li+ ion conductivity of PAN-based composite electrolyte is limited by the coordination of Li+ by CN sites of the PAN chain. However, conductivity of Li ion increases with increase in the dielectric constant of ceramic reinforcement. *From NMR studies*, three types of Li ion environments are observed in PAN-based composite electrolyte and they are (1) solvation shell, (2) vicinity of polymer chain in the presence of ceramic reinforcement, and (3) interactions of Li+ ions with both solvent and polymer.

24.1 Introduction

PAN-based composite gel electrolytes are the most investigated system among the composite electrolytes for application in lithium batteries (Jayathilaka et al. 2003). High lithium-ion conductivities of these electrolytes are due to their capacity to hold large volumes of liquid electrolyte (Song et al. 1999). It is also known that PAN gels with high solvent content are reported to suffer mechanical stability (Croce et al. 1994). However, there has been no complete investigation on the nature of the reinforcement and its role in the conductivity of the gels. Hence, it is necessary to study PAN gels with different reinforcements that can simultaneously offer high Li$^+$ ion conductivity and minimum level of mechanical stability. In this chapter, suitable liquid electrolytes for PAN gels were studied and with that various ceramic reinforcements such as barium titanate (BT), PZT, and alumina have been incorporated and analyzed by SEM, FTIR, differential scanning calorimetry (DSC), and impedance spectroscopy. ^{7}Li NMR has been used to study the behavior of Li$^+$ ion in the gel composites.

DSC and FTIR spectra confirm the miscibility of solvent occurs in liquid electrolyte. Vapor phase (VP) and solubility parameters suggest propylene carbonate and ethylene carbonate (PC-EC) is a suitable mixed solvent for poly acrylo nytrile (PAN)-based gel electrolytes.

24.2 PAN-Based Composite Gel Electrolyte

Figure 24.1 shows the X-ray diffractometry (XRD) pattern for PAN gel electrolyte and dielectric ceramic reinforced PAN-based gel electrolyte. In all the XRD patterns, a broad hump is present at 2θ angle 21°. The presence of a broad hump confirms amorphous in nature. In general, the presence of amorphous nature increases polymer flow and ionic diffusivity (Wesley 2003). The absence of the sharp peaks pertaining to LiClO$_4$ in the polymer complex indicates that complexation has taken place in the amorphous phase. On the other hand, peaks belonging to Al$_2$O$_3$, BT, and PZT are also not observed. The absence of these peaks is due to the dominance of the amorphous nature of the gel system.

24.3 FTIR Studies

Figure 24.2 shows the IR spectra of C≡N stretching of PAN_PC_EC gel with Li salt content along with pure PAN. The peak at 2243 cm^{-1} is due to the presence of polar group C≡N in the polymer chain. The shoulder seen at 2253 cm^{-1} for gels with LiClO$_4$ is attributed to the CN that is coordinated with the Li$^+$ (Durig et al. 1970, Yang et al. 1996). The combined presence of the original peak indicates that only part of polar groups is associated with Li$^+$ ions. The broadness of the peak increases with the increase in Li content in the gel electrolyte. It is expected that, in the polymer gel, if Li ion is coordinating with the backbone of the polymer matrix, then the conductivity of Li ion in the gel system is reduced. The polar group of the polymer strongly attracts the Li ion and hinders the mobility of the Li ion. Hence, it is required to have most of the Li ion in the liquid electrolyte rather than the polymer chain I order to increase the total Li ion conductivity of the gel electrolyte.

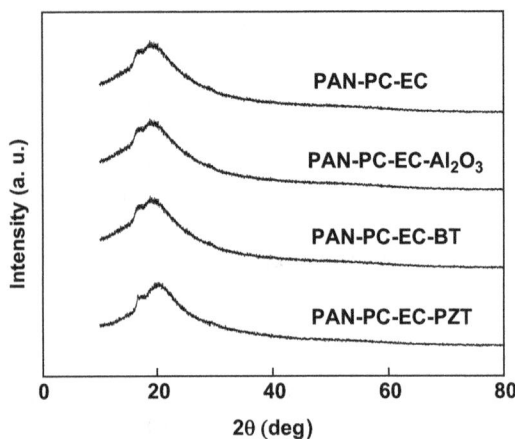

FIGURE 24.1 XRD pattern of PAN-based composite gel electrolytes.

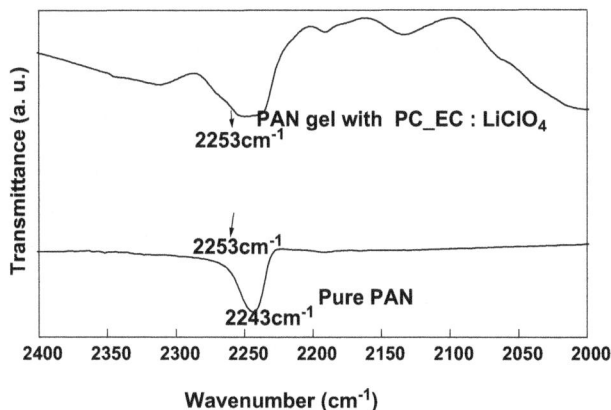

FIGURE 24.2   FTIR spectra of pure PAN and PAN composite gel electrolytes.

Figure 24.3 shows the IR spectra of C≡N stretching of PAN gel and PAN gel with Al_2O_3, BT, and PZT. The peak at 2243 cm^{-1} in all the composite gel electrolytes is attributed to the presence of polar group C≡N in the polymer chain. The broad peak is due to the merging of associated and unassociated CN-shouldered peaks. In order to identify the strength of the associated and unassociated C≡N in the polymer chain, the peaks are deconvoluted using Lorentzian distribution. It is clearly seen from the plot that PAN gel and Al_2O_3 reinforced PAN gel exhibit more Li ions that are associated with the polar group. In the case of PAN gel with BT, there is an equal proportion of associated and unassociated polar groups. Interestingly, in the presence of PZT, more unassociated polar groups are present when compared to the associated polar groups. The observation is very much similar to that of pure PAN. This indicates that, in the presence of PZT, more Li ions are dissolved in the liquid electrolyte rather than the polymer matrix. The presence of Al_2O_3 unalters the chemical nature of the gel system and contributes only to the mechanical property of the gel electrolyte. But in the case of BT and PZT, there is certainly a role in the gel electrolyte apart from contributing to the mechanical property of the gel electrolyte.

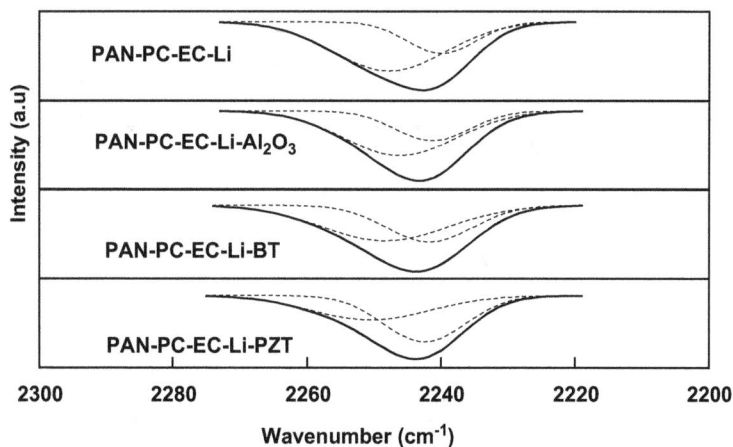

FIGURE 24.3   FTIR spectrum of PAN-based composite gel electrolytes.

24.4 Thermal Analysis

The DSC thermograms are shown in Figure 24.4, and the glass transition temperature Tg is shown in Table 24.1. The DSC curve for the PAN gels with Al_2O_3, BT, and PZT systems exhibits featureless except for the glass transition. Upon addition of PAN to the PC-EC liquid electrolyte, there is no significant change in the glass transition. In the presence of Al_2O_3 and BT, there is a slight change in the glass transition. But there is a considerable change in the glass transition temperature when the PZT is introduced. It is noteworthy that lower glass transition is required to achieve higher conductivity at ambient and sub-ambient temperatures.

24.5 Conductivity Behavior of PAN-Based PGE

Figure 24.5a, 24.6a, 24.7a and 24.8a shows typical impedance cole-cole plot for PAN-PC-EC gel and gel with Al_2O_3, PZT, and BT reinforcements from −80 to −40°C in the operated frequency range of 1Hz to 1MHz.

The DC conductivity of the PAN-PC-EC gel and gel with Al_2O_3, BT, and PZT reinforcements at room temperature are 4.01×10^{-3}, 0.61×10^{-3}, 0.89×10^{-3}, and 3.42×10^{-3} S/cm, respectively. The highest conductivity reported other than pure PAN-based gel electrolytes is for PAN-PZT system. It is clearly seen from Table 24.6 that as the dielectric constant of the reinforcement increases, the ionic conductivity of the composite increases. However, the conductivity of the PAN-PZT system is less than the pure PAN gel electrolyte. Microscopically, the presence of dielectric reinforcement increases the mobility of the ionic conductivity. To be precise, the dielectric field present in the medium increases Li ion to mobile in the liquid electrolyte and itself is not a conducting medium. Hence, more the volume of the dielectric

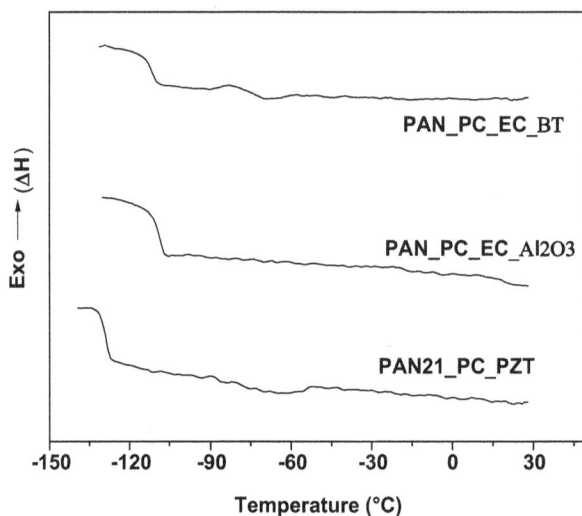

FIGURE 24.4 DSC pattern of PAN-based composite gel electrolytes.

TABLE 24.1 Thermal Data for PAN-based Gel Electrolytes

PAN Gels	$T_g(°C)$
PAN-PC-EC-Al_2O_3	−106
PAN-PC-EC-BT	−110
PAN- EC-GBL-PZT	−125

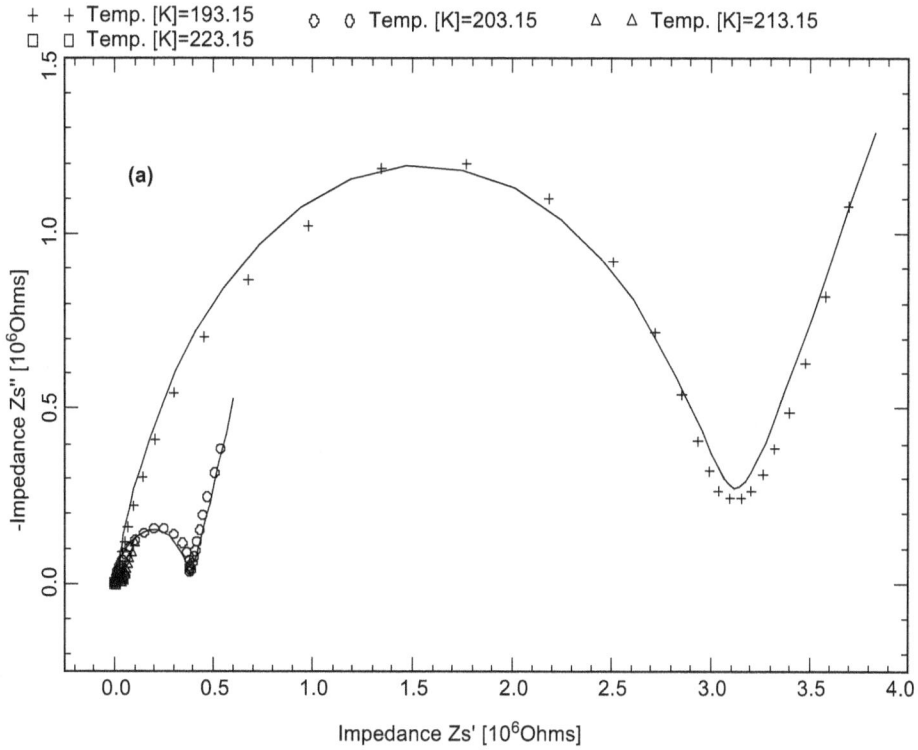

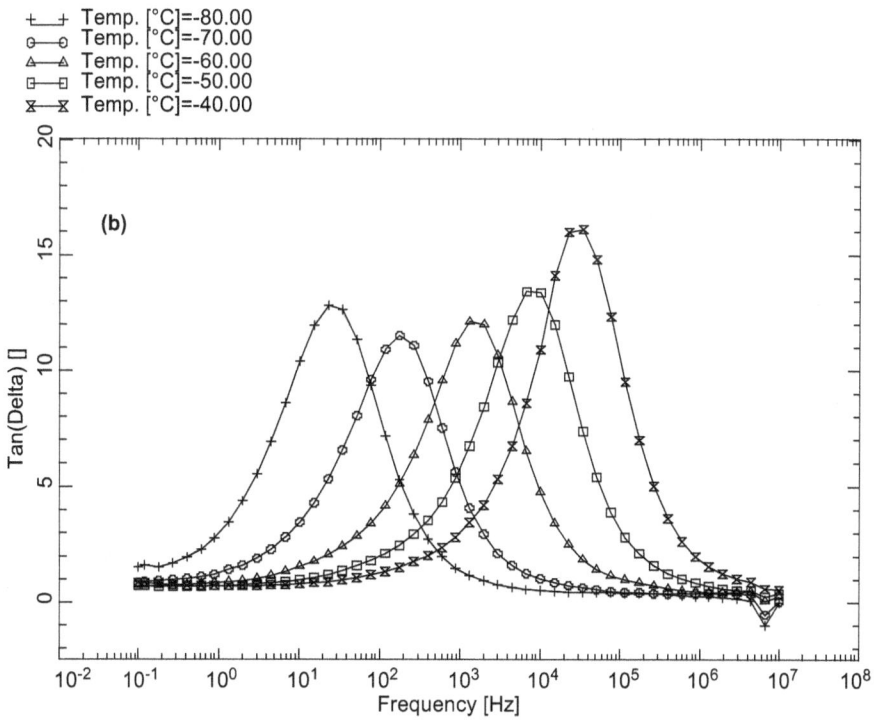

FIGURE 24.5 PAN-PC-EC system (a) Cole-Cole and (b) Tan δ plot.

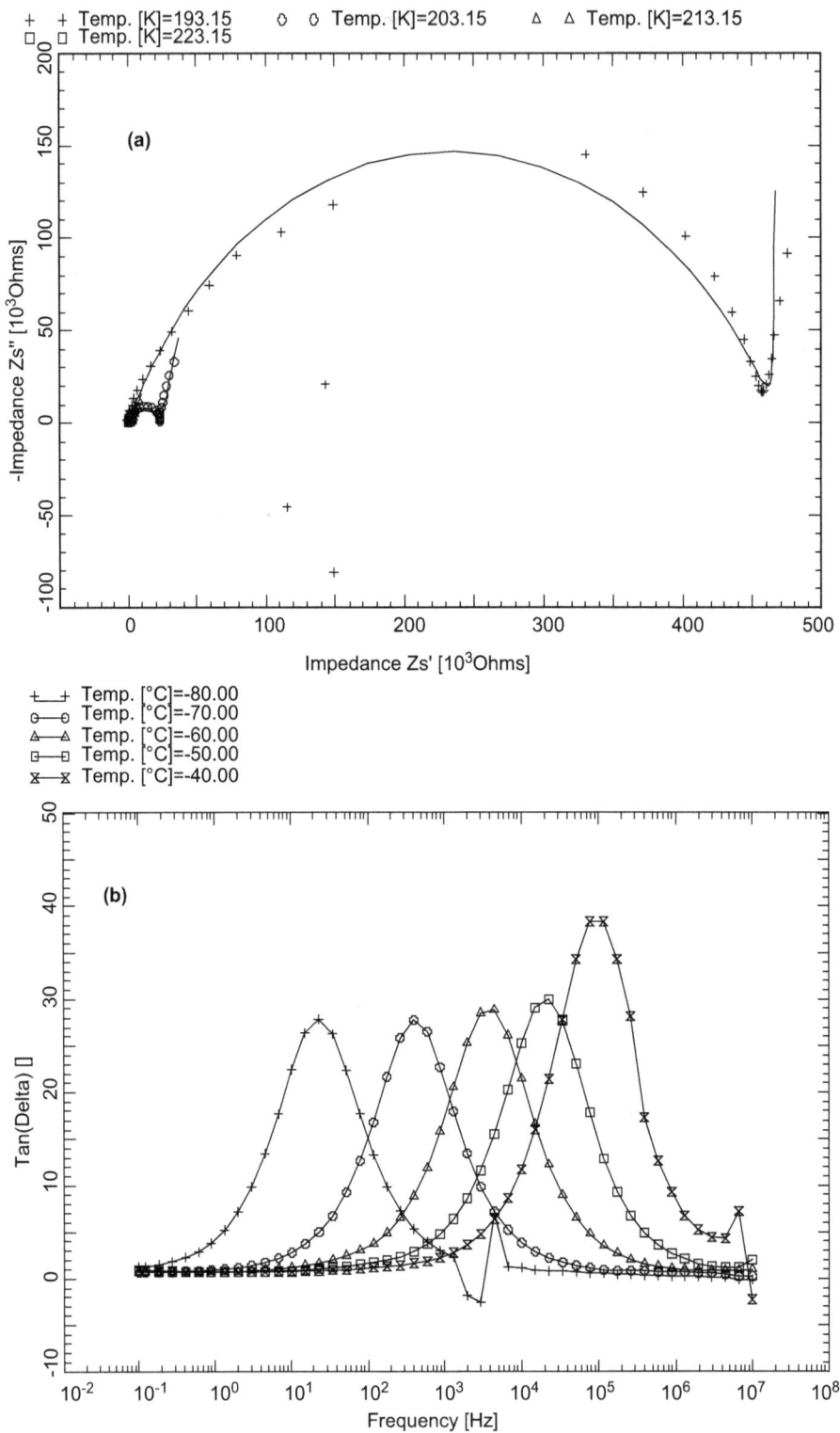

FIGURE 24.6   PAN-PC-EC-Al$_2$O$_3$ system (a) Cole-Cole and (b) Tan δ plot.

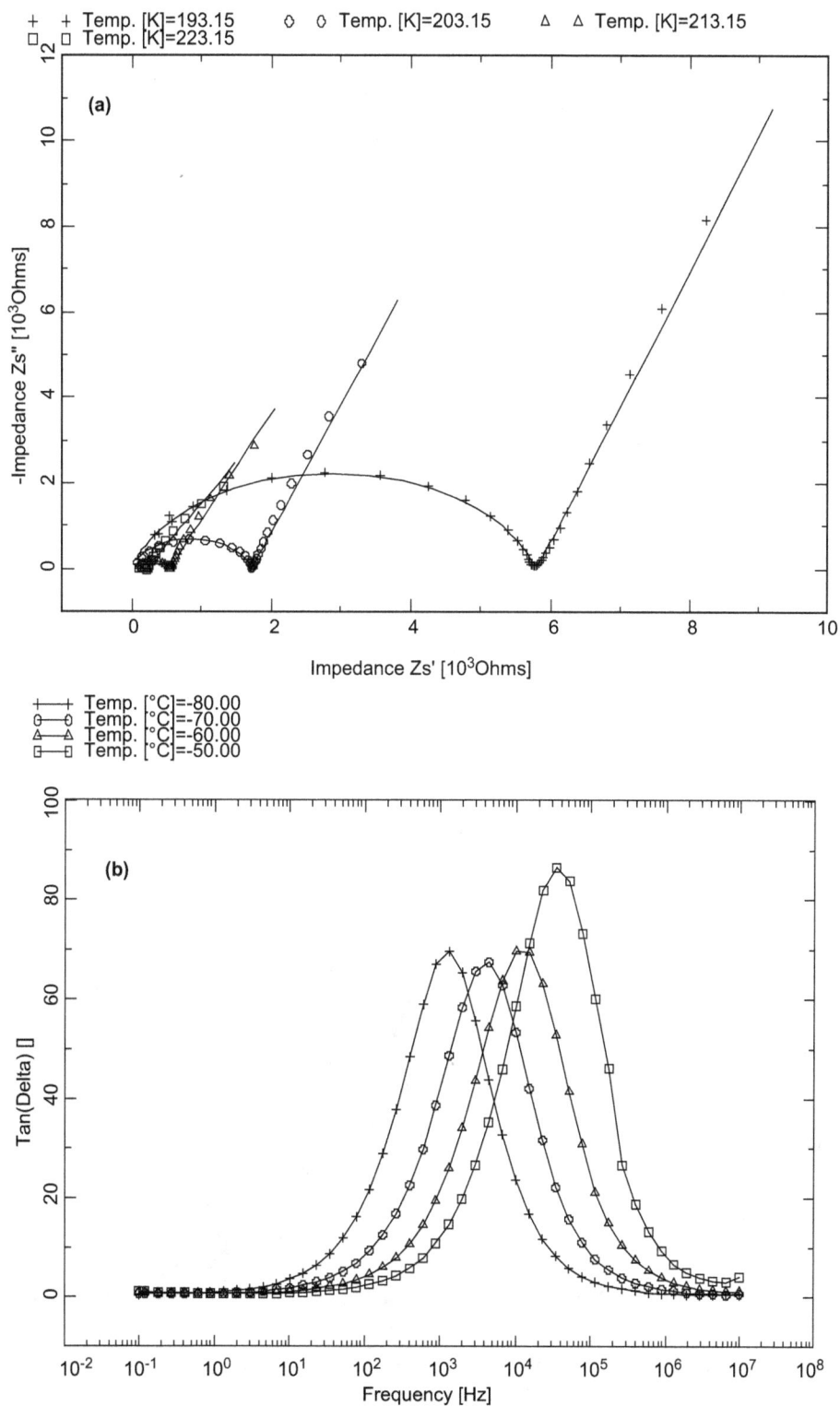

FIGURE 24.7 PAN-PC-EC-BT system (a) Cole-Cole and (b) Tan δ plot.

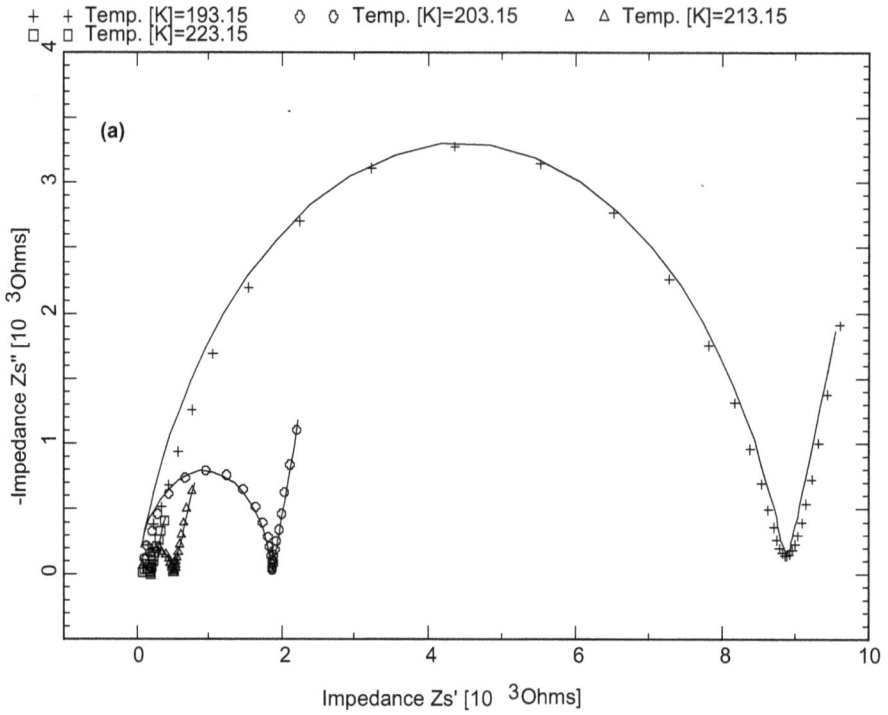

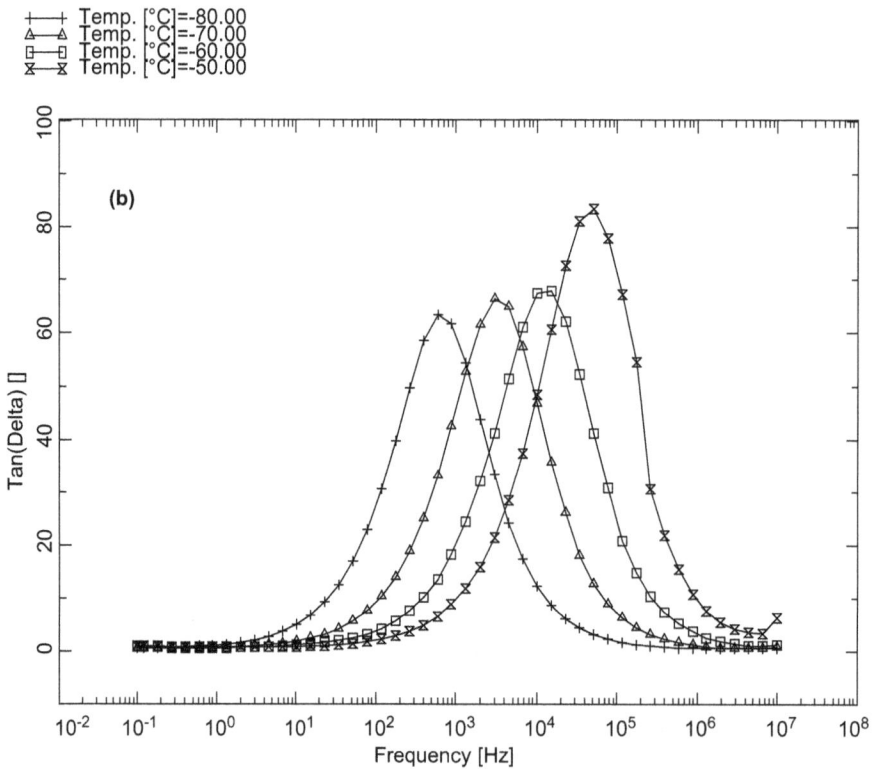

FIGURE 24.8 PAN-PC-EC-PZT system (a) Cole-Cole and (b) Tan δ plot.

reinforcement and less the number of charge carriers and subsequently reduces the total ionic conductivity of the composite gel electrolyte.

Figure 24.5b, 24.6b, 24.7b, and 24.8b shows Tan δ spectra for PAN-PC-EC gel and gel with Al_2O_3, PZT, and BT reinforcements from −80 to −40°C in the operated frequency range of 1Hz to 1MHz. The position of the peak is attributed to the relaxation frequency of the solvent molecules present in the gel electrolyte. The relaxation frequency increases with increase in temperature. It is well known that near T_g, the relaxation frequency decreases due to the higher viscosity of the solvent molecules at that temperature. The highest relaxation frequency at near T_g values is observed for the PAN gel with PZT reinforcement. As mentioned earlier, in the presence of high dielectric field, the solvent molecule easily relaxes when compared to the low electric field. The results are in excellent agreement with the DSC studies, where the PAN gels with the PZT system show very low glass transition temperature.

Arrhenius plot for all the systems obeys VTF relation (Souquet et al. 1996). Figure 24.9 shows Arrhenius plot for PAN-based composite gel electrolytes. The relation describes the transport properties in a viscous matrix and supports the fact that ion moves through the solvent-rich phase in a polymer gel electrolyte. Activation energy and glass transition temperature are obtained from the VTF relation which is shown in Table 24.2. As mentioned earlier the decrease in DC conductivity is shown in the Arrhenius plot.

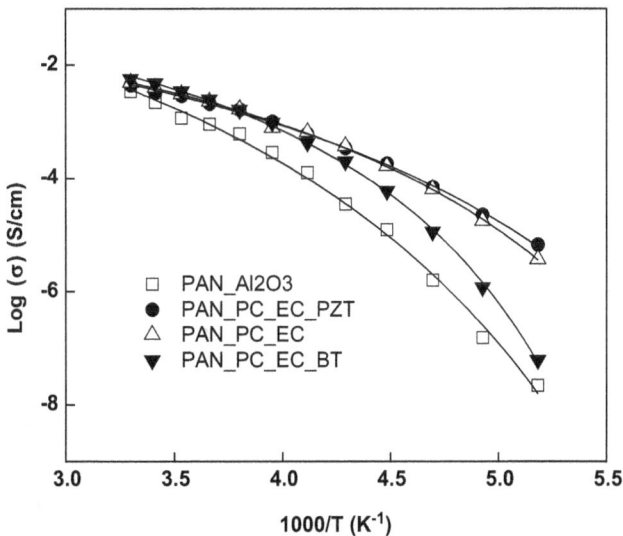

FIGURE 24.9 Arrhenius plot of the PAN-based gel electrolytes. Solid lines are Vogel –Tammann –Fulcher (VTF) fit to the data.

TABLE 24.2 DC Conductivity and VTF Fit Data for PAN-based Composite Gel Electrolytes

Code	σDC at RT (S/cm) ($\times 10^{-3}$)	Ea (kJ/mol)	To (K)
PAN_PC_EC	4.01	2.32	159
PAN_Al$_2$O$_3$	0.61	2.48	171
PAN_PC_BT	0.89	4.01	169
PAN_PC_PZT	3.42	4.30	162

24.6 NMR analysis of PAN-Based Composite Gel Electrolytes

The environment of the lithium ion in the gels and its mobility in the PAN-PC-EC gel and gel with Al_2O_3, PZT, and BT reinforcements were studied by ^7Li NMR spectroscopy. Figure 24.10–24.13 shows ^7Li NMR spectra at 303K for PAN-PC-EC gel and gel with Al_2O_3, PZT, and BT reinforcements. The NMR spectra were deconvoluted with Lorentzian functions. It can be seen from the fit for PAN gel that the spectra is composed of three components: a narrow and intense peak in upfield side, followed by a two broad low-intensity peaks at lower fields. The positions of the peaks at room temperature for PAN21_PC_EC

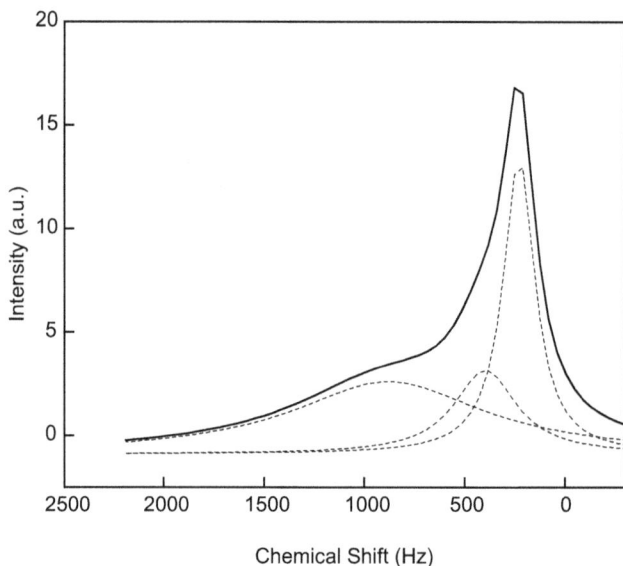

FIGURE 24.10   NMR spectra for PAN gel electrolytes.

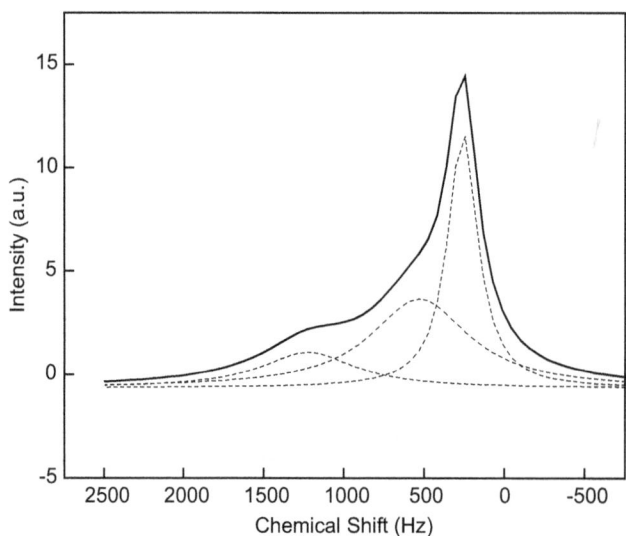

FIGURE 24.11   NMR spectra for PAN gel with Al_2O_3 reinforcement.

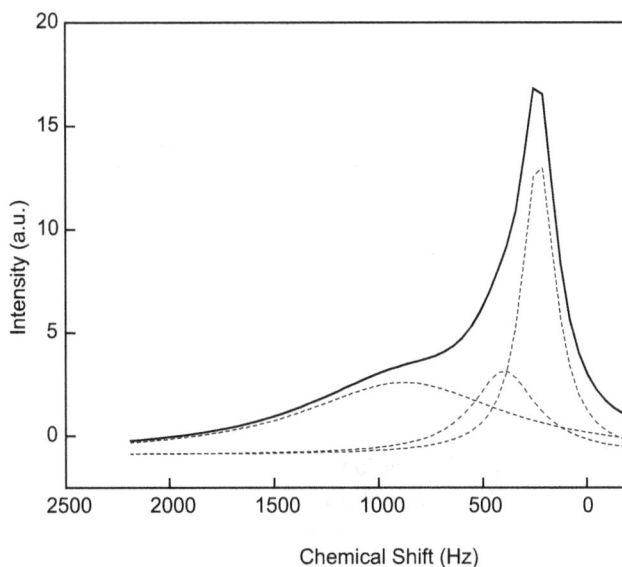

FIGURE 24.12 NMR spectra for PAN gel with BT reinforcement.

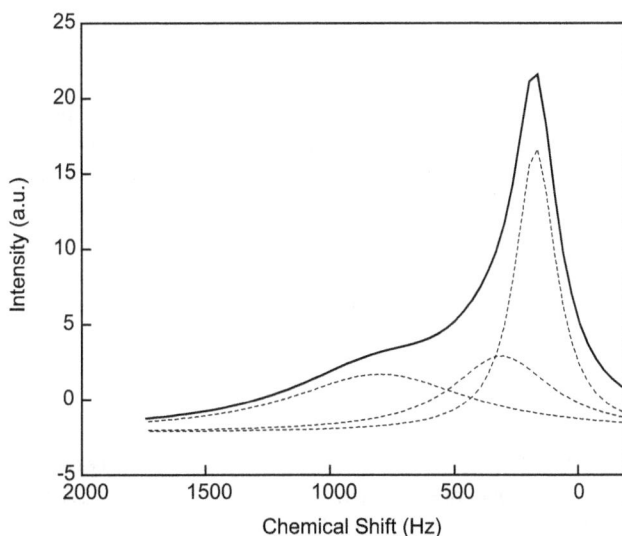

FIGURE 24.13 NMR spectra for PAN gel with PZT reinforcement.

solvent are shown in Table 24.3. Each of the components or resonance frequency in the spectrum represents Li^+ ions in different chemical environments. On the basis of comparison of these peaks with the reported for Li^+ ions (Stallworth et al. 1995), these environments were attributed as follows:

Type 1 Li^+: The most shielded, narrow, and high intense peaks seen near 200 Hz are attributed to the Li^+ ion in the solvent environment. Since Li^+ ions in the solvent environment are expected to move faster than in any other environment. It may be noted that the higher the mobility of ions, the narrower the peaks (Croce et al. 1993).

TABLE 24.3 Chemical Shifts and Line-widths of the Three Types of Li⁺ ions in PAN-PC-EC Gel and Gel with Al_2O_3, PZT, and BT Reinforcements

	PAN21_PC_EC		
Temp. (K)	Type 1	Type 2	Type 3
Chemical shift			
PAN gel	233	367	842
PAN-Al_2O_3	282	538	1215
PAN-BT	280	402	892
PAN-PZT	275	380	782

Type 2 Li⁺: The least shielded and low-intensity peak seen near 400 Hz is attributed to Li⁺ ions coordinated to solvent at the vicinity of the ceramic reinforcement. This peak is the broadest and this Li⁺ ion is expected to be the slower than the ion present in the solvent.

Type 3 Li⁺: The broad and low-intensity peak seen above 600 Hz is attributed to Li⁺ ions coordinated to the polymer at the vicinity of the ceramic reinforcement. The mobility of Li⁺ ion is expected to be slowest among the above-mentioned two types of Li⁺ ions.

It can also be seen that the type 1 Li⁺ ions being coordinated to most polar ligand are expected to be most shielded. This is reflected in its chemical shift that is lowest. In a similar manner, the chemical shifts of types 2 and 3 may be explained.

24.7 Diffusion Coefficient

The predominant lithium resonance peak was further characterized by pulsed-gradient spin-echo (PGSE)-NMR in order to get the diffusion coefficient. The conductivity value obtained from diffusion coefficient of Li in PAN-PC-EC gel and gel with Al_2O_3, PZT, and BT reinforcements is shown in Table 24.4. Table 24.4 shows comparison of conductivities obtained from NMR (calculated using Nernst–Einstein equation) (Williamson et al. 1999 & Dunst et al. 2014) and impedance spectroscopy. The conductivity values obtained from NMR agree well with the total conductivity obtained from impedance spectroscopy. The table clearly indicates that the Li ions mostly transported through the solvent domain of the gel electrolytes.

24.8 SEM Studies

Morphological studies on the PAN-based gels have been carried out on freeze-fractured gels in an environmental scanning electron microscope (ESEM). The pressure in the ESEM sample chamber was maintained at 1 mbar. The morphology of the fractured surface of PAN-PC-EC gel and gel with Al_2O_3, BT,

TABLE 24.4 Conductivity Obtained from NMR and Impedance Spectroscopy for PAN-PC-EC Gel and Gel with Al_2O_3, PZT, and BT Reinforcements

Code	σDC at RT (S/cm) ($\times 10^{-3}$)	σDC at RT (S/cm) ($\times 10^{-3}$) (Calculated from Diffusion Coefficient)
PAN_PC_EC	4.01	4.84
PAN_Al_2O_3	0.61	1.23
PAN_PC_BT	0.89	1.27
PAN_PC_PZT	3.42	3.98

and PZT reinforcements are shown in Figure 24.14, 24.15, 24.16 and 24.17, respectively. The morphology of PAN_PC_EC gel exhibits a plain structure without any pores. The pores in the gels are expected to exude from the gel and reduce the conduction of Li$^+$ ions. The morphology of PAN gel with ceramic reinforcement exhibits similar featureless behavior except for the reinforcement which appears as bright spot. The morphology clearly depicts the uniform distribution of reinforcement in polymer matrix.

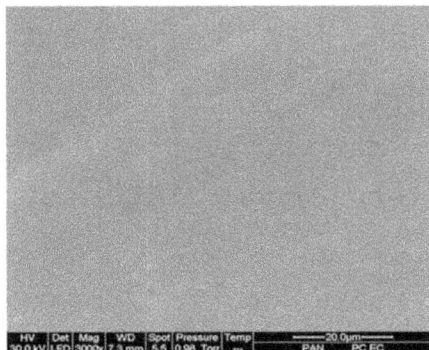

FIGURE 24.14   SEM morphology for PAN gel electrolyte.

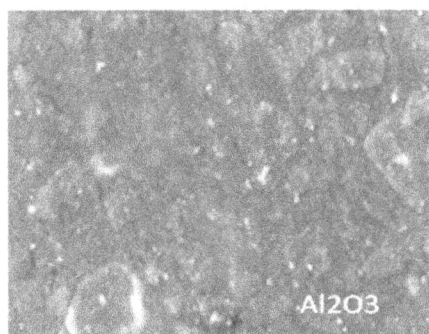

FIGURE 24.15   SEM morphology for PAN gel electrolyte with Al$_2$O$_3$.

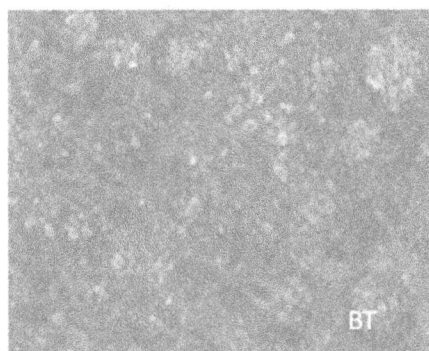

FIGURE 24.16   SEM morphology for PAN gel electrolyte with BT.

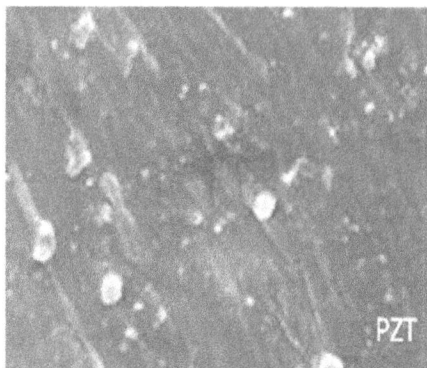

FIGURE 24.17   SEM morphology for PAN gel electrolyte with PZT.

24.9 Conclusion

PAN composite gel electrolytes based on (PZT)-12000, BT-1000 and Alumina (Al_2O_3) have been reported, with a special focus on their structural and conductivity studies. PAN composite electrolytes with various solvents are known to contain crystal domains. However, wide angle x-ray diffraction (WAXD) patterns of PAN composite with different solvents show amorphous nature in the presence of $LiClO_4$. IR spectra of PAN with solvents showed peaks at 715 cm^{-1} and 893 cm^{-1} attributed to ring bending and the ring breathing modes of the solvent, respectively. These peaks shift to higher frequency when Li salt is added to the solvent indicating the possible coordination. In the presence of PAN, these peaks further shift to a higher frequency and confirm complex formation of Li^+ ion with both polymer and solvent. The peak at 2243 cm^{-1} attributed to C°N in the polymer chain is seen to shift to higher frequencies with the addition of lithium salt. This indicates that Li^+ ion is coordinating with the PAN matrix. The association of Li^+ with PAN inhibits its mobility and leads to a decrease in its mobility and consequent decrease in the conductivity of the system.

ESEM results suggest the ceramic particles are homogeneously dispersed on the surface of PAN-based composite electrolyte. The morphology of PAN gel with ceramic reinforcement exhibits similar featureless behavior except for the reinforcement which appears as a bright spot. The morphology clearly depicts the uniform distribution of reinforcement in polymer matrix. The PAN composite with PC-EC solvent shows the highest conductivity (3.42×10^{-3} S/cm) at room temperature. The environment of the lithium ion and its mobility in the composite electrolytes were studied by ^7Li NMR. NMR spectra of composite electrolyte exhibit three peaks at 275, 380, and 782 Hz suggesting three different Li^+ environments. On the basis of comparison of these peaks with the reported for Li^+, these environments were identified to be (1) solvation shell, (2) vicinity of polymer chain in the presence of ceramic reinforcement, and (3) interactions of Li^+ ions with both solvent and polymer. The conductivity value calculated from diffusion coefficient (3.98×10^{-3} S/cm) for the Li^+ ion present in the solvent agrees well with the total conductivity value obtained from impedance spectroscopy (3.42×10^{-3} S/cm). The increase in dielectric constant of ceramic reinforcement increases the diffusion coefficient of Li ion. Conductivity and glass transition temperature of composite electrolytes are reported in Table 24.5.

TABLE 24.5 Conductivity and Glass Transition Temperature of Composite Electrolytes

Polymer	Reinforcement	σ at RT ($\times 10^{-3}$ S/cm)	T_g (K)
PAN	Al_2O_3	0.61	171
	BT	0.89	169
	PZT	3.42	162

References

Croce, F, Brown, SD, Greenbaum, SG, Slane, SM & Salomon, M 1993, *Chemistry of Materials*, vol. 5, pp. 1268.

Croce, F, Gerace, F, Dautzemberg, G, Passerini, S, Appetchchi, GB& Scrosati, B 1994, *Electrochimica Acta*, vol. 39, pp. 2187.

Dunst, A, Epp, V, Hanzu, I, Freunberger, SA & Wilkening, M 2014, *Energy & Environmental Science*, vol. 7, pp. 2739–2752.

Durig, JR, Clark, JW & Casper, JM 1970, *Journal of Molecular Structure*, vol. 5, pp. 67.

Jayathilaka, P, Dissnayake, ARD, Albinsson, MAKL & Mellander, IBE 2003, *Solid State Ionics*, vol. 156, pp. 179.

Song, JY, Wang, YY & Wan, CC 1999, *Journal of Power Sources*, vol. 77, pp. 183.

Souquet, JL, Duclot, M & Levy, M 1996, *Solid State Ionics*, vol. 85, pp. 149.

Stallworth, PE, Greenbaum, SG, Croce, F, Slane, S & Salomon, M 1995, *Electrochimica Acta*, vol. 40, no 13, pp. 2137.

Wesley, A 2003, *Henderson, Stefano Passerini*, vol. 5, no 7, pp. 575–578.

Williamson, MJ, Hubbard, A & Ward, IM 1999, *Polymer*, vol. 40, no 26, pp. 7177–7185.

Yang, CR, Perng, JT, Wang, YY & Wan, CC 1996, *Journal of Power Sources*, vol. 62, no 1, pp. 89–93.

25

High Gain Modified Luo Converter for Nano Capacitor Charging

Dr. S. Vijayalakshmi
and
Dr. M. Marimuthu

Saranathan College of Engineering, Trichy, India

Dr. N. Vengadachalam

Malla Reddy Engineering College for Women, Hyderabad, India

Dr. V. Subha Seethalakshmi

Sreyas Institute of Engineering and Technology, Hyderabad, India

Dr. M. Sangeetha

M.A.M. School of Engineering, Trichirappalli, India

J. Rekha

Sathyabama University, Chennai, India

The construction of a DC–DC converter is made possible by the rapid expansion of nano capacitors in the automotive industry. This chapter suggests a brand-new boosted gain Super Luo DC–DC converter. A modification topology is recommended for cascaded converters. Any scheme for a reliable capacitor can be modified using this suggestion. Before the load, the converter that is being described increases the voltage conversion ratio by including boost (voltage-multiplier) cells can be added. Exceptional qualities of this converter include an enhanced gain voltage, minimum stress voltages on power switches and capacitors, and a low component count. The proposed topology's operating theory and steady-state analysis are presented. Reduced capacitor stress is the suggested topology's key characteristic, which is assessed by a reliability analysis using the military manual. The correlation between minimum capacitor voltage stress and capacitor failure is explained in detail. The theoretical conclusions are validated by simulation of the resulting topology using MATLAB/Simulink. Additionally, the suggested topology is expanded to include dual output, which qualifies the modernized converter for use in supercapacitor charging electric vehicle applications. The simulation of the output circuit is carried out and evaluated. The experimental findings from the 50 W miniature model support the thorough steady-state evaluation done on the suggested circuit.

25.1 Introduction

The usage of petroleum fuels in automobiles has resulted in emissions of greenhouse gases and other hazardous substances that have significantly increased environmental risks, health problems, and price volatility in the current situation. Electric cars (EVs) and electric hybrid vehicles (EHVs) have recently become alternative options for intercity transportation due to a number of benefits including enhanced efficiency, easy and cheap maintenance, reduced noise and environmental pollution, and low cost-effectiveness. The most popular source of power for EVs and EHVs is now pollution-free battery power, which can be recharged at battery charging stations. However, there will likely be power quality issues in the current grid due to the exponential expansion of electric vehicles in the future [1–3].

Power quality issues, in general, harmonic pollution in the power system distribution system, are of growing concern owing to the widespread practice of the power electronics concept for battery charging stations. As more battery charging stations are installed in metropolitan areas, there is a heightened harmonics level risk in power transmission and distribution, and transformers in the distribution side whose lifespan is shortened. Traditionally, pulsed sinusoidal waveform is obtained from bridge rectifier with capacitor filter. Abridge rectifier circuit continued by a large filter capacitor draws a heavily distorted, non-sinusoidal, and has a low power factor (PF) of the order of 0.7–0.85 lagging and a high peak factor. This supply current has a high harmonic content of total harmonic distortion (THD) as 70–80% [4–6].

The ideal electronic energy storage system would have plenty of energy storage capacity, charge quickly, and deliver potent bursts when required. Sadly, modern electronics only have one of these capabilities: high power is provided by capacitors, while high storage is provided by batteries. The nano capacitor or the supercapacitor is the one that carries both works together. These kinds of capacitors require high voltage for charging, those high voltages can be supplied by a modified super-lift Luo converter.

25.2 PF Improvement Methodology for AC and DC

The AC voltage source power factor is the proportion of real power VICosϕ (measured in watts) passing into the load to apparent power VISinϕ measured as the sum of current and voltage. PF = VICosϕ/VISinϕ is how it is expressed (VA). It is an integer between 0 and 1, as the PF equation demonstrates. PF equals 1 when current and voltage are sinusoidal and in phase. The VISinϕ power is greater than the VICosϕ when both are sinusoidal but out of phase, and power factor is equal to the cosine angle of the current and voltage waveforms. When the load is linear and pure resistive in practice, PF = 1 represents the optimum scenario. Off-line AC/DC power sources that are seen in an electronic circuit are actually switched-mode devices that provide a nonlinear load [7–9].

The fact that switch-mode power supplies are the norm today causes them to draw a non-sinusoidal waveform, which causes a phase difference among an input current and input voltage. A PF lower than 1 is the outcome when the voltage waveform and the current waveform do not match. Along with power losses, harmonics produced by the neutral line interfere with other AC mains-associated devices. The harmonics content of the AC line increases with a decreasing PF number and decreases in the opposite direction.

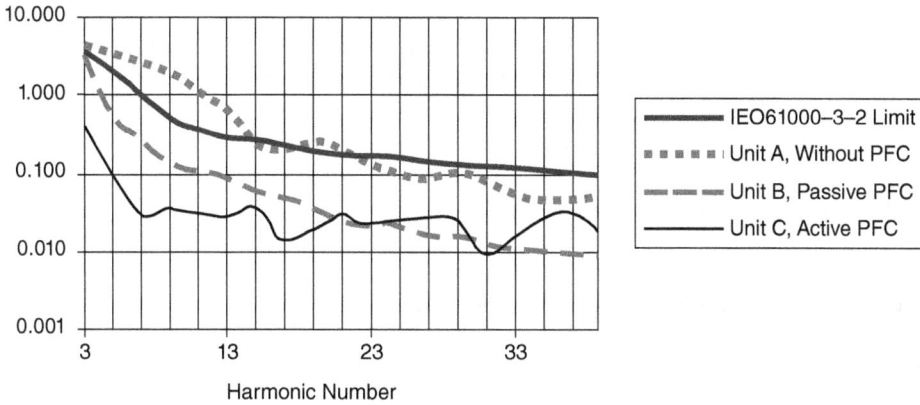

FIGURE 25.1 Power factor corrections graph.

25.2.1 Passive PFCs

Use of the passive filter can only permit ampere current at line frequency which is the simplest method of regulating the harmonic current (e.g., 50.0 or 60.0 Hz). Since the reduction of the current harmonic, the nonlinear device now resembles a linear load. Power factor can be brought close to unity by using filters constructed from capacitors and inductors. The filter's disadvantage is that it needs a high-voltage capacitor and an inductor with a large value and high current, both of which are cumbersome and expensive [10–12].

In contrast, a power supply with an active PFC controller outperforms a passive PFC supply to go above the mains line harmonics requirements outlined in IEC61000-3-2. Figure 25.1 relates the harmonics input for 3 different 250 W PC power supplies with the permissible levels. Conferring to these devices' input power, the harmonic amplitudes are proportional. As demonstrated in this graph, the passive PFC's performance just about meets the third harmonic limit. IEC-61000-3-2 specifications are met and exceeded by the active PFC unit.

Passive PFC circuits have some drawbacks despite being straightforward to design and operate. The inductor's usability in many applications is first and foremost limited by its size. The second need is a line-voltage range switch for global operation. The switch's inclusion creates the appliance or system vulnerable to user error if the switches are not appropriately chosen. At last, because the voltage is not controlled, the DC/DC converter that comes after the PFC step suffers in terms of cost and efficiency [13–15].

25.2.2 Existing System

Conversely, traditional DC–DC converter such as buck, boost, single-ended primary inductor converter (SEPIC), isolated, and Cuk converters have the capacity to support both power factor corrections functionality and step-up and step-down conversion capability.

The non-isolated converter output voltage can be either step-up or step-down than the input, and the buck-boost converter whose output is the inversion of the input. In contrast to the majority of other converter types, which use inductors as their primary energy storage device, it uses capacitors. Slobodan UK, a researcher at the California Institute of Technology, is honored by having his invention named after him [16–17].

The electrical potential drop at the output of a SEPIC, a type of DC-to-DC converter, may be greater, lower, or equal to that at the input. The control transistor's duty cycle regulates how the SEPIC outputs data.

In essence, a SEPIC is a buck-boost converter with non-inverted output, as a result, it is like a conventional buck-boost converter but has the benefits of having a non-inverted output (the output which follows the polarity of the input), by means of a capacitor connected in series to couple the input energy to send it to the output side (i.e., it could act perfect fully and transfer the energy from input side to output side), and being able to perform transferring on all conditions.

25.3 Existing System Block Diagram

A prominent methodology frequently used in the design of electrical circuits is the voltage lift technique. The impact of parasitic components restricts the efficiency and load voltage of the DC-to-DC converter is illustrated in Figure 25.2. The voltage-boost approach offers a useful means of enhancing the properties of the circuit. A group of novel DC–DC step-up converters called Luo converters were created using the voltage lift approach from prototypes [18].

In this paper, voltage-boost converters are examined. Luo converter with successful output. These converters have an enhanced voltage output with few ripples, a high power density, excellent efficiency, and a straightforward design. In particular, for high-output voltage projects, these converters are employed extensively in industrial appliances and computer peripheral tools. Numerous photovoltaic applications can make use of these converters. To emphasize the benefits of the converters, simulation data are provided.

25.3.1 Proposed System Block Diagram

One power switch, one inductor, and two diodes make up the Luo converter. Additionally, a filter is present to lower the output voltage's harmonic content. Inductor is activated when the power switch is flipped on, and voltage is then sent through diodes to the capacitor output is depicted in Figure 25.3.

The capacitors are used to deliver the load's voltage. Feedback to the peripheral interface controller is provided by the output voltage, along with set and real voltages. The output potential is identical to the set voltage since the set voltage is fixed. The voltage value that was set with the key operation. • Pulse generator: To create the pulse width modulation signal in this case, we used a NANO microcontroller [19].

- Driver circuit: This circuit amplifies pulses and uses an opto coupler to provide isolations. It serves two purposes:
- Amplification
- Isolation

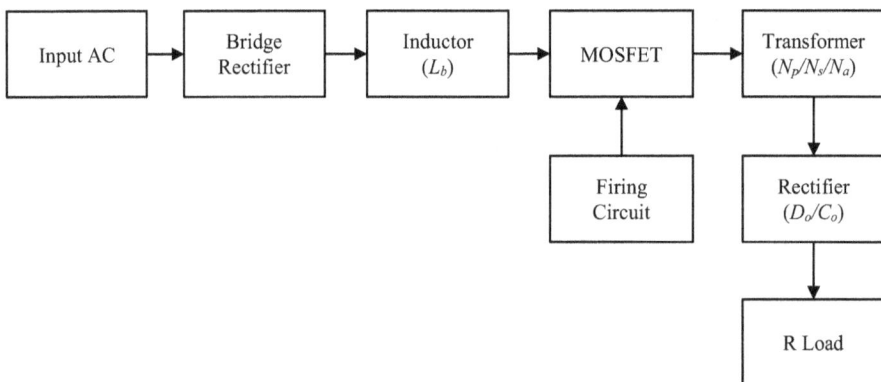

FIGURE 25.2 Proposed system block diagram of the existing system.

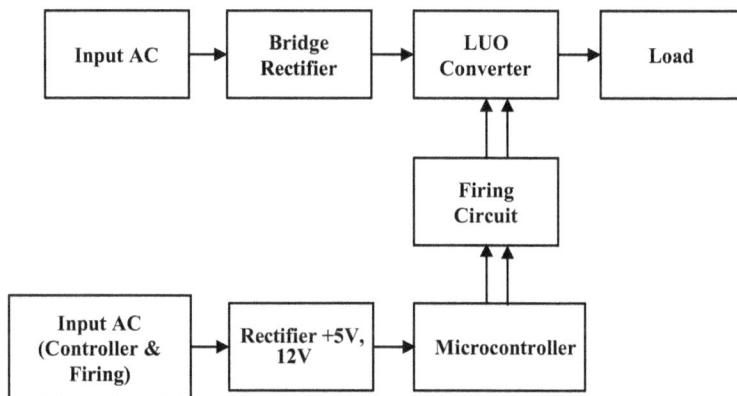

FIGURE 25.3 Proposed circuit block diagram.

- Bridge Rectifier: This device transforms AC power into DC power.
- The Luo converter transforms a low-potential DC supply into a high-potential DC source.

25.4 Proposed Circuit Topology

Section 3: The voltage lift approach is a well-liked and frequently used technique in electronic circuit design. With this method, the output voltage is significantly increased while parasitic elements are successfully overcome. As a result, these kinds of converters increase DC–DC voltage with a high power concentration, enhance efficiency, and high output potential with minimal ripples.

The voltage lift (VL) method was extensively applied to the design of electronic circuits. All DC-to-DC converters have a limited ability to transfer power and have a limited output potential owing to the influence of parasitic components. The output voltage increases arithmetically, which is one further drawback of the VL technique. The SuperLift (SL) scheme has emerged alongside the development of the conversion technique and has proven to be more potent than the VL technique. Super-Lift Luo converters, which are based on super-lift technology, are frequently used to generate high output voltages. The Super-Lift Luo Converter has a number of benefits [20]. It has a very high-voltage transmission gain, to start. A geometric progression can be used to raise the output voltage. Second, it has a high power density and efficiency. Additionally, the Super-Lift Luo Converter has the capacity to lessen ripple current and voltage.

Mode – I
When the switch is ON, an inductor (L) and a capacitor (C1) are receiving the dc output directly. When switch S – ON, the potential difference across capacitor C1 is charged to the input voltage. The current I_{L1} will rise with voltage V_{in} since an inductor L, and a capacitor C1 are c in parallel. To the voltage of Mode 2, the inductor charges.

Mode – II
In mode 2, the switch is in the OFF position, resulting in a potential difference across an inductor L is (Va-2Vin)*I= -(Va-2Vin)*(1-kT)/L. Thus, the present IL1 will drop. The switch-on and switch-off periods are presumed to be kT and (l-k) T, respectively.

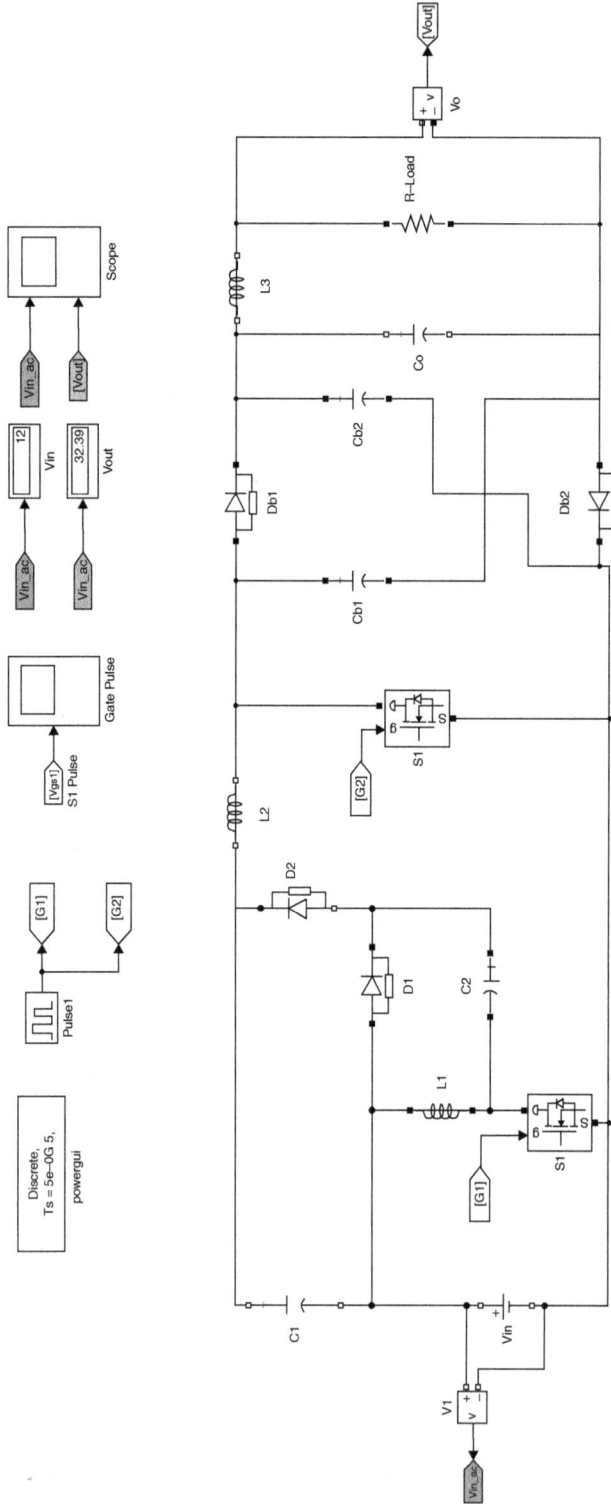

FIGURE 25.4 Proposed circuit topology.

The transfer functions obtained using the state space analysis techniqueand the circuit averaging technique are identical. The system's stability could be assessed using the derived transfer function. Additionally, it is employed to get the system's closed response. Bode plots created from these transfer functions illustrate the converter's operating frequency while the system's stability is indicated by the phase and gain margins.

25.5 Hardware Requirements

25.5.1 Microcontroller's Power Supply Section

Step-down Transformer
Depending on the required DC value, the transformer output can be either stepped up or stepped down the source AC voltage.

This circuit uses a transformer with a step-down voltage of 230V/12-0-12V to convert 230V AC into 12V AC across the secondary winding. The transformer's top becomes positive and its bottom becomes negative when the input is changed in any way. The next change will momentarily have the opposite effect. The transformer used in this project has a current rating of 500 mA. Along with providing isolation among the main circuit and power supply unit, it steps down AC voltages.

Diode
Rectifier diodes are used in power supplies to rectify AC into direct current (DC), which only allows electricity to flow in one direction. A bridge rectifier comprised of four diodes (4*1N4007) is required to achieve full wave rectification. Four diodes will operate in total: two during the negative cycle and two during the positive half cycle. The 1N4007 is suitable for the majority of low-voltage circuits with currents under 1A.

Filtering Unit
Following the rectifier unit are always has filter unit, it typically has a capacitor working as a surge arrester. This capacitor, which is also known as a decoupling capacitor or a bypassing capacitor, is used to leave the output's DC frequency unaffected while also shorting the ripple with a 130Hz frequency to ground.

- 1000 µF: to reduce pulsating's ripple effects;
- 100 µF: to avoid high-frequency disturbances

Voltage regulator
Voltage regulators are crucial components of any power supply unit. A regulator's main function is to assist the rectifier unit and filter unit in supplying the device with a consistent DC voltage. Without regulators, power supplies are prone to having their DC voltage values change as a result of changes in the load or variations in the AC linear voltage. The voltage can be kept near to the desired output with the help of a regulator linked to the DC output.

Fixed voltage regulators:
a. **Positive voltage regulator**:
   The three-terminal positive voltage regulators in the 78xx series come in seven different voltage configurations. These integrated circuits are fixed voltage regulators by design, and with sufficient heat dissipation, they can handle output currents of more than 1 A.

   We used the IC voltage regulators 7812, 7815, and 7805, which produce the voltages of +12 V, +15V, and +5 V, respectively, in the output.

b. **Negative voltage regulator**:

Devices from the 78xx family are supplemented by the 79xx series of fixed negative output voltage regulators. Additionally, three-terminal devices, these regulators are. Ground, input, and output are the three terminals. For an output voltage of -12 V, we used an IC voltage regulator called the 7912. Because voltage regulators are easy to use, dependable, affordable, and exist in a range of voltage and current ratings, almost all power supply employ some sort of voltage regulator IC.

These ICs have a fixed voltage regulator type and are made of monolithic silicon, which provides low cost, high reliability, size reduction, and great performance. In this regulator, the inductive effects brought on by lengthy distribution leads are often cancelled by connecting a capacitor between the input terminal and ground. The output capacitor enhances transient performance. 7805, 7812, and 7912 are used to adjust the filtered DC voltage as a result.

Resistor

Resistors are "Passive Devices," which means they don't have a power source or an amplifier inside of them and merely attenuate or weaken the voltage signal that passes through them. As a result of this attenuation, heat is produced as the resistor obstructs the flow of electrons through it.

25.5.2 Microcontroller – Arduino

Arduino is a user-friendly software-based open-source prototyping platform. It offers developers a versatile foundation on which to experiment while creating interactive environments. They can be programmed for particular uses to produce embedded systems that can sense and control real-time parameters. It is made up of a microcontroller, the ATmega328, which is programmed using the Arduino software.

25.5.3 Firing Circuit

25.5.3.1 6N137

An 850 nm AlGaAS - LED is optically linked to an integrated photodetector logic gates with a very high speed and a storable output in the 6N137, HCPL2601, and HCPL2611 single-channel opt couplers. A wired-OR output is possible thanks to the open collector of this output, as shown in Figure 25.5.

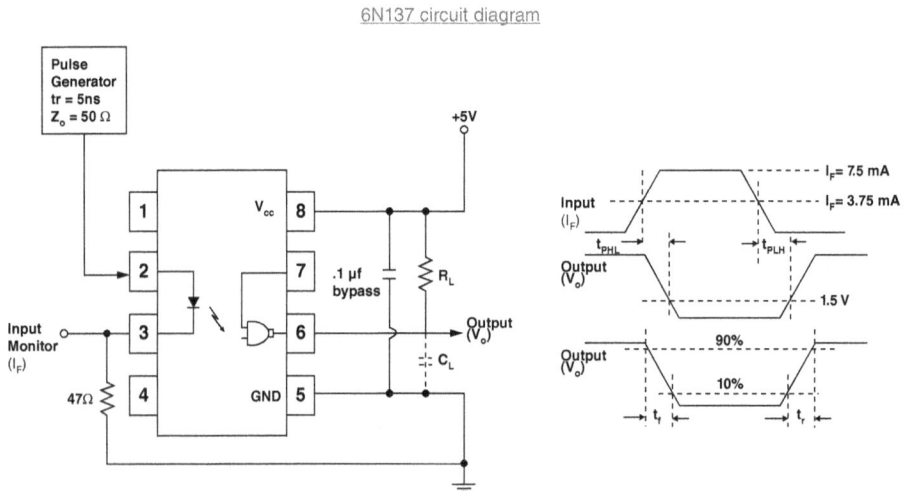

FIGURE 25.5　Opto-isolator interface circuit.

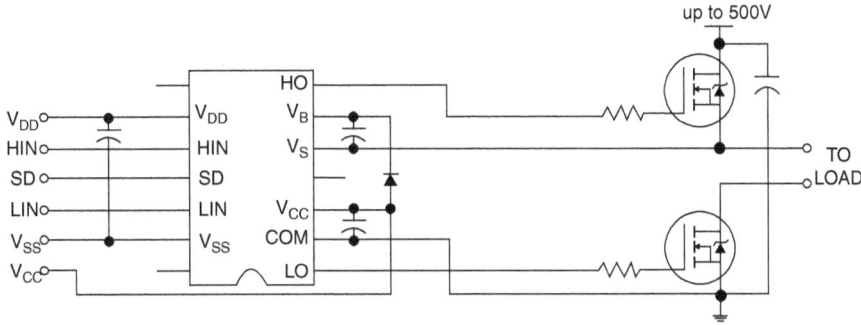

FIGURE 25.6 MoSFET driver circuit.

With the help of driver circuit, one can provide the gate pulse to the MoSFET. Because there is an impedance mismatching between the gate pulse circuit and MoSFET. If one gives the pulse to the MoSFET directly, either the MoSFET will damage or half of the pulse only reach to the gate of the MoSFET. Hence, driver circuit is most essential for the MoSFET. The MoSFET circuit is depicted in Figure 25.6.

25.6 Results and Discussion

25.6.1 Proposed Simulation

The proposed modified super-lift Luo converter was simulated using MATLAB software. The 12V input voltage provided to the input of the converter, 10KHz, 40% duty cycle of the gate pulse gave to the gate of MoSFET switch. The obtained output voltage measured through the scope of the software was 32V (Figures 25.7 and 25.8).

25.6.2 Measurements

25.6.3 Proposed Circuit Diagram

The converter circuit uses the controller to produce the PWM pulses. The driver circuit receives input from the controller pulses. The primary functions of a driver board are to separate and amplify the controller's input signals. The components of the primary power circuit are connected to the amplified driver output. Using a bridge rectifier, the ac supply is changed into dc. Luo converter is also used to enhance the dc voltage. The controller and key control are used to control the boosted dc output voltage.

For the AC-DC system controller circuit, the simulation was run on the positive output Luo converter with the parameters presented in Figure.

25.6.4 Input and Output Voltage Waveform

The modified Luo converter can be calculated and simulated using MATLAB software. The input voltage measured is 12V, and the output voltage measured is 32 V are depicted in Figure 25.9. The gate pulse of 40% duty cycle along with 10KHz provided to the MoSFET driver circuit of the modified Luo converter is illustrated in Figure 25.10.

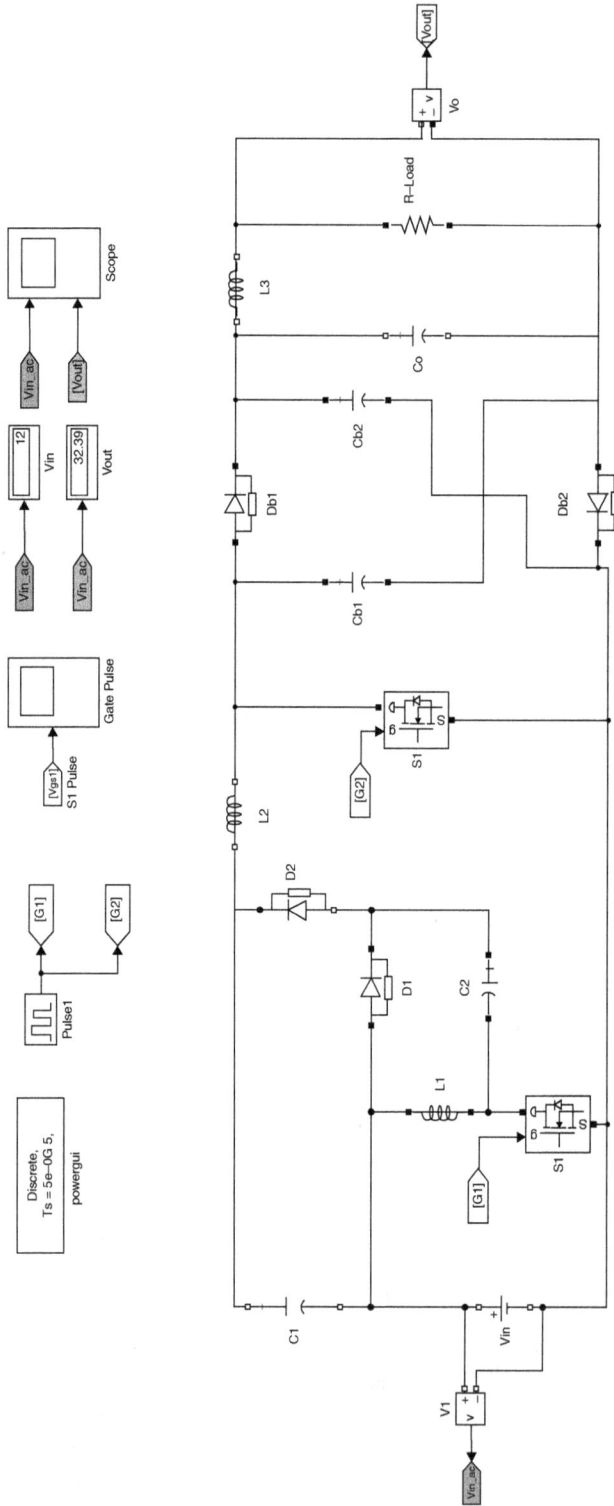

FIGURE 25.7 Simulation diagram of the modified Luo converter.

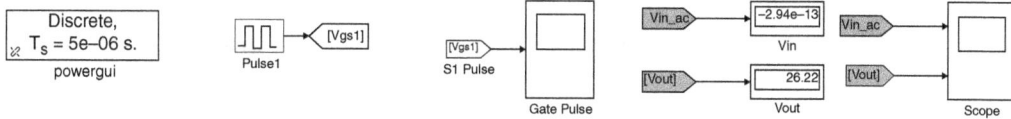

FIGURE 25.8 Output scope.

25.6.5 MoSFET Gate Pulse

25.6.6 Experimental Verification

Built and tested at 15V was the super-lift Luo converter. The proposed circuit's experimental configuration is displayed[5]. R0 = 50, L0 = 150, C0 = 1000, L = 100, and Cin = 0.94 are the circuit's parameters. Figure 25.11 depicts the Arduino NANO-based control circuit. Driving pulses for the MoSFET switches were produced by a NANO microprocessor. The IR2110 driver is used to amplify them. The port pin D9 is wired up to the gate signal.

25.6.7 Hardware

The schematic illustration indicates that the two outputs of this designed super-lift Luo converter could easily be linked to an electric vehicle. The DC power supply of the electric vehicle must be able to supply the same voltage to additional load locations as well as multiple loads with changing voltage levels. Switch S01 controls output voltage V01 whereas Switch S02 controls output voltage V02. The obtained simulation results validate the suggested converter's dual output capabilities. The converter displays two different voltage levels and two outputs with the exact same voltage in this example. In order to assess the performance of the designed topology, a prototype is used to test the MLHG converter. The topology's input and output voltages are 12 V with a 50 W power rating and a 0.25 duty cycle, respectively. Figure 25.13 shows that the actual outcomes are very similar to what was predicted.

25.6.8 Hardware Output

Input Voltage: The input voltage of 19.0 V is given in the hardware circuit. The input voltage is taken from the Data Storage Oscilloscope is given in Figure 25.12.

Output Voltage: The duty cycle of 30%, 10KHz gate pulse has been given to the MoSFET driver circuit, so that the generated output voltage is 38V, as illustrated in Figure 25.13.

25.7 Conclusion

The Luo converter's primary function is to transform low-voltage sources into usable voltage sources using the Luo converter to increase the low voltage for supercapacitor charging and converting the input AC voltage into DC voltage. Luo converter output voltage is thus used to regulate via Arduino NANO controller. Positive source voltage to positive load voltage is converted by the positive output super-lift Luo converter. For a 12V input voltage that has been rectified, it generates an output voltage of 38V. The converter has demonstrated its robustness around the working point, good dynamic performance when input voltage varies, and invariant dynamic performance when running under various situations.

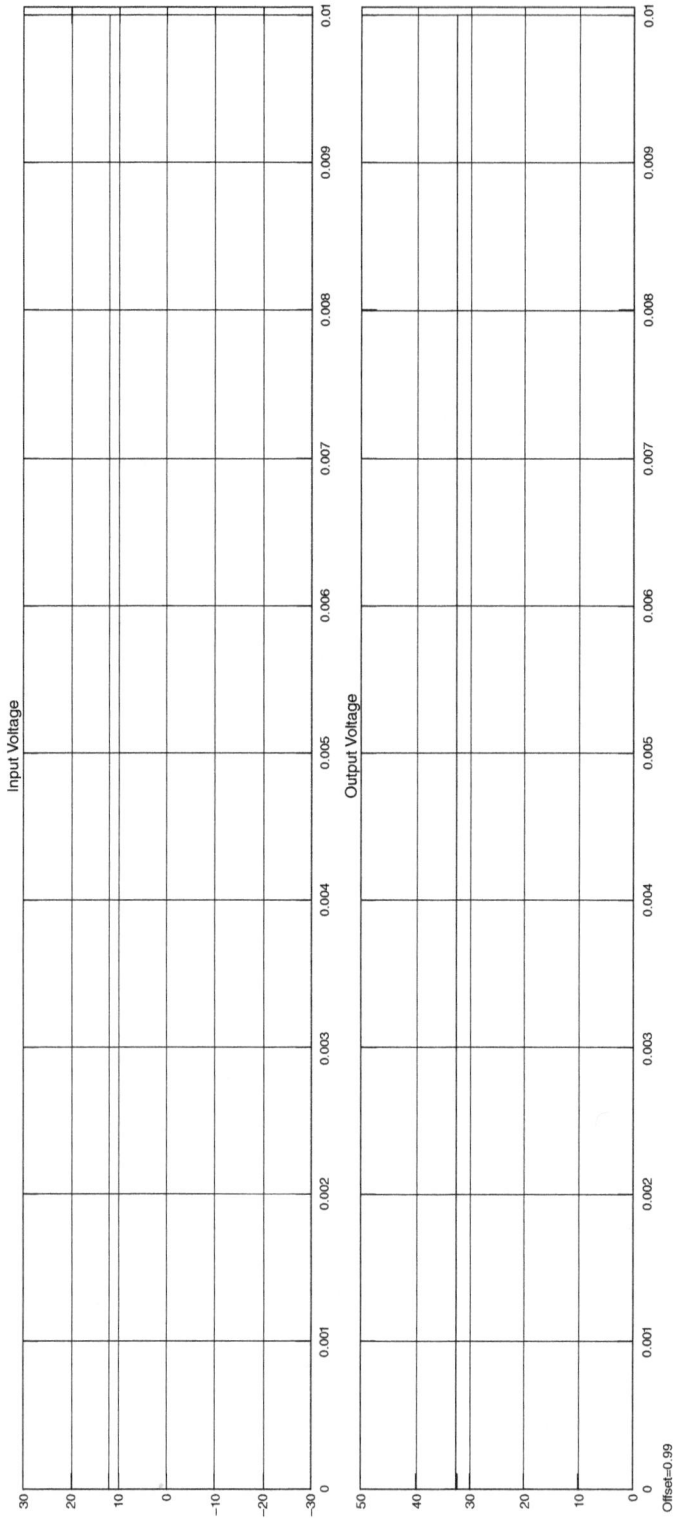

FIGURE 25.9 Input and output voltages.

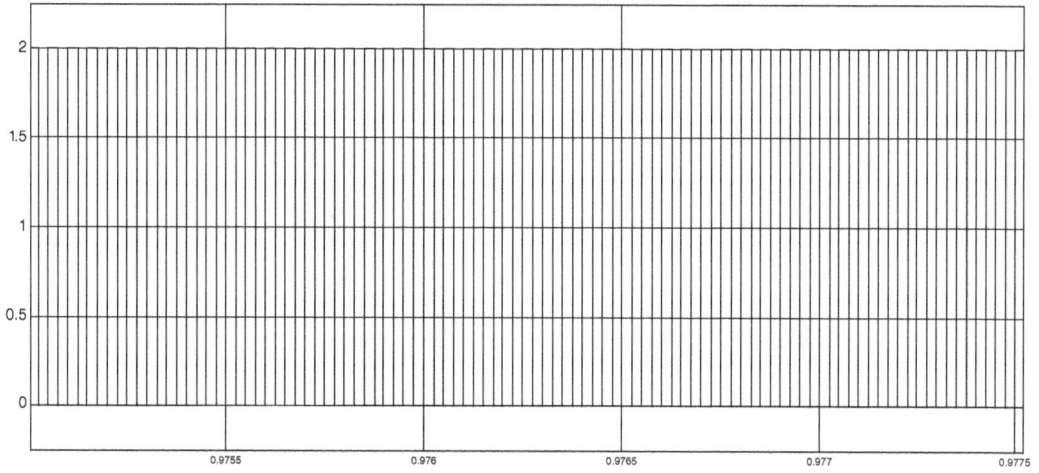

FIGURE 25.10   Gate pulse.

FIGURE 25.11   Hardware of the proposed circuit.

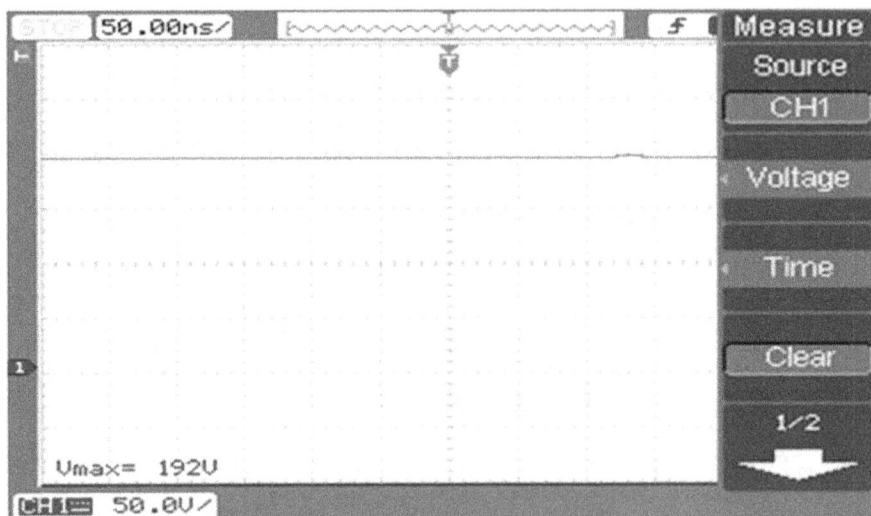

FIGURE 25.12   Input voltage.

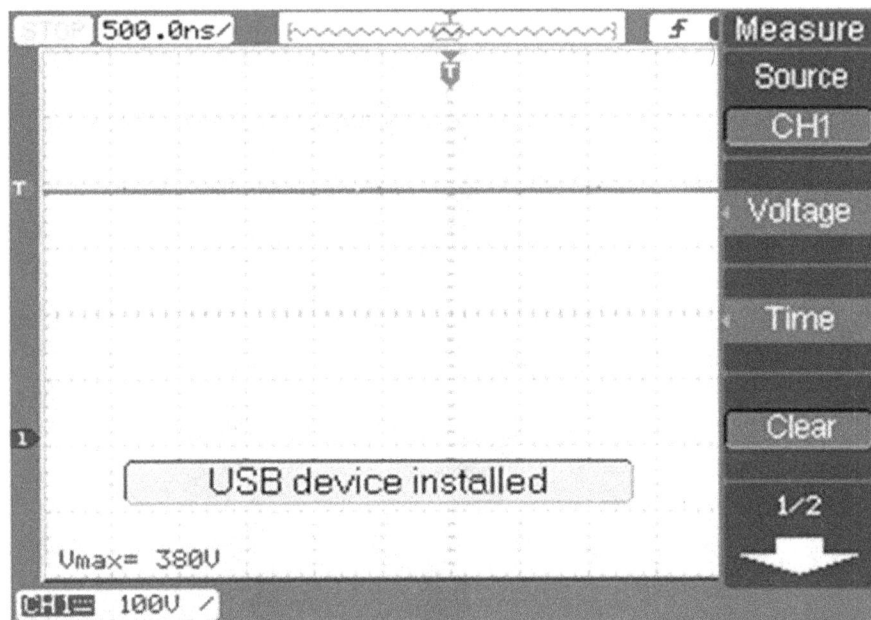

FIGURE 25.13   Output voltage.

References

1. S. Vijayalakshmi, T. Sree Renga Raja, "Robust discrete controller for double frequency buck con-verter," *Automatika*, Vol. 56, No. 3, 2015, pp. 303–317, DOI. 10.7305/automatika.2015.12.647.
2. Behzad Poorali, Ehsan Adib, "Analysis of the integrated SEPIC fly back converter as a single-stage single-switch power-factor-correction LED driver," *IEEE Transactions on Industrial Electronics*, Vol. 63, No. 6, 2016, pp. 3562–3570.

3. S. Vijayaakshmi, T. Sree Renga Raja, "Time domain based digital PWM controller for DC-DC converter," *Automatika*, Vol. 55, No. 4, 2014, pp. 434–445.

4. Dylan Dah-Chuan Lu, Herbert Ho-Ching Iu, Velibor Pjevalica, "Single-stage AC/DC boost–forward converter with high power factor and regulated bus and output voltages," *IEEE Transactions on Industrial Electronics*, Vol. 56, No. 6, 2009, pp. 2128–2132.

5. S. Vijayalakshmi, T. Sree Renga Raja, "Time domain based digital controller for buck-boost converter," *Journal of Electrical Engineering Technology*, Vol. 9, No. 5, 2014, pp. 1551–1561.

6. Rong-Tai Chen, Yung-Yaw Chen, "Single-stage push–pull boost converter with integrated magnetics and input current shaping technique," *IEEE Transactions on Power Electronics*, Vol. 21, No. 5, 2006, pp. 1193–1203.

7. Tomislav Pavlovic, Toni Bjazi, Zeljko Ban, "Simplified averaged models Of DC–DC power converters suitable for controller design and micro grid simulation," *IEEE Transactions on Power Electronics*, Vol. 28, No. 7, 2013, pp. 3266–3275.

8. S. Vijayalakshmi, T. Sree Renga Raja, "Design and implementation of a discrete controller for soft switching DC-DC converter," *Journal of Electrical Engineering*, Vol. 12, No. 25, 2012, p. 8.

9. M. Marimuthu, S. Habeebullah Sait, B. Paranthagan Vijayalakshmi, "Time domain based digital controller for Boost converter," *Journal of Electrical Engineering*, Vol. 18, No. 45, 2018, p. 10.

10. Xueshan Liu, Jianping Xu, Zhangyong Chen, et al., "Single-inductor dual output buck–boost power factor correction converter," *IEEE Transactions on Industrial Electronics*, Vol. 62, No. 2, 2015, pp. 943–952.

11. R. Shenbagalakshmi, S. Vijayalakshmi, K. Geetha, "Design and analysis PID controller for Luo Converter," *International Journal of Power Electronics*, Vol. 11, No. 3, 2020, pp. 283–298.

12. Koen De Gussemé, David M. Van de Sype, Alex P.M. Van den Bossche, et al., "Input-current distortion of CCM boost PFC converters operated in DCM," *IEEE Transactions on Industrial Electronics*, Vol. 54, No. 2, 2007, pp. 858–865.

13. M. Marimuthu, S. Vijayalakshmi, R. Shenbagalakshmi, "A novel non-isolated single switch multilevel cascaded DC–DC boost converter for multilevel inverter application," *Journal of Electrical Engineering and Technology (Springer)*, Vol. 15, 2020, pp. 2157–2166.

14. P. F. de Melo, R. Gules, E.F.R. Romaneli, et al., "A modified SEPIC converter for high-power-factor rectifier and universal input voltage applications," *IEEE Transactions on Power Electronics*, Vol. 25, No. 2, 2010, pp. 310–321.

15. M.G. Umamaheswari, G. Uma, S. Redline Vijitha, "Comparison of hysteresis control and reduced order linear quadric regulator control for power factor Correction using Dc-Dc cuk converters," *Journal of Circuits, Systems, and Computers*, Vol. 21, No. 1, 2012, p. 1250002.

16. M.G. Uma Maheswari, G. Uma, L. Annie Isabella, "Analysis and design of digital predictive controller for PFC Cuk converter", *Journal of Computational Electronics*, Vol. 13, 2014, pp. 142–154.

17. M. Marimuthu, S. Vijayalakshmi, "Symmetric multi-level boost inverter with single dc source using reduced number of switches," *Tehnički vjesnik*, Vol. 27, No. 5, 2020, pp. 1585–1591.

18. Ramprakash Ponraj, Titus Sigamani, Vijayalakshmi Subramanian, "A developed H-bridge cascaded multilevel inverter with reduced switch count," *Journal of Electrical Engineering Technology*, Vol. 16, 2021, pp. 1445–1455.

19. S. Vijayalakshmi, M. Marimuthu, N. Jayakumar, B. Devi Vighneshwari, B. Paranthagan, C. Nisha, R. Shenbagalakshmi, "A novel double frequency SEPIC converter with improved transient characteristics and efficiency," *Journal of Electrical Engineering & Technology*, Vol. 1, No. 17, 2022, pp. 1039–1050.

20. Wei Gu, Dongbing Zhang, "Designing a SEPIC converter," in *Excellent Design Guidelines, National Semiconductor in Application Note*, April, 2008, pp. 1–6.

26

Nanotechnology in Solar Energy

Dr. R. Shenbagalakshmi

*G.H. Raisoni College of
Engineering & Management,
Pune, India*

R. Femi

*Research Scholar, University
College of Engineering, Nagercoil*

Dr. M. Marimuthu

*Saranathan College of
Engineering, Trichy, India*

P. Rathidevi

*JJ College of Engineering and
Technology, Trichy, India*

Dr. S. Vijayalakshmi

*Saranathan College of
Engineering, Trichy, India*

Globally expanding energy demand, rising carbon emissions, global warming, and climate change all contribute to the need for modernization of the present energy industry. The globe is moving toward a more green and sustainable energy industry that will fulfill future energy demands by minimizing the use of raw and non-renewable minerals and resources, as well as lowering energy consumption and pollution. As one of the main technologies of the twenty-first century, nanotechnology has the potential to transform the whole energy sector, contributing to the development of more efficient and sustainable energy systems. Nanotechnology adoption in the energy sector is in various stages of study, development, and deployment. Nanotechnology solutions and techniques can aid in the development of novel methods of producing, changing, distributing, storing, and consuming energy. This chapter seeks to highlight some of the existing nanotechnology uses in the energy sector, emphasizing both the benefits and the dangers and concerns. In addition, this chapter discusses the development of solar cell technologies and the significance of nanotechnology in solar cell formation in detail.

26.1 Introduction

The creation and use of renewable energy technologies are becoming more crucial as the globe faces a severe energy crisis. Over the past two decades, solar energy utility has continuously increased as more

DOI: 10.1201/9781003355755-26

users understand the immense benefits of using solar panels. Sunlight is the source of solar energy. People can use the sun as a resource to generate electricity instead of expensive, polluting traditional electricity since it is inexpensive, renewable, and clean. We can employ solar energy to supply heat, illumination, and other electrically reliant needs for residential buildings. Solar energy is distinct from traditional energy since it does not depend on fossil fuel burning. It pollutes neither the air nor the water. It does not emit greenhouse gases. Due to the above reasons, many people favor solar energy over conventional energy sources. There seems to be an essential need to engage in alternative energy sources that can produce power effectively without contributing to the greenhouse effect. Since there is an increasing rate of extraction of traditional energy sources like oil, gas, and coal, the environmental degradation of ecosystems is being induced.

In addition to the challenge of worsening environmental circumstances brought on by the growth in the urban population and industrial activities, solar energy has gained more attention. It provides a solution to the growing demand for energy. Nowadays, photovoltaic plates are placed using fixed mounting techniques, which are not always parallel to the sunlight. As a result, maximum quantity of solar energy cannot be fully absorbed by photovoltaic plates. It further reduces photoelectric device performance.

The production of highly efficient photovoltaic panels and modules has been rising over the past few years in the photovoltaic cell industry. Furthermore, manufacturing solar panels using present technology for massive energy production is too costly and unreliable. Traditional solar panels have a few drawbacks, high production costs, and poor efficiency. The issue is that the usage of silicon cells lowers the efficiency. The intensity and band gap of the photons from sunshine is sufficient to push one electron out, which is the underlying cause. The excess energy is dissipated when the photon's energy is lower than the energy band gap. Nearly 70% of the solar irradiation on solar panels gets wasted due to the two significant above-said reasons.

On the other hand, many prospective technological advances might offer improved chances for producing solar cells that are both more affordable and effective. Quantum dots (QDs) also have the potential to transform the photovoltaic sector. Nanotechnology is one of the best technologies that can increase the efficiency of photovoltaic panels. At the same time, it significantly lowers the cost of production. Since the surface of modern solar cells will reflect 2% and 10% of incoming sunlight, these cells can lose as much as 10% of their obtained power due to direct optical losses. Using nanotechnology, we can resolve such issue [1–5].

Nanostructures, especially those made of silicon, are typically only a few hundred nanometers in size. They produce an interface between the air and the nanostructure, which causes it to become graded rather than planar. This modification to the solar cell's structure enables light to be precisely directed into the cell and absorbed within as opposed to being reflected away from it.

The thin-film photovoltaic cells' surface has been coated with layers of nanoparticles. Due to their enhanced and advanced physicochemical qualities and outstanding surface-to-area ratio, they provide several benefits. At first, the perceived optical channel for photon absorption is much wider than the real film width due to multiple reflections. Secondly, to reduce the loss due to recombination, the light generated electrons and holes must be transferred along a much shorter distance. Therefore, the absorbance layer of photovoltaic cells with nanostructures film could have a thickness of over 150 nanometers, as contrasted to many micrometers in traditional thin-film photovoltaic modules. The other benefit is that, by using different-sized nanoparticles, different layers of energy band gaps are constructed as per desired applications [20, 21].

There are numerous additional advantages to using nanomaterial in the production of solar modules. By creating flexible rolls rather than definite crystalline modules, then the implementation costs can be significantly decreased. Although nanotechnology-based solar cells are currently less effective than traditional ones, they are less expensive. Nanomaterial-used solar cells may eventually be more affordable. Moreover, if manufacturing uses QDs, there is a possibility that higher efficiencies are achievable. This chapter discusses the development of solar cell technologies and the significance of nanotechnology in solar cell formation in detail.

26.2 Generation of Solar Cell Technology

Photovoltaic cells are of different generations based on their categories and modifications. Wafer-based crystalline silicon (c-Si) technology, whether single-crystalline or multi-crystalline, is used in first-generation solar cells that are solely commercial. Thin-film photovoltaic technologies are the core of the second-generation solar panel for commercial applications. Following that, technologies like concentrating and organic solar cells, which are being tested or are not yet generally available, are included in third-generation solar cells. Some innovative ideas are in the early stages of research.

26.2.1 Overview of First-Generation Photovoltaic Cells

Photovoltaic cells of the first generation belong to crystalline film technology, which uses semiconductors like silicon and gallium arsenide (GaAs). In the first generation, the monocrystalline-based solar cell is the most popular, representing about 80% of the marketplace, and it will dominate till a more approachable and inexpensive photovoltaic cell technology is found. Under standard test conditions (STC), monocrystalline silicon cell efficiency is approximately 23%, while the highest measured efficiency was 24.7%. Although polycrystalline cells are an effective material for lowering solar panel operating costs, they have lower efficiency than monocrystalline cells and other emerging materials. Polycrystalline cells outperform monocrystalline cells by exhibiting fewer faults in metal pollution and crystalline structure. Although GaAs are the earliest material still being used to make solar cells due to their greater efficiency, silicon (Si) is the most widely utilized material for commercial applications, accounting for almost 90% of the photovoltaic cells industry.

A few micrometers of GaAs with a perfect band gap of 1.43 eV are required. GaAs have a 25–30% higher power conversion capability than crystalline Si. It is the smartest choice for concentrated solar systems and applications in outer space because of high resistance to thermal and radiation degradation. The next effective technique using silicon's crystalline structure in the first generation is emitter wrap-through cells, which improved the efficiency to 15–20% using improved cell layout instead of material advancements

26.2.1.1 Overview of Second-Generation Photovoltaic Cells

In the second generation, thin-film technology is utilized, which is less costly than solar cells made of crystalline silicon. It requires fewer resources and fewer production steps. Since it employs less material, the solar cells used in this technique are thin—between 35 and 260 nm. The earliest thin-film photovoltaic cell technologies were amorphous silicon (a-Si). The material's electrical characteristics are significantly impacted by this atomic structural unpredictability, resulting in a greater bandgap (1.7 eV) than crystalline silicon (1.1 eV). A-Si cells that capture more solar irradiance from the visible band of the spectrum than from the infrared, owing to the wider band gap. Multiple-junction a-Si devices are being developed to increase performance and address degradation issues [6]. This enhancement is related to the architecture of these cells, which allows the collection of various solar radiation wavelengths.

Following that, multiple-junction a-Si device development has been attempted to increase efficiency and address degradation issues [6]. This enhancement is related to the architecture of these cells, which allows the collection of various solar radiation wavelengths. The "stacked" or multi-crystalline (mc) junctions, also known as micro morph thin film, are one more way to increase the performance of solar panels and modules. The top layer of this technique layers has two or even more photovoltaic connectionsoverone another, with an extraordinarily thin layer of a-Si transforming the shorter wavelength of the visible solarspectrum. The next material, CdTe, is a polycrystalline semiconductor with higher light absorption characteristics. Despite being very thin at about 1 mm, 90% of the solarspectrum can be absorbed by this semiconductor. Another benefit is that its manufacturing process is simple and affordable. The performance instability of cells and modules is the primary issue with CdTe development in solar applications. Recently, [7] established a low-temperature way to solve cation exchange reactions to build cadmium sulfide core/copper sulfide shell nanowire solar panels. The open-circuit voltage and fill factor, which controls the highest amount of energy a solar cell can generate, support the low-cost and practical manufacturing process of nanowire solar cells. Additionally, the energy output of these novel nanowire photovoltaic cells is 5.4%.

Due to the great degree of flexibility in the compound creation having specific features in this material, significant advancements may be anticipated in the coming years. CIS is a functional material, but its manufacturing is challenging due to its complex nature. CIS is helpful in solar cells because of its composition, density, and adhesion characteristics. A multi-layered thin-film composite called CIGS (indium combined with gallium-enhanced band gap) has several layers. Research and experimental findings show that after an outdoor exposure of 130 kWh/m^2, CIS degrades only by 10% to thin-film materials.

26.2.1.2 Overview of Third-Generation Solar Cells

Third-generation solar cells solve the problems identified by second-generation approaches and improve efficiency through thin-layer fabrication. Although this advanced technology may be expensive, it reduces the cost per maximum watt. This technology is ideal for large-scale photovoltaic solar cell applications since it is non-toxic and employs ingredients that are easily accessible. Such materials could be organic or nanomaterial, and the employment of various charge carrier collection techniques allowed for high efficiency of more than 60%.

In the third generation, the capacity of QDs to increase absorptivity over various energy ranges and prolong the absorbance peak into the infrared region makes them interesting for photovoltaic applications. According to theoretical QDs solar cell modeling, the efficiency might rise to 64% for a well-adjusted intermediate band. Small, nanometer-sized QDs are composites of minute crystals that exhibit various unique semiconductor characteristics. Such nanoparticles typically consist of materials like zinc oxide (ZnO$_2$) and titanium oxide (TiO$_2$), and their sizes range from 1 to 20 nm [8]. Nowadays, the "energy transfer" solar panel is designed using QDs. Because QDs come in various sizes and volumes, they can absorb light of different wavelengths. QDs are arranged in layers of varied thickness in solar cells. QDs absorb solar radiation at various wavelengths because of their variable sizes. Hence, the cell produces more light. Electron holes are unable to hop because of the steady state at the p-n junction, increasing the cell's efficiency.

The subsequent technologies utilized in the third generation is a quantum well (QWs) that have only discrete energy values that are created in semiconductors by sandwiching gallium arsenide (GaAs) over two layers of a semiconductor having a large band gap, such as aluminum arsenide (AlAs) [9]. In solar panels, QWs regulate electron-hole pairs, which typically swing from three aspects to two aspects. The minimum number of electrons and holes is determined by the size of the semiconductor used, which typically ranges from 1 to 10 nm. ZnOnanorod arrays (NRAs) were used as the anti-reflection (AR) layer. Due to its enhanced optical communication, its efficiency is enhanced by 36%. The efficiency of InGaN-based multiple quanta well solar cells using quantized cheap ZnO NRAs improves, confirming promising nanofabrication of AR layers to make use of light absorption capability.

In recent years, nanostructured elements and nanoparticles have emerged as essential for electro-optical programming and system-improving performance. Recently created gasochromic windows have shown impressive outcomes. Due to the basic design of the product and the lack of transparent conductors, extremely high transmittance modulation ranges have been attained in comparison to the brief research study. This might also indicate that in the future, gasochromic commercial windows could develop into a competitive, high-performance replacement for present smart window methods. The next significant technique in the third generation is the distinct physicochemical characteristics against its bulk materials. The nanowires have lately received much interest in solar energy harvesting, transmission, and preservation. The smallest size for effective charge creation, isolation, and transmission, nanowires have grabbed the attention of researchers. One-dimensional nanostructures have unique chemical, structural, and physical features that make them perfect for capturing and converting solar energy. Short wavelengths reflection and improved light capture and absorbance are possible because of the unique geometry of arrays of nanowires.

26.3 Nanotechnology in Solar Cells

Nanoscale structures and nanotechnological solutions will improve the efficiency and performance of solar cell systems. Systems fabricated at the nanoscale have unique characteristics compared to their bulk or thin-film counterparts. Nanotechnology provides several merits, such as solar cells benefiting from high surface-to-volume ratio nanostructures, quantization effects at 1–20 nm scales, and different fabrication techniques. To outperform the Shockley–Queisser limits, multiple-junction solar cells use nanomaterials. Theoretically, a triple-junction solar cell made of III–V semiconductors can achieve up to 34.1% efficacy, while a solar cell with a finite number of junctions has an efficiency cap of 68% at one sun's intensity. These high-efficiency cells are costly. However, layers with various band gaps can be created using nanomaterial, and many junctions can be produced in solution at a much lower price. Additionally, optical losses and electrical performance are improved in solar energy systems by using the special features of nanostructures.

i. **Role of Nanostructures in optical losses reduction**:
   Broadening the solar spectrum to prevent transmission and heat losses or improving the probability of absorbance of incoming photons can reduce the optical losses of solar cells. For example, we can use QDs, resonators, and anti-reflective nanostructures as nanotechnology solutions to reduce optical losses. A variety of nanostructures are used to generate an anti-reflective surface and also to lower the reflection losses. These include nanowires, nanopillars, nanodomes, and nanocones. Gradually reducing the refractive index from the semiconductors to air requires proper optimization of the geometrical parameters. A constantly graded refractive index can aid this process. In addition, using light trapping and patterned nanostructures, even more light can be collected, requiring less material to absorb solar flux.

   The extensively employed phenomenon of resonance absorption restricts and re-directs light to active areas of solar cells to increase absorption. It is accomplished using photonic crystals, metallic or dielectric resonators, or both. Light scattering, enhanced light concentration (localized plasmonic resonance or LPRs), and activation of surface plasmonpolaritons are three ways that metallic resonators alter the absorption of light (surface plasmonic resonances or SPRs).

   The metallic nanoparticle resonators make the effect of metallic plasmonic resonance much more noticeable. Metallic nanoparticles cause LPRs and boost the semiconductors' absorption in the vicinity. Additionally, by adjusting the geometry and concentration of metal nanoparticles, LPRs can be improved. Placing metallic nanoparticles into the p-n junction makes their inclusion much more advantageous. Additionally, metallic nanoparticles are used to assist SPRs by boosting the semiconductor layer's absorption. The difficulty of positioning metallic nanoparticles in the solar cell optimally is a drawback. Metallic nanoparticles' reflecting qualities can result in optical losses if they are not positioned correctly. By coupling leaky optical modes into the active layer with the help of dielectric nanophotonic resonators that make use of Mie resonances, it is possible to circumvent the inherent losses that come with metallic nanoparticles and nanostructures. Utilizing semiconductor nanocrystals like QDs in solar cells is another application of nanotechnology. The most appealing characteristics of nanocrystals are their considerable size dependency on the bandgap and altered relaxation dynamics of photoexcited charge carriers. Narrower-bandgap materials absorb more solar photons because only light with an energy greater than the bandgap is absorbed, resulting in stronger photocurrents. On the otherhand, wider-bandgap materials enable greater voltages since the output voltage is linearly proportional to the bandgap. Shockley–Queisseranalysis shows that the ideal bandgap for solar devices lies between 1.2 and 1.4 eV. Due to the quantum confinement effect, it is possible to widen this bandgap using nanocrystals.

ii. **Role of Nanostructures in improving Electrical Performance**

The electrical characteristics of solar cells impact their device performance in addition to increasing light absorption. To achieve high power conversion efficiency, quick and effective charge separation is required in addition to transfer and collection. Changing the shape and arrangement of the donor and acceptor surfaces will enhance the carrier collection and transfer. The control of the junction interface and the control of carrier collection distance are two efficient ways to increase carrier collection and transmission.

Recombination during carrier transfer diminishes charge separation and transfer efficiency. Hence, the junction interface is crucial to prevent the same. Due to the irregular paths for charge transport and the disorganized donor/acceptor energy levels, a disordered junction interface enhances carrier recombination. As a result, the overall efficiency of solar cells is increased by an organized heterojunction at the nanoscale.

The carrier collection effectiveness of solar systems depends on managing the carrier collection distance. Utilizing nanostructures with core/shell architecture is one of the best ways to manage the carrier collection distance. These nanostructures' radial designs offer minimal separation distances that facilitate the efficient passage of carriers into depletion zones. Additionally, core/shell architectures offer vast surface areas in a small volume, improving the performance of the device.

26.3.1 Formulation Methodologies and Nanostructured Materials

Materials that have a minimum size of 100 nm or less are regarded as nanoscale materials. A nanometer is 100,000 times lower than that of the width of a human hair or one-millionth of a millimeter. Nanomaterials are incredibly tiny, with at least one dimension measuring no more than 100 nm. Nanoscale materials with nanostructures can have zero, one, two, or three dimensions (e.g., QDs), surface layers, strands, or fibers (e.g., particles). They come in spherical, tubular, and irregular shapes and can be found alone, fused, aggregated, or agglomerated. The most prevalent types of nanomaterials comprise fullerenes, QDs, dendrimers, and nanotubes. Nanomaterials are composed of structural elements that fall in between atoms and bulky substances. The majority of microstructured materials share identical attributes with their respective bulk materials. It is generally caused by the materials' nanoscale size, which further makes them: (i) increased surface energy; (ii) spatial detention; and (iii) minimal defects.

26.3.1.1 Nanostructured Materials Classification

Nanomaterials include a very high surface area to volume ratio due to its small size, which creates more surface or interfacial atoms and more "surface" dependent material features. Based on its dimensions, nanomaterials are of (a) 0D clusters and spheres, (b) 1D nanorods, nanofibers, and nanowires, (c) 2D networks, plates, and film (d) and (e) 3D nanomaterials, as illustrated in Figure 26.1.

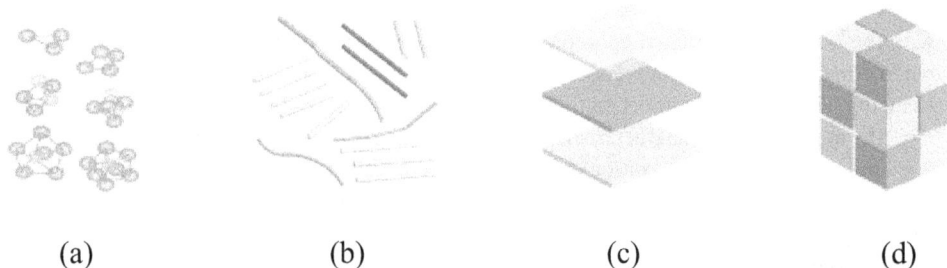

(a)   (b)   (c)   (d)

FIGURE 26.1 Nanostructured materials classification (a) 0D spheres and clusters, (b) 1D nanofibers, wires, and rods, (c) 2D films, plates, and networks, (d) 3D nanomaterials.

The surface properties of nanoparticles will impact the entire material, specifically when their sizes are equivalent to their lengths. The qualities of the bulk materials may then be improved upon or altered. Metallic nanoparticles, for instance, can be utilized as highly effective catalysts. Chemical sensors with nanowires and particles have improved sensitivity and selectivity. The electrical as well as optical properties of nanomaterials can be significantly altered from its bulk size, which will modify the materials' charge carrier densities and band structure of energy. The characteristics of the nanomaterials are also significantly influenced by reduced defects. For instance, some nanoparticles may have improved chemical stability, and their mechanical qualities will be superior to those of bulk materials. Nanomaterials, such as carbon nanotubes, fullerene, silver nano, and silica, have been used in the field of nanotechnology and differ from regular chemicals in respect of both chemical and physical composition.

26.3.1.2 Fabrication and Processing of Nanostructured Materials

To create nanomaterials, we can use both "bottom-up" and "top-down" methods. These methods involve assembling the atoms or dissociating bulk solids into smaller bits to reduce them into fewer atoms. Mechanical attrition is a classic example of a "top-down" technique of nanoparticle formation wherein the nanomaterial is created by structural disintegration of coarse ground structures as a result of extreme deformation of the plastic material rather than cluster formation. It is easy to use as it requires fewer number of equipment. It is highly capable of synthesizing all kinds of materials. It is one of the most popular techniques for producing nanocrystalline materials.

As demonstrated in Figure 26.2, mechanical milling is frequently accomplished using elevated shakers, planetary balls, or tumbler mills. Here, the compression effect during milling results development of nanoparticles. During processing in cryo solutions, the amount of crystallization of the powders can be greatly increased. This process can affect the breaking process. This synthesis method works well for creating elemental or complex powders and crystalline or nanocrystalline alloy particles.

Generally, there are two major categories into which we can divide the moist molecular techniques of nanomaterials: (i) The top-down approach involves, creating nanomaterials by dissolving single crystals in liquid solutions. For instance, porous silicon can be created by electrochemical etching. (ii) The

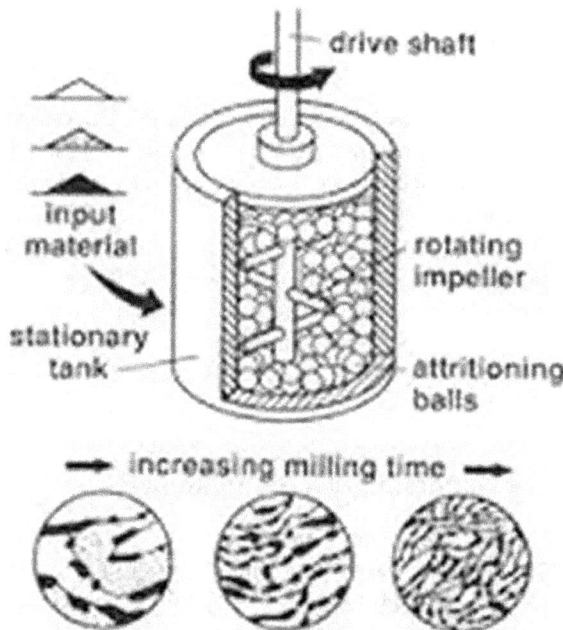

FIGURE 26.2 Mechanical milling process for the synthesis of nanoparticles.

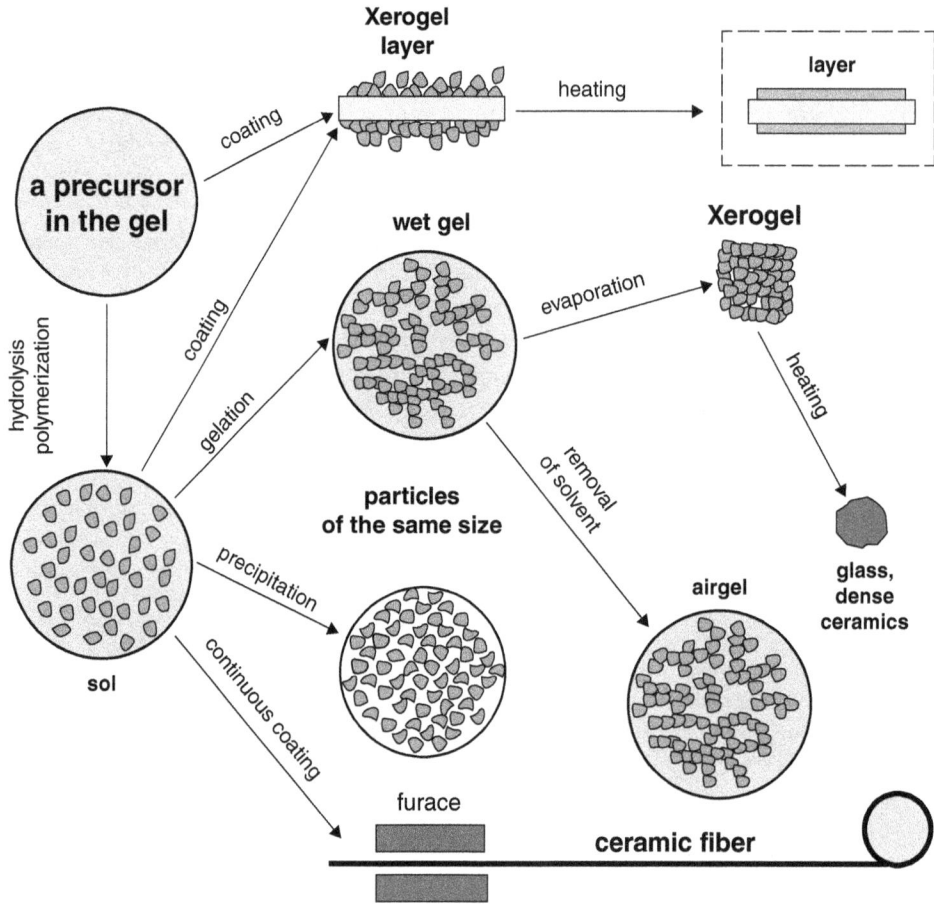

FIGURE 26.3 Schematic diagram of sol–gel nanomaterial production process.

bottom-up approach, which includes the sol–gel technique, precipitation, etc., entails carefully combining materials containing the required precursors to create a colloidal solution. Inorganic networks can be produced by forming a colloidal suspension (sol) and gelating the sol to form a chain in a continuous liquid phase (gel). The precursors used to create these colloids typically include a metallic or metalloid component encircled by several reactive ligands. If the beginning element gets in touch with liquid or dissolved acid, the processed starting material transforms into a sol. The gel is produced when the liquid is extracted from the sol. Then, the transition between the shapes and sizes of the particles will take place. Chemists frequently use the sol–gel process to create oxide materials. It is one of the very popularways to synthesize nanomaterials. As seen in Figure 26.3, several unique phases make up the sol–gel process.

26.3.2 Applications of Nanomaterials for Solar Cells

Coordination chemistry provides the necessary scientific method and is distinct from classical materials like Si, gallium arsenide (GaAs), and cadmium telluride. Coordination chemistry determines the surface characteristics and doping concentration chemistry of the transition metal compounds (CdTe). The ability to achieve homogenous photoactivity through quality control is anticipated to be a crucial component in the development of any future photosensitive materials. Thin films are defined as materials that are formed on supportive surfaces and comprise inorganic coatings, organic additives, and organic–polymer materials. The third group is set up as nanocrystals and employs QDs placed in a supporting matrix using a "bottom-up" methodology.

Only Si has been thoroughly studied in combining bulky and thin-film forms. Si photocells have several new competitors, including copper indium gallium selenide (CIGS), copper telluride (CdTe), dye-sensitized solar cells (DSSCs), and organic solar cells [8, 9]. Majority of photocells are intended for marking on commonly used packaging materials and then inexpensive flexible polymer films. Semiconducting polymers are one of these new materials that are receiving a lot of attention due to their broad parameter space and intrinsic ease of device construction. Therefore, they merit further study [10]. The degradation of the solar cells over time is a significant problem because they are designed to be used in conditions of prolonged sunlight exposure. For instance, it's important to carefully guard against oxidation when using thin-film materials, particularly chalcogenides, in solar modules. Nanomaterials and nanostructures get a great potential to enhance the performance of photovoltaic cells by enhancing either light absorption orphotocarrier collection. In the meantime, it is possible to construct these new materials and structures for photovoltaics affordably. Nanowires, nanopillars, nanocones, nanodomes, nanoparticles, and other characteristic materials make up the family of nano-materials. A PV device's performance depends on absorbing photons and collecting photocarriers.

In order to produce a photovoltaic system with an acceptable power conversion efficiency, these components must be modified. However, there may be a conflict between the criteria for maximizing optical absorbance and collecting carriers. For instance, a planar patterned pv cell, wider materials are needed to achieve the proper optical absorption, but it reduces carrier collection possibility since the minority carrier mobility path is extended, and vice versa. According to recent studies, 3-D nanostructures facilitate the collection of photocarriers by aligning the orientations of carrier collection and light dispersion.

26.3.2.1 PV Thin-Film Systems Using CdTe, CdSe, and CdS

Since the nanoparticle atoms are smaller in size, they are more likely to be found on their surfaces than inside them, which causes surface interactions to predominate in nanoparticle activity. As a result, they frequently have distinct traits and qualities that larger particles of the same material lack. Three important advantages exist for photovoltaic devices that have been constructed with nanostructured materials. First, result of multiple reflections, the actual visual path for absorbance is significantly more than the original film thickness. Second, because the distance that electrons and holes produced by light must travel over is shorter in length, recombination losses are significantly decreased. Third, by adjusting the size of the nanoparticles, energy band gap of several layers can be changed to the desired design value. It gives the absorber of solar cells an additional design latitude. A thin layer of the active ingredient is deposited on a cheap substrate in the thin-film method, which is a more affordable approach. As a result, expenses are reduced and substantially less element is used (1% or less in comparison to wafers). The majority of pv cells employ amorphous silicon, which has a substantially lower efficacy of 8% since, as its name suggests, it has a crystalline phase but is far more easily produced. A simple method for producing inexpensive and sizable areas of polycrystalline semiconductors, particularly for the PV solar power conversion and photoelectrochemical cells (PEC), involves the placement of an electrode and semiconducting particles are deposited chemically in an aqueous solution. The most desirable photovoltaic components for thin-film photovoltaic cells have been reported to be thin films of polycrystalline CdTe, CdSe, and CdS [11, 12].

26.3.2.2 Quantum Dot and Nanoparticle Solar Cells and PV Technology

PV applications have also undergone intensive research for QDs, yet another insightful class of nanomaterials. The motivation for the related research is based on either tiny nanoparticles or QDs, with distinctive physical characteristics like dimensions-specific band gap and multiple exciton generation (MEG), that enable modern PV mechanisms perhaps exceeding thermodynamic limitations. Additionally, the majority of QD syntheses are suitable using solution-based procedures, allowing for the utilization of high-capacity, short-term approaches for the manufacture of PVs relying on such nanoparticles.

The most common technique for producing colloidal QDs involves the regulated nucleation and growth of liquid components of precursor chemicals involving inorganic salts or organometallic compounds. The precursors are quickly injected into a heated, forcefully mixed solvent that contains organic surfactant compounds which can interact with the surfaces of the deposited QD nanoparticles so-called hot-injection method. Typically, this method is used to manufacture II–VI and I–VI semiconductor colloidal QDs. By preventing or limiting the development of particles via Ostwald ripening, the organic surfactant compounds play a crucial part in adjusting the growth and kinetics of nucleation. A variety of nanocrystals such as CDs, CdTe, CdSe, Copper-Indium-Selenide, etc., have been created by a growth technique resembling this one, and in photovoltaic cell, nanoparticles have been incorporated. According to Figure 26.4, the nanocrystals utilized in PV cells are rod-shaped CdSe, and CdTe nanocrystals made using air-free hot-injection methods [11, 13].

26.3.2.3 CuInS$_2$, Iron Disulfide Pyrite, and Cu$_2$ZnSnS$_4$

Due to its advantageous solid-state features, an attractive element for solar energy systems in photoelectrochemical and PV cells as well as sturdy solar cells is iron disulfide pyrite. There were major developments when Wohler [13] first created synthetic pyrite via combining Fe$_2$O$_3$ using liquid sulfur and NH$_4$Cl in an inclusive environment in the previous century and was successful in creating a tiny brass golden octahedron. Pyrite thin films and single crystals can be produced in a variety of ways, such as by metal-organic chemical vapor deposition (MOCVD), iron pent carbonyl, sulfur, or hydrogen sulfide in an organic solvent, and many others. Recently, one-dimension nanocrystalline substances likenanorods have been drawn a lot of attention and interest, and it has been shown that their morphology

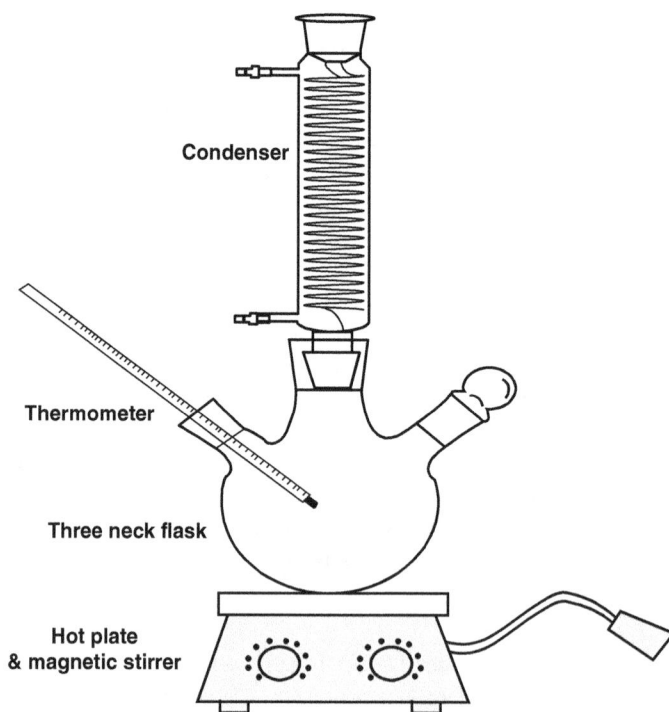

FIGURE 26.4 Airless hot-injection method.

can be controlled. Currently, copper-rich metal precursors are deposited, accompanied by tempering in a sulfur environment, to produce thin-film photovoltaic devices relying on CuInS2 absorbing layer by co-evaporation [14, 15]. Direct impeachment of magnetron sputtering could be advantageous for mass production in the future of solar cells because this method is simple to scale up to huge regions. Additionally, the ZnO/ZnO: Al window and interface surface, which are currently deposited by magnetron sputtering, provide the ability to create an uninterrupted vacuum manufacturing procedure of cells. Only a few publications published in the literature during the past ten years have discussed the triple composite semiconductors are deposited by magnetron sputtering [16].

Moreover, Thornton and colleagues at the University of Illinois (Urbana) made a significant effort in the 1980s to introduce CuInSe2-thin-film photovoltaic cells using a reactive magnetron. Sputtering is not appropriate for high-quality absorbers as low efficiencies are obtained for solar cells using sputtered absorbers. For the deposition process of carbides, oxides and nitrides, and carbides, reactive magnetron sputtering is a very well method. These elements fall under the category of reactive magnetron sputtering and are frequently employed as optical layers for coating materials and surface protection. Plastic solar cells on inexpensive polymer substrates provide the probability of quick and low-cost fabrication of large-area PV devices. Recently, in PV systems with absorber layers that are fewer than 100 nm thick and only absorb a small percentage of the sunlight, were able to reach over 2.5% solar efficacy. This was based on a set of p-type polymer networks with conducting C60 compounds that percolated electrons. CIS nanocrystals are the greatest solar absorber for photovoltaic devices.

Researchers are interested in iron disulfide (FeS2), which has a pyrite structure [17]. Regarding applications, FeS2, which makes up the majority of the sulfur in coal, has shown a considerable rise in photoelectrochemical activity. Iron-based materials have undergone substantial research as prospective replacements for silicon or gallium arsenide photovoltaic cells, which are commercially accessible owing to their excellent potential for use in photovoltaic applications. Binary FeS2 nanocrystals enable solution-processed solar cells in contrast to other multi-composition PV materials like Cu2ZnSnS4. To address the rising need for energy, recent pyrite FeS2 nanocrystal research and efficiency of low-cost photovoltaic cells has greatly improved as a result of development efforts. But this system's principal flaw is oxidation, which is caused by the orthorhombic metastable marcasite structure and is bad for PV characteristics.

26.3.2.4 Solar Cells Made of Nanowires and Organic Materials

A homogenous mixture of two organic materials or two organic layers is used to make organic solar cells. The other components serve as the electron acceptor, with one of the organic materials acting as the electron donor, such as semiconducting polymer or an organic additive. In PV devices, the transparency anode is made of substrates covered with Indium doped Tin Oxide. Other transparent conductive oxide (TCO) film, such as zinc oxide coated with aluminum has low device performances when employed asanode [18]. Science and engineering technology hot topics right now include research on nanowires and photovoltaic cells. Advances in nanoscience's synthetic control have enabled the development of high-performance electronic devices. Working nanowire photo voltaic has been made from a wide range of materials, including silicon, germanium, zinc oxide, zinc sulfide, cadmium telluride, cadmium selenide, copper oxide, titanium oxide, gallium nitride, indium gallium nitride, gallium arsenide, and a variety of polymer/nanowire combinations. But before using them commercially, research should explore several unanswered concerns. However, the output efficiency hasprogressively advanced to the extent in which most substances have efficiencies better than 1%, with certain approaching 10% [19].

26.3.2.5 Solar Cells Made of Polycrystalline Thin Films

Polycrystalline thin-film photovoltaic cells like CuInSe2 (CIS), Cu (In, Ga) Se2 (CIGS), and CdTe composite semiconductors are essential for powered by solar applications because of its high effectiveness, long-term durability, and prospects for less-cost production. Owing to the increased absorption coefficient, a covering as thinly as 2 mm is adequate to absorb the useful section of the spectral range (105 cm^{-1}). The greatest record performances of 19.2% and 16.5%, correspondingly, were achieved for CIGS and CdTe. According to various growing techniques, numerous groups worldwide have created CIGS photovoltaic cells that are approximately 15–19% effectiveness. Although glass is still the most popular substrate, significant work has lately been done to create flexible photovoltaic cells using metal and polyimide foils and its efficiencies were 12.8% and 17.6%, respectively. Flexible panels having 7.8% metal [91] and 11% polyimide efficiency have been produced. On the glass substrate, CdTe solar cells with efficiencies of 10–16% have been developed. CIGS and Because CdTe photovoltaic systems outperform Si or III—V photovoltaic cells in particle irradiation, there is currently a surge of interest in these polycrystalline composite semiconductor photovoltaic modules for usage in space applications. Also, solar cells that are lightweight and flexible have a high specific power (W/kg), which opens up a wide range of opportunities for many applications. The solar module encapsulation at a reasonable cost is made possible by the super-state arrangement.

26.3.3 Progressive Nanostructures for Technological Applications

Human space technology and advancement have relied heavily on pv systems. The development of unique device configurations and new element systems will be necessary to meet the demands for photovoltaic solar power in the future. We suggest a new device design that uses nanostructured materials in PV cells to increase device efficiencies while reducing weight and maintaining structural integrity. This strategy will enable us to enhance the best space solar cells currently on the market in terms of their efficiency and material characteristics, which are crucial for maximizing space use. Due to the advancement of nanomaterials, we will be able to create thin-film solar arrays for space that are flexible, lightweight, and made of polymer-based materials.

Scientists in the area of solar cells are concentrating more on polymer or plastic-based technologies. The researchers mainly focus on hybrid approaches that incorporate photoactive nanoparticles into PV systems using thin films made of polymers. In these hybrid photovoltaic cells, an organic–polymer matrix is used to disseminate inorganic semiconducting nanoparticles. This method offers effective, lightweight, strong, adaptable, and affordable solar energy. Numerous nanomaterials being considered for application in such polymeric photovoltaic panels have a variety of capabilities. Solar light with wavelengths above the conducting polymer-energy bandgap being converted is the only source of power for non-hybrid polymer devices (usually more than 2 eV, is notideal for the solar radiation). Due to the nanomaterials' absorption coefficient below the conductingpolymer bandgap, which they exhibit in the hybrid approaches, these hybrid devices can capture a sizable portion of the solar radiation. Composite photovoltaic cells may exploit additional quantum confinement effects that were demonstrated in other semiconductor materials QD systems.

26.3.3.1 Low-Cost Solar Cells Made of Nanocones

Nanoscale texturing is used in hybrid solar cells for two reasons: (i) it enhances light absorption, and (ii) it lowers the amount of silicon material required. Nanowires, nanodomes, and other structures have been used in the past to texturize solar cells at the nanoscale. Researchers have found that a nanocone shape including an aspect ratio of around one is the ideal form for boosting light absorption since it provides for both strong AR and refraction. Traditionally, nanoscale structuring has required the adoption of complete second layer because the spaces among structures were too minute to be covered with polymer. The polymer can be coated in open spaces without using additional materials. Processing costs can be decreased when the nanocone/polymer hybrid structure depicted in Figure 26.5 is formed using a straightforward, low-temperature approach.

The researchers developed a PV device with the highest efficiency of hybrid silicon/organic photovoltaic cells is 11.1%, after testing the solar cell and making minor changes. Additionally, the short-circuit

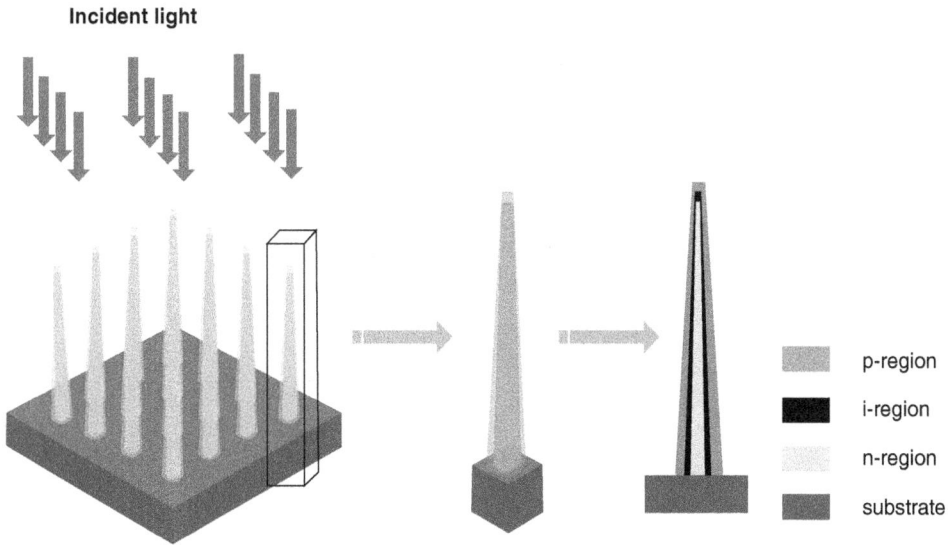

FIGURE 26.5 Nanocones for solar cells.

current density of photovoltaic cell is extremely close to the theoretical maximum and only slightly lower than the monocrystalline silicon solar cell world record. Researchers believe hybrid silicon nanocone-polymer solar cells could be utilized as inexpensively PV system because of their high performance and easy fabrication.

26.3.3.2 Nanoparticles with a Core and Shell for PV Applications

In the domains of information technology and microelectronics, the size-dependent features of semiconducting and magnetic nanocrystals are crucial. Smaller nanoparticles become unstable due to increased surface tension, which causes Ostwald ripening, in which smaller particles grow to bigger sizes. Thus, to prevent the nanoparticles from aggregating, and for the purpose of enveloping the nanoparticles with an organic chemical, core-shell nanoparticles are necessary. Hybrid solar cells, which combine inorganic nanoparticles or core/shell nanoparticles in the form of a photovoltaic layer in a semiconducting matrix material, are a rapidly expanding field. Due to its energy bandgap and material absorption coefficient, $CuInS2$ and $CuInSe2$ core/shell thin-film photovoltaic cells have presently the subject of studies and produced with conversion efficiencies of 18.8%. One can adjust the band gap for the creation of twin PV cells by changing one specific compound to obtain different band gaps. The development of an interpenetrating heterojunction network as a result of combining the inorganic nanoparticles in a polymer matrix is a crucial component in creating organic/inorganic hybrid solar cells.

26.3.3.3 Silicon Photovoltaic

The importance of silicon solar technology is growing worldwide, and amorphous silicon alloys are in extremely popular for huge-area andinexpensive PV. Si is a great contender for photovoltaic, which offer an energy source even in distant places. To address the energy demands of people in developing and developed nations, researchers are working on developing Si PV. Minimizing the contaminants, pressure as well asflaws in the silicon crystals will yield greater conversion efficiency, which is a significant issue in Si PV. In thick film solar cells, silicon or other indirect band gap semiconductors are employed. However, the transport characteristics of a semiconductor are influenced by the minority carrier diffusion length. A promising way for obtaining high-efficiency and inexpensive solar cells now appears to

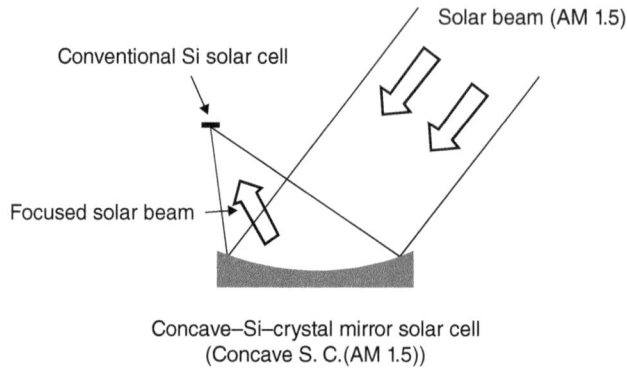

FIGURE 26.6   Concave mirror Si solar cell.

be hybrid solar cells, which consist of a thin-film photovoltaic cell mounted on a cheap micro-silicon photovoltaic cell. Solar cells frequently employ Si single crystals or multiple crystals.

The focused mirror solar cell system shown in Figure 26.6 consists of conventional photovoltaic cells and a focused PV system through a lens. This system can achieve high efficiency by effectively utilizing the photons that were reflected by the cell. At Bell Labs in the USA, the first Si p/n solar cells were created in 1954 with a 6% efficiency [19]. First photovoltaic cells usage was launched in 1958 by Soviet satellite Sputnik 3 and the US Vanguard employing n-type silicon with p-type boron as a dopant. Si solar cells are being used more frequently, which has led to a substantial study being conducted to make weightless modules and improved radiation-resistant machinery by boosting reliability and effectiveness.

26.3.4 Semiconductors (III–V)

Excellent-efficiency of solar cells iscomposed of Gallium Arsenide and InGaP due to its direct-bandgap energy and excellent durability. It hasa 30% efficiency when used as a source of power for satellite launches. Due to their increased physical qualities and higher efficiency rate compared to Si, Photovoltaics can benefit from GaAs-based III–V semiconductors that were produced on GaAs substrates. Because of flattened solar modules for aircraft applications owing to its low weight, increased irradiation resistance, and increased performance, photovoltaic system used III–V semiconductors have almost completely supplanted silicon solar devices in recent years. Operating at frequencies over 250 GHz, the PV device has a direct energy bandgap of 1.42 eV, high electron saturating velocity, and high electron mobility. GaAs is an III–V compound used in high-efficiency solar cells. High-frequency GaAs devices are less noisy and may operate at higher power levels than analogous Si devices because of their higher breakdown voltages, and they also have electronic properties that are generally better than Si.

Multi-junction photovoltaic cells are more efficient than single-junction photovoltaic cells when it comes to tuning individual junction to a light wavelength range collected. For satellite power sources, sophisticated heterostructures with multi-junction solar cells made of phosphides and arsenides and built on germanium substrates have been accomplished. Around the year 2000, the InGaP/GaAs/Ge three junction PV device allowed for an effectiveness rate of 30%. High-efficiency GaInP/GaAs/Ge is now used in space applications as well as terrestrial concentrators. To lessen the stress-induced flaws that affect solar systems' performance, III–V semiconductors with bandgaps similar to or lower than those of GaAs are preferred. For usage in high-efficiency solar cell applications, materials having abnormal direct-bandgap bending, like GaNAs and GaInNAs with preferred lattice matching to GaAs were shown to have short minority carrier diffusion lengths. Small diffusion duration of III–V semiconductors should be taken into account for moving forward because photovoltaic cells need long diffusion lengths to gather photogenerated carriers effectively.

26.4 Conclusion

The inclusion of nanotechnology into the films provides a unique opportunity to boost solar energy conservation effectiveness and lower manufacturing costs. Nanotechnology would have a huge impact on society even if it could merely provide low-power devices with enough energy. Although there are many potential uses for nanofluid, there are several obstacles that researchers must overcome to advance this subject, including

- The lack of consistency in experimental findings from various groups.
- The suspension of nanoparticles exhibits poor long-term stability.
- Lack of theoretical understanding of the basic energy transport mechanisms.
- The increased cost of nanofluids decreased specific heat and increased pressure drop and viscosity.
- Eco-friendly technology for massproduction of nanofluids.
- Need for experimental studies on nanofluid convective heat transfer.
- Additionally, when performing experimental research, certain factors such as particle migration, change in thermophysical properties about temperature, and Brownian motion of particles should be carefully taken into account.

Furthermore, the most challenging features of nanotechnologies are the properties of nanomaterials, like,

- Long-term nanoparticle dispersal stability.
- Increased stress decline.
- Enhanced pumping capacity.
- Temperature behavior of a nanofluid in a turbid zone.
- Strong viscosity, reduced heating value, heat capacity, and the profitability of nanofluids are all factors that contribute to the issue.

Although solar energy is an endless supply, technology for utilizing it has not advanced very far. By using nanofluids, the efficiency of straight absorption of solar collectors may be increased by up to 10%. Whereas the second generation of photovoltaic technology is currently in use in various areas, the third generation is in the testing phase. The first generation of photovoltaic devices is now commercially available. To harness the maximum amount of solar energy feasible, the features of nanofluids must be properly matched by technology.

In the future, it will be necessary to take advantage of unexpected possibilities as well as the aforementioned problems. For nanofluids to be used as planned, more study still needs to be done on their synthesis and uses. Nevertheless, numerous advancements and in the examined applications, observations regarding the characteristics of nanofluids have been made. Therefore, to assess the elevations in heat transport in lamina, transition, and turbulence zones, convective experiments using metallic nanoparticles of various geometries and concentrations must be carried out in the future.

References

[1] Z. Abdin, M.A. Alim, R. Saidur, M.R. Islam, W. Rashmi, S. Mekhilef, A. Wadi, Solar energy harvesting with the application of nanotechnology, *Renewable and Sustainable Energy Reviews* 26 (2013): 837–852.
[2] Suresh Sagadevan, Recent trends on nanostructures based solar energy applications: A review, *Reviews on Advanced Materials Science* 34 (2013): 44–61.
[3] Djamel Ghernaout, Abdulaziz Alghamdi, Mabrouk Touahmia, Mohamed Aichouni, Noureddine ait messaoudene, nanotechnology phenomena in the light of the solar energy, *Journal of Energy, Environmental & Chemical Engineering* 3, no. 1 (2018): 1–8.

[4] Mohammad H. Ahmadi, Mahyar Ghazvini, et al., Renewable energy harvesting with the application of nanotechnology: A review, *International Journal of Energy Research* 43, no. 4 (March 2019): 1387–1410.

[5] Murthy S. Chavali, Maria P. Nikolova, Metal oxide nanoparticles and their applications in nanotechnology, *SN Applied Sciences* 1 (2019): 607.

[6] C.N. Jardine, G.J. Conibeer, K. Lane. PV-compare: direct comparison of eleven PV technologies at two locations in northern and southern Europe. In *Seventeenth EU PVSEC* (2001).

[7] W. Li, C. Liang, W. Zhou, et al., Preparation and characterization of multiwalled carbon nanotube-supported platinum for cathode catalysts of direct methanol fuel cells, *The Journal of Physical Chemistry B* 107 (2003): 6292–6299.

[8] L. Etgar, M. Grätzel, Solid state PbS quantum dots/TiO_2 nanoparticles heterojunction solar cell, *MRS Online Proceedings Library* 1390 (2012): 00723.

[9] A. Scavennec, M. Sokolich, Y. Baeyens, Semiconductor technologies for higher frequencies, *IEEEMicrowave Magazine* 10 (2009): 77–87.

[8] V.M. Fthenakis, Life cycle impact analysis of cadmium in CdTe PV production, *Renewable and Sustainable Energy Reviews* 8 (2004): 303.

[9] H. Zervos, B. Kahn. Alternative Energy Press (2008).

[10] R. McConnell, Next-generation photovoltaic technologies in USA, *Semiconductors* 38 (2004): 5.

[11] T.K. Bandyopadhya, M.N. Majumdar and S.R. Chaudhari, *Journal of the Indian Chemical Society* 56 (1979): 167.

[12] T.K. Bandyopadhya, S.R. Chaudhuri, Influence of annealing ambients on thin film CdTe for solar energy conversion, *Physica Status Solidi A*92 (1985): 637–642.

[13] W. Wohler, *Liebigs Annalen* 17 (1836): 260.

[14] R. Klenk, U. Blieske, V. Dieterle, K. Ellmer, S. Fiechter, I. Hengel, A. Jager-Waldau, T. Kampschulte, C. Kaufmann, M. Saad, J. Klaer, M.C. Lux-Steiner, D. Braunger, D. Hariskos, M. Ruckh, H.W. Schock, *Solar Energy Materials & Solar Cells* 49 (1997): 349.

[15] J. Klaer, J. Bruns, R. Henninger, K. Siemer, R. Klenk, K. Ellmer, D. Braunig, *Semiconductor Science and Technology* 13 (1998): 1456.

[16] R.R. Arya, T. Lommasson, B. Fieselmann, L. Russell, L. Carr, A. Catalano, In: 6th International Photovoltaic Science Engineering Conference, vol. 10–14: pp. 1033.

[17] C. Wadia, Y. Wu, S. Gul, S. Volkman, J. Guo, A.P. Alivisatos, Surfactant-assisted hydrothermal synthesis of single phase pyrite FeS2 nanocrystals, *Chemistry of Materials* 21 (2009): 2568–2570.

[18] J. Owen, M.S. Son, K.H. Yoo, B.D. Ahn, S.L. Lee, *Applied Physics Letters* 90 (2007): 033512.

[19] M. Weyers, M. Sato, H. Ando, Red shift of photoluminescence and absorption in dilute GaAsN alloy layers, *Japanese Journal of Applied Physics* 31 (1992): 853.

[20] R. Femi, T. Sree Renga Raja, R. Shenbagalakshmi, Closed-loop control of solar fed high gain converter using optimized algorithm for BLDC drive application, *Intelligent Sustainable Systems* 1(2022): 245–256.

[21] R. Femi, T. Sree Renga Raja, R. Shenbagalakshmi, A positive output-super lift Luo converter fed brushless DC motor drive using alternative energy sources, *International Transactions on Electrical Energy Systems* 31, no. 2 (2020): 1–23.

27

Nanocomposites for Energy Storage

Sandip Sen
Guru Nanak Institutions Technical Campus, Hyderabad, India

Rachappa Jopate
University of Technology and Applied Sciences, Al Musannah, Sultanate of Oman

S. S. Kerur
KLE Technological University, KLE Society's Dr. M. S. Sheshagiri College of Engineering and Technology, Belagavi, India

L. H. Manjunatha
REVA University, Bangalore, India

Ayaz Ahmad
NIT P, Patna, India

Gitanjali Jothiprakash
Agricultural Engineering College and Research Institute, Tamil Nadu Agricultural University, Coimbatore, India

Because the world's energy requirements are growing and pollution is getting worse, renewable energy sources must substitute fossil fuels. Nanomaterials' rapid development has created new possibilities for storing and transforming renewable energy. Nanocomposites are presently developed to store energy due to the significant increase in efficiency. Many different types of nanocomposites are used to store energy in batteries and supercapacitors, including ionic nanocomposites, ferroelectric polymer nanocomposites, polymer nanocomposites, and electrochemically produced nanocomposites. This chapter covers about the various types of nanocomposites applications in energy storage devices.

DOI: 10.1201/9781003355755-27

27.1 Introduction

The foremost tasks to the sustainable growth of human society have been a lack of energy and environmental pollution (Sreesvarna et al., 2019). Verdict clean, renewable energy sources, and developing enormously effective energy storage methods are viable solutions to these issues (Gitanjali et al. 2015). One of the key difficulties facing both applied and central research in greener energy technologies is the creation of highly effective and inexpensive energy storage and conversion devices (Chen, Skordos, and Thakur 2020). When it comes to power density and energy density, supercapacitors might be a compromise between batteries and conventional capacitors. Because it can be charged and discharged quickly, as it has a great power density.

Supercapacitors can be divided into two types, electric double-layer capacitors and pseudo capacitors, according to their charge storage method. Polymers or metal oxides are used as electrode materials in pseudo capacitors that store charge by the Faraday mechanism. Electric double-layer capacitors energy storage devices can incorporate different electrode materials into the devices (Conway and Pell 2003).

With excellent cycle stability, high energy density, and appropriate capacity, lithium-ion batteries are regarded as the greatest promising power sources for portable devices and electric vehicles. Additionally, because sodium is more prevalent and far less expensive than lithium, it has been thought that sodium-ion batteries are a possible replacement for lithium-ion batteries. If it is eventually replaced with sodium-ion batteries, production costs will drop by around 30%. It should be mentioned that the characteristics of the electrode materials have a major role in the determination of their electrochemical properties for energy storage. Carbon-based materials, transition metal oxides/metal sulfides, and other materials are the primary anode materials for storing energy.

Commercial carbon-based compounds have good cycle stability and outstanding electrical conductivity. Though, their specific capacity is quite small owing to the intrinsic restraint of reversible ion absorption at the active material–electrolyte interface. Their low specific capacity as lithium/sodium storage materials also severely restricts large-scale applications. The most difficult task is to make materials for electrodes with high specific capacities for lithium/sodium storage materials and supercapacitors. Despite their impractically long cycle life, high electronic conductivity, and fast charging capacity, nanocomposites have been viewed as one of the best possibilities for lithium/sodium storage materials and supercapacitors (Roselin et al. 2019).

27.2 Electrochemically Synthesized Nanocomposites

Because the deposition process operates similarly to electrostatics, where the sample is deposited on the anode electrode by drawing in opposing charges, electrochemical deposition produces precise, high-quality products. If charge diffuses from the electrode-electrolyte–electrode during the cycle process, this strategy can also stop chemical expansion and contraction. The bandgap of the conducting polymer is the lowest photon energy needed to excite an electron from the highest occupied molecular orbital to the lowest unoccupied molecular orbital. In conducting polymer nanocomposites, high electrical conductivity causes free radical production, fragmentation, and free electron generation.

The hybrid electrode of polyaniline and nickel oxide produced electrochemically has 2000 cycles at 263 F/g. Electrochemical synthesis was used to create polyaniline-copper oxide nanocomposites, which exhibit remarkable cycling stability with a maximum specific capacitance value of 294 F/g. Therefore, nanocomposites produced electrochemically become good candidates for use in energy storage device applications (Ashokkumar et al. 2020).

27.3 Green Nanocomposites

Carbon dots are a particularly intriguing starting point for next-generation nanocomposite materials because of their multidimensionality, affordability, low toxicity, and great biocompatibility. According to the idea of green chemistry, a new research boom in materials science may also be sparked by the creation of carbon dots using biomass materials and low-energy, ecologically friendly synthesis techniques. One of the main design strategies for creating carbon dot/polymer nanocomposites is the synthesis of green nanocomposites. As a result of their availability, affordability, and renewable nature, a variety of biomass sources that fall under the categories of vegetable and animal waste have been used as carbon dot precursors (Gong et al. 2020). Carbon-derived biomass will be a good raw material for producing carbon-based nanocomposites (Jothiprakash and Palaniappan 2014). The three most widely used green synthesis techniques at the moment are sonication, microwave irradiation, and hydrothermal carbonization (Qin et al. 2021).

Biomass as a feedstock for carbon point synthesis has the benefits of availability, affordability, and environmental friendliness. However, it is challenging to accurately manage the flaws, surface conditions, and size distribution of the carbon dots since each biomass batch has a variable material composition. The primary method for creating carbon dot/polymer nanocomposites at the moment is the physical insertion of carbon dot into a polymer matrix. Although there are certain benefits to this process, particularly in the ease of the preparation step, there is a chance that carbon dots will migrate and disperse unevenly when the product is used (Wang et al. 2020).

Even though they are still in the early stages of research, emerging techniques like in-situ procedures may be able to address the aforementioned drawbacks. Today's polymers are still frequently made with toxic solvents, and high temperatures/pressures, and are frequently obtained from fossil fuels. Effective carbon dot electron transfer capacity must now meet higher standards due to the widespread use of carbon dot/polymer composites as electrodes and other components in energy storage devices. The extremely crystalline arrangement and little internal flaws of the carbon dot structure are what give it its effective electronic transfer capacity (Sreenath et al. 2017).

27.4 Graphene Nanocomposites

Graphene nanocomposites had exceptional mechanical qualities, high specific surface area, and outstanding conductivity, graphene has attracted a lot of attention. The nickel cobalt-coated graphene oxide electrode has a remarkable cycling performance and a high reversible specific capacity (Yu et al. 2020). The specific surface area and conductivity of the entire matrix are enhanced by the large-scale nanopores found in these nanosheets and the graphene framework (Yi et al. 2020).

27.5 Ionic Nanocomposites

Ionic nanocomposites have great potential for use in batteries, fuel cells, sensors, and displays, which is why electrochemical engineers and scientists are very interested in them. The longevity and use of these cutting-edge gadgets are restricted by the usage of liquid electrolytes, which are hazardous and caustic and break down quickly in the presence of high temperatures. Ionic nanocomposites have several benefits over liquid electrolytes, including ease of containment, chemical and thermal stability, and non-flammability. Nanocomposites are beneficial for electrochemical devices anticipated to operate across a wide temperature range (Moghimikheirabadi, Karatrantos, and Kröger 2021).

27.6 Polymer Nanocomposites

Due to their inexpensive cost, polymer nanocomposites have been widely used in industry. Polymer nanocomposites are materials made up of dispersed polymer matrixes and nanoadditives. Clay, spherical nanoparticles, nanotubes or fibers, and other nanomaterials are employed as nanosubstrates. These polymer nanocomposites contain a variety of characteristics, including greater elastic stiffness and rigidity. Admirable features of polymer nanocomposites include barrier resistance, magnetic effectiveness, and electrical qualities (Hossain and Hoque 2018). Numerous sectors use polymers because of their wide range of features, including their lightweight, high durability, and corrosion resistance, as well as their heat resistance, fire performance, and weak gas barrier capabilities. Melt mixing, in-situ polymerization, and filler substrate mixing can all be used to create polymer nanocomposites (Singh and Kumar 2020).

Alumina beads were combined with polystyrene, polypropylene, and styrene-acrylonitrile copolymer. Numerous techniques, including in-situ polymerization, polymer intercalation, melt intercalation, direct mixing of polymers and fillers, template synthesis, and sol–gel procedures, can be used to create polymer nanocomposites (Riggs et al. 2015). Electrical conductivity, processing simplicity, ionic conductivity, tensile strength, chemical, thermal, and mechanical stability are all significantly influenced by the type and qualities of the polymer employed as the matrix for polymer nanocomposites as well as the nanofiller. According to their method of polymerization, polymers can be categorized as addition polymers, condensation polymers, or thermoset or thermoplastic polymers (Hussain et al. 2020). However, two kinds of polymers ionically conductive polymers and electrically conductive polymers are of particular interest for use in electrochemical systems. A type of polymer known as conductive polymer exhibits high electrical conductivity in contrast to regular polymers. These primarily consist of organic monomers with double bonds that are conjugated. The conducting polymers polyaniline, polypyrrole, and polythiophene have received the greatest attention for a variety of real-world uses, including energy storage. These materials combine traditional polymers' low cost, lightweight, good processability, mechanical flexibility, and thermal stability with good electrical characteristics. Polyvinylidene fluoride, polyvinylidene fluoride-hexafluoropropylene, and poly vinylidene fluoride-trifluoro ethylene-chlorotrifluoroethylene are examples of ferroelectric polymers that are frequently utilized for polymer products (Saleh et al. 2020).

27.7 Ferroelectric Polymer Nanocomposites

Ferroelectrics made of polyvinylidene fluoride have the highest dielectric constant of any polymer. Ferroelectric polymer nanocomposites have shown the ability to further increase the dielectric constant while maintaining high breakdown strength and excellent flexibility by introducing high dielectric constant nanofillers into polyvinylidene fluoride-based polymer matrixes (Guo et al. 2019). Polymer nanocomposites have generated a great deal of academic interest recently due to their promise of superior energy storage capabilities. Numerous factors, including the characteristics of the nanofillers, the polymer/filler interfaces, and the spatial composite architectures, influence the dielectric and energy storage capabilities of nanocomposites (Yang et al. 2020).

27.8 Polymer–Ceramic Nanocomposites

Ionically conductive solids called polymer–ceramic nanocomposites are made of both polymer and ceramic phases. This kind of substance belongs to the category of solid electrolytes. Polymer–ceramic nanocomposites could be created by combining a greater variety of polymer and ceramic compounds. The conductivity, cation transport number, and stability of the electrode–electrolyte interface are all improved by the addition of a ceramic phase to an ionically conductive matrix. The ionic conductivity of nanocomposites is influenced by several variables, including temperature, chemistry, size and volume fraction of the ceramic phase, annealing parameters, physical and chemical characteristics of the polymer matrix, and reactivity between the polymer and ceramic phase effect (Kumar 2013). Films and bulk nanocomposite samples can be created using a variety of processing techniques, including solvent casting, melt casting, heat pressing, etc. The polymer, lithium, and ceramic ingredients are combined by crushing using a mortar and pestle or heated into pellets, and then pressed to the required thickness is the most appropriate method. This method can be used to create nanocomposite materials with a variety of ceramic phase concentrations in the polymer matrix.

27.9 Summary and Future Trends

Materials are needed in the developing field of renewable energy technologies to produce unique components with distinct electrical properties. Devices for efficient power generation and energy storage require superionic conductors. Collecting energy from acoustic vibrations with inexpensive mixed conductors may be simpler (Ibrahim et al. 2020). Energy conversion and other industrial applications seek catalyst materials that can enhance reaction kinetics at low temperatures. Generation, transmission, distribution, and storage are the fundamental building blocks aimed at the production of energy from renewable energy sources. Materials having a particular blend of electrical, mechanical, and thermal qualities are needed for each of these components. Interesting ionic and electronic conductivity is provided by nanocomposites. These materials can be used to store thermal energy storage (Bahari, Najafi, and Babapoor 2020). These requirements can be satisfied by nanocomposites with tailored electrical conductivity and the desired mechanical and thermal properties. The need for renewable energy technologies will be what influences future trends in this field.

References

Ashokkumar, S. P., H. Vijeth, L. Yesappa, M. Niranjana, M. Vandana, and H. Devendrappa. 2020. 'Electrochemically Synthesized Polyaniline/Copper Oxide Nano Composites: To Study Optical Band Gap and Electrochemical Performance for Energy Storage Devices'. *Inorganic Chemistry Communications* 115 (January): 107865.

Sreesvarna, B., S. Pugalendhi, P. Subramanian, and J. Gitanjali. 2019. 'Characterization of Rice Husk for Sustainable Applications'. *Madras Agricultural Journal* 106 (Spl): 279–283.

Bahari, Mehdi, Bahman Najafi, and Aziz Babapoor. 2020. 'Evaluation of α-AL2O3-PW Nanocomposites for Thermal Energy Storage in the Agro-Products Solar Dryer'. *Journal of Energy Storage* 28 (December): 101181.

Chen, S., A. Skordos, and V. K. Thakur. 2020. 'Functional Nanocomposites for Energy Storage: Chemistry and New Horizons'. *Materials Today Chemistry* 17: 100304.

Conway, B. E., and W. G. Pell. 2003. 'Double-Layer and Pseudocapacitance Types of Electrochemical Capacitors and Their Applications to the Development of Hybrid Devices'. *Journal of Solid State Electrochemistry* 7 (9): 637–644.

Gitanjali, J., S. Pugalendhi, S. Kamaraj, S. Karthikeyan, and V.J.F. Kumar. 2015. 'Feasibility Test of Agricultural Residues through Characterization for Utilization in Plasma Gasification'. *Indian Journal of Agricultural Sciences* 85 (12): 1534–1539.

Gong, Shengqin, Xiaomin Cheng, Yuanyuan Li, Xiuli Wang, Yanping Wang, and Hao Zhong. 2020. 'Effect of Nano-SiC on Thermal Properties of Expanded Graphite/1-Octadecanol Composite Materials for Thermal Energy Storage'. *Powder Technology* 367: 32–39.

Guo, Mengfan, Jianyong Jiang, Zhonghui Shen, Yuanhua Lin, Ce Wen Nan, and Yang Shen. 2019. 'High-Energy-Density Ferroelectric Polymer Nanocomposites for Capacitive Energy Storage: Enhanced Breakdown Strength and Improved Discharge Efficiency'. *Materials Today* 29: 49–67.

Hossain, S. K.S., and M. E. Hoque. 2018. *Polymer Nanocomposite Materials in Energy Storage: Properties and Applications. Polymer-Based Nanocomposites for Energy and Environmental Applications: A Volume in Woodhead Publishing Series in Composites Science and Engineering.* Elsevier Ltd, Woodhead Publishing India.

Hussain, Hafiza Vaneeza, Mateeb Ahmad, Muhammad Tamoor Ansar, Ghulam M. Mustafa, Saira Ishaq, Shahzad Naseem, Ghulam Murtaza, Farah Kanwal, and Shahid Atiq. 2020. 'Polymer Based Nickel Ferrite as Dielectric Composite for Energy Storage Applications'. *Synthetic Metals* 268 (May): 1–9.

Ibrahim, Idowu D., Emmanuel R. Sadiku, Tamba Jamiru, Yskandar Hamam, Yasser Alayli, and Azunna A. Eze. 2020. 'Prospects of Nanostructured Composite Materials for Energy Harvesting and Storage'. *Journal of King Saud University – Science* 32 (1): 758–764.

Jothiprakash, Gitanjali, and Venkatachalam Palaniappan. 2014. 'Development and Optimization of Pyrolysis Unit for Producing Charcoal'. *International Journal of Agriculture, Environment and Biotechnology* 7 (4): 863.

Kumar, B. 2013. 'Ceramic Nanocomposites for Energy Storage and Power Generation'. In *Ceramic Nanocomposites*, 509–529. Woodhead Publishing India.

Moghimikheirabadi, Ahmad, Argyrios V. Karatrantos, and Martin Kröger. 2021. 'Ionic Polymer Nanocomposites Subjected to Uniaxial Extension: A Nonequilibrium Molecular Dynamics Study'. *Polymers* 13 (22): 1–20.

Qin, Jingwen, Yunkang Chen, Changlu Xu, and Guiyin Fang. 2021. 'Synthesis and Thermal Properties of 1-Octadecanol/Nano-TiO2/Carbon Nanofiber Composite Phase Change Materials for Thermal Energy Storage'. *Materials Chemistry and Physics* 272 (July): 125041.

Riggs, Brian C., Shiva Adireddy, Carolyn H. Rehm, Venkata S. Puli, Ravinder Elupula, and Douglas B. Chrisey. 2015. 'Polymer Nanocomposites for Energy Storage Applications'. *Materials Today: Proceedings* 2 (6): 3853–3863.

Roselin, L. Selva, Ruey Shin Juang, Chien Te Hsieh, Suresh Sagadevan, Ahmad Umar, Rosilda Selvin, and Hosameldin H. Hegazy. 2019. 'Recent Advances and Perspectives of Carbon-Based Nanostructures as Anode Materials for Li-Ion Batteries'. *Materials* 12 (8): 1229.

Saleh, Tawfik A., Nagaraj P. Shetti, Mahesh M. Shanbhag, Kakarla Raghava Reddy, and Tejraj M. Aminabhavi. 2020. 'Recent Trends in Functionalized Nanoparticles Loaded Polymeric Composites: An Energy Application'. *Materials Science for Energy Technologies* 3: 515–525.

Singh, Rupinder, and Ranvijay Kumar. 2020. *Energy Storage Device From Polymeric Waste Based Nano-Composite by 3D Printing. Encyclopedia of Renewable and Sustainable Materials*. Elsevier Ltd. BV.

Sreenath, P. R., M. S. Seema Singh Prolay Das Satyanarayana, and K. Dinesh Kumar. 2017. 'Carbon Dot–Unique Reinforcing Filler for Polymer with Special Reference to Physico-Mechanical Properties'. *Polymer* 112: 189–200.

Wang, Yumeng, Jian Sun, Bin He, and Mi Feng. 2020. 'Synthesis and Modification of Biomass Derived Carbon Dots in Ionic Liquids and Their Application: A Mini Review'. *Green Chemical Engineering* 1 (2): 94–108.

Yang, Xin, Xiaoming Zhu, Liudi Ji, Peng Hu, and Zeyu Li. 2020. 'Largely Enhanced Energy Storage Performance in Multilayered Ferroelectric Polymer Nanocomposites with Optimized Spatial Arrangement of Ceramic Nanofillers'. *Composites Part A: Applied Science and Manufacturing* 139 (September): 106111.

Yi, Ting Feng, Jing Jing Pan, Ting Ting Wei, Yanwei Li, and Guozhong Cao. 2020. 'NiCo2S4-Based Nanocomposites for Energy Storage in Supercapacitors and Batteries'. *Nano Today* 33: 100894.

Yu, Qiang, Yuanwei Lu, Cancan Zhang, Xiaopan Zhang, Yuting Wu, and Adriano Sciacovelli. 2020. 'Preparation and Thermal Properties of Novel Eutectic Salt/Nano-SiO$_2$/Expanded Graphite Composite for Thermal Energy Storage'. *Solar Energy Materials and Solar Cells* 215 (May): 110590.

28

Development of Environmental Benign Nanomaterials for Energy and Environmental Applications

N. K. Anushkannan
Kathir College of Engineering, Coimbatore, India

Nikhat Farhana
YPCRC, Yenepoya (Deemed to be University), Naringana, Mangaluru, India

M. K. Valsakumari
Mookambigai College of Engineering, Pudukottai, India

N. Karthikeyan
Sri Srinivasa Matriculation Higher Secondary School, Thanjavur, India

Dr. V. Saravanan
Sri Meenakshi Vidiyal Arts and Science College, Trichy, India

Dr. M. Jayapriya
Pavendhar Bharathidasan Institute of Engineering and Technology, Trichy, India

DOI: 10.1201/9781003355755-28

Development of nanoparticles using environmental benign strategy is gaining more attention among the researchers for the fabrication of noble metal nanoparticles and metal oxide nanocomposites. The synthetic protocol using biological method is considered to be non-hazardous, simple, and lucrative does not produce any toxic byproducts. Biological synthetic protocol utilizes plants and their parts such as root, stem, flower, leaf, and microbes such as bacteria, fungi, algae, etc. Plant-based synthesis is more favorable than the microbe-mediated synthesis due to the maintenance of sterile environmental conditions. The plants are the renowned factories of bio constituents such as flavonoids, carbohydrates, tannins, anthocyanin, etc., which act as reducing and capping agents. The phytoconstituents standalone act as an antioxidant, antimicrobial, and antifungal agent while the nanomaterials synthesized using these biomolecules possess enhanced biocidal activities which act as a pivotal candidate in various fields, especially in environmental and biomedical applications. The utilization of plant materials for the synthesis of nanomaterials not only reduces the operation cost but also minimizes energy utilization in comparison to physiochemical method of synthesis, also the need of employing hazardous chemicals or the production of toxic byproducts, stimulates the usage of "green synthetic process". Implementing plant-mediated synthesis would be unquestionably a strong step toward sustainable development. Green nanotechnology seeks its application in almost all areas of science and technology, agriculture is no exception. Vast development in the field has to happen for the establishment of nanotechnology in various fields especially agro-industrial sector as agriculture is known as the backbone of our economy. Therefore, the present review is intended to recapitulate the ongoing developments and progress made in the green fabrication of nanomaterials and their various applications in several fields of biomedicine.

28.1 Introduction to Nanotechnology

In the past decade, nanotechnology has become the frontrunners and is defined as a multidisciplinary field with the implementation of science and technology to control matter at a molecular level [1]. Immense escalation in the field of nanotechnology paved the way in the field of materials science and engineering notably nanobiotechnology, surface-enhanced Raman scattering (SERS), applied microbiology, and quantum dots [2]. The word "nano" indicates small, which implies the dimensions of a particle that falls within the order of 10^{-9}m and is one billionth of a meter, which may be a thousand-fold smaller than a red blood cell, or about half the dimensions of the diameter of DNA. The smaller size and larger surface-to-volume ratio of the nanoparticles make them a potent candidate in a diverse field, which are significantly different from those bulk materials [3].

Nanotechnology includes measuring, imaging, sensing, modeling, and manipulating matter at nanoscale [4]. Nanoparticles exhibit higher structural integrity, unique optical, chemical, biological, electrical, and magnetic properties [5]. These unique properties may be due to its variations in certain characteristics notably size, morphology, and distribution of nanoparticles [6]. Owing to these astonishing properties, they present myriad applications in the field of biomedicine, food industries, pharmaceutical, and environmental bioremediation. Also, the synthesis of nanomaterials with controlled morphology plays a pivotal role in a diverse field such as optics, chemical industry, drug–gene delivery, energy science, non-linear optical device, and photoelectrochemical applications. Hence, the fabrication of nanomaterials with a novel application can be accomplished by controlling shape with various morphology and size at the nanometer scale [7].

28.2 Properties of Nanoparticles

The nanoparticles display pivotal properties, which are more different from their bulk counterparts. The particles at the nanometer scale exhibit low boiling and melting point and possess reduced lattice constants, owing to the presence of surface ions or atomic molecules. The higher surface energy plays a pivotal role in the thermal stability of the nanomaterials. The quantum confinement effect of nanomaterial and the electronic properties of the nanomaterials gets altered with respect to reduced particle size. At nanometric scale, the crystals are stable at higher temperature, which are more stable even at lower temperature. The bulk materials become insulator when the dimensions of the materials get reduced, i.e., in nanometric range.

28.3 Metal Nanoparticles

Metal nanoparticles gained greater attention due to their multifunctional properties in vast fields including biomedicine, catalysis, pharmaceutical industries, thin film formation, household products, optics, antimicrobial, antioxidant, anticancer and production of biomaterials [8]. Noble metal nanoparticles (NMNs) such as silver (Ag), palladium (Pd), gold (Au), and platinum (Pt) exhibit unique physiochemical properties that make them a versatile candidate in the promptly moving consumer products notably detergents, shampoos, shoes, toothpaste, cosmetics, and soaps along with their applications in pharmaceutical and medical products. Manipulation of these noble metals in the nanoscale dimensions exhibits a higher surface area-to-volume ratio and quantum confinement [9]. They are uniquely different from their atomic composition and bulk counterparts [10]. The physicochemical properties such as shape, size, crystalline nature, and composition of the materials are the key features that can be used to tailor the functionalities of NMNs and make them a potent material in various applications [11].

Among the aforesaid NMNs, silver NPs (AgNPs) have a lot of deliberation owing to their capacious spectrum in several fields such as catalytic activity, energy conversion, bactericidal and fungicidal activities, free radical quenching effect, anti-inflammatory activity, and the minimum propensity toward various disease resistance microbes. The astonishing feature of AgNPs is that at the least dosage, they inhibit the growth of microbes, especially more efficient against multidrug-resistant bacteria, but safer to human cells [12]. In the biomedical field, AgNPs are being added in antiseptic sprays, fabrics, wound dressing, and topical creams and exhibit a wide range of biocidal activity against various microorganisms by the distribution of their cellular membrane resulting in the inhibition of DNA replication ability [13]. Ag NPs integrate directly into the cellular membrane and create pores on the surface which eventually produce osmotic collapse and thus release intracellular materials [14].

Furthermore, Ag NPs possess pro-apoptotic, pro-inflammatory, and cytotoxic effects arbitrated by reactive oxygen species (ROS) produced in normal body cells and tumor cell lines [15]. Ag NPs with reduced size and morphology-controlled shapes such as rods, spherical shaped, and crystal-shaped structures exhibit fascinating applications in various fields owing to their increase in the higher surface-to-volume ratio [16]. While compared to bulk counterparts, the reduced size of AgNPs causes noteworthy change due to the exhibition of negative Fermi potential [17]. This is a main key factor that helps Ag NPs to transfer electrons from donor to acceptor molecules in various catalytic reactions. Also, excess production of ROS leads to various catastrophic diseases such as neurodegenerative disorder, Alzheimer, atherosclerosis, arthritis, diabetes, and cancer.

In the recent past, studies have demonstrated that Ag NPs is a promising candidate to quench free radical and act as a better antioxidant molecule [18]. Hence, the functionalization of nanomaterials with naturally available antioxidant could reduce ROS production and shield proteins present in the cell, and lipids that prevent adverse reactions [19]. However, there are certain shortcomings associated with oxidization and spontaneous aggregation of Ag NPs tends to reduce surface energy that limits the reaction efficiency and stability of nanoparticles. Also, ultrahigh cost of AgNPs hinders their further usage in practical applications. Hence, in order to satisfy the urgent demand for highly functional nanomaterials with multifunctional properties, the design of new generation nanocomposite by the combination of one or more dissimilar material with distinct properties has been employed, which offers synergetic multifunctional efficiency toward assorted applications [20]. Hence, in the realm of nanomaterial, considerable attention has been focused on the fabrication of noble metal-based metal oxide (NM-MOs), and metal oxide nanocomposite (MO-NCs) with multifunctional efficiency.

28.4 Metal Oxide Nanoparticles (MO-NPs)

Nanoparticles can be broadly organized into two categories specifically organic nanoparticles and inorganic nanoparticles. Organic nanoparticles comprise carbon nanoparticles (fullerenes), etc., whereas inorganic nanoparticles consist of NMNs, magnetic nanoparticles, and metal oxide nanoparticles. Among the aforesaid inorganic nanomaterials, metal oxide nanoparticles have garnered greater deliberation due its unique characteristic feature which makes them a potent candidate with enhanced optical, biological, electrical, and catalytic properties [21]. Metal oxides in their nanoform exhibit enhanced properties with a novel characteristic feature when compared with bulk counterparts.

Among the aforementioned nanomaterials, magnesium oxide nanoparticles have attracted greater interest due to their functional characteristic features which validate them as a waste remediation agent, catalyst, paints, an additive in refractory materials, superconductors, and as a biocidal agent in the pharmaceutical industry. The pivoting property of MgO NPs includes enhanced catalytic behavior, reduced electrical conductivity, and high thermal stability [22]. These properties can be further enhanced by synthesizing these materials at nanoscale. US Food and Drug Administration (USFDA) acknowledged that MgO NPs (21CFR184.1431) are non-hazardous, eco-friendly in nature that can be readily used in pharmaceutical industries.

Furthermore, zinc oxide (ZnO) is said to be biologically safe metal oxide that acts as an effective antimicrobial agent, i.e., it inhibits the growth of several disease-causing bacteria and fungi and has the ability to act as a better antioxidant agent. It also acts as a proficient nanocatalyst for the degradation of manmade pollutants released from several industries owing to the presence of higher surface-to-volume ratio, facile generation of active sites, and efficient electron transfer abilities to reduce the industrial pollutants [23]. The USFDA has identified (21 CFR 182.8991) zinc oxide as safe, and environment benign in nature. There isa wide range of techniques that are responsible for the production of nanostructures with several degrees of speed, quality, and cost-effective. The manufacturing approaches of nanoparticles fall under two broad categories "bottom-up" and "top-down" approaches.

28.5 Synthesis of Nanoparticles

Various strategies have been used for the synthesis and fabrication of nanoparticles; they are broadly classified into two classes they are "top-down" and "bottom-up" approaches. In top-down approach, suitable large material can be divided into fine particles with the reduction in size. It includes physical method with various lithographic techniques such as milling, sputtering, grinding, thermal, and laser ablation. There are several shortcomings that can be associated with these methods, which include heat liberation, high energy consumption, and expensive process.

In bottom-up approach, nanoparticles can be fabricated using chemical and biological methods. It is defined as the self-assembly of atoms into nuclei which resembles into nanoparticles. In bottom-up approach, chemical reduction is the most common technique used for the synthesis of metal and metal oxide nanoparticles. However, this method utilizes various organic and inorganic capping and reducing agents such as sodium citrate, sodium borohydride (NaBH$_4$), ascorbate, Tollen's reagent, elemental hydrogen, poly(ethylene glycol), and N,N-dimethylformamide (DMF). They produce toxic byproducts that are hazardous to the environment, making these nanomaterials unsuitable for biomedical, environmental, and energy applications [24]. The overview for the synthesis of nanomaterials is depicted in Figure 28.1.

In recent past, biological synthesis of metal and metal oxide nanoparticles has received greater interest because it is one-step, simple, lucrative, and environmental benign process [25–27]. Biological fabrication of nanoparticles utilizes assorted biological sources such as plants, fruit, seed extract, agro-waste products, and microorganisms such as bacteria, fungi, and yeast [28]. However, microbe-mediated synthesis of nanoparticles has a major drawback because it is not feasible for pilot-scale production due to the maintenance of cell culture under aseptic condition. In this regard, the syntheses of nanomaterials using plants are found to be better alternative in nanoparticle fabrication [29].

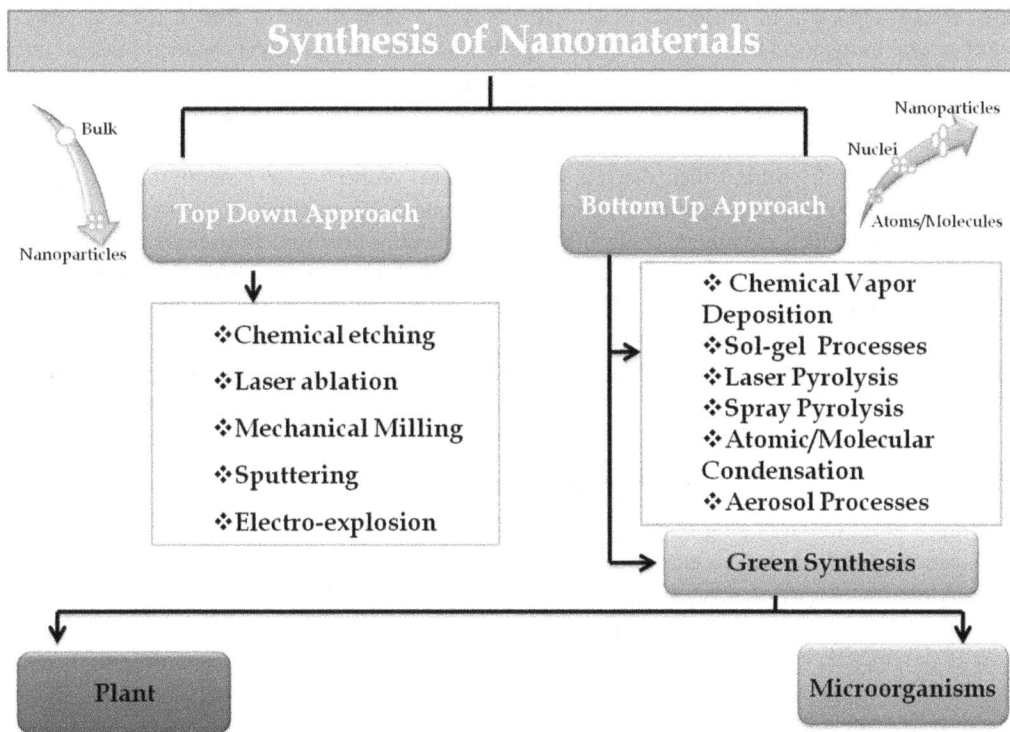

FIGURE 28.1 Overview for the synthesis of nanomaterials.

There are several advantages associated with this strategy which include low consumption of energy, moderate operational condition for pressure and temperature maintenance, without using any toxic chemicals. The plant extract acts as both reducing and capping mediators, and the nanoparticles thus formed are more stable with different shapes and dimensions, making the plant-mediated synthesis more efficient than other synthesis methods [30].

Generally, the nanoparticles with intended application mainly depend on its composition. If the nanoparticles interact with biological system, functional groups will be attached to its surface in order to prevent aggregation or agglomeration [31]. For biomedical applications, the nanomaterials with biocompatibility, i.e., metals with reduced cytotoxicity are required. This was achieved through the biogenic synthesis of nanomaterials which are free from toxic contamination of byproducts. The add-on advantage in biogenic synthesis of nanoparticles is that the phytoconstituents present in the plant extract are responsible for the reduction of nanomaterials itself having enhanced antimicrobial, anticancer, and antioxidant activity [32–35]. Hence, the nanomaterials fabricated via biogenic route have enhanced biomedical applications which are considered to be a novel one.

28.6 Application of Nanomaterials for Energy Utilization

In our day-to-day life, our world has been focusing on severe challenges to achieve energy utilization in a highly efficient way. Multifunctional nanomaterials gain greater attention in the field of energy conversion, energy production, storage of energy, and energy diffusion, which are mainly depend upon their optoelectrical properties, thermal, mechanical, and catalytic functions of the noble metals and metal-based materials at nanoscale [36]. The pivoting properties of nanomaterials such as thermoelectric, triboelectricity, devices based on photovoltaic fabrication, catalytic, and electrochromic substances have created important involvements to several energy-related applications. The distinctive properties of several inorganic nanomaterial include high thermal stability, electrical conductivity, and large surface-to-volume ratio which makes them a potent candidate in energy applications [37].

Multifunctional nanoplatforms are designed to perform various applications with the judicious combination of several metal and metal-based nanocomposites. They have high efficiency to deliver certain system-level capability behind their discrete physiochemical properties. In general, the functions of multifunctional nanomaterial may be concurrently or sequentially executed on the same scale at the nanometric range which are systematically organized on the same platform. The remarkable features of multifunctional nanomaterials include optical, electrical, and magnetic properties which have great impacts on various fields such as electronics, spintronics, and medical devices. Integration of several nanomaterials in single platform with various functionalities is a challenging one. Batteries require various electrodes with high electrochemical reactivity, lower toxicity, reversibility, and excellent chemical stabilities [38].

28.6.1 Energy Conversion Applications – Solar Cells

The crucial point in developing solar cells is the efficiency of photocurrent produced internally, the conversion of absorbed photons energy into electric current, the outer quantum level efficiency of the material, the fraction efficiency of incident photons on the device which are transferred into electric current, and the total efficiency of energy conversion. Several recent advances have been furnished in order to develop the conversion efficiency of solar cell-based devices [39–40]. The most widely used absorber is the silicon which governs the fabrication of PV devices in market. The first-generation solar cells were developed based on the silicon-based PV devices entrenched on p-n junctions said to have an overall conversion efficiency of about 25%. The design and the fabrication of second-generation photovoltaic devices on the surface of thin-film technologies to increase the efficiency of solar cell have been achieved. However, the second-generation efficiency was lower than the first-generation efficiency, i.e., silicon-based material. The third-generation cell was developed based on the exploitation of dye-sensitized solar cells fabrication, and solar cells fabrication on quantum dots. Based on the various materials utilized in solar cells application, perovskites-based applications have become more successful in the development of solar cells owing to its promising features such lucrative, flexibility, and higher efficiency toward solar cell applications [41–43]. Despite their processing advantages, the low stability of perovskites in accordance with their humidity, light, oxygen, and heat their industrial applications was constrained.

28.6.2 Fuel Cells

The up conversion of chemical energy obtained from the fuel into electrical energy using the conversion reaction mechanism undergone with the oxygen or several oxidizing mediators are achieved by fuel cells. The combination of fuel cells in the optoelectronic devices faces numerous provocations which include: (i) finding appropriate electrodes for the fabrication of malleable electronics. (ii) Utilization of alternative materials for high-cost plasmonic metals such as platinum, ruthenium, gold, silver, and their mixture of alloy compounds. (iii) Avoidance of metal electrode poisoning [44–46]. Hence, to overcome these hindrances, the development of new class of multifunctional materials with environmental benign strategy, lucrative, and high efficient material have been developed for the conversion of energy in electronic devices. The most widely utilized fuel cell production evolved from direct methanol-based fuel cells (DMFCs) and proton exchange membrane-based fuel cells (PEMFCs), which are considered to be the most pivotal candidate utilized in the portable electronic gadgets such as cellular phones, laptops, and personal electronics owing to the consumption of high power density ability and devices operating at low temperature [47–49]. The nanomaterials developed with platinum and its mixture of alloy-based materials have been utilized as a catalyst in fuel-based devices due to its enhanced catalytic activity but owing to high cost it hinders its industrial applications in order to commercialize the DMFCs. To overcome this drawback, the supporting material with low cost was achieved to improve the catalyst efficiency. The supporting nanomaterial must be more stable uniformly dispersed in solution. In the recent past, the carbon nanomaterials can be used as an excellent supporting material owing to their good electrical conductivity, easily available source, and low-cost materials [50].

28.7 Energy Storage Applications

The design and the fabrication of renewable and recyclable eco-friendly energy-storing devices is the best way to tackle the existing and the upcoming energy crisis all along with the various environmental problems. The exploitation and utilization of suitable nanomaterials are the pivotal properties for the fabrication of high-energy, efficient, stable, lucrative, and environmental benign energy-storing devices. Electrochemical repository devices especially supercapacitors and batteries are used for storing charges which are said to be limited through the process of diffusion which allows higher power that can be achieved [51–53]. The demand for higher performance of many rechargeable and reusable batteries which has become so much tangible and said to be ever-present in the current past which requires many functions that have been risen to the rank of very common knowledge.

28.8 Batteries

Batteries are the key constituent for energy-storing devices and it has the capability to store more electrical energy which is in the form of chemical energies. Through the process of reversible reactions, electrochemical reaction occurs and stored in batteries for future usage. In the recent past, many batteries have been utilized as a remarkable successful candidate in both research and the commercialization process in industries [54–55].

28.9 Conclusion

Nanotechnology has moved into modern biological and medicinal implications for the advancement of the biomedical field and environmental applications. Recent developments in nanoparticle synthesis have focused on safe and environmentally sustainable approaches. Plant-mediated nanoparticles are not only eco-friendly but also safe mode that utilizes biological resources for synthesis. The green approach to nanoparticle fabrication has opened up a new era of safe nanotechnology. The metal oxide and metal nanoparticles are well known for their unique outstanding behavior and properties. In the present view, we conceptualize the green synthesis of metal and metal oxide nanomaterials from bio-resources. Furthermore, we have documented the effect of reactants and other kinetic conditions regarding the size, shape, and charge of the NPS/NCs. Additionally, we broadly present the energy applications mediated by metal oxide nanoparticles and noble metal nanomaterials. Henceforth, the present review will benefit the researchers and applied scientists who attempt to produce the green synthesis and their subsequent applications in energy and environmental applications.

References

1. Naseer, B., Srivastava, G., Qadri, O.S., Faridi, S.A., Islam, R.U. and Younis, K. 2018. "Importance and health hazards of nanoparticles used in the food industry." *Nanotechnology Reviews* 7(6), 623–641. doi:10.1016/j.jafr.2022.100270
2. Huang, H.X. and Wang, X. 2019. "Biomimetic fabrication of micro-/nanostructure on polypropylene surfaces with high dynamic superhydrophobic stability." *Materials Today Communications* 19, 487–494. doi:10.1016/j.mtcomm.2019.04.005

3. Bayda, S., Adeel, M., Tuccinardi, T., Cordani, M. and Rizzolio, F. 2019. "The history of nanoscience and nanotechnology: From chemical–physical applications to nanomedicine." *Molecules* 25(1), 112. doi:10.3390/molecules25010112

4. Singh, R. and Nalwa, H.S. 2011. "Medical applications of nanoparticles in biological imaging, cell labeling, antimicrobial agents, and anticancer nanodrugs." *Journal of Biomedical Nanotechnology* 7(4), 489–503. doi:10.1166/jbn.2011.1324

5. Jeevanandam, J., Barhoum, A., Chan, Y.S., Dufresne, A. and Danquah, M.K. 2018. "Review on nanoparticles and nanostructured materials: History, sources, toxicity and regulations." *Beilstein Journal of Nanotechnology* 9(1), 1050–1074. doi:10.3762/bjnano.9.98

6. Ahmed, S., Ahmad, M., Swami, B.L. and Ikram, S.2016. "A review on plants extract mediated synthesis of silver nanoparticles for antimicrobial applications: A green expertise." *Journal of Advanced Research* 7(1), 17–28. doi:10.1016/j.jare.2015.02.007

7. Patil, S. and Chandrasekaran, R. 2020. "Biogenic nanoparticles: A comprehensive perspective in synthesis, characterization, application and its challenges." *Journal, Genetic Engineering & Biotechnology* 18(1), 1–23. doi:10.1186/s43141-020-00081-3

8. Roco, M.C. 1999. "Nanoparticles and nanotechnology research." *Journal of Nanoparticle Research* 1(1), 1. doi:10.1023/A:1010093308079

9. Aslam, B., Wang, W., Arshad, M.I., Khurshid, M., Muzammil, S., Rasool, M.H. and Baloch, Z. 2018. "Antibiotic resistance: A rundown of a global crisis." *Infection and Drug Resistance* 11, 1645. doi:10.2147/IDR.S173867

10. Obi, F.O., Ugwuishiwu, B.O. and Nwakaire, J.N. 2016. "Agricultural waste concept, generation, utilization and management." *Nigerian Journal of Technology* 35(4), 957–964.

11. Martins, D.A.B., do Prado, H.F.A., Leite, R.S.R., Ferreira, H., de Souza, M.M., Moretti, R.D.S. and Gomes, E. 2011. "Agroindustrial wastes as substrates for microbial enzymes production and source of sugar for bioethanol production." In *Integrated Waste Management-Volume II*. Intech Open. doi:10.5772/23377

12. Seetharaman, P., Chandrasekaran, R., Gnanasekar, S., Mani, I. and Sivaperumal, S. (2017). "Biogenic gold nanoparticles synthesized using Crescentia cujete L. and evaluation of their different biological activities." *Biocatalysis and Agricultural Biotechnology* 11, 75–82. doi:10.1016/j.bcab.2017.06.004

13. Ponarulselvam, S., Panneerselvam, C., Murugan, K., Aarthi, N., Kalimuthu, K. and Thangamani, S. 2012. "Synthesis of silver nanoparticles using leaves of Catharanthus roseus Linn. G. Don and their antiplasmodial activities." *Asian Pacific Journal of Tropical Biomedicine* 2(7), 574–580. doi:10.1016/S2221-1691(12)60100-2

14. Kumar, V., Yadav, S.C. and Yadav, S.K. (2010). "Syzygium cumini leaf and seed extract mediated biosynthesis of silver nanoparticles and their characterization." *Journal of Chemical Technology & Biotechnology* 85(10), 1301–1309. doi:10.1002/jctb.2427

15. Dubey, S.P., Lahtinen, M. and Sillanpa, M. 2010. "Tansy fruit mediated greener synthesis of silver and gold nanoparticles." *Process Biochemistry* 45(7), 1065–1071. doi:10.1016/j.procbio.2010.03.024

16. Krishnaraj, C., Jagan, E.G., Rajasekar, S., Selvakumar, P., Kalaichelvan, P.T. and Mohan, N.J.C.S.B.B. 2010. "Synthesis of silver nanoparticles using Acalypha indica leaf extracts and its antibacterial activity against water borne pathogens." *Colloids and Surfaces B: Biointerfaces* 76(1), 50–56. doi:10.1016/j.colsurfb.2009.10.008

17. Mustapha, T., Misni, N., Ithnin, N.R., Daskum, A.M. and Unyah, N.Z. 2022. "A Review on Plants and Microorganisms Mediated Synthesis of Silver Nanoparticles, Role of Plants Metabolites and Applications." *International Journal of Environmental Research and Public Health* 19(2), 674. doi:10.3390/ijerph19020674

18. Zu, H. and Gao, D. 2021. "Non-viral vectors in gene therapy: Recent development, challenges, and prospects." *The AAPS Journal* 23(4), 1–12. doi:10.1208/s12248-021-00608-7

19. Sorensen, C.M. 2001. *Nanoscale Materials in Chemistry*. New York: Wiley. Interscience.

20. Citakovic, N.M. 2019. "Physical properties of nanomaterials." *Vojnotehnički glasnik* 67(1), 159–171.

21. Reiss G, Hutten A. 2005. "Magnetic nanoparticles: Applications beyond data storage." *Nature Materials* 4(10): 725–6. doi:10.1038/nmat1494

22. Eastman, J.A., Choi, U.S., Li, S., Thompson, L.J. and Lee, S. 1996. "Enhanced thermal conductivity through the development of nanofluids." *MRS Online Proceedings Library (OPL)*, 457. doi:10.1557/PROC-457-3

23. Wong, T.S., Brough, B. and Ho, C.M. 2009. "Creation of functional micro/nano systems through top-down and bottom-up approaches." *Molecular & Cellular Biomechanics: MCB* 6(1), 1.

24. Khandel, P. and Shahi, S.K. 2016. "Microbes mediated synthesis of metal nanoparticles: Current status and future prospects." *International Journal of Nanomaterials and Biostructures* 6(1), 1.

25. Hinton, M., Hedges, A.J. and Linton, A.H. 1985. "The ecology of Escherichia coli in market claves fed a milk-substitute diet." *The Journal of Applied Bacteriology* 58, 27–35. doi:10.1111/j.1365-2672.1985.tb01426.x

26. Hosseini, S.M., Sarsari, I.A., Kameli, P. and Salamati, H. 2015. "Effect of Ag doping on structural, optical and photocatalytic properties of ZnO nanoparticles." *Journal of Alloys and Compounds* 640, 408–415. doi:10.1016/j.jallcom.2015.03.136.

27. Jeyabharathi, S., Kalishwaralal, K., Sundar, K. and Muthukumaran, A. 2017. "Synthesis of zinc oxide nanoparticles (ZnONPs) by aqueous extract of Amaranthus caudatus and evaluation of their toxicity and antimicrobial activity." *Materials Letters* 209, 295–298. doi:10.1016/j.matlet.2017.08.030

28. Chang, X., Li, Z., Zhai, X., Sun, S., Gu, D., Dong, L., Yin, Y. and Zhu, Y. 2016. "Efficient synthesis of sunlight-driven ZnO-based heterogeneous photocatalysts." *Materials and Design* 98, 324–332. doi:10.1016/j.matdes.2016.03.027

29. Brayner, R., Ferrari-Iliou, R., Brivois, N., Djediat, S., Benedetti, M.F. and Fievet, F. 2006. "Toxicological impact studies based on Escherichia coli bacteria in ultrafine ZnO nanoparticles colloidal medium." *Nano Letters* 6, 866–870. doi:10.1021/nl052326h

30. Azizi, S., Ahmad, M.B., Namvar, F. and Mohamad, R. 2014. "Green biosynthesis and characterization of zinc oxide nanoparticles using brown marine macroalga Sargassum muticum aqueous extract." *Materials Letters* 116, 275–277. doi:10.1016/j.matlet.2013.11.038

31. Harinee, S., Muthukumar, K., Abirami, A., Amrutha, K., Dhivyaprasath, K. and Ashok, M. 2019 "UV-light photocatalytic activity of biocompatible nanoparticles provide multiple effects." *Advanced Materials and Processes* 4(3), 115–118. doi:10.5185/amp.2019.0006

32. Subramanian, H., Krishnan, M. and Mahalingam, A.2022. "Photocatalytic dye degradation and photoexcited anti-microbial activities of green zinc oxide nanoparticles synthesized via Sargassum muticum extracts." *RSC Advances* 12, 985–997. doi:10.1039/D1RA08196A

33. Muthukumar, K., Vignesh, S., Dahms, H.U., Santhosh, G.M., Palanichamy, S., Subramanian, G. and James, R.A. 2015. "Antifouling assessments on biogenic nanoparticles: A field study from polluted offshore platform." *Marine Pollution Bulletin* 101, 816–825. doi:10.1016/j.marpolbul.2015.08.033.

34. Zhang, Z. and Feng, S.S. 2006. "The drug encapsulation efficiency, in vitro drug release, cellular uptake and cytotoxicity of paclitaxel-loaded poly (lactide)–tocopheryl polyethylene glycol succinate nanoparticles." *Biomaterials* 27, 4025–4033. doi:10.1016/j.biomaterials.2006.03.006

35. Shanmugam, N., Rajkamal, P., Cholan, S., Kannadasan, N., Sathishkumar, K., Viruthagiri, G. and Sundaramanickam, A. 2013. "Biosynthesis of silver nanoparticles from the marine seaweed Sargassum wightii and their antibacterial activity against some human pathogens." *Applied Nanoscience* 4, 881–888. doi:10.1007/s13204-013-0271-4

36. Bharathi, D., Vasantharaj, S. and Bhuvaneshwari, V. 2018. "Green synthesis of silver nanoparticles using Cordia dichotoma fruit extract and its enhanced antibacterial, anti-biofilm and photo catalytic activity." *Materials Research Express* 5(5), 1–8. doi:10.1088/2053-1591/aac2ef

37. Zare, M., Namratha, K., Alghamdi, S., Mohammad, Y.H.E., Hezam, A., Zare, M., Drmosh, Q.A., Byrappa, K., Chandrashekar, B.N., Ramakrishna, S. and Zhang, X. 2019. "Novel green biomimetic approach for synthesis of ZnO-Ag nanocomposite; Antimicrobial activity against food-borne

pathogen, biocompatibility and solar photocatalysis." *Scientific Reports* 9, 1–15. doi:10.1038/s41598-019-44309-w

38. Gao, S., Jia, X., Yang, S., Li, Z. and Jiang, K. 2011. "Hierarchical Ag/ZnO micro/nanostructure: Green synthesis and enhanced photocatalytic performance." *Journal of Solid State Chemistry* 184, 764–769. doi:10.1016/j.jssc.2011.01.025

39. Thareja, R.K. and Shukla, S. 2017. "Synthesis and characterization of zinc oxide nanoparticles by laser ablation of zinc in liquid." *Applied Surface Science* 253, 8889–8895.

40. Heiligtag, F.J. and Niederberger, M. 2013. "The fascinating world of nanoparticle research." *Materials Today* 16, 262–271. doi:10.1016/j.apsusc.2007.04.088

41. Hasnidawani, J.N., Azlina, H.N., Norita, N., Bonnia, N.N., Ratim, S. and Ali, E.S. 2016. "Synthesis of ZnO nanostructures using solgel method." *Progress in Chemistry* 19, 211–216. doi:10.1016/j.proche.2016.03.095

42. Li, G., He, D., Qian, Y., Guan, B., Gao, S., Cui, Y., Yokoyama, K. and Wang, L. 2012 "Fungus-mediated green synthesis of silver nanoparticles using Aspergillus terreus." *International Journal of Molecular Sciences* 13, 466–476. doi:10.3390/ijms13010466

43. Akhayere, E. and Kavaz, D. 2022. "Synthesis of silica nanoparticles from agricultural waste." *Nanobiotechnology for Plant Protection* 121–138. doi:10.1016/B978-0-12-823575-1.00028-7

44. Vasyliev, G. and Vorobyova, G.V. 2020. "Valorization of Food Waste to Produce Eco-Friendly Means of Corrosion Protection and "Green" Synthesis of Nanoparticles." *Advances in Materials Science and Engineering* 1–14. doi:10.1155/2020/6615118

45. Muhammad Idris, N., Li, Z., Ye, L., Sim, E.K.W., Mahendran, R., Chi-Lui Ho, P. and Zhang, Y. 2009. "Tracking transplanted cells in live animal using up conversion fluorescent nanoparticles." *Biomaterials* 30 (28), 5104–5113. doi:10.1016/j.biomaterials.2009.05.062

46. Ahmad, A., Wei, Y., Syed, F., Tahir, K., Rehman, A.U., Khan, A., Ullah, S. and Yuan, Q. 2017. "The effects of bacteria-nanoparticles interface on the antibacterial activity of green synthesized silver nanoparticles." *Microbial Pathogenesis* 102, 133–142. doi:10.1016/j.micpath.2016.11.030

47. Senthilkumar, P., Santhosh Kumar, D.S.R., Sudhagar, B., Vanthana, M., Parveen, M.H., Sarathkumar, S., Thomas, J.C., Sandhiya Mary, S. and Kannan, C. 2016. "Seagrass-mediated silver nanoparticles synthesis by Enhalus acoroides and its α-glucosidase inhibitory activity from the Gulf of Mannar." *Journal of Nanostructure in Chemistry* 6, 275–280. doi:10.1007/s40097-016-0200-7

48. Inbakandan, D., Kumar, C., Bavanilatha, M., Ravindra, D.N., Kirubagaran, R. and Khan, S.A. 2016. "Ultrasonic-assisted green synthesis of flower like silver nanocolloids using marine sponge extract and its effect on oral biofilm bacteria and oral cancer cell lines." *Microbial Pathogenesis* 99, 135–141. doi:10.1016/j.micpath.2016.08.018

49. Muthusamy, G., Thangasamy, S., Raja, M., Chinnappan, C. and Kandasamy, S. 2017. "Biosynthesis of silver nanoparticles from Spirulina microalgae and its antibacterial activity." *Environmental Science and Pollution Research* 24, 19459–19464. doi:10.1007/s11356-017-9772-0

50. Harinee, S., Muthukumar, K., Dahms, H.U., Koperuncholan, M., Vignesh, S., Banu, R.J., Ashok, M. and James, R.A. 2019 "Biocompatible nanoparticles with enhanced photocatalytic and anti-microfouling potential." *International Biodeterioration and Biodegradation* 145, 104790. doi:10.1016/j.ibiod.2019.104790

51. Nagarajan, S. and Kuppusamy, K.A. 2013 "Extracellular synthesis of zinc oxide nanoparticle using seaweeds of Gulf of Mannar, India." *Journal of Nanobiotechnology* 11, 1–11. doi:10.1186/1477-3155-11-39

52. Aragao, A.P., Oliveira, T.M., Quelemes, P.V., Perfeito, M.L.G., Araujo, M.C., Santiago, J.A.S., Cardoso, V.S., Quaresma, P., Leite, J.R. and Silva, D.A. 2019 "Green synthesis of silver nanoparticles using the seaweed Gracilaria birdiae and their antibacterial activity." *Arabian Journal of Chemistry* 12, 4182–4188. doi:10.1016/j.arabjc.2016.04.014

53. Ishwarya, R., Vaseeharan, B., Kalyani, S., Banumathi, B., Govindarajan, M., Alharbi, N.S., Kadaikunnan, S., Al-Anbr, M.N. and Khaled, G. 2018. "Facile green synthesis of zinc oxide

nanoparticles using Ulva lactuca seaweed extract and evaluation of their photocatalytic, antibiofilm and insecticidal activity." *Journal of Photochemistry and Photobiology B: Biology* 178, 249–258. doi:10.1016/j.jphotobiol.2017.11.006

54. Yadav, L.S.R., Pratibha, S., Manjunath, K., Shivanna, M., Ramakrishnappa, T., Dhananjaya, N. and Nagaraju, G. 2019. "Green synthesis of Ag-ZnO nanoparticles: Structural analysis, hydrogen generation, formulation, and biodiesel applications," *Journal of Science: Advanced Materials and Devices* 4, 425–431. doi:10.1016/j.jsamd.2019.03.001

55. Xiang, Y.H., Ju, P., Wang, Y., Sun, Y., Zhang, D. and Yu, J.Q. 2016. "Chemical etching preparation of the Bi2WO6/BiOI p-n heterojunction with enhanced photocatalytic antifouling activity under visible light irradiation." *Chemical Engineering Journal* 288, 264–275. doi:10.1016/j.cej.2015.11.103

29

ZnS Nanoparticles for High-Performance Supercapacitors

Ranjit Kumar Puse
Rabindranath Tagore University, Bhopal, India

T. Ch. Anil Kumar
Vignan's Foundation for Science Technology and Research, Guntur, India

Diip Mishra
ICFAI University, Raipur, India

Omprakash B. Pawar
Government Institute of Forensic Science, Aurangabad, India

M. K. Valsakumari
Mookambigai College of Engineering, Pudukottai, India

Dr. A. Arun kumar
Methodist College of Engineering and Technology, Hyderabad, India

A wet chemical technique has been used to synthesise pure ZnS nanoparticles. The synthesised ZnS nanoparticles were evaluated using X-ray diffraction (XRD), high-resolution transmission electron microscopy, UV-visible, photoluminescence (PL) spectra, permittivity tests, and photoconductivity experiments. XRD and high resolution transmission electron microscopy (HRTEM) analysis reveal the formation of zinc sulphide nanoparticles with a size of 40–60 nm. The ZnS nanoparticles UV-visible spectra revealed a band gap energy value of 3.80 eV, which corresponds to a semiconductor material. The dielectric constant and dielectric loss of ZnS nanoparticles at various temperatures are examined in the 50 Hz to 5 MHz frequencies.

DOI: 10.1201/9781003355755-29

29.1 Introduction

Zinc sulphide (ZnS) is a substantial semiconductor for the cubic phase at ambient temperature, with a linear and broad energy gap of 3.67 eV. It has historically demonstrated exceptional fundamental qualities, adaptability, and a promise for fresh, varied applications. It could be used in photovoltaic modules, thermal screens, laser beams, detectors, and screens, as well as in near-UV semiconductor applications [1, 2]. Because of its higher index of refraction (n 2.35) and it is often in the visible and near-IR regions of the spectrum, ZnS is well suited for use in interference filters, anti-reflection coatings, and reflectors [3]. ZnS nanoparticles have substantial potential applications as photodetectors [4, 5], catalysts [6, 7], light-emitting diode components [8, 9], and luminescence labels for biocontrol agents and cell imaging. Because of quantum confinement and their large surface area, their characteristics may vary from that of the solid matrix. This kind of substance has intriguing optical and catalytic features that make it appropriate for a variety of industrial uses. Numerous techniques, including microwave irradiation, reverse micelles, chemical vapour deposition, hydrothermal, and solid–liquid chemical reactions, have been used to create ZnS nanoparticles. However, the majority of these approaches rely on the employment of harmful compounds or reactions that produce toxic molecules. Because it is crucial to design a dependable methodology for making semiconductor nanoparticles, researchers are looking for ideas from environmentally benign techniques [9–11]. Therefore, it would be quite interesting to synthesise nanoparticles using biological components. The goal of the current thesis is to create ZnS nanoparticles employing various biological agents. Additionally, efforts are undertaken to correlate the synthesised ZnS nanocrystals' structural, optical, and electrical properties. It is anticipated that this study will expand the semiconducting ZnS nanomaterial used as an optoelectronic detector, emitter, and modulator material [12, 13]. High refractive index and visual transmittance make ZnS nanoparticles as a good reflector and dielectric filter [14]. ZnS is employed as a reflective pigment in reflective coatings because of its broad band gap and high refractive index [15]. Due to the grey tint caused by trace iron, zinc sulphide was not widely employed after its introduction as a synthetic white pigment in 1852; a purer form was introduced in 1927 [16, 17]. According to Heaton, when synthesised zinc sulphide was precipitated, it produced a white material with a high degree of opacity. Increasing energy consumption has sparked research into more efficient energy generation and storage technology. Supercapacitors or ultracapacitors have the potential to revolutionise energy storage in recent decades. This energy storage technology has gained a lot of attention and is regarded the best for future needs, due to its increase in power density, improved cycling longevity, and eco-friendliness [18, 19, 20–22]. This makes supercapacitors a promising solution for portable electronics, hybrid cars, backup energy systems, and space or military devices.

29.2 Experimental Methods

Zinc nitrate and sodium sulphide were used as the source ingredients in a wet chemical process for the manufacture of ZnS nanoparticles. Since every reagent was of analytical grade, no additional purification was necessary. Deionised water was employed throughout the method due to its inherent benefits of being straightforward and environmentally safe. The synthesis was carried out at ambient, low-temperature conditions throughout. In a typical preparation, distilled water in 100 ml was used to form a 1 M zinc nitrate solution. The solution was then added in drops with a mixture of 1 M Na2S while being agitated at 70°C with a magnetic stirrer, resulting in the formation of ZnS nanocolloid [23]. By centrifuging the nanoparticles for 15 minutes at 2500 rpm, they were extracted. An ultrasonic bath was used to further purify the area. The finished product was dried for two hours at 120°C. For characterization, the dry nanoparticles were ground into a fine powder.

29.3 Characterization Studies

The crystalline phases and grain size of nanoparticles were determined by XPERT-PRO X-ray diffractometer to CuKα radiation in the 2Θ ranges that ranged from 10 degrees to 70 degrees. The UV–vis spectrometer studies were performed by Systronics, India, for optical measurements. In a dc electric field, a Keithley-480 picoammeter measures photoconductivity. Spreading a thick layer of powdered materials between two Cu electrodes engraved on a Cu plate (PCB) with a 0.5-cm spacing allowed for the creation of a cell. The sample is attached to a dc power supply and picoammeter (Keithely-480). The photo current was measured from 200 to 240 volts. Transparent glass pressed the powdered layer. This glass plate features a 0.25-cm² slit.

29.4 Analysis of the Results

29.4.1 X-Ray Diffraction Studies

The specimen is placed in a mount that is at centre along the axis of the diffractometer in the plane. When different values of 2θ are used, the intensity of the radiation that has been diffracted is measured. When n is held constant at 1, it is possible to derive 2θ from the diffraction data. After that, the diffracted patterns were compared with the American Society for Testing and Materials (ASTM) data, which indexed the h, k, and l values of the nanoparticles. For the ZnS nanoparticles, the X-ray diffraction (XRD) patterns were captured and analysed. Figure 29.1 represents the diffraction pattern of ZnS nanoparticles and we found that there are distinct (111) (101) (220), (110) diffraction patterns Joint Committee on Powder Diffraction Standards (JCPDS) file no. 80-0020 [24]. It has been established through XRD research that the nature of the ZnS nanoparticles as they have been deposited is cubic zinc blende. The grain sizes are noticed to lie between the region of 40 and 60 nm.

The average grain size (D) was estimated using Scherer's formula and is given by,

$$D = \frac{k\lambda}{\beta \cos\theta}$$

where
  D is the average particle diameter (nm).
  θ is the diffraction angle,
  λ is the X-ray wavelength, and
  β is the full width half maximum.

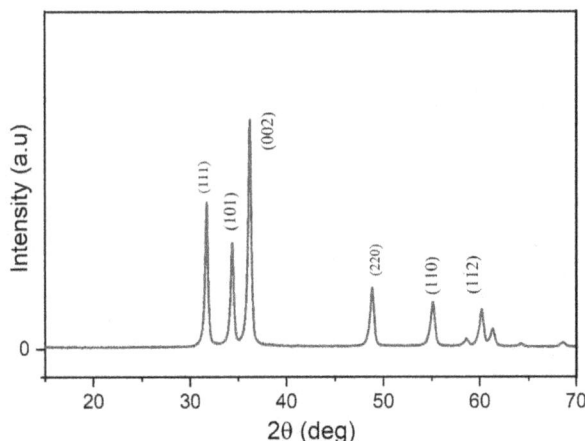

FIGURE 29.1 X-ray diffraction pattern of ZnS nanoparticles.

29.4.2 UV–Vis Studies

The UV–vis spectra at room temperature, as recorded by a Perkin Elmer spectrophotometer in the transmission mode, is shown in Figure 29.2. It displays the slightly blue-shifted ZnS nanoparticle absorption spectrum at 300 nm (3.7 eV), compared to bulk ZnS. The near-band-edge free excitons are responsible for the absorption peak's close proximity to the bulk ZnS crystals [25]. The quantum confinement of the nanoparticles may be the cause of the broadening of the absorption spectra.

The link between the optical absorption coefficient and photon energy aids study into electron band structure and changeover nature. Based on the following relationship,

$$\alpha = \frac{2.3026}{t}\log(1/T) \tag{29.1}$$

where T is transmittance, 't' is the thickness of the nanoparticles, 'α' is related to the extinction coefficient K by,

$$K = \frac{\alpha\lambda}{4\pi} \tag{29.2}$$

The radiance (R) in parameters of the absorbance rate and optical properties (n) can be calculated using the following relationships:

$$R = \frac{\exp(-\alpha t) \pm \sqrt{\exp(-\alpha t)T - \exp(-3\alpha t)T + \exp(-2\alpha t)T^2}}{\exp(-\alpha t) + \exp(-2\alpha t)T} \tag{29.3}$$

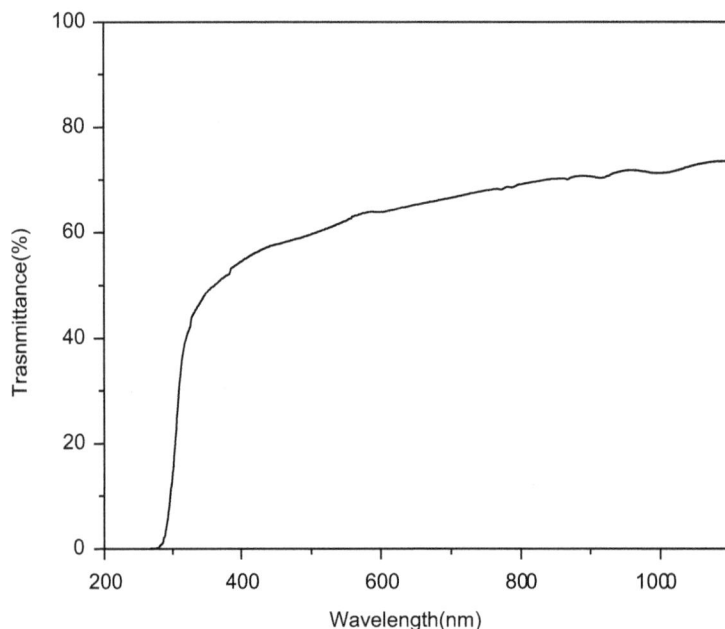

FIGURE 29.2 Transmittance spectra of ZnS nanoparticles.

$$n = \frac{-(R+1) \pm 2\sqrt{R}}{(R-1)} \tag{29.4}$$

The energy necessity of the absorption coefficient in the high photon energy area indicates the existence of a bandgap of ZnS nanoparticles and is given by [26],

$$(\alpha h v)^2 = A(E_g - h v) \tag{29.5}$$

where E_g is the optical band gap of the ZnS nanoparticles and A is a constant. The band gap of ZnS nanoparticles was estimated by plotting $(\alpha h v)^2$ vs $h v$ [27] as shown in Figure 29.3 and it was found to be 3.7 eV.

29.4.3 Photoluminescence Studies

The fluorescence emission spectrum of pure ZnS nanoparticles at room temperature is shown in Figure 29.4. At 300 nm, ZnS nanoparticles are stimulated, and emission maxima are detected at 440 and 480 nm, despite electron recombination between the donor and valence band being linked with the sulfur vacancy [28, 29]. Sharp exciton emission peak and broad trapped emission are observed [30–33] and most of the researchers have studied about blue UV emission of ZnS nanoparticles. Octahedral voids and pen surface states emission are usually found in semiconductor nanocrystals [34, 35].

29.4.4 Dielectric Studies

Ionic polarisation, which results from the comparative dislodgment between ions, is the next source of contribution. Molecules with a perpetual electric dipole moment can shift their alignment dipolar or orientation polarisation results from the application of an electric field. A molecule's dipole moment changes in alignment when an electric field is applied, causing space charge polarisation. The electrical

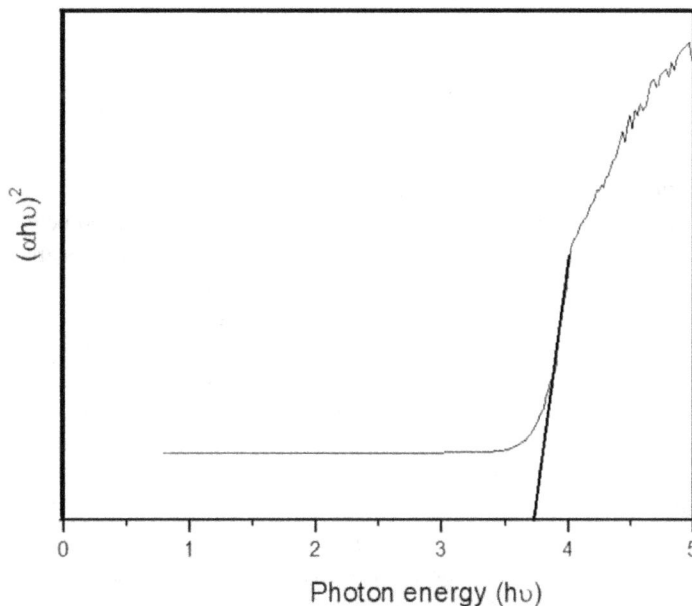

FIGURE 29.3 Graph $(\alpha h v)^2$ vs. $h v$ for ZnS nanoparticles.

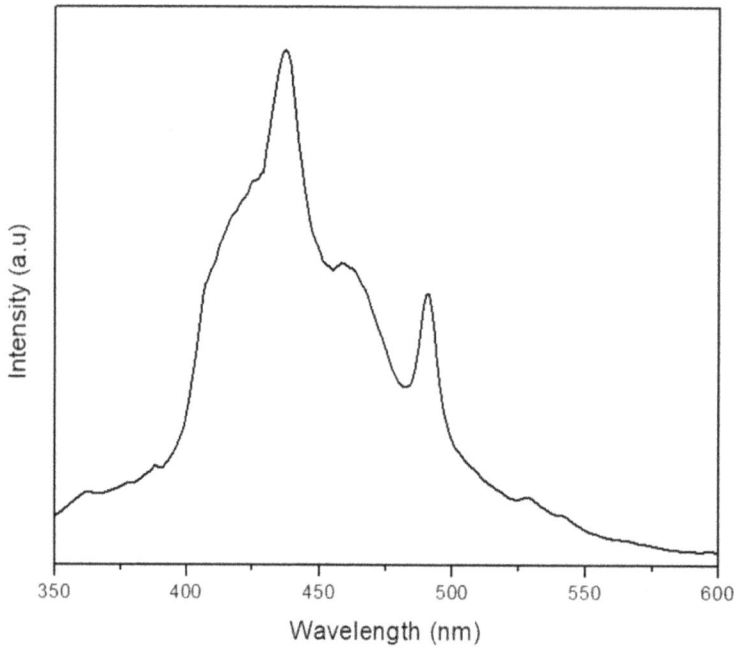

FIGURE 29.4   PL spectra of ZnS nanoparticles.

processes that occurred in ZnS nanoparticles can be determined by measuring dissipation and dielectric stability in relation to periodicity and different temperatures. Figures 29.5 and 29.6 illustrate the changes in the dielectric constant and dielectric loss of the permittivity was calculated using zinc sulphide nanomaterials at temperatures up from 40 to 120°C and frequency distribution from 50 Hz to 5 MHz which were utilised to compute the permittivity.

$$\varepsilon_r = \frac{Cd}{\varepsilon_o A}$$

where A is the sample's area, and d is the sample's width.

The chart illustrates that dielectric values are significant depending on periodicity of the alternating current signal and the temperatures of the ZnS nanoparticles. The dielectric constant is greater at 50 Hz and decreases as frequency increases (5 MHz). Because of electronic, ionic, dipolar, and space charge polarisations, ZnS nanoparticles have a relative permittivity at specific frequencies [36]. At lower frequencies, space charge polarisation occurs, implying the purity of nanoparticles. It has a strong influence at high- or low-temperature vibrations [37]. The nanocrystalline materials' grain boundaries become electrically active due to charge trapping. Low-frequency dipole moment changes quickly with electric field. Space charge and rotation polarisations raise the dielectric constant at interfaces. Therefore, nanostructured materials should have a higher dielectric constant [38]. The grain boundary interface structure contributes to the high dielectric constant in nanomaterials. Dielectric loss dominates the quick growth in permittivity at low frequencies and low temperatures at high enough temperatures. Space charge and ion jump polarisation decrease as material is heated, increasing the dielectric constant. Absorption current causes dielectric losses in dielectric materials. In polar dielectrics, molecular orientation requires power-driven liveliness to overwhelmed interior resistance. Alternative fraction of the electric energy is used for molecular variations and other energy-losing molecular transfers.

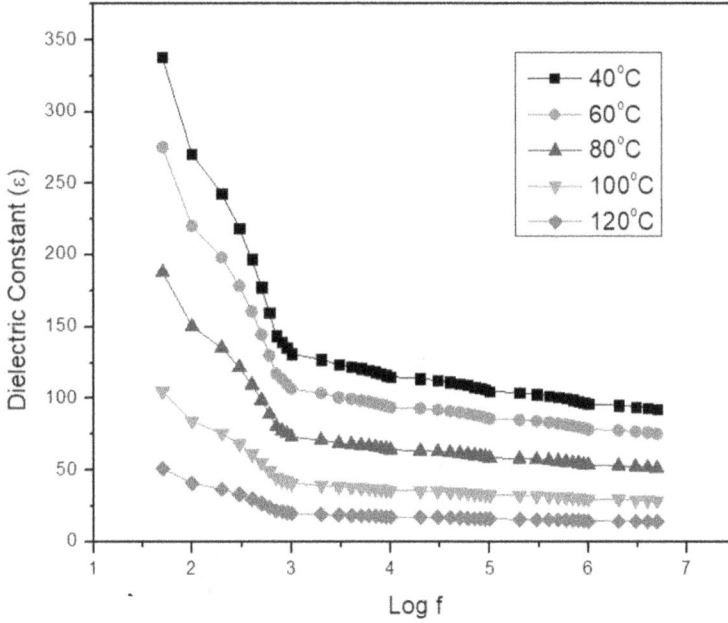

FIGURE 29.5 Dielectric constant vs log *f* of ZnS nanoparticles.

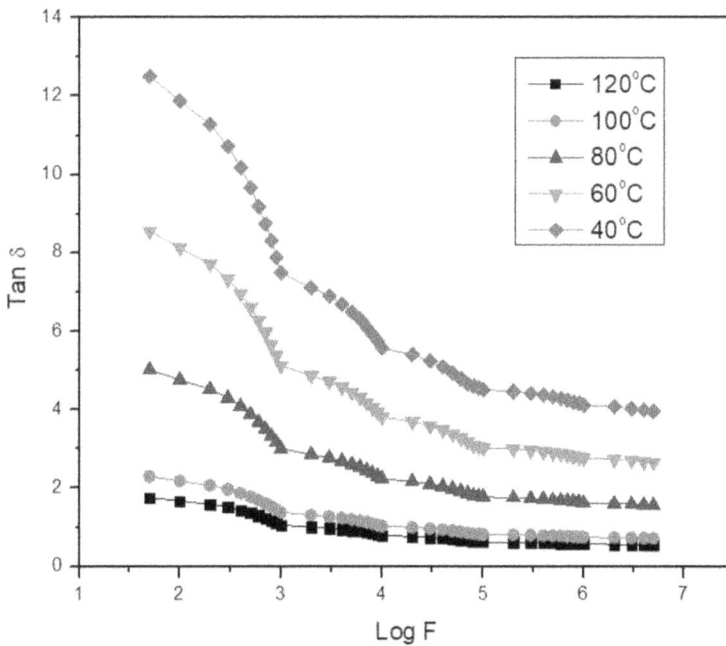

FIGURE 29.6 Dielectric loss vs log *f* of ZnS nanoparticles.

Nanophase materials exhibit glassy grain boundaries. All the inhomogeneities, flaws, and space charge creation cause dielectric losses. Due to dangling connections on the external coatings, nanoparticles are very reactive and may absorb oxygen or nitrogen. Adsorbed gases contribute to increased dielectric loss. Inhomogeneities such as defects and space charge production generate dielectric loss in nanophase materials. Figure 29.5 depicts dielectric loss vs. frequency at 40°C, 60°C, 80°C, 100°C, and 120°C. The trend of dielectric loss is similar to that of the dielectric constant. Dielectric loss appears to be frequency driven because it decreases with increasing frequency for all temperatures [39].

29.4.5 Photoconductivity Studies

The electrical energy applied was enlarged from 200 to 240 volts, and the associated photocurrent was recorded. In Figure 29.7, the photocurrent and dark current were shown as a function of applied voltage. The relationship between the photocurrent and dark current and the applied voltage was displayed. The graph shows that, for ZnS nanoparticles, both the photocurrent and the dark current develop steadily with increasing applied voltage, with the light current always being greater than the dark current. The photocurrent, however, always outweighs the dark current. The ZnS nanoparticles demonstrate positive photoconduction as a result. This may be because the absorption of photons generates mobile charge carriers, resulting in positive photoconductivity [40].

29.4.6 HRTEM Results

ZnS particle crystallinity was also confirmed by HRTEM investigation. The particles appear to be spherical in shape. Figure 29.8 shows a widefield image of a zinc sulphide nanoparticle. The TEM image reveals that the sizes of the made nanoparticles and agglomerates in smaller grains vary, but a large number of the particles show a size range between 40 and 60 nm [41]. The XRD data and these results match up well. The *inset* shows a charged–coupled detector (CCD) pattern that shows a uniform diffraction pattern, which also shows that the crystallinity is good. Electron diffraction also verified the crystallinity of the ZnS particles. The electron diffraction pattern of a ZnS layer in a selected location is displayed in Figure 29.9.

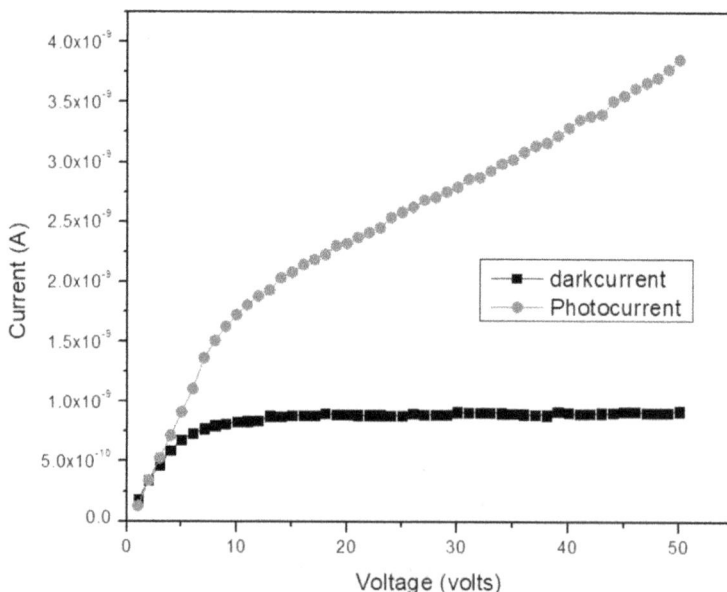

FIGURE 29.7　Photoconductivity studies of ZnS nanoparticles.

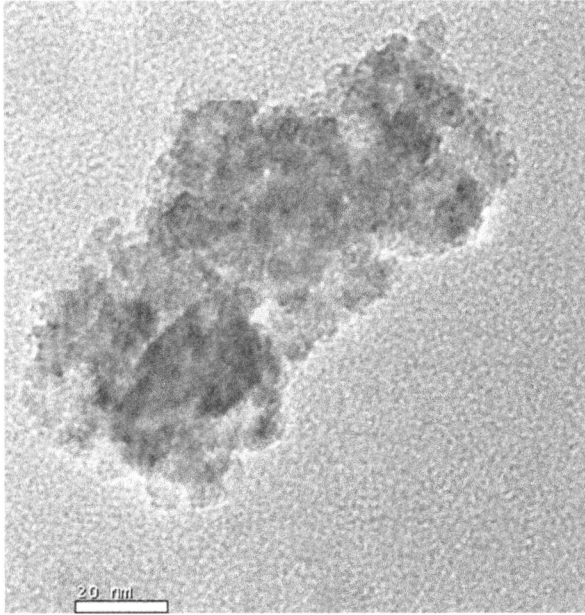

FIGURE 29.8 The bright field image of ZnS nanoparticles.

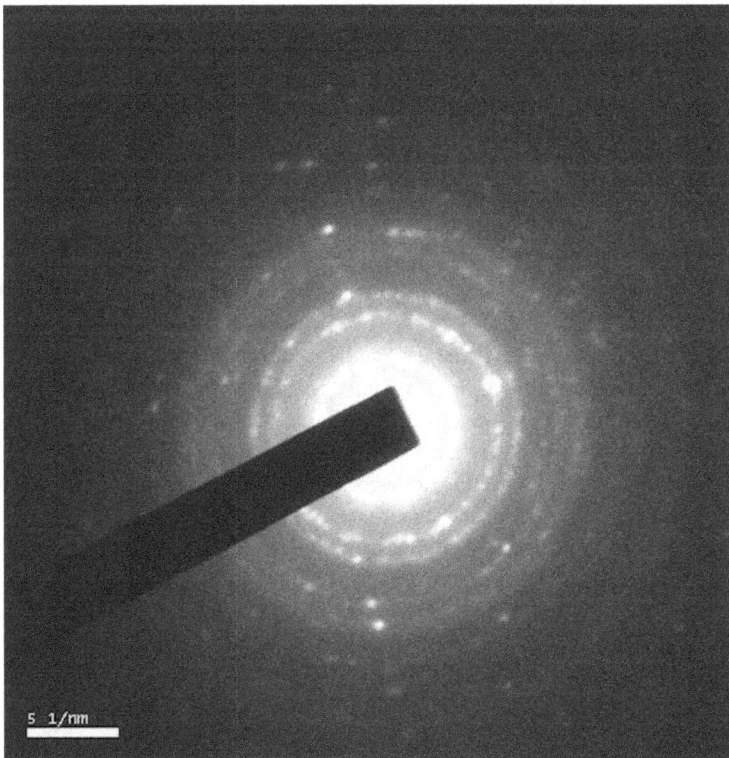

FIGURE 29.9 Electron diffraction pattern of ZnS nanoparticles in a certain location.

29.5 Conclusion

Wet chemical synthesis is used to create ZnS nanoparticles. Using XRD measurements to identify the particle's crystal structure and grain size, the ZnS nanoparticle's size is discovered to be 40–60 nm. Quantum confinement effect is responsible for the blue shift of 320 nm in the optical absorption spectrum relative to its bulk equivalent. The bandgap is determined to be 3.80 eV. ZnS emits at 440 and 480 nm, which is due to blue emission bands at room temperature. The dielectric studies reveal frequency-dependent changes in the low value of the dielectric constant at higher frequencies. The dielectric constant of ZnS nanoparticles is found to be significantly greater than that of conventional ZnS. The transmission electron microscopy (TEM) picture shows that the synthesised nanoparticles range in size from 40 to 60 nm. These results match with XRD data. The *inset* displays a CCD pattern showing good crystallinity and uniform diffraction. The cohort of mobile charge carters caused by the captivation of photon leads to encouraging photoconductivity.

References

1. K.W. Boer, *Survey of Semiconductor Physics*; Van Nostrand Reinhold: New York, 1990, p. 249.
2. G. Rothenberger, J. Moser, M. Grätzel, N. Serpone and D.K. Sharma, *J. Am. Chem. Soc.* 107, 8054 (1985).
3. M.R. Hoffmann, S.T. Martin, W.Y. Choi and D.W. Bahnemann, *Chem. Rev.* 95, 69 (1995).
4. Y.F. Hao, G.W. Meng, Z.L. Wang, C.H. Ye and L.D. Zhang, *Nano Lett.* 6, 1650 (2006).
5. S.I. Yanagiya, et al., *Mater. Chem. Phys.* 105, 250 (2007).
6. W. Tang and D.C. Cameron, *Thin Solid Films* 280, 221 (1996).
7. G.H. Motlan, K. Zhu, K. Drozdowicz-Tomsia, M. McBean, R. Phillips and E. M. Goldys, *Opt. Mater.* 29, 1579 (2007).
8. U. Gangopadhyay, K. Kim, D. Mangalaraj and J. Yi, *Appl. Surf. Sci.* 230, 364 (2004).
9. M. Bangal, S. Ashtaputre, S. Marathe, A. Ethiraj, N. Hebalkar, S.W. Gosavi, J. Urban and S.K. Kulkarni, *Hyperfine Interact.* 160, 81–94 (2005).
10. S. Gupta, J.S. McClure and V.P. Singh, *J. Am. Chem. Soc.* 104, 2977–2985 (1997).
11. A. Henglein, *Thin Solid Films.* 33, 299 (1989).
12. Y. Jiang, X.M. Meng, J. Liu, Z.Y. Xie and C.S. Lee, *Lee ST Chem. Rev.* 89, 1861–1873 (2003).
13. J. Meera, V. Sumithra, R. Seethu, J.M. Prajeshkumar, *J. Appl. Phys.* 73, 5237–5241 (2010).
14. W. Mingwen, S. Lingdong, F. Xuefeng, L. Chunsheng and Y. Chunhua, *Appl. Phys. Lett.* 83, 4241–4244 (2000).
15. C.S. Pathak, M.K. Mandal and V. Agarwala, *Solid State Comm.* 115, 493–496 (2013).
16. C.S. Pathak, M.K. Mandal and V. Agarwala, *Mater. Sci. Semicond. Process.* 16, 467–471 (2013).
17. C.S. Pathak, D.D. Mishra, V. Agarwala, M.K. Mandal, *Super Lattices Microst.* 58, 135–143 (2012).
18. C.S. Pathak, D.D.V. Mishra, V. Agarwala and M.K. Mandal, *Ceramics Int.* 38, 5497–5500 (2013).
19. Biswarup Chakraborty, Shweta Kalra, Rodrigo Beltrán-Suito, Chittaranjan Das, Tim Hellmann, Prashanth W. Menezes and Matthias Driess, *Chem Asian J.* 15, 852–859 (2020).
20. Biswarup Chakraborty, Rodrigo Beltran-Suito, Viktor Hlukhyy, Johannes Schmidt, Prashanth, W. Menezes and Matthias Driess, *ChemSusChem* 13, 1–9 (2020).
21. D. Brida, E. Fortunato, H. Águas, V. Silva, A. Marques, L. Pereira, I. Ferreira and R. Martins, *J. Non. Cryst. Solids* 299(302), 1272–1276 (2002).
22. Mukta Tripathi, *Anuj Kumar Ionics* 24, 3155–3165 (2018).
23. M. Suchea, S. Christoulakis, K. Moschovis, N. Katsarakis and G. Kiriakidis, *Thin Solid Films* 515(2), 551–554 (2006).
24. M.Y. Nadeem, Waqas Ahmed and M.F. Wasiq, *J. Res. (Sci.)* 16(2), 105–112 (2005).
25. H. Hu and W. Zhang, *Opt. Mater.* 28(5), 536–550 (2006).

26. M. Pal, N.R. Mathews, E.R. Morales, J.M. Gracia, Y. Jimenez and X. Mathew, *Opt. Mater.* 35(12), 2664–2669 (2013).

27. H.P. Klug and E.A. Leroy, *X-Ray Diffraction Procedures*, John Wiley & Sons, New York, NY, USA (1974).

28. F. Vetrone, J.C. Boyer and J.A. Capobianco, H.S. Nalwa, Ed., vol. 10, pp. 725–765, American Scientific Publishers, (2004).

29. B.S. Rema Devi, R. Raveendran and A.V. Vaidyan, *Pramana* 68(4), 679–687 (2007).

30. S. Lee, D. Song, D. Kim et al. *Mat. Lett.* 58(3–4), 342–346 (2004).

31. W.Q. Peng, G.W. Cong, S.C. Qu and Z.G. Wang, *Opt. Mater.* 29(2–3), 313–317 (2006).

32. N. Murase, R. Jagannathan, Y. Kanematsu et al., *J. Phys. Chem. B* 103(5), 754–760 (1999).

33. S. Yanagida, M. Yoshiya, T. Shiragami, C. Pac, H. Mori and H. Fujita, *J. Phys. Chem.* 94(7), 3104–3111 (1990).

34. B. Geng, J. Ma, and F. Zhan, *Mater. Chem. Phys.* 113(2–3), 534–538 (2009).

35. M. O'Neil, J. Marohn, and G. McLendon, *J. Phys. Chem.* 94(10), 4356–4363 (1990).

36. B. Thomas and M. Abdulkhadar, *XVI International Union of Crystallography*, Beijing, China (1993)

37. A.B. Cruz, Qing Shen and Taro Toyoda, *Mat. Sci. Eng. C* 25, 761–765 (2005).

38. S.E. Ahn, H.J. Ji, K. Kim, G.T. Kim, C.H. Bae, S.M. Park, Y.K. Kim and J.S. Ha, *Appl. Phys. Lett.* 90, 153106 (2007).

39. S.J. Nelson, A.M. Eppler, and I.M. Ballard, *J. Photochem. Photobiol, A1* 48, 23–28 (2002).

40. J. Carry, H. Carrere, M.L. Kahn, B. Chaudret, X. Marie and M. Respaud, *Semicond. Sci. Technol.* 23, 025003 (2008).

41. A. Bera and D. Basak, *Appl. Phys. Lett.* 93, 053102 (2008).

30

Cost-Effective-Mediated Fabrication of ZnO Nanomaterials and Its Multifaceted Perspective toward Energy Storage and Environmental Applications

Pradosh Kumar Sharma

Chinmaya Degree College, BHEL, Haridwar, India

Kalaivaani

Vivekananda College of Engineering for Women (Autonomous), Thiruchengodu, India

S. Gopakumar

Rohini College of Engineering and Technology, Nagarcoil, India

T. Ch. Anil Kumar

Vignan's Foundation for Science Technology and Research, Guntur, India

Shubhajit Halder

Hislop College, Nagpur, India

Dr. M. Jayapriya

Pavendar Bharthidasan College of Engineering and Technology, Trichy, India

DOI: 10.1201/9781003355755-30

In recent trends, nanotechnology is a multidisciplinary field of engineering that emphasizes the nano-particles with multidimensional properties. Nanoparticles aid to attain greater attention owing to their enhanced physical, chemical, and biological properties at the nanoscale size ranging from 1 to 100 nm. Owing to the development of a massive industrial arena, there is enormous scarcity in energy utilization. So to overcome this, energy crisis with viable technology is crucial for the conservation of ecosystem and environment. Nanotechnology plays a pivotal function in the energy storage devices. Hence, functionali-zation of nanomaterials with natural materials acts as an environment benign and safer technology to pro-vide versatile platform. Noble metal nanoparticles have received greater attention due to their fascinating physiochemical and enhanced biological characteristic features and they have been widely used in diverse fields. However, metal nanoparticles spontaneously aggregate and grow which tends to reduce surface energy that limits the reaction efficiency and stability of nanoparticles. Also, the ultra-high cost of metal nanoparticles hinders their further usage. Hence, the design of new generation of metal oxide nanomate-rial with distinct properties has been employed, which offers synergetic multifunctional efficiency toward assorted applications. Hence, in the realm of nanomaterial, considerable attention has been focused on the fabrication of metal oxide nanoparticles in various energy storage applications.

30.1 Introduction

30.1.1 Nanotechnology

Nowadays nanotechnology has been emerged as a prominent technology due to their unique physi-ochemical characteristic feature with enhanced catalytic activity in various chemical reactions, thermal stability, and optoelectronic characteristic features owing to the existence of a huge surface area to vol-ume ratio. The integration of nanomaterials in assorted industries includes food, health, energy storage applications, chemical, cosmetics, and pharmaceutical industries, which utilizes environmental benign strategy which is considered to be non-toxic, cost-effective methods [1].

30.1.2 Types of Nanoparticles

In general, the nanoparticles are divided into various categories based on the synthesis of materials with different morphology, various size, and different shape of the materials produced. Some significant classes of nanomaterials are mentioned below.

30.1.2.1 Organic Nanoparticles

Organic nanoparticles are non-hazardous and highly degradable, which include micelles, ferritin, den-drimers, and liposomes. Among them, liposomes and micelles are hollow spheres, otherwise known as nanocapsules, are highly light and heat sensitive in nature. Owing to the ideal characteristic features, organic nanoparticles were extensively utilized in targeted drug delivery [2]. These nanomaterials are otherwise called as polymeric nanocapsule, which are spherical in shape.

30.1.2.2 Inorganic Nanoparticles

In the development of inorganic nanomaterials, carbon is absent and they are not toxic in nature. They are biocompatible and hydrophilic in nature. These inorganic nanomaterials are developed with enhanced stability compared to organic nanomaterials. These nanomaterials are further classified into metal nanoparticles, for example, Ag, Au, etc., and metal oxide nanoparticles such as ZnO, MgO, etc.[3].

30.1.2.2.1 Metal Nanoparticles

The metal nanoparticles are fabricated via metal precursors. Owing to the renowned localized sur-face plasmon resonance (LSPR) characteristic feature, these nanoparticles acquire optoelectronic characteristics with several applications. Well-known examples of alkali and noble metals include Cu, Ag, and Au and have a broad spectrum in the visible region of the electromagnetic spectra. Various facets, shape, and morphology-directed fabrication of metal nanoparticles are more significant in the present area. Owing to the advanced optoelectrical properties, these metal nanoparticles find vast applications in many research areas [4]. In sophisticated analytical measurements such as SEM, gold nanoparticle-coated samples are used to enhance the electronic stream, which are very much useful in obtaining higher quality of SEM images.

30.1.2.2.2 Metal Oxide Nanoparticles

The synthesis of metal oxide nanoparticles is to change and modify the structural properties of respective nanomaterials in order to produce their oxidized form for instance iron nanomaterials are oxidized to form iron oxide nanomaterials. The reactive ions of iron oxide materials are enhanced while comparing the properties with iron nanoparticles. Owing to the reactivity with other materials and the enhanced efficiency of metal oxide nanoparticles, the nanomaterials are developed based on the materials we need. The best examples of metaloxide nanoparticles with various applications are ZnO, SiO_2, Fe_2O_3, CeO_2, MgO, and TiO_2 [5, 6].

30.1.3 Ceramics NPs

Ceramics nanomaterials are defined as inorganic materials with non-metallic solid particulates, fabricated through the process of successive heating and cooling. Ceramics nanomaterials originate in various forms which include amorphous, dense, porous, hollow forms, and polycrystalline in nature. Hence, they are utilized in various industrial applications such as catalysis, photocatalytic degradation of several textile dyes, and various imaging applications [7].

30.1.4 Semiconductor NPs

Semiconductor materials possess distinct difference between metal and non-metal nanoparticles and therefore they find versatile functions in a variety of fields such as photocatalysis, photo optics, and various electronic devices. These devices are potent candidates in water splitting owing to various bandgap and band edge positions [8].

30.1.5 Carbon-Based Nanomaterials

Carbon-based nanomaterials are further classified into fullerenes and carbon nanotubes. The globular hollow cage consists of allotropic form of carbon, which finally constitutes to form fullerenes nanomaterial. Pivotal properties which include electrical conductivity, electron affinity, high tensile strength, and various structures made them a versatile candidate in industrial applications. The arrangement of pentagonal and hexagonal carbon units with sp2 hybridized carbon materials showed that well-established fullerenes consist of C_{60} and C_{70} with diameter ranging from 7.114 and 7.648 nm, respectively. The extended tube-like structure with a diameter ranging from 1 to 2 nm is predicted as metallic or semiconducting materials based on their diameter. The carbon nanotubes are structurally resembled as a graphite sheet revolving upon itself with single-wall, double-wall, or layered as multi-walled carbon nanotubes. Main synthetic protocol for the development of carbon nanotubes is the deposition of precursor material, especially the atomic carbon, vaporization from graphite using laser or electric arc method through the process of chemical vapor deposition (CVD) technique. Owing to the unique physiochemical properties, carbon nanotubes are not only utilized in pristine form but they are used as nanocomposites for many industrial applications such as fillers, gas adsorbents for environmental remediation, and act as a support medium for various organic and inorganic catalysts [9, 10].

30.1.6 Nanoparticles Synthesis

The two main approaches for the fabrication of nanoparticles are bottom up and top down approaches. In the top down approach, the development of nanomaterials includes the process of milling or attrition of larger particles that are macroscopic in nature. The process of attrition initially involves the fabrication of particles with large-scale pattern and finally reduces it to nanoscale level through the process of plastic deformation. This technique cannot be used to scale-up for industries owing to high cost and very slow process of material formation. Hence, interferometric lithographic (IL) techniques are frequently used for the synthesis of nanomaterials. This technique employs the synthesis of nanomaterials from priorly miniaturized atomic components through the process of self-assembly. This process includes the formation of nanomaterials with both physical and chemical means. This synthetic methodology is comparatively cheap which employs both kinetic and thermodynamic approaches. The kinetic approach employs molecular beam epitaxy (MBE) for the fabrication of nanomaterial synthesis [11, 12]. The methods under bottom up approach are chemical and biological protocols. In the top down approach, the breakdown of huge particles results in the formation of imperfectly synthesized nanoparticles. It leads to disruption of the uniqueness of size and various shapes of the nanoparticles. High preference of bottom up approach is due to the synthesis of nanoparticles with homogeneous properties with almost similar shape and size. Therefore, nowadays bottom up approach is greatly preferred over top down approach.

30.1.6.1 Physical Methods

Physical methods include physical vapor deposition, thermolysis, pulsed laser method, high energy ball milling, microwave-assisted synthesis, melt mixing, sputter deposition of nanomaterials, laser ablation of particles, ion implantation technique, and electric arc deposition. These techniques rely on altering any one of the physical parameters. The important disadvantages of this method include costlier use of equipment and maintenance of elevated parameters than normal conditions. Also, this method is a time-consuming approach.

30.1.6.2 Chemical Methods

Chemical method of nanoparticle synthesis yields a greater amount than any other methods. The methods under this chemical approach include co-precipitation, the sol-gel process, CVD, solvo-thermal synthesis of nanoparticles, and radio-frequency plasma method. These fabrication methods involve the need for chemical reducing agents and other capping agents which are hazardous for stabilized production of nanoparticles [13–15]. The reducing agents involved are usually sodium citrate, N_2H_4, $NaBH_4$, H_2 gas, and other abundant sugar molecules such as sucrose, maltose, glucose, and fructose or dry ethanol is also used in case of metallic reagents. For the fabrication of metal nanoparticles, mesityl derivatives are also under use. Some common capping agents for size-controlled and stabilized production of nanoparticles are oleic acid, triethanolamine, and thioglycerol. Usually, the chemical synthesis of nanoparticles is a cost-effective approach for large-scale synthesis. But this method suffers from several other drawbacks like contamination from precursor chemicals, production of dangerous byproducts, and the usage of toxic chemicals [16]. For example, $NaBH_4$, which is popularly utilized as a reducing mediator for the synthesis of a wide variety of nanoparticles, is considered as a harmful agent on exposure to the environment as it is found to be corrosive and found to yield diborane and hydrogen which are depicted as flammable. Likewise, amines and thiols, even at moderate concentrations, were found to cause severe adverse effects to higher organisms like hemolytic effects, cytotoxic effects, tissue damage, and irritation in skin, eyes, and mucous membranes [17–19].

30.1.6.3 Biological Methods

To rectify these numerous problems, nanoparticle synthesis from biological sources had emerged. The need for the development of environment benign approach accelerated this method of nanoparticle synthesis over the past decade. The biological method of fabrication of nanoparticles mainly relies on the utilization of natural resources like plants and microorganisms. As the biological method is equipped with numerous advantages like low-cost production and obtaining non-toxic end products, this method is gaining a lot of interest by researchers and scientists. This way of synthesis is popularly known as the green fabrication of nanoparticles [20, 21].

The microbial development of nanoparticles involves a wide variety of organisms including most bacteria, virus, and fungal genera. Some examples of nanoparticle synthesis by various microbes include *Aspergillus fumigates* (Ag nanoparticle), *Fusarium oxysporum* (silver, zirconia, gold, cadmium sulfide, titanium particle, silica, and CdSe quantum dots), *Candida glabrata* (CdS nanoparticle), *Pseudomonas aeruginosa* (Au nanoparticle), and many others. The possible mechanism for the development of nanomaterials by this synthetic protocol is usually bioaccumulation, biosorption, efflux systems, alteration of solubility and toxicity as a result of reduction or oxidation, extra-cellular complexation, enzyme bio reduction or precipitation of metals and metal transport system, and cell wall reduction by ketones and aldehydes [22, 23]. The enzymes present in the microbes also play a pivotal role in this method of nanoparticle synthesis. Nowadays, it is also felt that nanoparticles by cellular proteins also play a prominent role in the reduction and stabilization of nanoparticles. The major disadvantage under the microbial synthesis of nanoparticles includes the long time needed for the synthesis of nanoparticles. Therefore, to overcome this disadvantage, plant-mediated fabrication of nanomaterials is gaining more importance than any other methods in use. The plant-mediated/plant extract-mediated synthesis of both metal and metaloxide nanomaterials is considered as the most rapid, non-hazardous, economical, and eco-friendly method [24].

The plant-mediated fabrication has turned out to be the most versatile technology than other technology in the synthesis of nanoparticles. This is one-step, lucrative method for the synthesis and fabrication of nanoparticles. It is because plants have so many metabolites for this purpose. These metabolites and other phytochemicals are capable of acting as both reducing mediators and capping agents that are non-toxic in nature [25]. Thus, it helps in the better fabrication of nanoparticles. Nowadays, research has been carried out with extracts from different plant parts like bark, stem, flower, pods, and fruits. The nanoparticles synthesized by this method are multifunctional in nature and they are found to have many applications in the field of medicine. Many parameters influence the synthesis of nanoparticles through this plant-mediated approach. They are concentration of the precursor, temperature, the presence of organic compounds such as phenols, flavonoids, anthocyanins, and other phytoconstituents in the leaf extract and pigments in the leaf extract. The predominant role played by the aldehydes, ketones, alkaloids, flavones, terpenoids, amides, and carboxylic acids present in the plant material is a vital function in the part of acting as capping agents. An added advantage is that a single plant species can produce a wide variety of different nanoparticles. Thus, plant extracts are considered as the most efficient reactors for the synthesis and fabrication of nanoparticles [26, 27].

Plant-mediated fabrication is considered to be an economical one and to add up this, the utilization of *Musa paradisiaca* bract is considered to be of less significance. It is generally considered to be an agro waste after the consumption of all other parts of the plants. The chemical composition of bract illustrates that it is a depository of many minerals (iron, calcium, potassium, lead, sodium, cadmium, and magnesium), vitamins (beta carotene, ascorbic acid, riboflavin, niacin, and thiamin), secondary metabolites (alkaloids, phenols, tannins, flavonoids, saponins, phylates, and oxalates), and various pigments. The bract itself is known to possess its innate antimicrobial, astringent, analgesic, and antispasmodic activity [28]. Therefore, it is obvious that the bract extract would contribute a lot to the medicinal effects of nanoparticles synthesized using this extract.

Nanoparticles are widely utilized as antimicrobial agents, which are emerged as a promising alternative method over the conventional method which employs chemical-based antimicrobial materials which are hazardous in the recent era, and it is commonly employed since their smart and the materials developed with multidimensional physical and chemical properties. The developed nanomaterials have high biological compatibility. The disadvantage of the chemically developed antimicrobial agents is the higher tendency to develop antimicrobial resistance toward multidisease-resistant bacteria [29].

The substances which have the ability to prevent the oxidation reaction especially used to stop the spoilage of various stored food products are called antioxidants. Many diseases are found to have the involvement of the formation of free radicals. Antioxidants are able to block the free radicals, prevent the cell from damage, or inhibit the diseases which are initiated due to free radicals. Antioxidants can protect the cells against the reactive oxygen species by scavenging them. Many plants are known to possess free radical scavenging molecules, which include several phytoconstituents, such as phenolic compounds, terpenoids, vitamins, and some other endogenous metabolites, that are rich in antioxidant activity. Therefore, plant extract-fabricated nanoparticles can better combat with antioxidant molecules [30, 31].

30.1.7 ZnO Nanoparticles

The dimensions of the semiconductor materials are reduced to the nanoscale range which is known as the semiconducting nanomaterial and these materials can be changing their properties. Because of this novel property, which includes semiconducting nanomaterials with versatile properties have attracted more and significant attention in various area of emerging advanced technologies which includes energy conversion, catalysis, non-linear optical imaging, imaging devices, sensors and detectors, photography, nanoelectronics, nanophotonics, solar cells, and biomedicine [32]. The semiconducting nanomaterials have interesting physical properties and chemical characteristics when compared to conventional bulk materials. The most attractive properties of these materials are continuous absorption bands, narrow and intense emission spectra, high chemical and surface functionality. The semiconducting nanomaterials such as TiO_2, SnO_2, GaN, CuO, GaAs, Si, and ZnO have been used in the different fields. Among this, ZnO semiconductor nanomaterials areone of the most promising materials for electronic and optoelectronic properties due to its band gap energy (3.37 eV) and it has very large excitation binding energy (60 meV) at room temperature. In addition, ZnO nanomaterials were applied in many fields such as photocatalysis, optical devices, and cosmetic products, such as sunscreens. Hence, recent researchers focused on the development of one-dimensional (1D), two-dimensional (2D), three-dimensional (3D) ZnO nanostructures such as rod, tube, sheet, and flower to enhance the physical, electrical, and photoluminance properties. In addition to that, researchers are focusing on ZnO-based core/shell nanocomposite (CSNC) materials due to their better physico-chemical properties [33–35].

30.1.8 1D Nanostructured Semiconductors

1D semiconducting materials are those which possess uni-directional confinement in atleast one of its three dimensions. In recent decades, 1D nanostructure has attracted many researchers due to its size and shape-related properties and found as fundamental building blocks for the new generation of electronic and photonic devices. Therefore, the growth of 1D nanostructure was applied in semiconducting metal oxides, metal sulfides, carbides, nitrides, etc. Similarly, in our studies, 1D nanorod-like growth was carried out for synthesizing ZnO. The growth of ZnO in 1D helps in boosting the performance of various applications such as optoelectronics, solar cells, gas sensors, and photocatalysts [36].

30.1.9 Core/Shell Nanocomposite Materials (CSNCs)

The development of core-shell nanomaterials has garnered greater attention in multifaceted fields owing to their unique optoelectronic properties and catalytic activity, and areapplied in various fields such as in energy storage devices. The innermost layer of the nanomaterial issaid to be the core and the outermost layer of the materials issaid to be shell; both the core and the shell are developed at nanoscale which are defined as core-shell nanoparticles [36, 37]. The development of core/shell material needs to select the core and the shell with the aim to optimize the structural defects and the lattice parameters. The application of CSNCs mainly depends on the composition of the core and the arrangement of the shell.

The CSNCs have greater attention in the new era with various applications such as nanobiotechnology, optoelectronic devices, imaging systems, storage of more energies in energy storage devices, such as supercapacitors, capacitors, genetic engineering, and many other applications. The development of CSNCs can be classified into various types which includes the development of core with dielectric and shell with metal nanoparticles, with dielectric-semiconductor materials, metal-semiconductor nanomaterials, etc. Among the aforementioned materials, the development of CSNCs with semiconductor as the core and metal as the shell, metal nanoparticles include Ag, Au, and Pt with chemical stability and the absorption of light from the infrared region to the visible region with unique optical, electrical, and catalytic properties [38].

30.2 ZnO Nanoparticles-Emerging Solar Cell Applications

30.2.1 ZnO Nanoparticles Utilized in Dye-Sensitized Solar Cells

The development of dye-sensitized solar cells (DSSCs) was first initiated by the scientist O'Regan and Gratzel in the year 1991, which have garnered greater attention owing to the alternative materials to traditional silicon solar cells. The fabrication of these solar cells iseasy, effective, and lucrative with very good and long-term stability with the conversion efficiency exceeding more than 10%. Ruthenium-based dyes are highly utilized as a photosensitizer which are broadly attached to the mesoporous metal oxide nanomaterials with a larger surface area and absorb solar light very efficiently. In this process, the electrons were injected to the optically excited molecules of dye and turn back into the metaloxide conduction band which diffuses across the semiconductor film layer and finally reached into the back contact. The redoxpairs diffuse in solution, which are reduced to the counter-electrode, and regenerate the oxidized dye. Due to sensitizers, the high cost of providing typical dyes (N3, N719) encouraged the alternative use of narrow-band semiconductor-based quantum dots due to their individually adjustable optical properties to match the solar spectrum [39].

30.2.2 Solar Cell Application

In organic solar cells, n-type inorganic semiconductor of ZnO has been widely used for hybrid solar cells owing to their prominent characteristics such as low cost, easy to synthesis, non-toxicity, high stability, and good electrical and optical properties. Also, various researchers focused on the development of different 1D ZnO nanostructures (nanorods, nanopillars, nanofibers, nanowires, and nanoarrays) for enhancing the performance of organic and organic/inorganic hybrid solar cells [40].

30.2.3 Photocatalytic Application

Metal oxide semiconductor materials can also be used for water purification treatment. Among the various metal oxide semiconductor materials, ZnO is one of the most prominent candidates for photocatalytic application due to its large specific surface area, wide band semiconductor, and large excitation binding energy. Also, ZnO nanostructures act as potential candidates for photocatalytic degradation of an organic dye pollutant. In organic dye degradation process, electrons in the valence band (VB) are excited to the conduction band (CB) at the same time holes will be generated in the VB in ZnO. From this reaction, electrons and hole pairs produce hydroxyl radicals and superoxide anion radicals which are responsible for dye degradation process [41].

30.2.4 Supercapacitor Application

ZnO nanostructures (nanorods, nanopillars, nanocones, nanowires, nanofibers, and nanoarrays) show enhanced supercapacitor characteristics due to theirhigher surface area. Various researchers focused on the development of combination of ZnO-based metal and metal oxide materials to improve the stability of supercapacitors. In the supercapacitor application, ZnO and ZnO-based materials act as the working electrode for improving the specific capacitance, galvanostatic charge discharge, and cycle stability [42].

30.3 Conclusion

The efficient storage of energy in the various electronic storage devices is the key factor to the future energy crisis problems. Among them, capacitors and supercapacitors play a pivotal role in various electronic devices such as electronic circuits, electronic vehicles, and telecommunication systems. In this regard, the applications of green nanocomposites prepared via various renewable feedstocks have garnered greater deliberation owing to the enhanced physiochemical, lucrative, and facile synthetic methodology. In order to attain enhanced storage applications, the developed environmental benign green nanomaterials must possess enhanced morphological characteristics, highly crystalline structure, enhanced electrical and thermal conductivity with better tensile strength. Also, green nanomaterials developed with unique optoelectronic and physiochemical characteristics are continuously employed in supercapacitors for the storage of energy in order to attain high reliability and charge storage performance. Also, the dielectric properties and capacitance enhancement may decrease the strength and physical properties of green nanomaterials. So, the interactions between the nanomaterials need to improve in order to attain higher energy storage of the devices and density of the materials. The chapter discusses the recent development of ZnO nanomaterials synthesized through facile green synthesized protocol for energy storage applications. This may be achieved by the inclusion of eco nanofillers with eco polymers which has led to enhancing energy storage performance through green synthesized nanocomposite. The parameters considered in order to attain high energy storage performance are morphology-controlled synthesis, matrix-filler interaction, high thermal and electrical conductivity, dielectric properties with enhanced charge density, dielectric properties, capacitance, charge/discharge ratio, etc. Hence, the chapter concludes with the fabrication of ZnO and ZnO-based nanomaterials applied successfully in several energy-storing devices including solar cells, electronics, nanogenerator capacitors, and supercapacitors.

References

1. Chauhan, K, Sharma, R, & Dharela, R, 2016, 'Chitosan-thiomer stabilized silver nano-composites for antimicrobial and antioxidant applications', *RSC Adv.*, vol. 6, pp. 75453–75464.
2. Moosavi, R, Ramanathan, S, & Lee, Y, 2015, 'Synthesis of antibacterial and magnetic nanocomposites by decorating graphene oxide surface with metal nanoparticles', *RSC Adv.*, vol. 5, pp. 76442–76450.
3. Moldovan, B, David, L, & Achim, M, 2016, 'A green approach to phytomediated synthesis of silver nanoparticles using Sambucusnigra L. Fruits extract and their antioxidant activity', *J. Mol. Liq.*, vol. 221, pp. 271–278.
4. Swaminathan, B, Muthukrishnan, S, & Sukumaran, M, 2016, 'Antimicrobial, antioxidant and anticancer activity of biogenic silver nanoparticles – An experimental report', *RSC Adv.*, vol. 6, pp. 81436–81446.
5. Ahmad, A, Syed, F, & Shah, A, 2015, 'Silver and gold nanoparticles from sargentodoxacuneata: Synthesis, characterization and antileishmanial activity', *RSC Adv.*, vol. 5, pp. 73793–73806.
6. Liu, J, Zhao, Z, & Feng, H, 2012, 'One-pot synthesis of Ag–Fe_3O_4 nanocomposites in the absence of additional reductant and its potent antibacterial properties', *J. Mater. Chem.*, vol. 22, pp. 13891–13894.
7. Moulton, MC, Braydich-Stolle, LK, & Nadagouda, MN, 2010, 'Synthesis, characterization and biocompatibility of 'green' synthesized silver nanoparticles using tea polyphenols', *Nanoscale*, vol. 2, pp. 763–770.
8. Sahni, G, Panwar, A, & Kaur, B, 2015, 'Controlled green synthesis of silver nanoparticles by Allium cepa and Musa acuminata with strong antimicrobial activity', *Int. Nano Lett.*, vol. 5, pp. 93–100.

9. Krishna, M, Bhagavanth Reddy, G, & Veerabhadram, G, 2016, 'Eco-friendly green synthesis of silver nanoparticles using salmaliamalabarica: Synthesis, characterization, antimicrobial, and catalytic activity studies', *Appl. Nanosci.*, vol. 6, no. 5, pp. 681–689.

10. Kumari, R, Brahma, G, & Rajak, S, 2016, 'Antimicrobial activity of green silver nanoparticles produced using aqueous leaf extract of hydrocotylerotundifolia', *Orient Pharm. Exp. Med.*, vol. 16, no. 3, pp. 195–201.

11. Yugandhar, P, Haribabu, R, & Savithramma, N, 2015, 'Synthesis, characterization and antimicrobial properties of green-synthesised silver nanoparticles from stem bark extract of syzygiumalternifolium (Wt.) walp', *3 Biotech*, vol. 5, no. 6, pp. 1031–1039.

12. Lee, KJ, Park, SH, & Govarthanan, M, 2013, 'Synthesis of silver nanoparticles using cow milk and their antifungal activity against phytopathogens', *Mater. Lett.*, vol. 105, pp. 128–131.

13. Kokila, T, Ramesh, PS, & Geetha, D, 2016, 'Biosynthesis of AgNPs using Carica Papaya peel extract and evaluation of its antioxidant and antimicrobial activities', *Ecotoxicol. Environ. Saf.*, vol. 134, pp. 467–473.

14. Ajmal, M, Demirci, S, & Siddiq, M, 2016, 'Simultaneous catalytic degradation/reduction of multiple organic compounds by modifiable p (methacrylic acid-co-acrylonitrile)–M (M: Cu, Co) microgel catalyst composites', *New J. Chem.*, vol. 40, no. 2, pp. 1485–1496.

15. Ji, T, Chen, L, & Schmitz, M, 2015, 'Hierarchical macrotube/mesopore carbon decorated with monodispersed Ag nanoparticles as a highly active catalyst', *Green Chem.*, vol. 17, no. 4, pp. 2515–2523.

16. Edison, TJI, & Sethuraman, MG, 2013, 'Biogenic robust synthesis of silver nanoparticles using punicagranatum peel and its application as a green catalyst for the reduction of an anthropogenic pollutant 4-nitrophenol', *Spectrochim Acta A: Mol. Biomol. Spectrosc.*, vol. 104, pp. 262–264.

17. Mahmood, A, Ngah, N, & Omar, MN, 2011, 'Phytochemicals constituent and antioxidant activities in Musa x Paradisiaca flower', *Eur. J. Sci. Res.*, vol. 66, no. 2, pp. 311–318.

18. Alexandra Pazmino-Duran, E, Monica Giusti, M, & Wrolstad, RE, 2001, 'Anthocyanins from banana bracts (Musa X paradisiaca) as potential food colorants', *Food Chem.*, vol. 73, no. 3, pp. 327–332.

19. Krishnaraj, C, Ramachandran, R, & Mohan, K, 2012, 'Optimization for rapid synthesis of silver nanoparticles and its effect on phytopathogenic fungi', *Spectrochim Acta A: Mol. Biomol. Spectrosc.*, vol. 93, pp. 95–99.

20. Balraj, B, Arulmozhi, M, & Siva, C, 2017, 'Cytotoxic potentials of biologically fabricated platinum nanoparticles from Streptomyces sp. on MCF-7 breast cancer cells', *IET Nanobiotechnol.*, vol. 11, no. 3, pp. 241–246.

21. Makarov, VV, Love, AJ, & Sinitsyna, OV, 2014, 'Green' nanotechnologies: Synthesis of metal nanoparticles using plants', *Acta Nat.*, vol. 6, pp. 35–44.

22. Kumar, B, Smita, K, & Cumbal, L, 2015, 'Fabrication of silver nanoplates using nepheliumlappaceum (Rambutan) peel: A sustainable approach', *J. Mol. Liq.*, vol. 211, pp. 476–480.

23. Anandalakshmi, K, Venugobal, J, & Ramasamy, V, 2016, 'Characterization of silver nanoparticles by green synthesis method using pedalium murex leaf extract and their antibacterial activity', *Appl. Nanosci.*, vol. 6, no. 3, pp. 399–408.

24. Paulkumar, K, Gnanajobitha, G, & Vanaja, M, 2014, 'Piper nigrumleaf and stem assisted green synthesis of silver nanoparticles and evaluation of its antibacterial activity against agricultural plant pathogens', *Sci. World J.*, vol. 2014, p. 9.

25. Medda, S, Hajra, A, & Dey, U, 2015, 'Biosynthesis of silver nanoparticles from Aloe vera leaf extract, and antifungal activity against rhizopus sp. and aspergillussp', *Appl. Nanosci.*, vol. 5, no. 7, pp. 875–880.

26. Baranwal, A, Mahato, K, & Srivastava, A, 2016, 'Phtyofabricated metallic nanoparticles and their clinical applications', *RSC Adv.*, vol. 6, no. 107, pp. 105996–106010.

27. Zheng, L-Q, Yu, X-D, & Xu, J-J, 2015, 'Reversible catalysis for the reaction between methyl orange and NaBH4 by silver nanoparticles', *Chem. Commun.*, vol. 51, no. 6, pp. 1050–1105.

28. Jyoti, K, & Singh, A, 2016, 'Green synthesis of nanostructured silver particles and their catalytic application in dye degradation', *J. Genet. Eng. Biotechnol.*, vol. 14, no. 2, pp. 311–317.

29. Baruah, B, Gabriel, GJ, & Akbashev, MJ, 2013, 'Facile synthesis of silver nanoparticles stabilized by cationic polynorbornenes and their catalytic activity in 4-nitrophenol reduction', *Langmuir*, vol. 29, no. 3, pp. 4225–4234.

30. Moosavi, R, Ramanathan, S & Lee, Y, 2015, 'Synthesis of antibacterial and magnetic nanocomposites by decorating graphene oxide surface with metal nanoparticles', *RSC Adv.*, vol. 5, pp. 76442–76450.

31. Moorthy, SK, Ashok, CH, Rao, KV & Viswanathan, C, 2015, 'Synthesis and characterization of MgO nanoparticles by Neem leaves through green method', *Mater. Today: Proc.*, vol. 2, no. 9, pp.4360–4368.

32. Moulton, MC, Braydich-Stolle, LK, & Nadagouda, MN, 2010, 'Synthesis, characterization and biocompatibility of 'green' synthesized silver nanoparticles using tea polyphenols', *Nanoscale*, vol. 2, pp. 763–770.

33. Musee, N, 2011, 'Nanowastes and the environment: Potential new waste management paradigm', *Environ. Int.*, vol. 37, no. 1, pp. 112–128.

34. Najafi, A, 2017, 'A novel synthesis method of hierarchical mesoporous MgO nanoflakes employing carbon nanoparticles as the hard templates for photocatalytic degradation', *Ceram. Int.*, vol. 43, no. 7, pp. 5813–5818.

35. Narayanan, KB & Sakthivel, N 2011, 'Green synthesis of biogenic metal nanoparticles by terrestrial and aquatic phototrophic and heterotrophic eukaryotes and biocompatible agents', *Adv. Colloid Interface Sci.*, vol. 2, no. 169, pp. 59–79.

36. Narayanan, RK & Devaki, SJ, 2015, 'Brawny silver-hydrogel based nanocatalyst for the reduction of nitrophenols. Studies on kinetics and mechanism', *Ind. Eng. Chem. Res.*, vol. 54, no. 4, pp. 1197–1203.

37. Nath, D & Banerjee, P, 2013, 'Green nanotechnology–a new hope for medical biology', *Environ. Toxicol. Pharmacol.*, vol. 3, no. 36, pp. 997–1014.

38. Naushad, M, Sharma, G, & Alothman, ZA, 2019, 'Photodegradation of toxic dye using Gum Arabic-crosslinked-poly (acrylamide)/Ni (OH)$_2$/FeOOH nanocomposites hydrogel', *J. Clean. Prod.*, vol. 241, pp. 118263.

39. Nguyen, NT, Yoo, J, Altomare, M, & Schmuki, P, 2014, 'Suspended Pt nanoparticles over TiO$_2$ nanotubes for enhanced photocatalytic H$_2$ evolution', *Chem. Commun.*, vol. 50, pp. 9653.

40. Nguyen, TB, Doong, RA, Huang, CP, Chen, CW, & Dong, CD, 2019, 'Activation of persulfate by CoO nanoparticles loaded on 3D mesoporous carbon nitride (CoO@ meso-CN) for the degradation of methylene blue (MB)', *Sci. Total Environ.*, vol. 675, pp. 531.

41. Nidheesh, PV, Zhou, M, & Oturan, MA, 2018, 'An overview on the removal of synthetic dyes from water by electrochemical advanced oxidation processes', *Chemosphere*, vol. 197, pp. 210–227.

42. Nigra, MM, Haz, JM, & Katz, A, 2013, 'Identification of site requirements for reduction of 4-nitrophenol using gold nanoparticle catalysts', *Catal. Sci. Technol.*, vol. 3, pp. 2976–2983.

31

Nanotechnology for Sustainable Energy Storage Devices in Medical Applications

Dr. V. R. Lenin

Anil Neerukonda Institute of Technology and Sciences, Visakhapatnam, India

Dr. V. Suresh Kannan

Madanapalle Institute of Technology & Science, Madanapalle, India

R. Muthukumaran

Kakinada Institute of Technological Sciences, Ramachandrapuram, India

Ayaz Ahmad

NIT P, Patna, India

Shubhajit Halder

Hislop College, Nagpur, India

Rama Krishna Yellapragada

Koneru Lakshmaiah Education Foundation, K L University, Vaddeswaram, India

Dr. S. Vijayalakshmi

Saranathan College of Engineering, Trichy, India

In this chapter, medical applications in nano energy storage devices are discussed. Nanomaterials have made an impact on cancer treatment; chemotherapy, radiation therapy, and heat therapy are also explained. This chapter also explains how nanomaterials are helpful to cure diabetes, kidney disease, heart disease, wound treatment, etc.

31.1 Introduction

In nanotechnology, the term "nanomaterial" refers to the study and utilization of structures with a size between 1 and 100 nanometers. In order to put it into perspective, consider how many particles, each 800 nanometers wide, would need to be arranged side by side in order to equal the width of a human hair. The ability to identify nanoscale materials has opened up a variety of options across many industries and scientific disciplines. Nanotechnology can be utilized for a variety of purposes, including those that are discussed below [1–5], as it is basically a set of techniques that enable the modification of properties at a very small scale.

31.2 Drug Delivery

The majority of negative side effects associated with therapies like chemotherapy today are brought on by medication delivery techniques that fail to precisely target the cells they are meant to cure. Scientists at Harvard and MIT have successfully attached unique RNA strands with a diameter of around 10 nm to nanoparticles and filled the nanoparticles with a chemotherapeutic medication. These RNA strands draw cancer cells to them. The nanoparticle attaches to a cancer cell when it comes into contact with it and delivers the medicine into the cancer cell. With fewer negative side effects than those caused by traditional chemotherapy, this targeted technique of drug delivery holds a great deal of potential for treating cancer patients [6, 7].

Utilizing nanoparticles to transport medications, heat, light, or other chemicals to particular types of cells is one current application of nanotechnology in medicine (such as cancer cells). The drug delivery system that uses nanotechnology is depicted in Figure 31.1. Engineered particles are drawn to damaged

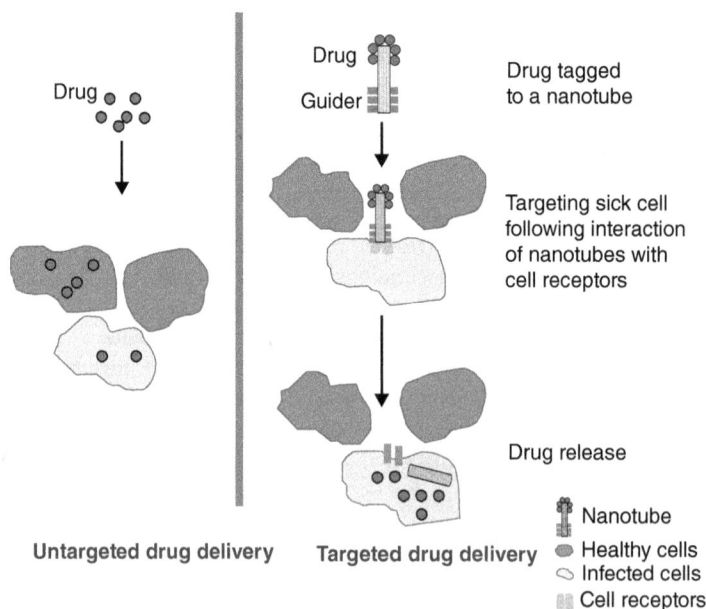

FIGURE 31.1 Untargeted drug delivery and targeted drug delivery.

cells, allowing for direct therapy of particular cells. By using this method, sickness can be detected earlier and less harm is done to the body's healthy cells.

An approach to transfer cardiac stem cells to injured heart tissue, for instance, is being developed by researchers at North Carolina State University. In order to transfer more stem cells to an injured tissue, they attach nanovesicles to the stem cells that are drawn to the injury. You might not think there is a substantial need for study on medicine delivery if all you ever take is the odd aspirin. But if you were a cancer patient suffering from crippling side effects from your medication or a diabetic who had to inject insulin many times each day, the advantages of enhanced drug delivery may completely alter your life. The use of nanoparticles to deliver medications to cancer cells has perhaps received the most attention in the field of drug delivery. Engineered particles are drawn to damaged cells, allowing for direct therapy of particular cells. This method lessens harm to the body's healthy cells.

But that's only the tip of the iceberg when it comes to drug administration; nanotechnology may also be used in a variety of other ways to improve drug delivery effectiveness and, perhaps most importantly, patient comfort. While some strategies are still just ideas, others are either in use right now or are in various phases of testing. A number of these methods are introduced in the review of drug delivery applications of nanomedicine that are discussed in the following sections.

31.3 Fabrics

In order to enhance performance, manufacturers are altering the properties of known materials by including nanoscale components. For instance, piezoelectric fibers are now being developed, and they may one day allow garments to produce electricity from regular motion [8].

31.4 Reactivity of Materials

When produced as nanoparticles, the characteristics of many conventional materials alter (nanoparticles). This is typically due to the fact that smaller particles, such as nanoparticles, have more surface area per weight than bigger particles, making them more reactive to certain chemicals. Studies have shown, for instance, that nanoparticles of iron can be useful in the removal of pollutants from groundwater because they react with those chemicals more effectively than bigger iron particles [9].

31.5 Strength of Materials

Buckyballs and nanotubes, two types of carbon nanoparticles, are very powerful. Only carbon makes up nanotubes and buckyballs, and the unique properties of the bonds between carbon atoms give these materials their strength. Making bulletproof vests consisting of carbon nanotubes that are the weight of a t-shirt is one proposed use that demonstrates the power of carbon nanoparticles [10].

31.6 Micro/Nano Electromechanical Systems

Manufacturing tiny sensors like those needed to activate your car's airbags is made possible by the capacity of silicon surfaces to produce gears, mirrors, sensor components, and electronic circuitry. MEMS is the name of this method (micro-electro mechanical systems). Similar to the process used to make computer chips, the MEMS methodology yields a close integration of the mechanical mechanism with the requisite electronic circuit on a single silicon chip. Comparing identical devices created using conventional techniques to those made using MEMS, the size and cost of the product are reduced. Nano-electro mechanical systems, or NEMS, are a stepping stone to MEMS. Few firms are currently producing NEMS products, but once manufacturers invest in the machinery required to develop nano-sized features, NEMS products will become the norm [11–13].

31.7 Molecular Manufacturing

The replicator, which could create anything from an advanced guitar to a cup of Earl Grey tea, will be familiar to Star Trek fans. Just by programming the replicator, your favorite character might make anything materialize. The Star Trek replicator might soon become a reality, thanks to a technique being developed by researchers called molecular manufacturing. The technology that these people imagine is a molecular fabricator; it would utilize tiny manipulators to place atoms and molecules in order to construct something as intricate as a desktop computer. Theoretically, practically any inanimate object can be replicated using this technique utilizing only raw materials [14–16]. Researchers are creating personalized nanoparticles the size of molecules that can carry medications directly to your body's sick cells. When developed, this technique ought to significantly lessen the harm that a patient's healthy cells endure during a treatment like chemotherapy.

31.8 Nanotechnology in Medicine – Nanoparticles in Medicine

Exciting opportunities exist for the application of nanotechnology in medicine. While some strategies are still just ideas, others are either in use right now or are in various phases of testing. Applications of currently being developed nanoparticles are part of nanotechnology in medicine, as are longer-term studies including the employment of produced nanorobots to perform cellular repairs. Whatever you want to call it, the application of nanotechnology to medicine has the potential to fundamentally alter how disease and bodily injury are identified and treated in the future. Many methods that were only a few years ago considered science fiction are now making remarkable strides toward becoming a reality [17].

31.8.1 Nanotechnology in Medicine Application: Diagnostic Techniques

A sensor that may quickly and easily be utilized with a hand-held testing instrument to identify viruses such as COVID-19 is being created by researchers at John Hopkins University using nanoimprint lithography. Antibodies coupled to carbon nanotubes in chips are being used by Worcester Polytechnic Institute researchers to identify cancer cells in the blood stream. The researchers think this approach might be applied to quick lab tests that could find cancer cells in the circulation early [18].

Figure 31.2 depicts the development of a test for the early diagnosis of kidney injury. The procedure employs gold nanorods that have been functionalized to attach to the specific protein produced by damaged kidneys. A change in color occurs as protein builds up on the nanorod. The test is made to be completed quickly and affordably in order to identify problems early.

31.8.2 Antibacterial Treatments Using Nanotechnology in Medicine

A method to kill bacteria using infrared light and gold nanoparticles is being developed by researchers at the University of Houston. In hospitals, using this technique might result in better instrument cleaning. Quantum dots may be used to treat illnesses that are resistant to antibiotics, according to research being done at the University of Colorado Boulder [19, 20].

31.8.3 Treatment of Wounds Using Nanotechnology in Medicine

A bandage that uses electricity generated by the patient's nanogenerators to apply electrical pulses to a wound has been shown by University of Wisconsin researchers.

Another strategy to stop blood loss is required for trauma patients who have internal bleeding. Polymer nanoparticles that mimic platelets are being created by scientists at Chase Western Reserve

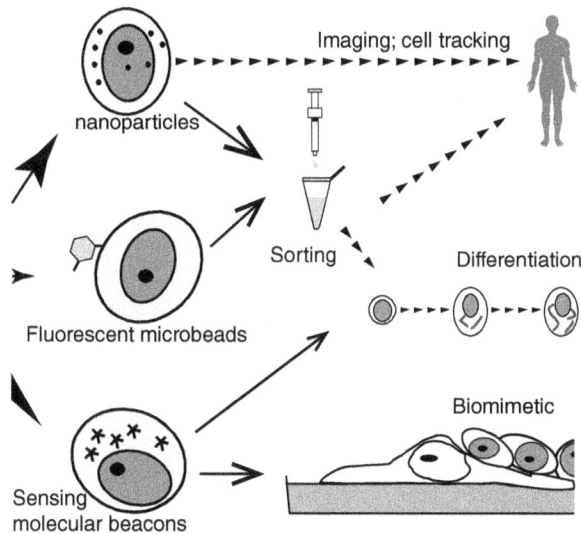

FIGURE 31.2   Nanotechnology in diagnosis.

University. According to laboratory testing, injecting these artificial platelets considerably lowers blood loss [21, 22].

Nanotechnology-based wound healing enables the utilization of

- Nanoparticles that cause blood to coagulate to stop bleeding
- Bandages that emit electric pulses to hasten the healing process
- Nanoparticles to lessen internal bleeding

31.8.4 Cell Repair Using Nanotechnology in Medicine

As with antibodies in our body's natural healing processes, nanorobots might be designed to mend particular sick cells. The following article discusses design analysis for one such cell-repairing nano-robot: What Is the Best Gene Delivery Vector? Chromosome repair therapy using chromallocytes and nanorobots for cell repair [23].

31.8.5 Medical Resources for Nanotechnology

Eight Centers of Cancer Nanotechnology Excellence and a Nanotechnology Characterization Lab are part of the National Cancer Institute Alliance for Nanotechnology in Cancer.

Alliance for NanoHealth: Eight research institutions engaged in cooperative research make up this alliance.

Nano robots in Medicine: Future applications of nanomedicine will be based on the ability to build nano robots. In the future, these nano robots could actually be programmed to repair specific diseased cells, functioning in a similar way to antibodies in our natural healing processes.

31.8.6 Nanotechnology vs. Covid-19

The elimination of Covid-19 is one of the world's largest problems right now. Coronavirus vaccines, enhanced safety goggles, more powerful disinfectants, and improved diagnostic tools are all developed using nanotechnology. Examples of the ongoing research and the potential of nanotechnology in this area are discussed. A number of the techniques mentioned have reached the stage of pre-clinical or clinical trials, and a few applications are now reported [24, 25].

Nanotechnology has been used in the field of immunology for a very long time. Since viruses are naturally occurring nanoparticles, turning to nano is a sensible approach when dealing with viruses like the coronavirus that causes Covid-19. In reality, nanomedicine has long attempted to imitate the traits of viruses for uses like targeted drug delivery. The development of vaccines and nanotechnology, however, have arrived in 2020.

31.8.6.1 How Nanotechnology Is Being Used to Fight Covid-19

An experimental vaccination being developed at Caltech may be able to protect against new coronavirus variations by producing antibodies for a variety of coronaviruses. A sensor that may quickly and easily be utilized with a hand-held testing instrument to identify viruses such as COVID-19 is being created by researchers at John Hopkins University using nanoimprint lithography.

A vaccine being developed by scientists at the Walter Reed Army Institute of Research may be able to defend against a variety of coronavirus species and strains by employing ferritin nanoparticles. Results from a vaccination trial using mRNA molecules that are enclosed in lipid nanoparticles have been released by Moderna. This vaccination has an Emergency Use Authorization from the FDA.

Novavax has begun the Phase 3 clinical trial for a coronavirus vaccine candidate that was created utilizing protein nanoparticles. A filter comprised of cellulose nanofibers can prevent particles the size of viruses, according to research from Queensland University of Technology. They think it is possible to produce the filters at a low cost and in large quantities required for single-use filter cartridges.

To create probes that adhere to Covid-19 RNA, scientists are utilizing gold nanoparticles. These probes are being used in the development of testing apparatus that will, in their opinion, result in quick turnaround and low error rates. To transfer peptide molecules to Covid-19 viral molecules, scientists at MIT and Northwestern University are developing nanostructures. The viral molecule may be rendered inactive by the peptide molecules' ability to bind to the Covid-19 spike protein. The peptide molecules will be carried by the nanostructures because peptide molecules don't last long in the bloodstream.

A nano-interferometric biosensor is being used by scientists at the Catalan Institute of Nanoscience and Nanotechnology to create a point-of-use testing tool for Covid-19. Gold nanorods are being used by Sona Nanotech to create a diagnostic test for Covid-19. Results from the test should be available in 5–15 minutes, and no lab workup is necessary. Orthogonal nanofibers were used by researchers at the Korea Advanced Institute of Science and Technology to create a filter mask that, according to them, is equivalent to the N95 masks required for coronaviruses, with the added benefit of continuing to filter small particles even after being washed numerous times.

To stop the coronavirus from spreading, several businesses treat surfaces including walls and ceilings with titanium dioxide nanocrystals. The nanocrystals function as a photocatalytic disinfection system when the surface is exposed to UV light, assisting in the killing of viruses on the surface. A Covid-19 test without chemicals has been created by researchers at the Norwegian University of Science and Technology (which are in limited supply). Silica-coated magnetic nanoparticles are used in the test. The nanoparticles attract virus RNA, which is subsequently drawn out of the sample by a magnetic field.

31.9 Nanotechnology in Cancer Treatment

With the help of nanotechnology, cancer treatments may be able to detect and remove cancer cells before they develop into tumors, as well as remove malignant tumors with little harm to good tissue and organs. The majority of initiatives to enhance cancer care with nanotechnology are in the research or development phase. However, a lot of institutions of higher learning and businesses operate in this field globally.

31.9.1 Nanotechnology Cancer Treatments: Nanoparticle Chemotherapy

A wearable patch that can deliver chemotherapeutic medications to the skin to treat melanoma is being developed by Purdue University researchers using silicon nanoneedles. A nanoparticle by the name of CRLX101 is used in an experimental targeted chemotherapy therapy. Cerulean Pharma is the firm creating this specific form of chemotherapy.

A strategy to combat prostate cancer is being developed by researchers at the University of Georgia. To transport a substance called IPA-3 to the cancer cells, they are making use of nanoparticles. In experiments using lab mice, IPA-3 seems to slow the growth of prostate cancer cells. To treat brain cancers, scientists are experimenting with nano diamonds that have chemotherapy medications linked to them. Because the nano diamond and chemotherapy medication combination persists in the tumor for a longer period of time than the chemotherapy agent alone, the effectiveness should increase.

31.9.1.1 A Survey of Nanoparticles in Chemotherapy

Targeted chemotherapy, which delivers the cancerous tumors with the tumor necrosis factor alpha (TNF) agent, is one form of treatment that is currently being researched. TNF and thiol-derivatized polyethylene glycol (PEG-THIOL), which conceals the TNF-carrying nanoparticle from the immune system, are linked to a gold nanoparticle. This makes it possible for the nanoparticle to pass through the bloodstream unharmed. Visit this link to read the article for additional information. The business known as CytImmune is working on this targeted chemotherapy technique to deliver TNF and other chemotherapy medicines to cancer tumors. Figure 31.3 depicts another targeted chemotherapy that is being developed that employs a nanoparticle called CRLX101. Cerulean Pharma is the firm creating this specific form of chemotherapy. In Phase 1 clinical trials, Cristal Therapeutics is delivering the medication docetaxel to tumors using nanoparticles dubbed CriPec.

The development of a wearable patch that can administer chemotherapeutic medications to the skin for the treatment of melanoma is being carried out by Purdue University researchers utilizing silicon nanoneedles. Scientists at MIT are creating nanoparticles that can deliver two chemotherapy medications to tumors of the glioblastoma form of brain cancer by breaching the blood–brain barrier and targeting the tumors. Doxorubicin's efficiency as a chemotherapeutic agent can be improved, according to research from the University of Toronto. This was made possible by the introduction of manganese dioxide nanoparticles that are intended to concentrate in a tumor and produce oxygen.

Irinotecan, a chemotherapeutic medication, has been successfully delivered to pancreatic cancer tumors by UCLA researchers using mesoporous silica nanoparticles. The toxicity of the treatment is lessened, according to tests done on mice, using this approach. Prostate cancer treatment is being developed by researchers at the University of Georgia. To deliver IPA-3 to the cancer cells, scientists are

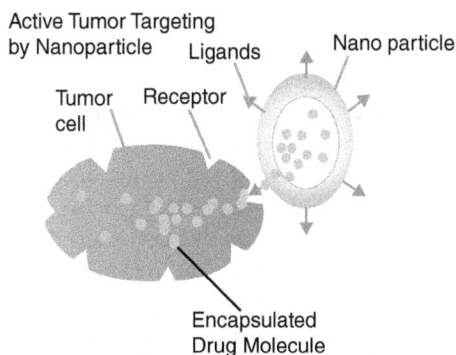

FIGURE 31.3 Nano particles targeting cancer cells.

employing nanoparticles. Prostate cancer cell proliferation seems to be inhibited by IPA-3 in investigations on laboratory animals.

Researchers at the University of Texas Southwestern Medical Center have used nanoparticles called dendrimers to deliver nucleic acids that suppress tumors to liver cancer tumors. The researchers have demonstrated, in lab tests, that this method can reduce tumor growth in mice. Researchers at the University of Leicester and three other universities are developing synthetic polymer nanoparticles, or nano-MIPs, which may result in improved delivery of chemotherapy drugs to cancer cells.

Researchers are developing graphene strips to deliver different drugs to specific regions of cancer cells. When the graphene strip reaches the cancer cell, one drug separates from the graphene and attacks the cell membrane while the graphene strip enters the cell and delivers the second drug to the cell nucleus. Researchers at MIT are developing nanoparticles that carry precise ratios of three different drugs. They are testing the effectiveness of this approach on ovarian cancer cells. Researchers at the Johannes Gutenberg University Mainz have developed a method to attach antibodies to nano capsules that they believe builds more stable nano capsules, which will improve the ability of the nano capsule to deliver drugs to cancer cells.

Researchers at UCLA are investigating a method to fight pancreatic cancer using two different nanoparticles. The first nanoparticle removes material on the exterior of the cancer cells that block the entry of chemotherapy drugs, the second nanoparticle carries the chemotherapy drug. Testing this method on laboratory mice showed significantly faster shrinkage of the tumors than other methods.

Researchers are testing the use of chemotherapy drugs attached to nano diamonds to treat brain tumors. The nano diamond/chemotherapy drug combination stays in the tumor longer than the chemotherapy drug by itself, which should increase the effectiveness. Researchers are also testing the use of chemotherapy drugs attached to nano diamonds to treat leukemia. It turns out that leukemia cancer cells can pump chemotherapy drugs out of the cancer cell, limiting the effectiveness of the drug. The cancer cell cannot pump the nano diamond out, so attaching the drug molecules to nano diamonds results in the drug staying in the cancer cell longer.

Researchers are connecting different DNA strands together into a structure they call a "nano train". They have demonstrated in lab studies that these nano trains are effective in delivering chemotherapy drugs to cancer cells, and that by using different DNA strands, they can customize which type of cancer cells the nanotrains target. Researchers are developing a nanoparticle that both delivers a chemotherapy drug and stimulates the immune system to attack cancer cells. They have tested the method on mice with positive results.

A method being developed to fight bladder cancer uses nanoparticles called micelles to deliver a chemotherapy drug called paclitaxel to bladder cancer cells. An improved way to shield nanoparticles delivering chemotherapy drugs from the immune system has been developed by forming the nanoparticles from the membranes of red blood cells. Researchers have demonstrated a method of delivering a protein to cancer cells that destroys the cancer cells. They use a polymer nanoshell to deliver the protein into the cancer cells. When the protein accumulates in the nucleus of the cancer cell, the protein causes the cancer cell to self-destruct.

JSI-124 is a chemical that is delivered to cancer tumours via polymer nanoparticles. This molecule degrades the ability of the cancer cells to suppress the immune system, possibly slowing the growth of cancer tumors. Another technique delivers chemotherapy drugs to cancer cells and also applies heat to the cells. Researchers are using gold nano rods to which DNA strands are attached. The DNA strands act as a scaffold, holding together the nano rod and the chemotherapy drug. When infrared light illuminates the cancer tumor, the gold nano rod absorbs the infrared light, turning it into heat. The heat both releases the chemotherapy drug and helps destroy the cancer cells.

Researchers have demonstrated a nanoparticle that kills lymphoma cancer cells. They use a nanoparticle which looks like high density cholesterol (HDL), but with a gold nanoparticle at it's core. When this nanoparticle attaches to a lymphoma cell, it blocks the cancer cell from attaching to real HDL cholesterol, starving the cancer cell. A method being developed to fight skin cancer uses gold nanoparticles to which RNA molecules are attached. The nanoparticles are in an ointment that is applied to the skin. The nanoparticles penetrate the skin and the RNA attaches to a cancer-related gene, stopping the gene from generating proteins that are used in the growth of skin cancer tumors.

31.9.2 Nanotechnology Cancer Treatments: Heat

Another technique being developed works on destroying cancer tumors by applying heat. Nanoparticles called Auro Shells absorb infrared light from a laser, turning the light into heat. The company developing this technique is called Nano spectra. Targeted heat therapy is being developed to destroy breast cancer tumors. In this method, antibodies that are strongly attracted to proteins produced in one type of breast cancer cell are attached to nanotubes, causing the nanotubes to accumulate at the tumor. Infrared light from a laser is absorbed by the nanotubes and produces heat that incinerates the tumor.

Another method that targets individual cancer cells inserts gold nanoparticles into the cells and then shines a laser on the nanoparticles. The heat explodes the cancer cells. The use of nanoparticles in cancer heat therapy, referred to as nanoparticle hyperthermia, involves applying heat to tumors to attack cancer cells. This type of treatment can destroy cancer tumors with minimal damage to the human body.

31.9.2.1 A Survey of Methods using Nanoparticles to Improve Cancer Hyperthermia

Using iron oxide nanoparticles and a magnetic field to heat up cancer tumors has been shown to stimulate the immune system to fight cancer cells in other parts of the body. Researchers believe that this methodology may be useful in preventing the spread of cancer cells, while other techniques are used to fight localized tumors. Another technique being developed works on destroying cancer tumors by applying heat. Nanoparticles called Auro Shells absorb infrared light from a laser, turning the light into heat. Researchers are reporting results from a clinical study using this technique to destroy tumors in prostate cancer patients. The company developing this technique is called Nano spectra.

Researchers at Massachusetts General Hospital are working on a method using infrared light to trigger the release of two anticancer drugs to tumors. Another technique delivers chemotherapy drugs to cancer cells and also applies heat to the cell. Researchers are using gold nano rods to which DNA strands are attached. The DNA strands act as a scaffold, holding together the nano rod and the chemotherapy drug. When infrared light illuminates the cancer tumor, the gold nanorod absorbs the infrared light, turning it into heat. The heat both releases the chemotherapy drug and helps destroy the cancer cells.

Researchers are using a photosensitizing agent to enhance the ability of drug-carrying nanoparticles to enter tumors. First, they let the photosensitizing agent accumulate in the tumor and then illuminate the tumor with infrared light. The photosensitizing agent causes the blood vessels in the tumor to be more porous, therefore more drug-carrying nanoparticles can enter the tumor. One heat therapy to destroy cancer tumors using nanoparticles is called Auro Shell™. The Auro Shell™ nanoparticles circulate through a patient's bloodstream, exiting where the blood vessels are leaking at the site of cancer tumors. Once the nanoparticles accumulate at the tumor, the Auro Shell™ nanoparticles are used to concentrate the heat from infrared light to destroy cancer cells with minimal damage to surrounding healthy cells.

Nano spectra Biosciences has developed such a treatment using Auro Shell™ that has been approved for a pilot trial with human patients. Targeted heat therapy is being developed to destroy breast cancer tumors. In this method, antibodies that are strongly attracted to proteins produced in one type of breast cancer cell are attached to nanotubes, causing the nanotubes to accumulate at the tumor. Infrared light from a laser is absorbed by the nanotubes and produces heat that incinerates the tumor.

An intriguing targeted chemotherapy method uses one nanoparticle to deliver the chemotherapy drug and a separate nanoparticle to guide the drug carrier to the tumor. First, gold nanorods circulating through the bloodstream exit where the blood vessels are leaking at the site of cancer tumors. Once the nanorods accumulate at the tumor, they are used to concentrate the heat from infrared light, heating up the tumor. This heat increases the level of a stress-related protein on the surface of the tumor.

The drug-carrying nanoparticle (a liposome) is attached to amino acids that bind to this protein, so the increased level of protein at the tumor speeds up the accumulation of the chemotherapy drug-carrying liposome at the tumor. Nanotechnology Cancer Treatments: Radiation

Researchers have developed nanoparticles containing a radioactive core with attached molecules that attach to lymphoma tumor cells. The researchers are designing this method to stop the spread of cancer from the primary tumor. Researchers are investigating the use of bismuth nanoparticles to concentrate radiation used in radiation therapy to treat cancer tumors. Initial results indicate that the bismuth nanoparticles would increase the radiation dose to the tumor by 90%.

A method to make radiation therapy more effective in fighting prostate cancer is using radioactive gold nanoparticles attached to a molecule that is attracted to prostate tumor cells. Researchers believe that this method will help concentrate the radioactive nanoparticles at the cancer tumors, allowing treatment of the tumors with minimal damage to healthy tissue.

The use of nanoparticles in radiation therapy offers some exciting possibilities. Radiation therapy, which has been used for years to treat cancer, can cause serious damage to the human body. Using nanoparticles, it may be possible to destroy cancer tumors with minimal damage to healthy tissue and without the serious side effects often caused by radiation therapy treatments.

31.9.2.2 A Survey of Methods Using Nanoparticles to Improve Radiation Therapy

Researchers have developed nanoparticles containing a radioactive core with attached molecules that attach to lymphoma tumor cells. The researchers are designing this method to stop the spread of cancer from the primary tumor. Researchers are investigating the use of bismuth nanoparticles to concentrate radiation used in radiation therapy to treat cancer tumors. Initial results indicate that the bismuth nanoparticles would increase the radiation dose to the tumor by 90%.

X-ray therapy may be able to destroy cancer tumors using a nanoparticle called nbtxr3. The nbtxr3 nanoparticles, when activated by x-rays, generate electrons that cause the destruction of cancer tumors to which they have attached themselves. This is intended to be used in place of radiation therapy with much less damage to healthy tissue. Nanobiotix and the University of Texas MD Anderson Cancer Center are conducting clinical trials for this technique.

Researchers at the University of Missouri are developing a nanoparticle that contains actinium, a radioactive element that emits alpha particles. A method to make radiation therapy more effective in fighting prostate cancer is using radioactive gold nanoparticles attached to a molecule that is attracted to prostate tumor cells. Researchers believe that this method will help concentrate the radioactive nanoparticles at the cancer tumors, allowing treatment of the tumors with minimal damage to healthy tissue.

31.9.3 Nanotechnology Cancer Treatments: Miscellaneous

Researchers at Tel Aviv University are developing a vaccine for melanoma using polymer nanoparticles to which melanoma-related peptides have been attached. Researchers have demonstrated a nanoparticle that kills lymphoma cancer cells. They use a nanoparticle which looks like HDL cholesterol, but with a gold nanoparticle at its core. When this nanoparticle attaches to a lymphoma cell, it blocks the cancer cell from attaching to real HLD cholesterol, starving the cancer cell.

Researchers have demonstrated a method of delivering a protein to cancer cells that destroys the cancer cells. They use a polymer nanoshell to deliver the protein into the cancer cells. When the protein accumulates in the nucleus of the cancer cell, the protein causes the cancer cell to self-destruct.

A method being developed to fight skin cancer uses gold nanoparticles to which RNA molecules are attached. The nanoparticles are in an ointment that is applied to the skin. The nanoparticles penetrate the skin and the RNA attaches to a cancer-related gene, stopping the gene from generating proteins that are used in the growth of skin cancer tumors. Delivery of short interfering RNAs (siRNA) is interesting because siRNA simply stops the cancer tumor from growing and there is the potential to tailor synthetic siRNA to the version of cancer in an individual patient.

A method to increase the number of cancer fighting immune cells in cancer tumors is interesting. Nanoparticles containing drug molecules called interleukins are attached to immune cells (T-cells). The idea is that when the T-cells reach a tumor, the nanoparticles release the drug molecules, which cause the T-cells to reproduce. If enough T-cells are reproduced in the cancer tumor, the cancer can be destroyed. This method has been tested on laboratory mice with very good results.

Magnetic nanoparticles that attach to cancer cells in the blood stream may allow the cancer cells to be removed before they establish new tumors. Researchers at the Institute of Bioengineering and Nanotechnology and IBM researchers have demonstrated sustained drug delivery using a hydrogel. The hydrogel is injected under the skin, allowing continuous drug release for weeks, with only one injection, rather than repeated injections. They demonstrated this method by injecting the hydrogel, containing the chemotherapy drug herceptin, under the skin of laboratory mice. The study showed significant reduction in tumor size.

Using gold nanoparticles to deliver platinum to cancer tumors may reduce the side effects of platinum cancer therapy. The key is that the toxicity level of platinum depends upon the molecule it is bonded to (for the tech types the toxicity depends upon the oxidation state of the platinum). So, the researchers chose a platinum-containing molecule that has low toxicity to attach to the gold nanoparticles. When the platinum bearing nanoparticle reaches a cancer tumor, it encounters an acidic solution which changes the platinum to its toxic state, in which it can kill cancer cells.

Other researchers are taking a different approach to delivering platinum to cancer tumors. Instead of attaching platinum to nanoparticles, they have used Iron oxide nanoparticles can be used to improve magnetic resonance imagining (MRI) images of cancer tumors. The nanoparticle is coated with a peptide that binds to a cancer tumor. Once the nanoparticles are attached to the tumor, the magnetic property of the iron oxide enhances the images from the MRI scan.

Sensors based upon nanoparticles or nanowires can detect proteins related to specific types of cancer cells in blood samples. This could allow early detection of cancer. T2 Biosystems uses superparamagnetic nanoparticles that bind to the cancer-indicating protein and cluster together. These clusters provide a magnetic resonance signal indicating the presence of the cancer-related protein. For another approach, researchers at John Hopkins University use quantum dots and molecules that emit a fluorescent glow to detect DNA strands that are early indicators of cancer.

The capacity to do surgery at the cellular level, eliminating specific sick cells and even fixing specific damaged cells. Significant lengthening of the human lifespan by repairing cellular level conditions that cause the body to age.

31.10 Nanotechnology vs. Heart Disease

The use of nanotechnology to treat heart disease offers some exciting possibilities, including the ability to:

- treat defective heart valves
- detect and treat arterial plaque
- understand at a sub-cellular level how heart tissue functions in both healthy and damaged organs, which can help researchers design better treatments

31.10.1 Nanotechnology in Medical Diagnostics

Nanotechnology-based diagnostic techniques currently under development may provide two major benefits:

- Rapid testing, potentially in a doctor's office, may allow complete diagnosis and start of treatment within one visit to the doctor.
- The detection of diseases at an earlier stage than possible with current techniques offers the potential of stopping a disease earlier, possibly with less damage to the patient.

31.11 Nanotechnology Treatments for Diabetes

The use of nanotechnology to treat diabetes offers some exciting possibilities:

- Glucose monitoring that does not require a blood sample.
- A system that would automatically release insulin when needed.
- A potential vaccine for Type 1 diabetes.

31.12 Nanotechnology Kidney Disease Treatments

The use of nanotechnology to treat kidney disease offers some exciting possibilities, including the ability to:

- target drug delivery to treat kidney disease
- develop an artificial kidney

31.13 Nanoparticles in Antibacterial Treatments

Nanoparticle antibacterial treatments offer exciting possibilities, including the ability for using nanoparticles to:

- treat antibiotic-resistant infections
- treat staph infections
- release antibiotics when an infection starts in a wound

31.13.1 Extending Life by Repairing Cells

Perhaps, the most exciting possibility exists in the potential for repairing our bodies at the cellular level. Techniques for building nanorobots are being developed that should make the repair of our cells possible. For example, as we age, DNA in our cells is damaged by radiation or chemicals in our bodies. Nanorobots would be able to repair the damaged DNA and allow our cells to function correctly.

This ability to repair DNA and other defective components in our cells goes beyond keeping us healthy: it has the potential to restore our bodies to a more youthful condition. This concept is discussed in Eric Drexler's *Engines of Creation*. Drexler states:

> Aging is fundamentally no different from any other physical disorder; it is no magical effect of calendar dates on a mysterious life-force. Brittle bones, wrinkled skin, low enzyme activities, slow wound healing, poor memory, and the rest all result from damaged molecular machinery, chemical imbalances, and mis-arranged structures. By restoring all the cells and tissues of the body to a youthful structure, repair machines will restore youthful health.

31.14 Conclusion

Nanomaterials have increased surface area and nanoscale effects, and hence used as a promising tool for the advancement of drug and gene delivery, biomedical imaging and diagnostic biosensors. Nanomaterials have unique physicochemical and biological properties as compared to their larger counterparts. The properties of nanomaterials can greatly influence their interactions with biomolecules and cells, due to their peculiar size, shape, chemical composition, surface structure, charge, solubility, and agglomeration. For example, nanoparticles can be used to produce exceptional images of tumor sites; single-walled carbon nanotubes have been used as high-efficiency delivery transporters for biomolecules into cells. There is a very bright future to nanotechnology by its merging with other technologies and the subsequent emergence of complex and innovative hybrid technologies. Biology-based technologies intertwined with nanotechnology are already used to manipulate genetic material, and nanomaterials are already being built using biological components. The ability of nanotechnology to engineer matter at the smallest scale is revolutionizing areas such as information technology cognitive science, and biotechnology are leading to new and interlinking these and other fields. By further research in nanotechnology, it can be useful for every aspect of human life. Medicine, regenerative medicine, stem cell research, and nutraceuticals are among the leading sectors that will be modified by nanotechnology innovations.

References

1. Yousaf SA, Salamat A (2008) *Effect of heating environment on fluorine doped tin oxide (f: SnO/sub 2/) thin films for solar cell applications.* Islamabad: Faculty of Engineering & Technology.
2. Wang Z, Ruan J, Cui D (2009) Advances and prospect of nanotechnology in stem cells. *Nanoscale Res Lett* 4: 593–605.
3. Ricardo PNE, Lino F (2010) Stem cell research meets nanotechnology. *Revista Da Sociedade Portuguesa D Bioquimica, Canal BQ* 7: 38–46.
4. Deb KD, Griffith M, Muinck ED, Rafat M (2012) Nanotechnology in stem cells research: advances and applications. *Front Biosci (Landmark Ed)* 17: 1747–1760.
5. Boisseau P, Loubaton B (2011) Nanomedicine, nanotechnology in medicine. *Comptes Rendus Physique* 12: 620–636.
6. Bertrand N, Leroux JC (2012) The journey of a drug-carrier in the body: an anatomo-physiological perspective. *J Control Release* 161: 152–163.
7. Nagy ZK, Balogh A, Vajna B, Farkas A, Patyi G, et al. (2012) Comparison of electrospun and extruded Soluplus Â®-based solid dosage forms of improved dissolution. *J Pharm Sci* 101: 322–332.
8. Hollmer M (2012) Carbon nanoparticles charge up old cancer treatment to powerful effect. *Fierce drug delivery.*
9. Garde D (2012) Chemo bomb' nanotechnology effective in halting tumors. *Fierce drug delivery.*
10. Peiris PM, Bauer L, Toy R, Tran E, Pansky J, et al. (2012) Enhanced delivery of chemotherapy to tumors using a multicomponent nanochain with radiofrequency-tunable drug release. *ACS Nano* 6: 4157–4168.
11. Radovic-Moreno AF, LuTK, Puscasu VA, Yoon CJ, Langer R, et al. (2012) Surface charge-switching polymeric nanoparticles for bacterial cell walltargeted delivery of antibiotics. *ACS Nano* 6: 4279–4287.
12. Wyss Institute (2012) *Harvard's wyss institute develops novel nano therapeutic that delivers clot-busting drugs directly to obstructed blood vessels.* Boston, MA: Wyss Institute.
13. Nourmohammadi N (2012) New Study Shows Promise in Using RNA Nanotechnology to Treat Cancers and Viral Infections. Nanomedicine: Notes, Fierce Drug Delivery.
14. Haque F, Shu D, Shu Y, Shlyakhtenko LS, Rychahou PG, et al. (2012) Ultrastable synergistic tetravalent RNA nanoparticles for targeting to cancers. *Nano Today* 7: 245–257.
15. Suzanne E (2012) Bacterial 'minicells' deliver cancer drugs straight to the target Fierce Drug Delivery.
16. Ahmed RZ, Patil G, Zaheer Z (2013) Nanosponges - a completely new nanohorizon: pharmaceutical applications and recent advances. *Drug Dev Ind Pharm* 39: 1263–1272.
17. Laurance J (2012) Scientists develop nanoparticle method to help tackle major diseases. *The Independent* 11: 12.
18. Getts DR, Martin AJ, McCarthy DP, Terry RL, et al. (2012) Micro particles bearing encephalitogenic peptides induce T-cell tolerance and ameliorate experimental autoimmune encephalomyelitis. *Nature Biotechnology* 30: 1217–1224.

19. Wong HL, WuXY, Bendayan R (2012) Nanotechnological advances for the delivery of CNS therapeutics. *Adv Drug Deliv Rev* 64: 686–700.

20. Davide B, Benjamin LD, Nicolas J, Hossein S, Lin-Ping Wu, et al. (2011) Nanotechnologies for Alzheimer's disease: diagnosis, therapy and safety issues. *Nano Medicine: Nanotechnology, Biology and Medicine* 7: 521–540.

21. Sivaramakrishnan SM, Neelakantan P (2014) Nanotechnology in dentistry - What does the future hold in store? *Dentistry* 4: 2.

22. Zarbin MA, Montemagno C, Leary JF, Ritch R (2013) Nanomedicine for the treatment of retinal and optic nerve diseases. *Curr Opin Pharmacol* 13: 134–148.

23. Zhang W, Wang Y, Lee BT, Liu C, Wei G, et al. (2014) A novel nanoscaledispersed eye ointment for the treatment of dry eye disease. *Nanotechnology* 25: 125101.

24. Sahoo SK, Dilnawaz F, Krishnakumar S (2008) Nanotechnology in ocular drug delivery. *Drug Discov Today* 13: 144–151.

25. Fouda MM, Abdel-Halim ES, Al-Deyab SS (2013) Antibacterial modification of cotton using nanotechnology. *Carbohydr Polym* 92: 943–954.

32

Nanotechnology on Energy Storage: An Overview

Shenbagalakshmi

*G.H. Raisoni College of
Engineering & Management,
Pune, India*

Rahul Singh

*N.B.G.S.M College,
Gurugram University,
Raipur, India*

Dr. N. Prakash

*Kumaraguru College of
Technology, Coimbatore,
India*

Dr. G. Raghu Babu

*VNR Vignana Jyothi
Institute of Engg. & Tech,
Bachupally, India*

Dr. A. Yasmine Begum

*Sree Vidhya Nikethan Engg.
College, Tirupati, India*

Ayaz Ahmad

NIT P, Patna, India

Dr. P. Janardhan
Saikumar

*Audisankara Institute of
Technology, Gudur, India*

Nanotechnology is receiving a lot of attention these days, which is raising hopes not only among academics but also among investors, governments, and businesses. Its exceptional capacity to create frameworks at the atomic scale has already resulted in the development of novel materials and gadgets with broad applications. Significant advances in the energy sector, in particular, are required to satisfy our growing thirst for energy, which grows in tandem with many people joining developed countries and our demand per capita. As we gather more evidence of human impact on the climate, biodiversity, and

DOI: 10.1201/9781003355755-32

air, water, and soil quality, this must be implemented in a manner, which includes the environment in the wealth-generating equation. This chapter gives an overall insight into the role of nanotechnology and nanomaterials in the area of sustainable energy storages such as batteries, ultracapacitors, fuel cells, thermal energy storage, and mechanical energy storage.

32.1 Introduction

One of humanity's most significant challenges in the twenty-first century is the increasing depletion of fossil resources [1]. There is a significant demand for clean, renewable energy resources which can replace fossil fuels to allow the economy and society to thrive sustainably. Renewable resources like solar energy, wind, tide, and geothermal energy are promising because they don't deplete over time and are safe for the environment [2–4]. The challenges associated with utilizing the electrical energy produced deterministically arise from the fact that these renewable energies are adaptive and alter depending on the season and location. Techniques for storing electrical energy that are effective are highly desired in order to avoid the sporadic nature of these flexible resources and on-demand power supplies. Storing the electrical energy is a somewhat challenging process. It first needs to be transformed into another type of energy that involves a loss. Even more energy can be lost during storage and conversion into power, depending on the type of storage used.

The high amount of energy they can store while maintaining a long lifespan makes lithium batteries, capacitors, and supercapacitors stand out among the many electrical energy storage (EES) products now on the market [8, 9]. Gadgets with improved energy storage capacity have been made possible in recent years thanks to the tremendous advancements in material science and nanotechnology [2, 6–10], and the technology has now taken over as the primary resources for movable gadgets in contemporary civilization. The performance of these EES devices still has to be enhanced despite the advancements made in order to fulfill the more stringent criteria of future systems, like microelectrochemical systems (MEMS) and electric vehicles. The advancement of high-performance EES systems depends on the development of electrode materials [11].

Currently, advances in nanotechnology are enhancing the conversion, storage, and transfer of energy. Future developments in the energy sector may heavily rely on nanotechnology (including the selective application of nanomaterial), particularly when it comes to the development of novel methods for energy storage. Rechargeable energy storage device development and enhancement are the focus of current research and development. The lithium-ion battery pack is seen to be the most feasible type of energy storage device due to the remarkable density of power and increased cell potential. By developing novel ceramic, thermal resistance, still supple separators, and greater electrochemical properties, nanotechnologies have the ability to significantly boost the storage and stability of lithium-ion batteries.

Even hydrogen seems to be a potential resource for a sustainable energy source in future. In addition to the critical nanostructure alterations, effective hydrogen storage is recognized as one of the key success criteria on the path to a potential hydrogen gas administration. The automotive industry requires chemical hydrogen storage materials to have the capacity to store hydrogen up to 10% by weight; however, the materials now available do not meet these requirements. Numerous nanomaterials, the ones based on nanosized metallic ions, among others, offer development opportunities that, at least in terms of the application of fuel cell technology to portable gadgets, appear to be economically feasible. Thermal energy storage is another significant area. Phase transition materials like latent heat stores, for instance, can drastically lower the energy demand in buildings. Economically speaking, desorption storage based on nanostructured materials, such as zeolites, is attractive since it might be employed as thermal storage in geothermal grids or in industries. The desorption of moisture in zeolite enables the bidirectional storing and heat release. Due to nanoparticles and composite nanostructures, we may be able to develop energy-storing systems with the power density of the most powerful battery packs but with the higher magnification, quick recharging, and sustained life properties of supercapacitors.

Batteries and electrochemical supercapacitors, two types of electrical storage devices, may also be significantly impacted by nanotechnology. Supercapacitors based on redox using nanocrystalline electrocatalysts have demonstrated that it is possible to combine the great electrical capabilities of electrostatic capacitors with the excessive lab-scale conventional battery energy density. The chemical and physical qualities of the electrocatalyst have a significant impact on the energy efficiencies and capabilities of reusable battery packs. In this context, nanomaterial of lesser size and higher floor location increases the electrode-electrolyte contact and the charge of electron transport, whereas the nanostructures themselves offer simple pressure reduction and resistance against fractures. A number of silicon nanowires and nanostructures made of graphene have all been investigated as potential hosts for conductive additives and high-capacity materials for anode applications [11]. This chapter gives an overall insight into the role of nanotechnology and nanomaterials in the area of sustainable energy storages such as batteries, ultracapacitors, fuel cells, thermal energy storage, and mechanical energy storage.

32.2 Nanotechnology in Batteries

Chemical energy is used to store electrical energy in batteries. Batteries consist of two electrodes connected to a circuit, which is how all batteries operate. The electrode that absorbs the "positive electrode" seems to be what releases electrons when a load is attached to a battery (cathode). The operating terminal that releases electrons is referred to as the "negative electrode" (anode). Electrons are produced during a chemical reaction on the anode, while they are consumed during a chemical reaction on the cathode. An electrical current is produced by the reactions. In the electrical circuit's direction, an external charge pushes the recharging of a battery by moving electrons in the opposite way. A battery is a combination of numerous related metal cells that are connected in series or parallel to convert chemical energy into electricity. Each metal cell consists of two electrodes and an electrolyte. Initially, only disposable batteries were meant by the name "battery" (primary batteries). Both linguistically and legally, the restrictive definition of the term is no longer valid. Today's batteries are known to consist of both primary and secondary cells that can be recharged. An accumulator, or accu, is another name for a secondary cell.

The terms "backup battery" or "rechargeable battery" are more frequently used than "accumulator." The electrode materials of the battery undergo structural changes as a result of the chemical processes that take place during charge-discharge cycles, reducing the number of recharge cycles. The following advantages come with making batteries with nanotechnology:

1. Extending a battery's power capacity and speeding up battery recharge. By applying nanoparticles to an electrode's surface, these advantages are realized. Between the electrodes, greater current can flow, and the battery's chemical components as a result of the electrode's surface area are being increased. By drastically lowering the weight of the batteries required supplying appropriate power, this technology could improve the efficiency of hybrid vehicles.
2. Separating the liquids from the solid electrodes in a battery pack using nanoparticle to prolong the battery's shelf life when the battery is not being used. The minimal effluent that takes place in a typical battery is avoided by this separation, thereby extending the battery's shelf life.

Nanomaterials such as carbon-coated silicon nanowires, carbon nanotubes (CNTs), layer-by-layer nanomaterials, vanadium oxide and manganese oxide, $LiMn_2O_4$ or $LiCo_2O_4$ nanoparticles, Li alloy/graphene foil, and phosphorene-graphene hybrid materials, among others are used as electrodes because of the aforementioned benefits. Similarly, solid polymer gel might be greatly improved in terms of conductivity and storage capacity by adding Al_2O_3, SiO_2, or ZrO_2 nanoparticles. Solid ceramics have also been researched because of their high-temperature resilience, which would be useful for demanding, high-stress applications like huge automobiles or renewable power plants. 2D MoS_2 is used as an effective protective layer for Li metal anodes in high-performance Li-S batteries. Now we will discuss in detail about the recently used batteries and their development:

32.2.1 Lithium-Ion Battery

The rechargeable lithium battery is a focus of current research and development. The following section will go into more depth about this kind of battery. For the creation of high-performance batteries, lithium is appropriate. The power density and the stableness of the batteries are raised by larger electrode surfaces and improved separators, expanding the range of potential uses like in either stationary power storage or electric and hybrid vehicles. The use of lithium in battery packs offers a variety of benefits, including an elevated specific charge; a typical potential that is extremely negative (−3.05 V); numerous organic (and some inorganic) electrolytes that have stability.

Consumers purchased more than 5 billion lithium-ion battery packs in 2013 for using in electric vehicles, computers, smartphones, and other rechargeable mobile energy storage devices. For the batteries to be suitable for such tasks, they must have a voltage level, a large capacity, and a longer cycle life while also being exceedingly safe and dependable. The power density of lithium battery pack is comparatively high. Within 10 years, it could nearly double thanks to the introduction of novel cathode materials. Power density for lithium battery packs currently range from about 80 Wh/kg of lithium iron phosphate to about 250 Wh/kg nickle cobalt aluminum oxide (NCA), depending on which subsystem is taken into account. In contrast, lead-acid batteries have power density of roughly 40 Wh/kg. The overall stability of lithium-ion batteries is good despite their higher voltages. Energy density could increase to 400 Wh/kg by 2017. Increased cell voltage values (operating voltage: 3.2–3.7 volts), a high cycle count (up to more than 5,000 sessions), sufficient electrode dimensional stability, and a simple assembly of the cell from air- and moisture-resistant materials are all still possible. An electrolyte, a cathode and anode, as well as a separator, are the components of lithium-ion batteries. Charge transmission between the electrodes is made feasible because the lithium in electrolytes is ionized.

Lithium ions go toward the anode as the battery is being charged. Graphite is generally used as the anode, and it has a 370 Ah/kg capacity. Doped carbon, fullerenes, and carbon black are additional materials that can be used as anodes. Additionally, nanocomposites (such as carbon matrix with nanoscale silicon or tin) and nanostructured materials (such as nanowires, nanorods, nanotubes, and nanoporous particles) are used or being developed. Typically, any substance that can store the crystal structure of lithium can be employed. Such electrodes enable a 700 Ah/kg capacity. Lithium ions go toward the cathode as part of the discharge process. The main application of $LiCoO_2$ called as the lithium cobalt oxide is its use as a cathode material. More and more people are using new materials due to their economic and environmental benefits. Nanomaterials constructed of transition metal phosphates, such as lithium manganese phosphate and lithium iron phosphate are among them. The number of materials required can be decreased if they are employed in nanoparticle form. Additionally, the cathode's surface can be increased through nanostructuring, improving energy storage and cycle stability.

All batteries depend heavily on separators. They safeguard the battery against internal short-circuits and overheating while enabling free ion passage between the electrodes. The separation is made of non-conductive-reinforced polymers that are covered with an organic substance that conducts ions on one or both sides. Preferably, the polymer nanofibers utilized in the separator materials are used. Due to the development of smaller pore diameters, these non-wovens have high porosity. The biggest diameter of the particles that make up the ion-conducting substance is less than 100 nm. Additionally, the separator's high tensile strength and good pliability enable it to adapt to changes in the electrodes' shape without being harmed.

32.2.2 Lithium-Air and Sodium-Air Batteries

Although research on air core lithium battery pack is still in its early stages, they should offer an elevated specific power density. Air-type lithium battery packs are capable of storing 10 times as much energy as current batteries, that is identical to the energy density of a petroleum engine. When a lithium-air battery is discharged, the lithium is converted to its oxide, which releases higher amount of energy; when the battery pack is again charged, it releases oxygen into the environment during the rejuvenation of the lithium anode. The benefit of such approach would be that the oxygen which acts as cathode does not get added to the battery pack's volume. These batteries' cathode (often carbon) and separator membrane structures, which permit the flow of oxygen while blocking the passage of moisture, depend on nanostructured materials [12]. Since charging an electric car would be prohibitively expensive, a lithium-air battery would almost certainly not be considered for this type of application [13]. As a result, experts believe that sodium-air batteries have better chances of developing since they are predicted to have power density that is very much greater than that of current lithium-ion batteries and because sodium is less expensive than lithium as a crucial component for batteries. However, these batteries have only been able to complete 100 charge cycles to date [14].

32.2.3 Lithium-Sulfur Battery/Sodium-Sulfur Battery

A promising type is the lithium-sulfur battery. These lithium-ion batteries will attain power densities that are 2–5 times higher than those of the current generation. Additionally, the parts are needed to make the battery pack as inexpensive. Batteries made of sulfur or lithium are soon to enter the market, and they should have a power density of roughly 500 Wh/kg [13]. The lithium-based anode in lithium-sulfur batteries functions as electrode as well as a source of lithium ions. The cathode is sulfur-based. Lithium on the anode melts during discharge, forming lithium sulfide with sulfur on the cathode. Lithium is generated on the anode, and the resultant lithium sulfide is decreased during charging. The electrode made of sulfur is covered with carbon-based nanoparticles, like graphene, to prevent early breakdown [15]. Additionally, there are initiatives to improve these batteries using CNTs.

Different electrolytes and anode compositions are used in other optimization variants, tin compound and silicon compound in particular are suggested as anode with such a potential to enhance cycle performance. The lithium-sulfur battery has an advantage over many other lithium-ion batteries in that it does not require expensive and rare heavy metals like Co or Ni, which are harmful to health of individuals and the surroundings. Sulfur is rather harmless for the nature, but the discharged sulfides of lithium are poisonous; when they interact with acids, they produce toxic hydrogen sulfides. The cells must be tightly sealed because of this. Since sodium is more widely available than lithium, other development work attempts to utilize sodium rather than lithium. However, batteries using sodium as the anode material have lower energy densities. They might be utilized in stationary applications because cycle stability and cost are key considerations in these settings.

32.2.4 Printed Battery

Another crucial technology for energy storage is the print of circuit boards using specialized inks on different materials (printed electronics). This process is a part of nanotechnology because the inks contain a dispersion of metallic nanoparticles. The inks are already used extensively as a conductor and have good conductive qualities. Metallo-organic decomposition inks only slightly block the spraying devices. Silver, copper, aluminum, and nickel are among the metals used. Metal oxide, typically zinc oxide, is used to create another sort of ink. These are stable with regard to oxidation and transparent. Electrodes made of organic conductors and metals can take the place of pricey indium tin oxide (ITO). The capacity of printed batteries is currently quite low. The usage of this technology for ultra-thin, flexible, and rechargeable batteries is predicted to rise in the near future by experts.

32.3 Nanotechnology Applications Being Developed for Batteries

Researchers from the University of Eastern Finland are working on developing mesoporous Si film electrodes for Li-ion battery packs. Anodes composed of silicon nanofibers are being tested in Li-ion batteries by a business by the name of TruSpin. Anode mechanical degradation at high power cycling may be prevented by using oxidized antimony nanocrystals in Li-ion batteries, according to researchers at Georgia Tech. A lithium metal battery with an automated construction process and thin coating of compounds with electrochemical activity to stop the growth of spikes formed over the lithium crystal which could lead to short-circuit the battery has been shown by Penn State researchers.

Anodes for solid-state batteries can now be produced via spray deposition of Si nanoparticles, according to Nizam's Institute of Medical sciences (NIMS) researchers. They anticipate that this procedure could result in a high-volume, low-cost technique of manufacturing anodes for solid-state batteries with high capacity. An antimony electrode in the form of what Purdue University researchers refer to as a nanochain has been demonstrated. They have demonstrated that these electrodes allow lithium-ion batteries to charge more quickly than electrodes made of graphite. A graphene oxide aerogel has been employed successfully by Chalmers University researchers as a lithium-sulfur battery electrode. Their research indicates that this technique might lengthen the life of lithium-sulfide batteries.

CNT films are being used by researchers at Rice University to halt the formation of neurons on lithium metal anodes. This process could aid in the development of lithium metal batteries, which might be significantly more powerful and quick to charge than lithium-ion batteries.

CNTs with silicon coatings have been used successfully as Li-ion battery anodes, according to research from North Carolina State University. The Li-ion battery's capacity might potentially grow by up to many times with the inclusion of silicon, according to their predictions. But during a battery's discharge cycle, silicon expands, which can harm silicon-based anodes. To protect the anode from harm when the silicon expands, the researchers plan to put silicon atop parallel-aligned nanotubes. Researchers at Stanford Linear Accelerator Center (SLAC) and Stanford University are working on methods to encase silicon nanoparticles in graphene cages. According to the theory, silicon that has expanded and caused cracks in the nanoparticles will stay in the cages of graphene and not harm the anode.

The need for CNT catalysts instead of platinum has been demonstrated by scientists at the Los Alamos research lab. According to scientists, such catalysts may be utilized with air-type lithium battery packs that have the ability to store up to 10 times more energy than the conventional ones. Scientists in University of Southern California (USC) are creating Li-ion batteries using silicons as nanoparticles in the electrode that can recharge in just 10 minutes. Using silicon nanoparticles as opposed to solid silicon protects the electrode from breaking, that happens with electrodes of silicon.

Nanotubes made of carbon can be used in electrodes employed in three-dimensional (3D) structures to boost the power densities of capacitances, according to research from the University of Delaware. Nanowire electrodes with a gel coating have been shown to have a considerably longer lifetime by University of California, Irvine researchers. Researchers from Rice University have developed CNT electrodes that have an extremely large surface and less impedance. The scientists primarily developed a film of graphene on a metallic base, and then they developed nanotubes of carbon on the sheets of graphene compound. The graphene sheet and every nanotube are joined atom for atom at their bases, making the graphene nanotube structures as a single larger molecule with more space. By growing nanowires of silicon on a substrate made of stainless steel, researchers at the Stanford University have illustrated that battery packs employing these positive electrodes will increase the energy densities of regular Li-ion batteries. By adopting such nanowires, the problem of cracking on electrodes made of bulk silicon is overcome.

As the battery pack discharges and the lithium ions flow out of the silicon, the silicon grows as it absorbs the ions of Li while charging. This swelling and contraction are what cause the cracks. Contrary to anodes made of bulk silicon, the researchers discovered that silicon nanowires do not shatter even if they swell and compress as lithium ions are absorbed during a battery's discharge and recharge, respectively. The positive as well as the negative electrodes of Li-ion batteries can be made by depositing nanotubes made of carbon on a base, according to a process developed by MIT researchers. The organic molecules that are linked to the CNTs aid in their alignment on the substrate and supply many oxygen atoms for lithium ions to connect to. As a result, lithium-ion batteries may have a large improvement in power density—up to 10 times, maybe. This technology has been licensed by Contour Systems, a battery manufacturer, and they want to incorporate it in their next Li-ion batteries. Carbon nanofibers were employed by MIT researchers to create lithium-ion battery electrodes that have a storage capacity that is 4 times greater than that of contemporary lithium-ion batteries.

To create Li-ion battery packs that restore more faster than other battery packs, scientists at Rensselaer employed graphene sheet on the anode. By means of cracks in the graphene sheet that were produced during heat treatment, lithium ions can adhere to the positive electrode base. MIT scientists had created CNT batteries that can produce electricity without using metals. Electricity will be generated while heat is delivered through the nanotubes, at the time of burning sugar. The scientists are believing that this technique might be used to produce incredibly small batteries, which might be needed for apparel technology. The recent advancement after Li-ion battery packs may be sulfur-based lithium battery packs, which may store energy at a rate that is several times greater than that of lithium-ion batteries (the sulfur is found in the cathode). While Liverpool University Munich and Waterloo University employ negative electrodes made of sulfur-containing mesoporous nanoparticles made of carbon, Stanford University researchers use cathodes built of sulfur-encapsulating carbon nanofibers.

Scientists at the Institute of Physical Chemistry of the Polish Academy of Sciences have built a negative electrode using nanoparticle made of carbon for using it in battery packs or fuel cells in medical applications. Cathodes made of nanocomposites were created to raise the power density of lithium-ion battery packs. Battery is used in the retina by placing it in the eye. Nanograss is used in long-lasting batteries to keep the solid electrode and liquid electrolytes apart until electricity is required.

Nanoparticle (Nanophosphate™) electrodes in lithium-ion batteries enhance performance while maintaining required safety levels for electric vehicles. In hybrid cars, ultracapacitors made from nanotubes may perform even better than batteries. Graphene sheets that are only one atom thick are used in ultracapacitors to store electrical charge. The charge/discharge rate of these batteries have improved at colder temperature range and raised the peak value at which the battery pack may operate without experiencing thermal runaway.

32.3.1 Nanotechnology in Supercapacitors

Nanotechnology has widened possibilities by introducing new enabling technologies and energy storage materials. CNTs and graphene sheets, two graphitic carbon nanomaterials that have gained increasing importance, have contributed to the development of high-performance supercapacitors [16].

A capacitor is composed of two conducting plates placed parallel to each other in between the dielectric material. Electric charges permeate the dielectric when a voltage is applied, which in turn leads to a progressive increment. The capacitance (C), which controls how well a capacitor can store charges, is dependent on both voltage (V) and the volume of stored electric charge (Q), as shown by the equation:

$$C = \frac{Q}{V}$$

With a vacuum between the plates of a parallel-plate electrostatic capacitor, the capacitance may be calculated using the following equation:

$$C = \varepsilon \frac{A}{d}$$

where the vacuum's permittivity (ε) is always constant and equals roughly 9×10^{-12} F/m. Through its energy and power, a capacitor's energy storage capacity can be assessed. One method for calculating the energy (E) is as follows:

$$E = \frac{1}{2}CV^2$$

Calculating power (*P*) is as follows:

$$P = \frac{1}{2} IV$$

The primary properties of a parallel-plate capacitors are their energies and power densities, that are calculated as the values previously stated as the unit mass of electrodes.

The three essential components of a typical supercapacitor are the electrodes, electrolytes, and separation. The total working of a supercapacitor is influenced by the physical characteristics of the materials employed for the electrolyte as well as electrodes. The positive and negative electrodes of a supercapacitor, however, are the significant components for charge distributions and storage since it has a considerable impact on the power density and energy density of the device. The performance of a supercapacitor's electrochemical system can be evaluated using galvanostatic charge-discharge experiments and cyclic voltammetry [17]. Figure 32.1 shows a schematic representation of the double two-layer electric capacitor hypothesis of storing energy in a supercapacitor. It may entail charge transfer through reversible (Faradaic) redox reactions to oxidant components (such as polymer nanocomposites and metal oxide nanoparticles) on the electrode surface, or it may include static electricity accumulation at the electrode/electrolyte interface (pseudo-capacitance). The two storage strategies frequently coexist in real supercapacitors [18].

Different charge transfer channels are used by the electric double layer capacitor (EDLC) and pseudo-capacitance [19]. The absence of charge transfer across the electrodes in EDLC and the energy storage mechanism at the electrode-electrolyte interface being solely ion adsorption, an electrostatic process, indicate a quasi-mechanism. In disparity, false capacitance is the result of reduction and oxidation interactions between reactive species at the surface and the electrolyte of electrodes. Although supercapacitors based on quasi-mechanism are capable of having pseudo-capacitance greater than EDLC capacitance, these devices typically exhibit poor electrical properties of the electroactive species, which results in low density and cycle stability. A supercapacitor's overall capacitance is increased by combining EDLC and pseudo-capacitance, which is a useful approach.

While pseudo-capacitance greater than EDLC capacitance is possible, supercapacitances based on false capacitance usually have the electroactive species with weak electrical conductivity, it results in cycle stability and low power density. Combining EDLC and pseudo-capacitance increases a supercapacitor's total capacitance, making it a beneficial technique. These tiny pores are unable to support an electrical double layer because they typically have little to no access to the electrolyte ions (particularly for organic electrolytes). The capacitance of an electrical double-layer capacitor, however, is predominantly caused by mesopores [20–23]. However, recent experimental and theoretical studies have demonstrated that charge storage is enhanced by reducing particle sizes since the ion center in the micropores was closer to the electrolyte interface [18, 24–26]. These pores are bigger than the smallest of solvated

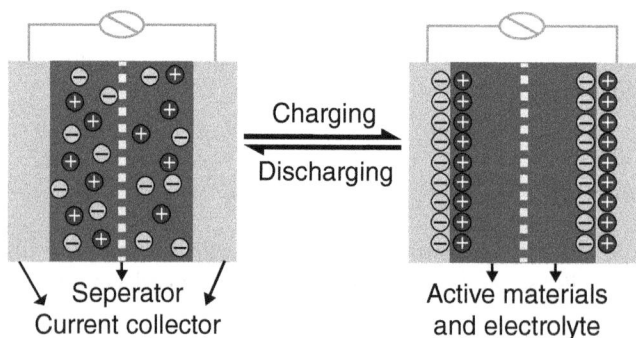

FIGURE 32.1   Two-layer electric capacitor.

electrolyte ions and range in size from 0.5 to 2 nm. Double layers cannot form in pores that are smaller than 0.5 nm in width [26]. The commercially available activated carbon materials have a high surface area but disappointingly a low mesoporosity, which results in a restricted capacitor since an electrolyte is not readily accessible [20].

As a result, superconductors built on active carbon electrodes that are now available have a constrained power density (4–5 Wh kg^{-1}) and a restricted specific power (1–2 kW kg^{-1}) [20]. To overcome the issues with active carbon electrode materials and improve the performance of supercapacitors, new resources are obviously needed. Because of their large surface area, high porous structure, accessibility to the electrolyte, and favorable electrical properties, carbon nanomaterials, in particular graphene-based materials (CNTs), are indeed very potential candidate to substitute porous carbon as the anode materials with high-performance supercapacitors [21].

Graphene is a single-atom thick sheet of sp2 carbon atoms packed tightly into a two-dimensional (2D) honeycomb lattice. It is the basic building block for graphene sheets of all other types of dimensions, including 0D fullerene, 1D nanotubes, and 3D graphite (Figure 32.2) [22] For example, a graphene that has been folded into a nanotube form can be used to conceptualize a human nanofibers (SWNT) or a multi-walled carbon nanotubes (MWNT) with additional graphene coaxial tubes encircling the SWNT core. CNTs with significant specific surface areas (SWNT >1600 m^2 g^{-1}, MWNT >430 m^2 g^{-1}) [23], high aspect ratios, and outstanding mechanical and electrical (5000 S cm^{-1}) properties [28, 32–37] have frequently been employed as the active electrodes in supercapacitors. Nanomaterials are optimal solutions

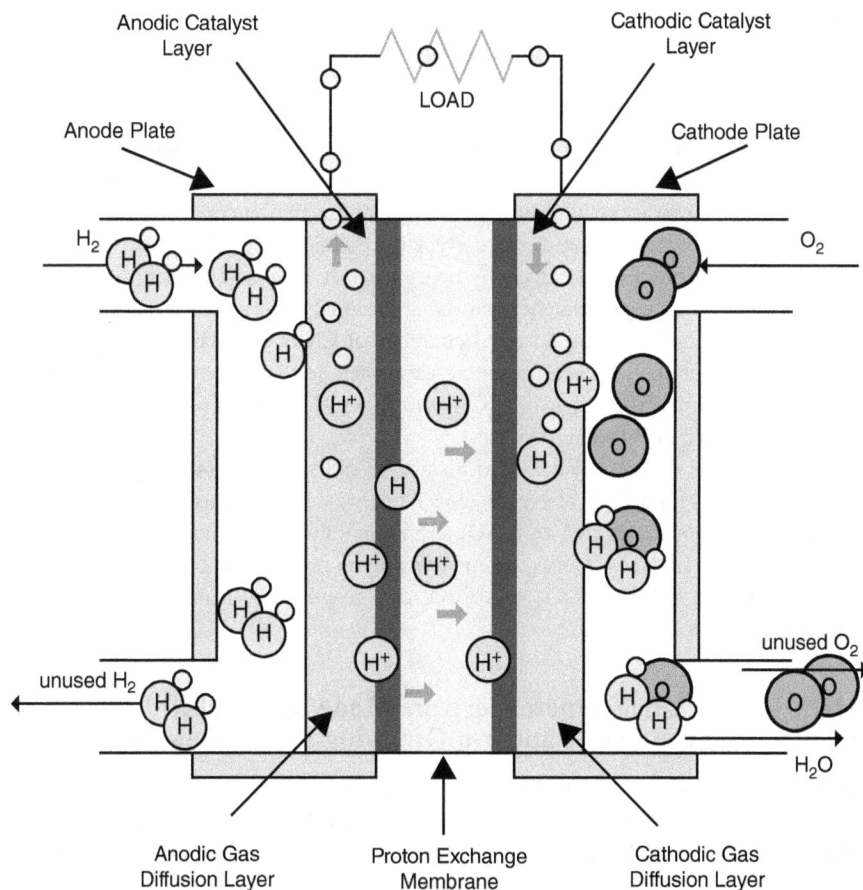

FIGURE 32.2 Proton exchange membrane fuel cell.

for a variety of potential applications that CNTs have previously been used for because it shares many structural and functional characteristics with nanostructures, including a high aspect ratio (the ratio of lateral size to thickness), a large surface area ($2630 \, m^2 \, g^{-1}$), outstanding carrier mobility ($15{,}000 \, cm^2 \, V^{-1} \, s^{-1}$ both for electrons and holes), and strong mechanical characteristics [24].

Each nanosheet with a 2D flat structure may be superior than CNTs and offer extra advantages as capacitive electrode materials [25]. As explained below, the availability of multiple rational materials' engineering approaches and a wide range of chemistries for controlled functionalization of both CNTs and graphene enable the creation of a variety of elevated supercapacitors based on carbon nanomaterials.

32.4 Managing the Alignment and Tip Formation of CNTs for Producing Highly Efficient Supercapacitors

As indicated earlier, the supercapacitors have frequently been made of CNTs with a greater aspect ratio, significant surface area, and good conductivity [26, 27–30]. For instance, by using electrodes constructed on standalone mats of entangled MWNTs in H_2SO_4 electrolytes, ultracapacitors with a specific capacity of $102 \, F \, g^{-1}$, a power density of $>8 \, kW \, kg^{-1}$, and a power density of $1 \, Wh \, kg^{-1}$ were produced [31]. A random SWNT network was used as the electrodes in a KOH electrolyte more recently (Figure 3a), which resulted in a greater capacitance of $180 \, F \, g^{-1}$ [32]. The maximum efficiency in this case was around $20 \, kW \, kg^{-1}$, as well as the ideal energy density was around $10 \, Wh \, kg^{-1}$. Even though this power density number is larger than that of activated carbon-based capacitors that are readily accessible in the market, the energy density can still be raised. The porous carbon tube specific area may have been greatly decreased by the arbitrary interlocking or bundling structure, as well as limiting the ability to store electrochemical energy [33].

It has been established that Vertically Alligned Carbon Nanotubes (VA-CNT) arrays are preferable to their randomly interlocked counterparts for application in supercapacitors. Compared to erratic porous structures of arbitary CNTs, the spaced evenly framework and the strictly delineated tube-spacing in a CNT array could allow more accessible surface for electrolytic structures. The joined materials must offer better charge-storing characteristics since every individual-oriented tube can participate in the charging and discharging process effectively by being directly coupled to a common electron beam. Additionally, it is possible to appropriately open the uppermost end-caps of the CNTs [44, 46], which makes the electrolyte accessible to the internal chamber of CNTs for charge storage even though it would normally be inaccessible. Indeed, recent research has demonstrated that randomly entangled CNTs cannot match the rate capabilities of CNTs, and that CNTs also have lower energy density and power density [47] ($148 \, Wh \, kg^{-1}$ and $315 \, kW \, kg^{-1}$).

A high capacitance was specifically achieved for a VA-CNT electrode ($440 \, F \, g^{-1}$) arranged using a pattern strategy [44, 49], and a CNT electrode ($365 \, F \, g^{-1}$) was compiled using a framework chemical vapor deposition (CVD) approach] in $1 \, M \, H_2SO_4$ and ionic liquid electrolytes, respectively. By adding pseudo-capacitance to the surface of CNTs, conductive polymers such as polyaniline and polypyrrole as well as metal oxides such asMnO_2, TiO_2, and RuO_2 can progress the functionality of carbon-based nanotubes supercapacitors.

32.4.1 High-Performance Supercapacitors Made by Carefully Controlling the Frame and p-p Grouping of Graphene Sheets

Due to the full contact tension of the CNT electrode with the conductive substrate and its ineffective connection to the electrolyte, superconductors built on CNTs electrodes do not perform as predicted (e.g., specific capacitance below $500 \, F \, g^{-1}$). Scaling up the production of VA-CNTs for the commercialization of supercapacitors is very challenging, if not impossible [34]. The ability to produce large quantities of graphene materials at competitive prices for a variety of device applications [36] has recently been made possible by advances in the manufacturing of rim-designed and synthesized nanosheet (EFG) plates such as through milling [35] as well as remedy graphene (GO) by removing the dead graphite powder utilizing solution oxidation [34].

Recent research has looked into a substitute carbon-based electrode for supercapacitors because of graphene's vast surface area, excellent charge carrier efficiency, and excellent mechanical and thermal stability [68]. Graphene is a carbon-based electrode with the highest intrinsic capacitance value and with a double-layer capacitance value that can theoretically exceed 550 F g^{-1}.

A supercapacitor with a phenomenally special energy density (136 Wh kg^{-1} at 80°C and 85.6 Wh kg^{-1} at room temperature) [37] and a capacitor with a specific capacitance of 341 F g^{-1} in organic and aqueous electrolytes, respectively, both have been developed utilizing molecularly reduced graphene oxide electrodes. Recently, by directly reducing graphene oxide films to graphene electrodes using a typical LightScribe DVD optical drive, capacitors with a high specific capacitance (276 F g^{-1}), specific power (20 W cm^{-3}, 20 times greater compared to the active carbon equivalents), energy density (1.36 mWh cm^{-3}), and excellent stability during bending from 08 to 1808 have been developed. Adding pseudo-capacitance through the use of polymer film or oxide coatings can further improve performance.

Recent research on the EFG with various edge groups (such as –H, –COOH, and –SO$_3$H) by efficient and environment-friendly ball milling of graphite) offers a practical way to synthesize nanocrystalline materials with custom chemical compositions and electronic properties appealing for applications in a variety of disciplines [38], including huge opacity, and conduct electrolyzer for electronic components and material electrocatalyst for oxygen reduction. EFGs haven't been used much for energy storage, but by using them as the electrode materials, it may be possible to create elevated supercapacitors with a higher percentage functionality due to a lot of active locations optimal conjugation (conductivity) at the periphery and on their surface. EFGs should make it easier to create hierarchically ordered sensing in superconductors for particular applications through programmable self-assembling.

32.4.2 Three-Dimensional CNT and Graphene Networks as the Basis for High-Performance Supercapacitors

Strong p-p interactions cause many of the single graphene sheets to be layered in the graphene electrode, blocking a sizable portion of their surface area. Graphene sheets with more surface areas make excellent materials for electrodes in storing energy. The increased surface region of 2D graphene sheets must be maintained; hence, it is vital to geographically separate them. One way to do this is by making 3D graphene foams using templates that have a huge surface area and a high porosity [39]. A sheet identity solution method was created to generate multi-layered hybrid carbon films comprised of differently changed graphene sheets and acid-oxidized MWNT layers in order to make supercapacitors. These capacitors have a significantly higher specific capacitance of 120 F g^{-1} (than those of upright direction [35] and random CNT electrodes [40]) and rate capability due to the strictly delineated interface region of CNT networks.

The pore size and porosity allocation inside the multi-layered CNTs/graphene hybrid film, however, are difficult to control because each of the individual CNT and graphene strands are randomly assembled, preventing it to further improve the device performance. Researchers have developed and studied 3D pillared VA-CNT/graphene architectures to address this issue. These structures alternate between sheets of CNTs and graphene, and it is simple to control how porous they are by adjusting the height and packing density of the VA-CNTs. A 3D pillared VA-CNT/graphene architecture was made by intercalating VA-CNT advances in heating expansion crystalline structure pyrolytic graphite (HOPG), which also demonstrated a specific capacitance of roughly 110 F g^{-1} inside a double-layer supercapacitor. The resultant 3D-columned architecture hybridization with nickel hydroxide coating displayed outstanding long-term electrochemical stability, a high specific capacitance (1065 F g^{-1}), and a remarkable rate capability with just 4%of overall capacity loss after 20,000 charge-discharge cycles.

This same value of 1065 F g^{-1} is within the range of 953–1335 F g^{-1} for various polymeric solitary nickel hypochlorite hexagonal nanoplates [90] and is roughly 10 times higher compared to large surface carbon materials (100 F g^{-1}). VA-CNTs have a greater conductivity and superior rate capability because they can function as mechanical supports and structures that create channels for electrons and ions inside the 3D-hierarchical structure.

32.4.3 Higher Efficiency Supercapacitors with Novel Structures

As mentioned earlier, typical supercapacitors that utilize liquid electrolytes have seen tremendous advances. However, they don't meet the requirements for some specific applications, like translucent, wearable, and portable devices. In this context, there has been considerable interest in recent work on the construction of supercapacitors with new structures, such as all-solid, fiber supercapacitors. These supercapacitors are translucent, lightweight, elastic, and/or flexible. By applying a thin nanotube film over a nature of the synthesized (PDMS) re-strained elastic substrate and then soothing the pre-strained substrate, it was also possible to create stretchable supercapacitors with crumpling nanotube macrofilm electrodes. Stretchable electrodes made of graphene or allied CNT films should really be especially promising for flexible supercapacitors with good precision that could be used as the source of power for wearable electronics because their electrical and chemical properties and stabilities remained largely unchanged even after being strained up to 30% discolored.

There has been a great deal of interest in Figure 6f–h. Highly porous supercapacitors based on buckling nanotube macrofilm conductors are built [11] by depositing a thin SWNT nanotube film over poly-dimethylsiloxane (PDMS) which was before elastomer substrate, and then relaxing the pre-strained substrate. Even after being elongated up to 30% discolored, these supercapacitors' electrical and chemical properties mostly remain constant and stable in comparison to their unmodified state. Stretchable supercapacitors with good precision, such as those based on aligned CNT sheets or graphene-based extensible electrode, should therefore be possible candidates for usage as the source of power for flexible electronics. Supercapacitors, on the other hand, are a good choice for powering portable electronics and energy windows because of their superior features. A range of many other energy techniques, such as self-powered light-emitting diodes, depend on them as essential components. The ability to incorporate supercapacitors into clothing, packs, and other fabrics has recently been shown to have a variety of special benefits, such as being breathable, light, and flexible [13–15]. For capacitors with electrodes composed of carbon-based composite fiber and a fiber-like shape, a particular conductance of up to 38 mF cm^2 has been attained [15].

Conventional supercapacitors are too big and heavy for potential uses in wearables and mobile optoelectronics. Some optically clear, physically flexible, and/or wearable wire-shaped supercapacitors with constrained electrocatalysts have been designed to address these issues. The progress of stretchy portable optoelectronics depends on the creation of coax cable superconductors and linked self-powering systems. Although research in this field is still in its early stages, oriented CNTs and/or graphene sheets with outstanding reference speed, endurance, and opacity are promising electrodes for opaque, elastic, and/or stretchy supercapacitors. Supercapacitor technology may be developed as a result of research in this developing field.

32.5 Nanotechnology in Fuel Cells for Energy Storage

It is amazing how fuel cells have the potential to be a power source that is amazingly efficient. Theoretically, they can run a variety of devices, from enormous additional equipment to power data centers to powering fuel cell portable laptops. While putting this concept practically, there are a variety of challenges that prevent the use of hydrogen in fuel cells. Platinum is a necessary and expensive component for electrode catalysts. Alternative fuels have the potential to result in electrode fouling, even though hydrogen is costly to produce and problematic to store. The significant types of fuel cells have a limited lifetime because hydrogen fuel parts degrade at such high temperatures. Many of these issues might be resolved through nanotechnology. Recent developments in nanotechnology have produced a variety of fascinating nanomaterials that could elevate the cost, size, and effectiveness of fuel cells [13].

32.5.1 Nanotechnology as a CNT or Fuel Cell Catalyst

The pricey metal platinum is being used in fuel cells as a catalyst. The fact that efficiency is increased and less metal is required when utilizing platinum nanoparticles as opposed to solid platinum surfaces is another well-known advantage. Another development in contemporary technology is the support of nanomaterials on porous surfaces, such as activated charcoal or nanomaterials like CNTs or nano-walls. Since platinum substrates are now more commonly available, less of the pricey metal is needed to create an effective catalytic electrode. Platinum may be entirely replaced by modified CNTs in fuel cells. The production of CNTs is developing quickly, and because the necessary raw materials are inexpensive and widely available, they may be produced quickly. The introduction of such catalyst will remove a major obstacle for so many hydrogen production because platinum currently accounts for at least 25% of the cost of commercial fuel cells. Nanotubes could be coated with an electron-attractive polymer or blasted with nitrogen to further boost efficiency. The nanotube electrodes also demonstrate greater durability. When methanol is used as the fuel, overall longevity of the cell is extended since, except in the case of platinum, neither carbon monoxide nor the result of crossover affects their catalytic activity [15].

32.5.2 Fuel Cells with Proton Exchange Membrane

Scientists from the University of Illinois created a proton-conducting membrane (PEM) as an illustration of how nanotechnology has enhanced fuel cell membranes in addition to catalysts. A coating of porous silica with a 5 nm diameter was added to the exchange membrane for protons that they developed. The intention is for moisture to accumulate in the silica layer's porous materials and interact with acid molecules. Due to the acidic medium in which the moisture in the silicon nanostructures develops, hydrogen atoms can pass through the membrane. Due to this, the fuel cell performs 10 times better than standard fuel cells, especially in low-humidity situations.

32.5.3 Fuel Storage for Hydrogen

The use of hydrogen fuel has been hindered by storage issues even if it meets all other criteria for a perfect substitute for conventional fuel. Extremely high pressures are necessary to store liquid fuels as a pressurized gas or liquid, which calls for pricey tanks and raises the risk of leakage or explosion. The only alternative that appears to work is to bind hydrogen into solid metals; however, even this strategy is unable to store a significant amount of hydrogen due to the hydride-producing substance's ability to handle hydrogen and securely store it at a high density. To solve this issue, researchers at Lawrence Berkeley National Laboratory created a composite material in 2011 that consists of magnesium nanoparticles enclosed in a matrix of stretchy organic polymer. As magnesium hydride, the molecule can safely store hydrogen gas at large levels and quickly release it when needed [15, 13].

32.6 Conclusion

For many years, storing sustainable energy has been a major concern. It can now be done with minimal effort thanks to technologies based on nanotechnology. This chapter went into great length about the use of nanotechnology in fuel cells, supercapacitors, and batteries. It has been discovered that nanotechnology has a number of benefits, including being very effective, light, flexible, and affordable.

References

[1] B.L. Allen, P.D. Kichambare, P. Gou, I. Vlasova, A.A. Kapralov, N. Konduru, V.E. Kagan, and A. Star, "Biodegradation of single-walled carbon nanotubes through enzymatic catalysis," *Nano Letters*, vol. 8, no. 11, pp. 3899–3903, 2008.

[2] B.L. Allen, G.P. Kotchey, Y. Chen, N.V. Yanamala, J. Klein-Seetharaman, V.E. Kagan, and A. Star, "Mechanistic investigations of horseradish peroxidase-catalyzed degradation of single-walled carbon nanotubes," *Journal of the American Chemical Society*, vol. 131, no. 47, pp. 17194–17205, 2009.

[3] O. Apul, G. Wang, Q.Y. Zhou, and T. Karanfil, "Adsorption of aromatic organic contaminants by graphene nanosheets: comparison with carbon nanotubes and activated carbon," *Water Research*, vol. 47, no. 4, pp. 1648–1654, 2013.

[4] R. Arvidsson, D. Kushnir, B.A. Sandén, and S. Molander, "Prospective life cycle assessment of graphene production ba ultrasonication and chemical reduction," *Environmental Science and Technology*, vol. 48, pp. 4529–4536, 2014.

[5] P. Banerjee, I. Perez, L. Henn-Lecordier, S.B. Lee, and G.W. Rubloff, "Nanotubular metal–insulator–metal capacitor arrays for energy storage," *Nature Nanotechnology*, vol. 4, pp. 292–296, 2009.

[6] B. Bellmann, O. Creutzenberg, A. Hackbarth, D. Schaudien, and A. Leonhardt, "Toxikologie von Nanomaterialien, Wirkmechanismen und Kanzerogenität CNT-Kinetik nach Kurzzeitinhalation," Umweltbundesamt Dessau-Roßlau TEXTE 77/2014, 2014.

[7] F. Ghasemzadeh, and M. Esmaeili Shayan, "Nanitechnology in the service of Solar energy systems," doi:10.5772/intechopen.93014. Available: https://www.intechopen.com/chapters/73145

[8] E.J. Fundam, Renewable Energy Appling Journal of Fundamentals of Renewable Energy Journal and Applications, 2016. Available: https://www.longdom.org/openaccess/nanotechnology-as-a-tool-forenhanced-renewable-energy-applicationin-developing-countries-2090-4541-1000e113.pdf

[9] Anonymous, *Green Hydrogen: Opportunities and Challenges for India*, 2021. Available: https://www.financialexpress.com/opinion/green-hydrogen-opportunitiesand-challenges-for-india/2341349/

[10] Z.M. Al-Azzawi, M. Al-Baidhani, A.R.N. Abed et al., "Influence of nano silicon carbide (SiC) embedded in Poly (Vinyl Alcohol)(PVA) lattice on the optical properties," *Silicon*, vol. 14, pp. 5719–5732, 2021. doi:10.1007/s12633-021-01325

[11] G. Richhariya, A. Kumar, and Samsher, *Solar Cell Technologies*. Elsevier BV; 2020. Available: https://www.sciencedirect.com/science/article/pii/B9780128196106000028

[12] H. Ahmed, "Applications of nanotechnology in renewable energies—A comprehensive overview and understanding," *Renewable and Sustainable Energy Reviews*, vol. 42, pp. 460–476, 2015. doi:10.1016/j.rser.2014.10.027. Available: https://www.researchgate.net/publication/267870574_Applications_of_nanotechnology_in_renewable_energiesA_comprehensive_overview_and_underst_anding

[13] H.K. Rai, and P. Rai, "Solar energy harvesting using nanotechnology," *International Journal of Applied Engineering Research*, vol. 13, no. 6, pp. 0973–4562, 2018. Available: https://www.ripublication.com/ijaspl2018/ijaerv13n6spl_62.pdf.

[14] Available: https://www.sciencedirect.com/journal/international-journal-of-hydrogenenergy/vol/45/issue/8

[15] P.S. Tom. *Nanofabrication. Encyclopedia Britannica*. 2021; 26. Available: https://www.britannica.com/technology/nanotechnology. Accessed 23 February 2022.

[16] R. Abed, A. Abed, and E. Yousif, "Carbon surfaces doped with (Co3O4-Cr2O3) nanocomposite for high-temperature photo thermal solar energy conversion via spectrally selective surfaces," *Progress in Color, Colorants and Coatings*, vol. 14, no. 4, pp. 301–315, 2021.

[17] F. Ghasemzadeh, and M.E. Shayan, "Nanotechnology in the Service of Solar Energy Systems," In (Ed.), Nanotechnology and the Environment Intech Open; 2020.

[18] W. Soutter, Nanotechnology in Fuel Cells. AZoNano. Retrieved; 2021.

[19] S. Satyapal, J. Petrovic, and G. Thomas, "Gassing up with hydrogen," *Scientific American*, vol. 296, no. 4, pp. 80–87, 2007.

[20] B. Sakintuna, F. Lamari-Darkrim, and M. Hirscher, "Metal hydride materials for solid hydrogen storage: a review," *International Journal of Hydrogen Energy*, vol. 32, no. 9, pp. 1121–1140, 2007.

[21] S.A. Sherif, F. Barbir, T.N. Vieziroglu, M. Mahishi, and S.S. Srinivasan, "Hydrogen energy technologies," in *Handbook of Energy Efficiency and Renewable Energy*, F. Kreith and D.Y. Goswami, Eds., chapter 27, CRC Press, Boca Raton, FL, USA, 2007.

[22] E. Fontes and E. Nilsson, "Modeling the fuel cell," *The Industrial Physicist*, vol. 7, no. 4, p. 14, 2001.

[23] R.H. Jones and G.J. Thomas, *Materials for the Hydrogen Economy*, CRC Press, Boca Raton, FL, USA, 2007, Catalog no. 5024.

[24] Report of the Basic Energy Science Workshop on Hydrogen Production, Storage and use prepared by Argonne National Laboratory, May 2003.

[25] L. Schlapbach, "Hydrogen as a fuel and its storage for mobility and transport," *MRS Bulletin*, vol. 27, no. 9, pp. 675–676, 2002.

[26] C. Read, G. Thomas, C. Ordaz, and S. Satyapal, "U.S. Department of Energy's system targets for on-board vehicular hydrogen storage," *Material Matters*, vol. 2, no. 2, p. 3, 2007.

[27] A. Züttel, "Materials for hydrogen storage," *Materials Today*, vol. 6, no. 9, pp. 24–33, 2003.

[28] A.M. Seayad and D.M. Antonell, "Recent advances in hydrogen storage in metal-containing inorganic nanostructures and related materials," *Advanced Materials*, vol. 16, no. 9–10, pp. 765–777, 2004.

[29] F.E. Pinkerton and B.G. Wicke, "Bottling the hydrogen genie," *The Industrial Physicist*, vol. 10, no. 1, pp. 20–23, 2004.

[30] F. Schüth, "Technology: hydrogen and hydrates," *Nature*, vol. 434, no. 7034, pp. 712–713, 2005.

[31] F. Schuth, B.C. Bogdanovi, and M. Felderho, "Light metal hydrides and complex hydrides for hydrogen storage," *Chemical Communications*, vol. 10, no. 20, pp. 2249–2258, 2004.

[32] N.B. McKeown, S. Makhseed, K.J. Msayib, L.-L. Ooi, M. Helliwell, and J.E. Warren, "A phthalocyanine clathrate of cubic symmetry containing interconnected solvent-filled voids of nanometer dimensions," *Angewandte Chemie International Edition*, vol. 44, no. 46, pp. 7546–7549, 2005.

[33] M. Fichtner, "Nanotechnological aspects in materials for hydrogen storage," *Advanced Engineering Materials*, vol. 7, no. 6, pp. 443–455, 2005.

[34] A.G. Wong-Foy, A.J. Matzger, and O.M. Yaghi, "Exceptional H2 saturation uptake in microporous metal-organic frameworks," *Journal of the American Chemical Society*, vol. 128, no. 11, pp. 3494–3495, 2006.

[35] V. Renugopalakrishnan, A.M. Kannan, S.S. Srinivasan, et al., "Nanomaterials for energy conversion applications," *Journal of Nanoscience and Nanotechnology*. In press.

[36] E.G. Baburaj, F.H. Froes, V. Shutthanandan, and S. Thevuthasan, "Low cost synthesis of nanocrystalline titanium aluminides," Interfacial Chemistry and Engineering Annual Report, Pacific Northwest National Laboratory, Oak Ridge, Tenn, USA, 2000.

[37] R. Schulz, S. Boily, L. Zaluski, A. Zaluka, P. Tessier, and J.O. Ström-Olsen, "Nanocrystalline materials for hydrogen storage," *Innovation in Metallic Materials*, pp. 529–535, 1995.

[38] L. Schlapbach, and A. Züttel, "Hydrogen-storage materials for mobile applications," *Nature*, vol. 414, no. 6861, pp. 353–358, 2001.

[39] W. Grochala, and P.P. Edwards, "Thermal decomposition of the non-interstitial hydrides for the storage and production of hydrogen," *Chemical Reviews*, vol. 104, no. 3, pp. 1283–1316, 2004.

[40] B. Bogdanovic, and M. Schwickardi, "Ti-doped alkali metal aluminium hydrides as potential novel reversible hydrogen storage materials," *Journal of Alloys and Compounds*, vol. 253–254, pp. 1–9, 1997.

[41] C.M. Jensen and R.A. Zidan, "Hydrogen storage materials and method of making by dry homogenation," US patent 6471935, 2002.

[42] Y.H. Hu and E. Ruckenstein, "H2 storage in Li3N. Temperature-programmed hydrogenation and dehydrogenation," *Industrial and Engineering Chemistry Research*, vol. 42, no. 21, pp. 5135–5139, 2003.

[43] J.J. Vajo, S.L. Skeith, and F. Mertens, "Reversible storage of hydrogen in destabilized LiBH4," *Journal of Physical Chemistry B*, vol. 109, no. 9, pp. 3719–3722, 2005.

[44] M. Au, "Destabilized and catalyzed alkali metal borohydrides for hydrogen storage with good reversibility," US patent Appl. Publ 0194695 A1, 2006.

[45] S.S. Srinivasan, D. Escobar, M. Jurczyk, Y. Goswami, and E.K. Stefanakos, "Nanocatalyst doping of Zn (BH4)2 for on-board hydrogen storage," *Journal of Alloys and Compounds*, vol. 462, no. 1–2, pp. 294–302, 2008.

[46] J. Yang, A. Sudik, D.J. Siegel, et al., "Hydrogen storage properties of 2LiNH2 + LiBH4 + MgH2," *Journal of Alloys and Compounds*, vol. 446–447, pp. 345–349, 2007.

[47] G.J. Lewis, J.W.A. Sachtler, J.J. Low, et al., "High throughput screening of the ternary LiNH2-MgH2-LiBH4 phase diagram," *Journal of Alloys and Compounds*, vol. 446–447, pp. 355–359, 2007.

[48] S.H. Joo, S.J. Choi, I. Oh, et al., "Ordered nanoporous arrays of carbon supporting high dispersions of platinum nanoparticles," *Nature*, vol. 412, no. 6843, pp. 169–172, 2001.

[49] A. Zaluska, L. Zaluski, and J.O. Ström-Olsen, "Structure, catalysis and atomic reactions on the nano-scale: a systematic approach to metal hydrides for hydrogen storage," *Applied Physics A*, vol. 72, no. 2, pp. 157–165, 2001.

[50] A. Bianco, "Graphene: Safe or Toxic? The Two Faces of the Medal (Minireview)," *Angewandte Chemie International Edition*, vol. 52, no. 19, pp. 2–14, 2013.

[51] Paulo Emílio V. de Miranda. *Hydrogen Energy*. Elsevier BV, 2020, 45.

[52] Aris Vourvoulias, "Pros and cons of solar energy, greenmatch, spetember," 2021. Available: https://www.greenmatch.co.uk/blog/2014/08/5-advantages-and-5-disadvantages-of-solar-energy

[53] M.S. Dresselhaus, and I.L. Thomas, "Alternative energy technologies," *Nature*, vol. 414, no. 6861, pp. 332–337, 2001.

[54] E.K. Stefanakos, D.Y. Goswami, S.S. Srinivasan, and J.T. Wolan, "Hydrogen energy," in *Environmentally Conscious Alternative Energy Production*, M. Kutz, Ed., vol. 4, chapter 7, pp. 165–206, John Wiley & Sons, New York, NY, USA, 2007.

[55] D. Chandra, J.J. Reilly, and R. Chellappa, "Metal hydrides for vehicular applications: the state of the art," *JOM*, vol. 58, no. 2, pp. 26–32, 2006.

[56] P. Chen, Z. Xiong, J. Luo, J. Lin, and K.L. Tan, "Interaction of hydrogen with metal nitrides and imides," *Nature*, vol. 420, no. 6913, pp. 302–304, 2002.

[57] A. Züttel, "Hydrogen storage methods," *Naturwissenschaften*, vol. 91, no. 4, pp. 157–172, 2004.

33

Biocompatible Nano-Electro-Mechanical System–Based Cantilever: An Overview

R. Rekha, M. V.
Suganya Devi and
M. Shanmugavalli

*Saranathan Engineering
College, Trichy, India*

P. V. Rajesh

*Saranathan Engineering
College, Trichy, India*

P. Jothi Palavesam

*Saranathan College of
Engineering, Trichy, India*

J. Femila Roseline

*Saveetha Institute of
Medical and Technical
Sciences, Saveetha
University, Chennai, India*

In the modern world, there is an inadvertent necessity to reduce the weight, increase power-to-weight ratio, and improve the performance of engineering materials without any compromise on quality. The field of nanotechnology has to be pondered upon exponentially to enhance the mechanical and electronic characteristics despite a reduction in weight, size, and structure. Miniaturization, the process of scaling down at nano (10^{-9} m) levels, is increasingly becoming inevitable due to its flexibility, adaptiveness, compatibility, and bio inertness. Nano-electro-mechanical systems (NEMSs) which are miniaturized devices consisting of built-in mechanical actuators interact with the immediate or local environment. Nanocantilever is one such NEMS sensor to measure mechanical forces generated by biological systems with improved static and dynamic characteristics. The integration of advanced simulation technique and additive manufacturing technology has realized the highly developed production of items through macro-, micro-, and nanoarchitecture control. The manufacturing methods are based on material, device boundaries, and the exact need for the last Nano scaffold. This chapter is a humble

DOI: 10.1201/9781003355755-33

attempt made by the authors for providing a deeper insight to the reader by throwing some light on the NEMSs, biocompatible nanomaterials, types and categories, design and analyze using COMSOL Multiphysics simulation software and fabrication methodologies, applications and advantages of nanometer scale–level systems developed so far and future road ahead.

33.1 Introduction

Research experts and academicians working in the field of micro-electro-mechanical system (MEMS) and nanotechnology-based materials have constantly calculated, developed, and analyzed a staggering variety of microsensors over the past few decades for practically every type of sensing modality imaginable [1, 2]. The phenomena include pressure, chemical species, radiation, magnetic fields, inertial forces, and temperature. Almost all the micro machined sensor and actuators proved to function better than their macro-scale peers. MEMS-based sensors and complementary metal oxide semiconductor (CMOS) circuitry are typically used in biocompatible medical devices to measure and transmit anatomical signals [3, 4]. Microcantilever is one type of microsensor, which is used to measure the mechanical signals originated by biological systems with large level of sensitivity. The mechanical deflection is the result of the stress and strain produced in the cantilever because of the vibration, element reaction, heat variation or audio frequency in the system [5]. In order to improve the performance of the sensor and to scale-down the device, nanotechnology is employed. According to researchers, precise positional control for a device assembly can be obtained only through miniaturization. This can be made possible only by nanotechnology that enables any structure or substance to be made in accordance with the rules of atomic physics, so that manufacturing costs and energy used in fabrication can be reduced. The recent progress of NEMSs and wireless technologies in medical devices and components particularly in wireless medical implants is astonishing. Novel minuscule medical devices aid in minimizing incision encumbrance, reduce risk to the life of the patient, augment magnetic resonance imaging (MRI) conformity, and enhance the standard of health services for chronic diseases [6]. NEMSs integrate mechanical and electronic components at nanolevel. The nanocantilever beam is used to measure many physical and biological parameters like temperature, pressure, mass, flow, and wetness. The signal-conditioning cantilever-based sensors can be electrostatic, piezoelectric or piezoresistive. It is fabricated using solid free-form manufacturing. This additive manufacturing (AM) technology is cost-effective as it reduces material wastage and improves precision and accuracy. The biomolecules applied on cantilever sensor change the mass of the sensor, which results in the resonant frequency of the beam to be changed. The piezoelectric material deposited on the one side fixed beam will detect the mass as the resonant frequency.

33.2 Nano-Electro-Mechanical Systems

Miniaturization, the process of scaling down prototypes is usually called as seeing the light of the day more often nowadays. The multi-disciplinary and multi-stream components and assemblies like nanosensor, nanoactuator, nanogear, and nanochip fabricated at nano (10^{-9} m) levels are increasingly becoming inevitable due to their flexibility, adaptiveness, compatibility, and bioinertness. These systems, which are often formed by various nanoscale manufacturing processes, can be found most sought after almost everywhere since the advent of 21st century. NEMSs which are miniaturized devices consisting of built-in mechanical actuators such as levers, vibrating structures, and deformable membranes and electrical components such as resistors, capacitors, and inductors that interact with the immediate or local environment. Nanocantilever is one such NEMS sensor that measures mechanical forces generated by biological systems with improved static and dynamic characteristics. The integration of advanced simulation technique and AM technology has realized the highly developed production of items through macro-, micro-, and nanoarchitecture control. The manufacturing methods are based on material, device boundaries, and the exact need for the last nanoscaffold. Their range of applications include NEMS pressure sensors, NEMS inertial sensors, nanofluidics for diagnostics, nanofluidics for drug delivery, nanosurgicalaccessories, sensing and imaging diagnostics, biomedical implants, and so on.

In the early 2000s, NEMS machine was found to have potential in mass sensing [7]. Later, a demonstration was carried out using top-down NEMS devices to obtain mass resolution at the unit of mass equal to 10^{-21} grams level [8]. Meanwhile, key contributions involving the detection of single molecules, atomic mass determination by shot noise statistics, and eventually yoctogram-level resolution have all been carried out using bottom-up NEMS systems. The researched part has been modeled in these studies as a slushy particle with a rigid stiffness. As a result, it is highly possible to connect the frequency shifts observed in the trials to the analyte's mass. However, the NEMS device has the location-reliant sensitivity which proved to be a persistent issue in the early trials. In NEMS, there are architectures like doubly clamped cantilever beams that have sporadic vibration profile of the flexural modes which causes positional sensitivity [9–11].

With the advent of microcantilevers, location-reliant sensitivity of various kind was discovered near the beginning [12]. Later, different modes were used to determine the analyte's mass and location [13]. The easiest way to comprehend this method is to understand that since mass and position are the two unknowns, it is required to find two independent equations. One such independent equation is provided by measuring the frequency in each mode. As the mass of the analyte is calculated when both modes have been employed in a beam with double clamp, this method can also be known as "two-mode technique". A substantial number of assumptions are to be made in this technique such as the analytes are considered as point loads, in mechanical modes both are orthogonal to each other, and particles dropped on the beam have no effect on the mode forms. With the help of multi-modes, the two-mode technique was further improved, and the analytes' mass distributions were measured. There is an emphasis that the diversity in geometry of the mode forms has to be taken advantage by the theory of two-mode along with inertial imaging; therefore, the concepts can be applied to transducers having different forms and operating in different field [14–16]. For example, in a microfluidic channel, sensing the position of droplets and single cell have been established by utilizing microwave resonant sensors operating in two modes concurrently [17]. Another common strategy for gathering bulk data is to completely avoid the problem of responsivity based on position-dependent by adopting various designs of contrivance [18].

Even though an analyte's location is crucial for measuring its mass, sufficient studies have not been done on how well NEMS sensors function in terms of position resolution. In a study, the association between modal sensitivity and location was investigated utilizing MEMS and a 0.9 micrometer radius gold particle [19]. In this study, a novel method has been used to gather a lot of statistics, and individual nanoparticles on the NEMS are sensed. The nanoparticles are delivered during the trials using laser desorption at room temperature. The two-mode method determined each particle's location in real time. The device was taken out of the experimental setup after each consecutive adsorption of multiple nanoparticles so that SEM could characterize it and identify the locations of each nanoparticle that had been adsorbed. To avoid any misinterpretation caused by crowding when the events detected by NEMS were plotted, the particles were viewed in SEM. In each cycle, the discovered particles count was normally restricted to 5 and maximum to 20.

Many researchers are quite apprehensive about the suboptimal characteristics of nanodevices. Nano NEMS/MEMS devices have subpar wear characteristics. These products are not ideal for electrical conducting applications and are very susceptible to poor fracture strength. Due to the tension placed on the material during the deposition process, NEMS/MEMS devices are prone to warping. Compared to semiconductor materials, metallic film production employing NEMS/MEMS devices has lesser strength [20]. The materials utilized to create nanodevices using NEMS/MEMS are more hazardous. Consequently, efficient device management is required. Due to their tiny size and special material qualities, nanolevel devices present handling challenges. Majority are non-biodegradable materials that require better solutions to prevent environmental damage. The materials used in the production of NEMS/MEMS nanodevices may have disastrous effects that are not foreseen. Better automation and increased production efficiency result in employment losses for many people. Figure 33.1 presents the classification of NEMS.

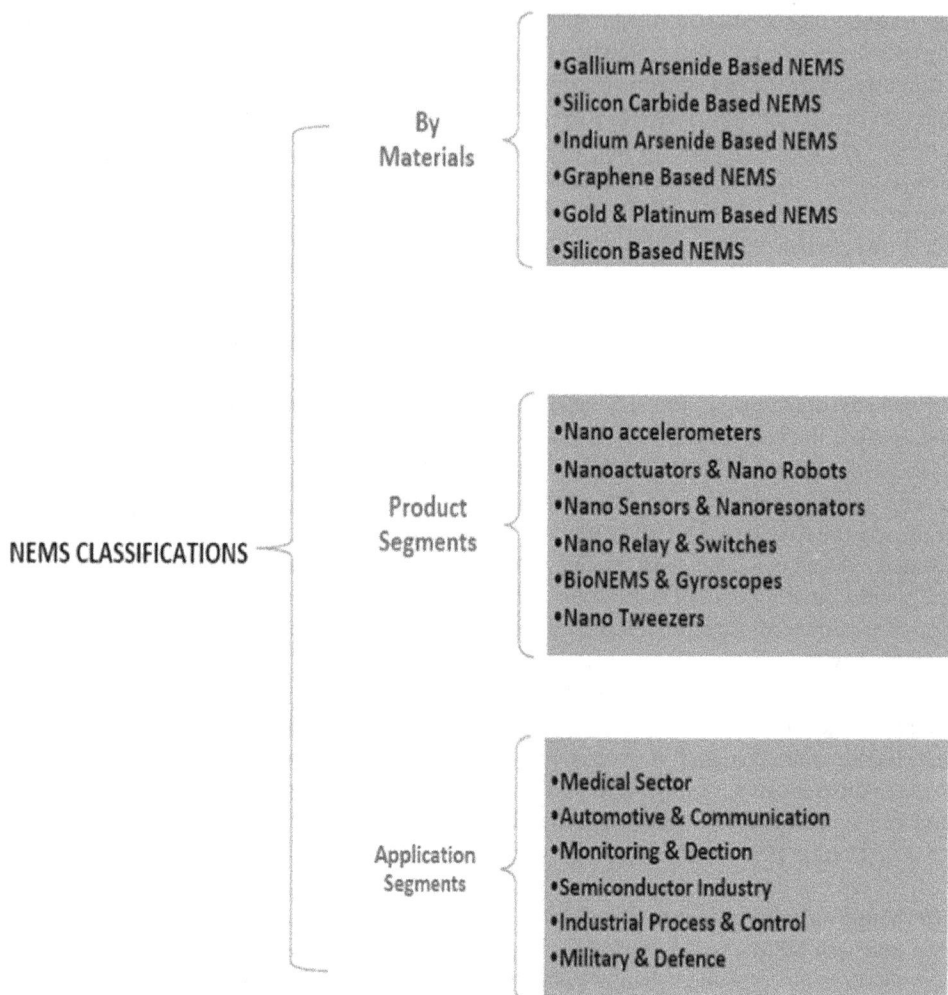

FIGURE 33.1 NEMS classifications.

The flexibility and universality of NEMS make their usage unrestrictive. They act as enabling technologies, fusing engineering, and biological sciences in ways that are currently impractical using microscale methods. Figure 33.2 enumerates the various benefits of NEMS. It will significantly affect a variety of industries, including:

Industry of semiconductors: The MOSFET is the semiconductor component that sees the most utilization. It is responsible for almost all transistors. The degree of essential length IC is lower than 50 nm when considering the transistors in CPU or DRAM devices into consideration. Modern silicon MOSFETs are built using 10 nm and 7 nm fin field-effect transistors.

Automobile industry: Nanomaterials like nanosheets, nanofibers, nanotubes, nanowires, and nanorods provide a number of advantages. For instance, nanoadditives can dramatically increase tire longevity, abrasion resistance, rolling resistance, and wet traction. Future generations of hydrogen-powered cars will benefit from improved fuel cell performance thanks in large part to NEMS.

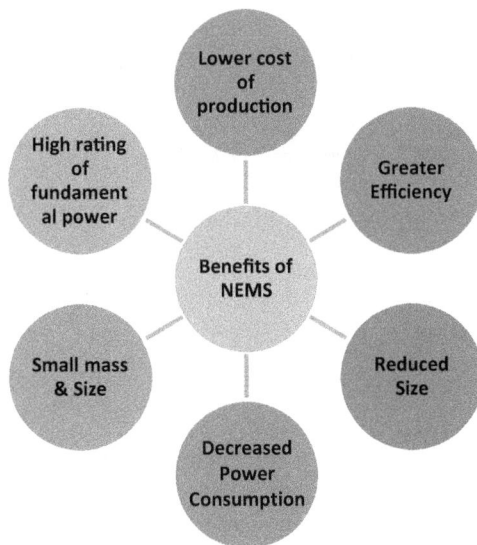

FIGURE 33.2   Benefits of NEMS.

Communication: NEMS resonators, particularly graphene resonators, offer a viable foundation for upcoming ultra-fast communication systems due to their distinctive mechanical features (which allow for high-resonance frequencies and high-frequency tunability). But right now, the majority of advancements in this area are restricted to theoretical models, computer simulations, and laboratory tests.

Medical sector: NEMS sensors identify and keep track of patient data such as blood glucose levels, hydration levels, and the existence of different ions and proteins. These components are possible to set up to sense certain proteins including beta-2-microglobulins and hypovolemia. They can diagnose and isolate cells of various sizes in addition to monitoring, which prevents clogging in a microfluidic system.

Power production and storage: Nanotech has a lot of potential to extend the lifetime and performance of lithium-ion batteries. Additionally, it is found to enhance power density, reduce recharging time, and decrease weight and size of the batteries while improving their stability and safety.

Additionally, work is being done on bionanogenerators which are electrochemical devices of nanoscale that get their energy from the blood glucose level of a living organism. Examples of bionanogenerators are fuel cells or galvanic with an objective of creating solar cells that are more effective and affordable than those made possible by conventional planar silicon-based compounds, research is also being done on a number of nanostructured materials, particularly nanowires.

33.3 Future of the Global Market

NEMS device sales are now in their formative stage. The features of NEMS like low energy consumption, elevated resonant frequency, size and cost reduction, and numerous frequencies on a single chip of integrated circuits are predicted to be responsible for their strong growth in the upcoming years [21]. Figure 33.3 presents the various applications of NEMS.

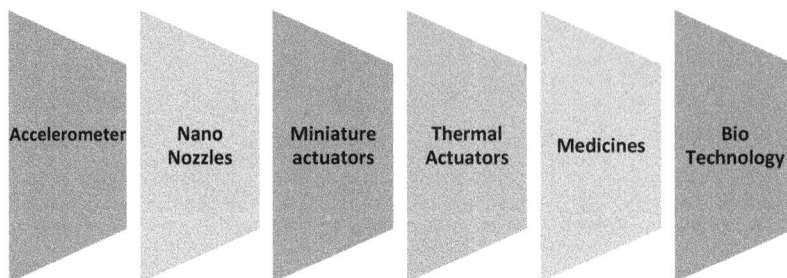

FIGURE 33.3 Applications of NEMS.

33.4 Working Principle

Nanocantilever is a NEMS sensor to measure mechanical forces generated by biological systems with improved static and dynamic characteristics. The cantilever is a beam, its one side is fixed and another end is suspended to rest. It can able to deflect both upwards and downwards due to the strain acting on the entire structure. The nanocantilever allows to measure the change in mass of the biological parameter as the change in the resonance frequency. Therefore, this cantilever has the ability to transducing a nanomechanical motion (i.e. deflection) into an electrical signal [22].

Nanomechanical resonators are used in NEMS sensing systems, and their frequency is electronically monitored. These devices can be made in wide range of shapes and have been applied to a number of sensing tasks. The mechanical resonator is electronically powered at resonant frequency. Once molecules touch the resonator, a small amount of the resonator's mass will change. The frequency of the resonator is altered linearly with an increase in mass. The temporal frequency shift can be calculated by monitoring the NEMS resonant frequency in a phase-locked loop in real time. To determine the mass of various molecules or even a single molecule, the sensor should be set up to detect stable frequencies using its Allan deviation parameters that statistically quantify the frequency noise during frequency as a function of time calculations. The relationship between this parameter and the resonator's consistency factor is inverse. This pattern illustrates how frequency drops as consistency factor changes.

In their study of the piezoresistive effect in semiconductors (such as silicon and germanium), Smith et al. made a substantial observation in the creation of tiny piezoresistive sensors. The employment of these sensors in sensing force, displacement, flow or pressure has made them one of the most well-known and sought-after techniques till date. A piezoresistive pressure sensor's basic operation is the transformation of the applied pressure stimulus into a recordable variation in resistance. A piezoresistive pressure sensor is typically made by sandwiching a layer of piezoresistive material between two parallel electrodes. For certain applications, the piezoresistive layer might be constructed as a cantilever beam or diaphragm and should provide exceptional electrical and mechanical qualities. Piezoresistive pressure sensors with this basic structure and operation enable quick reaction times, great sensitivity, and simple circuit interface. However, the performance of these sensors is restricted by the high temperature coefficient of piezoresistivity, necessitating temperature compensation methods [23, 24].

A piezoelectric material's two sides become oppositely charged when it experiences external stress. This phenomenon has been utilized to create piezoelectric pressure sensors, which directly translate pressure inputs into changes in electrical potential. Lead-Zirconium-Titanium oxide (PZT) thin films are typically utilized as active components in tiny piezoelectric pressure sensors, typically sandwiched between two electrodes. Additionally, ZnO has been touted as a promising material for piezoelectric pressure–sensing devices. These miniature sensors have characteristics akin to those of the microfabrication-based sensors previously mentioned. Indeed, due to their impulsive output signals, they are especially well suited for dynamic pressure–sensing applications.

33.5 Biocompatible Materials for NEMS

Since the development of glucose biosensors by the reference [25], biocompatible sensors are finding wide range of utility and increasingly becoming a crucial component among various diagnostic appliances like medical diagnostics, and ecological observance. Moreover, the first enzyme-based blood sugar monitoring sensor, created in reference [26], came only after that. Because of its adaptability, quick revealing time, and enhanced sensitivity to examine analytes such as proteins, cells, and DNA/ RNA in the nanoscale, biosensor design has since undergone substantial research. A transducer and a biologically active molecule are commonly used in biosensors to transform the biochemical reaction into a measureable signal. The detailed classification of nanoscale biocompatible materials is shown in Figure 33.4.

A biosensor consists of three fundamental parts: (i) a detector, (ii) a transducer, and (iii) a signal processor. Depending on the diagnostic and the physio-chemical characteristics of the analyte, the transducer may be of the optical, or calorimetric, electrochemical, or acoustic kind. A wide range of detecting principles, including conductometry, amperometry, potentiometry, and voltammetry, have been used to study biosensors. It's crucial to use the right biomaterial when creating a biosensing component. The development of most of the sensing systems has substantial use of enzymes, DNA/RNA, aptamers antibodies, receptors, organelles, and animal cells/tissues. The three main guidelines that should be followed when building a biosensor are (i) it must operate in a broad variety of pH and temperature settings, (ii) it must be easy to fabricate, and (iii) it must be highly sensitive with broad dynamic range. The capacity of a biomaterial to remain immobilized while maintaining organic action and detecting the objective

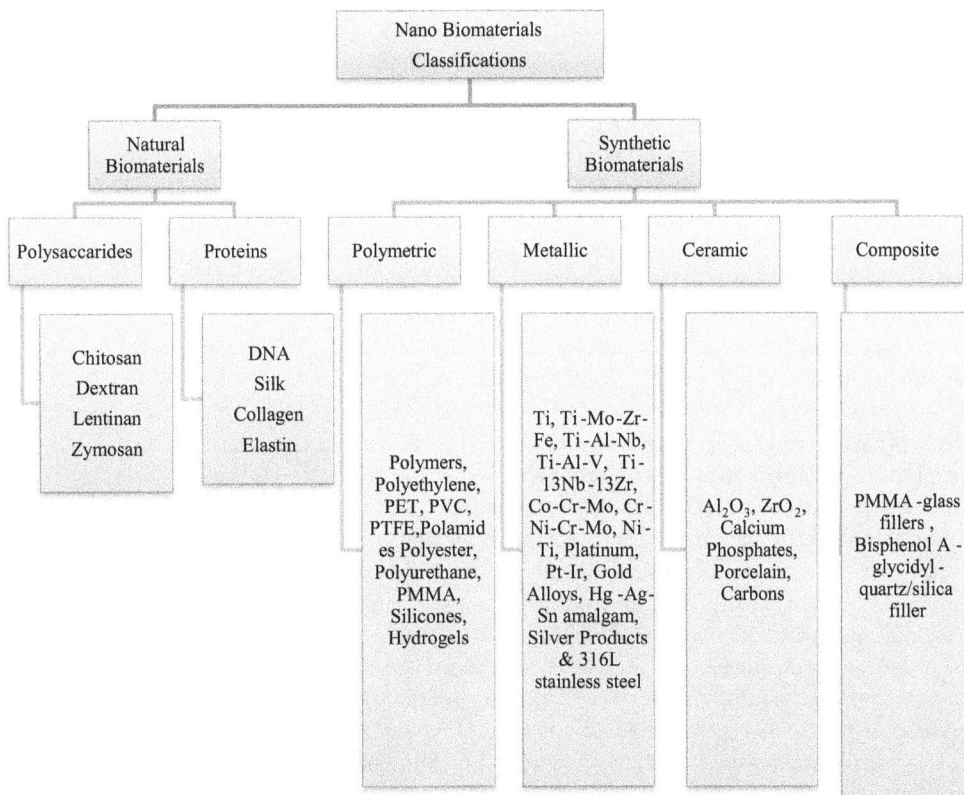

FIGURE 33.4　Classifications of nanobiomaterials.

is the successive step after choosing a biomaterial. A key benefit of biosensors is their ability to simplify problems for the average person by providing them with a bedside medical device for individualized diagnosis.

A lot of work has been put into creating numerous prototypes for dedicated biosensing over the years. These include synthetic materials like carbon nanotubes (CNTs) metal oxides and different polymer composites containing quantum dots and graphene [27] as well as natural biomaterials like chitin, chitosan, and collagen from plants. Designing biosensors with different biomaterials requires an interdisciplinary approach from many different scientific disciplines. This has led to extensive research employing various biomaterials over the past few decades to create even more effective diagnostic systems. Biosensors with protein immobilization have been developed using natural biomaterials like chitosan. The unique qualities of chitosan such as excellent biocompatibility and superior film-forming capacity have made this possible. Using the approach of one-step ball milling, a team of scientists from Zhejiang Foundation in China have revealed a quick but effective method for creating brand-new chitosan-graphene glucose biosensors which are water-dispersible, multi-functional, and modified version. By incorporating magnetic iron oxide (Fe_3O_4) nanoparticles, the multi-functional method was made possible. Similar work has been carried out to create a potentiometric urea biosensor utilizing cobalt oxide (Co_3O_4)-chitosan nanocomposites. These nanoparticles having potential properties like biocompatibility, high superparamagnetism, and low toxicity are incorporated for their usage in a variety of biomedical applications like hyperthermia therapy, medicine deliverance, and cell division. In order to create effective biosensing prototypes, biocompatible materials are widely exploited as polymer composites and polymeric fibers combined with conductive materials.

Using zirconia nanoparticles and collagen composite, a reagent less amperometric glucose biosensor is created by researchers. It was discovered that the collagen-grafted biosensor was more selective, extremely sensitive, and thermally stable. Research studies on biocompatible materials, such as collagen type I, are found to be outstanding and successful than employing natural biomaterials for building biosensors. This is because biosensors with great stability and sensitivity can be created using functionally graded synthetic materials. Among these important efforts is the development of biosensors using CNTs. For instance, by utilizing CNTs and electrospun collagen polymer, researchers have created an H_2O_2 biosensor [28]. The incorporation of CNTs, collagen fibers, and electrospunnanofibers assisted in the construction of biosensors with superior biocompatibility and high specificity. The abovementioned studies clearly show that sufficient research has been done and is currently being done on creating nanobiosensors by means of different biocompatible materials. Although existing biocompatible sensors are very potent, there is still room for improvement with regard to biocompatibility and for in vivo and in vitro investigations. Therefore, prospective research is on investigating different kinds of natural and synthetic biomaterial for a range of bioclinical appliances.

Allotropes of carbon or metallic nanoparticles are two main categories for the nanomaterials that researchers have recently used. Graphene, CNTs, and graphite are the three most popular carbon-based allotropes. The properties such as extremely high mechanical flexibility, high aspect ratio, high electrical conductivity, and exceptional optical transparency have made nanowires a popular choice for metallic nanomaterials. These nanomaterials have either been employed in their pure form or combined with a polymer matrix to create composites for the electrodes. Due to their biocompatibility, simplicity in processing and lack of harmful byproducts during their synthesis, carbon-based allotropes have an edge over metallic ones when compared to these two types of conductive elements. In fact, since its debut by Iijima in 1991, manufacturing and using CNT to produce sensors have become a routine because of its outstanding material features. In the construction of nanotubes, depending on the quantity of cylindrical walls existing, CNTs can be classified into several types. The prominent among them are single-walled carbon nanotubes (SWCNT), double-walled carbon nanotubes (DWCNT), and multi-walled carbon nanotubes (MWCNT). These types have different physio-chemical characteristics because of their structural peculiarities. SWCNTs and MWCNTs have received the most attention among the three classes because of their unique characteristics that are beneficial to the microelectronics industry.

While MWCNTs can be processed more easily and have greater electrical characteristics, SWCNTs have simpler architectures.

Additionally, when compared to SWCNTs, the functionalization process for MWCNTs is simpler and more efficient. This is a result of the MWCNTs improved interfacial interaction with their functionalized ligands. Despite these physio-chemical variations, the two types of CNTs are fabricated using comparable fabrication methods, which are discussed in the upcoming section. Both SWCNTs and MWCNTs have been widely used in the creation of sensors for healthcare-related applications due to their improved properties [29]. This study demonstrates how SWCNTs and MWCNTs are frequently used to create piezoresistive sensing technology. Different fabrication techniques were developed by experts in the academic and industry sectors to process the raw materials. Due to the nature of the processing stages, the majority of them fall under the category of printing techniques. These methods all provide sensors that use various operational mechanisms. The pressure that various body parts' movements exert on the piezoresistive sensors allows them to function. For rehabilitation purposes, these sensors are very crucial. In fact, to identify both sensitive and long-term irregularities in people, piezoresistive kind of sensors are employed. For instance, nowadays people are having chronic issues with their bones and muscles [30].

The functionalities of these piezoresistive sensors are also imitated by electronic skins and thin films which are adhered to the human body. Flexible MEMS sensors created using printing processes have two distinct advantages over traditional MEMS sensors in terms of electrochemical and strain-sensing applications. This chapter summarizes some of the important work on the creation, use, and functionality of flexible pressure sensors based on CNTs that use a piezoresistive sensing mechanism.

A group of solid materials known as piezoelectric materials, which respond to mechanical agitation, are having the ability to build up an electrical charge and making it easier to change mechanical energy into electrical energy and vice versa. Piezoelectricity has been exhibited by both organic and inorganic materials, and variation depends on the type of material. The inversion symmetry–deficient dielectric materials that make up inorganic piezoelectric materials' piezoelectric action undergo an ion rearrangement to produce the piezoelectric effect. In contrast, when subjected to mechanical stress, polarization in organic piezoelectric materials occurs due to the reorientation of the molecular dipole. The market for electromechanical devices, including sensors, actuators, energy harvesting, and storage, has been completely dominated by these materials. Medically implanted and mountable devices are the most recent uses for piezoelectric materials that have garnered a lot of attention. In comparison to inorganic piezoelectric materials, organic piezoelectric biomaterials have a number of advantages, such as great biocompatibility, outstanding processability, ecological balance, and a good flexibility. Researchers have attempted to design molecular structure and adding dopants in order to improve the physical and chemical properties of the materials. Recent studies suggest that, even though organic piezoelectric biomaterials exhibit weak piezoelectricity compared to their inorganic counterparts, when properly prepared, biocompatible piezoelectric materials that connect with humans' biological systems can serve as functional materials in the field of medically implantable and mountable applications. Due to damage in locality and non-local elasticity, organic piezoelectric materials may pose certain manufacturing issues for a variety of devices, from nano- to millimeter-scaled devices.

33.6 COMSOL Multiphysics Simulation Tool for NEMS

To theoretically characterize devices and systems before manufacturing or even before prototyping, modeling and simulation have been used in all engineering fields and disciplines for decades. This is done for a variety of reasons, among them being the reduction in manufacturing time and cost. However, modeling and simulating NEMS components are an uphill task due to the heterogeneous nature of NEMS structures and the multi physics events that are involved in their operation. Since NEMS are too small, direct experimentation to ascertain certain of their physical attributes is challenging, and metrological mistakes happen when working with these micro-level systems. Before prototyping, it is necessary to evaluate how the design behaves under various experimental phenomena and levels of parameterization in order to save production costs. NEMS are typically combined with other micromachined components, and systems are incorporated within other systems. As a result, modeling and simulation are required at both device level and nano level. Simulation and modeling can produce a quicker design cycle that enables thorough scoping and accurate decision-making. It provides room for operating system and device optimization and enables comprehensive work. Designers and system developers can observe and further explore system behavior that would not have been possible without modeling and simulation. Figure 33.5 portrays the V-shaped model of the design process.

A ubiquitous computer tool for engineering simulation, COMSOL Multiphysics software is based on finite element analysis (FEA) and provides a wider range of capabilities for experimental investigations and solutions. This package contains the modules for fluid flow, heat transport, structural mechanics, acoustics, chemical engineering, and electromagnetic. Additionally, it has a sizable number of pre- and post-processing tools, which foster a conducive working atmosphere for resolving both technical difficulties and intricate scientific conundrums. It enables individuals to do away with pointless and dull finite element programming. The ability to simultaneously solve connected multi physics phenomena is COMSOL Multiphysics' second benefit. It is recognized as the top software package for any number of connected multi physics areas as a result of these benefits.

There are two different types of operation modes offered by COMSOL Multiphysics: command style using scripts and graphical user interface style. Most users find both approaches convenient [5]. For COMSOL Multiphysics, script mode is mostly used for optimal design and second development. COMSOL Multiphysics is divided into three parts: pre-, post-, and solution processes. Pre-processing includes the creation of finite element models and determining load parameters. Division by meshes and resolving equations are both included in the solution section.

COMSOL's user interface lessens the need for complex computer coding which is a benefit of modeling. COMSOL's versatility in modeling complex geometries and linking various physics is another benefit. COMSOL also has a large number of built-in capabilities for post-processing data, making it simple to examine different quantities. To avoid time-consuming and expensive clinical trials, biophysical problems are simulated using the computational finite element method (FEM). Various tools in software for using finite element models are part of the emerging technologies. The use of these models helps clinicians better plan the ablation process, which improves patient outcomes. COMSOL Multiphysics simulation tools for biomedical challenges play a vital function in the safe design, optimization, quick

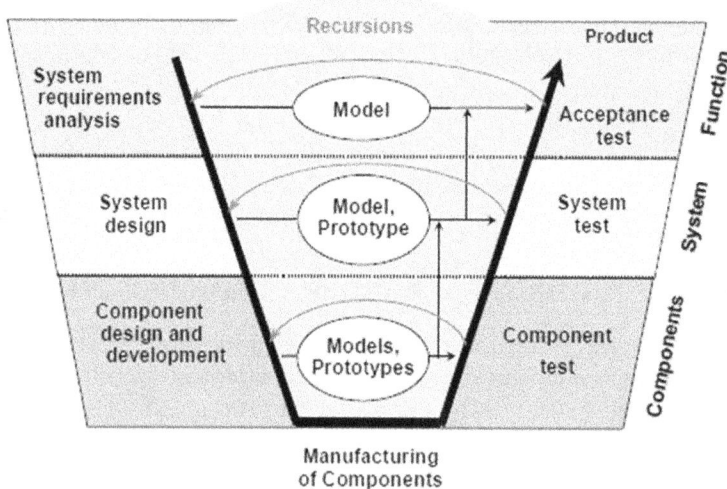

FIGURE 33.5   V-Shaped model of the design process.

development, evaluation of the virtual model, and archetyping for such a complicated ablation system. The simulation tools play a vital role in replicating the ablation system being studied which includes tumor tissues, healthy tissues, energy source, and applicator, in order to establish the optimal parameter selection and production method for finite element models. The constructed simulation model can then be used with a for-profit finite element program to identify the ablation applicator with optimal design. As a result, depending on where the tumor is located, tumors in various tissues, such as the bone, liver, lung, and skin, may be considered. COMSOL Multiphysics software is reported to be one of the best simulation tools for developing test design models and the energy radiator. To control and determine the best radiometric performance under diverse circumstances, it mimics the electromagnetic and temperature effects of the radiative applicator. For the purpose of shortening the assessment of the safety and sturdiness of therapeutic ablation equipment and obtaining regulatory approval, the COMSOL model can simulate the tissue ablation procedures involving lasers, effects of bioheating, low- and high-frequency EM radiation, and focused high-intensity sound.

The finite element approach adds few arithmetic errors, like any distinct numerical approach for solving continuous partial differential equations (PDEs); to reduce this, a suitable mesh must be used. The mesh size and expansion rate between elements can directly be changed in COMSOL Multiphysics at any location, on any limit, or on any quantity inside the model. The claimed inability to resolve small objects is no longer a worry because the mesh has to adhere to all geometric features like faces and edges. An "adaptive" mesher is also provided by COMSOL Multiphysics to automatically refine the mesh where it is needed. Despite taking longer to complete than a user-controlled mesh, the authors discovered that 20 rounds of adaptive mesh refinement on the two-dimensional microdisc problem produced a 0.1% inaccuracy when compared to the analytical solution. Klymenko and team have critically explored the application of adaptive meshing in the context of reaction fronts [6]. The necessary numerical time steps are automatically selected by the COMSOL Multiphysics solver. These are set to be as lengthy as they can be while still being between a specific relative or absolute tolerance for the integration's accuracy, which is calculated at runtime. By taking into account a rectangular MoS_2 sheet in the nanoscale span, Mandar Maitra et al. did the finite element (FE) simulation in longitudinal vibration of piezoelectric monolayer MoS_2 using the program of COMSOL Multiphysics. Fixed-fixed and free-free boundary conditions were taken into consideration. For different sizes of the MoS_2 sheet, the frequency of longitudinal oscillation has been determined using the eigenmode analysis. The longitudinal vibration's resonance frequency is discovered by eigenmode study with COMSOL Multiphysics. The computed speed was discovered to be highly near to the mean value obtained in FE simulations when employing an analytical formula for longitudinal mode.

33.7 Fabrication of NEMS Sensor

The traditional machining processes such as drilling, turning, milling, and laser machining are not applicable for nanomachining processes since the size is very minimum. Micromachining is the technology to fabricate microsensors and actuators of sizes ranging 10^{-6} meters. Nanomachining is the technology to fabricate nanosensors and actuators of sizes ranging 10^{-9} meters. The sensors and actuator materials fabricated on a micro- and nanometer-scale possess unique properties. NEMS devices, MEMS devices, and integrated circuits follow the same fabrication methods. Micro-and nanomachining are batch processes where thousands of identical sensors or actuators are fabricated at a time on the same wafer.

Microsensors are the components which convert a non-electrical input into equivalent electrical output, e.g. strain gauge and inertia sensor.

Few Sensing Mechanisms:
- Piezoresistive: Resistivity changes with the applied stress and is converted to electrical voltage.
- Piezoelectric: Applied strain/force results in a potential difference across the surfaces and vice versa.

- Capacitive: Includes one or more fixed and moving conducting plates.
- Tunneling: A small change in separation between tip and surface results in exponential change in tunneling current.

Microactuators are the components that respond to an electrical input signal, e.g. air bag deployment system.

Few Actuation Mechanisms:
- Piezoelectric: The applied potential difference is proportional to the displacement which is electrically induced.
- Electrostatic: Parallel plates with dielectric in between cause motion.
- Thermal: Temperature difference causing motion in lower and higher coefficient material bond.
- Shape memory alloy: Some materials (e.g. Ti/Ni alloys) regain their original shape when heated to a predefined value of heat.

The fabrication of miniaturized devices is done using huge micromachining, plane micromachining, Lithographie Galvanoformung Abformung (LIGA) technique and microstereolithography (MSL). NEMSs are mechanical and electrical devices at nanometer scale. While miniaturizing, the size reduces, space occupied is low, power consumption is low, higher resonant frequency, and high responsivity. It finds huge application in the fields of automobile, medicine, aerospace, and biomedical. The fabrication of miniaturized system started with the invention of transistors at 1947. Many biomedical and automobile and aerospace sensors and actuators are fabricated and commercialized. The time to market (TTM) was 10 years before. FEA and computer-aided design (CAD) tool plays a major role in reducing TTM. Nowadays due to the FEA CAD tools availability, the design and simulation of sensors and actuators are accomplished in advance, and so the TTM is reduced from 10 years to below.

Silicon as an electronic material already has an advanced microfabrication technology. Similar to fabrication of ICs, it is suitable for batch processing. It is a viable semiconductor sensor material with high performance/cost ratio. Miniaturized machines can be realized on silicon with high precision and can be electronically integrated. It has excellent mechanical properties. Figure 33.6 shows how easy to make semiconductor to act as either conductor or insulator.

Few basic process steps in fabrication are
- Select regions on wafer to do a particular processing (photolithography).
- Define n/p type doping (ion implantation).
- Deposit or grow insulator regions (oxidation/deposition).
- Remove material from unwanted areas (etching/micromachining).
- Make contacts and interconnect devices (metallization).

Thin-Film deposition
- High temperature of thermal oxidations makes it unattractive for already processed wafers.
- Thin-film deposition (physical and chemical).
- Physical (sputtering and evaporation – done mostly for metals).

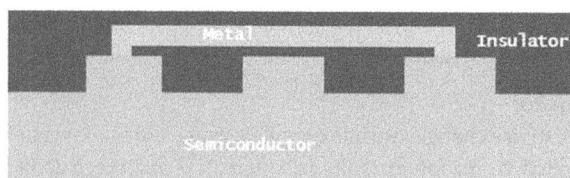

FIGURE 33.6 Semiconductor as conductor and insulator.

- Chemical vapor deposition (molecular beam epitaxy [MBE], MOCVD (metal-organic CVD), APCVD (atmospheric pressure CVD), LPCVD (low-pressure CVD), PECVD (plasma-enhanced CVD), high density plasma chemical vapor deposition [HDP]).

Sputtering

Sputtering is one way to make contacts (Figure 33.7). It is achieved by immersing an Al target in Ar plasma. Ar ions bombard target. The ejected Al atoms deposit on target wafer.

Chemical vapor deposition is one of the deposition processes to form a rigid layer on the substrate by chemical reaction. The reactant gases are given into the place of reaction and are decomposed and reacted at heated areas to form the thin layer. Few chemical vapor sediments are APCVD, MOCVD, PECVD, and LPCVD. Chemical vapor deposition (CVD) system for atmospheric and low pressure is given in Figure 33.8.

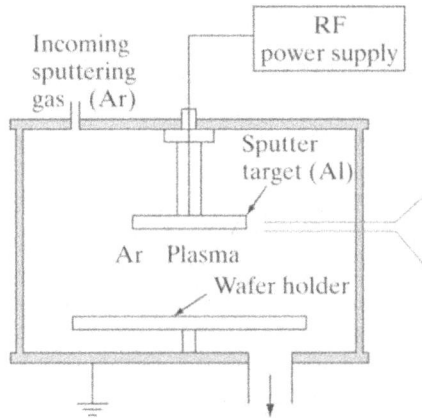

FIGURE 33.7 Sputtering process.

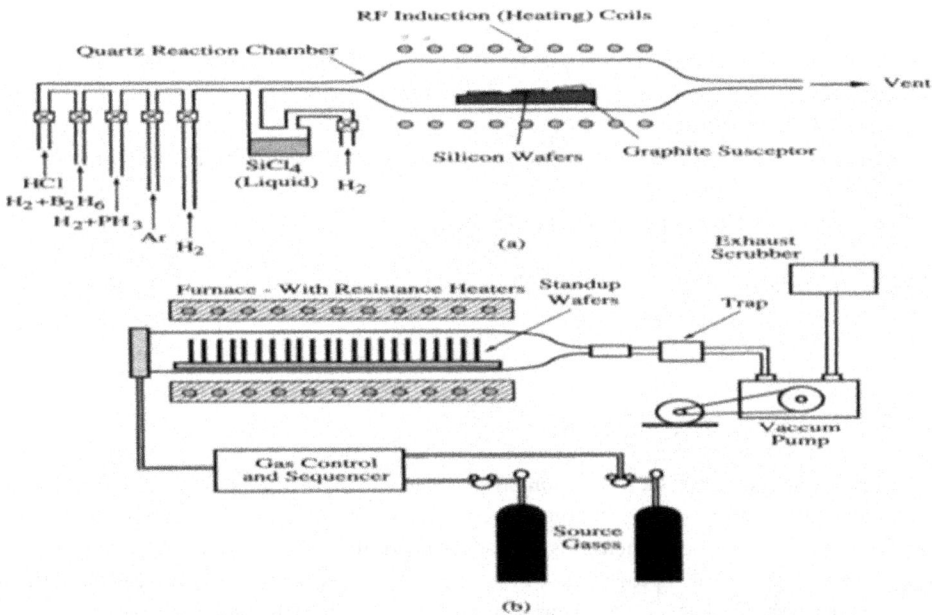

FIGURE 33.8 CVD operation for atmospheric pressure and low pressure.

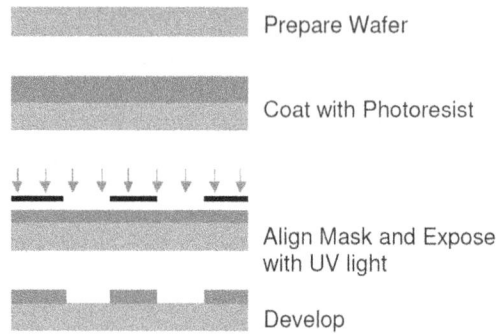

Prepare Wafer

Coat with Photoresist

Align Mask and Expose
with UV light

Develop

FIGURE 33.9   Photolithography process.

APCVD produces solid films at atmospheric pressure with mass transfer–limited regime. LPCVD is a batch processing surface reaction–dominated regime. PECVD is a low-temperature process (e.g. Al deposition already done). Extra energy from plasma is required. Pressure requirement of P » 50 mtorr–5 torr is possible.

Photolithography (Figure 33.9) is the most critical process in fabrication. The pattern is transferred from a mask to a substrate. Masks are opaque plates with transparencies that allow UV light to shine through a pattern. Initially, the substrate is coated with photoresist, the substrate needs to be aligned with mask, and then light needs to be exposed and finally developed to release structures. This determines how small and closely packed the individual devices/structures can be made. Patterning is the costliest and most complex process in fabrication.

Wet etch, isotropic, and selective are the etch process available. Physical sputtering is anisotropic, not selective etching type. Reactive ion etching is anisotropic and selective etching type.

The process steps followed in photolithography are

1. Surface preparation – Promoter like HMDS to offer good adhesion to photoresist.
2. Spin coating of photoresist.
3. Post-apply bake to remove excess solvent and hence stabilize the resist film (<70°C to prevent decomposition and cross-linkage) followed by chilling.
4. Alignment and exposure.
5. Post-exposure bake to reduce standing waves and needed for chemical amplification of resists (since ≥ 48 nm).
6. Post-develop bake to harden the final resist image.
7. Measure/inspect/rework.
8. Pattern transfer.
9. Removal of resist using plasma.

33.8 Bulk Micromachining

Bulk micromachining (BMM) is a fabrication process that is used to produce devices into the substrate by etching selectively. Fabrication of micromechanical devices by etching deeply into the silicon wafer is taking advantage of all its three dimensions. Well-developed silicon-sensor technology (from the 1970s) generally involves double-sided processing – one side exposed to the measured variables and the other side enclosed in a clean package. Device fabrication using IC batch fabrication processes includes deposition, lithography, etching, and bonding. The main steps in BMM are etching, etch stop technique, and bonding. There are two ways the engineer can stop wet etching in silicon substrates: dopant-based etch stop and electrochemical etch stop. Wafer bonding techniques are commonly used in sealing microsensors and in the construction of bulk micromachined sensors. Few bonding methods used in BMM are fusion bonding, anodic bonding, metallic seals, and low-temperature glass bonding.

33.9 Surface Micromachining

Surface micromachining (SMM) is a fabrication operation that is used for producing devices onto the substrate by etching selectively after step-by-step deposition (Figure 33.10). Normally SMM is two and a half dimensions at structures that is located onto the silicon wafer that exists as a thin film. The main advantage of surface micromachined devices is that the mechanical structures can be integrated easily with electronics. Features built up on a silicon wafer by SMM integrate well with electronics. In 1983, Howe and Muller demonstrated this technique using polysilicon with a sacrificial silicon dioxide. Polysilicon cantilever is 25 μm long, 230 nm thick, and 3.5 μm beam-substrate separation. Steps in SMM are shown in Figure 33.11.

Small gaps and smooth surfaces in surface micromachined structures cause the surfaces to stick when they come in contact is called as stiction, which needs to be addressed in SMM.

LIGA is the acronym for lithigraphy galvanoformung, abformung technique. This process is used for fabricating high aspect ratio devices. Aspect ratio is defined as the height to lateral dimensions of etched microstructures. The three steps followed in LIGA are LI = Lithographie (using X-ray), G = Galvanik (electroforming), and A = Abformung (molding).

Few key features of LIGA process are

- Wide variety of materials possible
- Extreme structure heights up to some millimeters (high aspect ratio)
- Sub-micrometer structural details

FIGURE 33.10 Difference between BMM and SMM.

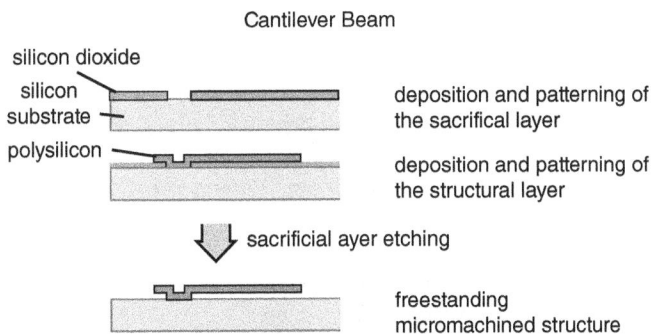

FIGURE 33.11 NEMS fabrication using surface micromachining.

- Smallest structures of some micrometer
- Arbitrary cross-sectional shape

X-rays provide high penetrating power because of their short wavelength, high resolution, and high aspect ratio. X-rays are provided by synchrotron radiation source high throughput and high flux of collimated rays shortens shorter time.

33.10 Microstereolithography

Stereolithography is suitable for fabricating complex 3D microstructures with high aspect ratios. The complex structures were fabricated from the shape saved through CAD files and transferring the information to the monomer on the vat to polymerize to generate polymer structures. These models were used for visualization of odd shapes that are not easily legible on traditional drawings. MSL is similar to rapid prototyping manufacturing technology, which helps to generate any complex polymeric structures. The light exposure solidifies a liquid resin (layer by layer) into a desired 3D shape. SL is used for microparts, so it is named as MSL. SL is a photopolymerization process, so it is called as microphoto forming. SL and MSL are the same but except for the resolution of MSL is low. Classical MSL, IH process, mass-IH process, and super-IH process are the different types of MSL.

33.11 Conclusion

The nanocantilever beam can be used to measure many physical and biological parameters like temperature, pressure, mass, flow, and moisture. The AM technology is a cost-effective method for fabricating nanocantilever as it reduces material wastage and enables precise manufacturing. The nanocantilever sensor with piezoelectric signal conditioning finds huge application in biomedical. The fabrication of biocompatible NEMS cantilever using 3D printing is a novel idea which is new initiative. 3D printing of biocompatible nanomaterial will be a challenging task, which may need a lot of trials and also modifications or additions of accessories in the existing 3D printing machine. Developing the cantilever in a nanoscale will require the applications of the latest technology which is in infant stage. It is stated in the literature that it is very difficult to deposit thin films with controlled material properties. Possibilities of runaway external and residual (intrinsic) film stress, induced cracks, voids, hillock formation, and film lifting should be avoided which may lead to yield loss and poor reliability issues. It may be difficult to get all the individual processing steps to work together and may need many cycles to solve process integration issues and get working devices. This work is a multi-disciplinary research and involves the integration of manufacturing, instrumentation, process control, and information tools (ICTs). However, the developed device will largely contribute for biomedical applications thereby enhancing healthcare.

References

1. Jiang, S.; Shi, T.; Zhan, X.; Xi, S.; Long, H.; Gong, B.; Li, J.; Cheng, S.; Huang, Y.; Tang, Z. Scalable fabrication of carbon-based MEMS/NEMS and their applications: A review. *J Micromech Microeng* 2015, 25, 113001.
2. Katz, E.; Willner, I. Biomolecule-functionalized carbon nanotubes: Applications in nano-bioelectronics. *Chem Phys Chem* 2004, 5, 1084–1104.
3. Kis, A.; Zettl, A. Nanomechanics of carbon nanotubes. *Philos Trans R Soc A* 2008, 366, 1591–1611.
4. Janas, D.; Koziol, K.K. A review of production methods of carbon nanotube and graphene thin films for electrothermal applications. *Nanoscale* 2014, 6, 3037.
5. Zang, X.; Zhou, Q.; Chang, J.; Liu, Y.; Lin, L. Graphene and CNT in MEMS/NEMS applications. *Microelectron Eng* 2015, 132, 192–206.

6. Sharma, S.; Sharma, A.; Cho, Y.-K.; Madou, M. Increased graphitization in electrospun single suspended carbon nanowires integrated with carbon-MEMS and carbon-NEMS platforms. *ACS Appl Mater Interfaces* 2012, 4, 34–39.

7. Varney, M.W.; Aslam, D.M.; Janoudi, A.; Chan, H.Y.; Wang, D.H. Polycrystalline-diamond MEMS biosensors including neural microelectrode-arrays. *Biosensors* 2011, 1, 118–133.

8. Choi, J.; Eun, Y.; Pyo, S.; Sim, J.; Kim, J. Vertically aligned carbon nanotube arrays as vertical comb structures for electrostatic torsional actuator. *Microelectron Eng* 2012, 98, 405–408.

9. Kim, M.-O.; Lee, K.; Na, H.; Kwon, D.-S.; Choi, J.; Lee, J.-I.; Baek, D.-H.; Kim, J. Highly sensitive cantilever type chemo-mechanical hydrogen sensor based on contact resistance of self-adjusted carbon nanotube arrays. *Sens Actuators B Chem* 2014, 197, 414–421.

10. Choi, J.; Kim, J. Defective carbon nanotube-silicon heterojunctions for photodetector and chemical sensor with improved responses. *J Micromech Microeng* 2015, 25, 115004.

11. Agrawal, V.K.; Patel, R.; Boolchandani, D.; Varma, T.; Rangra, K. Sensitivity and reliability enhancement of a MEMS based wind speed sensor. *Microelectron Reliab* 2020, 104, 113513.

12. Algamili, A.S.; Khir, M.H.M.; Dennis, J.O. et al. A review of actuation and sensing mechanisms in MEMS-based sensor devices. *Nanoscale Res Lett* 2021, 16, 16. https://doi.org/10.1186/s11671-021-03481-7

13. Su, Y. et al. Printable, highly sensitive flexible temperature sensors for human body temperature monitoring: A review. *Nanoscale Res Lett* 2020, 15(1), 200.

14. de Oliveira Hansen, R. et al. Magnetic films for electromagnetic actuation in MEMS switches. *Microsyst Technol* 2018, 24(4), 1987–1994.

15. Faudzi, A.A.M.; Sabzehmeidani, Y.; Suzumori, K. Application of micro-electro-mechanical systems (MEMS) as sensors: A review. *J Robot Mechatron* 2020, 32(2), 281–288.

16. Singh, A.D.; Patrikar, R.M. Development of nonlinear electromechanical coupled macro model for electrostatic MEMS cantilever beam. *IEEE Access* 2019, 7, 140596–140605.

17. Mohd Ghazali, F.A.; Hasan, M.N.; Rehman, T.; Nafea, M.; Mohamed Ali, M.S.; Takahata, K. MEMS actuators for biomedical applications: A review. *J Micromech Microeng* 2020, 30(7), 073001.

18. Joshi, P.; Kumar, S.; Jain, V.; Akhtar, J.; Singh, J. Distributed MEMS mass-sensor based on piezoelectric resonant micro-cantilevers. *J Microelectromech Syst* 2019, 28(3), 382–389.

19. Sathya, S.; Muruganand, S.; Manikandan, N.; Karuppasamy, K. Design of capacitance based on interdigitated electrode for Bio MEMS sensor application. *Mater Sci Semicond Process* 2019, 101, 206–213.

20. Agrawal, V.K.; Patel, R.; Boolchandani, D.; Varma, T.; Rangra, K. Sensitivity and reliability enhancement of a MEMS based wind speed sensor. *Microelectron Reliab*, 2020, 104, 113513.

21. Algamili, A.S.; Khir, M.H.M.; Dennis, J.O. et al.A review of actuation and sensing mechanisms in MEMS-based sensor devices. *Nanoscale Res Lett* 2021, 16, 1–21. https://doi.org/10.1186/s11671-021-03481-7

22. Su, Y. et al. Printable, highly sensitive flexible temperature sensors for human body temperature monitoring: A review. *Nanoscale Res Lett* 2020, 15(1), 200

23. Chen, K.-Y.; Xu, Y.-T.; Zhao, Y. Recent progress in graphene-based wearable piezoresistive sensors: From 1D to 3D device geometries. *Nano Mater Sci*, https://doi.org/10.1016/j.nanoms.2021.11.003

24. Faudzi, A.A.M.; Sabzehmeidani, Y.; Suzumori, K. Application of micro-electro-mechanical systems (MEMS) as sensors: A review. *J Robot Mechatron* 2020, 32(2), 281–288.

25. Singh, A.D.; Patrikar, R.M. Development of nonlinear electromechanical coupled macro model for electrostatic MEMS cantilever beam. *IEEE Access* 2019, 7, 140596–140605.

26. Mohd Ghazali, F.A.; Hasan, M.N.; Rehman, T.; Nafea, M.; Mohamed Ali, M.S.; Takahata, K. MEMS actuators for biomedical applications: A review. *J Micromech Microeng* 2020, 30(7), 073001.

27. Chen, K.-Y.; Xu, Y.-T.; Zhao, Y.; Li, J.K.; Wang, X.P.; Qu, L. Recent progress in graphene-based wearable piezoresistive sensors: From 1D to 3D device geometries. *Nano Mater Sci* 2022, https://doi.org/10.1016/j.nanoms.2021.11.003

28. de Oliveira Hansen, R. et al. Magnetic films for electromagnetic actuation in MEMS switches. *Microsyst Technol* 2018, 24(4), 1987–1994.

29. Joshi, P.; Kumar, S.; Jain, V.; Akhtar, J.; Singh J, Distributed MEMS mass-sensor based on piezoelectric resonant micro-cantilevers. *J Microelectromech Syst* 2019, 28(3), 382–389.

30. Sathya, S.; Muruganand, S.; Manikandan, N.; Karuppasamy, K. Design of capacitance based on interdigitated electrode for Bio MEMS sensor application. *Mater Sci Semicond Process* 2019, 101, 206–213.

34

Eco-friendly for Sustainable Nanomaterials for Renewable Energy Storage

J. Prakash Arul Jose
Paavai Engineering College (Autonomous), Namakkal, India

Sumanta Bhattacharya
MAKAUT, Kolkata, India

Vishwanath V. Hokrani
KLS Vdit, Haliyal, India

Gururaj Hatti
BLDEA'S V.P. Dr. P. G. Halakatti College of Engineering and Technologym, Vijayapur, India

Ayaz Ahmad
NIT P, Patna, India

Deepa Jaganathan
Agricultural Engineering College and Research Institute, Tamil Nadu Agricultural University, Coimbatore, India

The development of nanomaterials using sustainable raw materials is a developing technology since they are a natural resource with high carbon content. Due to the world's increasing demand for energy and rising pollution, fossil fuels must be replaced by environmentally friendly alternative energy sources. Nanomaterials' rapid growth has opened up new possibilities for the storage and transformation of renewable energy. With the current advancement of nanomaterials, it's feasible that in the future a higher amount of our daily energy storage will come from electricity produced by solar and wind energy. On the other hand, sustainable nanomaterials have a number of definite advantages, such as their low cost, recyclability, and relative lack of toxicity, as well as their simple and environmentally friendly synthesis. This chapter discusses various production processes of sustainable nanomaterials and utilization in renewable energy storage.

34.1 Introduction

The foundation of nanoscience and nanotechnology is nanomaterials. In the past several years, research and development activity in the vast and interdisciplinary subject of nanostructure science and technology has expanded quickly across the globe. It has the potential to fundamentally alter the processes used to make things and materials, as well as the variety and type of functionality that can be accessible. It already has a large commercial influence, and this impact will undoubtedly grow in the future. The Nanomaterials with the special optical, magnetic, electrical, and other capabilities at this size are fascinating to explore in various applications. Electronics, medicine, and other industries stand to benefit greatly from these new features. Nanomaterials are already being used in industry, and some of them have been around for years or even decades. The variety of commercial goods on the market today is very broad and includes electronics, paints, varnishes, cosmetics, and fabrics that are resistant to wrinkle and stain. Many consumer goods, including windows, sporting goods, bicycles, and automobiles, use nanocoatings and nanocomposites (Akmal and Lah 2021).

There are new butyl-rubber/nano-clay composite tennis balls that last longer, as well as UV-blocking coatings on glass bottles to protect beverages from sun damage. For instance, silica nanoparticles are utilized as fillers in a number of items, including cosmetics and dental fillings, while nanoscale titanium dioxide is used in sunscreens, self-cleaning windows, and cosmetics. Michael Faraday's creation of colloidal gold particles is considered one of the earliest scientific reports. In order to replace ultrafine carbon black for rubber reinforcements, nanoparticles of precipitated and fumed silica were created. Today, nanophase engineering is being applied to an ever-increasing range of organic and inorganic structural and functional materials, enabling the control of mechanical, catalytic, electrical, magnetic, optical, and electronic capabilities.

The development of discrete tiny clusters that are later melted into a bulk material or cast into compact liquid or solid matrix materials is atypical method for producing nanophase or cluster-assembled materials. For instance, nanophase silicon, which has different physical and electrical properties from regular silicon, could be used in macroscopic semiconductor processes to make new devices. In the creation of nanomaterials, current green nanotechnology approaches frequently make use of renewable resources, health-friendly solvents, and energy-saving techniques.

34.2 Sustainable Nanomaterial

Electrical energy storage has recently been regarded as a major area and these nanomaterials changed the electronic device sector with these sustainable energy storage devices. High-capacity energy storage devices with greater power ratings are also used in stationary applications like load balancing in modern power plants and in non-stationary applications such as electric vehicles. Charge accretion and reversible redox processes are two methods that are used to store charge in super capacitors. Electric double-layer capacitors and hybrid pseudo super capacitors are the two subcategories in super capacitors (Strauss 2017).

Ion migration causes an electrical double layer to form on the electrode material's outer surface, which results in a capacitance through the balanced charged particle. Faraday capacitors typically use metal oxides (sulfides, hydroxides, etc.) or high-conductivity polymers. Higher storage capacities require the reversible redox reaction to occur at or near the surface in order to store electrical energy. Electric double-layer capacitor electrodes and pseudo (plutonium) capacitor electrodes can be employed as a combination of the benefits of both capacitors to construct and use hybrid super capacitors for the storage process by means of the power of physical and chemical processes (Chen et al. 2021).

As a crucial component that directly affects the performance of the super capacitor, the electrode's material primarily consists of the following types of materials: carbon-based materials (graphene, carbon nanotubes, activated carbon, carbon fibers), oxides derived from metals (cobalt oxide, oxide manganese oxide, oxide nickel oxide, oxide ruthenium oxide), associated composites, and polymers with high cohesion (Li et al. 2021).

Metal oxides are unique among electrode materials in that they are readily available, have great electrochemical stability, and are extremely simple to produce. Due to their low conductivity values, metal oxides unfortunately have very slow reaction rates, which restrict their use in practical applications and slow down their growth. Highly conductive polymers, which have stayed in the spotlight for many years because of their low rates, high conductivity, environmental sustainability, better charge storage capacity, and excellent performance, are a crucial study area for materials used as electrodes for superconductors (Akmal and Lah 2021).

As a result of the undesired behavior of conducting polymers, the mechanical stability and strength of the electrodes will deteriorate and the capacity will alter during cycling which expands and shrinks during charging and discharging. Comparing conducting polymers to carbonaceous materials, they have poor cycle stability, which is an issue even for modern conducting polymers.

34.3 Sustainable Nanomaterials Production Methods

There are three ways to make nanomaterials: the first way is physically, using up–bottom techniques like spray pyrolysis, sonication, arc discharge, pulsed wire discharge, pulsed laser ablation, radiation, electrodeposition, evaporation–condensation, vapor and gas phase, ball milling, and lithography; and the second way is chemically, using bottom-up techniques like sol–gel technology, co-precipitation, redox processes, pyrolysis. The third way is biological approach with the use of microorganisms. In recent years, green synthesis has gained popularity in current nanomaterials research and development as a means of addressing the sustainability challenges of conventional approaches. A subfield of green chemistry, green synthesis aims to create safer chemical products and processes by reducing or eliminating the production and use of hazardous materials. Green chemistry originated as a result of the recognition of the need for sustainable procedures in the chemical industry. As a result, green synthesis borrows some of its fundamental ideas from green chemistry, including waste minimization or prevention, derivative use mitigation, the use of non-toxic (or safer) solvents/excipients and renewable raw materials, and pollution reduction (Thunugunta, Reddy, and Lakshmana Reddy 2015).

Through redox reactions, living things or their byproducts convert metal ions into stable nanomaterials by employing their nicotinamide adenine dinucleotide phosphate hydrogen and dependent reductase enzymes, which is known as green synthesis or biosynthesis. Biogenic nanomaterials are collectively referred to as nanomaterials produced by biological processes. Due to their excellent biocompatibility and sensitivity, these nanoparticles find use in a variety of fields, including biomedicine, agriculture, electronics, and environmental cleanup. Additionally, due to their distinct features, metal oxide nanoparticle synthesis employing biological components has drawn a lot of interest.

Although the use of microorganisms for the synthesis of nanomaterials has been extensively studied, the synthesis of nanomaterials from plants is gaining more attention and also the biosynthesis of nanomaterials without solvents is also being investigated and it is still in its early stages, to further strengthen and enforce one of the fundamental principles of green chemistry. The development of ecologically friendly and sustainable procedures is the second side of green nanotechnology, which also involves employing natural starting and processing materials to synthesize nanoparticles. As alternatives to organic solvents in this regard, water and supercritical carbon dioxide have been researched. It's very intriguing to consider many alternative heating methods. The hydrothermal technique using water as the reaction medium is the most well-liked green tactic (Samuel et al. 2022).

Alternative heat sources for the creation of nanoparticles like microwave energy and focused sunlight can also be used. Even if these methods work well in the lab, putting them to use in commercial production presents a number of formidable obstacles. For instance, microwave heating's quick reactions are only possible in small-scale processes. The changing intensity of the light and requirement for a lens restrict the usage of concentrated sunlight. The cost-effectiveness of green chemistry has recently been enhanced by the development of flow chemistry techniques.

34.3.1 Plant Sources

Using naturally available renewable resources, such as plant extracts and polyphenolic antioxidants from tea and coffee, biodegradable polymers like carboxymethyl cellulose, reducing sugars and agricultural residual waste like red grape pomace from winery waste allow for the green synthesis of nanometal/nanometal oxides/nanostructured polymers and subsequent stabilization (including dispersants, biodegradable polymers).

34.3.2 Vitamins

Non-harmful alternatives that decrease or eliminate the use and manufacture of hazardous compounds are part of our sustainable synthetic activity for the preparation of nanoparticles. For the synthesis of bulk nanospheres, nanorods, nanowires, nanosphere-aligned nanobelts, and nanoplates, one-pot green synthetic procedures using vitamins as well as biomass-derived polyphenols are quite straightforward.

34.3.3 Microwave Heating

A new alternative energy source that is potent enough to carry out chemical changes in minutes as opposed to hours or days is microwave technology. Control over the synthetic processes is crucial when the manifestation of material attributes is based primarily on size and form in the context of nanomaterial synthesis. This is so because surface reconstructions, vacancies, and defects all have an impact on the formation of nanoscale materials and is highly dependent on the thermodynamic and kinetic reaction barriers as specified by the reaction trajectory. The reaction vessel serves as a conduit for the energy transmission from the heating mantle to the solvent and, finally, to the reactant molecules in conventional thermal procedures, which are based on blackbody radiation conduction.

Sharp temperature gradients across the solution volume and ineffective reaction conditions can result from this. In the synthesis of nanomaterials, the necessity of uniform nucleation and growth rates for material quality seems to be a challenging problem. Through the use of increased reaction kinetics, quick initial heating, and consequently increased reaction rates, microwave heating techniques can address the issues of heating inhomogeneity in conventional thermal techniques, leading to clean reaction products with quick starting material consumption and higher yields. The methodology is also relevant to biological and enzymatic systems under various circumstances.

Magnetic nanoferrites, micropin-structured catalysts, and metal oxides with 3D nanostructures can all be produced using the microwave hydrothermal technique. Without the aid of any reducing or capping agents, these materials were easily made from inexpensive starting components in water. The ability to fine-tune a material's responses to magnetic, electrical, optical, and mechanical stimuli may eventually be made possible by this environmentally and economically sound synthetic idea (Al-Hazmi et al. 2015).

34.3.4 Magnetic Nanocatalysts

Green chemical pathways have greatly benefited from the discovery of sustainable paths to nanoparticles and their diverse ecological uses in catalysis via magnetically recoverable and recyclable nanocatalysts for reduction, oxidation, and condensation processes. It has been discovered that an innovative magnetically recoverable and nanoparticle-supported organocatalyst may efficiently catalyze the Paal-Knorr reaction without the need of harmful organic solvents, even throughout the work-up stage.

Wider applications for the highly efficient use of these nanocatalysts are made possible by the post-synthetic alteration of nanoferrites with dopamine and the subsequent anchoring of metal particles. For instance, benzonitrile on magnetic nanoferrites is converted to benzamide in water by hydration with ruthenium hydroxide. Due to the paramagnetic properties of nano-$Ru(OH)x$, the reaction mixture was purified after the reaction was finished and the catalyst was placed on the magnetic rod. By employing an external magnet to extract the catalyst, the filtration stage was skipped. After the catalyst was separated, the transparent reaction liquid was progressively cooled, and satisfactory purity benzamide crystals precipitated.

No organic solvents were used during the reaction or the processing steps, and the entire operation was carried out in a clean aqueous environment. The fact that the reaction proceeded with a high turnover number and at high turnover frequencies is an impressive aspect of these protocols. The catalyst worked properly multiple times with no traces of metal in the finished product. In the upcoming years, various C–C asymmetric bond formation reactions, asymmetric hydrogenation, and asymmetric cycloaddition reactions, among others, will find numerous applications for the new design of magnetically recoverable heterogeneous asymmetric catalysts supported on nanoparticle systems.

34.3.5 Hydrothermal Methods

One of the most well-known and often utilized processes for creating nanostructured materials is the hydrothermal process. By performing a heterogeneous reaction in an aqueous environment at high pressure and temperature near the critical point in a sealed container, the hydrothermal process produces nanostructured materials. The use of a non-aqueous medium is the only distinction. The microwave-assisted hydrothermal approach, which combines the benefits of the hydrothermal and microwave processes, has recently attracted a lot of interest in the field of engineering nanomaterials (Szabó et al. 2010).

34.4 Advantages of Sustainable Nanomaterials

The nanomaterials make use of resources that are naturally replenishable and won't deplete. This implies that it can be used to benefit future generations without endangering the environment. The energy industry may also profit from nanotechnology. This technology allows for the development of smaller, more effective devices that can produce, absorb, and store energy more efficiently. Smaller versions of things like batteries, fuel cells, and solar cells are possible, thanks to this technology, which also increases efficiency. By reducing CO_2 emissions, waste management lessens the impact of global warming and offers alternatives for trash treatment and recycling.

Additionally, it boosts productivity, which favors specific regions (agricultural) economically. The advantage of green synthesis is that it offers an environmentally friendly alternative to chemical and physical processes. Green synthesis additionally offers the benefits of cheap production costs, energy efficiency, secure procedures and products, decreased waste production, and increased application potential in the pharmaceutical and biomedical sectors (Choi et al. 2021).

34.5 Use of Nanomaterials

The use of nanomaterials in sustainable development and renewable energy technologies continues to be a significant area of academic and industrial research. The incorporation of nanoparticles can enhance device performance via a variety of processes. These include, for instance, supporting better energy storage capabilities, streamlined and quick manufacturing procedures for novel device architectures, and greater harvesting and conversion efficiency. Authors are encouraged to submit original research articles or summaries of reviews addressing the most recent advancements and unique uses of nanomaterials for highly effective, novel devices pertinent to applications in renewable energy and sustainability (Prasanna and Deshmukh 2020).

In addition to the extremely effective conversion of solar energy, it is also highly desirable to store the converted energy because solar energy can become unstable at night or when it is overcast outside. In order to use solar energy when there is no sunlight, we should collect and store it. As a result, energy storage is crucial for the effective use of energy resources. Nanomaterials are strongly tied to energy conversion and storage as one of the most crucial components. Energy storage nanotechnologies, in particular research on hydrogen storage, have advanced significantly over the years as a result of breakthroughs and advancements in materials science (Bin et al. 2021).

34.5.1 Batteries for the Accumulation of Renewable Energy

Batteries for solar panels come in wide varieties. Lithium-ion batteries are the most typical kind, although other types including lead acid, gel, deep cycle, and more might also come up. Electrochemical accumulators are made up of a grouping of cells that are essentially formed of electrodes, an electrolyte, and an electrical insulator due to the physical makeup of the device. When power is applied to the battery, energy is stored in chemical form. Modern electricity is produced by connecting the load to the electrodes using wind energy and energy storage technology. Due to the transfer of electrons, reversible chemical reactions take place that charge and discharge the battery.

Lead-acid and lithium-ion batteries stand out among the many different electrochemical batteries available. The positive and negative electrodes in lead-acid batteries are made of lead oxide and sponge lead, respectively. The microporous substance serves as electrical insulation while the two electrodes are submerged in a sulfuric acid solution that serves as an electrolyte. The Li-ion kind of lithium battery is currently the most well-known, with a sizable consumer base in the field of portable devices. Lithium salts are used as the electrolyte in Li-ion batteries, which have a positive electrode made of lithium oxide and a negative electrode made of carbon (Packiyalakshmi, Chandrasekhar, and Kalaiselvi 2019).

34.5.1.1 Lithium-Ion Batteries

Lithium-ion batteries are currently the most popular type of battery for solar panels. The solar industry quickly discovered the advanced lithium-ion technology's promise for household energy storage, yet it was originally created for electric vehicle batteries (Ruan et al. 2021). Because of their high, extended lifespan, and high density, lithium-ion batteries are appealing. All of the top solar batteries on our list incorporate lithium-ion technology in some capacity. It's important to remember that lithium-ion technology includes lithium iron phosphate batteries (Ruan et al. 2021).

34.5.1.2 Lead-Acid Batteries

The most rechargeable battery is made with lead-acid batteries. This technology has been around for a very long time. Since lead-acid batteries have a lower density, a shorter lifespan, and a slower charging rate than their lithium-ion equivalents, they provide less power. Lead-acid batteries come in a variety of forms, including absorbent glass mat batteries, sealed lead-acid batteries, flooded lead-acid batteries, and deep cycle batteries (Piao 2016).

34.5.1.3 Flow Batteries

An innovative new technology called flow batteries has the potential to replace lithium-ion batteries in the near future. A chemical reaction that occurs when the liquid is energized by solar panels results in a flow of energy between two different chambers in the battery. Because of their relative simplicity, and low precious metal requirement, flow batteries are appealing. Flow batteries are still being developed and are currently too big for household use (Wolfgang Dubbert et al. 2014).

34.5.1.4 Redox Flow Battery

Its operating method is distinct from that of electrochemical batteries. The flow batteries are another name for redox flow batteries. Two tanks make up the system, where the electrolytes are kept apart. Both electrolytes are pumped into the cell when necessary, where the ion-selective membrane and electrodes enable ion exchange between the two electrolytes. The ability of this technology to decouple power and energy capabilities is one of its most noticeable characteristics. Energy delivery is dependent on the volume of the reservoir tanks, while performance rates are constrained by the velocity of ion transfer over the selective membrane (Conway and Pell 2003).

34.5.1.5 Sodium–Sulfur Batteries

An exterior layer of molten sulfur surrounds a solid ceramic electrolyte of beta aluminum oxide, which serves as the positive electrode in sodium–sulfur batteries. Molten sodium serves as the negative electrode on the interior surface of the electrolyte. To maintain the sulfur and sodium in a liquid state, a high temperature of between 300 and 350°C is necessary. In order to start the reaction, heat must be supplied; but, once it has started, the heat generated by the constant charge and discharge cycles is sufficient to keep the temperature within the appropriate range, neglecting the need for an external heat source (Dubbert et al. 2014).

34.5.2 Supercapacitor

An electric field is produced between two electrodes of a supercapacitor, which is separated from the electrodes by a very thin layer of liquid electrolyte that is only a few in thickness. Supercapacitors therefore adhere to the same principles as regular capacitors. High rates of capacitance per unit volume are, however, made possible by particular electrode designs and materials. Since no chemical interactions are required to store or release energy, they have a longer lifespan than batteries with frequent charge–discharge cycles (Chen et al. 2016).

34.5.3 Superconducting Magnetic Energy Storage

This technology can release the energy that is stored in enormous magnetic fields as electricity. Magnetic fields that can reach up to several Tesla must be produced and maintained using sophisticated procedures and machinery. The system is based on electromagnetism, producing magnetic fields by passing direct current through a superconducting coil. To keep the superconducting material at a temperature of roughly −270°C, a cooling system is necessary. To meet this need, liquid helium is employed in a vacuum enclosure. The cooling system uses a lot of energy during this operation, raising the whole system's running costs and decreasing its ability to compete economically with alternative technologies (Dubbert et al. 2014).

34.5.4 Hydrogen Technology

There are three distinct phases. These include the creation of hydrogen, hydrogen storage, and power. These stages can be created in various locations and call for particular tools. While delivering electricity, an electrolyzer separates hydrogen and oxygen from water. As a result, in the second step of the process, oxygen is released into the atmosphere and hydrogen is deposited. There are now four storage technologies available: While hydrogen liquefaction and adsorption on carbon nanofibers are currently being studied, hydrogen pressurization and adsorption in metal hydrides are more advanced processes. In the last step, the fuel cell uses inverse electrolysis to produce power from the reaction of stored hydrogen and oxygen. Water is squandered as a result, with no negative effects on the environment. Two electrode-electrolyte configurations make up fuel cells. There are different types depending on the materials used (Mao, Shen, and Guo 2013).

34.6 Challenges and Outlook for the Future

New fields of contemporary study that are widely categorized as nanoscale science and technology have evolved over the past ten years. Financial and regulatory hurdles brought on by the murky toxicity of nanoparticles may be the main challenge for the majority of nanotechnologies. Nanotechnology is a complicated platform for the existing scientific disciplines of chemistry, physics, biology, medicine, neurology, information technology, and engineering. It is not a distinct field of study in and of itself. Recent years have seen a significant increase in interest in both synthesis techniques and broad applications of nanotechnology in areas including as electronics, energy, and the environment. Nanoscience and technology have made considerable strides, and despite the proliferation of commercial nanomaterial products, they continue to face numerous hurdles, particularly in the fields of energy and health, which are of particular public interest.

Due to their extraordinary ability to control energy flow and alter light at the near-atomic level, nanostructured materials, such as quantum dot solar cells, nanowires, mesoscopic nanostructures, etc., have generated a lot of interest as adaptable components of optoelectronic devices. In fact, the majority of the so-called next-generation solar cells are based on nanostructured materials. They have a lot of promise for fresh ideas or methods to transform solar energy into different forms of energy, such electronic or chemical fuels. Before they can be completely employed in real applications, there are still a number of issues that need to be resolved.

Some characteristics of nanotechnology that are advantageous to solar cells might also bring additional difficulties. Since nanocrystals have a high surface-to-volume ratio, their impact on recombination pathways is more significant. Nanocrystalline grains may prevent photoexcited charges from moving through them, and an increase in surface states may render the charge carriers inactive. Therefore, charge transport is dominated by the interparticle medium in a nanocrystalline solid device where each individual nanocrystal contains the size-dependent features of nanomaterials. To (i) provide colloid stabilization, (ii) ensure simple and steady charge transport between nanocrystals, and (iii) complement the features of the solid nanocrystalline substance, the surface ligands of colloidal nanostructures must be carefully developed.

34.7 Conclusion

Numerous nanostructures have been developed during the past ten years to solve important material and application issues in the fields of energy, environment, and health. Although there are several specific needs for nanomaterials in various applications, structure–activity link between nanomaterials and applications, controllable targeted synthesis is another requirement for better services for constructing the structure–activity link in nanoscience research. In addition, synthetic methods for low-cost mass production of nanomaterials and systematic evaluation of the toxicity and environmental concerns of nanomaterials are crucial.

References

Akmal, Nurul, and Che Lah. 2021. 'Late Transition Metal Nanocomplexes: Applications for Renewable Energy Conversion and Storage'. *Renewable and Sustainable Energy Reviews* 145 (March): 111103.

Al-Hazmi, F. S., Ghada H. Al-Harbi, Gary W. Beall, A. A. Al-Ghamdi, A. Y. Obaid, and Waleed E. Mahmoud. 2015. 'One Pot Synthesis of Graphene Based on Microwave Assisted Solvothermal Technique'. *Synthetic Metals* 200: 54–57.

Bin, Safat, M. A. Hannan, M. S. Reza, Pin Jern, R. A. Begum, M. S. Abd Rahman, and M. Mansor. 2021. 'Battery Storage Systems Integrated Renewable Energy Sources: A Biblio Metric Analysis towards Future Directions'. *Journal of Energy Storage* 35 (November 2020): 102296.

Chen, Chong, Dengfeng Yu, Gongyuan Zhao, Baosheng Du, Wei Tang, Lei Sun, Ye Sun, Flemming Besenbacher, and Miao Yu. 2016. 'Three-Dimensional Scaffolding Framework of Porous Carbon Nanosheets Derived from Plant Wastes for High-Performance Supercapacitors'. *Nano Energy* 27: 377–389.

Chen, Shengrui, Runming Tao, Chi Guo, Wang Zhang, Xiaolang Liu, Guang Yang, Pingmei Guo, Gengzhi Sun, Jiyuan Liang, and Shih-yuan Lu. 2021. 'A New Trick for an Old Technology : Ion Exchange Syntheses of Advanced Energy Storage and Conversion Nanomaterials'. *Energy Storage Materials* 41 (July): 758–790.

Choi, Jonghyun, Taylor Wixson, Adam Worsley, Surendra Dhungana, Sanjay R. Mishra, Felio Perez, and Ram K. Gupta. 2021. 'Pomegranate: An Eco-Friendly Source for Energy Storage Devices'. *Surface and Coatings Technology* 421 (June): 127405.

Conway, B. E., and W. G. Pell. 2003. 'Double-Layer and Pseudocapacitance Types of Electrochemical Capacitors and Their Applications to the Development of Hybrid Devices'. *Journal of Solid State Electrochemistry* 7 (9): 637–644.

Dubbert, Wolfgang, Kathrin Schwirn, Doris Völker, and Petra Apel. 2014. 'Use of Nanomaterials in Energy Storage', no. October 2011: 1–19.

Li, Ke-ke, Guo-yang Liu, Li-si Zheng, Jia Jia, You-yu Zhu, and Ya-ting Zhang. 2021. 'Coal-Derived Carbon Nanomaterials for Sustainable Energy Storage Applications'. *New Carbon Materials* 36 (1): 133–154.

Mao, Samuel S., Shaohua Shen, and Liejin Guo. 2013. 'Progress in Natural Science : Materials International Nanomaterials for Renewable Hydrogen Production, Storage and Utilization'. *Progress in Natural Science: Materials International* 22 (6): 522–534.

Packiyalakshmi, Parameswaran, Bongu Chandrasekhar, and Nallathamby Kalaiselvi. 2019. 'Domestic Food Waste Derived Porous Carbon for Energy Storage Applications'. *Chemistry Select* 4 (27): 8007–8014.

Piao, Yuanzhe. 2016. *Preparation of Porous Graphene-Based Nanomaterials for Electrochemical Energy Storage Devices*, Nano Devices and Circuit Techniques for Low-Energy Applications and Energy Harvesting. Springer Science, 229–252.

Prasanna, Y. S., and Sandip S. Deshmukh. 2020. 'Materials Today: Proceedings Significance of Nanomaterials in Solar Energy Storage Applications'. *Materials Today: Proceedings* 38, 2633–2638.

Ruan, Songju, Dan Luo, Matthew Li, Jitong Wang, Licheng Ling, Aiping Yu, and Zhongwei Chen. 2021. 'Synthesis and Functionalization of 2D Nanomaterials for Application in Lithium-Based Energy Storage Systems'. *Energy Storage Materials* 38 (March): 200–230.

Samuel, Melvin S., Madhumita Ravikumar, Ashwini John, Ethiraj Selvarajan, Himanshu Patel, P. Sharath Chander, J. Soundarya, Srikanth Vuppala, Ramachandran Balaji, and Narendhar Chandrasekar. 2022. 'A Review on Green Synthesis of Nanoparticles and Their Diverse Biomedical and Environmental Applications'. *Catalysts* 12 (5): 459.

Strauss, Volker. 2017. 'Ecofriendly Carbon Nanomaterials for Future Electronic Applications'. *CHEMPR* 2 (3): 319–321.

Szabó, Andrea, Caterina Perri, Anita Csató, Girolamo Giordano, Danilo Vuono, and János B. Nagy. 2010. 'Synthesis Methods of Carbon Nanotubes and Related Materials'. *Materials* 3 (5): 3092–3140.

Thunugunta, Tejaswi, Anand C. Reddy, and D. C. Lakshmana Reddy. 2015. 'Green Synthesis of Nanoparticles: Current Prospectus'. *Nanotechnology Reviews* 4 (4): 303–323.

35

Nanomaterials in Solar Energy Applications

K. Anandan

Academy of Maritime Education and Training (Deemed to be University), Chennai, India

Sarvani Jowhar Khanam

University of Hyderabad, Hyderabad, India

K. Suneeta

Pallavi Engineering College, Hyderabad, India

Kodanda Rama Rao Chebattina

GITAM (Deemed to be University), Visakhapatnam, India

R. G. Padmanabhan

Arasu Engineering College, Kumbakonam, India

Fabian I. Ezema

Africa Centre of Excellence for Sustainable Power and Energy Development (ACE-SPED), University of Nigeria, Nsukka, Nigeria

K. Raju Yadav

Agricultural Engineering College and Research Institute, Tamil Nadu Agricultural University, Coimbatore, India

DOI: 10.1201/9781003355755-35

429

The discussion of recent advancements in flexible solar cells of carbon nanomaterials is mostly centered with a focus on material synthesis and structural design. With the development of flexible solar cells of carbon nanomaterials, enduring problems and future directions are finally resolved. The sun's energy generates electricity, heating, or cooling for the residential, commercial, and industrial sectors at affordable, low- or no-emission costs. Phase-change materials, nanocomposites, and nanofluids have created a brand-new area of study that has been specifically designed to improve solar collector efficiency.

35.1 Introduction

Nanotechnology, a term regularly used to depict materials and peculiarities at a nanoscale, has been generally utilized in different designing and logical applications. A lot of interest has been aimed at the utilization of nanotechnology and nanomaterials in the energy area. Nanotechnology can possibly foster new enterprises that add to feasible monetary development. Additionally, nanotechnology has been utilized in numerous applications expected to give cleaner and more proficient energy supplies and uses. Nanomaterials can assume a significant part in different areas of the energy area, specifically, energy change, energy capacity, and energy-saving frameworks. Ongoing advances in nanotechnology have prompted the improvement of an imaginative class of intensity-move nanofluids made by scattering nanoparticles (10–50 nm) in customary nanofluids. Nanofluids show the possibility to essentially increment heat-move rates in different regions like modern cooling applications, atomic reactors, the transportation business (cars, trucks, and planes), microelectromechanical frameworks, hardware and instrumentation, and biomedical applications (nano-drug conveyance, disease therapeutics, and cryo-preservation). Conceivable superior warm conductivity converts into higher energy effectiveness, better execution, and lower working expenses. The nanofluids both tentatively and hypothetically shows the quick development of nanofluid research as of late (Ahmed et al. 2017).

The energy request is expanded to a great extent because of both populace development and innovative headways. Because of the absence of energy assets, innovative exploration studies and creations are expected to ensure the coherence and productive utilization of these sources while likewise settling financial issues. Giving minimal expense, emanations free energy for family, modern, and enlistment heaters is fundamental. Sunlight-based energy is a limitless, safe, and bountiful power source. There are two principal techniques for changing over sunlight-based energy into power: those that utilize the aggregate electromagnetic nature of light, where light is caught by radio wires and corrected, and those that influence the existing molecule idea of light in ordinary photovoltaic cells. In the two cases, designed nanomaterials are significant. Models incorporate varieties of semiconductor nanostructures as a transitional band (band sunlight-based cells), semiconductor nanocrystals for different exciton ages, or nanomaterials in receiving wire rectifier cells for high optical recurrence amendment. Structures use a surprising measure of energy, comprising up to 45% of all energy consumed around the world. PCM is utilized in structures as to restrict indoor intensity gain, advance temperature balance, and limit heat aggregation. Imaginative liquids incorporate nanofluids. To accomplish wanted warm actual boundaries, cross-breed liquids are made by blending a few nanoparticles to the right extent with customary liquid. Perhaps, the best issue confronting the world is presently freshwater creation. Desalination utilizes a ton of energy despite the fact that it might supply clean water from a source free of the environment. The demonstrated sun-powered refining innovation has inhaled new life through the advancement of photothermal nanomaterials and the pressing need to change to environmentally friendly power energy. Efficiencies are vigorously improvement of photothermal vaporization, which generally overlooks appropriate water assortment for power acquired and interfacial warming. One more serious issue with seawater sun-powered driven vaporization is salt collection. Initially, a concise prologue to photovoltaic gadgets is given, covering color-sharpened sun-based cells, natural sun-oriented cells, and adaptable perovskite sun-powered cells. The requirement for carbon nanomaterials like fullerene, carbon nanotubes, and graphene for photovoltaic applications is then framed.

35.2 Nanomaterials in Solar Cells

The key parts of the latest age of sun-based cells are designed nanomaterials. The point behind the IBSCs, for example, is to embed at least one energy level inside the bandgap so they can ingest photons simultaneously with a solitary bandgap cell's regular activity. Exceptionally confused semiconductor compounds that normally show a middle-of-the-road band and nanostructures, specifically varieties of quantum dab that structure bionanomaterials in sunlight-based cells, have been the principal focuses on the chase after the best sub-bandgap safeguards (Deshmukh et al. 2021).

Nanomaterials made in compound ways present a high chance for effectiveness improvement by expanding light catching and photocarrier assortment without extra expense in sun-oriented cell manufacture. The physical and compound properties change from the mass material to the nanomaterial. For instance, the dissolving point is least for the nanomaterial contrasted with its mass one. This can be because of the great surface-to-volume proportion of iotas in a nanoparticle.

35.3 Nanomaterials as Solar Cell Electrodes

Sun-oriented cell innovations have progressed extraordinarily lately, and their cost has dropped fundamentally. Analyzing the three ages of PV, which are recognized by the harmony between cost decrease and proficiency development, might be the best method for checking mechanical headway. Sun-oriented cells with semiconducting p-n intersections, like a single gem and poly-grain Si, are utilized in original PV. Slender film advances were presented for second-age PV, which diminished costs. Third-age PV currently being used ought to increment effectiveness. The essential activities of thermoplastics are analyzed, going from the material science of intensity move and photothermal transformation to warm impacted processes. The essential activities and instruments of various structure advances, for example, self-tunable plasmons, plasmon coupling systems, and dynamic plasmons with alterable hole distances, are then completely analyzed. The advancement of creative thermoplastic designs utilized for such materials incorporated in solar radiators, thermo-photovoltaic, solar desalination and cleansing, sunlight-based decay, and catalysis (Shamshirgaran, Assadi, and Sharma 2018).

The use of carbon nanotubes, graphite particles, and graphene as added substances toward the superior proficiency of the electrolyte in these sun-based cells is additionally examined. At last, a concise viewpoint is given on the future improvement of carbon nanomaterial composites as imminent materials for DSSCs, especially as parts for printable sun-oriented cells, which are supposed to assume a significant part later on sun-powered cell market.

35.4 Nanomaterials in Perovskite Solar Cells

The photophysical properties of semiconducting filmsre associated with their crystallinity and surface shape. To make profoundly proficient sun-powered cells, perovskite film creation with huge, great precious stone areas is expected to empower photo excited charge detachment and extraction. Since there are a lot of charge traps at grain limits, little grains advance charge recombination in perovskite films. Subsequently, charge extraction is hampered and sun-oriented cells perform more regrettably (Liu and Bashir 2015). Perovskite film improvement occurs in two phases. The initial step is perovskite nucleation on the substrate. Following that, perovskite seed inter diffusion happens during the drying and warm strengthening processes, and the response between the seeds causes the perovskite grain to develop. To expand the pore size, an extensive distance between perovskite precious stone seeds and their high porous limit versatility is fundamental. The greater part of the examination remembered for this study exhibited how influenced perovskite morphology. The presence of two terminal gatherings in a particle's design is a huge component. The other connects artificially with the perovskite film, while one of their capabilities is a mooring bunch (Kim et al. 2020).

Perovskite sunlight-based cells have been viewed as one of the best options in contrast to silicon sun-oriented cells due to their profitable elements like their enormous light retention coefficient, high charge transporter portability, and high change proficiency. Nonetheless, there are a few basic difficulties in working on the photovoltaic execution and steadiness of PSCs, particularly with respect to transformation effectiveness. The perovskite sensitizer is restricted by its bandgap (1.55 eV), which brings about its retention range ascending to 780 nm. In any case, around 52% of the entire sun-oriented energy is in the close infrared (NIR) area ($\lambda > 700$ nm). Thus, the energy loss of close infrared (NIR) light prompted the breaking point for the PCE of PSCs. Then again, the thermalization of charge transporters brought about by retaining high-energy photons from which energy is bigger than the bandgap of the perovskite sensitizer additionally restricts the exhibition of PSCs. Just a single electron-opening pair matching the bandgap of the perovskite sensitizer is produced by engrossing a high-energy photon and the overabundance energy of the great-energy photon is changed into heat, which is destructive to the strength of PSCs. Subsequently, understanding how to decrease thermalization misfortune is a vital component of the elite presentation of PSC gadgets.

35.5 Nanomaterial-Based Phase Change Materials

Among various latent cooling building procedures, PCM innovation has accumulated far-reaching interest in planning prevalent energy-execution green structures. Stage change innovation gives a viable procedure to work on the warm mass of structures, which could diminish temperature varieties, eliminate heat energy from inside and dispose of it to the outside climate bringing about better tenants' warm solace. Studies have shown PCMs could save cooling energy utilization by around 10–30% in various climatic circumstances in the Joined States.

Stage change innovation gives a great deal of potential to cutting-edge mechanical advances in light of nanomaterials. There is a developing interest in consolidating nanomaterial-scaled warm guides into PCM as nanoparticles, nanofibers, nanotubes, nanosheets, and nanofoams. Adding high warm conductivity metallic nanostructures or nanoparticles of Ag, Al, or Cu to the PCM will build its warm conductivity, as well as microencapsulating the PCM or implanting it in a permeable medium like permeable carbon and metallic froths. The warm conductivity and wall elasticity of microcapsules can both be superior by silver nanoparticles. Because of their high conductivity and minimal expense, copper particles are utilized as a compelling PCM-added substance for expanded warm execution. Metals (silver, aluminum, copper), carbon-based nanostructures (carbon nanotubes CNT, graphene nanoflakes, nanoplatelets, and nanofibers), and metallic oxide nanoparticles are explored as warm conductivity advertisers (Mlinar 2013).

The trial business related to the use of nanoparticles was tended to. The instance of cleared tube sun-powered gather frameworks, with an illustrative concentrator, was researched (graphene-based nanofluid was utilized). Graphene was chosen as a nanomaterial because of its high warm conductivity and the focal point of the examination was to dissect the warm effectiveness of the inspected sun-oriented based framework. As indicated by the trial information, the warm productivity of the sun-powered authority can be improved from 31% to 76% when contrasted with base liquid, and for the considered nanoparticle focuses. The execution of nanoparticles in photovoltaic cell applications was explored. The focal point of the work was to investigate the cooling impact on photovoltaics (PV) with the utilization of the compound improvement approach and the use of stage change materials (PCM). The water was upgraded with an Al_2O_3/PCM blend with various nanoparticle fixations. The effect of the occupation proportion in the channels was likewise analyzed. The outcomes uncovered that it was feasible to accomplish execution improvement in the conveyed power yield from the PV board. Appropriate nanoparticle focus was recognized (1 wt%) and connected with the most elevated improvement of force yield from the PV board. The impact of microwaves on the Al_2O_3 nanofluid was inspected. For the instance of pool bubbling. The trial results uncovered an expansion in the intensity move for the Al_2O_3 nanofluid, with the most elevated revealed heat move proficiency of 128%. It was likewise found that the expansion in heat move effectiveness causes a descending pattern for the situation when intensity motion is expanded. The impact of SiO_2 and Al_2O_3 nanoparticle expansion in low-saltiness high temp water was examined in Ref. [9]. The principal objective of the work was to actually take a look at the possibility to further develop weighty oil recuperation. The mathematical reproductions uncovered that it ought to be feasible to build the weighty oil recuperation rate in a scope of 2.4%–7.2% with the utilization of the referenced nanoparticles when contrasted with ordinary water infusion. It was likewise announced in the very concentrate on that it would be feasible to further develop oil recuperation by around 40%. Also, legitimate improvement would permit low energy utilization overall. A novel thermomechanical fluid comprising of Ni-Br-IL fluid (nickel-bromine-ionic fluid) upgraded with VO2 nanoparticles (vanadium dioxide) was proposed i.e., to shape a novel nanocomposite film. The expansion of nanoparticles demonstrated the increment of ultrahigh optical properties in a measure of 27% when contrasted with unadulterated VO2 film (which is around 14.1%). The created film could be a compelling potential thermochromic material for shrewd window applications. Acquired results could empower an effectiveness improvement in superior execution structures. An exploratory examination of mineral ointment enhanced with TiO_2 nanoparticles was talked about. On account of condensed petrol gas refrigerant. The exploratory outcomes uncovered that the expansion of the recently referenced nanoparticles in the mineral oil worked on the productivity of the fridge. The mean drew in power was decreased from 8% to practically 15%, i.e., contingent upon the grouping of the additional TiO_2 nanoparticles.

35.6 Nanofluids as a Working Fluid in Solar Collectors

The discoveries layout that nanofluids are superior to ordinary liquids with regard to the improvement of warm and vigorous productivity. They consistently upheld, in light of the exploration result, that nanofluids push pressure drop and entropy misfortunes yet this is very small contrasted with the addition in heat move rate, warm conductivity, and intensity move coefficient. The utilization of nanofluids' enthusiastic proficiency increments with a little punishment of tension drop. It is a superior method for contrasting the exhibition of sun-based gatherers and furthermore gives the fundamental information to plan gatherers for ideal (Ebrazeh and Sheikholeslami 2020).

Power tower sun-oriented gatherers could profit from the potential productivity upgrades that emerge from utilizing a nanofluid working liquid. A notional plan for this kind of nanofluid recipient is introduced. Utilizing this plan, we show a hypothetical nanofluid improvement in the proficiency of up to 10% when contrasted with surface-based gatherers when sun-powered focus proportions are in the scope of 100–1000. Moreover, our examination shows that graphite nanofluids with volume parts on the request for 0.001% or less are reasonable for 10–100 Mwepower plants. Probes a research center scale nanofluid dish collector proposes that up to a 10% increment in effectiveness is conceivable (comparative with a regular liquid) assuming working circumstances are selected cautiously. Finally, we utilize these discoveries to look at the energy and income produced in a regular sun-oriented warm plant to a nanofluid-based one. It is found that a 100 Mwelimit sunlight-based nuclear energy tower working in a sun-oriented asset like Tucson, AZ, could produce ~$3.5 million more each year by consolidating a nanofluid beneficiary.

35.7 Nanomaterials in Solar Photothermal Collection

As 97% of the World's surface water is in the sea, desalination which transforms seawater into freshwater assumes a basic part in handling the issue of exhausting freshwater supplies. Be that as it may, desalination and water therapy frequently include serious energy utilization, which additionally expands the emanation of ozone-depleting substances. Sun-based-driven refining i.e., fueled by the generally accessible and bountiful sun-oriented energy is a suitable answer for energy-effective and sans CO_2 freshwater creation and thus accommodates the water-energy nexus. Moreover, the licentious advancement of photothermal nanomaterials with proficient sun-based-to-warm transformation reinforces sun oriented helped refining to become one of the most skilled and alluring innovations for freshwater making in the standpoint of manageability. Conventional sun-powered vaporization framework includes volumetric warming with enormous warm mass and huge intensity misfortune. The advances in nanostructured photothermal materials and interfacial evaporator plans understand a profoundly limited photothermal warming at the vanishing surface, which boosts sun-powered use and stifles heat misfortune to the mass water (Gao et al. 2021).

Power tower sun-based gatherers could profit from the potential productivity enhancements that emerge from utilizing a nanofluid working liquid. A notional plan of this sort of Photothermal change is a proficient method for using sun-based energy that permits the change of sun-powered light into nuclear power, consequently empowering MXenes to be applied in different fields, like sun-oriented steam age and biomedical. Notwithstanding, the light-to-warm capacity of MXenes has been given considerably less consideration now. Late advancement in photothermal MXenes is checked on to give a far-reaching comprehension of their photothermal change component and applications. To start with, engineered methodologies of MXenes and their nanocomposites will be momentarily summed up, and the conversation of the photothermal transformation component and, above all, current advances in their photothermal applications will follow. It is exceptionally guessed that 2D MXenes, through the intricate material plan and interdisciplinary methodology, will become one of the standard photothermal materials and their application fields will likewise be extended sooner rather than later.

35.8 Nanomaterials in Solar Photothermal Collection

Adaptable photovoltaic arrangements have drawn much thought due to their propitious applications in movable or wearable gadgets, power-produced materials, building-coordinated photovoltaic frameworks, battery-powered cars, automated in-flight vehicles, and space robots. Also, because of the benefits of overflow, long haul consistency, great photography, high conductivity, and mechanical versatility, carbon nanomaterials have been broadly utilized for understanding the adaptability and superior execution of sunlight-based cells. In this audit, photovoltaic gadgets including color-sharpened sun-oriented cells, natural sun-powered cells, and perovskite sun-based cells, which can be made adaptable, are first presented momentarily. The need for carbon nanomaterials including fullerene, carbon nanotube, and graphene is then summed up for the photovoltaic applications (Khanafer and Vafai 2013).

The notability behind this innovation guarantees a fundamental product profoundly open through the execution of limited warming standards to disintegrate seawater at the water/air interface, rather than inordinate energy-drinking mass water vanishing desalination frameworks. The meaning of any innovation relies upon its part, and photothermal desalination innovation lies in its material. In this survey, the attention on ultramodern nanomembranes, Janus nanoarchitectures, and reticular materials as promising sunlight-based warm nanostructures is featured. This report means to move novel progressions in nanomaterials for photothermal desalination by giving perception into the components of photothermal change and standards of slender activity to build imaginative water pathways to consistently siphon water with the phenomenal salt-dismissing highlight, subsequently improving PTIE execution. With desalination as the concentration, bits of knowledge into fume buildup for the assortment of desalinated water are talked about. Moreover, it provides details regarding creative sun-oriented driven water purging activities embraced by various nations to be in real life soon are consolidated. In the long run, this survey reveals insight into a few functional viewpoints toward the enhancement of reasonable nanomaterials for a maintainable tomorrow.

35.9 Flexible Solar Cells Based on Carbon Nanomaterials

Solar based nuclear power is capable in view of three principal ideas: reasonable intensity, dormant intensity, and thermochemical energy As such, intensity or cold as a sort of nuclear power can be put away by the modification of the inside energy of a material through reasonable intensity, inactive intensity, and thermochemical implies. Sun-based warm innovation delivers the Sun's energy, generally, more than petroleum derivatives, to create a minimal expense, low/zero-discharge warm, i.e., warming or cooling, or electrical energy for use in possessed and attractive areas or industry. For example, the created heat in sun-based authorities during summer can be used in winter, and on the other hand, the virus air common during winter can give agreeable cooling during summer. The capacity medium which can be water, ice, borehole, spring, rock, and water-filled pit, eutectic, stage change material, etc., assumes a significant part in the capacity of nuclear power. A few investigations are embraced to work on the proficiency of capacity materials. A few key variables influence the warm stockpiling of sun-powered energy e.g., capacity temperature, term of capacity, the greatness of energy put away, charging and releasing rates, activity and control, and monetary improvement. By and by, the accessibility of sun-powered energy is vital and should be thought of. Many environments all over the planet are related to everyday, month-to-month, and occasional varieties that might impact sunlight-based radiation (Moore and Wei 2021).

35.10 Solar Thermal Energy Storage Using Nanomaterials

Current innovation in the field of flimsy film testimony might permit the advancement of new items for sunlight-based energy change. The new items are expected to have exceptionally severe properties to fulfill clients, be savvy, rival ordinary sources and consider natural worries. Just in this manner, the field of sunlight-based energy change will enter the traditional energy framework at an extended rate (Prasanna and Deshmukh 2020).

The most customary and conventional intensity stockpiling component is water. Be that as it may, because of low warm conductivity (TC) in the fume express, its applications as an intensity stockpiling medium are restricted. An elective choice is to use natural and inorganic TES materials as the two of them work at low and medium temperature ranges. Natural TES materials, for example, paraffin are non-destructive and have a high dormant intensity limit. Going against the norm, inorganic TES materials have high thicknesses and considerable explicit intensity limits (SHC). Because of quick advancement and headway in nanotechnology, assortments of nanomaterials were scattered in different base fluid(s) to improve thermo-actual properties. This survey paper presents the ongoing status and future advancement patterns of TES materials. Besides, a broad examination of the improvement of TC and SHC of different TES materials doped with nanomaterials has been talked about.

35.11 Conclusion and Future Trends

Sunlight-based energy capacity utilizing stage change materials (PCMs) implanted with nanomaterials, the primary spotlight depended exclusively on the improvement of warm conductivity and explicit intensity of the material to store a lot of energy with high thickness. Likewise, the dissolving portion during the stage change assumes a significant part in assessing the best reasonable nanomaterial to be added for a PCM. From the consequences of various exploration articles, it is seen that the paraffin wax installed with graphene-based nanomaterials retains more energy for capacity because of its higher warm conductivity. There is an extraordinary need of making nanostructured stage change material i.e., solid to working expense and for the most part, centers around transformation and capacity of huge idle intensity in the nuclear power stockpiling framework.

Designed nanomaterials are the key structure blocks of sunlight-based cells, regardless of the components for the transformation of sun-oriented energy to power—those taking advantage of the molecule idea of light and those utilizing the aggregate electromagnetic nature, where light is caught by receiving wires and amended. Understanding how to design nanomaterials for designated sun-oriented cell applications is the way to work on their proficiency and could prompt leap forwards in the plan. The plan of nanomaterial-based sun-oriented cells is a long way from being direct. Conventional methodologies to improve the plan of sun-powered cells, or the plan of sunlight-based cells in light of new components, have depended on the instinct of scientists, zeroing in on experimentation search and unplanned disclosure.

References

Ahmed, Sumair Faisal, M. Khalid, W. Rashmi, A. Chan, and Kaveh Shahbaz. 2017. 'Recent Progress in Solar Thermal Energy Storage Using Nanomaterials'. *Renewable and Sustainable Energy Reviews* 67: 450–460.

Deshmukh, Megha A., Sang Joon Park, Bhavna S. Hedau, and Tae Jun Ha. 2021. 'Recent Progress in Solar Cells Based on Carbon Nanomaterials'. *Solar Energy* 220 (November 2020): 953–990.

Ebrazeh, Sh, and M. Sheikholeslami. 2020. 'Applications of Nanomaterial for Parabolic Trough Collector'. *Powder Technology* 375: 472–492.

Gao, Minmin, Connor Kangnuo Peh, Fan Lu Meng, and Ghim Wei Ho. 2021. 'Photothermal Membrane Distillation toward Solar Water Production'. *Small Methods* 5 (5): 1–17.

Khanafer, Khalil, and Kambiz Vafai. 2013. Applications of Nanomaterials in Solar Energy and Desalination Sectors. In *Advances in Heat Transfer*. 1st ed. Vol. 45. Elsevier. Copyright © 2013 Elsevier Inc. All rights reserved.

Kim, Seo Yeon, Soo Jin Cho, Seo Eun Byeon, Xin He, and Hyo Jae Yoon. 2020. 'Self-Assembled Monolayers as Interface Engineering Nanomaterials in Perovskite Solar Cells'. *Advanced Energy Materials* 10 (44): 1–21.

Liu, Jingbo, and Sajid Bashir. 2015. Sustainable Energy Application: Nanomaterials Applied in Solar Cells. In *Nanomaterials Applied in Solar Cells. Advanced Nanomaterials and Their Applications in Renewable Energy*. Elsevier Inc.

Mlinar, Vladan. 2013. 'Engineered Nanomaterials for Solar Energy Conversion'. *Nanotechnology* 24 (4).

Moore, Katherine, and Wei Wei. 2021. 'Applications of Carbon Nanomaterials in Perovskite Solar Cells for Solar Energy Conversion'. *Nano Materials Science* 3 (3): 276–290.

Prasanna, Y. S., and Sandip S. Deshmukh. 2020. 'Materials Today: Proceedings Significance of Nanomaterials in Solar Energy Storage Applications'. *Materials Today: Proceedings* 38: 2633–2638.

Shamshirgaran, Seyed Reza, Morteza Khalaji Assadi, and Korada Viswanatha Sharma. 2018. 'Application of Nanomaterials in Solar Thermal Energy Storage'. *Heat and Mass Transfer/Waerme- Und Stoffuebertragung* 54(6): 1555–1577.

36

Carbon Nanomaterials for Energy Storage

K. Rajesh

Academy of Maritime Education and Training, Kanathur, India

P. Krishnan

St. Joseph's college of Engineering, Chennai, India

R. Selvam

St. Joseph's College of Engineering, Chennai, India

Shanthala Kollur

RV Institute of Technology and Management, Bengaluru, India

A. Joseph Arockiam

Arasu Engineering College, Kumbakonam, India

Fabian I. Ezema

Africa Centre of Excellence for Sustainable Power and Energy Development (ACE-SPED), University of Nigeria, Nsukka, Nigeria

SreeSvarna BhaskaraMohan

Tamil Nadu Agricultural University, Coimbatore, India

With various environmental issues and a rapidly changing global climate, people have a strong desire for sustainability. In recent decades, a great deal of state-of-the-art research has been carried out to achieve converting and sustainably storing energy. However, alternative materials and advanced manufacturing techniques remain major obstacles to the commercialization of energy conversion and storage systems. The development of nanostructures and device architectures is also important for storing energy in planar form. In today's society, energy conversion and storage are essential to meet the growing energy demand. Energy sustainability, energy conversion, energy storage, environmental protection, and greenhouse gas

DOI: 10.1201/9781003355755-36

reduction all rely on electrochemical and chemical energy storage technologies. Developing innovative materials for energy conversion and energy storage applications is critical to meeting today's energy challenges. Novel materials need to be developed that can improve energy conversion and storage.

36.1 Introduction

Recent developments in several key technologies rely heavily on nanomaterials and nanostructures. The unique physical properties and characteristic geometric dimensions that distinguish nanomaterials from micro- and bulk materials can open up new possibilities for a variety of technological applications. The band gap can be changed simply by changing the size of the material, allowing the optical absorption and emission spectra to be precisely tailored to the needs of the intended application. Electrical energy storage is the main research priority of this era. Energy storage has already completely transformed the electronics industry, but more than that, higher capacity and higher power energy storage would be very useful for other stationary applications such as electric vehicles (Zhong, Basu, and Sun 2021).

Nanostructured materials are advantageous in solar cells, catalyst materials, thermoelectric devices, batteries, supercapacitors, and H_2 storage systems to nanoscale dimensions. This chapter focuses on some of these topics in detail, pointing out that nanostructured materials possess large surface areas that support electrochemical processes or molecular adsorption in solid–liquid or solid–gas mixtures. It increases the efficiency of electrochemical reactions and creates highly crystalline and/or porous structures, allowing electron or ion migration and electrolyte diffusion. It also improves the light absorption of solar cells. The supercapacitor uses ion adsorption or redox processes to store energy. In this reaction, most of the charge is transported to or near the surface of the electrode material (Bai et al. 2020). Batteries are important components of energy storage because they can store electrical energy into chemical energy through an electrochemical reaction and release it again through a reverse reaction. Low cost and access to abundant sodium resources make sodium-ion batteries an attractive option for grid energy storage. Lithium-sulfur (Li-S) batteries have fascinated consideration owing to their high theoretical capacity of 1675 mAh/g, abundant materials, low cost, and environmental friendliness. Owing to their very high theoretic energy density, high safety, cheaper, and long lifetime, zinc-air batteries have the potential to become large-scale energy storage solutions (Singh and Kumar 2020).

36.2 Composite-Based Nanomaterials for Energy Storage

Composites of three-phase structured conducting carbon materials offer interesting prospects for better-quality dielectric properties. Significant surface modification of the cavity allowed the filler to be miscible with the base polymer. It also reduced voids in the composite, reduced leakage current, and significantly increased dielectric loss (Ibrahim et al. 2020). Nanocomposites of insulating polymers and conductive fillers have great potential as future energy generation and storage systems. Good microwave absorbers include conducting polymers and carbon-based compounds such as silicon carbide and graphene (Zou et al. 2020).

Nanocomposites rapidly synthesize nanostructured materials by absorbing microwaves and converting them into heat. The addition of 10% vanadium or alumina improved the dielectric loss properties of the composites. The microwave operated at 2.45 GHz and reached a temperature of 1750°C after 2 minutes. The formation of the silica-vanadate glass phase or amorphous alumina-silica phase could be responsible for the temperature reached. Using the resulting combination as a heating element led to the rapid generation of oxide nanostructures (Hussain et al. 2020). Nanostructure-based composites contain improved UV shields and high electrical conductivity. A wearable technology was developed that stores energy in your pocket at a textile factory. Nanocomposites can be prepared by simply adding solvothermal-generated ZnO to existing composites. By incorporating nanostructured materials, cotton cloth (an insulator) can be easily turned into a conductive cloth (Guo et al. 2019).

36.3 Silicon-Based Nanomaterials for Energy Storage

Owing to its increased theoretical specific capacity and more elements, silicon is a feasible anode for high-performance lithium-ion batteries (LIBs). Owing to their low electrical conductivity and low volume change during lithiation cycling, Si-based anodes currently have few practical applications. Silicon is a favorable anode material for rechargeable LIBs due to its abundance, increased theoretical capacity of 4200 mmH/g, and low operating potential of less than 0.5 V which is one-tenth that of graphite. Si-based anodes, which have inherently poor electronic conductivity, undergo massive volume expansions of over 300% during the lithiation/delithiation process of the cell, fragmenting the silicon particles and dissolving the electrode structure (Barua et al. 2018).

As a result, the thick and unstable solid–electrolyte interface (SEI) layer caused rapid capacity loss during cycling performance. In high-energy LIB applications, Si has been post-treated with various types of coatings and structural designs to reduce silicon expansion. Silicon incorporated in carbon nanostructured materials is widely used in energy storage systems. The reasons for this are the large surface area-to-volume ratio, good transport properties, varying physical properties, and nanoscale dimensions (Pomerantseva et al. 2019).

These nanostructured materials have large surface areas that facilitate electrochemical reactions or molecular adsorption at solid–liquid or solid–gas interfaces. They are also highly crystalline, produce optical effects that improve light absorption in solar cells, and/or have porous structures that are advantageous for these applications. It increases the efficiency of electrochemical processes by facilitating electron or ion passage and electrolyte diffusion. Silicon-based nanomaterials and related polymer–nanocomposite polymers are attracted to layered silicates because of the high surface energy that aids the migration and expansion of the nanomaterials. We tested the energy harvesting efficiency of mesoporous silica, organosilicon-based chromophores, melt-blended polyamide 11/Kroisite 20A NCs, laponite nanoclay/poly (vinylidene fluoride) NCs, and conjugated polymer mesoporous silica (Gong et al. 2020).

36.4 Protein/Peptide-Based Nanomaterials for Energy Application

New charge transport materials, often fabricated using toxic organic solvents that are incompatible with the biological environment, continue to drive the development of organic electronics. The burgeoning field of organic bioelectronics paved other alternative materials to transport electrons by protein combined with carbon nanomaterials. Field-effect transistors and piezoelectric devices are just two of the examples of important applications made possible by these intriguing bio-inspired structures. In addition to the artificial self-assembly of oligopeptides and peptides, natural protein biopolymers also have the scope for electrical applications.

The peptides and proteins in carbon nanomaterials are novel techniques among a variety of biocompatible and biodegradable building blocks because they have great potential to achieve desired energy storage properties. Self-assembling proteins exhibit an extensive variety of normal structural multiplicity, ranging from nanoscale collagen protein formation to macroscale cytoskeleton formation. Other important properties of peptide biomaterials include abundance, self-repair ability, non-toxicity, and antimicrobic ability. These structured biomaterials have been transformed into energetic materials. Field-effect transistors are used in energy storage materials. Recently, the emerging field of bioelectronics has developed biocompatible nanomaterials for electronic components. The properties viz. electron, hole, and proton transport are increased in biomaterials (Panda, Katz, and Tovar 2018).

Polypeptides are electrically polarizable with properties that helps to control the charge density of near materials, non-volatile electric fields as electromechanical energy, and the charge transport subunits in current conducting devices. The diverse functions of biomaterials allow their natural properties to be tuned to enhance their use in artificial applications. Solid-state protein conduction studies typically use one of two strategies. One is to creating electrical contacts over single molecules using molecular array-based techniques or scanning microscopy or conducting probe atomic force microscopy. Another strategy is to create protein monolayers. Self-assembled monolayers on gold substrates using protein cysteine groups as anchor points and conducts a photocurrent of 10 μA. Biocompatible optoelectronic switches and photoconductivity of protein samples can also be produced. The protein was tagged with histidine (6xHN), which increased film self-assembly and ultimately increased conductivity (Lee et al. 2013). Due to their incompatibility with biological environments, inorganic piezos and ferroelectrics are not suitable for practical bioelectrical applications. Ferroelectric materials need enduring dipoles along each axis of the particle and are realigned by the external electric field applications, causing hysteresis loops in the electrical polarization, whereas piezoelectric materials typically exhibit a hysteresis loop under mechanical stress and indicate the electric polarization. Though ferroelectric and piezoelectric materials need non-centrosymmetric molecules for polarization.

36.5 Carbon-Based Nanomaterials for Energy Storage

Three types of carbon have been known since ancient times: amorphous carbon, graphite, and diamond. It has different properties depending on how the carbon atoms are organized. A common and soft form of carbon is graphite, which has weak lateral van der Waals interactions between layers but strong in-plane covalent bonds. Each carbon atom in a diamond is bonded with the remaining carbon atoms in an arranged lattice, making the material hard and transparent. The carbon hexagons are arranged concentrically to form elongated nanotubes that are often end-capped with pentagonal structures resembling fullerenes. The newest member of the carbon family is graphene (Chen et al. 2016).

Graphene is a potential alternative material with several probable applications, such as energy conversion and storage. It serves as a building block material for carbon nanotubes (CNTs) and other carbon nanomaterials. The controlled development and functionalization of nanotubes have unlocked novel perspectives in energy study, as C60 is extensively used as an electron acceptor in solar cells. Similar to corn stalks, carbon nanotubes can be packed very densely when supported vertically by electrodes. The exposed "top" of these cylindrical nanotubes provides the same amount of usable surface area as regular unaligned nanotubes. The tube spacing creates additional surface area, including the distinct "sides" of the cylinder. This innovative structure can increase the surface area of solar cells by a factor of 1000. This increased surface area and special orientation improve solar absorption and charge collection, resulting in significantly improved efficiency and performance. By layering photosensitive dyes and titanium dioxide onto individual nanotubes arranged invertical arrays, it is possible to fabricate advanced polymer and dye solar cells. Fullerenes and other charge transport materials have replaced CNTs and graphene in polymer solar cells due to their superior electron mobility and low percolation threshold, but their use in organic solar cells is currently under investigation. Research on this is just beginning. The field of photovoltaic technology has the potential to thrive if continued research efforts are made (Dai et al. 2012).

Nanotechnology will have a major impact on energy conversion technologies other than solar cells. Fuel cells are one of the most actual and eco-friendly skills to change the need for substitute energy sources. Instead of burning coal to generate heat, it converts chemical energy directly into electricity. The oxygen reduction reaction that occurs in fuel cells requires a catalyst to break the bonds between oxygen molecules. Owing to the situation of high catalytic efficiency in fuel cells, platinum is measured as a pioneering catalyst in this field. However, it is expensive and its properties deteriorate over time. This poses a challenge to the mass production of industrial fuel cells. The main hurdles for fuel cells are still lowered valuable metal loading, improved catalyst activity, and longer catalyst life. Graphene and nitrogen-doped CNTs have emerged as promising materials to solve these problems. CNTs and graphene have the highest value in surface area, mesoporosity, electrical conductivity, mechanical strength, and corrosion resistance than traditional carbon black supports. These excellent properties support CNT and graphene catalysts with high activity and longevity. Compared with random CNTs, they have better three-phase boundaries, higher charge transfer ability, higher catalytic activity, and better catalyst dispersion (Wang et al. 2020).

Nitrogen doping and catalyst binding to the CNT surface compared with bare CNTs and graphene thereby improves catalytic activity and catalyst lifetime. Carbon nanomaterials with or without nitrogen doping are formed by intra- and intermolecular charges. It has been identified as an alternative to platinum electrochemical catalysts and may work well as a metal-free catalyst.

Furthermore, these novel catalysts were shown to be superior to platinum-based electrodes in terms of mechanical stability and electrochemical robustness. This catalyst technology has important implications for the fuel cell industry and the power sector to produce effective catalysts without metal ions for reducing oxygen in the fuel cell at a reasonable cost and to develop new catalyst materials for use in applications. It suggests that there is still plenty of room. Other than fuel cells, super capacitors have long been sought for higher energy and power densities. The unique properties of CNTs make them suitable for supercapacitors. The higher value in the properties such as electrical conductivity, charge transport capacity, mesoporosity, and electrolyte accessibility are achieved CNTs are superior to conventional activated carbon electrodes in terms of capacitance and rate capability due to their well-balanced surface area and mesoporosity. Compared with the casual counterpart, VA-CNT has a larger electrolytically accessible surface area due to well-defined tube orientation, tube spacing, and end cap holes. This causes capacitive behavior. It will be improved. CNT-based supercapacitors have been shown to have superior power densities over existing super capacitor technologies (Yi et al. 2020).

CNT can be combined with activated carbon, nanosheets, polymers, or metal oxides with a higher surface area to create higher-capacity composite anodes/cathodes or combined with electrolytes which is anionic liquid to achieve high capacity and energy density in CNT supercapacitors. Meanwhile, carbon nanomaterials are being researched for energy storage applications to reduce destructive emissions and dependence on fossil fuels. The most advanced battery technology today is the lithium-ion battery, but there is still room for improvement in the performance of modern lithium-ion batteries. To solve these problems, research on carbon and its derivatives as new electrode materials or electrode additives is needed. As CNTs are considered novel electrode materials, they are also used in lithium-ion battery designs to advance rate performance, safety, and cyclability in combination with traditional graphite anodes. CNTs are excellent materials for electrode materials in energy storage systems (Roselin et al. 2019).

The utilization of carbon nanofibers for the improvement of energy storage devices, including both supercapacitors and lithium-ion batteries, is explored. Following a short conversation about the manufacturing cycle and portrayal techniques for ultrafine electro-turned carbon filaments, late advances in their presentation as supercapacitors and lithium-ion batteries anode materials are summed up. Further difficulties will be in sustaining the conductivity, surface region, and mechanical properties of the carbon nanofiber lattice, as well as the scale-up capacity of the production method Enhancement of large electrochemical properties of these materials through the decision of warm treatment conditions, the fuse of extra dynamic parts, and the age of novel stringy designs are featured. Further difficulties will be in sustaining the conductivity, conductivity, surface region, and mechanical properties of the carbon nanofiber lattice, as well as the scale-up capacity of the production method (Jiang, Sheng, and Fan 2017).

Carbon materials are an amazing contender for energy-related applications, from batteries and supercapacitors to energy components and electrocatalysis. A progression of carbon nanostructures is through adaptable amalgamation procedures, to empower the take-up of these materials in applications. A progression of carbon nanostructures is through adaptable amalgamation procedures, to empower the take-up of these materials in applications. Carbon is a straightforward, stable, and famous component with numerous allotropes. The carbon relatives incorporate carbon specks, carbon nanotubes, carbon filaments, graphene, graphite, diagramming, hard carbon, and so forth. They can be partitioned into various aspects, and their designs can be open and permeable. Besides, it is exceptionally fascinating to dope them with different components or hybridize them with different materials to shape composites. The essential and underlying qualities offer us to investigate their applications in energy, climate, bioscience, medication, hardware, and others. Among them, energy capacity and change are very alluring, as advances in this space might further develop our life quality and climate. Some energy gadgets will be incorporated into this, for example, lithium-ion batteries, lithium-sulfur batteries, sodium-ion batteries, potassium-ion batteries, double-ion batteries, electrochemical capacitors, and others. Furthermore, carbon-based electrocatalysts are additionally concentrated on hydrogen advancement responses and carbon dioxide decrease responses. Nonetheless, there are as yet many difficulties in the plan and readiness of cathode and electrocatalytic materials (Liu et al. 2010).

Carbon-based materials have been generally utilized as energy stockpiling materials due to their enormous explicit surface region, high electrical conductivity, as well as magnificent warm and compound stabilities. The conventional manufactured strategies, such as pyrolysis of natural atoms or biomass materials, fume stage disintegration techniques, high-temperature solvothermal and aqueous strategies have experienced specific limits of morphology, explicit surface region, and size controls. It also prevents the swindling of their electrochemical presentation and furthermore the investigation of their response mechanisms. To overcome the above limits, the metal-organic-derived carbon materials have become research areas of interest. Metal-organic compound is another kind of permeable organic–inorganic crossover material with metal particles and natural ligands. Metal-organic-derived carbon materials claim the special benefit of having the option to control their creation and design by choosing reasonable natural ligands and metals. Subsequently, these materials are described by rich arrangement and construction, various pore channels, and high porosity and antecedents for the preparation of permeable carbon materials. These materials hold the pore design and huge explicit surface area and have solid electrical conductivity and high stability. Furthermore, the morphology, explicit surface region, and molecule size of these carbon materials can likewise be tuned through planned manufactured control, making them as a serious kind of carbon materials, particularly for energy applications. In this manner, determined permeable carbon materials normally show an unrivaled presentation in many fields like energy stockpiling gadgets, and oxygen decrease responses prompt fast advancement and improvement of combination and applications. The different sorts of carbon materials have a lower thickness, more uncovered dynamic destinations, and simpler to be completely reached with the response medium. On the whole, the as-arranged empty carbon materials have rich underlying variety and controllable morphology, pore channels, and cavity size, which shows extraordinary benefits, particularly in the fields of energy stockpiling applications Moreover, the inward depression of the empty carbon materials unblock the dispersion pathway yet additionally give a support space to the volume development of the material. On the whole, the as-arranged empty carbon materials have rich underlying variety and controllable morphology, pore channels, and cavity size, which shows extraordinary benefits, particularly in the fields of energy stockpiling applications (Ren et al. 2020).

Energy storage devices using carbon nanomaterials are viewed as one of humankind's generally practical, bountiful, boundless, and clean energy assets, fit for meeting our advanced society's overall energy requests in a harmless way to the ecosystem. Sun-oriented cells have extraordinary potential as energy transformation gadgets; however, their utilization is restricted because of discontinuous daylight. Self-fueling coordinated sun-based cells and electrical energy stockpiling gadgets can be an answer to this issue. For this reason, sun-based energy change and electrical energy stockpiling should be incorporated into one framework. Joining sun-oriented cells with an electrical-energy-capacity unit takes into consideration sun-based energy capacity as well as decreases the fluctuation of sun-powered light as a result of power source. Somewhat recently, the improvement of green carbon materials or biomaterials, created from biomass, has drawn exceptional consideration. The fundamental benefits of those biomass materials are their overflow, minimal expense, and waste alleviation. Lignocellulose is a bountiful and inexhaustible biomass material with rich surface utilitarian gatherings which shows amazing potential for creating electroactive carbon nanomaterials. Interest in lignocellulosic materials has expanded in the line-up with research in terminal materials (Wu et al. 2020).

Normally plentiful and a green elective carbon source, they have as of late shown broad applications in electrochemical applications. Carbon materials got from lignocellulosic antecedents display phenomenal conductivity and rich porosity making them likely contenders for electrochemical energy stockpiling. These materials have been utilized as folios, anodes, and electrolytes for energy-capacity gadgets. In this exploration, the electrochemical exhibition of biomass-based anodes will be examined through trial draws near. Minimal expense combination cycles of such materials will be created. Electrochemical portrayal incorporates current-likely assessment and administration life tests for disappointment investigation. One of the vital uses of this examination lies in energy capacity. Proficient energy stockpiling frameworks are a fundamental prerequisite today. Batteries, as well as electrochemical capacitors, are the fundamental advancements as of now being used. High requests in energy-capacity gadgets require minimal expense manufacture and are harmless to the ecosystem materials. The ongoing energy stockpiling advancements are either excessively costly or destructive to the climate. This study should give a minimal expense answer for superior execution materials (Zhu et al. 2014).

Recently, various examinations on carbon-based anodes have been broadly detailed. The expected use of carbon-based anode materials in batteries is basically determined by two key boundaries, energy thickness, and power thickness Carbon-based materials are used as graphitic carbon anodes, leaning toward high energy thickness, and nebulous carbon anodes with huge pseudo capacitance, leaning toward high power thickness. . Graphitic carbon has a generally low and stable release level, which has prompted the improvement of high-energy-thickness anodes. Carbon has an irregular arrangement of the graphene layer and can oblige more volume variety contrasted with graphitic carbon. It can likewise offer little measured pores in its construction for putting away over-the-top potassium particles, and besides, the permeable design could likewise work with particle transportation to upgrade the rated capacity. Its lower thickness and moderately enormous surface-explicit region, in any case, would consume electrolytes to frame its strong electrolyte interphase layer, which would bring down the Coulombic effectiveness (Mao, Hatton, and Rutledge 2013).

36.6 Conclusions

The carbon materials with high energy capacity have persuaded us to continue the work with carbon materials in energy storage rather than adopting novel nanomaterials.

Further research in this carbon nanomaterials area will result in the appearance of novel, affordable, and strong materials that are practical for producing/harvesting, converting, and storing energy. The creation of novel and advanced materials from nanostructures and nanocomposites will give the energy sector a huge boost. Nanostructured materials have the potential to meet the requirements of the existing materials that have been in use for decades. The energy and power densities of energy storage devices are expected to increase light absorption, reduce charge recombination and other energy losses associated with electron transport in solar cells and improve light absorption based on novel nanostructures. The latest technological developments in nanostructured materials enable the creation of new materials.

References

Bai, Xiaoxia, Zhe Wang, Jingying Luo, Weiwei Wu, Yanping Liang, Xin Tong, and Zhenhuan Zhao. 2020. 'Hierarchical Porous Carbon with Interconnected Ordered Pores from Biowaste for High-Performance Supercapacitor Electrodes'. *Nanoscale Research Letters* 15 (1): 1–10.

Barua, Shaswat, Satyabrat Gogoi, Raju Khan, and Niranjan Karak. 2018. *Silicon-Based Nanomaterials and Their Polymer Nanocomposites. Nanomaterials and Polymer Nanocomposites: Raw Materials to Applications.* Elsevier Inc.

Chen, Chong, Dengfeng Yu, Gongyuan Zhao, Baosheng Du, Wei Tang, Lei Sun, Ye Sun, Flemming Besenbacher, and Miao Yu. 2016. 'Three-Dimensional Scaffolding Framework of Porous Carbon Nanosheets Derived from Plant Wastes for High-Performance Supercapacitors'. *Nano Energy* 27: 377–389.

Dai, Liming, Dong Wook Chang, Jong Beom Baek, and Wen Lu. 2012. 'Carbon Nanomaterials for Advanced Energy Conversion and Storage'. *Small* 8 (8): 1130–1166.

Gong, Shengqin, Xiaomin Cheng, Yuanyuan Li, Xiuli Wang, Yanping Wang, and Hao Zhong. 2020. 'Effect of Nano-SiC on Thermal Properties of Expanded Graphite/1-Octadecanol Composite Materials for Thermal Energy Storage'. *Powder Technology* 367: 32–39.

Guo, Mengfan, Jianyong Jiang, Zhonghui Shen, Yuanhua Lin, Ce Wen Nan, and Yang Shen. 2019. 'High-Energy-Density Ferroelectric Polymer Nanocomposites for Capacitive Energy Storage: Enhanced Breakdown Strength and Improved Discharge Efficiency'. *Materials Today* 29: 49–67.

Hussain, Hafiza Vaneeza, Mateeb Ahmad, Muhammad Tamoor Ansar, Ghulam M. Mustafa, Saira Ishaq, Shahzad Naseem, Ghulam Murtaza, Farah Kanwal, and Shahid Atiq. 2020. 'Polymer Based Nickel Ferrite as Dielectric Composite for Energy Storage Applications'. *Synthetic Metals* 268 (May): 1–9.

Ibrahim, Idowu D., Emmanuel R. Sadiku, Tamba Jamiru, Yskandar Hamam, Yasser Alayli, and Azunna A. Eze. 2020. 'Prospects of Nanostructured Composite Materials for Energy Harvesting and Storage'. *Journal of King Saud University-Science* 32 (1): 758–764.

Jiang, Lili, Lizhi Sheng, and Zhuangjun Fan. 2017. '% LRPDVV GHULYHG FDUERQ PDWHULDOV ZLWK VWUXFWXUDO GLYHUVLWLHV DQG WKHLU DSSOLFDWLRQV LQ HQHUJ \ VWRUDJH' 5 (December).

Lee, Jung Ho, Jae Hun Lee, Yun Jung Lee, and Ki Tae Nam. 2013. 'Protein/Peptide Based Nanomaterials for Energy Application'. *Current Opinion in Biotechnology* 24 (4): 599–605.

Liu, Chang, Feng Li, Ma Lai-Peng, and Hui Mmg Cheng. 2010. 'Advanced Materials for Energy Storage'. *Advanced Materials* 22 (8): 28–62.

Mao, Xianwen, T. Hatton, and Gregory Rutledge. 2013. 'A Review of Electrospun Carbon Fibers as Electrode Materials for Energy Storage'. *Current Organic Chemistry* 17 (13): 1390–1401.

Panda, Sayak Subhra, Howard E. Katz, and John D. Tovar. 2018. 'Solid-State Electrical Applications of Protein and Peptide Based Nanomaterials'. *Chemical Society Reviews* 47 (10): 3640–3658.

Pomerantseva, Ekaterina, Francesco Bonaccorso, Xinliang Feng, Yi Cui, and Yury Gogotsi. 2019. 'Energy Storage: The Future Enabled by Nanomaterials'. *Science* 366 (6468): 1–12.

Ren, Jincan, Yalan Huang, He Zhu, Binghao Zhang, Hekang Zhu, Shenghui Shen, Guoqiang Tan, et al. 2020. 'Recent Progress on MOF-Derived Carbon Materials for Energy Storage'. *Carbon Energy* 2 (2): 176–202.

Roselin, L. Selva, Ruey Shin Juang, Chien Te Hsieh, Suresh Sagadevan, Ahmad Umar, Rosilda Selvin, and Hosameldin H. Hegazy. 2019. 'Recent Advances and Perspectives of Carbon-Based Nanostructures as Anode Materials for Li-Ion Batteries'. *Materials* 12 (8): 1229.

Singh, Rupinder, and Ranvijay Kumar. 2020. *Energy Storage Device From Polymeric Waste Based Nano-Composite by 3D Printing*. Encyclopedia of Renewable and Sustainable Materials. Elsevier Ltd.

Wang, Huilin, Xitong Liang, Jiutian Wang, Shengjian Jiao, and Dongfeng Xue. 2020. 'Multifunctional Inorganic Nanomaterials for Energy Applications'. *Nanoscale* 12 (1): 14–42.

Wu, Mingguang, Jiaqin Liao, Lingxiao Yu, Ruitao Lv, Peng Li, Wenping Sun, Rou Tan, et al. 2020. '2020 Roadmap on Carbon Materials for Energy Storage and Conversion'. *Chemistry-An Asian Journal* 15 (7): 995–1013.

Yi, Ting Feng, Jing Jing Pan, Ting Ting Wei, Yanwei Li, and Guozhong Cao. 2020. 'NiCo2S4-Based Nanocomposites for Energy Storage in Supercapacitors and Batteries'. *Nano Today* 33: 100894.

Zhong, Yu Lin, Soumendra N. Basu, and Ziqi Sun. 2021. 'Nanomaterials and Composites for Energy Conversion and Storage'. *Jom* 73 (9): 2752–2753.

Zhu, Jixin, Dan Yang, Zongyou Yin, Qingyu Yan, and Hua Zhang. 2014. 'Graphene and Graphene-Based Materials for Energy Storage Applications', *Small* 10(17): 3480–3498.

Zou, Kailun, Zhenhao Fan, Chaohui He, Yinmei Yuanwei Lu, Haitao Huang, Qingfeng Zhang, Yunbin He, et al. 2020. 'Functional Nanocomposites for Energy Storage: Chemistry and New Horizons'. *Inorganic Chemistry Communications* 3 (September): 110590.

37

Green Energy Storage Devices Using Nanocellulose

K. Gayathri

Academy of Maritime Education and Training, Kanathur, India

M. Muralidhar Singh

RV Institute of Technology and Management, Bengaluru, India

E. Rajkumar

Agricultural Engineering College and Research Institute, Tamil Nadu Agricultural University, Coimbatore, India

S. Vijayaraj

Vels Institute of Science, Technology & Advanced Studies, Chennai, India

Ashish Kumar Srivastava

Galgotias University, Greater Noida, India

Dr. V. Saravanan

Sri Meenakshi Vidiyal Arts and Science College, Trichy, India

The creation of pure chemicals and functional materials from natural sources is of great public interest because using chemicals and products derived from petroleum-based resources causes environmental and ecological problems. Nanocellulose has the potential to become one of the most hopeful green materials of the modern era because of its qualities, renewability, and abundance. In recent years, nanocellulose and its derivatives have arisen as some of the most alluring spare components utilized in energy storage systems. To extract cellulose for commercial applications, numerous plants, algae, and other

DOI: 10.1201/9781003355755-37

445

animals can be utilized. This is the most predominant natural biopolymer on the planet. Furthermore, the easily oxidizable carbon compounds made from nanocellulose have excellent electrical conductivity and are frequently used as electrode materials or current collectors in sustainable energy storage. Owing to the cheaper, ease of production, nanocellulose and its derivatives, will compete with the next generation of green electronics.

37.1 Introduction

Nanocellulose is not only a typical natural polymer but also exists in nanoform. There are several different types of nanocellulose, known as nanocellulose, which, like other nano polymers, is a nanoscale cellulosic material. The two most common types are cellulose nanocrystals and cellulose nanofibers, both of which are readily available in the market. Not only is nanocellulose transparent, conductive, and has high tensile strength, but the surface is easily functionalized, making it highly adaptable by making nanocellulose a viable choice for a variety of applications (Chen et al. 2018). In addition, its biological properties, product richness, and ease of mass production make it a cheaper and greener alternative to many synthetic polymers. Commercial products of nanocellulose are available. It is used in a variety of applications, most commonly in food packaging, where it is used as a biological and ecological alternative to traditional packaging materials. Its applicability to energy storage applications is currently a new topic (Chen and Hu 2018).

Modern energy storage devices show tremendous promise for managing practical electronic devices and electric cars. Reducing overall costs through the use of maintainable technologies and eco-friendly materials is one of the essential problems for the development of cutting-edge energy capacity devices. As a result, conventional and environmentally friendly cellulose-based materials have attracted extensive consideration from experts in the field of electrochemical energy storage (Guo et al. 2020).

Massive efforts are currently being made to develop electrochemical energy storage devices in light of green economic resources due to the considerable challenges in energy improvement and environmental issues. The possibilities of utilization of nanocellulose in energy storage are in supercapacitors, electrodes, batteries, and solar cells. Some of the uses and applications are detailed in this chapter.

37.2 Nanocellulose for Supercapacitors

Supercapacitors or ultracapacitors or electrochemical capacitors improve performance due to their high-power density, excellent speed, fast charge/discharge, long life, simple operating principle, fast charge propagation dynamics, and low maintenance costs. Nanocellulose is a promising solution for electric capacitors, to energy storage by the charge accumulated at the electrode/electrolyte interface. Nanocellulose created from waste biomass is a superb alternative while paired with disparate nanoconductive substances along with carbon, conductive polymers, and metallic oxides as lively substances or nanocomposites for supercapacitors.

Pseudocapacitors, which rely on fast redox reactions at the electrodes to obtain high pseudocapacitances, fall into the latter category. Typical components of supercapacitors are castings, current collectors, electrodes (anode and cathode), and separators. Above all, the electrode material has a great influence on the performance of supercapacitors. Even though nanocellulose is not electrically conductive, using this biopolymer material to fabricate nanocellulose-based electrode materials has several advantages. First, depending on the cellulose source, nanocellulose can have different chemical structures and surface qualities that other materials containing hydroxyl groups can accommodate reactive surfaces. Moreover, nanocellulose is a potential substrate for loading other conductive materials due to its increased surface area, outstanding mechanical properties, and outstanding thermal stability. Moreover, carbonized or activated forms of nanocellulose can be directly used as conductive and capacitive electrodes due to its one-dimensional fibrous structure and high electrical conductivity. These supercapacitors have increased cycle stability, energy density, and power density (Durairaj et al. 2022).

With this development, supercapacitors can fill the gap amid batteries and fuel cells and traditional capacitors. A huge specific surface area is advantageous for electrochemical double-layer supercapacitors because it provides more adsorption space for electrolyte ions. Nanocellulose can be used to create extremely porous materials with increased mechanical strength and surface areas. These porous structures have great potential as flexible supercapacitor substrates and can be doped with carbon-conducting resources viz. carbon nanotubes and graphene oxide for enhancing efficiency (Vilela et al. 2019). Cellulose materials are environmentally friendly, but the chemical processes used to manufacture them pollute the environment. Therefore, further efforts are needed to minimize the impact of chemical treatments in constructing ecological energy storage systems. Using a combination of papermaking and flash reduction processes, Koga et al. successfully combined recycled pulp waste fibers with single-layer graphene oxide sheets to create a cellulose paper-reduced graphene oxide composite. After exposure to intense pulsed light, the composites underwent a millisecond reduction of graphene oxide at ambient temperature without the need for additional additives (Jose et al. 2019).

37.3 Carbon Materials Derived from Nanocellulose

Most of the carbonaceous materials used for energy storage come from precursors that are fossil fuels. As a sustainable precursor for carbonaceous materials, nanocellulose has recently attracted a great deal of interest. Nanocellulose is pyrolyzed at high temperatures in an inert atmosphere to form conductive carbon compounds. Porous, heteroatom-doped, and carbon composites are different types of carbon produced from nanocellulose (Osong, Norgren, and Engstrand 2016). Carbon nanofibers and carbon from biomass are often used as building blocks for fabricating porous carbon aerogels. The specific charge-matching surface area can be effectively improved by activating the carbon aerogel with catalysts namely potassium hydroxide or potassium citrate. In the carbonization process, macropores can be generated by a cross-linking reaction of biomass carbon. Figure 37.1 depicts the

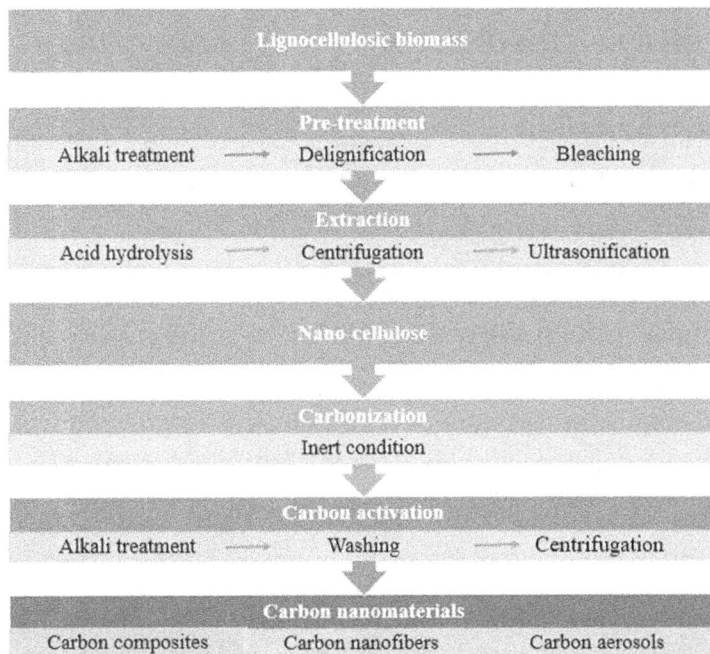

FIGURE 37.1 Carbon nanomaterial synthesis from nanocellulose derived from biomass.

process to synthesize carbon nanomaterials from nanocellulose derived from biomass. The resulting carbon nanomaterials from nanocellulose featured a honeycomb, three-dimensional, interconnected, hierarchical network structure. In the growth of anode materials for nickel metal hydride batteries, nanocellulose-derived carbon materials show rewards such as chemical and thermal stability, electrical conductivity, and sodium storage capacity. Pyrolysis is a method that can be used to produce carbon nanofibers directly. The characteristic sound structure of carbon nanofibers enabled extremely reversible ion transfer and large contact space amid electrodes and electrolytes (Nguyen Dang and Seppälä 2015).

37.4 Nanocellulose for Batteries

At present, lithium-ion batteries rule the versatile battery-powered gadgets market. Lithium-ion batteries have gone through significant review to further develop them on account of their high energy thickness and long life expectancy. Embedding/separating lithium ion between the cathode and the anode permits lithium-particle batteries to be charged and released. Lithium ions might diffuse (relocate) from the cathode to the anode affected by outside energy during the charging system. Lithium ions are taken out from the anode during release, where they return to the cathode and convey the current. Despite the fact that lithium-particle battery terminals' energy thickness and limit have expanded essentially as of late, most of commercial lithium-ion battery cathodes actually have various disadvantages, including significant expense, non-inexhaustibility, disappointing mechanical characteristics, and charging and discharging. The use of lithium-ion batteries in applications like battery electric vehicles is seriously compelled by these issues. For the making of terminals for battery-powered lithium-ion batteries that empower further developed energy thickness, very high charge and release rates, and better cycling, the utilization of nanostructured materials opens up additional opportunities. The unmistakable construction of nanocellulose, a pristine naturally harmless nanomaterial, goes about as a platform to charge conductive components, making stable composite anode materials effectively (Szabó et al. 2010).

Coordination of the part materials is essential for an exhaustive energy storage framework to operate well. Terminal materials, electrolytes, and separators make up the heft of a typical optional battery's major parts. To increase gadget explicit capacitance, rate execution, coulombic productivity, and cycle life, these utilitarian materials especially terminal materials are pivotal. Nanocellulose can be used as an option in contrast to conventional materials in the production of cathodes for different electrochemical energy stockpiling gadgets due to its exceptional highlights. One ongoing subject of enormous interest is adaptable lithium-ion batteries. As a mechanical help, or recently, cellulose has been used to create permeable nanostructures in adaptable batteries.

Lithium-ion batteries have attracted a lot of attention owing to their higher values of operating voltage, energy density, and lifetime and lower values of memory effect, and self-discharge. Nanocellulose is a probable component of hard composite electrolytes, porous separator membranes, electrode substrates, and carbon material precursors as a novelfactual approach to support the sustainable growth of lithium-ion batteries. Conventional lithium-ion battery electrodes consist of heavy metal current collectors, plus active materials, conductive additives, and polymer binders (Santos et al. 2020). The representative polymer binder most commonly used for lithium-ion battery electrodes is polyvinylidene fluoride. However, polyvinylidene fluoride-based electrode fabrication needs the usage of hazardous chemical solvents viz., N-methyl-2-pyrrolidinone, which needs expensive and complex dryers. Furthermore, the mechanical properties of polyvinylidene fluorideremaininadequate to endure external distortion when used in flexible power supplies. A flexible graphite anode created by a water evaporation process can be bonded with carbon nanofibers. Graphite anodes formed by this method exhibited a well-known porous structure and retained considerable mechanical flexibility. Liquid electrolytes are the most commonly used electrolytes in conventional lithium-ion batteries due to their strong ionic conductivity and suitable electrochemical properties (Kim et al. 2019).

Due to the inherent limitations of liquid electrolytes, newer options such as gels and solid electrolytes should be explored. The use of nanocellulose as a component of solid electrolytes has been explored as one of many strategies to achieve this goal. Lithium-sulfur, sodium, polyvalent (magnesium, calcium, aluminum), and the new energy storage system metal-air enable the use of nanocellulose. This indicates that nanocellulose can be a promising building block to overcome difficult post-lithium-ion battery problems that conventional battery materials and chemistries have not yet resolved. Lithium-sulfur batteries require a lot of care due to their high theoretic energy density, little cost, and ample supply of naturally friendly sulfur-active materials. Sulfur's inherent low electrical conductivity and the pendulum phenomenon caused by polysulfides leading to capacity loss, self-discharge, and low coulombic efficiency remain major obstacles to practical application. Much attention has been paid to these topics, focusing on separators, conductive interlayers, electrolytes, and sulfur electrodes (Chen et al. 2018). In lithium-sulfur batteries, nanocellulose can be processed into carbon compounds or used as electrode binders and separators. Nanocellulose fibers can be used to make very slender and strong paper terminals since they have a more modest fiber measurement than conventional cellulose fibers. Nanocellulose-based materials show great underlying and electrochemical capacities in lithium-particle batteries by surface enactment or blend with different materials to produce microporous designs. To forestall the oxygen-rich practical gatherings on the outer layer of nanocellulose from dissolving in the electrolyte, it is favorable to associate them with polysulfides utilizing sulfur oxygen bonds.

With a semi-open design that devours an infinite quantity of oxygen at the air electrode and produces a highertheoretic energy density, metal-air batteries are one of the most efficient energy sources of the new generation. Achieving electrochemical reliability requires energetic and vigorous electrocatalysts that can quicken the oxygen reduction reaction and development process in metal-air batteries. Precious metals (platinum and radium) and their alloys have been used to overwhelm the slow reaction rate and quicken the electrochemical conversion development due to their relatively slow electrocatalytic reaction rates (Lasrado, Ahankari, and Kar 2020).

37.5 Nanocellulose as Conductive Materials

Carbon-based compounds, metal drops, and conductive polymers make up most of customary conductive materials. It is feasible to make novel cross-breed materials by blending nanocellulose with these customary conductive materials to make conductive composites. These composite materials consolidate nanocellulose's extra abilities with the first conductive material highlights. To blend nanocellulose with conductive materials, there are two fundamental methodologies. One of these methods is the statement of conductive materials on the nanocellulose lattice, which can be used to fabricate composites through covering, affidavit, and different strategies. The subsequent technique, which generally involves in-situ polymerization and blending, includes straightforwardly joining nanocellulose with conductive components.

Any object that conducts electricity must be made of a conductive material including metal particles with different conductivities, conducting polymers, and conducting carbon compounds (carbon nanotubes, graphene, soot, etc.). Electrically conductive materials are required for the fabrication of energy storage devices (Hsu and Zhong 2019). The advantages of both components can be integrated into novel combinations of nanocellulose and these conducting compounds. The number of conductive materials that can be used to create conductive nanocellulose is theoretically limitless (Li et al. 2015). In practice, however, other factors such as cost, conductivity level, chemical and physical stability, colonization, toxicity, ecology, biodegradability, and ease of manufacture must be considered. Ternary hybrids based on nanocellulose were created for special applications requiring the necessary electrical conductivity and mechanical strength. A similar design approach can be used to fabricate ternary hybrid structures involving coating, in-situ polymerization, mixing, and doping. Systems typically consist of conductive materials (conductive polymers, metal particles, activated carbon, etc.). Other foreign objects may occasionally be added to the system to accommodate specific characteristics (Du et al. 2017).

37.6 Nanocellulose/Metal Oxide Composites

In addition to conducting polymers, nanocellulose is often combined with transition metal oxides to increase capacity and energy density over wild and rescindable redox processes. Transition metal oxides such as NiO, MnO_2, and Mn_2O_3, in addition to being affordable, non-toxic, and widely available, make them suitable for supercapacitors. Suitable for use in supercapacitor applications is a composite aerogel combining carbonized nanocellulosefibers and strong manganese oxide (MnOx). The porosity of carbonized nanocellulosefibers makes them perfect substrates for electrolyte ions and MnOx (Wesling et al. 2020).

37.7 Composites Based on Nanocellulose for Solar Energy Applications

Various materials are utilized in solar cells to competently convert solar energy. However, engineering solar cells from inorganic resourcesare expensive. Biomaterials are currently being investigated for use in solar cells because they are abundant in nature, renewable, and sustainable (Maros and Juniar 2016). Devices used for conversion must have large surface areas and excellent charge transfer capabilities to efficiently convert solar energy into electricity and thermal energy storage. The various applications of nanocellulose composites in solar energy are illustrated in Figure 37.2. Due to its high optical turbidity, carbon nanocellulose is ideal for use in high-efficiency conversion devices. The carbon nanocellulose can be combined with silver to fabricate reusable solar cells with semi-transparent electrodes. Using a 35-film transfer laminate, the carbon nanocellulose-based solar cell was further improved and found to have a conversion efficiency of 4%. Materials such as polyethylene terephthalate and polyethylene naphthalate are commonly used as substrate materials in organic photovoltaic systems. Perovskite solar cells (also called pseudo-supercapacitors) have the potential for use in wearable electronics due to their cheap and higher energy conversion efficiency. Cells made of crystalline silicon closely resembled metal hydride perovskite cells. Pseudo-supercapacitors are typically made from petroleum-derived

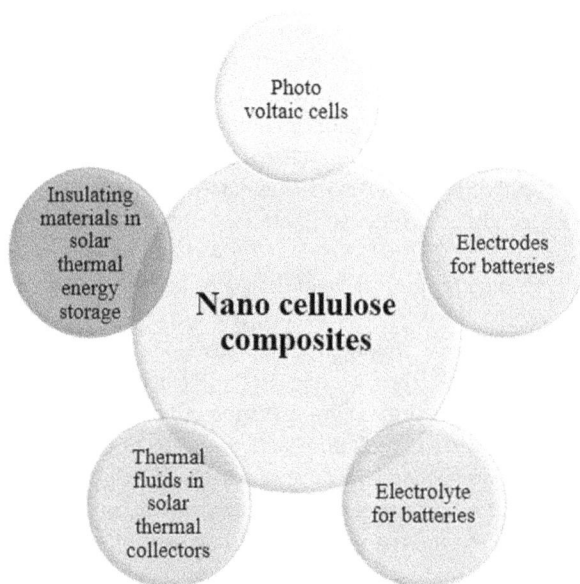

FIGURE 37.2   Nanocellulose-based composites for solar energy applications.

polymer substrates and are a significant source of pollution. The crude cellulose based nanocrystals can be used as a source of carbon nanodots to fabricate photosensitizers in dye-sensitized solar cells (Thomas et al. 2018).

37.8 Nanocellulose Composites for Piezoelectric Applications

Hybrid materials made of cellulose and carbon nanoparticles are known as nanocellulose/nanocarbon composites. Nanocarbon usefulness is delivered by joining nanocarbon materials with nanocellulose in a lattice or transporter that is non-metallic, non-dissolving, non-poisonous, solid, and correspondingly steady. Nanocellulose is engaging for biomedical applications because of its tunable compound properties, non-creature beginning, and likenesses to biomolecules regarding size, substance arrangement, and viscoelastic abilities, notwithstanding the valuable coupling with nanocarbon materials. In contrast with materials that simply hold back carbon nanoparticles or cellulose nanoparticles, nanocellulose/nanocarbon composites have various useful highlights.

The generation of electric charges in crystalline materials by pressure, tension, bending, and torsion is called piezoelectricity. There are commercial uses for polyvinylidene difluoride piezoelectric films. Polyvinylidene fluoride is used in the manufacture of commercially available flexible piezoelectric films. These films cost a lot of money and are not biodegradable. Nanocellulose has just been used in composite films, a recent development (Xing et al. 2019). The piezoelectric responses of carbon nanofiber green piezoelectric materials were measured utilizing the solvent-castingtechnique. Plain chitosan foil, used as a biodegradable sensor material, showed twice the piezoelectric sensitivity than quartz crystal. The mixture of chitosan and carbon nanofibers showed lower sensitivity compared to pure chitosan and carbon nanofiber film (Trache and Thakur 2020). The microcrystalline chitosan and nanocellulose form equivalent piezoelectric films. Carbon nanocellulose was arranged on an ultra-thin sheet. There are different degrees of orientation: the more highly oriented the film, the greater the piezoelectric effect with a similar carbon nanocellulose film with a highly organized crystalline structure (Lv et al. 2019).

The formation of an electric charge in a glasslike substance because of mechanical upgrades like strain, pressure, bowing, or bending is known as piezoelectricity. Commercially, piezoelectric films made of polyvinylidene fluoride are utilized. Films made of polyvinylidene fluoride are pricy and not biodegradable. Nanocellulose has quite recently been utilized into composite films, which is another turn of events. Utilizing the dissolvable projecting method, green piezoelectric materials are analyzed for their piezoelectric reaction. A contender for a biodegradable sensor material, plain chitosan film shows twofold the piezoelectric responsiveness of quartz. Contrasted with plain film, the mix of polyvinylidene fluoride films was less delicate. Using nanocellulose and microcrystalline chitosan, a similar piezoelectric film was made (Lasrado, Ahankari, and Kar 2020).

37.9 Conclusion

For cutting-edge electrochemical change or energy-saving gadgets, nanocellulose is recognized as an extremely encouraging feasible and sustainable nanomaterial for green and inexhaustible electronics. Since cellulose has a strong capacity to make light, permeable, and confounded networks, it is brilliant as a substrate or film material. In summary, the arena of nanocellulose in energy storage systems has made great strides recently but is still in the early stages of development. Shortly, cellulose-based devices may be the best choices for flexible electronic applications that require high performance but are inexpensive. Fabrication of thick and mass-stressed electrodes is a hopeful way to additionally increase the device-level energy density for use in applied applications.

Furthermore, aperture engineering to increase ionic conductivity and storage capacity is preferred for near-future applications. Furthermore, due to the comparative expense of nanocellulose formed by advanced production technology, the practical application of new production technology with high scalability and low cost is desperately needed. It is obviously difficult to ensure such environmentally friendly materials for energy applications, and additional efforts are required. The manufacture of nanocellulose continues to be labor-intensive and expensive. There should be a new generation of low-cost, easily scalable ways for generating nanocellulose. Nanocellulose will be a potential material to be used in all kinds of energy storage devices in order to increase the efficiency and operating life.

References

Chen, Chaoji, and Liangbing Hu. 2018. 'Nanocellulose toward Advanced Energy Storage Devices: Structure and Electrochemistry'. *Accounts of Chemical Research* 51 (12): 3154–3165.

Chen, Wenshuai, Haipeng Yu, Sang Young Lee, Tong Wei, Jian Li, and Zhuangjun Fan. 2018. 'Nanocellulose: A Promising Nanomaterial for Advanced Electrochemical Energy Storage'. *Chemical Society Reviews* 47 (8): 2837–2872.

Du, Xu, Zhe Zhang, Wei Liu, and Yulin Deng. 2017. 'Nanocellulose-Based Conductive Materials and Their Emerging Applications in Energy Devices - A Review'. *Nano Energy* 35: 299–320.

Durairaj, Arulppan, Moorthy Maruthapandi, Arumugam Saravanan, John H.T. Luong, and Aharon Gedanken. 2022. 'Cellulose Nanocrystals (CNC)-Based Functional Materials for Supercapacitor Applications'. *Nanomaterials* 12 (11): 1–25.

Guo, Ruiqi, Lixue Zhang, Yun Lu, Xiaoli Zhang, and Dongjiang Yang. 2020. 'Research Progress of Nanocellulose for Electrochemical Energy Storage: A Review'. *Journal of Energy Chemistry* 51: 342–361.

Hsu, Helen H., and Wen Zhong. 2019. 'Nanocellulose-Based Conductive Membranes for Free-Standing Supercapacitors: A Review'. *Membranes* 9 (6): 74.

Jose, Jasmine, Vinoy Thomas, Vrinda Vinod, Rani Abraham, and Susan Abraham. 2019. 'Nanocellulose Based Functional Materials for Supercapacitor Applications'. *Journal of Science: Advanced Materials and Devices* 4 (3): 333–340.

Kim, Jung Hwan, Donggue Lee, Yong Hyeok Lee, Wenshuai Chen, and Sang Young Lee. 2019. 'Nanocellulose for Energy Storage Systems: Beyond the Limits of Synthetic Materials'. *Advanced Materials* 31 (20): 1–16.

Lasrado, Dylan, Sandeep Ahankari, and Kamal Kar. 2020. 'Nanocellulose-Based Polymer Composites for Energy Applications—A Review'. *Journal of Applied Polymer Science* 137 (27): 1–14.

Li, Yuanyuan, Hongli Zhu, Fei Shen, Jiayu Wan, Steven Lacey, Zhiqiang Fang, Hongqi Dai, and Liangbing Hu. 2015. 'Nanocellulose as Green Dispersant for Two-Dimensional Energy Materials'. *Nano Energy* 13: 346–354.

Lv, Yanyan, Yi Zhou, Ziqiang Shao, Yanhua Liu, Jie Wei, and Zhengqing Ye. 2019. 'Nanocellulose-Derived Carbon Nanosphere Fibers-Based Nanohybrid Aerogel for High-Performance All-Solid-State Flexible Supercapacitors'. *Journal of Materials Science: Materials in Electronics* 30 (9): 8585–8594.

Maros, Hikmah, and Sarah Juniar. 2016. '済無 No Title No Title No Title', 1–23.

Nguyen Dang, Luong, and Jukka Seppälä. 2015. 'Electrically Conductive Nanocellulose/Graphene Composites Exhibiting Improved Mechanical Properties in High-Moisture Condition'. *Cellulose* 22 (3): 1799–1812.

Osong, Sinke H., Sven Norgren, and Per Engstrand. 2016. 'Processing of Wood-Based Microfibrillated Cellulose and Nanofibrillated Cellulose, and Applications Relating to Papermaking: A Review'. *Cellulose* 23 (1): 93–123.

Santos, Mayara C.G., Débora R. da Silva, Paula S. Pinto, Andre S. Ferlauto, Rodrigo G. Lacerda, Wander P. Jesus, Thiago H.R. da Cunha, Paulo F.R. Ortega, and Rodrigo L. Lavall. 2020. 'Buckypapers of Carbon Nanotubes and Cellulose Nanofibrils: Foldable and Flexible Electrodes for Redox Supercapacitors'. *Electrochimica Acta* 349: 1–9.

Szabó, Andrea, Caterina Perri, Anita Csató, Girolamo Giordano, Danilo Vuono, and János B. Nagy. 2010. 'Synthesis Methods of Carbon Nanotubes and Related Materials'. *Materials* 3 (5): 3092–3140.

Thomas, Bejoy, Midhun C. Raj, B. K. Athira, H. M. Rubiyah, Jithin Joy, Audrey Moores, Glenna L. Drisko, and Clément Sanchez. 2018. 'Nanocellulose, a Versatile Green Platform: From Biosources to Materials and Their Applications'. *Chemical Reviews* 118 (24): 11575–11625.

Trache, Djalal, and Vijay Kumar Thakur. 2020. 'Nanocellulose and Nanocarbons Based Hybrid Materials: Synthesis, Characterization and Applications'. *Nanomaterials* 10 (9): 1–5.

Vilela, Carla, Armando J.D. Silvestre, Filipe M.L. Figueiredo, and Carmen S.R. Freire. 2019. 'Nanocellulose-Based Materials as Components of Polymer Electrolyte Fuel Cells'. *Journal of Materials Chemistry A* 7 (35): 20045–20074.

Wesling, Bruno N., Gabriella M.V. Dias, Daliana Müller, Rafael Bento, Serpa Dachamir Hotza, and Carlos R. Rambo. 2020. 'Enhanced Electrochemical Performance of Nanocellulose/PPy·CuCl2 Electrodes for All-Cellulose-Based Supercapacitors'. *Journal of Electronic Materials* 49 (2): 1036–1042.

Xing, Jinghao, Peng Tao, Zhengmei Wu, Chuyue Xing, Xiaoping Liao, and Shuangxi Nie. 2019. 'Nanocellulose-Graphene Composites: A Promising Nanomaterial for Flexible Supercapacitors'. *Carbohydrate Polymers* 207: 447–459.

38

Synthesis of Graphene Nanomaterials for Energy Storage Applications

Anitha Rexalin
Devaraj
*Academy of Maritime
Education and Training
(AMET), Kanathur, India*

Balkeshwar Singh
*Adama Science and
Technology University,
Adama City, Ethiopia*

Thirumalvalavan
*Arunai Engineering College,
Tiruvannamalai, India*

A. Pandian
*Koneru Lakshmaiah
Education Foundation (K L
Deemed to be University),
Guntur, India*

Ankush Balajirao
Khansole
*Shreeyash College of
Engineering and Technology,
Aurangabad, India*

A.V.K. Shanthi
*AIMAN College of Arts and
Science for Women, Trichy,
India*

In a recent trend, nanotechnology is an interdisciplinary engineering field that emphasizes nanoparticles with multidimensional properties. Nanoparticles are attracting interest owing to their enhanced physiochemical and biological characteristic feature at nanoscale from 1 to 100 nm. Due to the development of large industrial fields, there is a significant shortage of energy usage. Overcoming this energy crisis with viable technologies is therefore essential to protect ecosystems and the environment. Nanotechnology plays a central role in energy storage devices. Therefore, the functionalization of nanomaterials with natural materials serves as an environmentally friendly and safe technology and provides

DOI: 10.1201/9781003355755-38

a versatile platform. Graphene oxide nanomaterials and their composites have garnered greater intention owing to attractive physicochemical and enhanced biological properties and are widely used in various fields. However, metal nanoparticles tend to aggregate and grow spontaneously, which tends to lower the surface energy, limiting the reaction efficiency and stability of the nanoparticles. The very high cost of metal nanoparticles also prevents their further use. Therefore, the design of a new generation of metal oxide nanomaterials with different properties that offer synergistic multifunctional efficiencies for different applications has been adopted. Therefore, in the field of nanomaterials, keen interest has been endowed to the synthesis of graphene oxide nanosheets using synthetic protocol that have myriad application in energy storage devices.

38.1 Introduction to Nanotechnology

In scientific research, nanoscience and technology plays a significant role in meeting out the modern generation needs. The word "Nano" originated from the greek word "Dwarf" which denotes that the reduction in size of nanomaterials are smaller in order of nanometer (nm) range. One nanometer of the particle is corresponding to one billionth of meter in metric scale (i.e. 1 nm = 10–9 m) [1].

Nanoscience and technology is defined as the broad area which includes all the scientific communities (chemists, physicists, engineers, biologists etc.) working together to transfer science into technology (Pal et al. 2011). The word "nanotechnology" was first discovered by Japanese scientist Dr. Taniguchi in 1974. Then, Prof. Eric Drexler developed the concept of nanotechnology by utilizing the nanomaterials for his product [2–4].

Generally, nanomaterials can be synthesized by two methods: (i) bottom-up approach and (ii) top-down approach. The materials are shaped by splitting (or) breaking down from bulk into smaller-sized materials either by means of chemical (or) physical route in top-down approach [5]. Whereas in the bottom-up approach, the growth of materials starts from nucleation by means of molecule-by-molecule (or) atom-by-atom process. Nanomaterials differ from bulk materials by means of their interesting properties such as optical, electronic, magnetic, thermal, etc. and thereby find utility in a wide range of applications like modern electronics, medical, chemical, pharmaceutical, and agricultural. Based on the synthetic approach, concentration of the material, morphology of the surface varies which in turn gets reflected on its surface properties [6, 7].

38.2 Graphene Oxide Nanomaterials

In the past several decades, several endeavors had been made to enhance the performance of devices that store energy. This chapter mainly focuses on the current development made in areas of energy storing appliances using GO nanomaterials amalgamated with several inorganic metal oxide nanomaterials. For example, GO coupled with hex a cyanoferrate material, graphene coupled with quantum dots, graphene materials with metal organic frameworks (MOFs) and its Mechanistic investigation for the development of devices with high energy storing capacity [8]. In the year 2004, graphene was initially invented by the cluster of research personnel's from the University of Manchester, United Kingdom, and GO was considered as a marvelous nanomaterial of the 21st century. GO nanomaterial is soluble in water fabricated by addition of oxygen ions in stack material of graphite through the process of chemical method. Further complete development of GO solid materials into the thin sheet of atomic thickness is made possible through the process of chemical, mechanical, or thermal means. GO have the tendency to act as transparent conductors of electricity with reduced level of resistance and transparency ability when compared with that of carbon nanotubes. The formation of single layer of graphene sheets or stacks of graphitic carbon materials insisted to form quantum dots which in turn possess unique physical and chemical properties. This chapter mainly focuses on the application of GO-based nanomaterials for multifaceted applications such as sensing, energy storage material in various devices, and bio imaging [9–11].

Development of graphene or graphene-based nanomaterials clusters to form nanomaterials with similar morphology and different nomenclature which in turn indicates formation of carbon materials with single or multiple graphene sheets [12]. In the current scenario, various methods were readily developed for fabrication of graphene materials and derivatives which in turn produce the products with various size and surface functionalities, notably C, O, H or functional group present on the surface includes carboxyl, epoxy, carbonyl and hydroxyl materials. The development of brainstorming allotropes of carbon nanomaterials has garnered greater attention owing to their unique physiochemical properties from all branches of science and technology. Carbon nanomaterials in their graphitic nanoforms are classified into several types which includes zero dimension, 2D nanomaterials, 3D nanomaterials, and 1D materials of carbon [13, 14].

In the last decade, researchers have developed GO nanosheets with various dimensional structures in the development of zero dimensions, one-dimension material, and two-dimension material. These materials were developed with the monolayer or multilayer sheets with the confinement edge of materials. The development of GO with various dimension offers robust physical and chemical properties which includes chemically inert, enhanced photo stability, high fluorescence capability, and reduced cytotoxicity. These unique physicochemical properties offer novel structure with enhanced optoelectronic properties that are not available in other nanomaterials. The tuning of electronic behavior as well as quantum confinement effect of zero dimensional GO has become more delightful when compared with graphene materials. These enhanced properties made GO a potent candidate for various industrial applications, notably sensors, imaging devices, and energy storage capacitors [15]. Zero dimensional GO nanomaterials exhibit better solubility in various organic solvents mainly acetone, ethanol, dimethyl sulfoxide, hydro furan, form amide, etc. Although the upgraded dissolving ability in aqueous solvents have been enhanced due to their multifaceted applications in assorted arena of biological imaging and drug delivery at targeted site into the host, the aqueous dissolving capacity of GO originates from the hydroxyl and carboxyl groups that adhere to the surface of zero-dimensional GO nanosheets. The surface moieties of GO are responsible for the hydrophilic nature of zero-dimensional GO and can be tuned through the synthetic strategy, mainly chemical method [16]. Applications of GO nanomaterials were clearly explained in Figure 38.1

Modified Hummer's method is the most widely used method for the fabrication of GO. The primary treatment for the fabrication of GO involves the amalgamation of sodium nitrate, sulfuric acid, and

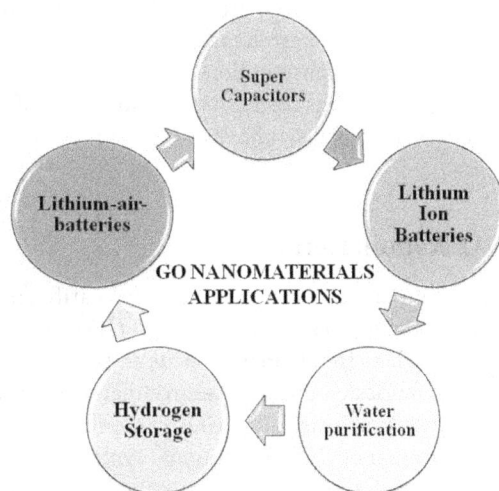

FIGURE 38.1 Application of GO nanomaterials in various energy storage devices.

potassium permanganate. However, GO is readily soluble in aqueous media and it will also disperse well in organic solvents. Through the process of electrostatic repulsion, GO sheets are stable in colloidal solution as they possess negative charge. GO was unloaded easily through the process of drop-casting method; filtration, dip coating, and spray coating were done by spray pyrolysis method, spin coating, and Langmuir- Blodgett techniques [17]. Although the GO fabricated using electrochemical deposition method was a widely used strategy in order to remove the functional moieties when compared to that of chemical synthetic strategy that utilizes toxic chemicals such as N_2H_4, KOH while electrochemical reduction method is easy, lucrative and environmental benign method to fabricate reduced graphene oxide for energy storage thin films. The composite obtained with the combination of several metal oxide nanomaterials such as SnO_2 and CO_3O_4 has been developed in order to enhance the electrochemical action of energy storing devices. While comparing the synthesized rGO nanocomposites with several other materials, it has been found that they possess greater energy and power consumption efficiency compared to traditional carbon compounds [18, 19].

The rechargeable batteries developed in the presence of Ni-metal, for instance development of Ni-Fe ions in the electrode, and combination of Ni-Co ions have outstanding application, i.e., it acts as an alternative form to store energies. . Further the commonly used method in order to increase the action of Ni-Fe ions-based batteries is hybridizing the nano Fe ionic materials with highly conducting carbon ions-based nanostructures such as networks of carbon nanotubes, graphene thin flims, and conducting polymers [20].

38.3 Synthesis of Nanoparticles

There are several synthetic strategies involved in the fabrication of nanoparticles which include "top-down" and "bottom-up" approaches. Bulk material splits into fine smaller particles by reduction in size with various lithographic techniques in "top-down approach". The synthesis of nanoparticles by physical technique has great disadvantages like liberation of heat, energy wastage, and need of more labor power [21]. Nanomaterials can be fabricated using chemical and biological means through the mechanism of self-assembly of atoms which in turn grow into nuclei further forms into a nanoparticle. In this method, chemical synthesis of metal and metal oxide nanoparticles is the key component. The chemical synthetic method for the fabrication of metal and metal oxide nanoparticles may in turn produce toxic byproducts which can be absorbed on the outer surface of nanomaterials, which leads to toxicity issues [22].

Plant extract and its parts such as stem and leaf are found to be better alternative methods for the fabrication of nanoparticles. The utilization of green materials possesses greater advantages over other synthetic protocols which include very low consumption of energy and mild environmental conditions and doesn't lead to the generation of hazardous chemicals. The fabrication of nanomaterials using plant-based synthesis is more useful over microbial synthesis as the maintenance of cell culture cost is high [23–25].

38.3.1 Synthesis of GO Nanomaterials

In general, variability in application and the results depends primarily on controlled morphological structure, and physiochemical characteristics of the fabricated GO sheets are mainly determined by synthetic strategy. Therefore, researchers should mainly focus and pay keen interest to synthetic strategy materials. GO nanosheets are fabricated using the source that contains carbon source containing materials notably fullerenes, glucose, graphene oxide (GO), carbon nanotubes, and carbon fibers used as a starting material for the fabrication of GO. Conventional synthetic methods are very hard for the synthesis of semiconductor quantum dots. Carbonization is the best synthetic method and isused to fabricate GO from suitable organic moieties or polymeric materials [26].

38.4 Top-Down Methods for the Fabrication of GO

38.4.1 GO Preparation by Liquid Exfoliation (LE)

Exfoliation of liquid for fabrication of two-dimensional GO nanosheets has attracted much interest owing to its scalability. Ultrasonic LE is the most capable operation to obtain nanosheets with many advantages such aslucrative fabrication and easy manipulation, and minimize the impact on the environment. In this process, if the graphite, which is the precursor of the graphene sheets, is exfoliated, it is possible to fabricate GO with good crystallinity by the LE method [27–29]. In this process, graphite and acetylene carbon powders were used as precursors with low and high defects (edge and surface defects), respectively, to create GOs. GOs by utilizing LE using probe sonication of graphite powder are a better fabrication technique. The synthetic process involves ultrasound waves with elevated energy to incise the GO sheets into ultrafine tiny particles. The GOs formed using water is very tiny than the particle extracted using the dimethylformamide protocol. The size and height of the fabricated GOs are ~4.5 nm and 1.8 nm, respectively, and they contained two to three layers of graphene with an interlayer spacing of 0.353 nm through this method while comparing with other methods [30]. The overview of synthetic approach for the fabrication of GO nanomaterials is shown in Figure 38.2.

38.4.2 Hydrothermal Method

A promising technique for the development of GO from carbon-based feedstock using strong oxidizing agents such as sulfuric acid, nitric acid, and hydrogen peroxide breaks the carbon materials into GO nanosheets. Main advantage over hydrothermal synthetic strategy for the fabrication of GO is that the application of varying hydrothermal temperature results in the reduction of GO particle size [31].

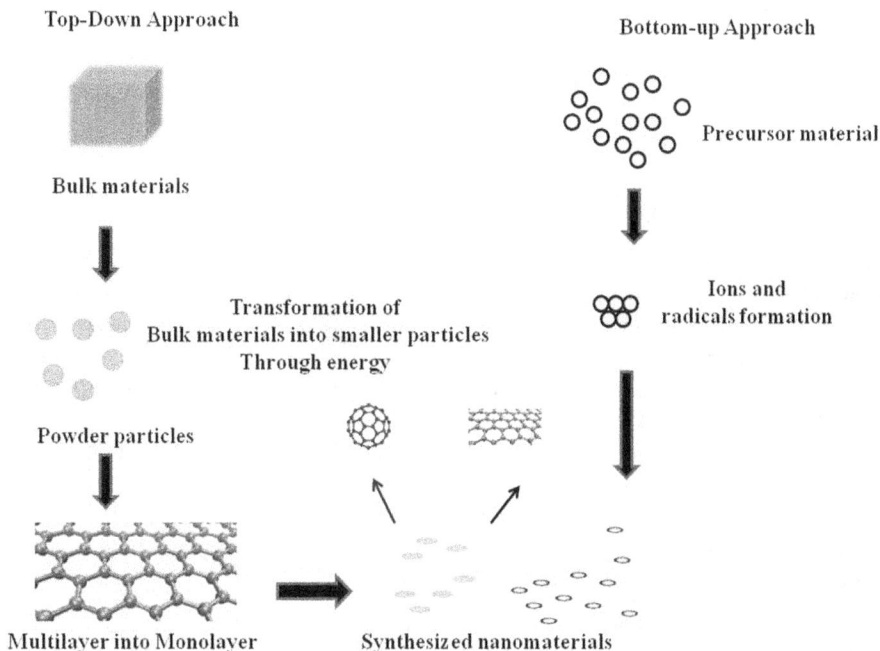

FIGURE 38.2 Overview of bottom-up and top-down approach for the fabrication of GO nanomaterials.

The mechanism behind the formation of GO from carbon source and H_2O_2 as a reagent that dissociated into OH radicals in the presence of high temperature results in the formation of graphite sheets [32].

38.4.3 Sol–Gel Method

In the sol–gel method, change of a system occurs from a colloidal sol (liquid state) into gel (solid state) phase. Sol–gel process is used to make the glass and ceramic materials. In addition, this method is the wet-chemical-based self-assembly process for the formation of nanomaterials [33]. The sol–gel process consists of four steps such as (i) hydrolysis, (ii) condensation, (iii) growth, and (iv) agglomeration of particles. The benefit of the sol–gel method is simplicity, reliability, repeatability, lucrative and enhanced thermal stability, elevated mechanical strength, and low temperature. The sol–gel method is used to synthesize nanowires, nanorods, and nanotubes [34–36].

38.4.4 Co-precipitation Method

The co-precipitation technique tends to propose higher advantages owing to its simplicity and lucrative experimental set up. Also, the shape and size of particles can be controlled easily by adjustment of medium pH [37]. The starting material will be reduced by a reducing agent, and by maintaining optimized pH value, the size and shape of the particles will be controlled. After this process, precipitations will be obtained and then the precipitation is centrifuged for some more times. After centrifugation, the achieved product will be dried to get NPs [38–40].

38.4.5 Spray Pyrolysis Method

The spray pyrolysis approach is a powerful technique for the synthesis of high-purity ceramic nanopowders. In this method, a large quantity of the oxide powders can be produced. Homogeneous particles with a crystalline size of less than 100 nm will be synthesized. In this method, homogeneous solution is the starting phase for obtaining the nanopowders. In spray pyrolysis process, precursor's material solution droplets takes place in the heated substrate. After that, solvent was evaporated by thermal decomposition [41–43]. The main advantage of this method is high production rate and high flexibility of the material with respect to other methods.

38.4.6 Hydrothermal Method

Compared with the sol–gel, co-precipitation, and spray pyrolysis methods, the hydrothermal method has more advantages due to reduced agglomeration rate, lucrative, simple, and single-step synthetic process. Also, elevated purity and various morphology of the particles (tubes, dots, sheets, rods, cubes, and flower) and controlled size can be achieved by this method. Hence, the hydrothermal method is used in the present investigation. Various structures are obtained by the hydrothermal method. In addition, the hydrothermal process showed an extraordinary ability in the fabrication of the inorganic semiconducting materials [44]. The inorganic nanostructures will be formed in the hydrothermal process by the following process. Due to the ambient pressure and temperature condition, the nanostructures can be grown. Also, the nanostructures will be formed due to their concentration, pH value, pressure, time, templates (or) additives. Under the above experimental conditions, the inorganic semiconducting nanostructures can be easily developed [45].

38.4.7 Microwave-Assisted Hydrothermal Method

The microwave-assisted hydrothermal method has greater attention and extensive interests than the hydrothermal method due to its rapid heating, low reaction time, high reaction rate, reduced temperature and higher phase purity with better yields. Also, nanostructures can be easily formed in this method [46].

38.5 Potential and Emerging Applications of GO Nanosheets

38.5.1 Sensor

Sensor is a technical detection device that can be used to measure the concentration of the molecules in environment and it converts into electrochemical signals that are received and displayed in the monitor. Researchers have developed several nanosensors using GO owing to the presence of several unique characteristic features, notably fixed bang gap energy, that have applications in assorted fields. GO acts as an excellent sensing material owing to the distinct properties of high electron motion with very high-speed reactions which act as excellent sensing material [47]. The widely used sensor, notably glucose monitoring device, for detecting the glucose level by the utilization of assorted fluorescent GO nanosheets contains the surface functionalities of carboxyl group and hydroxyl moieties [48].

38.5.2 Solar Cell Application

In organic solar cells, N-type inorganic semiconductor based GO had been utilized for the formation of solar cells due to enhanced properties such as good electrical and optical conductivity, simple, lucrative, low toxicity, and high stability. Also, various researchers have focused on developing different 1D nanostructures (nanorods, nanopillars, nanofibers, and nanowires and nano arrays) for enhancing the performance of organic and organic/inorganic hybrid solar cells [49].

38.5.3 Photoluminescence Sensor

GO exhibits excellent properties obtained from graphene nanomaterials and quantum dots. GO have a vast range of scope in several branches of science including biology and chemistry. In recent past, the photosensitive technique was utilized as the most recognized method for detection of metal ions by investigating the fluorescence characteristic features of both raw and modified GO nanosheets. There are several advantages achieved by keeping GO as sensing material in order to detect the metal ion concentration but due to the poor product yield, GO limited its detection to the industrial scale. To overcome this hindrance, the newly designed materials by doping heteroatom in GO was achieved. The heteroatoms such as nitrogen, boron, sulfur, and phosphorous was doped with GO in order to increase the quantum yield of the product [50].

38.5.4 Electrochemiluminescence Sensor

The technique that amalgamates the electrochemistry and chemiluminescence in terms results in the formation of electrochemiluminescence sensor. The mechanism behind the electro chemo luminescence sensor is mainly based on the transfer of electron from the middle radical cations and anions of the luminophore of the material. This sensor converts the electrochemical signals from the device into radiative energy by means of applied potential energy on the outer surface of the electrode. Throughout the reaction, the signal from the luminescence can be found to be in the excited form from the electrode material [51–53].

38.6 Energy Storing Devices

38.6.1 Supercapacitor

Vast consumption of energy and the development of industries and technology are two main factors which utilize high energy performance fordevices that store energy. Similarly, the conversion and the electrochemical storage of energy serve as an alternating option and it has been an emerging technology for the industrial and academic industries. Electrical energy storage system (EESS) can be used to store the energies which convert chemical energy into electrical energy. These devices have garnered greater

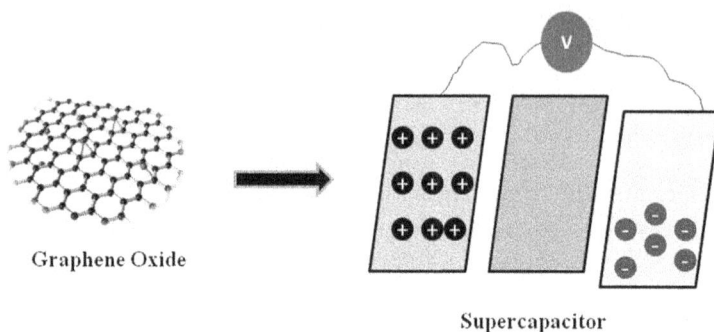

FIGURE 38.3 Application of fabricated GO nanomaterials in super capacitor devices.

attention due to elevated charge, discharge rate, and lifetime expectancy which are key points for energy storing devices. The electrochemical performance of supercapacitors, along with capacitance and cycle life, depends on the composition and structure of the probe material. Limited transition metal oxides, conjugated polymers, and carbon sources have been established as energy storage media. Recently, numerous studies have been conducted on the EESS properties of graphene quantum dots and its potential application as an electrode material [54]. The application of GO in super capacitors is shown in Figure 38.3.

38.6.2 Lithium-Ion Batteries

The depletion of fossil fuel resources and growing concerns about environmental issues are making people aware of the need for energy, which is of great importance for the improvement of sustainable energy technologies, renewable energy sources such as tidal, wind and solar energy in human society. Therefore, the development of efficient energy storage devices (EES) has attracted the attention of many researchers. Reliable energy storage systems such as batteries and super capacitors are key enablers for the development of these power structures. Undoubtedly, it is a leading alternative to other energy storage devices. Lightweight, high energy, and excellent performance characteristics can reduce existing demand and promote the application of important energy sources [55].

38.7 Conclusion

Efficient storage of energy in various electronic storage devices is a key factor in solving the problems of the future energy crisis. Among them, capacitors and super capacitors play a central role in various electronic devices such as electronic circuits, electric vehicles, and communication systems. In this regard, the application of green nanocomposites made from various renewable raw materials is gaining more attention due to their improved physicochemical, favorable, and simple synthetic methodology. To achieve improved storage applications, the developed ecofriendly green nanomaterials have improved morphological properties, highly crystalline structure, improved electrical and thermal conductivity, and better tensile strength. Green nanomaterials were also developed and their unique optoelectronic and physicochemical properties were continuously applied to energy storage super capacitors to achieve high reliability and charge storage performance. Also, the enhanced dielectric properties and capacitance may reduce the strength and physical properties of green nanomaterials. Therefore, there is a need to improve the interactions between nanomaterials to achieve higher energy storage in devices and higher density of materials. This chapter described the recent development of GO nanomaterials synthesized by a simple green synthesis protocol for energy storage applications. This could be achieved by encapsulating the eco-nanofiller leading to improved energy storage performance with synthetic environmentally friendly nanocomposites. Parameters believed to achieve high energy storage performance are morphology-driven synthesis, matrix–filler interaction, high thermal and electrical conductivity, dielectric properties with improved charge density, dielectric properties, capacitance, charge-discharge ratio, etc. We thus conclude the chapter on the fabrication of GO and GO-based nanomaterials, which are successfully used in multiple energy storage devices such as solar cells, electronics, nanogenerators, capacitors, and super capacitors.

References

[1] J. Zhao, L. Liu, F. Li, *Graphene Oxide: Physics and Applications*, Springer, 2015.

[2] H.A. Becerril, J. Mao, Z. Liu, R.M. Stoltenberg, Z. Bao, Y. Chen, *ACS Nano* 2 (2008) 463–470.

[3] D. Chen, H. Feng, J. Li, *Chem. Rev.* 112 (2012) 6027–6053.

[4] D.W. Boukhvalov, M.I. Katsnelson, *J. Am. Chem. Soc.* 130 (2008) 10697–10701.

[5] R. Lahaye, H. Jeong, C. Park, Y. Lee, *Phys. Rev. B* 79 (2009) 125435.

[6] J.-A. Yan, L. Xian, M.Y. Chou, *Phys. Rev. Lett.* 103 (2009) 086802.

[7] J.-A. Yan, M.Y. Chou, *Phys. Rev. B* 82 (2010) 125403.

[8] L. Liu, L. Wang, J. Gao, J. Zhao, X. Gao, Z. Chen, *Carbon* 50 (2012) 1690–1698.

[9] G. Eda, G. Fanchini, M. Chhowalla, *Nat. Nanotechnol.* 3 (2008) 270–274.

[10] G. Eda, M. Chhowalla, *Adv. Mater.* 22 (2010) 2392–2415.

[11] J. Robertson, *Mater. Sci. Eng., R* 37 (2002) 129–281.

[12] H. Chen, M.B. Müller, K.J. Gilmore, G.G. Wallace, D. Li, *Adv. Mater.* 20 (2008) 3557–3561.

[13] C. Chen, Q.-H. Yang, Y. Yang, W. Lv, Y. Wen, P.-X. Hou, M. Wang, H.-M. Cheng, *Adv. Mater.* 21 (2009) 3007–3011.

[14] Q. Zheng, Y. Geng, S. Wang, Z. Li, J.-K. Kim, *Carbon* 48 (2010) 4315–4322.

[15] L. Liu, J. Zhang, J. Zhao, F. Liu, *Nanoscale* 4 (2012) 5910–5916.

[16] Y. Gao, L.-Q. Liu, S.-Z. Zu, K. Peng, D. Zhou, B.-H. Han, Z. Zhang, *ACS Nano* 5 (2011) 2134–2141.

[17] F. Perrozzi, S. Prezioso, L. Ottaviano, *J. Phys. Condens. Matter* 27 (2015) 013002.

[18] K.S. Novoselov, V. Fal, L. Colombo, P. Gellert, M. Schwab, K. Kim, *Nature* 490 (2012) 192–200.

[19] K.P. Loh, Q. Bao, G. Eda, M. Chhowalla, *Nat. Chem.* 2 (2010) 1015–1024.

[20] C. Gómez-Navarro, J.C. Meyer, R.S. Sundaram, A. Chuvilin, S. Kurasch, M. Burghard, K. Kern, U. Kaiser, *Nano Lett.* 10 (2010) 1144–1148.

[21] A. Fujishima, *Nature* 238 (1972) 37–38.

[22] A. Kudo, Y. Miseki, *Chem. Soc. Rev.* 38 (2009) 253–278.

[23] S. Park, R.S. Ruoff, *Nat. Nanotechnol.* 4 (2009) 217–224.

[24] T.-F. Yeh, S.-J. Chen, C.-S. Yeh, H. Teng, *J. Phys. Chem. C* 117 (2013) 6516–6524.

[25] T.F. Yeh, J.M. Syu, C. Cheng, T.H. Chang, H. Teng, *Adv. Funct. Mater.* 20 (2010) 2255–2262.

[26] K. Krishnamoorthy, R. Mohan, S.-J. Kim, *Appl. Phys. Lett.* 98 (2011) 244101.

[27] T.-F. Yeh, H. Teng, *ECS Trans.* 41 (2012) 7–26.

[28] T.-F. Yeh, F.-F. Chan, C.-T. Hsieh, H. Teng, *J. Phys. Chem. C* 115 (2011) 22587–22597.

[29] T.-F. Yeh, C.-Y. Teng, S.-J. Chen, H. Teng, *Adv. Mater.* 26 (2014) 3297–3303.

[30] Y. Matsumoto, M. Koinuma, S. Ida, S. Hayami, T. Taniguchi, K. Hatakeyama, H. Tateishi, Y. Watanabe, S. Amano, *J. Phys. Chem. C* 115 (2011) 19280–19286.

[31] X. Jiang, J. Nisar, B. Pathak, J. Zhao, R. Ahuja, *J. Catal.* 299 (2013) 204–209.

[32] A.K. Agegnehu, C.-J. Pan, J. Rick, J.-F. Lee, W.-N. Su, B.-J. Hwang, *J. Mater. Chem.* 22 (2012) 13849–13854.

[33] X. An, C.Y. Jimmy, *RSC Adv.* 1 (2011) 1426–1434.

[34] Q. Xiang, J. Yu, *J. Phys. Chem. Lett.* 4 (2013) 753–759.

[35] A. Iwase, Y.H. Ng, Y. Ishiguro, A. Kudo, R. Amal, *J. Am. Chem. Soc.* 133 (2011) 11054–11057.

[36] N. Zhang, Y. Zhang, Y.-J. Xu, *Nanoscale* 4 (2012) 5792–5813.

[37] J. Zhang, J. Yu, M. Jaroniec, J.R. Gong, *Nano Lett.* 12 (2012) 4584–4589.

[38] Q. Li, B. Guo, J. Yu, J. Ran, B. Zhang, H. Yan, J.R. Gong, *J. Am. Chem. Soc.* 133 (2011) 10878–10884.

[39] C. Dong, X. Li, P. Jin, W. Zhao, J. Chu, J. Qi, *J. Phys. Chem.* C116 (2012) 15833–15838.

[40] Q. Xiang, J. Yu, M. Jaroniec, *Nanoscale* 3 (2011) 3670–3678.

[41] X.-J. Lv, W.-F. Fu, H.-X. Chang, H. Zhang, J.-S. Cheng, G.-J. Zhang, Y. Song, C.-Y. Hu, J.-H. Li, *J. Mater. Chem.* 22 (2012) 1539–1546.

[42] J. Wang, C. An, J. Liu, G. Xi, W. Jiang, S. Wang, Q.-H. Zhang, *J. Mater. Chem. A* 1 (2013) 2827–2832.

[43] Q. Xiang, J. Yu, M. Jaroniec, *J. Phys. Chem. C* 115 (2011) 7355–7363.

[44] J. Yang, X. Zeng, L. Chen, W. Yuan, *Appl. Phys. Lett.* 102 (2013) 083101.

[45] P.D. Tran, S.K. Batabyal, S.S. Pramana, J. Barber, L.H. Wong, S.C.J. Loo, *Nanoscale* 4 (2012) 3875–3878.

[46] A. Mukherji, B. Seger, G.Q. Lu, L. Wang, *ACS Nano* 5 (2011) 3483–3492.

[47] J. Hou, Z. Wang, W. Kan, S. Jiao, H. Zhu, R. Kumar, *J. Mater. Chem.* 22 (2012) 7291–7299.

[48] Z. Khan, T.R. Chetia, A.K. Vardhaman, D. Barpuzary, C.V. Sastri, M. Qureshi, *RSC Adv.* 2 (2012) 12122–12128.

[49] Q. Xiang, J. Yu, M. Jaroniec, *J. Am. Chem. Soc.* 134 (2012) 6575–6578.

[50] S. Patchkovskii, S.T. John, S.N. Yurchenko, L. Zhechkov, T. Heine, G. Seifert, *Proc. Natl. Acad. Sci. U.S.A.* 102 (2005) 10439–10444.

[51] G. Srinivas, Y. Zhu, R. Piner, N. Skipper, M. Ellerby, R. Ruoff, *Carbon* 48 (2010) 630–635.

[52] C.X. Guo, Y. Wang, C.M. Li, *ACS Sustainable Chem. Eng.* 1 (2013) 14–18.

[53] B.H. Kim, W.G. Hong, H.Y. Yu, Y.-K. Han, S.M. Lee, S.J. Chang, H.R. Moon, Y. Jun, H.J. Kim, *Phys. Chem. Chem. Phys.* 14 (2012) 1480–1484.

[54] J.M. Kim, W.G. Hong, S.M. Lee, S.J. Chang, Y. Jun, B.H. Kim, H.J. Kim, *Int. J. Hydrogen Energy* 39 (2014) 3799–3804.

[55] Q. Sun, Q. Wang, P. Jena, Y. Kawazoe, *J. Am. Chem. Soc.* 127 (2005) 14582–14583.

39

Electrical Energy Storage Analysis of $Li_4Ti_2O_6$ Nanomaterials by Sol–Gel Method

Radhika G. Deshmukh

Shri Shivaji Science College, Amravati, India

M. Sudha

Paavai Engineering College (Autonomous), Namakkal, India

Anamika Gupta

Aligarh Muslim University, Aligarh, India

D. S. Vijayan

Aarupadai Veedu institute of Technology, Paiyanur, India

Kumari Manisha

Gokaraju Rangaraju Institute of Engineering and Technology, Hyderabad, India

Fabian I. Ezema

Africa Centre of Excellence for Sustainable Power and Energy Development (ACE-SPED), University of Nigeria, Nsukka, Nigeria

C. Pavithra

Marudhar Kesari Jain College for Women, Vaniyambadi, India

DOI: 10.1201/9781003355755-39

Lithium titanate (LT) is an anode material used for storage devices. In this chapter, the composition of $Li_4Ti_2O_6$ synthesized by sol–gel method and processed in microwave is presented. The microwave-processed $Li_4Ti_2O_6$ is characterized by structure and microstructure by X-ray powder diffraction and scanning electron microscope confirming the single-phase formation and uniform morphology with a particle size varying from 23 nm to 30 nm. AC conductivity confirms that the materials use three different conduction mechanisms. In the high-temperature region, the conduction is purely due to hopping of electrons. The impedance analysis of LT is confirmed by the negative temperature coefficient of resistance because the resistance value is decreased by increasing the temperature. The conductivity, relaxation time, and capacitance values are measured at different temperatures from 300°C to 480°C.

39.1 Introduction

Lithium-based compounds are important materials in charge storage devices [1]. Lithium-based ceramics also find their applications as test blanket. In lithium-based compounds, lithium releases its outer electron easily, and ceramics are mostly dielectric in nature and titanates are known for their very high dielectric constant [2, 3]. This chapter provides an insight on the electrical behavior of lithium in the double perovskite structure. Partial replacement of lithium with Ni^{2+} resulted in no significant behavioral change.

Lithium is used as a tritium breeding material in fusion reactor. Lithium titanate (LT) is an attractive property for most stable, fast recharging; low activation, good density and tritium released in low temperature, because of many researchers, are attracted in LT [4–8]. Different methods have successfully been employed to synthesize LT such as sol–gel, solid state, high energy ball milling, wet chemical method, solution combustion, polymer solution and hydrothermal. Compared to these methods, sol–gel technique is a good method because of low cost, easy to process, and good purity achieved [9–10].

In recent researchers interested in microwave sintering, because of uniform sintering, and less time and low sintering temperature. Many researchers have successfully synthesized LT (Li_2TiO_3), and we report the double perovskite $Li_4Ti_2O_6$ ceramic and microwave processing [11–12]. In this chapter, we focus on $Li_4Ti_2O_6$ synthesized by sol–gel technique and microwave processing and its structural, microstructural, AC conductivity, and impedance studies are analyzed.

39.2 Materials and Methods of Synthesis

The origin materials of lithium monohydrate (Sigma-Aldrich, 98%, $LiOH.H_2O$), nickel nitrate (Sigma-Aldrich, 98.0%, $Ni(NO_3)_2. 6H_2O$), and titanium butoxide (Sigma-Aldrich, 98.0%, $Ti(OC_4H_9)_4$) are taken in stoichiometric ratio. The starting materials of $LiOH.H_2O$ and $Ni(NO_3)_2. 6H_2O$ materials are dissolved in glacial acetic acid. After dissolving, the solution is dried at 100°C and the solution is cooled to ambient temperature and then $Ti(OC_4H_9)_4$ is supplemented to that solution by drop. After that, the solution of ethanol and water is added for fast gelation. At the end, the gel is heated at 100°C for 6 hr to get the powder sample. Prepared sample is calcinated at 950°C for 50 min at a rate of 40°C per min using microwave furnace. The calcined powder is sintered at 1150°C for 35 min using microwave furnace to make a dense ceramic sample. Then, the prepared sample is made into a pellet for dielectric measurements.

39.3 Results and Discussion

X-ray powder diffraction pattern is shown in Figure 39.1. It confirms single-phase formation of LT. No extra additional peak was found and all the peaks match with those in JCPDS file no 71-2348 and are reported in Reference [9]. The LT is a monoclinic crystal structure and belongs to C2/c space group. The lattice parameters evaluated using powder X software are a = 5.06Å, b = 8.79Å, and c = 7.65Å. The size

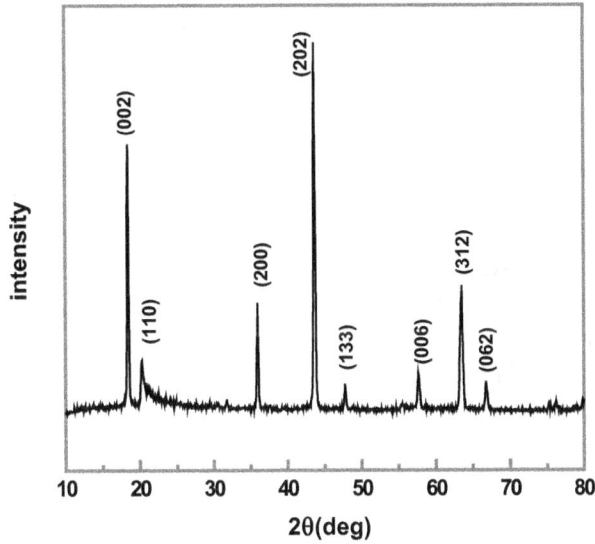

FIGURE 39.1 XRD of LT sample.

of the crystal structure is calculated using the Scherrer's equation (39.1) and the average crystal size is between 23 nm and 30 nm. The density and porosity calculated using relation (39.2) and (39.3) are 4.16 g/cm^3 and 0.023%, respectively [12].

$$G = \frac{K\lambda}{\beta \cos\theta} \tag{39.1}$$

$$\rho_B = \frac{m_{air}}{m_{air} - m_{xylene}} \times \rho_{xylene} \tag{39.2}$$

$$P = 1 - \left(\frac{\rho_B}{\rho_X} \right) \tag{39.3}$$

where k is the constant value of 0.9, the wavelength of X-rays used Cu $K\alpha$ radiation ($\lambda \approx 1.54$Å) the full width half maximum of the maximum intensity peaks is β and θ is the angle of the diffraction.

The single-phase formation of double perovskite LT is confirmed using the powder X-ray diffraction. The details of the structure and microstructural properties are discussed elsewhere [1]. The perovskite formation is confirmed by the Goldschmidt tolerance factor (t) given in relation (39.4) [5].

$$t = \frac{R_A + R_o}{\sqrt{2}\left(R_B + R_o\right)} \tag{39.4}$$

where the ionic radii of the A-site and B-site cations are R_A and R_B, and R_o is the ionic radius of O^{2-} ions. Perovskite structure is said to be stable if 't' is between 0.8 and 1.1. The calculated 't' value for LT is found to be 0.795.

39.4 Infrared Analysis

Fourier transformation infrared spectrum used to study the metal–oxygen bond in LT is shown in Figure 39.2. In the lattice transmittance spectrum, in general form, it can be split into two regions, one less than 1000 cm^{-1} is called fingerprint region for metal oxides and the other above 1000 cm^{-1} from 4000 to 1000 cm^{-1} is called the functional group region. The absorption peak sat 1431 cm^{-1} and 867 cm^{-1} are the stretching vibration of the Ti-O bond and MO$_6$ (TiO$_6$). An Oxo bond is found to be below 800 cm^{-1}. Oxo bond is Ti-O-Ti bond TiO$_6$ octahedron stretching. The peak at 1215 cm^{-1} is antisymmetric and at 867 cm^{-1} is symmetric stretching vibrations of CO$_3^{2-}$ anions. The bond at 459 cm^{-1} is the vibration bond of metal with organic group, as it can be Li-O-R or Ti-O-R [13–15].

39.5 Scanning Electron Microscope

Figure 39.3 shows the high-resolution scanning electron microscopy images of LT. It clearly shows the uniform morphology and close packing of LT. The particle size varies from 23 nm to 30 nm [16].

39.6 AC Conductivity

An AC electrical conductivity with the function of 1000/T with different nickel concentration is shown in Figure 39.4 at 1 kHz and Figure 39.5 at 10 kHz. The temperature-dependent conductivity will explain the intrinsic and extrinsic behavior of the material. It has been observed that there are three different regions of conductivity of LT. The decrease in conductivity with temperature in the low temperature region III (<80°C) suggests metallic behavior in this region. However, on further increase of temperature (region II), the conductivity increases linearly, and then at very high temperatures (region I), a rapid steep rise is observed. This is semiconducting type of behavior with temperature. At this temperature, the material is used as a thermal switch. This change of conduction behavior at around 80°C can act as an instrument safety and health measure to control the electrical supply through the system or device.

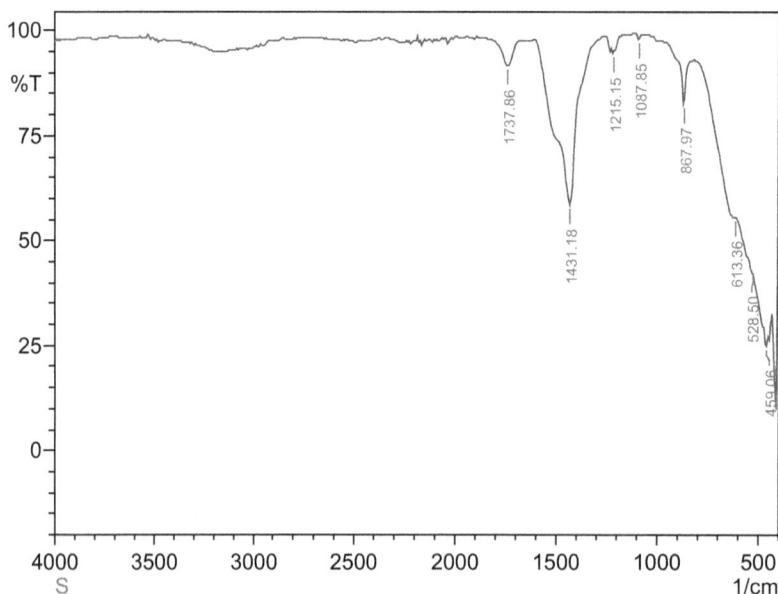

FIGURE 39.2 FTIR analysis of LT.

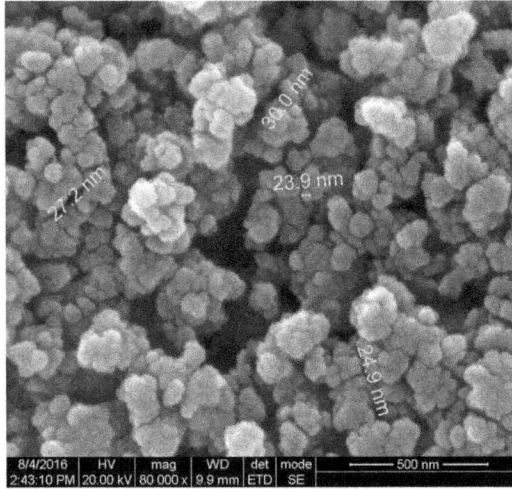

FIGURE 39.3   SEM image of LT.

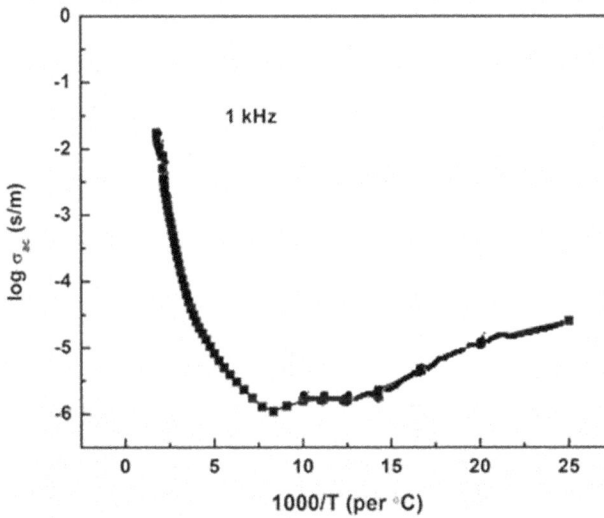

FIGURE 39.4   AC conductivity of LT at 1 kHz.

The sudden change in conduction mechanism might be due to thermally generated charge carriers [17–19]. Furthermore, at high temperature, the conductivity due to might be double perovskite of LT are merging with each other at both the frequencies. This suggests that at high temperatures the thermally generated charge carriers reached saturation. These plots clearly show the multiple conduction and different activation energies [20].

The activation energies, evaluated from Figures 39.4 and 39.5, and the measurements are shown in Table 39.1. Activation energies are calculated from the linear variation of AC conduction with 1000/T using Arrhenius relation (39.5):

$$\sigma_{ac} = \sigma_o \exp\left(-E_a / kT\right) \tag{39.5}$$

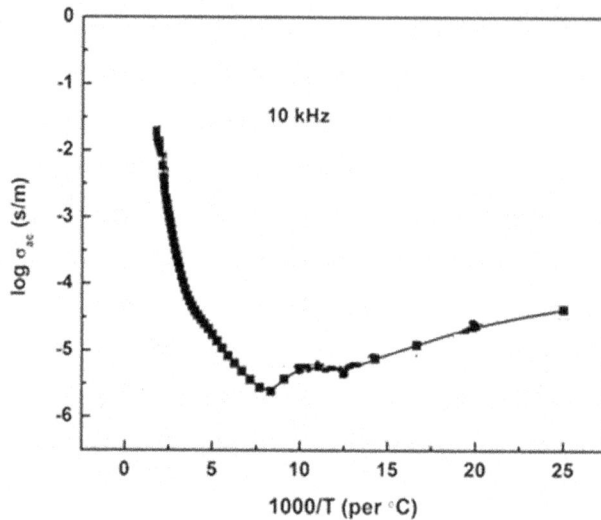

FIGURE 39.5   AC conductivity of LT at 10 kHz.

TABLE 39.1   Activation Energy at 1 kHz and 10 kHz Frequency

	1 kHz			10 kHz		
Sample	E_1	E_2	E_3	E_1	E_2	E_3
LT	0.137	0.029	0.007	0.131	0.026	0.005

where σ_o is the AC conductivity pre-exponential factor and E_a is the activation energy of AC conductivity and k is the Boltzmann constant [7]. The donor or acceptor ionization energies are usually 0.1eV as observed in region I. The activation energies $\leq$0.1 eV suggest band type of conduction in the temperature range of region I and II. Very low values of E_a in the low temperature regions III and II once again confirm metallic and semiconducting band nature as observed from the plots [8]. However, in the high-temperature region I, the steep rise and activation energy $\geq$0.1 eV suggests electron hopping between multivalent cations [21]. Thus, at very high temperatures, the conduction is purely due to electron hopping and the mechanism shifts from band type to hopping. This property is used for electrical switch.

39.7 Impedance Analysis

Figure 39.6 (a, b) shows the temperature-dependent Nyquist plots of LT. In general, Nyquist plot exhibits parabola (semicircular) when plotted between the complex terms of impedance. The parabola represents the relaxation behavior of LT at that temperature. The relaxation time and equivalent RC circuit elements can be estimated from relaxation equation (39.6) Nyquist plots facilitate the understanding of grain and the grain boundary effects of the electrical property of system [22].

$$\tau = \frac{1}{\omega} = R_b C_b \tag{39.6}$$

where $\omega = 2\pi\nu_{max}$. ν_{max} is an applied frequency corresponding to the maximum arc, R_b is bulk resistance, and C_b is bulk capacitance.

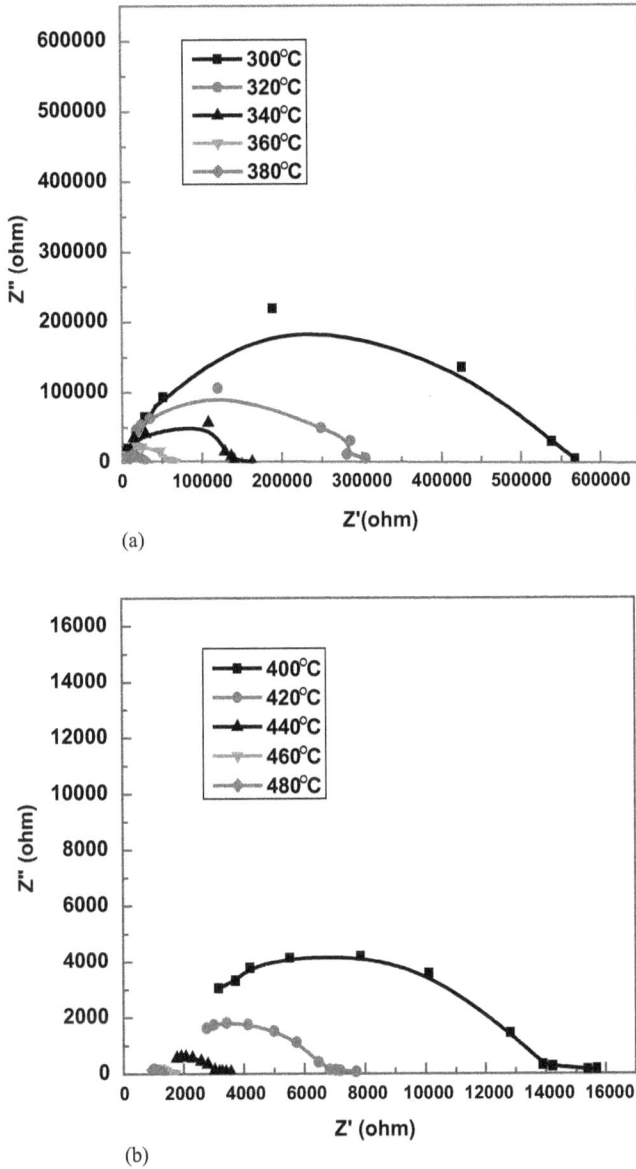

FIGURE 39.6 (a) Nyquist Plot of LT at different temperatures and (b) Nyquist Plot of LT at different temperatures.

The conductance of the materials is calculated using relation (39.7):

$$\sigma_{dc} = \frac{l}{R_b A} \tag{39.7}$$

where L is thickness and A is the cross-sectional area of the pellet sample.

Nyquist plots presented in Figure 39.6 (a, b) are in the temperature range of 300°C–480°C. It has been observed from the plots that at all temperatures the system exhibits single semicircle arc indicating grain (bulk) effect and no grain boundary effects exist in the system. The lower frequency arc represents

TABLE 39.2

Temperature (°C)	Resistance (ohm)	Capacitance (pF)	$\tau(\mu$ sec)	σ_{dc} (μS/cm)
300	566980.0	17.63	1.592	62.89
320	261490.3	6.08	1.592	136.3
340	133875.2	2.37	0.318	266.3
360	58106.9	2.74	0.159	613.7
380	26771.6	5.94	0.159	1332.0
400	13060.5	12.19	0.159	2730.5
420	6501.6	8.16	0.053	5485.0
440	3124.5	16.98	0.053	11413.5
460	1456.0	36.45	0.053	24493.0
480	1205.9	33.01	0.039	29572.7

grain boundaries conduction and a high frequency arc is due to grains conduction [23–25]. Moreover, it has been noticed that as the temperature increases the bulk resistance decreases, indicating LT has a negative temperature coefficient of resistance (NTCR). Similar results are observed by Umasankar and Fehr [16, 17]. The relaxations are Debye type from temperature 380°C and up. Below this temperature, the center for the semicircle is below X-axis, suggesting non-Debye type of relaxations below 400°C. The equivalent RC parallel circuit is shown in the *inset* in Figure 39.6 (a, b). The values of bulk resistance (R_b) bulk capacitance (C_b), relaxation time (τ), and conductance (σ) at different temperatures are tabulated in Table 39.2. Different capacitance (electrical energy storage) values are from 300°C to 480°C at 460°C and the capacitance value is 36.45 pF.

39.8 Conclusion

The double perovskite $Li_4Ti_2O_6$ has been synthesized by sol–gel method using microwave sintering method. The XRD pattern of $Li_4Ti_2O_6$ confirms the single-phase formation and monoclinic crystal structure with C2/c space group. The SEM images show the uniform morphology and purity of the sample; the particle size is around 23–30 nm. The AC conductivity confirms that the materials has three different conduction mechanisms. The material is used as a thermal switch. This change of conduction behavior at around 80°C can act as an instrument safety and health measure to control the electrical energy supply through the system or device. In the high-temperature region, the conduction is purely due to hopping of electrons. In double perovskite $Li_4Ti_2O_6$, the NTCR and non-Debye type of relaxation behaviors are observed. Impedance analysis reveals that the energy storage property of the prepared sample at 460°C is found to be 36 pF.

References

1. C. Pavithra, W. Madhuri, *J. Mat Sci. Mat Ele.*, 29 (2018), 2259–2266.
2. Th. Fehr, E. Schmidbauer, *Solid State Ionics*, 178 (2007), 35–41.
3. Jun Liang, Wen-Zhong Lu, Jia-Min Wu, Jian-Guo Guan, *Materials Science and Engineering B*, 176 (2011), 99–102.
4. C. Pavithra, W. Madhuri, *Mechanics, Materials Science & Engineering* (2017), DOI:10.2412/mmse.86.65.206.
5. S. Shanmuga Sundari, Binay Kumar, R. Dhanasekaran, *IOP Conf. Ser.: Mater. Sci. Eng.*, 43 (2013), 012010.
6. Chonghe Li, Kitty Chi Kwan Soh, Ping Wu, *Journal of Alloys and Compounds*, 372 (2004), 40–48.
7. M. Krimi, K. Karoui, A. Ben Rhaiem, *Journal of Alloys and Compounds*, 698 (2017), 515170.

8. W. Madhuri, M. Penchal Reddy, Il Gon Kim, N. Rama Manohar Reddy, K.V. Siva Kumar, V.R.K. Murthy, *Materials Science and Engineering B*, 178 (2013), 843–850.

9. Xiangwei Wn, Zhaoyin Wen, Bin Lin, Xiaogang Xu, *Materials Letters*, 62 (2008), 837–839.

10. Choong Hwan Jung, *Journal of Nuclear Materials*, 341 (2005), 148–152.

11. M. Venkateswarlu, C.H. Chen, J.S. Do, C.W. Lin, T.C. Chou, B.J. Hwang, *Journal of Power Sources*, 146 (2005), 204–208.

12. Ionela Carazeanu Popovici, Elisabeta Chirila, Viorica Popescu, Victor Ciupina, Gabriel Prodan, *Journal of Material Science*, 42 (2007), 3373–3377.

13. D. Mandal, D. Sathiyamoorthy, V. Govardhanan Rao, *Fusion Engineering and Design*, 87 (2012), 7–12.

14. Wn Xiangwei, Zhaoyin Wen, Xiaoxiong Xn, Zhonghua Gu, Xiaohe Xu, *Journal of Nuclear Materials*, 373 (2008), 206–211.

15. Choong-Hwan Jung, Sang Jin Lee, Waltraud M. Kriven, Ji-Yean Park, Woo-Seog Ryu, *Journal of Nuclear Materials*, 373 (2008), 194–198.

16. Bambang Priyono, Anne Zulfia Syahrial, Akhmad Herman Yuwono, Evvy Kartini, Mario Marfelly, *International Journal of Technology*, 4 (2015), 555–565.

17. R. Ramaraghavulu, S. Buddhudu, G. Bhaskar Kumar, *Ceramics International*, 37 (2011), 1245–1249.

18. Choong-Hwan Jung, Ji-Yeon Park, Weon-Ju Kim, Woo-Seog Ryn, Sang-Jin Lee, *Fusion Engineering and Design*, 81 (2006), 1039–1044.

19. M. Penchal Reddy, G. Balakrishnaiah, W. Madhuri, M. Venkata Ramana, N. Ramamanohar Reddy, K.V. Siva Kumar, V.R.K. Murthy, R. Ramakrishna Reddy, *Journal of Physics and Chemistry of Solids*, 21 (2010), 1373–1380.

20. W. Madhuri, M. Penchal Reddy, Il Gon Kim, N. Rama Manohar Reddy, K.V. Siva Kumar, V.R.K. Murthy, *Materials Science and Engineering B*, 178 (2013), 843–850.

21. Ao Mei, Xiao-Liang Wang, Jin-Le Lan, Yu-Chuan Feng, Hong-Xia Geng, Yuan-Hua Lin, Ce-Wen Nan, *Electrochimica Acta*, 55 (2010), 2958–2963.

22. Qianyu Zhang, Chengli Zhang, Bo Li, Shifei Kang, Xi Li, Yangang Wang, *Electrochimica Acta*, 98 (2013), 146–152.

23. Shweta Thakur, Radheshyam Rai, Igor Bdikin, *Materials Research*, 9 (2016), 1–8.

24. Umasankar Dash, Subhanarayan Sahoo, Paritosh Chaudhuri, S.K.S. Parashar, Kajal Parashar, *Journal of Advanced Ceramics*, 3 (2014), 89–97.

25. Th. Fehr, E. Schmidbauer, *Solid State Ionics*, 178 (2007), 35–41.

Index

Pages in *italics* refer to figures and pages in **bold** refer to tables.

O

For Product Safety Concerns and Information please contact our EU
representative GPSR@taylorandfrancis.com
Taylor & Francis Verlag GmbH, Kaufingerstraße 24, 80331 München, Germany

www.ingramcontent.com/pod-product-compliance
Lightning Source LLC
Chambersburg PA
CBHW080120220326
41598CB00032B/4903